FUNDAMENTALS OF
ELECTRONIC DEVICES

Dr. SANJAY SHARMA
B.E., M.Tech, Ph.D, MIETE

Edited By
DEEKSHA SHARMA
and
V. SHARMA

S.K. KATARIA & SONS®
Educational Publishers
4885/109, Prakash Mahal, Dr. Subhash Bhargav Lane,
Opposite Delhi Medical Association, Daryaganj, New Delhi-110002 (INDIA)
Phone: +91-11-43551243, Mobile +91-9871775858, 7678425209
e-mail: katariabook@gmail.com, katariabooks@yahoo.com
Website: www.skkatariaandsons.com
(e-books, subscriptions and bulk buying also available on our website)

Published by :

S.K. Kataria & Sons®

4885/109, Prakash Mahal, Dr. Subhash Bhargav Lane,
Opposite Delhi Medical Association, Daryaganj, New Delhi-110002 (INDIA)
Phone: +91-11-43551243; *Mobile:* +91-9871775858, 7678425209
e-mail: katariabook@gmail.com, katariabooks@yahoo.com
Website: www.skkatariaandsons.com

Send all correspondence to : **M/s S.K. Kataria & Sons,** New Delhi
e-mail: katariabook@gmail.com

First Edition: August, 2012, Reprint: 2026

Price: ₹ 595/-

Laser Typesetting by: **PLUS COMPUTERS**, Shahdara, Delhi-110093

Printed at: **Arora Printing Press**, Delhi–110051

Preface to the First Edition

Fundamentals of Electronic Devices is a core subject for undergraduate students in Electronics & Communication Engineering, Electrical Engineering and Electronics and Instrumentation Engineering. This book is primarily intended to serve as a textbook in accordance with the syllabus of Fundamentals of Electronic Devices offered by various universities in India as well as abroad.

In this book, a significant effort has been made to find simple ways to develop theoretical aspects of Fundamentals of Electronic Devices. A strong emphasis has been given on numerical examples. Neat and Clear diagrams have been used for explanation.

Every effort has been made to make the treatments simple and comprehensive. The whole text has been divided in Nine chapters :

Chapter 1 introduces Semiconductor Materials and Crystal Properties.

Chapter 2 introduces the Charge Carriers in Semiconductors.

Chapter 3 introduces Excess Carriers in Semiconductors.

Chapter 4 explains about Junction Properties.

Chapter 5 introduces BJTs.

Chapter 6 discusses FETs

Chapter 7 deals with Photonic Devices.

Chapter 8 discuss Negative Conductance Microwave Devices.

Chapter 9 explains Some Special Power Devices.

The author hopes that the book will fulfil the need of readers and would welcome any suggestions towards the improvement of the book.

I would like to express my sincere thanks to **Mr. Vibhav Sachan** and **Mr. Manish Gupta** for constant inspiration and encouragement.

Special thanks to **Dr. B.P. Singh** (Former Professor, Electronics & Comm. Engg., MMMEC, Gorakhpur), **Dr. R.K. Chauhan** (Professor, Electronics & Comm. Engg., MMMEC, Gorakhpur), **Dr. G.S. Tripathi** (Electronics & Comm. Engg., MMMEC, Gorakhpur), **Dr. Arun Kumar** (Former Professor, Electronics & Comm. Engg., MMMEC, Gorakhpur) and **Dr. C.B. Tripathi** (Former Professor Electronics & Comm. Engg., MMMEC Gorakhpur) and **Dr. Vipin Kumar** for their guidance and valuable suggestions.

I am grateful to my younger brothers **Rajeev** and **Rakesh,** daughter **Ritika**, for inspiring me for this project.

I deeply express my heartful thanks to the publishers **S.K. Kataria & Sons** for publishing this book in such a beautiful get-up and well in time.

Lastly, I am also grateful to **Mr. Manoj Tyagi** and his staff from **Plus Computer's** for their hard work.

Dr. Sanjay Sharma

August, 2012

DEDICATED

TO

MY LOVING SON

MRIDUL

Contents

UNIT-I

Syllabus

Crystal Properties and Charge Carriers in Semiconductors: Elemental and Compound semiconductor materials, Crystal lattice structure.
Bonding Forces and Energy Bands in Solids, Charge Carriers in Semiconductors, Carriers Concentrations, Drift of Carriers in Electric and Magnetic Fields.

Above syllabus has been covered in following chapters :

Chapter 1

Semiconductor Materials and Crystal Properties

Syllabus

Crystal Properties and charge carriers in semiconductors: Elemental and compound semiconductor materials, Crystal lattice structure.

Chapter Outline

- Introduction
- Semiconductoor Materials
- Elemental Semiconductor Materials
- Compound Semiconductor Materials
- Classification of Crystalline Solids Based upon Bonding
- Concept of Crystallography
- What is Space Lattice?
- Basic Aspects of Unit Cell
- Various Laws of Crystallography
- Crystal Systems
- Number of Atoms Per Unit Cell in a Cubic Lattice
- Lattice Planes or Net Planes
- Atomic Radius (i.e., Cubic Crystal)
- Packing of Atoms in Metals
- Atomic Packing Factor (APF)
- Co-ordination Number
- Miller Indices
- X-Ray Diffraction
- Laue Spots
- Bragg's Law
- Numericals on Radius-Density-Molar Mass
- Defects or Imperfections in Crystals
- Line Defects (i.e., Dislocations)
- Plane Defects (i.e., Surface Imperfections)

1.1. INTRODUCTION

As a matter of fact, solids are the most important materials handled by man for the designs, development and in day to day life. They are used as engineering materials for conduction, engines, refrigeration, abrasives, machine designs, installations, insulation, fabrication, construction, abrasives, chemical applications, etc. Various solids differ in different parameters such as strength, conductivity, magnetic character, density melting point, hardness, etc. properties. Solids like diamond, silica, silicon carbide are very hard whereas graphite, molybdenum disulphide, talc, magnesium silicate are quite soft. Various metals, alloys exhibit good thermal and electrical conductivities whereas ceramics, asbestos, many fibrous materials, thermocol etc. are bad conductors. The solids may be used as per their characteristic properties in diversified applications. Hence, it is essential to study and understand the solid materials regarding their structure, bonding in them so as to use them in a proper manner.

1. Forces of Attraction in Solids

The strong primary forces of attraction in solids are in the form of:

(i) ionic bonds

(ii) covalent bonds - co-ordinate bonds

(iii) metallic bonding.

Relatively weaker secondary forces of attraction are in the form of

(i) Vander Waals intermolecular forces

(ii) Hydrogen-bonding

2. Characterization of Solids*

Solids may be characterised by their low compressibility, ordered arrangement of their constituent particles (atoms/ions/molecules), high mechanical strength, rigidity, specific geometrical size and shape, stronger primary and / or secondary forces of attraction in them and close packed structure.

3. Few aspects abouts Solids and Liquids

(i) Two important factors, *i.e.*, their volume and shape are much less dependent on external conditions of temperature and pressure.

(ii) The constituent particles in solid are systematically arranged and do not move in a random manner with respect to each other. However, they vibrate about their mean positions with certain amplitude that depends upon the temperature.

(iii) This means that the constituent particles in solid have **confined positions**.

(iv) In the liquids, groups of constituent particles move freely with respect to each other and hence there is high extent of randomisation.

(v) Further, The particles in liquids not only vibrate with greater amptitude but also have some freedom of rotation and translation. An X-ray photograph of liquid shows various concentric circles.

(vi) With the help of X-rays diffractions, we can do the study of crystals regarding arrangement of constituent particles in the solid, the interatomic distances, interplanar distances and crystal angles between various intersecting planes X-rays have wave length of about 1 A° that is almost of the same order as interatomic distances in the crystal.

(vii) The multielectron atoms in the crystals have ability to reflect X-rays from the successive planes of atoms in the crystal. When the angle of incidence of X-rays is 90° to the plane of atoms, we obtain a photograph containing spots which shows the pattern of arrangement of atoms in the crystal.

(viii) When the angle of incidence to the parallel planes is q, then, we get diffraction patterns of first order (i.e., n = 1), second order (i.e., n = 2) and so on.

4. Bragg's Expression

We can calculate the 'd' interplanar distance in the parallel planes by Bragg's expression as under:

$$n\lambda = 2d \sin \theta$$

where n is the order of diffraction.

λ is wavelength of X-rays and

θ is the angle of incidence to the plane of atoms

The atoms having fewer electrons cannot reflect X-rays and for the study of solids consisting atoms of such type, the costlier 'neutron diffraction' technique is used. Sometimes electron diffraction technique is also used for the study of crystals but the electron beams have low penetration power and hence only surface of the crystal. However, low penetration power and hence only the surface of the crystal can be studied by electron diffraction techniques. In studying solid state electronic devices, we are interested primarily in the electrical behaviour of solids.

* The atoms in solids are held together by bonding forces among them.

1.2. SEMICONDUCTOR MATERIALS

1. Basic Aspects about Conducting and Insulating Materials

(i) We know something about conducting and insulating materials. Conducting materials are good conductors of electricity and hence are characterised by a large electrical conductivity and small electrical resistivity.

(ii) Basic, examples of good conductors are gold, silver, copper, aluminium etc.

(iii) On the other hand. Insulating materials are characterized by poor electrical conductivity and are used to block current from flowing where it is not to pass.

(iv) Basic, examples of insulators are mica, porcelain, glass, quartz, rubber, bakelite etc.

2. Basic Aspects about Semiconductor

(i) There is another group of materials that is called as semiconductors.

(ii) Semiconductors have electrical conductivities intermediate between metals and insulators.

(iii) The resistivity of various semiconductor materials falls in a very wide range from 10^{-4} W-m to about 0.5 W-m whereas the resistivities of conductors and insulators are of the order of 10^{-8} W-m and 10^{9} W-m respectively.

(iv) It is important that the conductivity of semiconducting materials can may be varied over orders of magnitude by the changes in temperature; optical excitation and impurity content. This variability of electrical properties make the semiconductor materials natural choices for electronic device investigations.

(v) Semiconductor materials are found in column IV and neighbouring columns of the periodic table as shown in Table 1.1.

(vi) The column IV semiconductors, silicon and germanium are called elemental semiconductors because they are composed of single species of atoms.

(vii) In addition to the elemental materials, compound of column III and column V atoms, as well as certain combinations from II and VI, build up the inter-metallic or compound semiconductors.

Table 1.1. Common Semiconductor materials

(a) The portion of the periodic table where semiconductors occur

II	III	IV	V	VI
	B	C		
	Al	Si	P	S
Zn	Ga	Ge	As	Se
Cd	In	Sn	Sb	Te

Table 1.1(b). Elemental and compound semiconductors

Elemental	IV compounds	III - V compounds	II - VI compounds
Si	SiC	AIP	Zns
Ge		AlAs	ZnSe
		AlSb	ZnTe
		GaP	CdS
		GaAs	CdSe
		GaSb	CdTe
		InP	
		InAs	
		InSb	

(viii) The resistance of conductor increases with the increase in temperature. Also, the resistance of insulator remains almost constant upto a certain temperature. Further, The resistance of semiconductor decreases with the increase in temperature as depicted in figure 1.1. This means that the temperature co-efficient of resistance of semiconductors is negative.

(ix) Semiconductors can be used as a insulator at very low temperature and as a conductor at high temperature*. The resistivity of semiconductors decreases in active state i.e., under illumination, under application of electric field. The current flowing in semiconductor under the application of electric field is not proportional to voltage like conductors which obey Ohm's law.

(x) However, in semiconductors current increases by far more than the applied voltge. This means that semiconductors are non-linear resistors.

3. Impurities and their Importance

The electronic and optical properties of semiconductor change considerably when even minute amounts of certain other substances known as the impurities, are added to them. In addition, the introduction of certain impurities like arsenic, antimony, gallium, indium may even change the mechanism of electrical conductivity in the semiconductor and certain physical phenomena will start to occur at the junctions between sections of the semiconductor having different impurities.

4. Doping

This process of controlled addition of impurities into the semiconductor materials is called as doping. The devices like diodes, transistors fabricated from semiconductors materials are called as solid state devices. The solid state devices have many advantages like low cost, light weight, compact size, more reliable, low operating voltage, instantaneous operation (no heating required), high efficiency and long life (if operated within the specified limit of temperature and frequency).

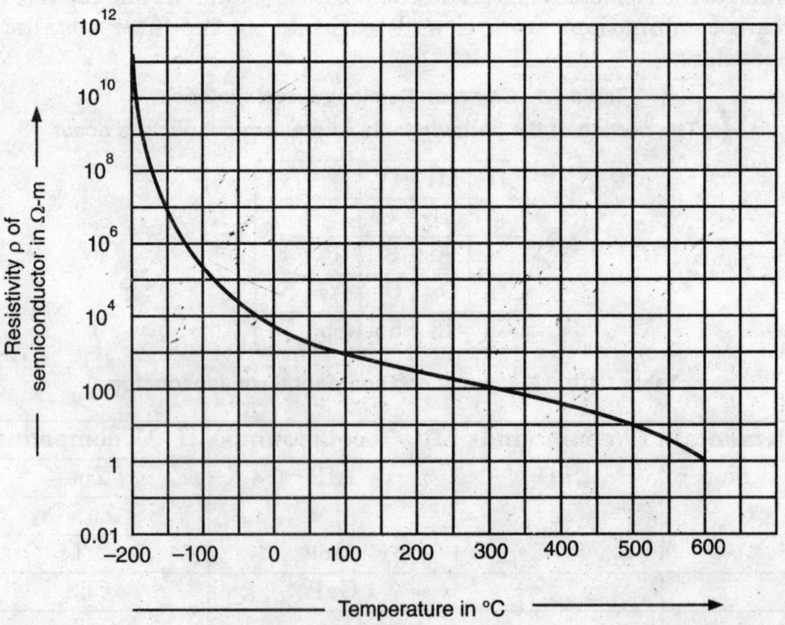

Fig. 1.1.

* Semiconductors are basically insulators, however, one important property that differentiates between the two is that the conductivity of semiconductors can be varied over orders of magnitude by changes in temperature, photonic excitation, irradiation, and most importantly, by doping.

1.3. ELEMENTAL SEMICONDUCTOR MATERIALS

1. Basic Aspects

(i) As depicted in table 1.1(a), Group IV of the periodic table consists of five elements like carbon, silicon, germanium, tin, and lead.

(ii) The electronic configuration of these elements has been shown in table 1.2.

(iii) It is obvious from the electronic configuration that all these elements have four electrons in their valence shell, two of which are in s-orbital where as the remaining two are in p-orbital.

Table 1.2. Electronic configuration in Group IV (A)

S.No.	Element	Atomic Number	Electronic configuration
1.	Carbon (C)	6	$1s^2\ 2s^2\ 2p^2$
2.	Silicon (Si)	14	$1s^2\ 2s^2\ 2p^6\ 3s^2\ 3p^2$
3.	germanium (Ge)	32	$1s^2\ 2s^2\ 2p^6\ 3s^2\ 3p^6\ 3d^{10}\ 4s^2\ 4p^2$
4.	Tin (Sn)	50	$1s^2\ 2s^2\ 2p6\ 3s^2\ 3p^6\ 3d^{10}\ 4s^2\ 4p^6\ 4d^{10}\ 5s^2\ 5p^2$
5.	Lead (Pb)	82	$1s^2\ 2s^2\ 2p^6\ 3s^2\ 3p^2\ 3d^{10}\ 4s^2\ 4p^6\ 4d^{10}\ 4f^{14}\ 5s^2\ 5p^2\ 5d^{10}\ 6s^2\ 6p^2$

(iv) Hence, all these elements have s^2p^2 configuration in the outer shell i.e., valence shell. The penultimate shell of C possess s^2 electrons, of Si contains s^2p^6 electrons, of Ge contains $s^2p^6d^{10}$ (saturated) electrons while Sn and Pb contain $s^2p^6d^{10}$ (unsaturated) electrons.

(v) This implies why C differs from Si in properties and both from the other members of this subgroup like carbon behaves as an insulator in crystalline form (diamond), Si and Ge behave as semiconductors and tin, lead behave as metal.

2. Physical Properties

(i) It may be observed in Group IV that there is a gradual change from non-metallic to metallic nature with the increase of atomic number. So, carbon is a non-metal, silicon and germanium are metalloid and tin, lead are metallic in nature.

Non - metal	Metalloid	Metal
Carbon	Silicon, Germanium	Tin, Lead

(ii) All the elements are tetravelent elements.

(iii) The size of atoms also increases with increase in atomic number.

(iv) Ionisation potential of these elements is high and decreases gradually with the increase of atomic number.

(v) The volume occupied in the solid state by one gram-atom of the element i.e., atomic volume increases with the increase in atomic number (C = 5, Pb = 18).

(vi) The most electronegative element of this subgroup is carbon and electronegativities decrease from carbon to lead but not systematically.

(vii) The tendency of an element to form long chains of identical atoms is called as catenation and decreases with the decrease in the value of bond energy. Also, carbon has the maximum tendency to form chains (C – C = 85 kcal/mole), silicon has less tendency (Si – Si = 54 kcal/mole), germanium has lesser tendency (Ge – Ge = 40 kcal/mole) and tin has very small tendency (Sn – Sn = 37 kcal/mole).

3. Chemical Properties

(i) All the elements of Group IV form covalent hydrides. The number of stable hydrides and the ease with which these elements form the hydride, decreases with the increase in atomic number.

(ii) All the elements form acids H_2MO_3 type like H_2CO_3 (carbonic acid), H_2SiO_3 (silic acid), H_2SnO_3 (meta stannic acid), H_2PbO_3 (meta plumbic acid). Sodium salts of all these acids are stable.

(iii) All these elements of Group IV form oxy-chlorides like $COCl_2$ (carbonyl chloride), $SiOCl_6$ (silicyl chloride), $SnOCl_2$ (tin oxy-chloride), except lead.

(iv) All elements in this Group form dioxides of MO_2 type. The acidity of these oxides decreases with the increase in atomic number. Therefore, CO_2 and SiO_2 are acidic and GeO_2, SnO_2 and PbO_2 are amphoteric. All these oxides react in alkalies to form carbonates, silicates, germanate, stannates and plumbates.

(v) All these elements react with caustic alkalies with the evolution of hydrogen except carbon.

4. Occurrence

Elemental carbon is available in widely varied forms like extreme hardness and transparent form like diamond, and soft and grey form like graphite. Another forms of carbon is coal and electrial carbon materials are manufactured from graphite. Silicon is not available free in nature. It is always available in form of compound. It is most abundant element after oxygen in the earth's crust. In fact, it is available in combination with oxygen as silica (SiO_2) and silicates. The naturally available minerals having silicon are feldspar [K_2O, Al_2O_3, $6SiO_2$] kaolimite[Al_2O_3, $2SiO_2$, $2H_2O$], Asbestos [$CaMg_3$ $(SiO3)_4$], and mica [KH_2 Al_3 $(SiO_4)_3$].

Free silicon available in three different form as under:

(i) Amorphous form

(ii) Polycrystalline form and

(iii) Crystalline form

These three forms of silicon are prepared by different techniques at different temperature. Germanium is basically an earth element and it is extracted from the ash of certain coals or from the flue dust of zinc smelters. Usually, extracted germanium appear in the form of germanium dioxide powder that is converted to pure germanium by reduction method. Tin is not available free in nature. It is available in the form of ore like, (i) cassiterite or tinstone (ii) SnO_2 (iii) Wolframite. Lead is available in the form of ore mainly galena, Pbs, which is associated with zinc blend, iron pyrites and traces of silver. Lead is mainly recovered from the ore galena.

5. Primary Applications

(i) Crystalline form of carbon (i.e., diamond) has very high melting point (above 3000°C). Pure carbon behaves as a semiconductor with negative temperature coefficient of resistance. Conductivity of carbon is slightly less than that of metals and their alloys.

(ii) Carbon elements are widely used in electrical engineering as (i) carbon electrodes for electric arc furnaces, electrolytic baths and arc welding, (ii) battery cell elements (iii) brushes for electrical machines (iv) non-wire resistors (v) arc-light (vi) microphone powders and other components of telecommunication equipments etc.

(iii) Silicon is used in the manufacture of alloy carborundum (SiC) and ferro-silicon. Silicon is also used as a deoxidiser in the manufacture of steel. Ultra pure silicon is used to design electronic, solid state devices and to fabricate integrated circuit. Germanium is mostly used for manufacturing semiconductor devices.

(iv) Tin metal is used in manufacturing of alloys like white metal, bell metal, solder, babbit metal etc. which has different applications. Tin metal is used in chemical technology like in tinning of household utensils and in tin plating of iron sheets.

(v) Tin amalgam is used for making mirrors and tin compounds are used as mordants in dyeing-calico printing.

(vi) Lead metal is used to manufacture lead-acid accumulator plates, lead shots, fuse wire and compounds like red lead, litharge, lead tetraethyl and white lead.

(vii) Lead metal is also used in the manufacturing water pipes, protective sheets for roofs and drains, cable coverings, for lining the chambers in H_2SO_4 preparation, etc. Lead sheets are used in X-ray generating machine to prevent hazardous effect.

1.4. COMPOUND SEMICONDUCTOR MATERIALS*

1. Definition and Basic Concepts

Basically, Compound semiconductor materials are mostly prepared either from the combinations of Group III and Group V elements or from the combinations of Group II and Group VI elements, as illustrated in table 1.1(b). Compounds prepared from two elements are called as binary compounds where as prepared from three elements are called as ternary compounds. The wide variety of electronic and optical properties of semiconductors enable greater flexibility in the design and manufacturing of electronic and optoelectronic devices. As discussed in table 1.1, there are numerous semiconductor materials. Among these, Si is used for the majority of semiconductor devices like rectifiers, transistors, SCRs, UJTs, resistors, thermistors and integrated circuits are usually made of Si.

2. Special Characteristics

(i) Compound semiconductors are most widely used in devices that require the emission or absorption of light. As an example, light emitting diodes (LEDs) are generally fabricated from compounds like GaAs, GaP and mixed compounds like GaAsP.

(ii) For television screen, fluorescent materials are used which are generally II-VI compound semiconductors like ZnS. Light sensors are commonly made with InSb, CdSe, or other compounds of lead like PbTe, PbSe.

(iii) Si and Ge are also used widely as nuclear radiation and infrared detectors.

(iv) Microwave device, gunn diode, semiconductor laser are commonly fabricated from compound semiconductor like GaAs, InP.

(v) Therefore, the wide range of semiconductor materials offers considerable variety in properties and provides engineers with much flexibility in the design of electronic functions.

(vi) Compound semiconductors are most widely used in high speed semiconductor devices.

(vii) The electronic and optical properties of semiconductor materials are strongly affected because of doping of impurities in precisely controlled amounts. This process of doping impurities causes wide variation in the conductivities of the semiconductor and even after the conduction processes from conduction by negative charge carrier to positive charge carriers.

(viii) An impurity concentration variation in ppm (parts per million) order can change the property of silicon from poor conductor to a good conductor of electricity.

(ix) As even slight variation of impurities produce dramatic change in electrical and optical properties of semiconductors, so details study about the specific arrangement of atoms in each semiconductor is required to use it properly for semiconductor device applications.

(x) The most important feature which distinguishes semiconductor from conductor and insulator, is its energy band gap which is of the order (1 – 2) eV. This energy band gap can easily overcome by using visible IR light as and when required.

(xi) As an example, the energy band gap of GaAs is about 1.43 eV that corresponds to light of wavelength in the near infrared and GaP has a energy gap 2 – 3 eV which corresponds to wavelength in the green portion of the visible spectrum.

(xii) As the semiconductors (compound) are available in the wide range of energy band gap so light emitting diodes (LEDs) and semiconductor laser can be designed with wavelengths over a broad range of spectrum i.e., infrared to visible range of the spectrum. In addition, compound semiconductors are used to design efficient solar cells.

---

* Two element components are called the binary compounds whereas three element compounds are called ternary compounds.

1.5. CLASSIFICATION OF CRYSTALLINE SOLIDS BASED UPON BONDING

Based upon the **nature of forces binding** the constituent particles in the crystalline solids, the solids may be classified into following four types:

1. Ionic solids
2. Molecular solids
3. Covalent solids
4. Metallic solids

1. Ionic solids

(i) The constituent particles in the ionic solids are basically positive and negative ions. It is known, that a positive ion is formed by loss of electrons by an electropositive element and a negative ion is formed by gain of electrons by element that is having high electron affinity.

(ii) There is a **strong electrostatic force** of attraction between the positive and negative ions.

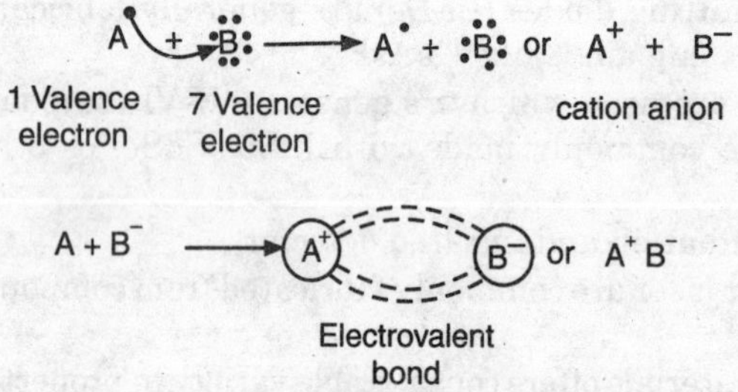

Fig. 1.2.

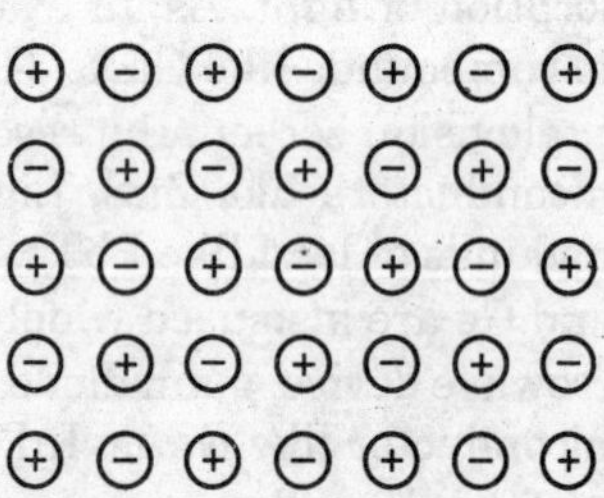

Fig. 1.3. A two dimensional array of positive and negative ions in an ionic solid like NaCl, $CuSO_4$, etc

(iii) An ionic solid contains of an array of positive and negative ions in a characteristic fashion throughout the crystal. Every positive ion is attracted by a definite number of negative ions in neighbouring from all directions.

(iv) Hence, the ionic bond is understood as **nondirectional bond.** The number of oppositely charged ions in the neighbouring, around an ion, is known as its co-ordination number. This means that each Na^+ ion is surrounded by six Cl^- ions (and vice versa). Hence, the co-ordination number is six.

(v) The ions are compactly packed due to strong bonds.

(vi) The co-ordination number depends upon the relative sizes of positive and negative ions as shown in figure 1.3.

(vii) The complete crystal can be regarded as one molecule and consists of a very large number of ions. We can express chemical formula of an ionic solid in the form of simple ratio of positive and negative ions i.e., NaCl (1 : 1 ratio of Na^+ and Cl^-), CaF_2 (1 : 2 ratio of Ca^{++} and F^-).

(viii) The ionic solids possess:

(a) high rigidity or hardness

(b) high melting point

(c) electrical conductivity in molten state or in solution

(d) more or less solubility in polar solvents like water, alcohol

(e) high lattice energy

(f) brittleness

Illustrative Examples of ionic solids

(i) Formation of NaCl

When sodium atom (with atomic number 11) and a chlorine atom (with atomic number 17) approach each other, an electron is transferred from sodium to chlorine atom and thus yielding oppositely charged sodium ions (Na^+) and chloride ions (Cl^-) respectively as depicted in figure 1.4(a) These ions are held together by the forces of electrostatic attraction between them.

(ii) Formation of Magnesium Chloride

Just to attain stable configuration, Mg atom (atomic number 12) loses its 2 electron of last orbit. While two chlorine atoms (atomic number 17) gain 1 electron each to form stable configuration of 2, 8, 8. Mg by losing 2 electrons acquired + 2 charge; while chlorine atoms by gaining one electron each acquire –1 charge each, as depicted in figure 1.4(b).

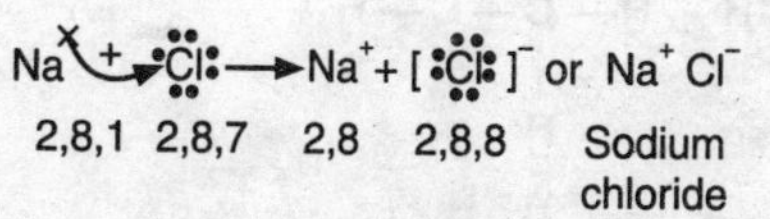

Fig. 1.4(a)

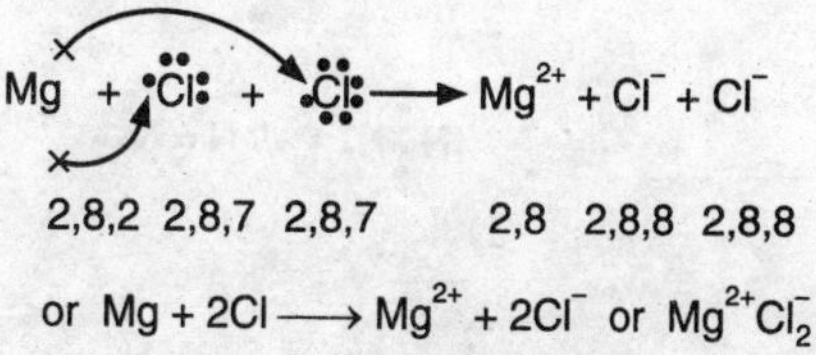

Fig. 1.4(b)

Mg^{2+} and two Cl^- ions unite by forces of electrostatic attraction to form $MgCl_2$ molecule.

2. Molecular solids

(i) A small number of atoms are joined by covalent bonds to form molecules. These neutral molecules are the lattice points in the crystal lattice.

(ii) The covalent bonds as primary binding forces are the bonds between particular atoms. Hence, they are called as directional bonds.

(iii) A molecular solid is made up of large number of the molecules. These molecules are bound by much weaker secondary forces of attraction i.e., by dipole-dipole attraction, hydrogen bonding or Vander Waals force of attraction.

(iv) Due to weak force of attraction between the molecules as constituent particles of molecular solid, these solids have

(a) softness

(b) low melting and boiling points

(c) volatility

(d) low density

(e) compressibility

(f) bad conductors of heat and electricity

(g) low heat of fusion

(h) low heat of vapourisation

(v) Few examples of the molecular solids are, various organic solids, sulphur (S_8), phosphorous (P_4) iodine, ice, PCl_5, $AlCl_3$ etc.

3. Covalent solids

(i) Covalent bond is formed by sharing of electrons in pairs between two atoms, both of which make equal contribution of electrons. When each atom contributes one, two or three electrons such that two, four or six electrons are shared in all, the union so set-up, is called as single or double or triple bond respectively.

(ii) Examples of covalent bond formation has been depicted in figure 1.5.

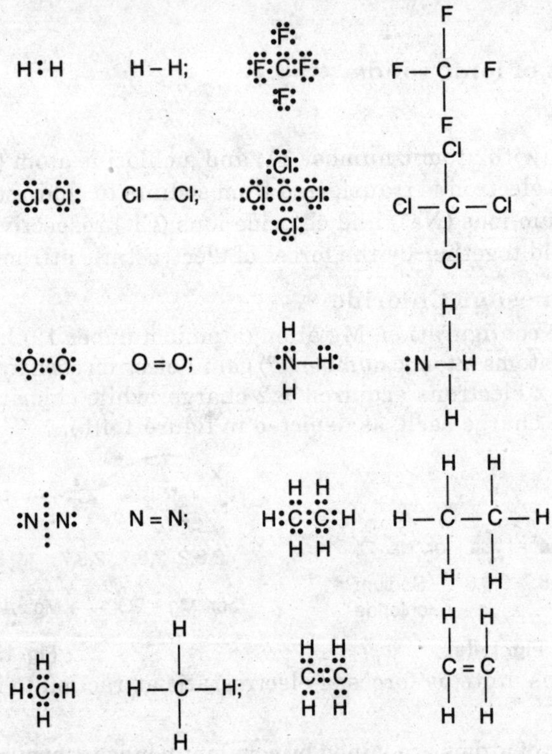

Fig. 1.5.

(iii) In covalent solids, atom is the constituent particle and it is represented by lattice point in a crystal lattice.

(iv) Each atom in the covalent solid is linked with definite number of neighbouring atoms by directional covalent bonds as per its valency. The bonding is continuous throughout the crystal. The whole crystal can be regarded as a giant molecule.

(v) Although, the covalent bond is weaker than ionic bond, however, the continuous network of the covalent bonds makes the covalent solid as a very strong, hard material.

(vi) In case, the covalent solid consists of large number of the giant molecules, the secondary force of attraction between the molecules may be quite weak.

(vii) Generally, these solids are hard, strong, and having very high melting point.

(viii) Few examples of covalent solids are, silica (SiO_2), diamond, graphite, silicon, silicon carbide (SiC), ZnS, AgI, etc.

(ix) Structures of diamond and graphite have been shown in figures 1.6 shown 1.7 to understand covalent solids.

(x) Each carbon atom sp^3 hybridised joined to four other such carbons by covalent bonds, bond angles 109.5°, bond lengths 1.54 A°, rigid three dimensional network; FCC crystal lattice.

(xi) Each carbon sp^2 hybridised, joined to three other such carbons by two single bonds and one pi bond, bond angles 120°, bond length 1.39 A°, hexagonal two dimensional or flat molecules, weak Vander Waal force of attraction between molecules, soft conductor, Hexagonal crystal lattice.

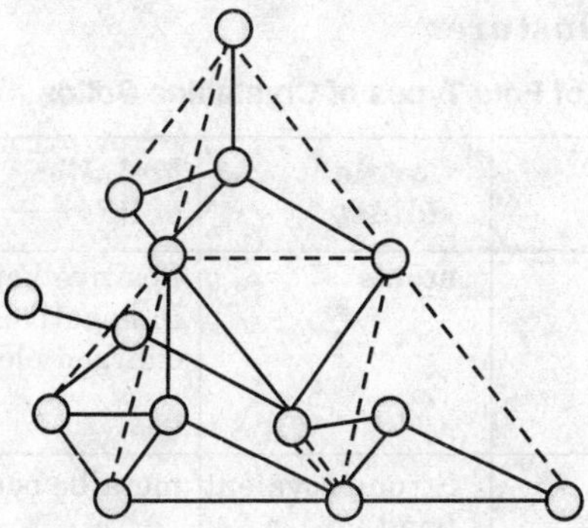

Fig. 1.6. Structure of diamond

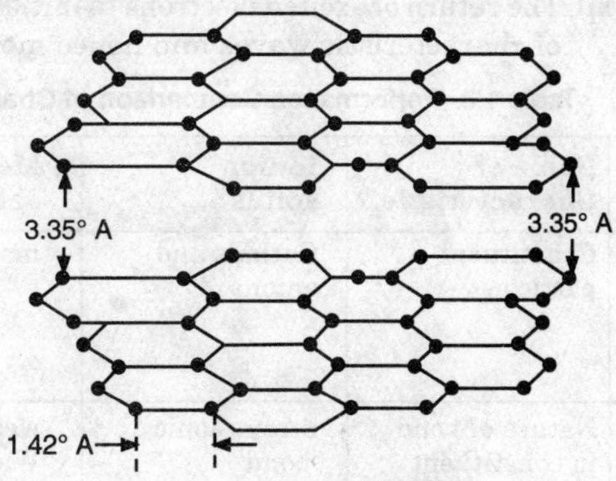

Fig. 1.7. Structure of graphite

4. Metallic solids*

(i) The constituent particles of metallic solids, are the metal atoms. All metals and their alloys have metallic bonding between the atoms.

(ii) All the metals have high electropositive character or low ionization potential. This means that the valence electrons are loosely bound with nucleus and become free electrons or mobile electrons.

(iii) The positively charged part of the metal atom excepting the valency electrons is called as **'Kernel'**.

(iv) According to Lorentz theory the Kernels are compactly arranged in the sea of free valence electrons.

(v) Each Kernel is attracted by the valence electrons in its sphere of influence and each valence electron is attracted by different Kernels in its sphere of influence, from all directions.

(vi) This force of attraction is called as metallic bond and it is non-directional bond.

(vii) The metals having greater nuclear charge and more number of valency electrons, have stronger metallic bond as depicted in figure 1.8.

(viii) Meatallic bond is weaker than ionic and covalent bonds but because of larger number of metallic bonds throughout the metal crystal, the metals have

(a) high mechanical strength

(b) hard,

(c) high melting points,

(d) high density,

(e) high tensile strength.

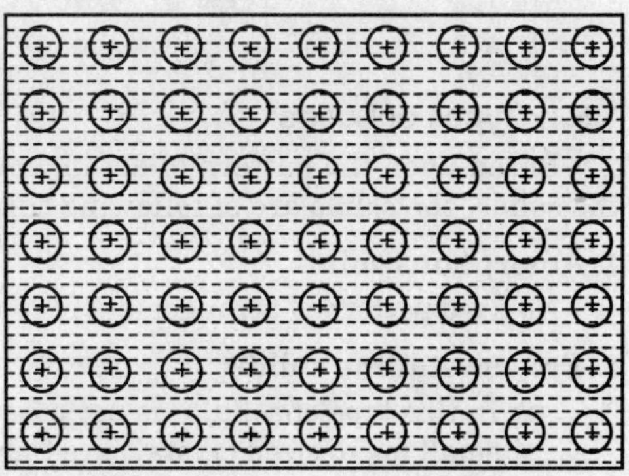

Fig. 1.8. Structure of metallic crystal

(ix) Further, since free or mobile electrons, metals are good conductors of heat and electricity. In addition, the Kernel layers can slide and rearrange in new fashion, on application of tensile force or compression, the metal have got ductility and malleability.

(x) The free or mobile electrons, absorb the light rays and hence get excited or oscillate to and fro.

---

* Solids are characterized by incompressibility, regidity and mechanical strength.

(xi) The return of excited electrons to initial energy levels or the oscillation, produces emission of characteristic waves and hence metals show **lusture**.

Table 1.3. Performance Comparison of Characteristics of Four Types of Crystalline Solids

S. No.	Name of Characteristic	Ionic solids	Molecular solids	Covalent solids	Metallic solids
1.	Constituent particles	Cations and anions	molecules	atoms	Positive kernel in negatively charged electron gas.
2.	Nature of bond in constituent particles	Strong ionic bond	weak Vander Waal force	Strong covalent bond	metallic bond
3.	Directional nature	Non-directional bonds	Directional bonds between atoms in molecules	Directional bond	Non-directional bond
4.	Bond strength	700 – 1000 kJ/mole	3 – 20 kJ/mole	50 – 700 kJ/mole	150 – 800 kJ/mole
5.	Melting point	high	Low	very high	Low to very high
6.	Feature properties	Hard, rigid soluble in water, conduct electricity in molten or solution state	soft, compressible, low density, bad conductors	Very hard insulators except graphite	Good conductors lusture, ductile malleable, high density, soft to very hard, high tensile strength strong
7.	Examples	NaCl, CuSO4, CaF_2, $Al(NO_3)_3$ etc.	glucose, sugars, organic solids, Urea ice, P_4, S_8 etc.	SiC, SiO_2, diamond, graphite, Zns, Al_2O_3	Various metals and their alloys

1.5.1. Types of Solids*

Based upon the pattern of arrangement of constituents particles (atoms/ions/molecules) in the solids, the solids are of following three types:

1. Amorphous solids

No fixed pattern of arrangement of constituent particles in solid i.e., glass, plastics, rubbers etc.

2. Polycrystalline solids

(i) In this type of solid, it is not possible to see clearly, the shape of the crystals, since many small micro sized crystals are tightly packed together without any specific order.

(ii) Such substances, which are basically crystalline, however are superfine to be seen as crystals, are termed polycrystalline. Few examples of such substances are various metals, alloys and semiconductors prepared at particular temperature.

3. Crystalline solids

Regular three dimensional arrangement of constituent particles in solid. i.e., metals, salts, orgnic solids, silicon, silica, talc, mica, ice, graphite etc. All above mentioned structures of solids have been depicted in figure 1.9.

* Some solids, like sodium chloride, sulphur and sugar, besides being incompressible and rigid, have also characteristic geometrical forms.

The details of amorphous and crystalline solids can be explained as under:

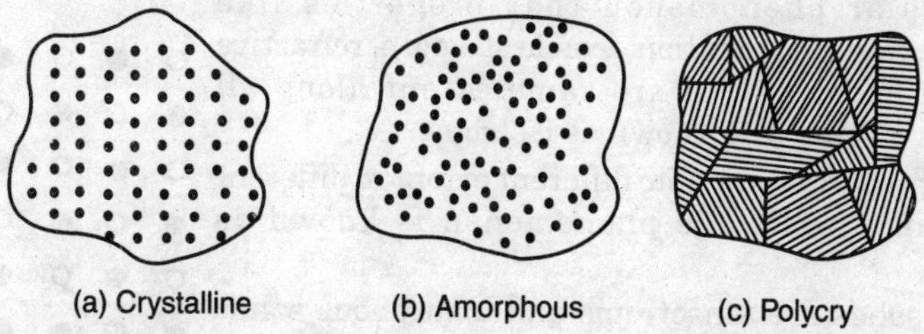

(a) Crystalline and (b) amorphous materials are illustrated by microscopic views of the atoms, whereas (c) Polycrystalline structure is illustrated by a more macroscopic view of adjacent single-crystalline region, such as (a)

Fig. 1.9. Three types of solids, classified according to atomic arrangement

(i) Characteristic Geometry

(a) Because of regularly arranged molecules or atoms in three dimensional space, crystalline solids have definite geometrical shape, faces.

(b) On the other hand, amorphous solids have no fixed pattern of arrangement of atoms or molecules and therefore do not have any definite geometrical shape.

(c) The definite and orderly arrangement of atoms/ions/molecules in crystalline solid extends over a long range from one edge or face of the crystal to the opposite edge or face and hence the crystalline solids are known as **long range ordered** solids.

(d) In the amorphous solids the orderly arrangement does not extend long and at the most it may be few angstroms only.

(ii) Melting Points

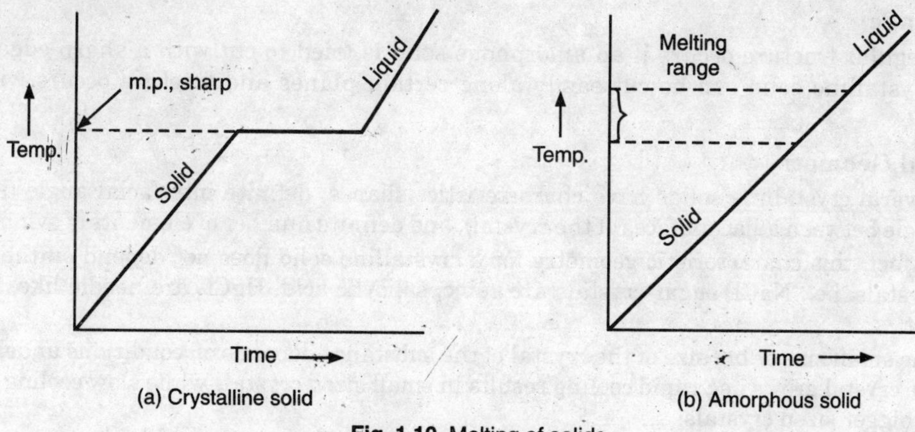

Fig. 1.10. Melting of solids

(a) With the increase in the temperature, the constituent particles in the solid vibrate with greater amplitude and at certain high temperature they may not return to their respective mean positions. This results in randomisation.

(b) In addition the translated atoms or molecules are free to have rotational degree of freedom.

(c) Melting point may be defined as the minimum temperature at which the atoms or molecules in a solid become sufficiently disordered and the solid changes to liquid state.

(d) A crystalline solid has a sharp melting point where as an amorphous solid has no sharp melting point. However, the apparent change of solid to liquid takes place over a range of temperature.

(e) Let us consider figure 1.10(a). Now, at the melting point, the temperature remains constant for certain time in the case of crystalline solids, although heating is going on.

(f) The amorphous solids do not show such feature. Let us consider figure 1.10(b).

(iii) Isotropy and anisotropy

(a) The particular phenomenon that properties like conductivity, rate of dissolution, tensile strength, refractive index, brittleness, etc, if are same extent along all directions in a solid, it is known as isotropy.

(b) If the above properties are upto different extent in different directions in a solid, the phenomenon is known as anisotropy.

(c) Crystalline solids are anisotropic and amorphous solids are isotropic like liquids.

(d) In figure 1.11 a simple two dimensional arrangement of a crystalline solid consisting of two kinds of atoms has been shown. Along direction AB and direction CD, the velocity of light would be different.

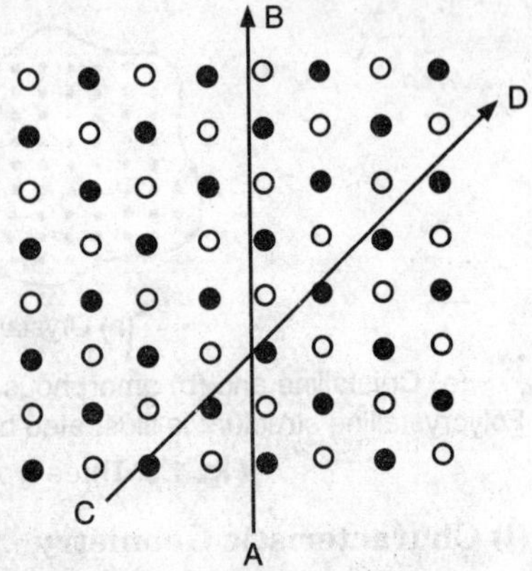

Fig. 1.11. Anisotropic behaviour of crystalline solid

(e) In a similar manner, various other properties like tensile strength, rate of dissolution of the crystal, conductivities, refractive index, etc. would also be different in the two directions Along direction CD, the alternately arranged atoms are in the row while along AB, atoms of one kind are in the row.

(f) In amorphous solids, the atoms are usually arranged at random, and hence, the magnitude of the properties is same in all the directions. Therefore, amorphous solids are isotropic and behave in the way the liquids do.

(g) Therefore, the crystalline solids are known as **true solids** and amorphous solids are called as **supercooled liquids**.

(iv) Cleavage

An irregular fracture occurs if an amorphous solid is tried to cut with a sharp edged tool. While a crystalline solid can be cut easily along certain planes and fracture occurs with flat faces.

(v) Crystal Geometry

(a) Several crystalline solids have characteristics shapes, definite interfacial angle (i.e. the angle between adjacent faces of the crystal), and definite number of elements of symmetries.

(b) In fact, this characteristic geometry for a crystalline solid does not depend on the size of crystals. i.e., NaCl, sugar crystals are cubic, salicylic acid, $HgCl_2$ are needle like, mica is flat.

(c) The small size or big size of the crystal of the substance depends on conditions under which the crystal grows i.e., rapid cooling results in small sized crystals while slow cooling results in bigger sized crystals.

(d) Some crystalline solids have microscopic crystals and are called as microcrystalline solids i.e., crystals of talc, graphite powder, $CaCO_3$ powder etc.

(e) Also, In case of amorphous solids, there is no such characteristic shapes, angles.

Table 1.4. Performance Comparison of Crystalline and Amorphous Solids

S.No.	Crystalline solid	Amorphous solid
1.	Arrangement of atoms/ions/molecules is regular or ordered in three dimensions.	Arrangement of the constituent particles is random.
2.	Crystalline solids are anisotropic.	Amorphous solids are isotropic.

3.	These have sharp melting points.	These have certain melting temperature range
4.	Crystals of a crystalline solid have definite geometry regarding shape, interfacial angle, elements of symmetries.	These solids do not have any definite geometry.
5.	These solids are cut much easily along certain planes and the fracture occurs with flat surfaces	Fracture on cutting is irregular.
6.	These are true solids.	Apparent physical state is solid but they are supercooled liquids.
7.	X-ray or neutron photograph shows regular pattern of spots.	Photograph of these with X-rays shows various irregular spots, highly diffused.
8.	Examples: various salts, various organic solids, solids, silica, diamond, graphite, metals, alloys, quartz, sulphur, silicon etc.	Examples: wood, plastic, coal, rubber, glass, stone, many impure solids etc.

1.6. CONCEPT OF CRYSTALLOGRAPHY

(i) As a matter of fact, crystallography is branch of science that deals with the study of crystalline solids regarding crystal planes, interatomic distances, interfacial angles, packing of constituent particles.

(ii) The internal structure of crystal may be studied with the help of X-ray spectrometer, goniometer, neutron diffractometer etc. Several crystalline solids have visible big sized crystals having flats faces and sharp edges.

(iii) They may be easily identified as crystalline solids, i.e., sugar, NaCl, quartz, silica, mica etc.

(iv) However, some crystalline solids are powders or agglomerates of fine particles and their crystals can be visualised with microscope.

(v) The size of crystals mainly depends upon the conditions under which crystallization occurs. The crystallization of a solid may be done from the molten state of the solid or from solution of the solid or during formation of a solid by chemical reactions. For obtaining crystals of bigger size, the rate of cooling should be very slow and if the crystallization is done from a solution, it should be of lower concentration.

(vi) The substances that exist in the form of very small sized crystals are called as microcrystalline solids i.e., graphite, metals, talc, some organic solids, $CaCO_3$, etc. Whether crystals of a solid are big or small, it has same internal structure.

1.7. WHAT IS SPACE LATTICE? (U.P. Tech, Sem. Exam; 2007-08, 2008-09)

1. Crystal Lattice*

Basically, **crystal lattice** is a higly ordered set of constituent particles in three dimensions. If the constituent particles are represented by points, then the crystal lattice may be regarded as a set of points regularly arranged in thee dimensions.

2. Space Lattice

Also, **space lattice** is an array of points showing how the constituent particles of a solid, are arranged in three dimensional space. Each point in the space lattice is the atom or molecule or ion.

* A crystal lattice is an array of points showing how molecules, atoms or ions are arranged at different sites in three-dimensional space.

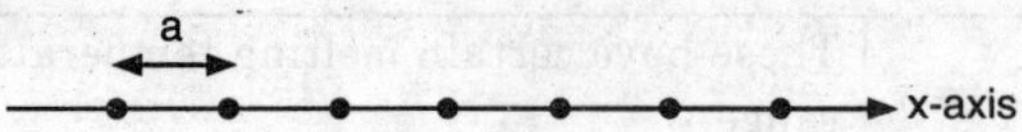

Fig. 1.12. (a) One Dimensional Lattice

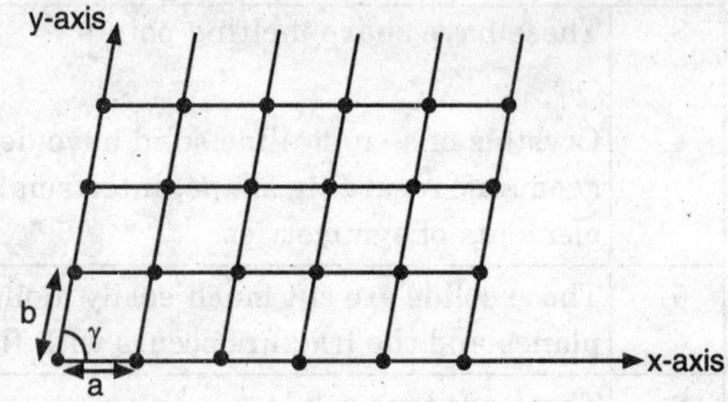

Fig. 1.12. (b) A two dimensional lattice

3. One Dimensional Lattice

Figure 1.12. (a) illustrates an the one dimensional lattice (along one axis) showing how the constituent particles are arranged regularly along X-axis. The distance between any two sucessive points is same, noted by 'a'.

4. Two Dimensional Lattice

Figure 1.12 (b) illustrates a two dimensional showing how the constituent particles are regularly arranged in the plane (X and Y axis). The interatomic distance along X-axis is 'a' and along Y-axis is 'b'. The angle between the X and Y axes is noted by 'λ' (not necessarily 90°).

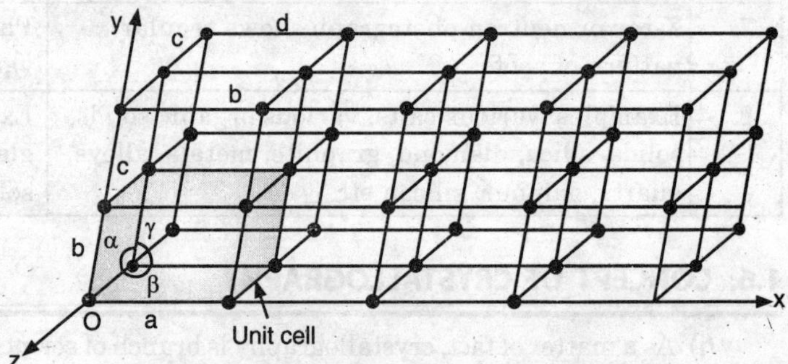

Fig. 1.13. Space lattice and unit cell

a = interatomic distance along x-axis
b = interatomic distance along y-axis
c = interatomic distance along z-axis
α = angle between b and c or y and z axis
β = angle between a and c or x-z axis
γ = angle between a and b or x-y axis

1.8. BASIC ASPECTS OF UNIT CELL*

(U.P.Tech. Sem. Exam, 2003-04, 2008-09)

1. Definition

Basically, unit cell is the smallest three dimensional unit of a space lattice which forms crystal on repeating in all directions. It may also be defined as the minimum volume cell of a space lattice. It is the fundamental building unit of a lattice or crystal. A unit cell of a crystal represents all the features of entire crystal.

2. Length Parameters

Let us consider figure 1.14(a). The unit cell lengths along X, Y and Z axis are noted by a, b, c respectively a, b, c are also known as **length parameters** of a unit cell. A unit cell also has angle parameters as α (Y-Z axis), β (X-Z axis) and γ (X-Y axis). These angles are not necessarily at right angle but can be more or lesser than 90°. Further, they all may be equal (90°) or two equal and third

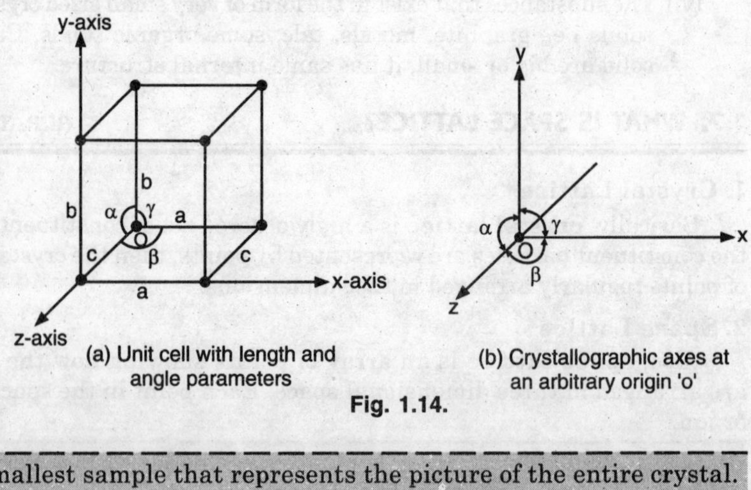

(a) Unit cell with length and angle parameters

(b) Crystallographic axes at an arbitrary origin 'o'

Fig. 1.14.

* The unit cell, infact, is the smallest sample that represents the picture of the entire crystal.

different or all the three angles may be different. The angles are characteristic of all crystals of the same solid. Goniometer helps measure them, at certain arbitary fixed position or origin. The length parameters a, b, c may be all different in dimensions, certain two same but third different or all the three same in length. These length parameters have characteristics values for crystals of a solid.

1.9. VARIOUS LAWS OF CRYSTALLOGRAPHY*

A crystal of a solid has definite geometrical shape with flat faces and sharp edges. Study of crystals may be concluded in the form of three fundamental laws. All crystals (i.e., small or big) of the same substance, are similar regarding the following three laws:

1. Law of constant interfacial angles.
2. Law of constant elements of symmetries.
3. Law of rational indices.

1. Law of constant interfacial angle

(i) A crystalline solid may crystallize under different conditions to produce crystals varying in size.

(ii) However, the angle between the two corresponding faces is always same, for all crystal of the same solid. i.e., the interfacial angle for all crystals of NaCl is 90°.

(iii) Angle between two intersecting faces of a crystal, is called as interfacial angle.

(iv) The law states that the interfacial angle between the two corresponding faces of crystals is same and independent of the size of crystals of that solid substance as shown in figures 1.15(a) and (b).

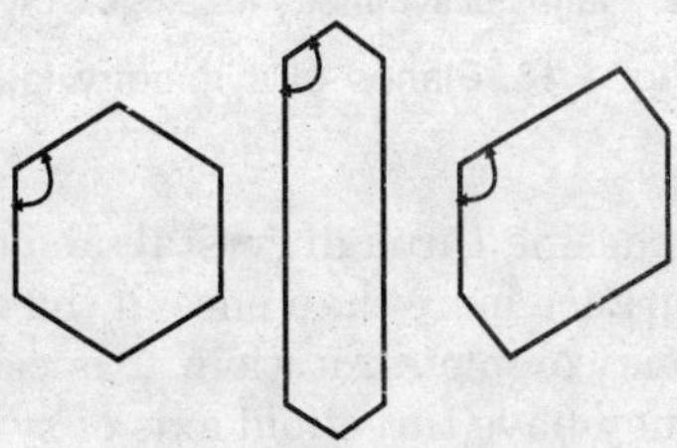

(a) Interfacial angle of crystals of a solid (θ – 120°) e.g. graphite

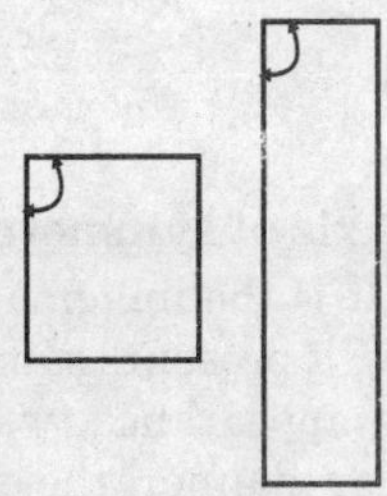

(b) Interfacial angle of crystals of solid (θ – 120°) e.g. sugar, NaCl

Fig. 1.15.

2. Law of constant elements of symmetries

The symmetries in crystal may be of following three types:

(i) Planes of symmetry
(ii) Axis of symmetry
(iii) Centre of symmetry.

The law states that all crystals of a substance possess same number of elements of symmetries. i.e., for one crystal if planes of symmetry are 9, then for other crystals of the same substance, the planes of symmetry are 9. In a similar manner, if one crystal of a solid has 7 axis of symmetry, then for all other crystals of the same solid, the axis of symmetry are 7. Let us consider cubic crystals and discuss the elements of symmetries for it.

(i) Plane of symmetry

It is the imaginary plane passing through a crystal which divides it into two parts such that one part is exact mirror image of the other. A cubic crystal has three rectangular planes of symmetry and six diagonal planes of symmetry as shown in figures 1.16(a), b and (c).

* The crystal may be considered to consist of infinite number of unit cells.

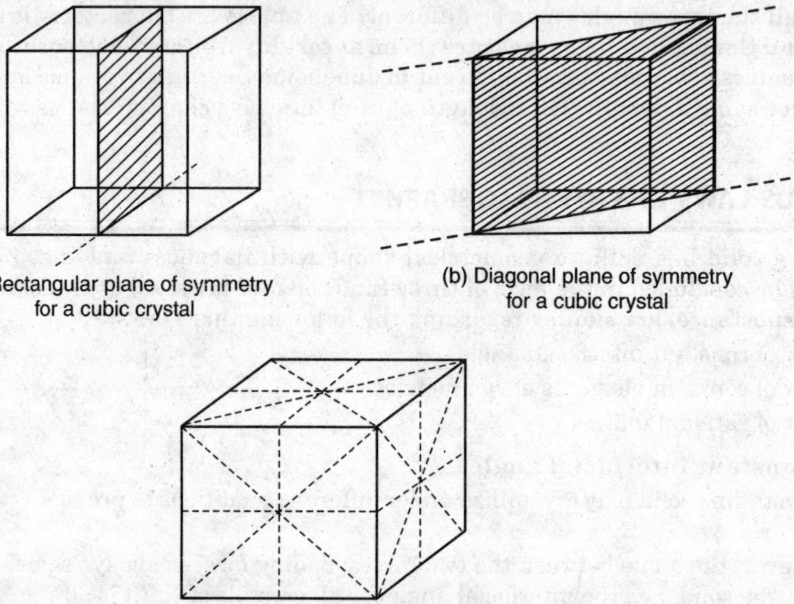

(a) Rectangular plane of symmetry for a cubic crystal

(b) Diagonal plane of symmetry for a cubic crystal

(c) Planes of symmetry for a cubic crystal

Fig. 1.16. Planes of symmetry for cubic crystal

(ii) Axis of symmetry

It is the imaginary line passing through crystal; about which if the crystal is rotated by 360° angle, it presents same appearance more than once. if the crystal is rotated about a line and shows same appearance twice, in one complete rotation, it is called as two-fold axis of symmetry. In a similar manner, the crystal may have three-fold axis of symmetry, four-fold axis of symmetry, six-fold axis of symmetry. In general, during one complete rotation of crystal about a line, if same appearance is repeated at angle $\frac{360°}{n}$, then the axis is called as n-fold axis of symmetry. For a cubic crystal, there are three-four-fold axis of symmetry, six two-fold axes of symmetry and four-three-fold axis of symmetry as shown figures 1.17(a), (b) and (c).

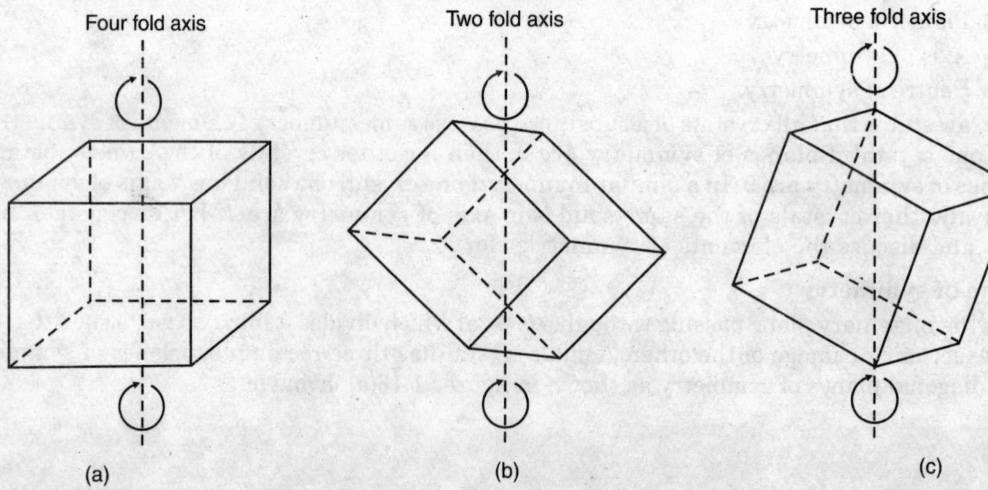

Fig. 1.17. Axis of symmetry for cubic crystal

(iii) Centre of symmetry

It is the imaginary point inside the crystal such that if a line is drawn through the point, it cuts the opposite faces of the crystal at equal distances. It should be noted here tht crystal of any substance has only one centre of symmetry. Refer figure 1.18

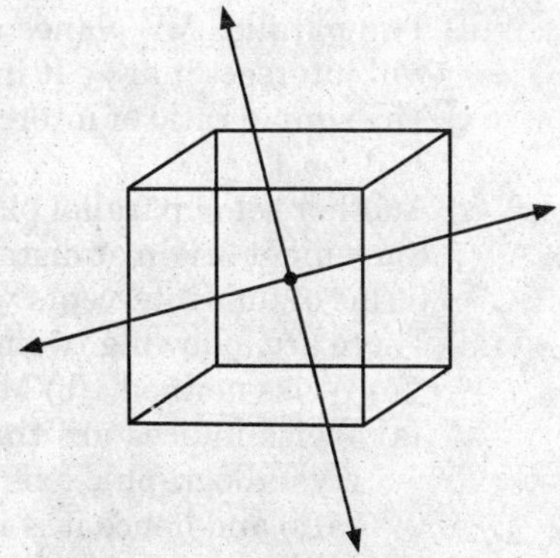

Fig. 1.18. Centre of symmetry for cubic crystal

Thus, the total elements of symmetries for a cubic crystal are 23 as under.

Planes of symmetry (9)	: 03 rectangular 06 diagonal
Axis of symmetry (13)	: 06 two-fold axis 03 four-fold axis 04 three-fold axis
Centre of symmetry (1)	: $\frac{01}{23}$

For hexagonal crystals, the elements of symmetry are as under,

Planes of symmetry (7)	: 04 rectangular planes : 03 diagonal planes
Axis of symmetry (7)	: 01 six fold axis 06 two-fold axis
Centre of symmetry	: $\frac{01}{15}$

3. Law of rational indices (Hauy's law)

(i) A space lattice contains a set of parallel horizontal planes, a set of parallel diagonal planes, a set of vertical planes, sets of inclined parallel planes.

(ii) Such planes contain the two dimensional regularly arranged lattice points and they are known as lattice planes.

(iii) The sets of parallel lattice planes can be designated by rational numbers.

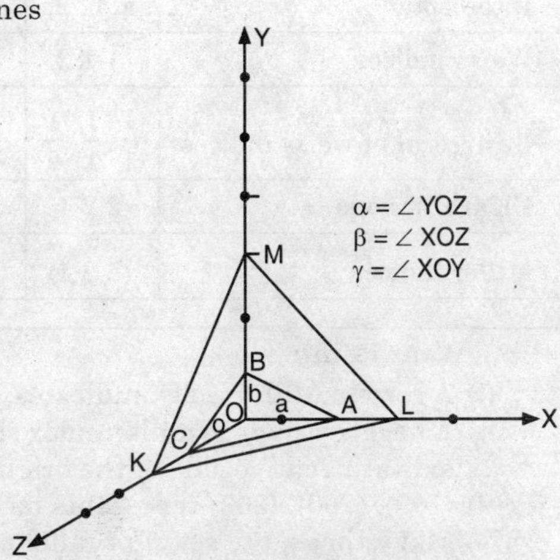

Fig. 1.19. Intersection of a lattice plane on crystallographic axes

(iv) A lattice plane cuts the three crystallographic axis at certain distances from the conveniently selected origin (sometimes the plane may be parallel to certain axis and it is supposed to intersect the axis at infinite distance).

(v) The intercepts of a plane on X, Y and Z axis can expressed in terms of integral multiples of unit intercepts on the axis. The law of rational indices states that the ratio of intercepts of a lattice plane on the crystllographic axis, can be expressed in terms of simple rational numbers.

(vi) Plane ABC intersects the crystallographic axis at distances 2a, b, 2c and the simple ratio of intercepts is 2 : 1 : 2, ABC plane is 212.

(vii) In figure 1.19, plane LMK intersects the crystallographic axis at distances 3a, 3b, 3c from the origin 'O'. The simple ratio of the intercepts is 1 : 1 : 1.

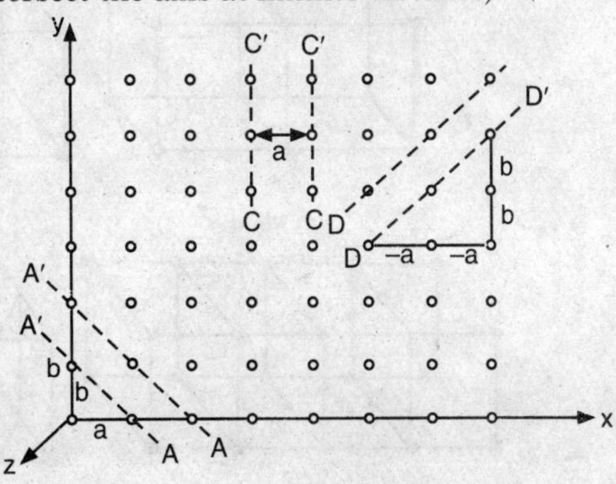

Fig. 1.20. Few sets of parallel planes in a two dimensional lattice

(viii) Two parallel AA′ planes are shown in figure 1.20. These planes are parallel to Z-axis and will intersect it at ∞. It intersects the X and Y axis at distances a and b.

(ix) The simple ratio of intercepts of the plane AA′ in terms of unit intercepts will be a : b : ∞ = 1 : 1 : ∞.

(x) Another set of parallel planes DD′ are shown. The DD′ intersects the X and Y axis from the convenient origin at distances – 2a and 2b. Being parallel to Z-axis, the ratio of intercept in terms of unit intercepts will be – 2a : 2b : ∞ = – 2 : 2 : ∞.

(xi) There are following two methods of designating or indexing the lattice planes in crystals:
(a) Weiss method, (b) Miller method.

(a) Weiss indices are the ratio of coefficients of unit intercepts of a lattice plane on the crystallographic axis from an arbitrary origin. However, sometimes ∞ appears in the ratio and hence it is incovenient.

(b) Miller indices of a lattice plane are the simple ratio of reciprocals of Weiss indices i.e., the simple ratio of reciprocals of coefficients of unit intercepts of a lattice plane on the X, Y, Z axis. Miller indices are written in parenthesis e.g. (121), (110), (200) ($\bar{1}$01). The digits are known in hkl older.

Following table shows the calculations of Miller indices.

Intercepts	a, b, 2c	– a, b, ∞	2a, 2b, 2c	∞, b, c	a, – 2b, c	3a, – b, c
Weiss indices	1 1 2	– 1 1 ∞	2 2 2	∞ 1 1	1 – 2 1	3 – 1 1
Reciprocal of weiss indices	$\frac{1}{1}\ \frac{1}{1}\ \frac{1}{2}$	$\frac{1}{-1}\ \frac{1}{1}\ \frac{1}{\infty}$	$\frac{1}{2}\ \frac{1}{2}\ \frac{1}{2}$	$\frac{1}{\infty}\ \frac{1}{1}\ \frac{1}{1}$	$\frac{1}{1}\ \frac{1}{-2}\ \frac{1}{1}$	$\frac{1}{3}\ \frac{1}{-1}\ \frac{1}{1}$
Clear fractions	2 2 1	– 1 1 0	1 1 1	0 1 1	2 – 1 2	1 – 3 3
Miller indices	(2 2 1)	($\bar{1}$ 1 0)	(1 1 1)	(0 1 1)	(2 $\bar{1}$ 2)	(1 $\bar{3}$ 3)

Important Point:

(i) A zero in Miller index indicates, the plane is parallel to corresponding axis.

(ii) A negative digit in miller index shows that the plane (e.g. $\bar{h}$ k l) cuts the corresponding axis at negative side of the origin.

(iii) (100), (200), (300) represents parallel planes.

(iv) Many times, the set of parallel planes have same Miller index or designation.

(v) (110), (1 $\bar{1}$ 0), ($\bar{1}$ 1 0) are the parallel planes. Refer figure 1.21.

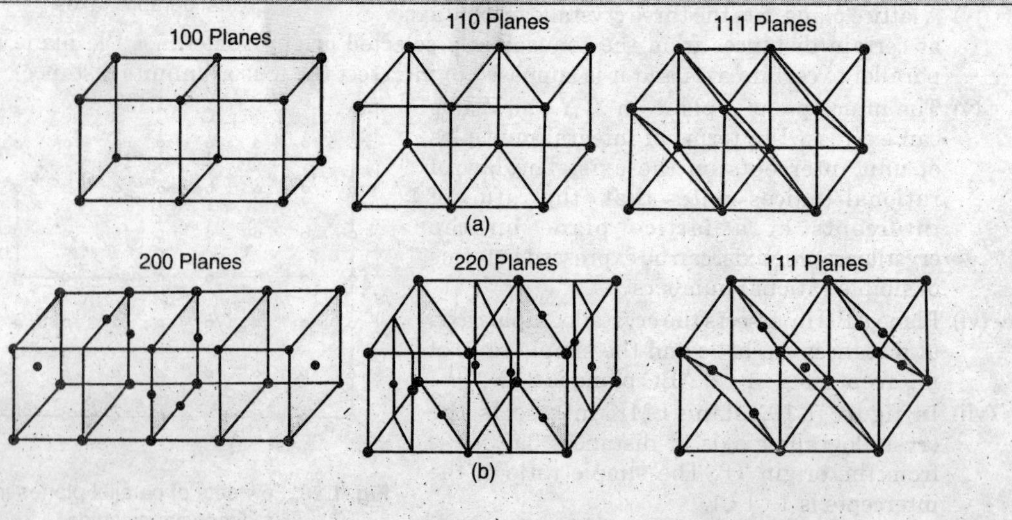

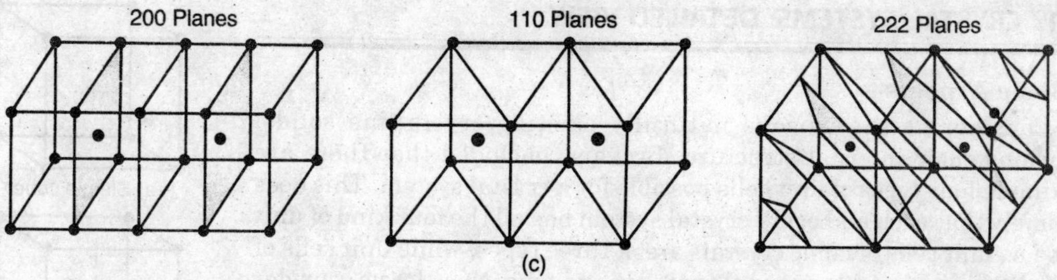

Fig. 1.21. Designation of planes of cubic lattices (with two unit cells)

1.10. CRYSTAL SYSTEMS

(U.P. Tech. Sem. Exam. 2005-06, 2006-07)

1. Formation of Seven Crystal Lattices

A unit cell in a space lattice is characterised by edges a, b, c and angles α, β, γ. Theses are known as lattice parameters. From these parameters, a total of following seven crystal lattices can be formed.

(i) cubic
(ii) orthorhombic
(iii) tetragonal
(iv) monoclinic
(v) triclinic
(vi) hexagonal
(vii) trigonal

These are called as the seven **basic crystal systems**. From the geometrical considerations and the different combinations of elements of symmetry, there are 32 crystals systems possible. But some of the systems have been grouped together so that we have only seven basic crystal systems. These systems with their maximum number of planes of symmetry and axis of symmetry with examples belonging to them are given in following table. (Each crystal system has one centre of symmetry).

S. No.	Crystal system	Parameters of unit cell		Elements of symmetry	Examples
		lengths	angles		
1.	Cubic	$a = b = c$	$\alpha = \beta = \gamma = 90°$	9 planes, 13 axis	NaCl, KCl, ZnS, Fe diamond, alums, Pb.
2.	Orthorhombic	$a \neq b \neq c$	$\alpha = \beta = \gamma = 90°$	3 planes, 3 axis	KNO_3, $BaSO_4$, Rhombic sulphur
3.	Tetragonal	$a = b \neq c$	$\alpha = \beta = \gamma = 90°$	5 planes, 5 axis	Sn, SiO_2, SnO_2
4.	Monoclinic	$a \neq b \neq c$	$\alpha = \gamma = 90°$ $\beta \neq 90°$	1 planes, 1 axis	Na_2, SO_4, 10 H_2 O, monoclinic sulphur
5.	Triclinic	$a \neq b \neq c$	$\alpha \neq \beta \neq \gamma \neq 90°$	no planes, no axis	$CuSO_4$, $5H_2O$ $K_2Cr_2O_7$
6.	Hexagonal	$a = b \neq c$	$\alpha = \beta = 90°$ $\gamma = 120°$	7 planes, 7 axis	graphite, ice, Zn, Cd
7.	Trigonal	$a = b = c$	$\alpha = \gamma = 90°$ $\beta \neq 90°$	7 planes, 7 axis	quartz, Sb, NaNO3, Calcite

1.11. CRYSTAL SYSTEMS: DETAILED VIEW

1. Basic Aspects

Considering the various naturally existing crystalline solids, regarding their internal structure, Bravais concluded that there are maximum four types of unit cells possible for a crystal system. This does not imply that each and every crystal system has all the four kind of unit cells i.e., unit cells of cubic crystals are of three types, while unit cells of triclinic, trigonal and hexagonal is of only one type. Now, let us consider how lattice point in a unit cell of cubic crystal are present according to Bravais,

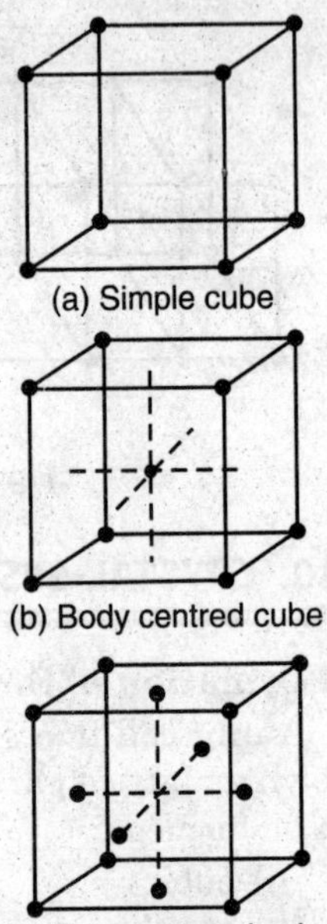

Fig. 1.22. Unit cells of cubic lattice

(i) Simple or primitive cube (P)

In this type of cubic unit cell, the lattice points are at 8 corners of the cube.

(ii) Face centered cubic cell (F)

In this type of cubic unit cell, the lattice points are at 8 corners as well as at the centres of 6 faces of the cube.

(iii) Body centered cubic cell (I)

In this unit cell lattice points are at the 8 corners of the cube and one at the centre.

2. Seven basic Crystals Systems and Fourteen Bravais Lattices

Although the unit cells of a cubic lattice, are of three types, the cubic lattice of a solid possesses just one type of unit cell i.e., copper, Al cubic crystal systems have FCC unit cells, sodium metal cubic crystal system has SC type of unit cells, molybdenum, iron, chromium cubic crystal systems have BCC type of unit cells. The seven basic crystals systems and the 14 Bravais lattices may be listed as under:

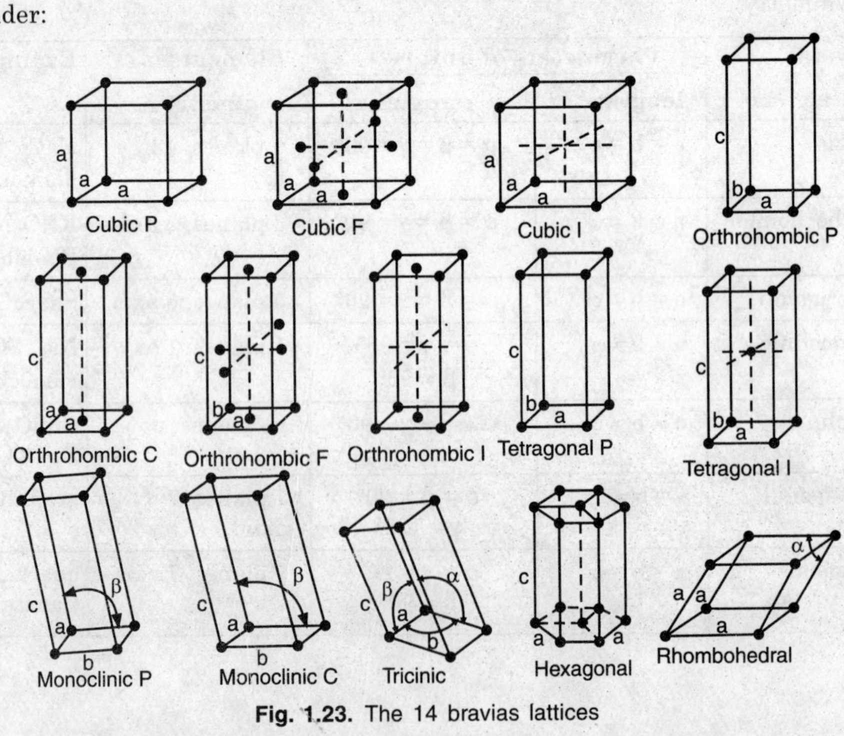

Fig. 1.23. The 14 bravias lattices

Crystal system	Bravais Lattice	Symbol
1. Cubic	1. Primitive (simple) 2. Face centered 3. Body centered	P F I
2. Orthorhombic	4. Primitive 5. Face centered 6. Body centered 7. End centered	P F I C
3. Tetrogonal	8. Primitive 9. Body centered	P I
4. Monoclinic	10. Primitive 11. End centered	P C
5. Triclinic	12. Primitive	P
6. Hexagonal	13. Primitive	P
7. Trigonal	14. Primitive	P

1.12. NUMBER OF ATOMS PER UNIT CELL IN A CUBIC LATTICE

1. Simple-cubic cell (SC)

(i) In SC unit cell, an atom situated at any corner of the cube, is shared by eight adjoining cells. Thus it belongs 1/8 only to each unit cell.

(ii) There are 8 corners to the cube.

(iii) Therefore, the number of atoms per unit cell of the SC crystal would be $= \frac{1}{8} \times 8 = 1$.

2. Face centered cubic cell (FCC)

(i) Unit cell of FCC has 8 atoms at the corners of cube and one each at the centre of each of the six faces of the cube.

(ii) Each corner atom is common to 8 adjoining unit cells and belongs only 1/8 to each cell.

(iii) Further, each atom at the centre of the face of cube is common to two unit cells and belongs only 1/2 to each unit cell.

Therefore, Number of atoms per unit cell of FCC will be $= \left(\frac{1}{8} \times 8\right) + \left(\frac{1}{2} \times 6\right) = 1 + 3 = 4$.

3. Body centered cubic cell (BCC)

(i) Unit cell of BCC has 8 atoms at the 8 corners of the cube and one atom at the centre of the unit cell.

(ii) Each corner atom belongs only 1/8 to one cell as the atom at the corner is common to 8 adjoining unit cells.

(iii) However, the atom at the centre of the cube completely belongs to one BCC unit cell.

Hence, number of atoms per unit cell in BCC crystal will be $= \left(\frac{1}{8} \times 8\right) + 1 = 2$.

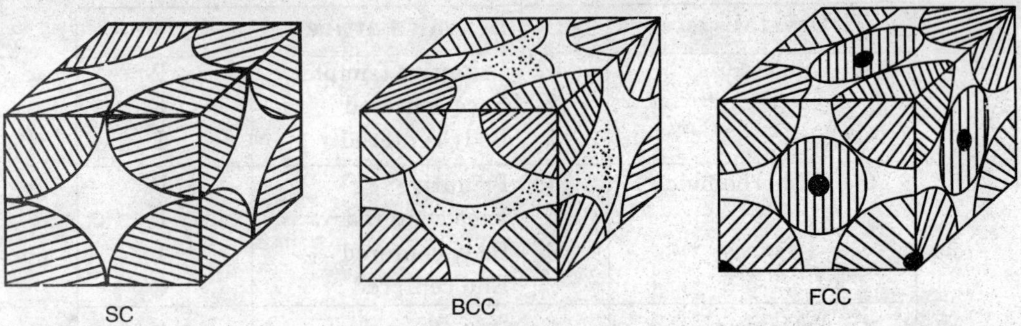

Fig. 1.24. Illustrations of Atoms per unit cell in cubic crystals

1.13. LATTICE PLANES OR NET PLANES

1. Lattice Planes

Basically, in a space lattice, all the points may be included in a set of parallel and equally spaced planes, known as **lattice planes.**

2. Separation (Spacing) between two Successive Lattice Planes

The separation between two successive lattice planes of a cubic, tetragonal and orthorhombic crystal system, in which edges are mutually perpendicular, may be expressed by following expression:

$$d_{hkl} = \left[\frac{h^2}{a^2} + \frac{k^2}{b^2} + \frac{l^2}{c^2}\right]^{-1/2}$$

Here, d_{hkl} = separation between two successive planes

and h, k, l = miller indices.

For a cubic lattice a = b = c, and the above expression will reduce to

$$d_{hkl} = \left[\frac{h^2}{a^2} + \frac{k^2}{b^2} + \frac{l^2}{c^2}\right]^{-1/2} = \left[\frac{h^2 + k^2 + l^2}{a^2}\right]^{-1/2}$$

or
$$d_{hkl} = \frac{a}{[h^2 + k^2 + l^2]^{1/2}} \quad ...(1.1)$$

here, a = unit cell edge length

3. Interplanar Spacing for Simple Cubic Lattice

The different planes of a simple cubic (sc) lattice have been shown in figure 1.25(a). The separation between two successive lattice planes may be calculated as under:

(i) Spacing between two successive 100 lattice planes

In simple cubic lattice, 100 planes cut the X-axis and the parallel to the Y and Z-axis as shown in figure 1.25 (a). The perpendicular distance between two successive 100 lattice planes may be calculated by substituting h = 1, k, = 0, 1 = 0 in equation (1.1).

Hence, we write,
$$d_{100} = \frac{a}{(1^2 + 0^2 + 0^2)^{1/2}} = a \quad ...(1.2)$$

(ii) Spacing between two successive 110 lattice planes

The 110 planes cut obliquely across the X and Y-axis, however, cut the Z-axis at infinity i.e. parallel to Z-axis as shown in figure 1.25(b). The perpendicular distance between two successive (110) planes may be calculated by substituting h = k = 1 and 1 = 0 in equation (1.1).

Hence, we write, $$d_{110} = \frac{a}{(1^2 + 1^2 + 0^2)^{1/2}} \quad \text{...(1.3)}$$

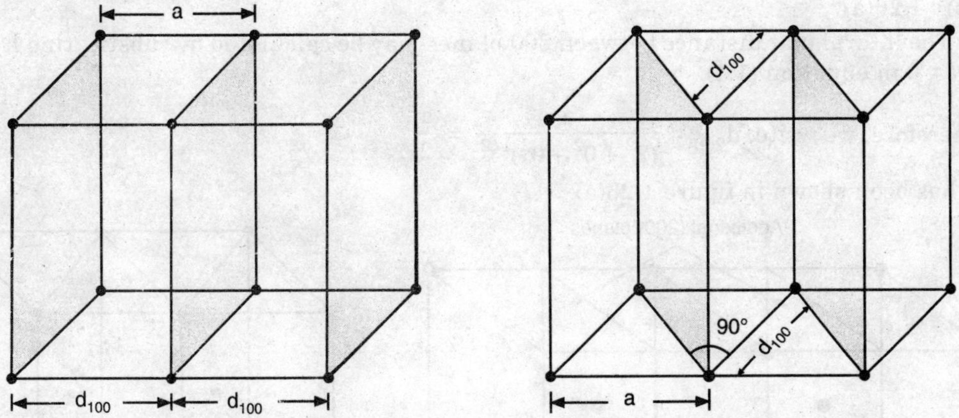

(a) The shaded lattice planes are (100) planes.in the simple cubic lattice (i.e., point are situated at the corners)

(b) The shaded lattice planes are (110) in the simple cubic lattice and the points are situated at the corners

Fig. 1.25.

(iii) Spacing between two successive 111 lattice planes

As illustrated in figure 1.25(c) lattice planess 111 cut all the three axis (X, Y, Z) obliquely and intercept each axis at the same distance from the origin. The interplanar distance between 111 planes may be calculated by substituting h = 1, k = 1 and l = 1 in equation (1.1).

Hence, we write, $$d_{111} = \frac{a}{(1^2 + 1^2 + 1^2)^{1/2}} = \frac{a}{\sqrt{3}} \quad \text{...(1.4)}$$

Therefore, the ratio of separation between successive (100), (110) and (111) lattice planes will be as under:

$$d_{100} : d_{110} : d_{111} = a : \frac{a}{\sqrt{2}} : \frac{a}{\sqrt{3}}$$

or $$\frac{1}{d_{100}} : \frac{1}{d_{110}} : \frac{1}{d_{111}} = 1 : \sqrt{2} : \sqrt{3}$$

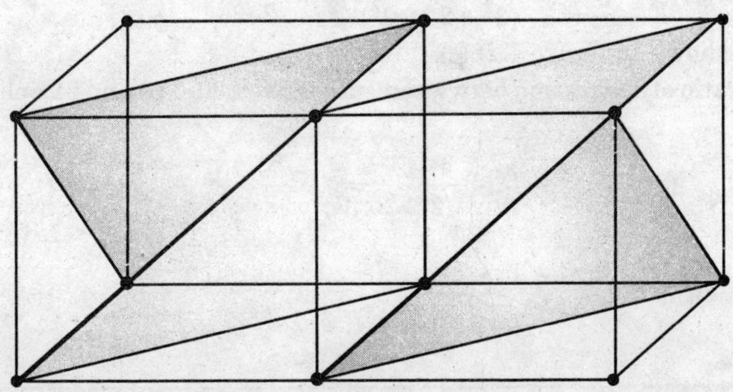

Fig. 1.25 (c). The shaded lattice planes are (111) planes and interplanar distance is d_{111} = a in simple cubic lattice. The points are situated at the corners

4. Body centered cubic lattice*

In this crystal, one unit is situated at each corner and one unit at the centre of the cube. Further, this lattice, additional planes may be passed through the structural units as depicted in figure 1.26(a).

(i) The interplanar distance between 200 planes may be calculated by substituting h = 2, k = 0 and l = 0 in equation (1.1).

$$\text{Therefore, we write, } d_{200} = \frac{a}{(2^2 + 0^2 + 0^2)^{1/2}} = \frac{a}{2}$$

It has been shown in figure 1.26(a).

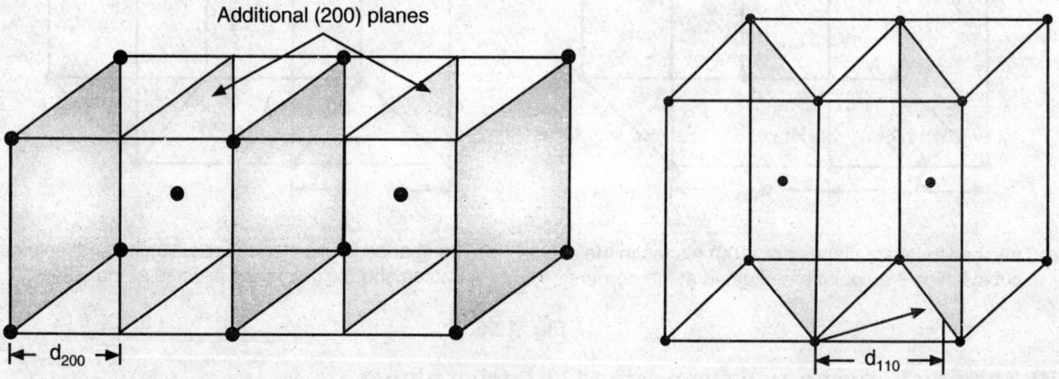

(a) Illustrating (200) planes and d_{200} between two successive planes in body-centered cubic lattice. Here, dark shaded planes are (200) and dotted planes are additional (200) planes.

(b) Illustrating (110) planes in body centered cubic lattice and interplanar distance (d_{110})

Fig. 1.26.

(ii) Also, Interplanar spacing between 110 planes may be calculated by substituting h = 1, k = 1, and l = 0 in equation (1.1).

$$\text{Hence, we write,} \quad d_{110} = \frac{a}{(1^2 + 1^2 + 0^2)^{1/2}} = \frac{a}{\sqrt{2}}$$

It has been shown in figure 1.26(b).

(iii) Interplanar spacing between 222 planes may be calculated by substituting h = 2, k = 2 and l = 2 in equation (1.1).

$$\text{Hence, we write,} \quad d_{222} = \frac{a}{(2^2 + 2^2 + 2^2)^{1/2}} = \frac{a}{2\sqrt{3}}$$

This has been shown in figure 1.26(c)

Therefore the ratio of separation betwween successive 200, 110 and 222 planes may be written as under:

$$d_{200} : d_{110} : d_{222} = \frac{a}{2} : \frac{a}{\sqrt{2}} : \frac{a}{2\sqrt{3}}$$

or $$\frac{1}{d_{200}} : \frac{1}{d_{110}} : \frac{1}{d_{222}} = 1 : \frac{1}{\sqrt{2}} : \sqrt{3}$$

* The coordination number of any given sphere in a crystalline structure is the number of other spheres by which it is immediately surrounded.

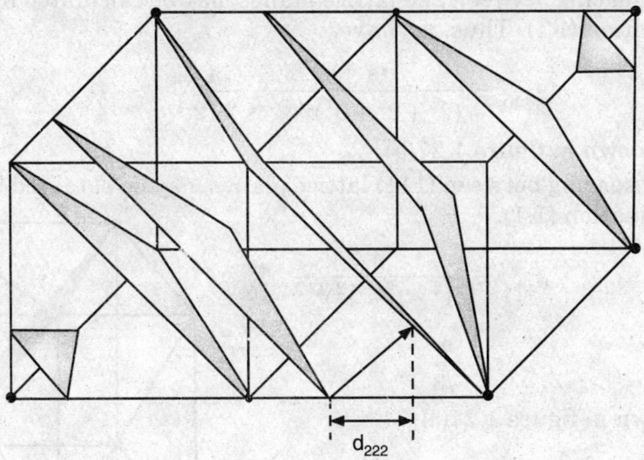

Fig. 1.26(c). Illustration of showing (222) planes and interplanar distance d_{222} in body centered cubic lattice

5. Face centered cubic lattice

In this crystal, one unit is situated at each corner and one at the centre of each of faces. Also, additional planes can be passed through the black dots as depicted in figure 1.27(a). They pass through the centre of front and back faces vertically.

(i) The interplanar spacing between 200 lattice planes may be calculated by substituting h = 2, k = 0 and l = 0 in the equation (1.1). Hence, we can write,

$$d_{200} = \frac{a}{(2^2 + 0^2 + 0^2)^{1/2}} = \frac{a}{2}$$

It has been shown in figure 1.27(a).

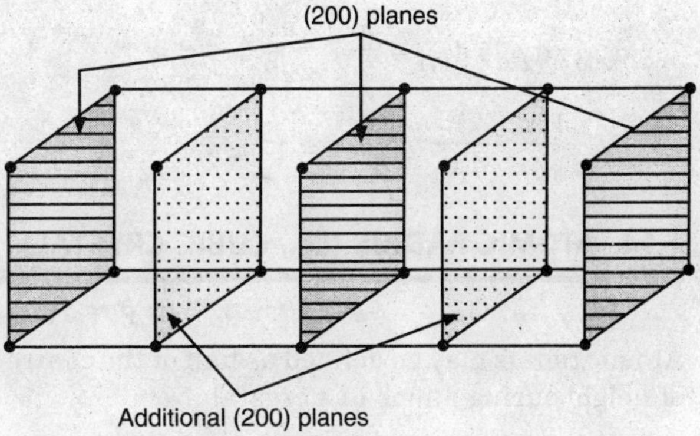

Fig. 1.27(a) The dark shaded lattice planes are (200) planes in face centered cubic lattice. The dotted lattice planes are additional (200) planes. Here, d_{200} = a/2

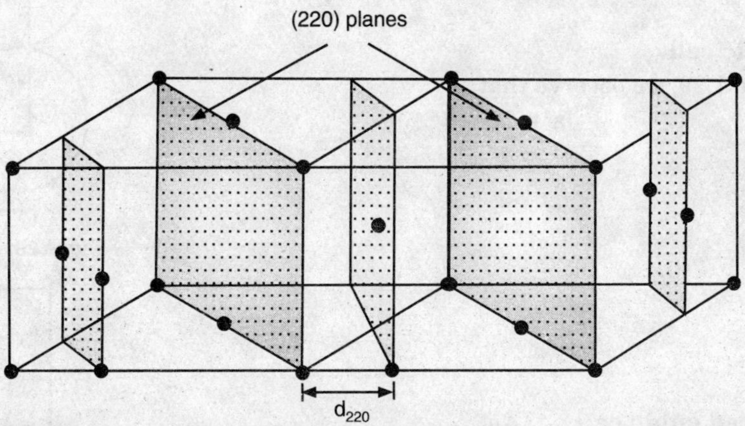

Fig. 1.27(b). The dark shaded lattice planes are (220) planes and dotted lattice planes are additional (220) planes. The $d_{220} = a/2\sqrt{2}$

(ii) Interplanar spacing between 220 lattice planes may be calculated by substituting h =2, k = 2 and l = 0 in equation (1.1). Thus, we have,

$$d_{220} = \frac{a}{(2^2+2^2+0^2)^{1/2}} = \frac{a}{2\sqrt{2}}$$

This has been shown in figure 1.27(b).

(iii) Interplanar spacing between (111) lattice planes may be obtained by substituting h = 1, k = 1 and *l* = 1 in equation (1.1).

Hence, we get,
$$d_{111} = \frac{a}{(1^2+1^2+1^2)^{1/2}}$$
$$= \frac{a}{\sqrt{3}}$$

It has been shown in figure 1.27(c)

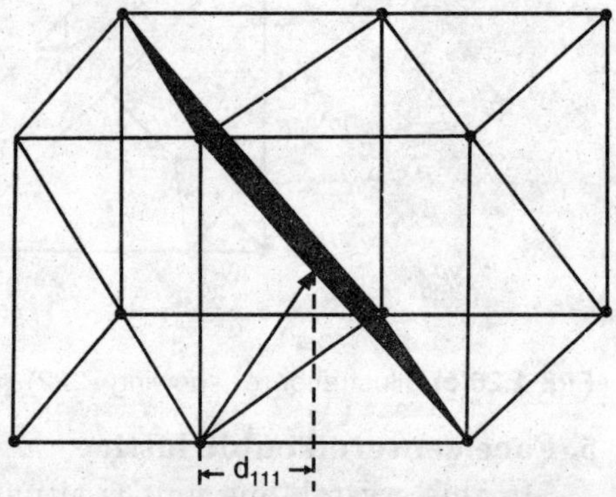

Fig. 1.27(c). Showing the (111) planes and d_{111} between two successive. Planes in face centered cubic lattice. The $d_{111} = \frac{a}{\sqrt{3}}$

Therefore, the ratio of separation between successive (200), (220) and (111) planes would be,

$$d_{200} : d_{220} : d_{111} = \frac{a}{2} : \frac{a}{2\sqrt{2}} : \frac{a}{\sqrt{3}}$$

$$\text{or} \quad \frac{1}{d_{200}} : \frac{1}{d_{220}} ; \frac{1}{d_{111}} = 1 : \sqrt{2} : \frac{\sqrt{3}}{2}$$

1.14. ATOMIC RADIUS (i.e., CUBIC CRYSTAL)

(U.P. Tech, Sem. Exam; 2009-10)

Atomic radius may be defined as half of the centre to centre distance of neighbouring atoms in a crystal.

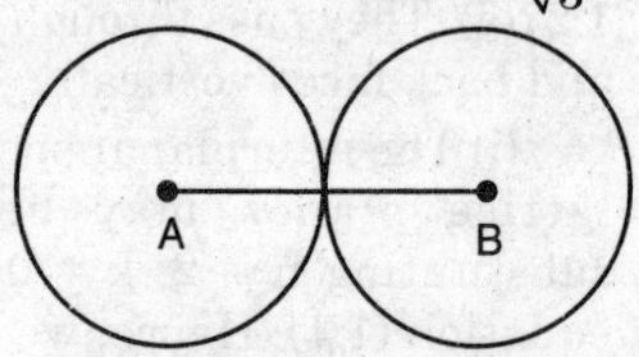

Fig. 1.28. Atomic radius

Thus, AB = a and atomic radius = $r = \frac{a}{2}$

The atomic radius may be expressed in terms of edge length of cubic unit cell, 'a'.

1. Simple-cubic cell

From figure 1.29, we observe that

$$a = 2r$$

$$\text{or} \quad r = \frac{a}{2}$$

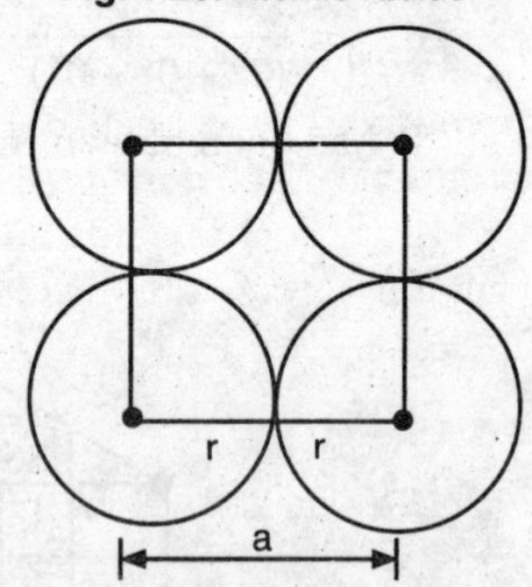

Fig. 1.29. Atomic radius in SC

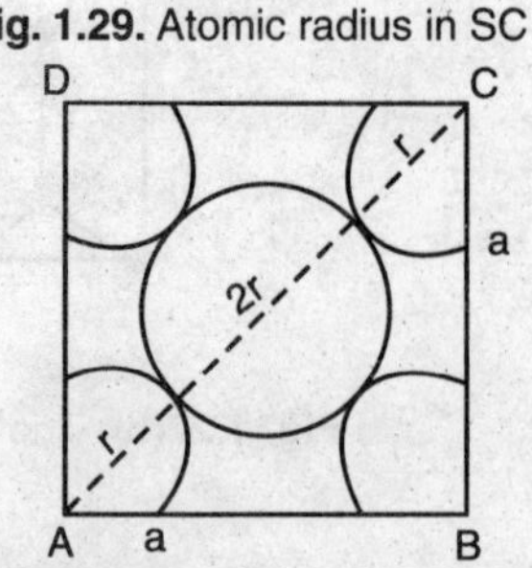

Fig. 1.30. Atomic radius in FCC

2. Face centered cubic cell

In triangle ABC, we have

$$BC^2 + AB^2 = AC^2$$

$$\text{or} \quad a^2 + a^2 = (4r)^2$$

or $$2a^2 = 16r^2$$

or $$r = \frac{a\sqrt{2}}{4} \text{ (FCC)}$$

3. Body centered cubic cell

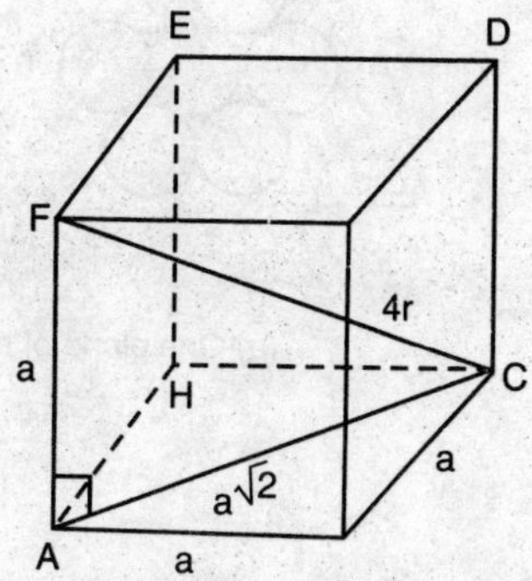

Fig. 1.31. Atomic radius in BCC

From figure 1.31, diagonal AC of ABCH square is $a\sqrt{2}$.

Now, let us consider right angled triangle FAC,

We have, $$FC^2 = FA^2 + AC^2$$

or $$(4r)^2 = a^2 + (a\sqrt{2})^2$$

or $$16\, r^2 = 3a^2$$

or $$r = \frac{a\sqrt{3}}{4}$$

1.15. PACKING OF ATOMS IN METALS

(U.P. Tech, Sem. Exam; 2009-10)

1. Basic Concept

Metal atoms may be considered as hard spheres. The metal atoms pack in such a manner that each atom has maximum possible number of neighbouring atoms. Further, in the closest packing, maximum available space can be filled by metal atoms.

2. Three Different Ways

But, depending upon the nature of metal, the packing of metal atoms is in different three ways:

(i) Hexagonal close packing (HCP)

(ii) Cubic close packing (CCP or FCC)

(iii) Body centered close packing (BCP)

All the metal atoms in a metal, are identical and the closest packing involves one atom surrounded by six other atoms of same size, in a plane, as in figure 1.32(a). Let this layer be noted by 'A' as first layer. Now the second layer of atoms must be placed below the first year such that the atoms in the second layer fit into the hollows between every three atoms. Let the second layer be called as 'B'.

The third layer of metal atoms arranges in two different ways.

(i) HCP

Let us consider the three atoms in layer B. The atoms in the third layer below B arrange in such a manner that each atom in layer B is fitted in the hollows of every three atoms in third layer. Also, the third layer is similar to the first layer. Hence, this type of packing form ABABABAB ... alternate similar layers and it is called as **hexagonal close packing (HCP)**.

In HCP, each atom is surrounded by twelve atoms of same size i.e., 6 in the same plane; 3 in upper plane and 3 in lower plane. Therefore, co-ordination number is 12. Further it is the closest packing and the packing factor is 0.74 or 74% of the available space is filled with atoms.

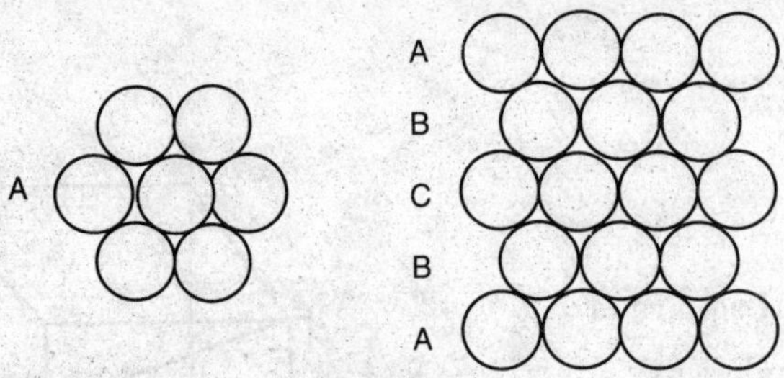

(a) One plane of metal atoms

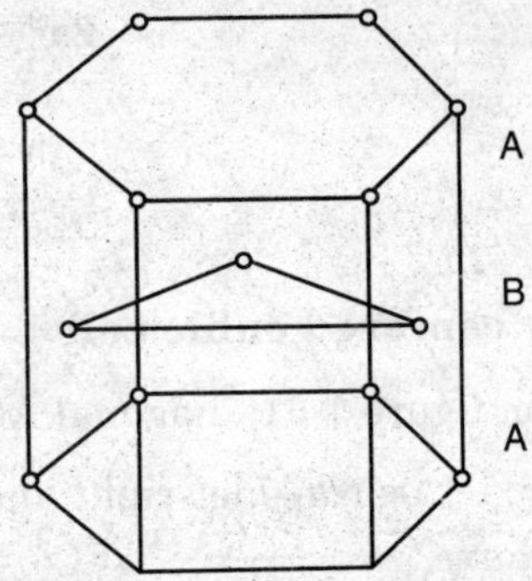

(b) Hexagonal close packing

Fig. 1.32.

(ii) CCP

Let us, consider the three atoms in layer B. The atoms in the third layer below layer B, arrange in such a manner that the every third layer atom fits in the valley between every two atoms of layer B. Let the third layer be 'C'. The fourth layer would be arrangement of atoms below layer C and it is similar to layer A. Hence, the layer repeat in the pattern ABCABCABC... This way of packing of atoms is known as **cubic close packing (CCP or FCC).***

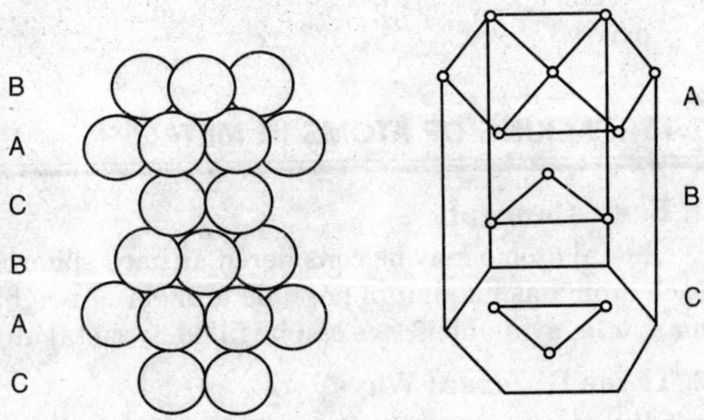

Fig. 1.33. Cubic close packing (CCP)

In the CCP or FCC close packing, each atom is surrounded by touching 12 atoms of same size i.e., 6 in the same plane, 3 in upper and 3 in lower plane. Therefore, the coordination number is 12. This is also closest packing with packing factor of 0.74.

(a) HCP is observed in : Zn, Be, Cd, Mg, etc.

(b) CCP is observed in : Al, Cu, Ag, Au, Co, Ni, Pd, etc.

(c) BCP is observed in : alkali metals, iron, Cr, Ba, V, W, etc.

(iii) Body centered cubic packing (BCP)

The body centered cubic packing is not closest packing. An atom is surrounded by eight other atoms in slightly open manner, and it is not directly touching them. Let this layer of atoms be known as A-layer. The second layer of atoms (B) may be placed on top of the first layer so that each atom of the second layer is in contact with 4 atoms of the layer below it. Successive building of third layer will be exactly like the first layer.

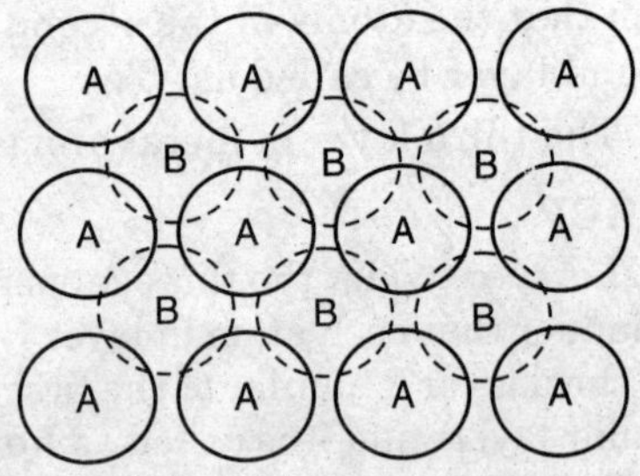

Fig. 1.34. BCP packing of metal atoms

A = Layer of atoms in one plane

B = Layer of atoms in another plane below A layer.

Out of the over 75 metals, about 60% metals have closest packing of HCP or CCP type.

---

* The basic lattice structure for many important semiconductors is the diamond lattice, which is characteristic of Si and Ge.

Important Point: It must be noted that some metals show packing of more than one type e.g., Mn, Fe, Cr possess BCP and CCP; calcium can have all the three types of packing of atoms. In the case of ionic crystals, the packing depends upon relative sizes of cation and anions and also upon the ratio of number of cations to anions in the solid.

1.16. ATOMIC PACKING FACTOR (APF)

(U.P. Tech. Sem. Exam 2003-04, 2009-10)

1. Definition

The efficiency of filling the available space in a crystal is understood from the atomic packing factor. Atomic packing factor may be defined as the ratio of volume of atoms per unit cell to volume of the unit cell.

$$\text{APF} = \frac{\text{Volume of atoms per unit cell}}{\text{Volume of unit cell}}$$

Knowing the length parameters of the unit cell (a, b, c) and type of unit cell, it is possible to calculate density of the crystalline solid and atomic radii. Let us discuss atomic packing factor regarding cubic crystal systems.

2. APF for SC crystals

The SC unit cell contains 1 atom/unit cell.

Therefore,
$$\text{APF} = \frac{\text{Volume of atoms one unit cell}}{\text{Volume of unit cell}}$$

Let 'r' be the atomic radius and 'a' be the edge length of the cube. For a simple cube, we have r = a/2.

Thus, we have
$$\text{APF} = \frac{1\times\frac{4}{3}\pi r^3}{a^3} = \frac{\frac{4}{3}\times\pi\left(\frac{1}{2}\right)^3}{a^3} = \frac{4\times\pi\times a^3}{3a\times 8} = \frac{\pi}{6} = 0.52$$

Important Point: The APF for a simple cube is 0.52 or the packing efficiency is 52%. This means that only 52% of the available space is occupied by atoms out of the total volume, in simple cubic crystals. This is a loose packing.

3. APF for BCC structure

For BCC structure, there are 2 atoms per unit cell and atomic radius in terms of lattice parameter 'a' is,

$$r = \frac{a\sqrt{3}}{4}$$

We have
$$\text{APF} = \frac{\text{Volume of atoms in one unit cell}}{\text{Volume of unit cell}}$$

or
$$\text{APF} = \frac{2\times\frac{4}{3}\pi r^2}{a^3} = \frac{2\times 4\times\pi\left(\frac{a\sqrt{3}}{4}\right)^2}{3a^3} = \frac{\pi\sqrt{3}}{8} = 0.68$$

Important Point: APF for a BCC structure is 0.68 or the packing efficiency is 68%. This is meant by only 68% of total volume of crystal is occupied by atoms. Hence, this is also a loose packing.

4. APF for FCC structure

In FCC structure, the number of atoms per unit cell is 4 and the atomic radius $r = \frac{a\sqrt{2}}{4}$

Therefore, $$APF = \frac{\text{Volume of atoms in unit cell}}{\text{Volume of unit cell}}$$

$$= \frac{2 \times \frac{4}{3}\pi r^3}{a^3} = \frac{16\pi}{3a^3}\left(\frac{a\sqrt{2}}{4}\right)^3 = \frac{\pi}{3\sqrt{2}} = 0.74$$

APF for FCC structure is 0.74 and the packing efficiency will be 74%. Hence, FCC structure has closest packing.

(The APFs calculations for HCP structure comes 0.74 and the hexagonal close packing structure is also a closest packing.)

Density (ρ) of a cubic crystal structure and the edge length of cube, are related by following expression:

$$\rho = \frac{nM}{a^3 N}$$

where n = number of atoms per unit cell,

N = Average number (6.023×10^{23})

and M = molar mass of the solid.

1.17. CO-ORDINATION NUMBER

(U.P. Tech. Sem. Exam 2007-08)

(i) Co-ordination number may be defined as the number of nearest atoms of a particular atom, in crystal lattice. In the case of ionic crystals, the co-ordination number is the number of nearest oppositely charged ions around an ion.

(ii) For covalent solids, the co-ordination number is the number of atoms joined directly to a particular atom covalently. e.g., in NaCl, each Na^+ is surrounded by six Cl^- from all side and hence Co-ordination number of NaCl is 6.

(iii) In diamond (covalent solid) each carbon is covalently bonded to 4 other carbons.

(iv) Hence, Co-ordination number of carbon in diamond is 4.

(v) If simple cubic cell considered, the co-ordination number is 6 (4 atoms in the plane, one below the plane and 1 above the plane, in an octahedral way).

(vi) There are two nearest neighbours on 2 ± X-axis, 2 on ± Y-axis and 2 on ± Z-axis. Hence, an SC unit cell cubic crystal has co-ordination number 6.

(vii) In the body centered cubic cell, the co-ordination number is 8 as the atom in the centre has all eight corner atoms to its close.

(viii) In the FCC structure the co-ordination number is 12. An atom is in contact with 6 atoms in the plane, 3 from layer above and 3 from layer below.

S.No.	Type of cubic cell	Simple cubic	Face centered cubic	Body centered cubic
1.	Co-ordination number	6	12	8
2.	Atomic radius	a/2	$a\sqrt{2}/4$	$a\sqrt{3}/4$
3.	Atoms per unit cell	1	4	2
4.	APF	0.52	0.74	0.68

1.18. MILLER INDICES

(U.P. Tech. Sem. Exam, 2003-04, 2005-06, 2009-10, 2011-12)

1. Definition and Basic Concepts

(i) Crystal lattice contains unit cells arranged in parallel planes. Hence, each crystal plane lies parallel to the crystal face as well as to the unit cell face.

(ii) These planes intersect the three axes along the three crystallographic axes (OX, OY, OZ).

(iii) According to Hauy's law a given crystal plane should be described in terms of intercepts along the axes. The reciprocals of these intercepts are small whole numbers.

(iv) These numbers are denoted by h, k and 1 is known as Miller indices according to the name of the British Scientist, W. H. Miller.

(v) Therefore Miller indices of a plane is defined as the reciprocals of the intercepts which the plane makes with the axes.

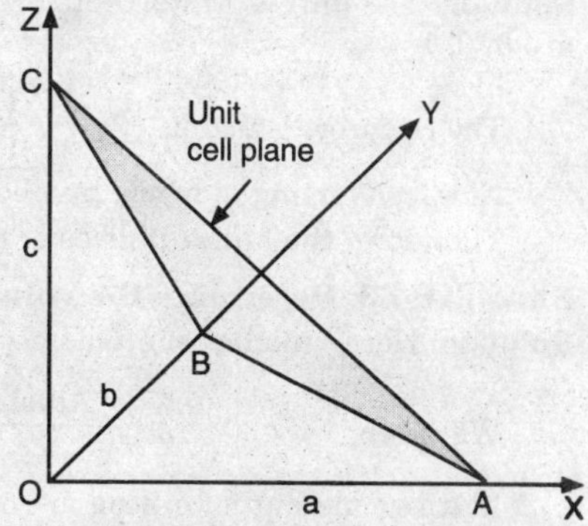

Fig. 1.35. A unit cell plane intercepts the axes to give intercepts a, b, and c

2. Illustrated Example

(i) As an example, let us assume a crystal system with the axis OX, OY and OZ. As depicted in figure. 1.35, ABC represents a unit cell surface and in figure 1.36 LMN represents another crystal plane under study.

(ii) The intercepts of the unit cell plane are OA, OB and OC which have the lengths a, b and c respectively.

(iii) The intercepts of the plane under study are OL, OM and ON. These can be written as multiples of the intercepts a, b, c i.e., la, mb and nc. Hence, 1, m and n are either integral whole numbers or ractions of whole numbers.

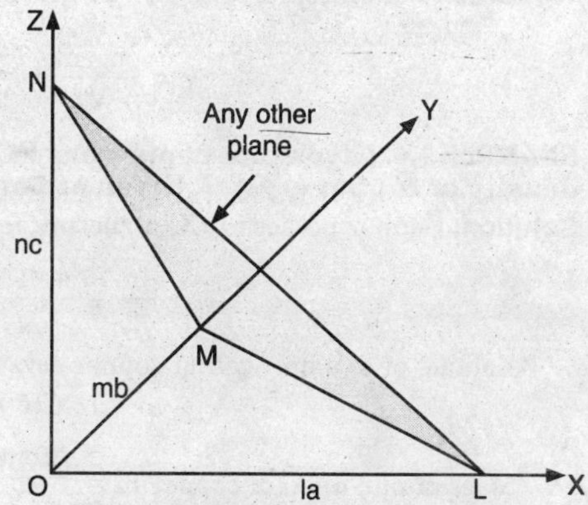

Fig. 1.36. Any other plane intersects the axes to give intercepts equal to la, mb and nc.

(iv) The reciprocals of these numbers are represented together in brackets (h, k, l) to provide the miller indices of the plane under study.

3. Determination of Miller indices

(i) Write the intercepts of the axes as multiple of a, b, c i.e., la, mb and nc.

(ii) Take the reciprocals of l, m and n.

(iii) Clear fractions to get whole numbers h, k, l.

(iv) So Miller indices of the plane are denoted as (h, k, l).

EXAMPLE 1.1. Determine the Miller indices of a plane that makes intercept 1 on a-axis, 2 on b-axis and is parallel to c-axis. **(U.P. Tech, Sem. Exam, 2006-07)**

Solution: Here, The unit cell intercepts are a, b, c

The intercepts of the given plane are 1a, 2b, αc. The lengths of the intercepts in terms of unit cell intercepts are 1, 2, α.

Hence, the reciprocals are $\frac{1}{1}, \frac{1}{2}, \frac{1}{\alpha}$

Now, converting to whole number i.e., multiplying by 2, we obtain 2, 1, 0.

Therefore, Miller indices of the given plane will be (2, 1, 0). **Ans.**

EXAMPLE 1.2. Determine the Miller indices for a plane when the intercepts along the axes are 3a, 2b and 1c.

Solution: The unit cell intercepts are a, b, c. The lengths of the intercepts in terms of unit cell intercepts are 3, 2, 1.

The reciprocals will be $\frac{1}{3}, \frac{1}{2}, \frac{1}{1}$

Now, converting to whole numbers i.e. multiplying by 6, we obtain 2, 3, 6.

Therefore, the Miller indices of the given plane are (2, 3, 6). **Ans.**

EXAMPLE 1.3. Determine the volume density of Si atoms with its lattice constant of 5.3 Å.

Solution: Here, number of atoms per unit cell of silicon, n = 4, mass of one silicon atom

We have, $$m = \frac{\text{Atomic weight of Si}}{N_A} = \frac{28}{6.023\times10^{23}}\text{ gm} = 4.65\times10^{-26}\text{ kg}$$

Lattice constant for silicon a = 5.3 A° = 5.3×10^{-10} m

Volume of one unit cell of Si will be

$$V = a^3 = (5.3\times10^{-10})^3 = 148.877\times10^{-3}\text{ m}^3$$

Density of silicon crystal will be given by

$$\rho = \frac{m\times n}{V} = \frac{4.65\times10^{-26}\times4}{148.877\times10^{-30}} = 1.249\text{ kg/m}^3$$ **Ans.**

EXAMPLE 1.4. Given that copper has FCC structure and its atomic radius is 1.278 Å. Evaluate the density of copper crystal. Given atomic weight of copper = 63.5, and $N_A = 6.023 \times 10^{23}$.

Solution: Copper possess FCC structure for which, lattice constant is given by

$$a = \frac{4r}{\sqrt{2}} = \frac{4\times1.278}{\sqrt{2}} = 3.615\text{ A}° = 3.615\times10^{-10}\text{ m}$$

Volume of one unit cell of copper crystal will be

$$V = a^3 = (3.615\times10^{-10})^3 = 47.24\times10^{-30}\text{ m}^3$$

Mass of one atom of copper $m = \frac{\text{Atomic weight}}{N_A}$

or $$m = \frac{63.5}{6.023\times10^{23}} = 10.543\times10^{-23}\text{ gm} = 10.543\times10^{-26}\text{ kg} = 10.543\times10^{-26}\text{ kg}$$

Number of atoms present per unit cell of copper for FCC structure will be n = 4

Therefore, Density of crystal,

$$\rho = \frac{m\times n}{V}$$

or $$\rho = \frac{10.543\times10^{-26}\times4}{47.24\times10^{-30}} = 8.927\text{ kg/m}^3$$ **Ans.**

1.19 X-RAY DIFFRACTION

(i) X-rays were discovered by Roentgen in 1895 when he was studying the phenomenon of discharge of electricity through rarefied gases.

(ii) According to his experiment when a beam of fast moving electrons strikes a solid of high atomic weight (like tungsten) kept in vacuum, an invisible radiation having high penetrating power is produced.

(iii) Because of the unknown nature of these radiations, Roentgen called them X-rays.

(iv) The physical nature of X-rays was not exactly known for many years after their discovery.

(v) Two theories were put forward regarding the nature of these rays. One theory regards X-rays as h gh speed particles having a high penetrating power.

(vi) The other theory regards X-rays as electromagnetic waves of very high frequency or very short wavelength.

(vii) A number of experiments were performed to test the wave nature of X-rays but success could not be obtained for a long period. In order to establish that X-rays have the same electromagnetic nature as ordinary light it was essential to show that they can produce interference and diffraction effects.

(viii) To observe the diffraction of light waves (l = 6000 A°) the spacing between the consecutive ruled lines on the grating should be of the order of the wavelength of light wave.

(ix) Hence, if X-ray diffraction is to be observed, the X-ray grating would require-40 million rulings per cm which is practically not possible.

(x) Max Van Laue was the first physicist to suggest that a crystal can effectively act as a grating for X-rays. The wavelength of X-rays is of the order of 1 A°. The spacing between the regularly arranged atoms is also of the same order of magnitude.

(xi) Hence, Laue suggested that a crystal consisting of three dimensional array of regularly spaced atoms can serve as a grating for X-rays.

(xii) The crystal acts as a three dimensional space grating rather than a two dimensional plane grating like optical grating.

1.20. LAUE SPOTS

1. Basic Concepts

(i) The use of crystal to produce X-ray diffraction was put to an experimental test by Laue and his associates. The experimental arrangement has been illustrated in figure 1.37.

(ii) The X-ray beam from a Coolidge tube is collimated into a fine pencil beam by the pinholes S_1 and S_2 in the lead screens L_1 and L_2.

(iii) This beam is then allowed to pass through a thin crystal of zinc sulphide. The transmitted beam is received on a photographic plate P placed perpendicular to the beam. the plate was developed after exposure of many hours.

(iv) The pattern on the plate consists of a central dark spot at O and a series of many faint spots arranged in a definite pattern about O.

(v) This symmetrical pattern of spots is known as Laue pattern or Laue spots. A simple interpretation of the Laue pattern was given by Bragg. The central spot is due to the X-rays that strike the plate directly through the crystal.

(vi) The surrounding spots are produced due to the reflections of incident X-rays from various sets of parallel crystal planes containing a large number of atoms.

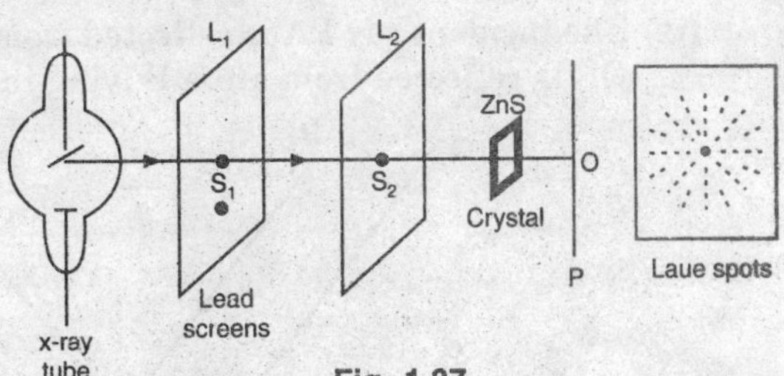

Fig. 1.37.

(vii) These equally spaced layers of atoms were supposed to act like the rulings of diffraction grating and give weak diffracted beams in different directions; that produce a series of dark spots on the photographic plate.

2. Confirmation from Laue Spot Experiment

Thus, Laue spot experiment confirms that:

(i) X-rays are electromagnetic waves of very short wavelength.

(ii) Atoms in the crystal are arranged in a regular three dimensional lattice.

Important Point: The Laue spots may be used for the determination of the crystal structure.

1.21. BRAGG'S LAW*

(i) Laue's experiment consider the crystal as a transmission grating. Further study of X-ray diffraction by crystals was carried out by Prof. Bragg in which the crystal was considered as a reflection grating.

(ii) According to him, through any crystal a set of equidistant parallel planes can be drawn which are rich in atoms.

(iii) Actually a large number of such families of planes can be drawn; the planes of each family being separated from each other by a characteristic distance.

(iv) Such planes are called Bragg planes and their separations called as Bragg spacings. Some of such planes have been illustrated in figure 1.38.

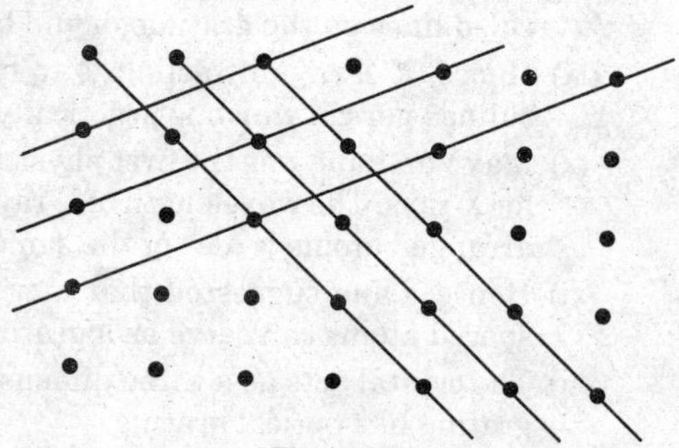

Fig. 1.38.

(v) When a beam of monochromatic X-rays falls on a crystal, each atom becomes a source of scattering radiation. The combined scattering of X-rays from Bragg's planes can be looked upon as reflections from these planes.

(vi) If the beam is incident at some particular angles on the planes then the reflections from the set of parallel planes are in phase with each other. Hence, they reinforce each other giving maximum intensity along a particular direction.

(vii) For other angles, the reflections from different planes are out of phase and hence they give zero intensity. To find the condition under which maximum intensity is obtained consider a set of parallel planes of a crystal as shown in figure 1.39. Let d be the interplaner spacing. The figure 1.39 shows a narrow beam of X-rays of wavelength l incident upon the crystal planes.

(viii) Let θ be the angle between the incident beam and the crystal plane θ is called angle of glancing. The beam is partially reflected in all directions by the atoms of various planes.

(ix) The incident ray PA is reflected from atom A along AR from plane 1 while the other ray QB is reflected from atom B lying in plane 2 immediately below atom A.

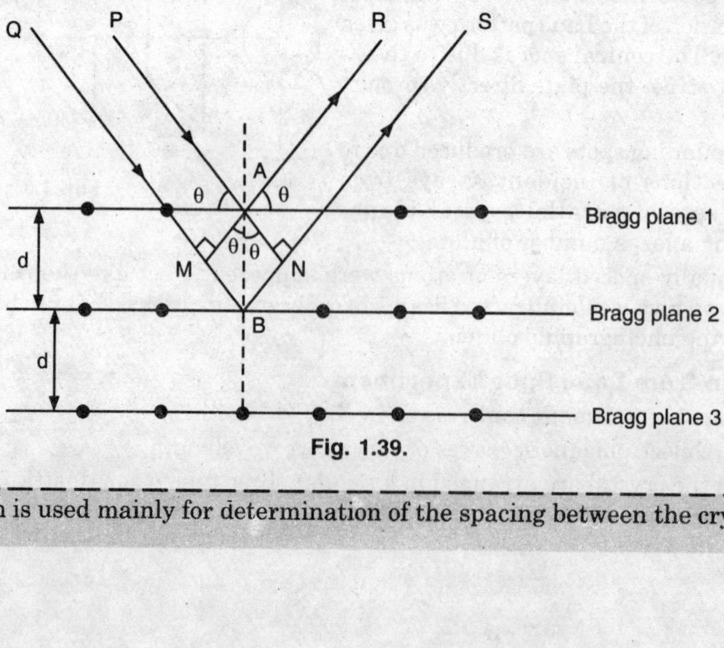

Fig. 1.39.

* Bragg equation is used mainly for determination of the spacing between the crystal planes.

(x) Whether these two reflected rays will be in phase or not, will depend upon the path difference between them. To find the path difference, draw AM perpendicular to QB and AN perpendicular to BS. Therefore, the path difference between the two reflected rays AR and BS is

$$\delta = MB + BN = d \,.\, \sin\theta + d \,.\, \sin\theta$$

or
$$\delta = 2d \sin\theta$$

(xi) The two reflected rays will be in phase with each other if the path difference is an integral multiple of wavelength 'λ'. Therefore, the condition for producing principal maxima is

$$\delta = 2\,d \sin\theta = n\lambda$$

where n = 1, 2, 3, 4 ... giving the first order, second order, third order ... etc. maxima respectively. This relation is known as **Bragg's law.**

(xii) Different directions in which maximum intensity will be produced can be obtained for different values of n.

i.e., For first maxima, n = 1, we have

$$\therefore \quad \sin\theta_1 = \frac{\lambda}{2d}$$

For second order maxima, n = 2, we have

$$\therefore \quad \sin\theta_2 = \frac{2\lambda}{2d}$$

For the third order maxima, n = 3, we have

$$\therefore \quad \sin\theta_3 = \frac{3\lambda}{2d} \text{ and so on.}$$

Important Point: As the order of maximum increases, the intensity decreases.

1.21.1. Bragg's Spectrometer

1. Basic Aspects

X-ray diffraction by crystals was studied by Bragg for which X-ray spectrometer was devised. The crystal is used here not as a transmission grating but as a reflection grating. Bragg's relation 2d sin θ = nλ can be verified by the X-ray spectrometer.

2. Schematic Arrangement

The schematic arrangement of Bragg's spectrometer is shown in figure 1.40. It consists of three parts:

(i) A source of X-rays suitably collimated.

(ii) A crystal held on a graduated circular table provided wit h a vernier.

(iii) A detecting device.

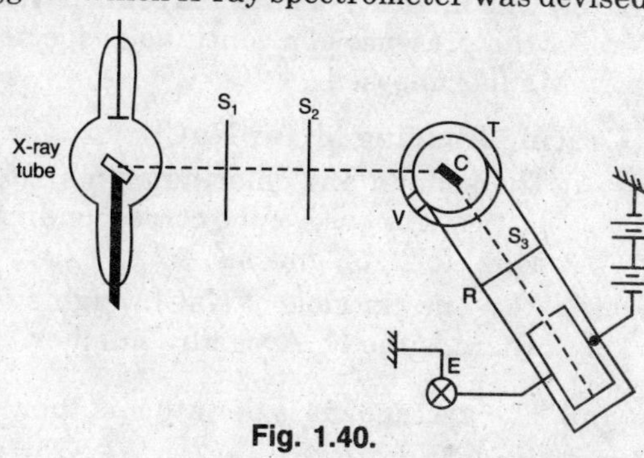

Fig. 1.40.

X-rays from an X-ray tube are made to pass through two slits S_1 and S_2 of lead and thereby collimated into a fine beam. This beam is then allowed to fall on a crystal C mounted on a circular table T known as the turn-table. The table T can rotate about the vertical axis and its position be read by the vernier V. A side-arm R, rotating about the same axis as the crystal table, carriers an ionization chamber I and is provided with a vernier of its own to fix its position accurately. The intensity of X-ray beam that enters the chamber is measured by the electrometer E. The third slit S_3 protects the chamber against stray scattered radiations. The turntable T and the side-arm R are so linked together that when the turntable (and hence the crystal) rotates through an angle θ, the arm R (and hence the ionization chamber) turns through 2θ. In this way the beam is always reflected into the ionization chamber whatever be the glancing angle at the crystal surface. The ionization current produced by the reflected beam of X-rays can be measured by the sensitive electrometer E. According to Bragg's law, $2d \sin \theta = n\lambda$.

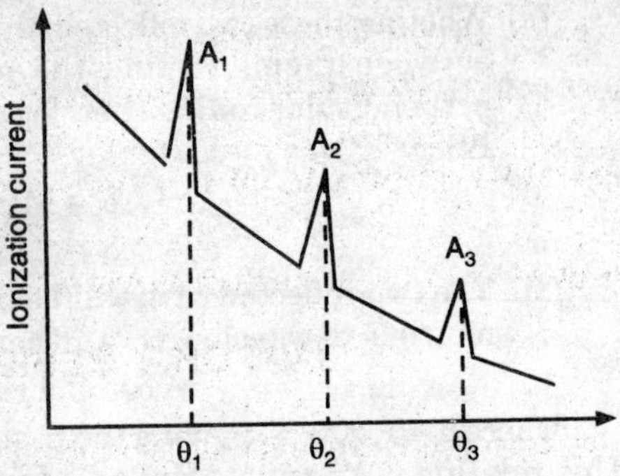

(a) Incident angle θ

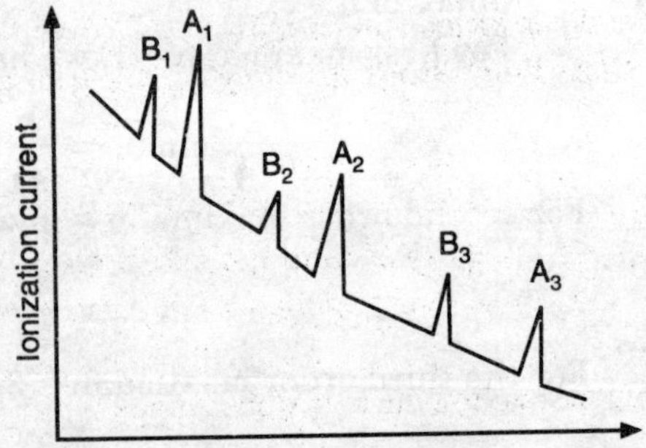

(b) Incident angle θ

Fig. 1.41.

3. Curves

Using X-ray beam of single wavelength λ and starting with a small glancing angle θ for the incident beam, the ionization chamber is set at angle 2θ to the incident beam direction. The intensity of the beam reflected from the crystal face is measured by the ionization current shown by electrometer. The angle θ is then increased in steps and the corresponding observations are recorded. The ionization current (and hence the intensity of reflected beam) is then plotted as a function of glancing angle θ. The graph as shown in figure 1.41(a) is obtained. If the incident beam contains two wavelengths, we get the curve as in figure 1.41(b).

For X-ray beam of wavelength λ, ionization current shows maxima A_1, A_2, A_3 ... at only certain glancing angles $\theta_1, \theta_2, \theta_3$ ... This indicates that the peaks A_1, A_2, A_3 ... refer to different order maxima 1, 2, 3 ... respectiely.

4. Conclusion from Curves

It is also observed from the curves that:

(i) The intensity of reflected beam decreases with the increase in order of spectrum.

(ii) The position of the peaks depend on the metal used as the target in X-ray tube, on the crystal used and on the particular crystal face reflecting the X-rays.

(iii) The intensity of reflected beam never falls to zero, but only to a minimum value, indicating the presence of a continuous spectrum over which the characteristics line spectrum is superimposed.

5. Grating Spacing 'd' for NaCl

(i) The sodium and chlorine atoms alternately the corners of the elementary cube of the lattice. For every cube corresponding volume is d^3 with mass ρd^3 where ρ is the density, equal to 2.165 gm/cm³.

(ii) Also one gm-mole of NaCl weights 58.454 gm and contains N_o molecules or $2N_o$ atoms, where is the N_o Avogadro number.

(iii) The average mass per atom is then $\frac{58.454}{2\,N_o}$ gm.

and $$\rho d^3 = \frac{58.454}{2\,N_o}$$

or $$d^3 = \frac{58.454}{2\,(2.165)\,(6.025)\,10^{23}}$$

and $$d = 2.82 \times 10^{-8} \text{ cm.}$$

Calcite ($CaCO_3$) is now used as the standard crystal : It has a lattice spacing of 3.03567 × 10^{-8} cm.

(iv) The wavelengths of X-ray lines can now be measured accurately with ruled grating and with the known wavelengths, the lattice spacing of crystals can be determined experimentally.

1.21.2. Bragg's Law and Crystal Structure

(i) In case of KCl crystal, Bragg obtained the maxima of reflected X-rays at glancing angles 5.22°, 7.30° and 9.05° respectively using three different reflecting planes.

(ii) According to Bragg's law of X-ray diffraction, we have

$$2d \sin \theta = n\lambda$$

(iii) Now for the first order maximum, n = 1 and hence

$$2d \sin \theta = \lambda$$

or $$\frac{1}{d} = \frac{2 \sin \theta}{\lambda}$$

(iv) Therefore, for the above values of the glancing angles we have,

$$\frac{1}{d_1} : \frac{1}{d_2} : \frac{1}{d_3} :: \sin 5.22° : \sin 7.30° : \sin 9.05°$$

or $$\frac{1}{d_1} : \frac{1}{d_2} : \frac{1}{d_3} :: 0.091 : 0.127 : 0.157$$

or $$\frac{1}{d_1} : \frac{1}{d_2} : \frac{1}{d_3} :: 1 : 1.4 : 1.73$$

This result implies that KCl is simple cubic crystal.

Important Point: Hence, Bragg's law can be utilized to analyse different types of crystal structure.

EXAMPLE 1.5. The spacing between the principal planes of NaCl/crystal is 2.82 Å. What will be the wavelength of X-ray when the first order Brangg's reflection is observed at an angle 10°?

Solution: Given that, $d = 2.82$ A° $= 2.82 \times 10–10$ m $\quad n = 1 \quad \theta_1 = 10°$

We have, $2d \sin \theta = n\lambda = 2d \sin \theta_1 = \lambda$

or $\lambda = 2 \times 2.82 \times 10–10 \times \sin(10) = 9.79 \times 10^{-10}$ m $= 0.979 \times 10^{-10}$ m

or $\lambda = 0.979$ A° **Ans.**

EXAMPLE 1.6. X-ray of wavelength 1.6 Å are diffracted by a Bragg's crystal spectrometer at an angle 14.2° in the first order. What will be the spacing of atomic layer in the crystal?

Solution: Given that, $\lambda = 1.6$ A° $= 1.6 \times 10^{-10}$ m $\quad \theta = 14.2° \quad n = 1 \quad d = ?$

We have $2d \sin \theta = n\theta =$ or $d = \dfrac{n\lambda}{2 \sin \theta}$

$n = 1$

or $$d = \frac{1 \times 1.6 \times 10^{-10}}{2 \times \sin (14.2)} = \frac{1.6 \times 10^{-10}}{2 \times 0.2453} = 3.26 \times 10^{-10} \text{ m}$$

or $d = 3.26°$ **Ans.**

EXAMPLE 1.7. The Bragg's angle corresponding to the first order reflection from the plane of crystal is 30° when the X-rays of wavelength 1.78° are used. Evaluate the interplaner spacing.

Solution: Given that, $n = 1 \quad \theta_1 = 30° \quad \lambda = 1.78$ A° $= 1.78 \times 10^{-10}$ m

We know that $2d \sin \theta = n\lambda$

$$d = \frac{1 \times 1.78 \times 10^{-10}}{2 \times \sin 30} = \frac{1.78 \times 10^{-10}}{2 \times 0.5}$$

or $d = 1.78 \times 10^{-10}$

$d = 1.78$ A° **Ans.**

EXAMPLE 1.8. An X-ray analysis of a crystal is made with monochromatic X-rays of wavelength 0.58 Å. Bragg's reflections are obtained at angles of:

(i) 6.45° (ii) 9.15° and (iii) 13°

Evaluate the interplaner spacing of the crystal.

Solution: According to Bragg's law, we have 2d sin θ = nλ

Therefore, $$\frac{d}{n} = \frac{\lambda}{2\sin\theta}$$

(i) When θ = 6.45°, we have

$$\frac{d}{n} = \frac{0.58\ \text{A}^\circ}{2\sin 6.45^\circ} = 2.568\ \text{Å} \quad \text{...(i)}$$

(ii) When θ = 9.15°, we have

$$\frac{d}{n} = \frac{0.58\ \text{A}^\circ}{2\sin 9.15^\circ} = 1.817\ \text{Å} \quad \text{...(ii)}$$

(iii) When θ = 13°, we have

$$\frac{d}{n} = \frac{0.58\ \text{A}^\circ}{2\sin 13^\circ} = 1.288\ \text{Å} \quad \text{...(iii)}$$

Here, let us that the value of $\frac{d}{n}$ given by Equation (i) is almost double the value given by Equation (iv). Therefore, the angles 6.45° and 13° represent the first and second order reflection maxima from one set of parallel planes. The interplaner spacing d can be obtained by substituting n = 1 in equation (i) or n = 2 in equation (iii).

Therefore, d = 2.568 Å [n = 1 for equation (i)]

The glancing angle 9.15° used in equation (ii) refers to some other set of reflecting planes for which the interplanar spacing is

d = 1.817 Å [n = 1 in equation (ii)]

EXAMPLE 1.9. Determine the glancing angle for a cubic (100) crystal of rocksalt (a = 2.814 Å) corresponding to second order diffraction maximum for X-rays of wavelength 0.710 Å.

Solution: According to Bragg's Law, we have 2 d sin θ = nλ

The interplaner distance d for a set of parallel planes in a cubic crystal is given in terms of the Miller indices as under

$$d_{hkl} = \frac{a}{\sqrt{h^2 + k^2 + l^2}}$$

a = 2.814 A°, h = l,

k = 0, l = 0

Therefore, d_{hkl} = 2.814 A°

n = 2 and λ = 0.710 Å

Thus, Bragg's law yields

$$\sin\theta = \frac{n\lambda}{2d} = \frac{2\times 0.710\ \text{A}^\circ}{2\times 2.814\ \text{A}^\circ} = 0.2523$$

Therefore, $\theta = \sin^{-1}(0.2523) = 14^\circ\ 36'\ 40''$

EXAMPLE 1.10. X-rays of unknown wavelength gives first order Bragg reflection at glancing angle of 60° with (100) planes of copper having FCC structure. Determine the wavelength of this X-ray if the lattice constant for copper is 3.65 Å.

Solution: Using nλ = 2d sin θ

and $$d = \frac{a}{\sqrt{h^2 + k^2 + l^2}} = \frac{3.65\times 10^{-10}}{\sqrt{1^2 + 0^2 + 0^2}} = 3.65\times 10^{-10}\ \text{m}$$

Therefore, for 1st order n = 1

Hence, $\lambda = 2 \times 3.65 \times 10^{-10} \times \sin 60° = 2 \times 3.65 \times 10^{-10} \times 0.8660$

$= 6.3218 \times 10^{-10} = 6.3218$ Å

EXAMPLE 1.11. The radiation of wavelength 1.54 Å from a X-ray tube is diffracted by a F.C.C. copper crystal of atomic mass 63.54 and density 9.024 gm/cc. Evaluate the glancing angle for first order diffraction, from the principle plane (100) of the crystal.

Solution: We have, λ = 1.54 Å; Copper crystal (FCC) of atomic weight = 63.54

Density = 9.024 gm/cc; θ = ?, n =1 ; M.I. = (100)

$$\text{Mass of copper crystal} = \frac{\text{Atomic Weight}}{N} = \frac{63.54}{6.023 \times 10^{23}}$$

Therefore, Volume of a crystal = Density × Mass

i.e.,
$$a^3 = 9.024 \times \frac{63.54}{6.023 \times 10^{23}}$$

or
$$a^3 = 95.1992 \times 10^{-23}/\text{cm}^3 = \frac{951990}{10^{27}}$$

or
$$a = 98.37 \times 10^{-9} = 9.837 \times 10^{-8} \text{ cm}$$

Now, we have
$$d = \frac{a}{\sqrt{h^2 + k^2 + l^2}} = \frac{9.837 \times 10^{-8}}{\sqrt{1^2 + 0^2 + 0^2}} = 9.837 \times 10^{-8} \text{ cm}$$

Thus, we write $n\lambda = 2d \sin \theta$ If, n = 1

Hence,
$$\sin \theta = \frac{\lambda}{2d} = \frac{1.54 \times 10^{-8}}{2 \times 9.837 \times 10^{-8}} = 0.07828$$

or
$$\theta = 4.4895°$$

EXAMPLE 1.12. X-rays of unknown wavelength give first order Bragg's reflection at glancing angle 22° with (111) planes of Cu.

(i) Determine the wavelength of the X-ray if the lattice constant for Cu is 3.615 Å.

(ii) Now, the crystal is rotated by a certain angle and the detector is also rotated by the corresponding angle to observe the 2nd order spectrum. What would be the position of the detector in degree to get 2nd order?

Solution: Given that, a = 3.615 A°; θ = 22°, n = 1

The interplaner spacing is given by

$$d = \frac{a}{\sqrt{h^2 + k^2 + l^2}}$$

$$d = \frac{3.615 \text{ Å}}{\sqrt{1^2 + 1^2 + 1^2}} = \frac{3.615 \text{ Å}}{\sqrt{3}} = 2.087 \text{ A°}$$

(i) According to Bragg's law, we have

$$2d \sin \theta = n\lambda$$

or
$$\lambda = 2d \sin \theta \ (\because n = 1)$$
$$= 2 \times 2.087 \text{ A°} \times \sin 22°$$

or
$$\lambda = 1.563 \text{ Å}$$

(ii) For n = 1

$$\theta_1 = 22° \quad \text{and} \quad \sin \theta_1 = \frac{\lambda}{2d}$$

Therefore, Position of the detector is 44°.

Now for 2nd order spectrum, n = 2

Therefore, f θ_2 is the glancing angle for second order, then, we have

$$2d \sin \theta_2 = 2 \times \lambda$$

$$\text{or} \quad \sin \theta_2 = \frac{\lambda}{d} = \frac{1.563 \text{ Å}}{2.087 \text{ Å}} = 0.7489$$

Hence, we have $\theta_2 = \sin^{-1} (0.7489)$

or $\theta_2 = 48.49°$

Thus, to get the 2nd order spectrum the position of the detector must be 2 × 48.49° i.e., 96.99°.

EXAMPLE 1.13. Bragg's reflection for the first order is observed at 21.7° for (111) planes for a crystal under test. If the wavelength of X-rays used is 1.54 Å, determine lattice constant of the crystal.

Solution: We know that the condition for nth order Bragg's reflection is $2d \sin \theta = n\lambda$

For first order reflection we have, n = 1

Therefore, $2d \sin \theta = \lambda$ ...(i)

The interplaner spacing d for the FCC crystal is given by,

$$d = \frac{a}{\sqrt{h^2 + k^2 + l^2}} \qquad ...(ii)$$

From equation (i), we write

$$d = \frac{\lambda}{2 \sin \theta} = \frac{1.54 \text{ Å}}{2 \times \sin 21.7°} = \frac{1.54 \text{ Å}}{2 \times 0.3697} = 2.0827$$

From equation (ii), the lattice constant of crystal we have

$$a = d\sqrt{h^2 + k^2 + l^2} = 2.0827 \text{ A°} \left(\sqrt{1^2 + 1^2 + 1^2}\right) = 2.087 \times \sqrt{3} \text{ A°}$$

Therefore, a = 3.607 A°

or, Lattice constant = 3.607 A°

1.22. NUMERICALS ON RADIUS-DENSITY-MOLAR MASS

We can use following information and expressions for calculations:

1. Number of atoms per unit cell for SC, BCC and FCC are 1, 2 and 4 respectively.
2. Packing factor for SC, BCC and FCC structures are 0.52, 0.68 and 0.74 respectively.
3. Co-ordination number for SC, BCC and FCC are 6, 8 and 12 respectively.
4. Density and molecular weight (atomic weight) are related by

$$\rho = \frac{nM}{a^3 N}$$

where n = Number of atoms per unit cell
M = Molar mass or atomic weight
a = Lattice parameter for cubic unit cell
ρ = Density of the solid
Nq = Avagadro Number
= 6.023×10^{23}

5. Interplanar distance in planes of a crystal, d_{hkl}

$$d_{hkl} = \frac{n\lambda}{2 \sin \theta} \text{ (Bragg's equation)}$$

λ = wavelength of X-ray,
θ = glancing angle of X-ray on face of a crystal,
n = integer, (hkl) = miller indices

6. Atomic radius for SC structure = $\frac{a}{2}$

Atomic radius for BCC structure = $\frac{a\sqrt{3}}{4}$

Atomic radius for FCC structure = $\frac{a\sqrt{2}}{4}$

7. Distance between neighbouring parallel planes in terms of miller indices for a cubic crystal.

$$d_{hkl} = \frac{a}{\sqrt{h^2 + k^2 + l^2}}$$

where h, k, l are the intercepts of a plane on the X, Y, Z axis in terms of lattice parameters.

EXAMPLE 1.14. Evaluate the distance between d_{211} planes; if the cubic lattice has a = 4.8 Å.

Solution: h = 2, k = 1, l = 1

Therefore, $$d_{211} = \frac{a}{\sqrt{h^2 + k^2 + l^2}} = \frac{4.8}{\sqrt{2^2 + 1^2 + l^2}} = \frac{4.8}{\sqrt{6}} = 1.96 \text{ Å}$$

EXAMPLE 1.15. Copper has FCC structure and atomic radius 1.28 A°. Determine its density if its atomic weight is 63.5 gm/mole.

Solution: For FCC, we have $r = \frac{a\sqrt{2}}{4}$

$$a = \frac{4r}{\sqrt{2}} = \frac{4 \times 1.28}{\sqrt{2}} = 3.62 \text{ A°} = 3.62 \times 10^{-8} \text{ cm}$$

Density is given by $\rho = \frac{nM}{N a^3}$

Since n = 4 atoms per unit cell for FCC and M = 63.5.

Therefore, $$\rho = \frac{4 \times 63.5}{6.023 \times 10^{23} \times (3.62 \times 10^{-8})^3} = 8.89 \text{ gm/c.c.}$$

EXAMPLE 1.16. Determine the number of atoms per unit cell in iron cubic lattice of lattice parameter 2.9 Å and density 7.87 gm/c.c. Atomic weight of metal = 55.85.

Solution: We have, M = 55.85 a = 2.9 A° = 2.9×10^{-8} cm, ρ = 7.87 gm/c.c

We know that $\rho = \frac{nM}{N a^3}$

or $$7.87 = \frac{n \times 55.85}{6.023 \times 10^{23} \times (2.9 \times 10^{-8})^3}$$

or n = 2.069 = 2 atoms per unit cell

Therefore, a lattice having 2 atom per unit cell is a BCC structure.

EXAMPLE 1.17. An element of atomic weight 60, has density 6.23gm/c.c. What will be the radius of its atom if it is a FCC structure?

Solution: Given that, M = 60 gm/mole, ρ = 6.23 gm/c.c

n = 4 as it is FCC structure

We know that $\rho = \frac{nM}{N a^3}$

or $$6.23 = \frac{4 \times 60}{6.023 \times 10^{23} \times a^3}$$

or a = edge length of cube = 4×10^{-8} cm

Radius of atom in a FCC structure will be given by = $\frac{a\sqrt{2}}{4} = \frac{4 \times 10^{-8}\sqrt{2}}{4} = 1.414 \times 10^{-8}$ cm or 1.414 Å

EXAMPLE 1.18. Determine packing factor for chromium metal having BCC structure if its density = 5.96 gm/c.c and atomic weight 50.

Solution: Given that, ρ = 5.96 gm/c.c, M = 50, n = 2 for BCC

We know that $$\rho = \frac{nM}{N a^3}$$

$$5.96 = \frac{2\times 50}{6.023\times 10^{23}\times a^3}$$

We have, a^3 = Volume of one cubic unit cell = 1.66×10^{-22} cm^3

or $a = 5.496 \times 10^{-8}$ cm

We know that, Radius of an atom in BCC = $\frac{a\sqrt{3}}{4} = \frac{5.496\times 10^{-8}\sqrt{3}}{4} = 2.38 \times 10^{-8}$ cm

Also, $$\text{Packing factor} = \frac{\text{Volume of atoms in the unit cell}}{\text{Volume of unit cell}}$$

or $$\text{Packing factor} = \frac{2\times\frac{4}{3}\pi r^3}{a^3} = \frac{2\times\frac{4}{3}\pi(2.38\times 10^{-8})^3}{1.66\times 10^{-22}} = 0.68$$

EXAMPLE 1.19. An element has atomic mass 120 and BCC structure. Density of the element is 5.2 gm/c.c. Evaluate number of unit cells in its 20 gm.

Solution: For BCC, number of atoms per unit cell, n = 2

M = 120, = 5.2 gm/c.c

We have, Mass of 1 atom = $\frac{\text{atomic mass}}{6.023\times 10^{23}} = \frac{120}{6.023\times 10^{23}} = 1.99 \times 10^{-22}$ gm

Therefore, Mass of unit cell = $2 \times 1.99 \times 10^{-22}$ gm

$2 \times 1.99 \times 10^{-22}$ gm mass corresponds to 1 unit cell.

Hence, 20 gm mass corresponds to $\frac{20}{2\times 1.99\times 10^{-22}}$ unit cells = 5.025×10^{22} unit cells

EXAMPLE 1.20. Determine the distance between K^+ and F^- ions in its FCC structure. Given, density of KF = 2.48 gm/c.c and atomic weights of K and F are 39 and 19 respectively.

Solution: Molecular weight of KF = 39 + 19 = 58

ρ = 2.48 gm/c.c and for FCC, n = 4

We know that $$a^3 = \frac{nM}{\rho N} = \frac{4\times 58}{2.48\times 6.023\times 10^{23}} = 1.55 \times 10^{-22} \text{ cm}^3$$

$\sqrt[3]{a^3}$ = a = Edge length of FCC cell = 5.375×10^{-8} cm = 5.375 Å

For FCC, radius of ion is given by = $\frac{a\sqrt{2}}{4} = \frac{5.375\sqrt{2}}{4}$

The centre to centre distance between ions

$2r = 2 \times 1.9 = 3.8$ Å

EXAMPLE 1.21. Referring to a unit cell, draw the following Miller planes and directions:

Planes: (1 0 0), (1 $\bar{1}$ 0), (2 3 1)

Directions: [1 0 0], [1 1 1], [1 2 3]

Solution: The planes (1 0 0), (1 $\bar{1}$ 0), (2 3 1).

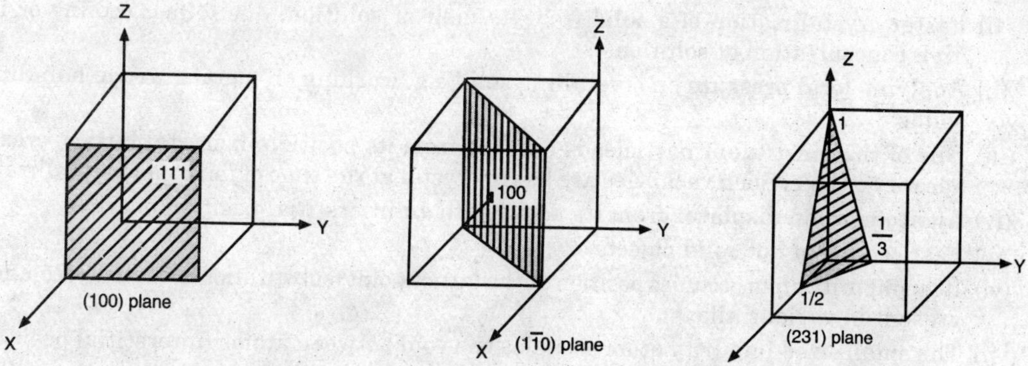

Fig. 1.42.

and directions $[\bar{1}\,0\,0]$, [1 1 1] have been illustrated figure 1.42.

Direction [1 2 3] is normal to plane (1 2 3).

EXAMPLE 1.22. With reference to a cubic unit cell, draw the following:

Planes: $(1\,\bar{2}\,0)$, $(1\,2\,\bar{3})$

Directions: [1 0 1], [1 2 3].

Solution: The planes $(1\,\bar{2}\,0)$, $(1\,2\,\bar{3})$ and directions [1 0 1], [1 2 3] have been illustrated in figure 1.43.

The direction [1 2 3] is normal to the plane (1 2 3).

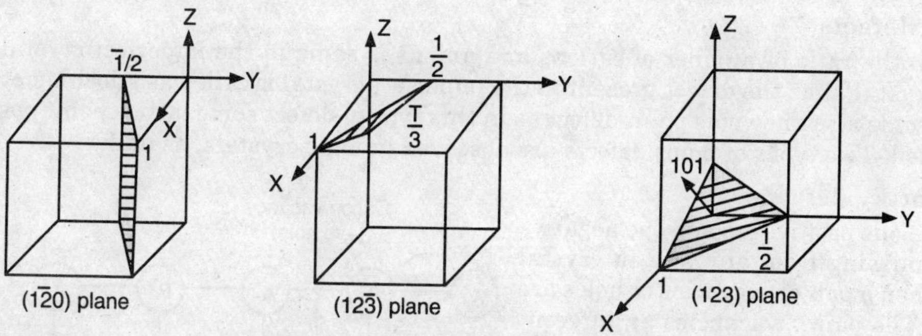

Fig. 1.43.

1.23. DEFECTS OR IMPERFECTIONS IN CRYSTALS

(U.P. Tech. Sem. Exam, 2003-04, 2007-08, 2008-09)

1. Definition and Basic Concepts

(i) An ideal or a perfect crystal is the one which has a regular arrangement of constituent particles of solid in all directions, throughout the crystal.

(ii) A perfect crystal at 0 K temperature is truly perfect and it has no disorder. But, as temperature increases, the atoms start vibrating about their mean positions and disorder increases. In addition to this, in practice, it is quite difficult to get perfectly crystalline solid.

(iii) Therefore practically, all crystals suffer from defects of one kind or other and the defects increase with increase of temperature.*

2. Defects or Imperfections

Any departure from complete order of arrangement of atoms in a crystal, is known as defect.

The defects or imperfections in crystal arise due to following reasons:

* At absolute zero, crystals tend to have perfectly ordered arrnagement of ions and there are no defects.

(i) Faster crystallization of a solid from its melt or solution, due to fast cooling or from high concentration of solution.
(ii) Applying local pressure on crystalline solid e.g. working on metals, wrong handling of solids.
(iii) One of the constituent particles is missing from its position in crystal lattice, creating vacancy or hole. Such vacancies are left in crystal at the time of improper crystallization.
(iv) An atom or ion displaces from its position to an interstitial position nearby.
(v) Fast cooling of hot solid objects.
(vi) If an impurity atom occupies position at the lattice point (substitutional impurity, commonly present in various alloys).
(vii) The small sized impurity atom occupies the void between atoms (interstitial position of impurity).

3. Effect of defects on properties of crystal

(i) Decreases melting point
(ii) Changes refractive index
(iii) Increase in electrical and thermal conductivities because of holes.
(iv) Changes magnetic strength.
(v) Decreases mechanical properties of the solid like tensile strength, hardness, toughness.
(vi) Increase in brittleness character.
(vii) Density decreases.

4. Point defects

When the ratio of number of cations and anions is same in the imperfect crystal as in the perfect crystal, then the defect present in the imperfect crystal is called as stiochiometric defect. This defect is also known as point defect as in this type of defect some lattice point positions are unoccupied. Two types of point defects are observed in ionic crystals, as under.

(A) Schottky defects

When one positive ion and one negative ion are missing from any site in crystal lattice, then a pair of vacancies or holes are created. The paired vacancies at different sites in a crystal due to missing of pairs of cation-anion, is known as schottky defect. The schottky defect is generated in ionic crystals when

(i) Co-ordination number of an ion is high,
(ii) When the size of anions and cations are almost equal.

Crystals of KCl, KBr, CsCl, etc. shows schottky defect. Crystallographic studies with X-rays on NaCl, shows that over 10 lakh paired vacancies are present in each gm-mole of NaCl.

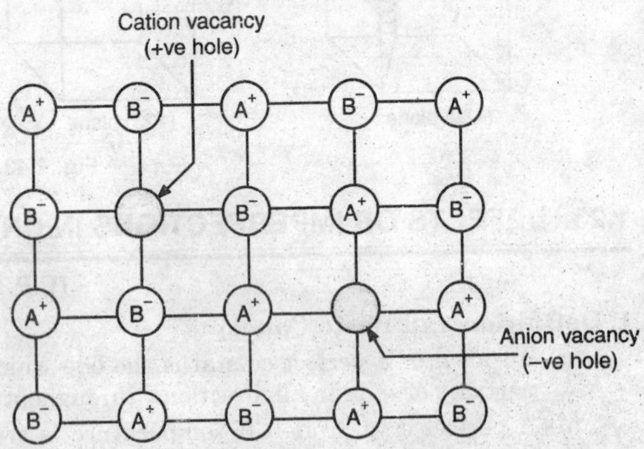

Fig. 1.44. Schottky defect

(B) Frenkel defect*

When an ion (generally cation) shifts from its position to interstitial position, in the crystal lattice then a vacancy or hole is created. The defect is known as Frenkel defect. Generally cations

* The Frankel defect is usually favoured by the compounds (i) with low coordination numbers (ii) having ions of different sizes, and (iii) having highly polarising action and an easily polarisable anion.

are smaller than anions and a cation may occupy the void (empty space) between the anions i.e. a cation occupies the interstitial position between the anions.

This defect can occur in ionic crystals when:

(i) The anion is much larger than size of cation

(ii) The ion has low co-ordination number.

In such cases, the energy required to shift a small cation to interstitital site is small. The Frenkel effect is observed in LiCl, ZnS, AgBr, AgCl, CaF_2 etc.

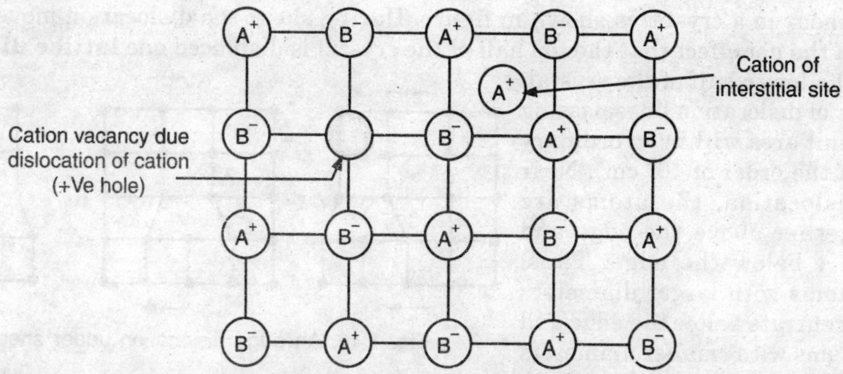

Fig. 1.45. Frenkel defect

5. Performance Comparison of Defects in Schottky and Frenkel defects

S.No.	Schottky defect	Frenkel defect
1.	In this defect, the ion pair (cation and anion) is missing from their respective position in crystal lattice.	In this defect generally the small sized cation gets dislocated from its normal position to interstitial position somewhere in nearby, in lattice.
2.	It occurs in highly ionic compounds with high co-ordination number of ion.	It occurs when an ion is smaller sized than the oppositely charged ion and has low co-ordination number.
3.	An ion pairs are missing in lattice, this defect causes decrease in density of compound.	As in this defect, just position of an ion is changed and not the total number of ions, hence density of crystal remains same.
4.	This defect generally occurs when cations and anions are of nearly same size.	This defect occurs when an ion is much smaller than its oppositely charged ion.
5.	Electrical conductivity increases because of holes.	Electrical conductivity increases slightly but dielectric constant also increases.
6.	Holes are of both negative and positive types.	Holes are of only positive type.

1.24. LINE DEFECTS (i.e., DISLOCATIONS)

1. Definition

Study of stress-strain curve for metal crystals shows that the practically observed shear force to deform a crystal permanently is at least 100 times lower than the theoretically predicted force for a perfect metal crystal. Therefore, line defects are proposed to be present in crystals.

2. Types of Line Defects

There are two types of line defects:

(i) Edge dislocations

(ii) Screw dislocations

(i) Edge dislocations

An edge dislocation is named for the edge of an atomic plane that terminates within the crystal instead of passing all the way through the crystal. The way in which an edge dislocation facilitates shear in a crystal is shown in figure. During shear the dislocation moves across the crystal with the net effect that the top half of the crystal is displaced one **lattice distance** with respect to the lower half of the crystal.

The number of dislocation lines passing through a unit area within an ordinary crystal is of the order of 10^6 cm^2. Near an edge dislocation, the atoms are pushed together above the edge and pulled apart below the edge. Thus, impurity atoms with larger diameters tend to concentrate below the edge and impurity atoms with smaller diameters tend to concentrate above edge. Such binding of impurity at the dislocation tends to make it more difficult to move a dislocation in an alloy than in pure metal. Therefore alloys are mechanically stronger than pure metals.

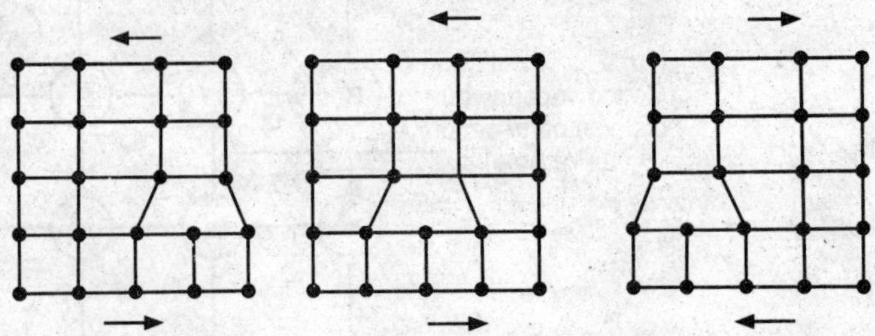

Fig. 1.46. An edge dislocation under shear force

(ii) Screw dislocations

A screw dislocation forms a continuous helical ramp of one set of atomic planes about the dislocation line. A screw dislocation provides site for easy crystal growth because atoms can be added to it. Screw dislocation also gives ease during permanent deformation. In real practice there are both the line defects (edge and screw) in a crystal. Refer figure 1.47.

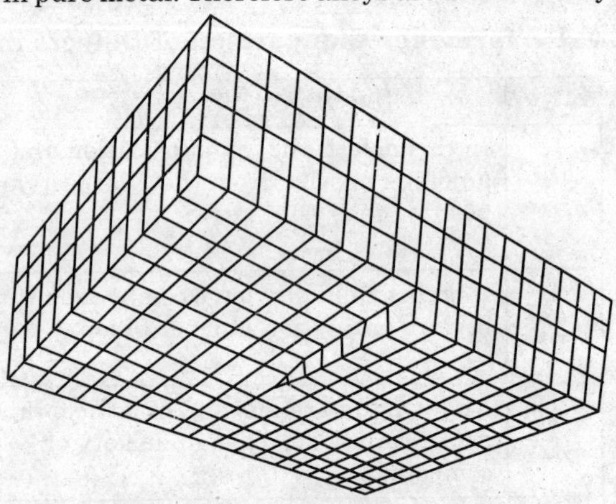

Fig. 1.47. Screw dislocation

3. Effects of line defects

(i) Pure solid becomes easy to deform its shape permanently

(ii) Alloyed metals become stronger than pure metals.

(iii) A deformation of metal does not only involve sliding dislocations already present but it produces more dislocations.

(iv) The line defect site on surface of crystal is also the site of more chemical reactivity.

(v) When a crystal is rubbed (etched), the easiest place to remove atoms from the surface is the site of dislocation.

(vi) When a crystal grows in size, from a melt or vapour, the easiest place for deposition is the line defect site.

1.25. PLANE DEFECTS (i.e., SURFACE IMPERFECTIONS)

(i) When a metal starts crystalling, its crystals grow to form grains. The neighbouring grains to a grain have different crystallographic orientations and there is strain existing between them.

(ii) Therefore, the grain boundaries are at excited state of energy and show more chemical reactivity, corrosion. Further the boundaries between the grains act as site for deposition of impurities as shown in figure 1.48.

(iii) Another type of the plane defect is regarding irregular stacking of atomic planes, when the crystals grow e.g, in CCP kind of atomic packing, the normal patterns of stacking planes is ABC ABC ABC.

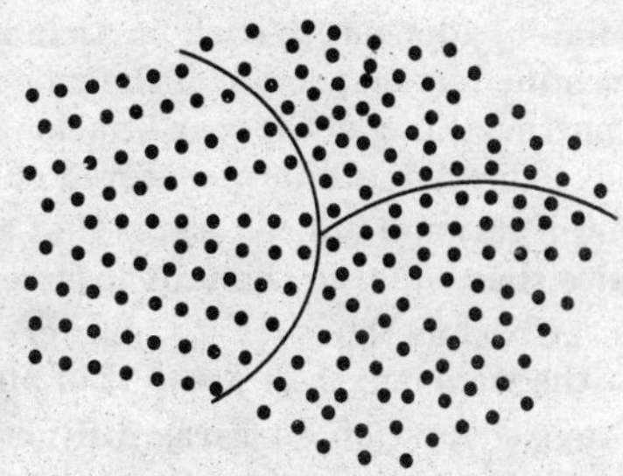

Fig. 1.48. Grain boundaries with different orientations

(iv) If a B layer is accidentally formed on the top of a A layer then the normal repeating pattern of planes is disturbed and changed to ABCBACBA ...

SUMMARY

- Resistivity of metal directly proportional to temperature.
- Resistivity of semiconductor indirectly proportional to temperature.
- Based on bonding there are four types of solids.
  (i) Ionic solids (ii) Molecular solids
  (iii) Covalent solids (iv) Metallic solids
- Unit cell is the smalest three dimensional unit of a space lattice which forms crystal on repeating in all directions.
- There are seven basic crystal systems.
  (i) cubic (ii) orthorhombic
  (iii) tetragonal (iv) monoclinic
  (v) triclinic (vi) hexagonal
  (vii) trigonal
- There are mainly three Bravais lattice i.e. simple cubic, face centered, body centered.
- Density (ρ) of a cubic crystal structure and the lattice constant are related by Equation,

$$\rho = \frac{nM}{a^3 N_A}$$

- The Bragg's law for X-ray diffraction is given by $2\,d \sin\theta = n\lambda$
- When one positive and one negative ion are missing from any site in crystal lattice, then a pair of vacancies or holes are created, this type defect is known as Schottky defects.
- Frenkel defects: When an ion (generally cation) shifts from its position to interstitial position, in the crystal lattice then a vacancy or hole is created. The defect is known as Frenkel defect.
- An edge dislocation is named for the edge of an atomic plane that terminates within the crystal instead of passing all the way through the crystal.
- A screw dislocations forms a continuous helical ramp of one set of atomic planes about the dislocation line.
- A crystal may be considered as a regular array of units-smallest groupings that repeat themselves in all directions in space. The repeating unit obeys the translation operation.
- A lattice is a regular periodic arrangement of imaginary points in space. The structure of all crystals is described in terms of lattice with a group of atoms attached to each lattice point. The groups of atoms is called basis, it is repeated in space to form the crystal structure.
- Lattice points are an array of points generated by the translation operations.
- A unit cell is the smallest repeating unit in the space lattice which, when repeated over and over again, results in the crystal of the given substance.
- The coordination number of a crystal structure is the number of particles (ions, atoms or molecules)

immediately adjacent ιo each particle in the lattice. It is a characteristics of a given space lattice and is determined by inspection of the model. In a simple cubic e.g. NaCl, each particle is surrounded by 6 other particles and so the coordination number is 6. It will be observed that the coordination numbers for body-centered and face-centered cubic lattices are 8 and 12 respectively.

- The law of rational indices states that the intercepts of any face of a crystal along the crystallographic axes ar either equal to the unit intercepts (a, b, c) or some multiples of a, b, c.
- One of the best methods for studying crystal structure by X-ray diffraction. Ths basis of X-ray diffraction is that the wavelength of X-ray is of about the same order as the interplanar distance in the crystal.
- The Bragg's equation for X-ray diffraction is $2d_{hkl} \sin \theta = n\lambda$, where n is an integer.
- An ideally perfect crystal is one which has the same unit cell and contains the same lattice points throughout the crystal. An imperfection as a defect is generally used to describe any deviation of the ideally perfect crystal from the periodic arrangement of its constitutes.
- Lattice constant, $\alpha = \left(\dfrac{nM}{\rho N_A}\right)^{1/3}$.

SHORT QUESTIONS WITH ANSWERS

Q.1. What is a crystal?

Ans. A crystal may be considered as a regular array of units – smallest groupings that repeat themselves in all directions in space. The repeating unit obeys the translation operation.

Q.2. What is crystal lattice?

Ans. A lattice is a regular periodic arrangement of imaginary points in space. The structure of all crystals is described in terms of lattice with a group of atoms attached to each lattice point. The groups of atoms is called basis, it is repeated in space to form the crystal structure.

Q.3. What are the lattice points?

Ans. Lattice points are an array of points generated by the translation operations.

Q.4. How many atoms are assigned to the unit cell of a (a) primitive (b) face centered and (c) body centered cubic lattice?

Ans. (a) In a primitive unit cell of a cubic lattice, one particle is present at each corner of the cibe. Thus, there are in all eight particles. But each corner atom is contributing 1/8 to the unit cell. Therefore, the number of particles per unit cell of this type,

$$= 8 \times \frac{1}{8} = 1$$

(b) In a face-centered unit cell, there are eight corner atoms (one on each corner) and six face particles (one on each face). Therefore, the total number of particles per unit cell of this type

$$= \left(8 \times \frac{1}{8}\right) + \left(6 \times \frac{1}{2}\right) = 1 + 3 = 4$$

(c) In a body centered unit cell, there are eight corner atoms and one at the body centre. Therefore, the total number of particles in a such unit cell

$$= \left(8 \times \frac{1}{8}\right) + 1 = 1 + 1 = 2$$

Q.5. What is unit cell?

Ans. A unit cell is the smallest repeating unit in the space lattice which, when repeated over and over again, results, in the crystal of the given substance.

Q.6. What is the law of rational indices?

Ans. The law of rational indices states that the intercepts of any face of a crystal along the crystallographhic axes are either equal to the unit intercepts (a, b, c) or some multiples of a, b, c.

Q.7. What are Miller indices?

Ans. Miller indices are a set of integers (h, k, l) which are used to describe a plane in a crystal. The interplanar distance d_{hkl} in a crystal is given by

$$\frac{1}{d_{hkl}^2} = \left(\frac{h}{a}\right)^2 + \left(\frac{k}{b}\right)^2 + \left(\frac{l}{c}\right)^2$$

Q.8. Enlist the elements included in Group - IV.

Ans. The elements in group IV are (i) Carbon (ii) Silicon (iii) Germanium (iv) Tin and (v) Lead

Q.9. Give examples of good conductors and semiconductors.

Ans. Conductors - Silver, copper, aluminium etc.

Semiconductors – Germanium, silicon, grey (crystalline) tin, selenium, tellurium etc.

Q.10. Classify the following as conductor or semiconductor.
(i) Gold, (ii) Germanium, (iii) Silver, (iv) Silicon

(U.P. Tech. Sem. Exam., 2007-08, 2009-10)

Ans. Gold and Silver – Conductors.

Germanium and Silicon – Semiconductors

Q.11. What is coordination number of a crystal lattice?

Ans. The coordination number of a crystal structure is the number of particles (ions, atoms or molecules) immediately adjacent to each particle in the lattice. It is a characteristic of a given space lattice and is determined by inspection of the model. In a simple cubic lattice e.g., NaCl, each particle is surrounded by 6 other particles and so the coordination number is 6. It will be observed that the coordination numbers for body-centered and face-centered cubic lattices are 8 and 12 respectively.

Q.12. What is the use of X-ray diffraction?

Ans. One of the best methods for studying crystal structure is by X-diffraction. The basis of X-ray diffraction is that the wavelength of X-ray is of about the same order as the interplanar distance in the crystal.

Q.13. What is Bragg's equation?

Ans. The Bragg's equation for X-ray diffraction is $2d_{hkl} \sin \theta = n\lambda$, where n is an integer.

Q.14. What do you understand by imperfections of the crystal?

Ans. An ideally perfect crystal is one which has the same unit cell and contains the same lattice points throughout the crystal. An imperfection as a defect is generally used to describe any deviation of the ideally perfect crystal from the periodic arrangement of its constitutes.

REVIEW QUESTIONS

1. Explain the classification of solids on the basis of bonding.
2. Define metal and semiconductor and distinguish between them.
3. Explain the properties which changes gradually from one element to the other in the series carbon, silicon, germanium and lead, tin.
4. What do you understand by co-ordination number. Determine the co-ordination number for
   (i) simple cubic (ii) face centered and
   (iii) body centered unit cell.
5. Explain in brief
   (i) unit cell
   (ii) space lattice and co-ordination number
6. How crystal systems have been classified? Write different types of crystal systems.
7. Write short notes on the following:
   1. Elemental semiconductor materials 2. Compound semiconductors materials
   3. Ionic solids 4. Covalent solids
   5. Miller indices 6. Defects in crystal
8. What are the types of defect observed in crystals. Explain briefly the following types of defects.
   (i) Schottky (ii) Frenkel defects
   (iii) Edge dislocations (iv) Screw dislocations

9. Define elemental and compound semiconductor.
10. Distinguish between crystalline, polycrystalline and amorphous solids.
11. Define unit cells and lattice constant.
12. Define lattice point.
13. Explain significance of Miller indices.
14. What do you understand by Symmetry of crystal?
15. How many types of cubic unit cells are there?
16. How many atoms are present in the unit cell
    (a) Primitive (b) Face centered and
    (c) Body centered lattice.
17. Write the name of elements present in Group-IV.
18. Give examples of good conductors and semiconductors.
19. What is co-ordination number of a crystal lattice?
20. Classify the following as conductor or semi-conductor.
    (i) Gold (ii) Germanium
    (iii) Silver and (iv) Silicon
21. Why silver is a good conductor?
22. Explain the significance of Bragg's law.
23. Write the applications of X-ray diffraction.
24. What is Bragg's law?
25. What do you mean by atomic packing factor?
26. What are the different types of imperfections may present in the crystal?
27. How do the following properties vary when we proceed from carbon to lead in group IV?
    (i) Metallic character (ii) Ionization potentials (iii) Electronegativities and (iv) Size of atom.
28. Mention a property which changes gradually from one element to the other in the series carbon, silicon and lead.
29. Give the atomic number and electronic configurations of the elements falling under group IV.
30. Sketch the lattice structure of:
    (i) simple cubic
    (ii) body centered cube, and
    (iii) face centered cube. **(U.P. Tech., Sem. Exam., 2005-06)**
31. Explain in brief:
    (i) unit cell
    (ii) space lattice and coordination number. **(U.P. Tech., Sem. Exam., 2007-08)**
32. What is meant by coordination number. Calculate the coordination number of (i) cubic centered unit cell (ii) face centered unit cell and (iii) body centered unit cell.
33. What is atomic radius? Calculate for (i) simple cubic structure (ii) body centered cubic structure and face centered cubic structure. **(U.P. Tech., Sem. Exam., 2005-06)**
34. What is atomic radius and atomic packing factor. Calculate it for simple cubic structure. **(U.P. Tech., Sem. 2009-10)**
35. How simple cubic, centered cubic, and face-centered cubic structures differ from one another. **(U.P. Tech. Sem. Exam., 2007-08)**
36. Describe with examples (a) unit cell (b) atomic packing factor (c) coordination number (d) Miller indices. **(U.P. Tech. Sem. Exam., 2008-09)**
37. How crystal systems have been classified? Name different types of crystal systems.
38. Name and draw the seven systems of crystals. **(U.P. Tech. Sem. Exam., 2008-09)**
39. What is Miller indices? Explain the procedure for finding Miller indices. **(U.P. Tech. Sem. Exam., 2007-08)**
40. What do you understand by Miller indices? How this is obtained that describes a plane in a crystal? **(U.P. Tech. Sem. Exam., 2005-06)**
41. Derive the formula for density and lattice constant. **(U.P. Tech. Sem. Exam., 2009-10)**
42. Explain Bragg's law. How Braggs equation can be used to determine the lattice parameters? **(U.P. Tech. Sem. Exam., 2006-07)**

43. State and explain Bragg's law. How can it be used to determine the lattice parameters? **(U.P. Tech. Sem. Exam., 2007-08)**
44. Derive an expression for Bragg's law. **(U.P. Tech. Sem. Exam., 2007-08)**
45. Explain with diagrams point, line and surface imperfections. **(U.P. Tech. Sem. Exam., 2008-09)**
46. What is meant by point defects in a crystal lattice? What are different types of point defects? How are they caused? **(U.P. Tech. Sem. Exam., 2008-09)**
47. What types of defects are commonly produced in crystals? Discuss briefly the following types of defects:
    (i) Schottky defects
    (ii) Frenkel defects
    (iii) Metal deficiency defects.
48. Explain with diagrams point imperfection, line imperfection and surface imperfection. **(U.P. Tech. Sem. Exam., 2007-08)**
49. Explain the basic crystal growth techniques. **(U.P. Tech. Sem. Exam., 2009-10)**
50. Write short notes on the following:
    (i) Compound semiconductor materials
    (ii) Chemical bonding
    (iii) Miller indices
    (iv) Imperfection crystals.

NUMERICAL PROBLEMS

1. Determine lattice constant of silicon if its density is 1,249 kg/m^3. **(Ans. 5.3 Å)**
2. A crystal plane intercepts the three crystallographic axes of the following multiples of unit distances, 3/2, 2 and 1. What are the Miller indices of the plane? **(Ans. 4, 3, 6)**
3. Copper crystal has density of 8,900 kg/m^3 and its atomic weight is 63.5. Determine its lattice constant. **(Ans. 3.62 Å)**
4. The crystal structure of copper is FCC mono-atomic. Its density is 8.9 gm/cm^3 and atomic weight is 63.55. Calculate the lattice parameter of copper and the effeective diameter of copper atom. Avcgadro numbers is N = 6.023 × 10^{26}/kg.mole.. **(Ans. a = 3.619 A°, d = 2.55 Å)**
5. Copper has FCC structure. If the interplanar spacing d = 2.08 Å for a set of (111) planes, find the density of copper. Given, atomic weight of Cu = 63.54 and N_A = 6.023 × 10^{23}/gm-mole. **(Ans. ρ = 9.03 gm/cm³)**
6. Calcium chloride (CaCl) has simple cubic structure. The atomic weight of calcium is 132.9 and that of chlorine is 35.5. If the lattice constant of CaCl is 4.12 Å, determine its density. Given Avogadros number = 6.023 × 10^{26}/kg-mole. **(Ans. 3.997 × 10³ kg/m³)**
7. Aluminium has FCC structure (mono-atomic). Its density is 2700 kg/m^3. Calculate the lattice constant and the diameter of the aluminium atom. Given atomic weight of Al = 26.98 and N_A = 6.023 × 10^{26}/kg-mole. **(Ans. 4.05 Å, d = 2.86 Å)**
8. Molybdenum has BCC structure. Calculate atomic radius and mass density. Given that lattice constant a = 3.15 Å, atomic weight of Mo is 95.94 and NA = 6.023 × 10^{26}/kg-mole. **(Ans. r = 1.364 Å, ρ = 10,192.6 kg/m³)**
9. Calcualate the glancing angle on the cube (100) of a rock salt (a = 2.814 Å) corresponding to 2nd order diffraction maximum for X-rays of wavelength 0.714 Å. **(Ans. θ = 14.69°)**
10. The wavelength of X-rays is 7.1 × 10^{-11} m. They are reflected from (110) planes of rock salt crystal of atomic distance 2.8 × 10^{-10} m. Calculate the glancing angle for second order reflection. State the formulae used. **(Ans. θ = 20.79°)**
11. Copper crystal has density of 8,900 kg/m^3 and its atomic weight is 63.5. Determine its lattice constant. **(Ans. 3.62 Å)**
12. What will be wavelength of X-rays which gives a diffraction angle, 2θ equal to 16.8° for a crystal, if the interplanar distance in the crystal is 0.4 nm and only second order diffraction is observed? **(Ans. λ = 0.584 Å)**

□□□

Chapter 2

Energy Bands and Charge Carriers in Semiconductors

Syllabus

Bonding forces and energy bands in solids, charge carriers in semiconductors, carrier concentration drift of carriers in electric and magnetic fields.

Chapter Outline

2.1. INTRODUCTION

As a matter of fact, it is essential to have the knowledge of the behaviour of materials regarding their magnetic, electric optical, dielectric thermal, mechanical properties. We know that quantum mechanics plays a fundamental role in determining all these properties of a material. From electrical point of view, the materials may be classified according to their resistivity range as under:

(i) Conductors

(ii) Semiconductors

(iii) Insulators

(i) Conductors have low resistivity (hence high conductivity) in the range 1.6×10^{-8} to 10^{-6} ohm-m.

(ii) For insulators, the range of resistivity is very high about 10^{7} to 10^{16} ohm-m. Therefore, they are very poor conductors of electricity.

(iii) Semiconductors are the materials whose resistivity (and hence the conductivity) lies between those of conductors and insulators, and has a range of about 10^{-4} to 10^{6} ohm-m.

In this chapter, we shall discuss about semiconductors. The semiconductors devices are highly compact, efficient, more reliable low power consuming, free from mechanical noise and are quito cheap. Therefore, they have replaced the vacuum tubes to a great extent. We can understand the properties of semiconductors on the basis of the band theory of solids. For this, let us first consider the electron energy states of an isolated atom that can be described with the quantum numbers.

In this chapter, we shall also discuss the specific mechanisms by which current flows in a solid. We shall also discuss how the conductivity of a semiconductor varies with change of doping and changes of temperatue.

2.2. ELECTRON ENERGY STATES OF AN ISOLATED ATOM

The four quantum numbers required to specify the electron state in an atom may be listed as under;

(i) The principal quantum number 'n' determines the size and the energy of electron orbit. It can have only integral values n = 1, 2, 3, 4, ... and the corresponding electron orbits are represented by K, L, M, N respectively.

(ii) An orbital quantum number 'l' that determines the shape of an electron orbit. It can have values l = 0, 1, 2, ... (n – l) and the corresponding subshells are denoted by s, p, d, f respectively. The number of electrons contained in a subshell are $2(2l + 1)$.

(iii) An orbital magnetic quantum number determines the oprientation of an orbit in space. It can have values form + l to – l including zero. For a given, l, m_l has $(2l + 1)$ values.

(iv) The magnetic spin quantum number m_s that gives the direction of spin or self rotation. It can have values $+\frac{1}{2}$ and $-\frac{1}{2}$.

These four quantum numbers n, l, m_l, m_s describe completely the state of an electron in an atom. Different physical and chemical properties of various elements are because of different configuration of electrons in their atoms.

(v) According to exclusion principle, no two electrons in an atom can have the same set of quantum numbers n, l, m_l, m_s. Therefore, an energy level cannot have more than two electrons. Hence, the energy level occupied by an electron is discrete in an atom.

2.3. FORMATION OF ENERGY BANDS IN SOLIDS (BAND THEORY OF SOLIDS)

(i) A single isolated atom has discrete energy levels. However, if two atoms come closer to each other, they interact and significant changes in their energy levels can be observed.

(ii) A solid, is an aggregate of atoms in very close proximity.

(iii) To form a solid, a large number of atoms are to be brought very close to each other. Every atom is affected by the presence of the neighbouring atoms.

(iv) During the formation of a solid, it is found that the energy levels of the inner shell electrons are not much affected by the presence of neighbouring atoms.

(v) However, the energy levels of outer shell electrons are changed considerably. These electrons are shared by more than one atoms in the crystal.

(vi) According to Pauli's exclusion principle, not more than two interacting electrons can have the same energy level and therefore new energy levels must be established; which are discrete but infinitesimally different. This group of discrete but closely spaced energy levels is called an energy band.

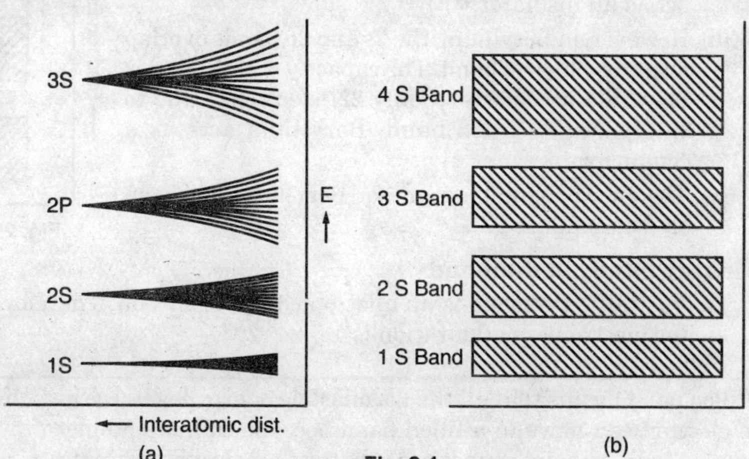

Fig. 2.1.

(vii) Hence, in solids, the allowed energy levels of an atom are modified by the proximity of other atoms. Every discrete level of an individual atom gives rise to a band in solid. Each band contains as many discrete levels as the number of atoms in the solid, i.e., in a solid containing N atoms, there are N possible levels in each band and they can be occupied by 2N electrons.

(viii) Figure 2.1(a) shows the energy level splitting in a solid as a function of interatomic distance.

(ix) The splitting of energy levels is the greatest for the outermost electrons and least for inner electrons. The individual energies in the band are so close together that each band can be considered as continuous. Figure 2.1(b) illustrates the bands fcrmed.

2.4. ENERGY BAND STRUCTURES OF CERTAIN SOLIDS*

The formation of energy bands in solids may be explained for some typical solids as under:

1. Energy bands in Lithium metal

(i) The atomic number of Lithium is 3. Hence, Lithium atom has 3 electrons ($1s^2\ 2s^1$). Consider that Lithium solid has been formed from N Lithium atoms.

(ii) For these N atoms, there will be 3N electrons. 2N electrons will occupy N different energy levels in K shell i.e., Is band.

(iii) This energy band would be completely filled inner band. The remaining N electrons shall occupy the higher energy L shell i.e., 2s band with N energy levels and capacity of 2N electrons.

(iv) Thus the 2s band of Lithium is only half filled and all other higher bands are completely empty. The energy band picture of Lithium has been shown in figure 2.2. Because of the partially filled band, Lithium acts as conductor.

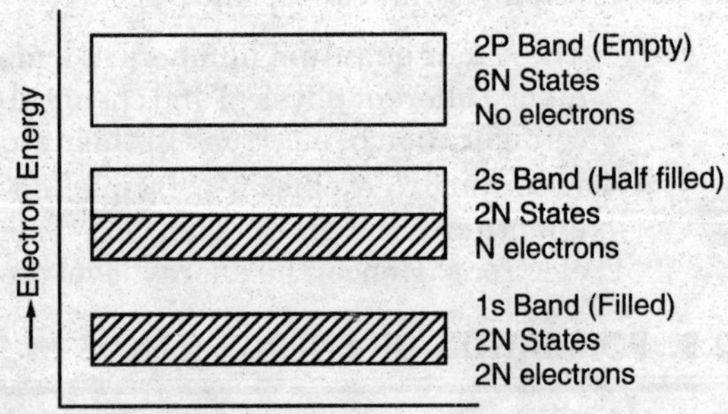

Fig. 2.2. Energy bands in Lithium

2. Energy bands in Beryllium metal

(i) The atomic number of Beryllium is 4. Hence, the electronic configuration is $1s^2\ 2s^2$. Consider a Beryllium crystal with N atoms.

(ii) Out of total 4N electrons, 2N electrons will fill 1s band completely and remaining 2N electrons will fill 2s band completely. Hence, Beryllium seems to act as an insulator.

(iii) However, in Beryllium, the 2s and 2p bands overlap forming a hybrid band. The capacity of this band is 8N while it consists of only 2N electrons. Due to this partially filled band, Beryllium acts as a conductor.

(iv) The energy band diagram for Beryllium has been shown in figure 2.3.

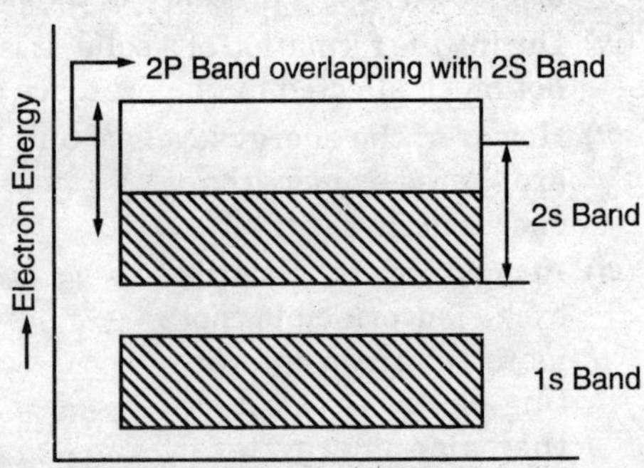

Fig. 2.3. Energy bands in Beryllium

3. Energy bands in diamond

(i) Basically, Diamond is an allotropic form of carbon. Therefore, the atomic number is 6 and the electronic configuration is $1s^2\ 2s^2\ 2p^2$.

* Filled band means that all the permissible energy levels in the band are occupied with electrons. No electron can move in a filled band because there is no blace to move. Thus, an electron in a completely filled band cannot contribute to an electric current.

(ii) In a Diamond crystal with N atoms, the 6N electrons will be distributed as 2N electrons in 1s band, 2N electrons in 2s band and 2N electrons in 2p band. The 2P level has three subshells $2p_x$, $2p_y$ and $2p_z$.

(iii) All the 2N electrons will occupy the $2p_x$ bands will be empty.

Important Point: Hund's rules for pairing of electrons is applicable only to fill the electrons in discrete energy levels of an isolated atom. However, in solids, there are energy bands instead of discrete energy levels. The band of lowest energy is filled first, then the next lowest and so on.

(iv) The interatomic distance in Diamond is such that the unfilled 2Py band is separated from the filled 2Px band by 6eV energy gap. No free electron is available for conduction. Therefore, Diamond acts as an insulator. The energy band diagram for Diamond has been shown in figure 2.4.

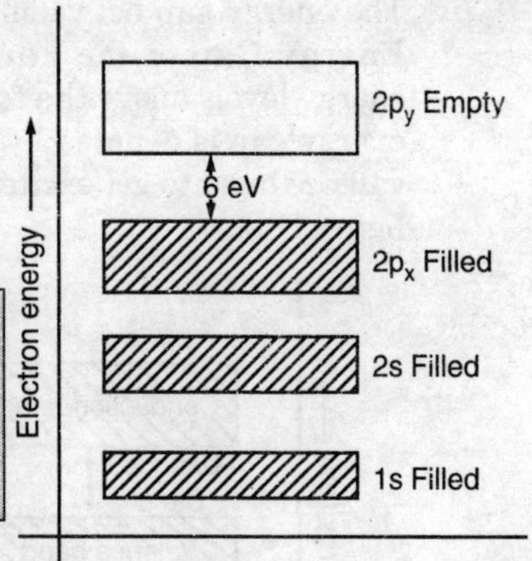

Fig. 2.4 Energy bands in Diamond

4. Energy bands in silicon

(i) Silicon has atomic number 14. Therefore, the electronic configuration is $Is^2\ 2s^2\ 2p^6\ 3s^2\ 3p^6\ 3s^2\ 3p^2$.

(ii) For a silicon crystal with N atoms, the 14 N electrons will be distributed as under:
2N electrons in 1s band
2N electrons in 2s band
6N electrons in 2p band
2N electrons in 3s band
2N electrons in $3p_x$ band

(iii) Thus, $3p_x$ band is completely filled and the unfilled $3p_y$ band is separated from it by, an energy gap 1.1 eV. All bands above $3P_y$ are empty.

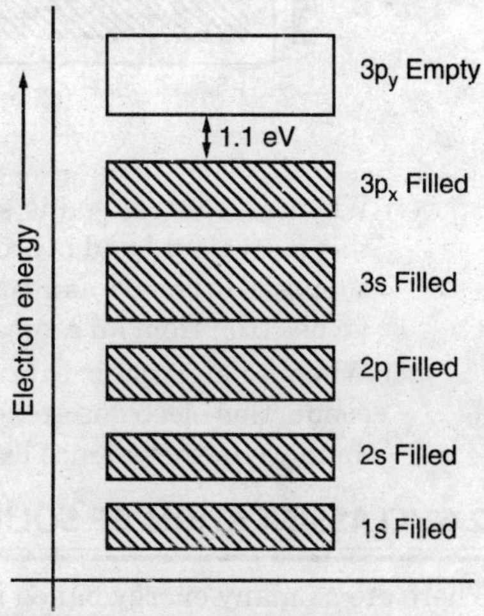

Fig. 2.5. Energy bands in Silicon

(iv) The energy band diagram of Silicon is shown in figure 2.5.

(v) With enough thermal energy some electrons in $3p_x$ band can overcome the gap of 1.1 eV and occupy the empty states in $3P_y$ band. Hence, Silicon acts as a semiconductor.

2.5. VALENCE BAND, CONDUCTION BAND AND FORBIDDEN ENERGY GAP

(i) We know that in an atom, the electrons in inner shells are tightly bound to the nucleus whereas the electrons in the outermost shell (i.e., the valence electrons) are loosely bound to the nucleus. During the formation of a solid, a large number of atoms are brought very close to each other, the energy levels of these valence electrons are affected most. The energies of inner shell electrons are not affected much.

(ii) The band formed by a series of energy levels containing the valence electrons is called the **Valence Band**. It is the highest occupied energy band. It may be completely filled or partially filled with electrons.

(iii) Also, The next higher permitted energy band is called the **Conduction Band**. It may also be defined as the lowest unfilled permitted energy band. it may be empty or partially filled with electrons. The electrons can move freely in the conduction band and hence the electrons in conduction band are known as conduction electrons.

(iv) The energy gap between the valence band and conduction band is called the **Forbidden Energy Gap** or the **Forbidden Band**. This band is formed by series of non-permitted energy levels above the top of valence band and below the bottom of conduction band. This energy gap is denoted by E_g and is the amount of energy to be supplied to the electron in valence band to get excited into the conduction band. The band diagram has been shown in figure 2.6(a).

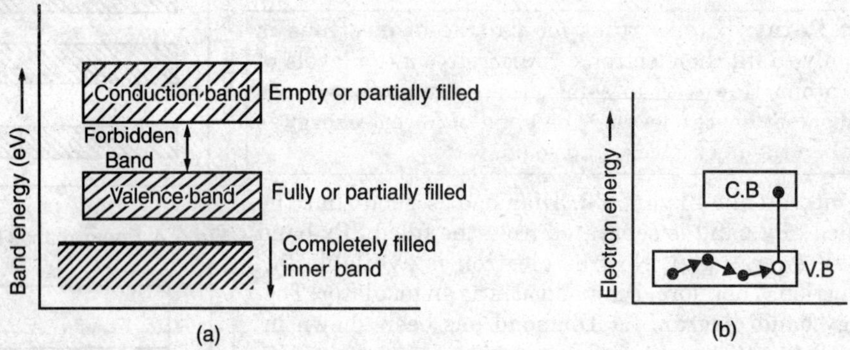

Fig. 2.6.

(v) When an electron gains sufficient energy, it is ejected from the valence band. Because of this a covalent bond is broken and a vacancy for electron, known as hole, is generated. It is supposed to behave as a positive charge. This hole can travel to the adjacent atom by acquiring an electron from an atom. This has been shown in figure 2.6(b).

(vi) When an electron is captured by a hole, the covalent bond is again re-established. Hence, conduction electrons are found in and move freely in the conduction band. The holes exist in an move in the valence band.

2.6. CLASSIFICATION OF SOLIDS BASED UPON BAND THEORY*

There are as many energy bands in a solid as there are energy levels in the parent atom. As far as the electrical conduction properties of solids are concerned, only the valence band and conduction band energies of the electrons are considered. Completely filled bands and completely empty bands do not contribute to electrical conduction. The electrons in the innermost filled shells do not take part in conduction process. Electrons in the conduction band are free and move easily under an electric field. The electrons in valence band are attached to the lattice and are not free to move. If they acquire sufficient energy to cross the forbidden gap E_g, they can occupy the conduction band states and are available for conduction.

Therefore, E_g is the parameter whch decides whether the material acts as conductor insulator or semiconductor. Based upon band theory, solids may be classified into following three categories:

1. Insulators
2. Semiconductors
3. Conductors

1. Insulators

Insulators are those materials in which the valence electrons are bound very tightly to their parent atoms. Thus, very large electric field is needed to remove them from attraction of nuclei. The band picture of insulator has been shown in figure 2.7(a).

* For any given material, the forbidden energy gap may be large, small or non-existent.

The characteristic features of insulators may be listed as under:

(i) A full valence band

(ii) An empty conduction band

(iii) A large energy gap Eg ($\approx$ 5 to 10 eV)

(iv) Very high resistivity (of the order or 10^9 Ω cm).

(v) For some materials at high temperature and under sufficient electric field, some electrons may move from valence band to conduction band. Therefore, insulators show small conductivity at high temperatures.

Illustrative Examples: Diamond, glass

2. Semiconductors

A semiconducting material has electrical properties between those of insulator and good conductors. The band picture of semiconductor has been illustrated in figure 2.7(b).

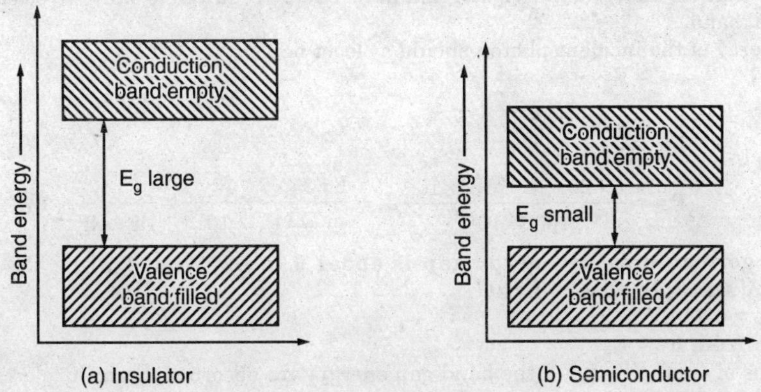

Fig. 2.7.

The characteristic features of semiconductors may be described as under:

(i) Almost empty conduction band and a filled valence band with a very narrow energy gap E_g ($E_g \cong 1$ eV).

(ii) At 0° K, the valence band is completely filled and the conduction band is empty. As temperature increases the electrons in valence band acquire enough energy to cross the small energy gap and move to conduction band. Hence, the conductivity of semiconductors increases with temperature. The electrons moving to conduction band leave behind positive holes in the valence band. Therefore, semiconductor current is the sum of electron and hole currents flowing in opposite directions.

(iii) Resistivity varies from 10^{-12} to 10^9 Ω cm.

Illustrative Examples : Germanium (E_g = 0.7 eV), Silicon (E_g = 1.1 ev).

3. Conductors

The conductors have a large number of free electrons available for electrical conduction. Conductors may be defined as solids characterised by a single energy band called conduction band, which is partially filled at any temperature.*

Alternatively we can also define a conductor as a solid, in which the conduction and valence bands overlap and there is no energy gap between the two bands. These two situations have been shown in figures 2.7(c) and (d). Since there is no forbidden gap, there is no structure to establish holes. The total current is simply due to the flow of electrons.

Illustrative Examples: Lithium, Beryllium, Sodium etc.

* Due to overlapping of valence and conduction bands, a slight potential difference across the conductor causes the free electrons to constitute electric current.

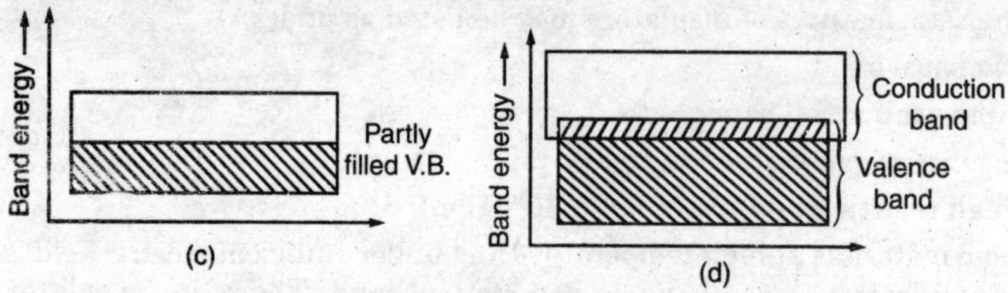

Fig. 2.7. Conductor

EXAMPLE 2.1. Evaluate the energy gap in Si given that it is transparent to radiation wavelength greater than 11,000 Å.

Data: λ = 11000 Å

Formula required: $h\nu = E_g$

Solution: We know that the energy gap E_g is the minimum energy required to shift the electron from valence band to conduction band.

Therefore, Energy of the incident photon should at least be

$$h\nu = E_g$$

Therefore $$E_g = h\nu = h \cdot \frac{c}{\lambda}$$

or $$E_g = \frac{6.625\times10^{-34}\times3\times10^{8}}{11000\times10^{-10}}\,J = \frac{6.625\times3}{111}\times\frac{10^{-26}}{10^{-7}}\times\frac{1}{1.6\times10^{-19}}\,eV = 1.29\ eV$$

EXAMPLE 2.2. In germanium, the energy gap is about 0.75 eV. What will be the wavelength at which germanium starts to absorb light?

Data: $E_C - E_F$ = 0.75 eV

Formula required: $h\nu = E_g$

Solution: If photons of energy $h\nu = E_g$ (the band gap energy) are absorbed, then the electrons can go from valence band to conduction band.

Therefore, $$h\nu = E_g$$

or $$\frac{hc}{\lambda} = E_g$$ where λ is the corresponding wavelength

Hence, $$\lambda = \frac{hc}{E_g}$$

Now, we have $$E_g = 0.75\ eV = 0.75 \times 1.6 \times 10^{-19}\ J$$

or $$\lambda = \frac{6.63\times10^{-34}\ J\times3\times10^{8}\ m/s}{0.75\times1.6\times10^{-19}\ J} = 16.575\times10^{-7}\ m$$

or $$\lambda = 16575\ Å$$

2.7. INTRINSIC AND EXTRINSIC SEMICONDUCTORS (U.P. Tech. Sem. Exam, 2005-06, 2008-09)

The semiconductors may be classified into following two categories as:

1. Intrinsic semiconductors and
2. Extrinsic semiconductors

1. Intrinsic Semiconductors

(i) Intrinsic means pure, so intrinsic semiconductors are the semiconductors in their purest possible form. The presence of impurity (i.e., atoms of other material) is as low as 1 part per 100 million parts of the semiconductors.

(ii) The intrinsic semiconductors are insulators or very poor conductors, at room temperature.

(iii) The increase in temperature or application of increased voltage does not increase their conductivity significantly.

(iv) Therefore, the intrinsic semiconductors are not practically used for manufacturing of devices.

2. Extrinsic Semiconductors

(i) Extrinsic means impure, so we can obtain the extrinsic semiconductors from intrinsic ones by adding impurities to them. Impurity is nothing but some other material.

(ii) The process of adding impurities is called as **Doping**. Due to doping, the conductivity of the semiconductors increase.

(iii) Thus extrinsic semiconductors have a better conductivity than the intrinsic semiconductors. Therefore, they are used in manufacturing of all the electronic components such as diodes, transistors etc.

(iv) Extrinsic semiconductors are of two types namely:

(a) N-type semiconductors

(b) P-type semiconductors

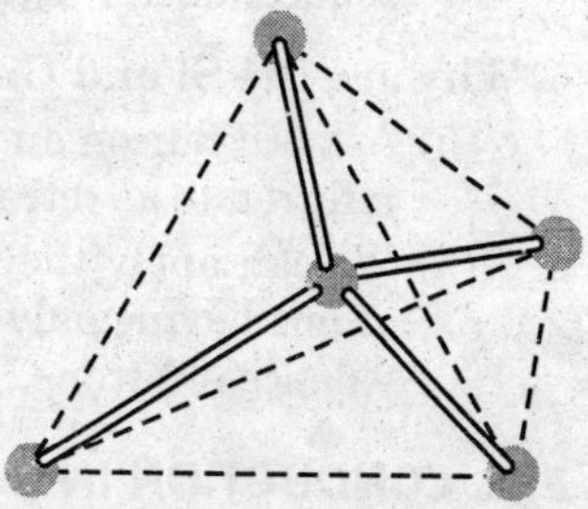

Fig. 2.8. Silicon and Germanium crystal structure

The type of extrinsic semiconductor (N or P) depends on the type of impurity (or dopant) being used.

2.8. CRYSTALLINE STRUCTURE OF INTRINSIC SEMICONDUCTOR

1. Crystal Structure

(i) Some of the unique qualities of Silicon and Germanium are due to their special type of atomic structure.

(ii) The atoms of both these materials form a very definite pattern which is periodic in nature *i.e. it continuously repeats itself.*

(iii) One such complete pattern is called as crystal and the periodic arrangement of atoms is called as lattice.

(iv) For the Silicon and Germanium the crystal has a three dimensional diamond shape as depicted in figure 2.8.

2. Two dimensional representation of a silicon crystal

(i) We know that in a Silicon or Germanium atom there are four valence electrons.

(ii) In an intrinsic Silicon or Germanium crystal these four valence electrons are bounded to four adjoining atoms as depicted in the two dimensional representation of Silicon or Germanium crystal of figure 2.8.

3. Covalent bonds

(i) Each one of the four valence electrons in each atoms forms a bond with a valence electron from the adjoining atom as shown in figure 2.9. This bond is nothing but sharing of electrons.

(ii) These bonds are known as the covalent bonds.

(iii) The covalent bonds result in a stronger bond between the valence electrons an their parent atoms.

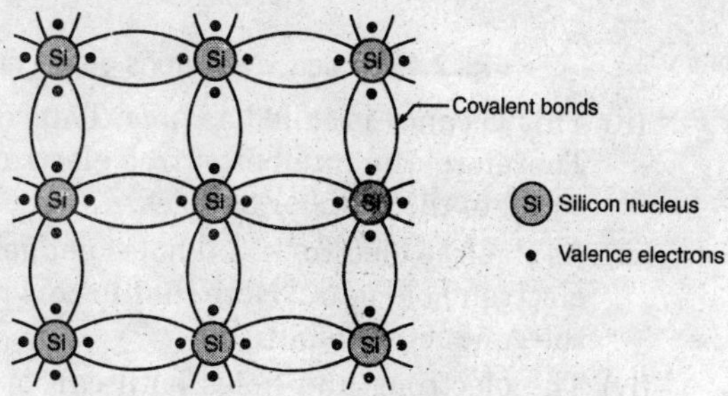

Fig. 2.9. Two dimensional representation of a silicon crystal

4. Why intrinsic semiconductors behave like insulators at low temperatures?

(i) In each covalent bond, both the electrons involved are shared by both the parent atom. Thus, with four covalent bonds associated with each atom, all the silicon atoms with effectively have eight electrons in their valence shells.

(ii) Four of these electrons are their own and remaining four are due to the four covalent bonds. Due to this, the outermost shells of all the atoms are completely filled.

(iii) This makes the outermost shells of all atoms stable. Therefore not a single free electron is available at the absolute zero temperature i.e. at 0° K or – 273°C. Hence the intrinsic semiconductor acts like an insulator at the absolute zero temperature.

5. Why are the Si and Ge referred to as tetravalent atoms?

(i) As both Silicon and Germanium atoms have four electrons in their valence shells they are referred to as tetravalent atoms.

(ii) We can apply the same logic to define *trivalent* and *pentavalent* atoms. Trivalent means atoms having only three valence electrons while pentavalent refers to an atom having five valence electrons.

2.9. CONDUCTION IN INTRINSIC SEMICONDUCTORS (U.P. Tech, Sem. Exam; 2011-12)

As a matter of fact, the intrinsic semiconductors behave like perfect insulators at the absolute zero temperature. However, the behaviour changes with increase in temperature. At around the room temperature, electrons become available for conduction and current can flow. The generation of free electrons takes place as explained below in following subsection.

2.9.1. Thermal Generation of Carriers in Intrinsic Semiconductors

With increase in temperature, many valence electrons absorb the thermal energy, break the covalent bonds and go into the conduction band. Therefore, they become free for conduction. In fact, these electrons are called **conduction electrons.**

1. Generation of a hole

(i) When an electron breaks a covalent bond and becomes free, a vacancy is created in the broken covalent bond.*

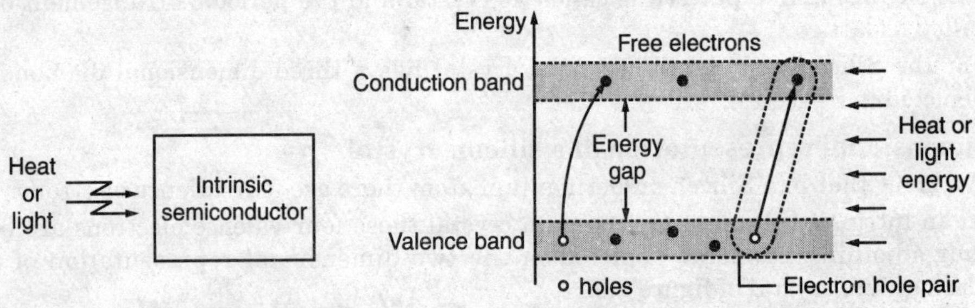

Fig. 2.10. Valence electrons go to conduction band due to absorbed thermal energy

(ii) This vacancy is called as *hole*. Thus corresponding to every free electron, a hole is created. Therefore, the number of free electrons generated due to increased temperature is exactly equal to the number of holes.

(iii) As the free electrons and holes get generated in pairs they are called as thermal generated electron hole pairs. Note that in this pair the hole is in the valence band when electron is in the conduction band.

(iv) The electrons and holes both can operate as charge carriers. The hole is said to have a positive charges as it is nothing but absence of an electron.

(v) The generation of electron hole pairs is shown in figues 2.10 and 2.11. The conductivity of intrinsic semiconductor thus increase due to the increase in temperature.

2. Effect of light on semiconductor

The effect of light incident on the semiconductor is the same as that of the heat.

* The liberated electron and the hole left behind is called an electron-hole pair.

(i) When light of adequate intensity and frequency is focussed on the semiconductor, photons will impart energy to the electrons.

(ii) Due to this energy, electrons will break their covalent bonds and enter into conduction band and become free for conduction.

(iii) The conductivity of a semiconductor thus increases due to incident light. In other words the resistivity of the material decreases due to the light incident on it.

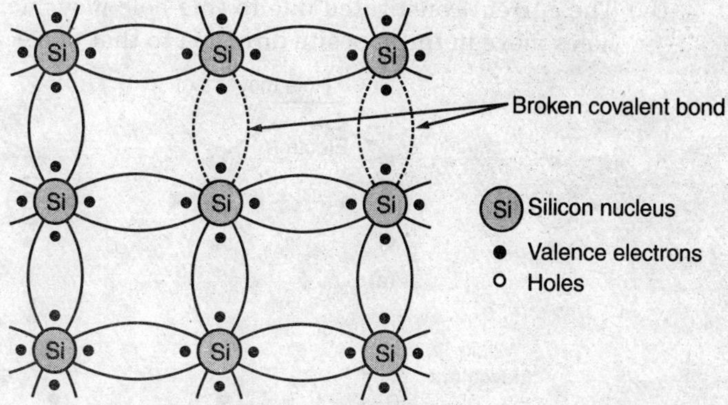

Fig. 2.11. Generation of electron-hole pairs due to increased temperature

2.9.2. Recombination*

(i) The free electrons in the conduction band, when come across the holes will jump in the holes. This process is called as the recombination process.

(ii) After the electron-hole pairs are generated due to external thermal energy, to conduction band electrons tend to return back to valence band.

(iii) The recombination was take place when a free electron returns to valence band.

(iv) When a conduction band electron returns back to valence band, and recombines with hole, it releases energy equal to the energy gap between conduction and valence bands. This principle is used in the light emitting diode (LED).

2.9.3. Electron and Hole Current

1. Electron current

(i) As we apply voltages across a piece of intrinsic semiconductor material, the thermally generated free electrons in the conduction band are attracted towards the positive end.

(ii) The current constituted due to movement of free electrons is called as electron current as depicted in figure 2.12.

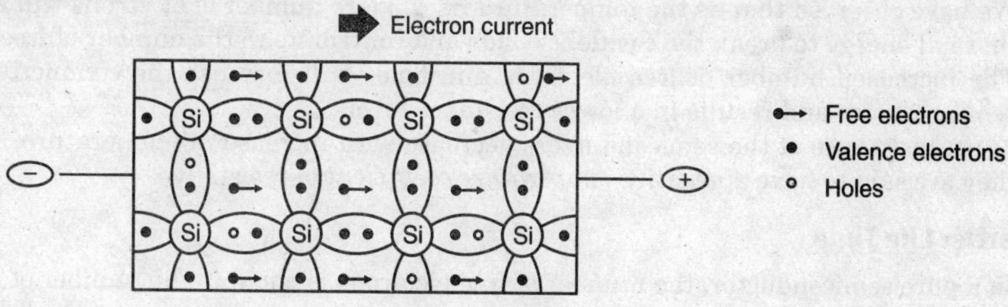

Fig. 2.12. Electron current

2. Hole current

(i) Let us consider to figure 2.13(a) that shows three neighbouring silicon atoms A, B, and C. The covalent bonds do exist but for convenience they have not been shown. A hole exists in a broken covalent bond with atom A as shown in figure 2.13 (a).

(ii) A valence electron may jump from another atom B to fill up this hole as shown in figure 2.13 (b).

* When an electron is freed from valence band, certain amount of energy is absorbed. The same amount of energy is released when an electron falls into a hole.

(iii) When it jumps, the electron leaves a hole behind it and thus the hole has moved in the direction opposite to that of the electron. In a similar manner, the hole passes from atom B to atom C as shown in figure 2.13 (c).

(iv) The current constituted due to this hole movement is known as the hole current. Hence, holes move in the opposite direction to that of electrons.

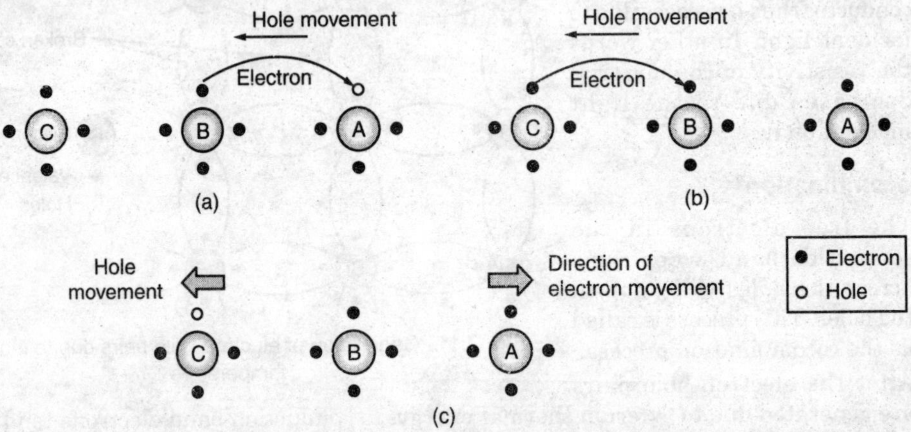

Fig. 2.13. Movement of electrons and holes

Important point: Here, it is important to note that the hole current flows due to the valence electrons and not due to the free electrons. These electrons move due to the presence of holes. Therefore, *holes* also are considered as charge carriers like the electrons, and the 'hole current' will always flow in opposite direction to that of the "electron" current.

3. Total current

The flow of an electric current is due to the movement of electrons in the conduction band and movement of holes in the valence band.

Total current = Electron current + Hole current

4. Semiconductors have a negative temperature coefficient of resistivity

(i) We have observed that as the temperature rises, more number of electrons will absorb the thermal energy to break the covalent bonds and contribute to the number of free carriers.

(ii) The increased number of free electrons and holes will increase the conductivity of the semiconductor and results in a lower resistance level.

(iii) Thus, resistance of the semiconductors decrease with increase in temperature. Therefore, they are said to have a *negative temperature coefficient* of resistivity.

2.9.4. Carrier Life Time

(i) In a pure semiconductor, the number of free electrons is equal to the number of holes. The thermal energy keeps producing new electron-hole pairs whereas some of the hole electron pairs will disappear due to the process of recombination.

(ii) Thus, thermal agitation gives rise to production of electron-hole pairs while the recombination will make them disappear. The time duration between these two processes is called as the carrier lifetime.

(iii) On an average, a hole or an electron will exist for t_p (for holes) or t_n (electron) respectively. This time is called as *mean lifetime* of holes and electrons respectively. The value of carrier lifetime ranges from a few nanoseconds to few microseconds.

(iv) The carrier lifetime is very important because it indicates the time taken by the electron or hole concentrations, to return to their equilibrium value.

(v) The carrier lifetime can be precisely adjusted to a desired value by introducing gold as recombination agent into the silicon under controlled condition.

2.10. CURRENT FLOW IN INTRINSIC SEMICONDUCTORS

1. Basic Concepts

Let us consider that a battery is connected across a slab of an intrinsic semiconductor as shown in figure 2.14. The free electrons will be pulled towards the positive end of the external battery and the holes will experience a pull towards the negative end of the battery. Hence, there will be two types of current flowing namely the conventional current and electron current flow.

2. Conventional Current

The current flow from positive to negative is referred to as conventional current. Every graphical symbol used to represent an electronic device has an arrowhead which indicates conventional current direction. Therefore, the electronic circuits are most explained by using the conventional current flow.

3. Electron Flow (Electron Current)

Under the influence of the external dc source or battery the free electrons in the semiconductor slab which are negatively charged will get attracted towards the positive terminal and holes being positively charged will be attracted towards the negative terminal of the external battery. Thus as depicted in figure 2.14, the electrons flow exactly in the opposite direction to that of the conventional current. The study of electron movement is essential in understanding the operation of any electronic device.

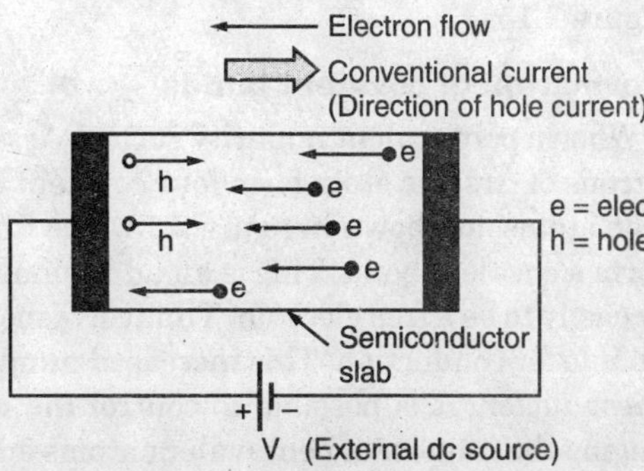

Fig. 2.14. Conventional current and electron flow in semiconductors

2.11. EXTRINSIC SEMICONDUCTORS

1. Doping Process

In the previous section we have learnt that to increase the conductivity of intrinsic semiconductors we have to supply the heat energy or light energy externally. These methods of increasing the conduction are not practically usable. Hence, the conductivity is increased by means of the Doping process. In the process of doping, impurities are added to the pure Silicon or Germanium. The impurities are the materials used to dope the intrinsic semiconductor materials.

These materials can be of two types:

(i) Donor impurity

(ii) Acceptor imputity

(i) Donor impurity: The material which is being used as impurity in the process of doping is called as dopant. When the dopant is a pentavalent atom i.e., the atom containing five valence electrons then it is called as the *donor impurity* and the doping is called as *donor doping*. Donor doping is used to manufacture n-type extrinsic semiconductor.*

(ii) Acceptor impurity: When the dopant is a trivalent atom i.e., the atom consisting of only three valence electrons, then it is called as the *acceptor impurity* and the doping is called as *acceptor doping*. Acceptor doping is used to manufacture p-type extrinsic semiconductors.

* Though each impurity provides only one free electron yet an extremely small amount of impurity provides enough atoms to supply millions of free electrons.

2.11.1. N-type Semiconductors

1. Basic Concepts

The n-type semiconductor is formed by adding s small amount of *pentavalent* impurity to the pure Silicon or Germanium material which acts as a base material. The *pentavalent* element is the one which has five valence electrons. The examples of pentavalent materials are: Antimony, Arsenic and Phosphorus. The effect of addition of such pentavalent impurity atoms to the silicon base has been shown in figure 2.15.

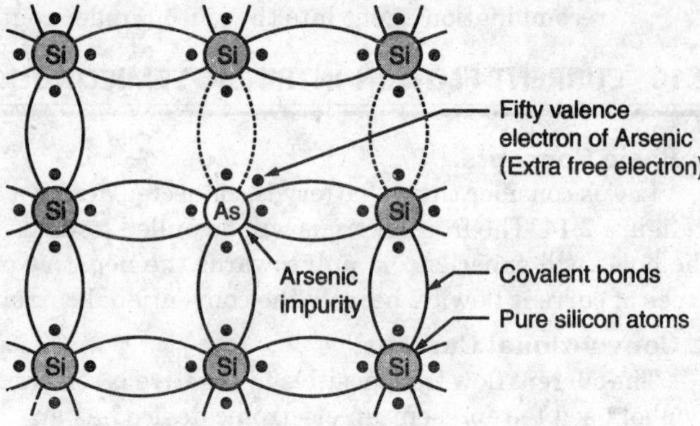

Fig. 2.15. Formation of n-type semiconductor

2. Formation of covalent bonds

When a pentavalent impurity such as Arsenic is added to the intrinsic semiconductor four valence electrons of Arsenic atom form four covalent bonds with four valence electrons of the neighbouring silicon atoms, as shown in figure 2.15. The fifth electron of the Arsenic atom does not have a chance to form a covalent bond. This is an additional valence electron which can enter the conduction band very easily to be a free electron. Thus corresponding to each impurity atom an extra electron becomes available for conduction. This increased number of electrons will make the semiconductor an *N-type semiconductor*. It is possible to control the number of free electrons available for conduction but changing the number of pentavalent atoms added to silicon material. Thus, conductivity is increased in a controlled manner with the help of the doping process.

3. Donor impurities

Each pentavalent impurity atom generates a free electron as discussed earlier. The number of free electrons will be equal to the number of impurity atoms added. As each impurity atom 'donates' one free electron, the pentavalent impurities are known as the *donor* impurities.

4. Majority and minority carriers in the n-type material

A large number of free electrons are present alongwith a small number of thermally generated holes in an N-type semiconductor. Hence, the conduction largely takes place due to the free electrons. Therefore, the free electrons are called as *majority carriers* and holes are known as *minority carriers*.

5. Conduction in n-type material

When an external DC voltage is applied to the N-type semiconductor material, the free electrons move towards the positive terminal of the source and holes move towards the negative end as depicted in figure 2.16.

As electrons outnumber the holes, the conduction in N-type material is mainly due to the majority carriers *electrons*.

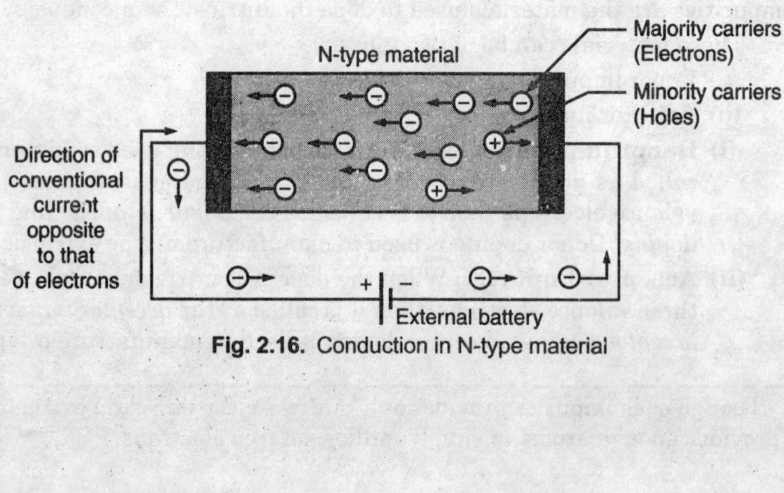

Fig. 2.16. Conduction in N-type material

2.11.2. P-type Semiconductors

1. Basic Concepts

The P-type semiconductors are formed by doping a pure germanium or silicon crystal with the *trivalent* impurity atoms. The materials having three valence electrons per atom are known as the **trivalent materials.** Examples of *trivalent* materials are : Boron, Gallium and Indium. The effect of adding one of these elements *Gallium* on the base of silicon has been shown in figure 2.17.

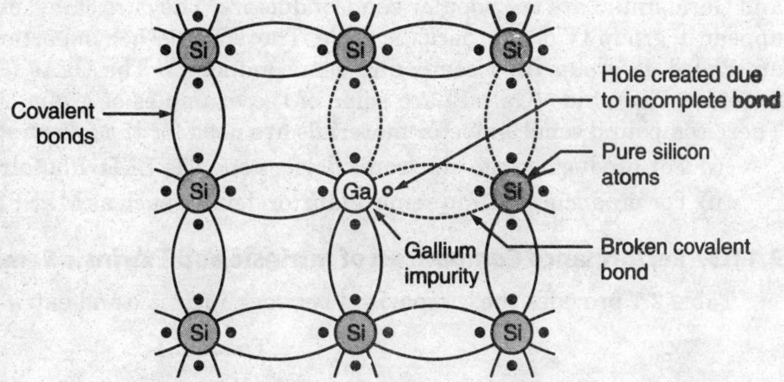

Fig. 2.17. Formation of P-type semiconductor

2. Covalent bonds and creation of a hole

As shown in figure 2.17, when a Gallium atom is added to the silicon base, its three valence electrons will form covalent bonds with the valence electrons of three neighbouring silicon atoms. The fourth covalent bond however, remains incomplete as the Gallium atom has only three valence electrons. The resulting vacancy is known as a *hole* and it is represented by a small circle in figure 2.17. A hole is positively charged as it represents the absence of a negative charge. Hence, corresponding to each trivalent impurity atom added, a hole is created. This increased number of holes make the semiconductor a P-type semiconductor. It is possible to control the number of holes by controlling the *doping concentration* of trivalent impurity atoms.

3. Acceptor impurities

Since the vacancy resulting due to the incomplete covalent bond will readily accept a free electron, the trivalent impurities are known as acceptor impurities.

4. Majority and minority carriers in p-type material

A large number of holes are present in a P-type semiconductor along with a small number of thermally generated electrons. hence, the conduction largely takes place due to the holes. Therefore, holes are called as the *majority carriers* and electrons are *minority carriers* for a P-type semiconductor. This is exactly opposite to what we learnt for the N-type semiconductor. The conduction inside a P-type semiconductor has been shown in figure 2.18.

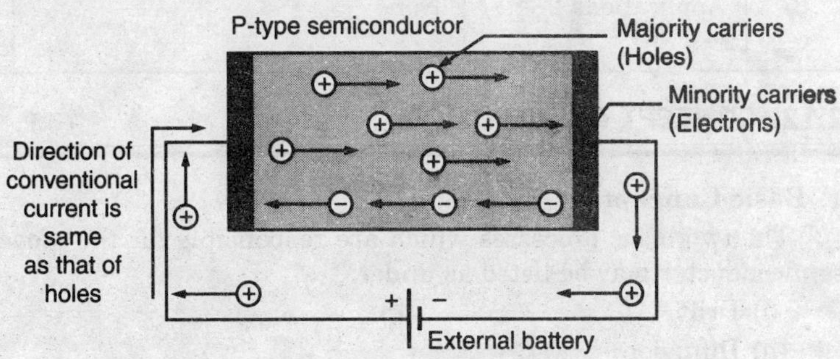

Fig. 2.18. Conduction in a P-type semiconductor

When an external dc source is applied across a P-type material, the holes move inthe valence band. The movement of holes is towards the negative terminal of the supply. The free electrons move towards the positive terminal of the battery, as shown in figure 2.18. Here, it may be noted.

Important Point: Here, it may be noted that the P and N type semiconductors are electrically neutral. P or N does not mean that they are positively or negatively charged.

5. Semiconductor materials used for opto devices, modern semiconductor devices and ICs

Elements in the periodic table are grouped according to the number of valence electrons. Silicon and Germanium are the popular semiconductors. They are called as elemental semiconductors and appear in group IV of the periodic table. There is another important type of semiconductors. They are called as group III-V compound semiconductors. The GaAs (Gallium arsenide) and Gallium Arsenide Phosphide (GaAsP) are some of the examples of group III-V compound semiconductors. These compound semiconductor materials are used for the following applications:

(i) For producing opto electronic devices such as LED, Phototransistors, Photodiodes etc.

(ii) For producing certain semiconductor devices such as MESFET and integrated circuits (ICs).

2.11.3. Performance Comparison of Intrinsic and Extrinsic Semiconductors

Table 2.1 provides the comparison between intrinsic and extrinsic semiconductors.

Table 2.1

S.No.	Parameter of comparison	Intrinsic semiconductors	Extrinsic semiconductor
1.	Purity	Extremely pure. No impurity added	Not pure as it contains impurity atoms
2.	Doping	Not done	Is done
3.	Type	—	P or N type
4.	Number of electrons and holes	Number of electrons is always equal to number of holes	They are never equal to each other
5.	Conductivity	Poor	Much higher
6.	Current conduction	Due to electrons and holes	Mainly due to electrons or holes depending on type of semiconductor
7.	Position of Fermi level	At the centre of the forbidden	It is either near the conduction band or near valence band
8.	Applications	None	For manufacturing of electronic components

2.12. CONCEPT OF DIFFUSION

(U.P. Tech, Sem. Exam; 2008-09, 2010-11)

1. Basic Concepts

The two basic processes which are responsible for the movement of electrons and holes in a semiconductor may be listed as under:

(i) Drift

(ii) Diffusion

The transport of charges takes place in a semiconductors which can be explained by a mechanism called diffusion. The diffusion can be encountered only in semiconductors. It is absent in the metals.

2. Non-uniform concentration of particles

It is possible to have a non uniform concentration of particles in a semiconductor as shown in figure 2.19 (a). The concentration of holes i.e., p varies with distance 'x' in the semiconductor due to which there exists a concentration gradient dp/dx, in the density of carriers as depicted in figure 2.19 (b).

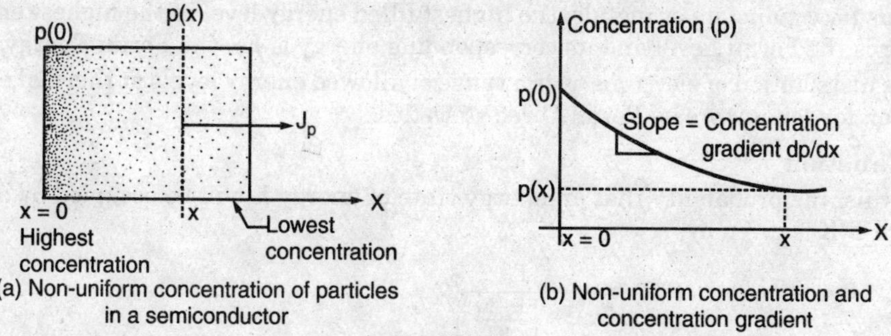

Fig. 2.19.

3. Transport of holes

Let us consider figure 2.19(a) which shows that the concentration of holes on the left of the dashed line is more than the concentration on its right hand side. The holes are always in random motion due to their thermal energy. Due to this random motion, they will move back and forth across the dashed line shown in figure 2.19(a). In a given time, due to this back and forth motion, we can expect that more number of holes will across the dashed line from left to right than in the reverse direction. That means more number of holes will move from higher concentration side to the lower concentration side.

2.12.1. Diffusion Current

This net transport of holes across the surface constitutes a current in the positive X direction which is called as the *diffusion current.** Here, it may be noted that the diffusion current does not flow due to the repulsive force that exists between the holes.

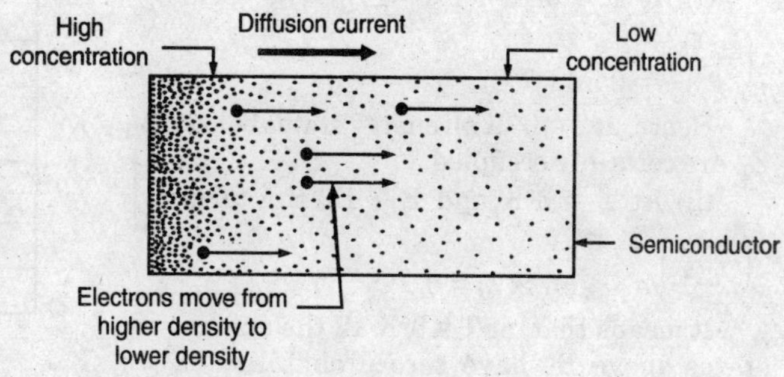

Fig. 2.20. Diffusion current

It exists because of the non-uniform concentration i.e. due to the existence of concentration gradient. Such a diffusion current will be present even in the n-type semiconductor, due to the non-uniform concentration of electrons, as depicted in figure 2.20.

2.13. FERMI LEVEL AND FERMI FUNCTION: DETAILED ASPECTS

(U.P. Tech, Sem. Exam; 2010-11)

1. Definition and Basic Concept

(i) A solid is formed by bringing a large number of atoms together. Each atomic level splits into a large number of closely packed levels and energy band are formed. The filling of the bands takes place such that the lowest energy bands are filled first then the next lowest and so on; till all the electrons are accommodated.

(ii) In case of metals i.e. conductors, only electrons are the current carriers. The highest occupied energy level at 0°K is called the Fermi Level and the energy corresponding to it is called the **Fermi-Energy E_F**. At 0°K, all energy states upto E_F are occupied and all states above E_F are empty.

(iii) At higher temperatures the random thermal energy can excite the electron to higher energy states and there may be some empty states below E_F.

* Diffusion current occurs when charge carriers diffuse from a point of concentration, to spread uniformly throughout the volume of a piece of material.

(iv) Thus for conductors or metals, the highest filled energy level in the highest energy band at 0° k is the Fermi Level and its corresponding energy is E_F; the Fermi Energy.

(v) The distribution of electrons over a range of allowed energy levels at thermal equilibrium is given for conductors by Fermi-Dirac statistics.

2. Fermi Function

As per this, the probability that an energy state of energy E will be occupied by an electron at temperature T°K is given by,

$$P(E) = \frac{1}{1 + e^{\left(\frac{E - E_F}{KT}\right)}}$$

where, K = Boltzmann's constant

E_F = Fermi energy

P(E) = Distribution function known as **Fermi function.**

3. Observation from fermi curves

This function as plotted for different temperatures is shown in figure 2.21. From these curves, it is observed that:

(i) At T = 0°K, for $E < E_F$, the term $e^{\left(\frac{E-E_F}{KT}\right)} = 0$ and P(E) = 1

Hence, at T = 0° K all energy states below E_F are certainly occupied.

(ii) At T = 0°K and $E > E_F$ the term $e^{\left(\frac{E-E_F}{KT}\right)} = \infty$ and P(E) = 0

It means that, at T = 0°K all the energy states above E_F have zero probability of occupancy and hence are empty.

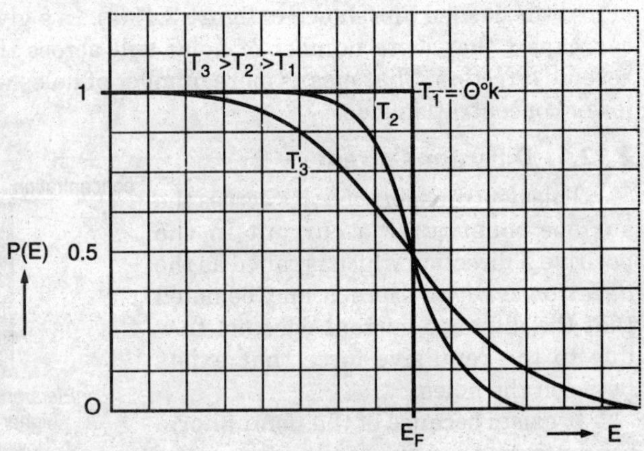

Fig. 2.21.

(iii) At T > 0°K and $E = E_F$, the term $e^{\left(\frac{E-E_F}{KT}\right)} = 1$ and $P(E) = \frac{1}{2}$ i.e., the energy state at the Fermi Level has 50% probability of being occupied by an electron at any finite temperature T > 0°K.

2.14. FERMI LEVEL AND FERMI FUNCTION

(U.P. Tech, Sem. Exam; 2010-11)

1. Basic Concepts

(i) In semiconductors, we have two types of current carriers. These are electrons and holes. In semiconductors, Fermi Level is the reference level that describes the probability of occupancy of states in conduction band as well as the unoccupied states in valence band.

(ii) Fermi level in semiconductors, may be defined as the energy which corresponds to the centre of gravity of conduction electrons and holes weighted according to their energies.

(iii) Fermi level is only an abstraction. The Fermi level lies in between the conduction band and valence band. For the case of an intrinsic semiconductor,the Fermi level lies at the centre of the forbidden band implying that the states occupied in conduction band is equal to the states unoccupied in valence band.

(iv) This means that for every electron in conduction band, there is a hole in valence band.

2. Location of Fermi energy level in semiconductors

(i) Intrinsic semiconductor:

In pure or intrinsic semiconductor the concentration of free electron in conduction band, the concentration of free holes in valence band is exactly same. Therefore, the Fermi level in intrinsic semiconductor lies exactly at the middle forbidden band as depicted in figure 2.22.

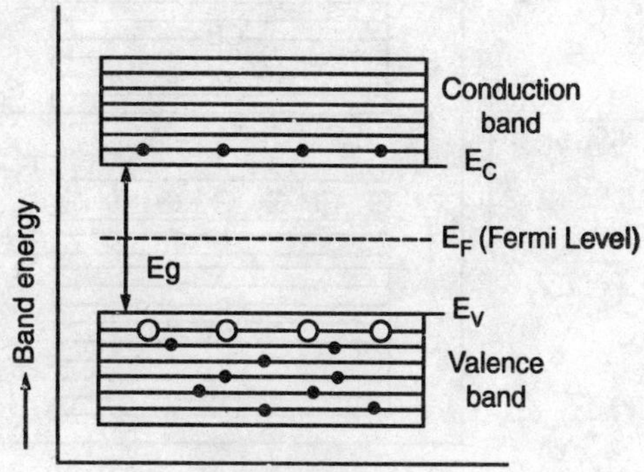

Fig. 2.22.

(ii) Extrinsic semiconductor

The position of Fermi level in extrinsic semiconductor depends both upon doping and upon the temperature. If the intrinsic semiconductor is doped, the Fermi level is always found to towards the energy band, that has higher concentration of charge carriers.

(a) N-type semiconductor

In N-type semiconductor the concentration of free electrons in conduction band is greater than that of free holes in valence band. Therefore, the Fermi level gets shifted towards the conduction band.

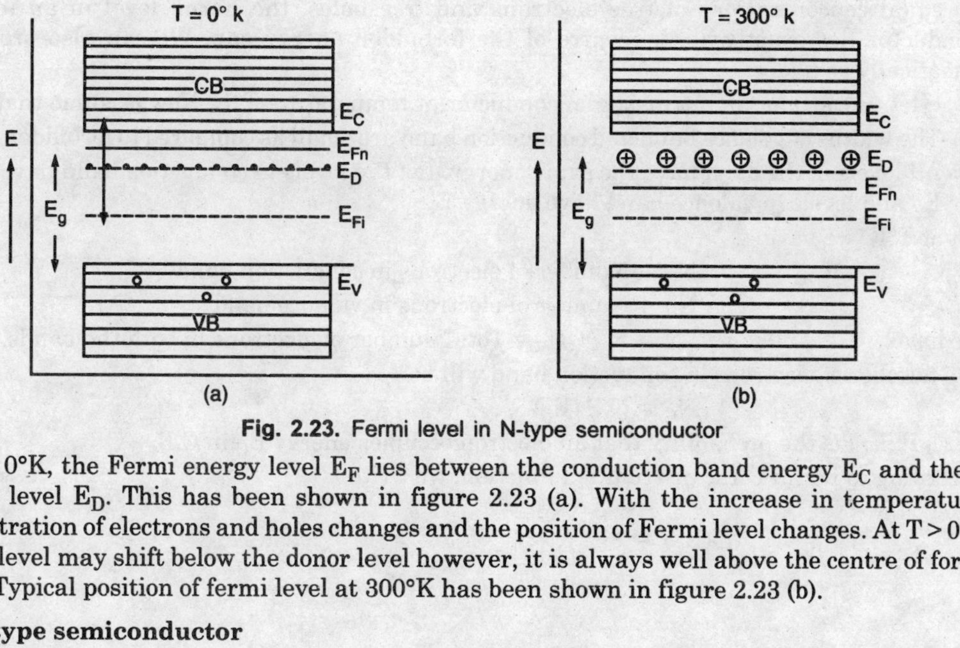

Fig. 2.23. Fermi level in N-type semiconductor

At 0°K, the Fermi energy level E_F lies between the conduction band energy E_C and the donor energy level E_D. This has been shown in figure 2.23 (a). With the increase in temperature, the concentration of electrons and holes changes and the position of Fermi level changes. At T > 0°K, the Fermi level may shift below the donor level however, it is always well above the centre of forbidden band. Typical position of fermi level at 300°K has been shown in figure 2.23 (b).

(b) P-type semiconductor

In P-type semiconductor, the hole concentration in valence band is much more than the electron concentration in conduction band. Therefore, the Fermi level lies above the top of the valence band. At 0°K, the Fermi level lies between the valence band energy E_v and the acceptor energy level E_A. This has been shown in figure 2.24 (a). As temperature rises electrons are intrinsically available in conduction band. Therefore, Fermi level may shift above the acceptor level, however it is always well below the centre of forbidden band. Typical position of Fermi level at 300°K has been shown in figure 2.24 (b).

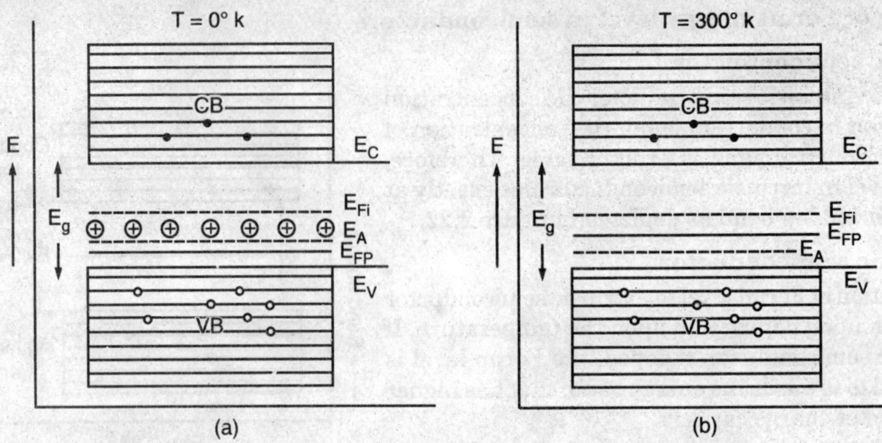

Fig. 2.24. Fermi level in P-type semiconductor

2.15. POSITION OF FERMI LEVEL IN INTRINSIC SEMICONDUCTOR **(Important)**

Due to equal concentrations of free electrons and free holes, the Fermi level in an intrinsic semiconductor lies exactly at the centre of the forbidden energy gap. We can also prove this mathematically as under:

First, let us consider an intrinsic semiconductor at temperature T°K. Now, assume that:

(i) The widths of valence band and conduction band are small as compared to forbidden gap E_g.

(ii) All levels in the band have the same energy. Let the levels in conduction band have energy E_C and levels in valence band have energy E_v.

Let at T°K, we have

$$N_C = \text{Number of electrons in conduction band}$$

$$N_V = \text{Number of electrons in valence band}$$

We have, $N = N_C + N_V$ = Total number of electrons in both the bands.

The number of electrons in conduction band will be

$$N_C = N.P\,(E_C) \qquad ...(2.1)$$

Here, $P\,(E_C)$ is the probability that an electron occupies energy E_C in C.B.

According to Fermi Dirac distribution function, we write

$$P\,(E_C) = \frac{N}{1 + e^{\left(\frac{E_C - E_F}{KT}\right)}} \qquad ...(2.2)$$

Therefore, from equation (2.1) and (2.2), we have $N_C = \dfrac{N}{1 + e^{\left(\frac{E_C - E_F}{KT}\right)}}$ ...(2.3)

In a simialar manner, we have $N_v = \dfrac{N}{1 + e^{\left(\frac{E_V - E_F}{KT}\right)}}$ ...(2.4)

Now, we have $N = N_C + N_V$

Therefore, we write

$$N = \frac{N}{1 + e^{\left(\frac{E_C - E_F}{KT}\right)}} + \frac{N}{1 + e^{\left(\frac{E_V - E_F}{KT}\right)}}$$

$$\text{or}\quad \left[1+e^{\left(\frac{E_C - E_F}{KT}\right)}\right]\left[1+e^{\left(\frac{E_V - E_F}{KT}\right)}\right] = \left[1+e^{\left(\frac{E_C - E_F}{KT}\right)}\right] + \left[1+e^{\left(\frac{E_V - E_F}{KT}\right)}\right]$$

This yields, $e^{\frac{E_C + E_V - 2E_F}{KT}} = 1 = e^0$

$$E_C + E_V - 2E_F = 0$$

Hence, we have $\quad E_C = \dfrac{E_C + E_V}{2} = 0$

Important Point: Therefore, Fermi level in intrinsic semiconductor lies exactly in the middle of the forbidden. gap.

2.15.1. Electron-Hole Concentration at Thermal Equilibrium Condition

(i) We have already discussed that the Fermi energy for an intrinsic semiconductor is near midgap. Later, We will observe that, in some particular situations, the Fermi energy may deviate from this midgap energy. We will assume initially, however, that the Fermi level remains within the bandgap energy.

(ii) The expression for the thermal-equilibrium concentration of electrons may be found by integrating following equation over the conduction band energy, or

$$n_0 = \int g_c(E)\, f_F(E)\, dE \qquad ...(2.5)$$

(iii) The lower limit of integration is E_c and the upper limit of integration should be the top of the allowed conduction band energy. But, since the Fermi probability function rapidly approaches zero with increasing energy so, we can take the upper limit of integration to be infinity.

(iv) Let us assume that the Fermi energy is within the forbidden-energy bandgap. For electrons in the conduction band, we have $E > E_c$. If $(E_c - E_F) >> kT$, then $(E - E_F) >> KT$, so that the Fermi probability function reduces to the Boltzmann approximation, i.e.

$$f_F(E) = \frac{1}{1 + \exp\dfrac{E - E_F}{KT}} \cong \exp\frac{[-(E - E_F)]}{KT} \qquad ...(2.6)$$

(v) Applying the Boltzmann approximation to equation (2.5), the thermal-equilibrium density of electrons in the conduction band can be determined from

$$n_0 = \int_{E_C}^{\infty} \frac{4\pi\,(2m\,n)^{3/2}}{h^3} \sqrt{E - E_c}\, \exp\left[\frac{-(E - E_F)}{KT}\right] dE \qquad ...(2.7)$$

(vi) The integral of equation (2.7) may be evaluated quite easily by making a change of variable. If we let

$$\eta = \frac{E - E_c}{KT} \qquad ...(2.8)$$

then equation (2.7) will become,

$$n_0 = \frac{4\pi\,(2m\,n\,.\,kt)}{h^3} \exp\left[\frac{-(E_c - E_F)}{KT}\right] \int_0^{\infty} \eta^{1/2} \exp(-\eta)\, d\eta \qquad ...(2.9)$$

(vii) The integral is the gamma function, with a value of

$$\int_0^{\infty} \eta^{1/2} \exp(-\eta)\, d\eta = \frac{1}{2}\sqrt{\pi} \quad ...(2.10)$$

Now, the equation (2.9) will become

$$n_0 = 2\left(\frac{2\pi m_n{}^* n\, kT}{h^2}\right)^{3/2} \exp\left[\frac{-(E_C - E_F)}{KT}\right] \quad ...(2.11)$$

(viii) Let us now define a parameter N_c as under:

$$N_c = 2\left(\frac{2\pi m_n{}^* n\, kT}{h^2}\right)^{3/2} \quad ...(2.12)$$

so that the thermal-equilibrium electron concentration in the conduction band may be written as under:

$$n_0 = N_c \exp\left[\frac{-(E_c - E_F)}{KT}\right] \quad ...(2.13)$$

(ix) The parameter N_c is known as the *effective density of states function in the conduction band.* If we were to assume that $m_n{}^* = m_0$, then the value of the effective density of states function at T = 300 K is $N_c = 2.5 \times 10^{19}\ cm^{-3}$, which is the order of magnitude of N_c for most semiconductors. If the effective mass of the electron is larger or smaller than m_0, then the value of the effective density of states function changes accordingly, but is still of the same order of magnitude.

(x) The thermal-equilibrium concentration of holes in the valence band may be evaluated by integrating equation over the valence band energy, or

$$p_0 = \int g_v(E)\,[1 - f_F(E)]\, dE \quad ...(2.14)$$

We may note that

$$1 - f_F(E) = \frac{1}{1 + \exp\left(\frac{E_F - E}{KT}\right)} \quad ...(2.15)$$

(xi) For energy states in the valence band, $E < E_v$. If $(E_F - E_V) >> KT$ (the Fermi function is still assumed to be within the bandgap), then we have a slightly different form of the Boltzmann approximation. Equation (2.15) may be written as under:

$$1 - f_F(E) = \frac{1}{1 + \exp\left(\frac{E_F - E}{KT}\right)} \approx \exp\left[\frac{-(E_F - E)}{KT}\right] \quad ...(2.16)$$

(xii) Applying the Boltzmann approximation of equation (2.16) to equation (2.14), we obtain the thermal-equilibrium concentration of holes in the valence band as under:

$$p_0 = \int_{-\infty}^{E_V} \frac{4\pi\,(2m_p{}^*)^{3/2}}{h^3} \sqrt{E_V - E}\, \exp\left[\frac{-(E_F - E)}{KT}\right] dE \quad ...(2.17)$$

where the lower limit of integration is taken as minus infinity instead of the bottom of the valence band. The exponential term decays fast enough so that this approximation is valid.

(xiii) Equation (2.17) may be solved quite easily by again making a change of variable. If we let

$$n' = \frac{E_V - E}{KT} \quad ...(2.18)$$

then equation (2.17) will become

$$p_0 = \frac{-4\pi(2m_p{}^*KT)^{3/2}}{h^3}\exp\left[\frac{-(E_F-E_V)}{KT}\right]\int_{+\infty}^{0}(\eta')^{1/2}\exp(-\eta)\,d\eta' \quad ...(2.19)$$

where the negative sign comes from the differential $dE = -kT\,d\eta'$.

(xiv) Here, it may be noted that the lower limit of η' becomes $+\infty$ when $E = -\infty$. If we change the order of integration, we introduce another minus sign. From equation (2.10), equation (2.19) becomes,

$$p_0 = 2\left(\frac{2\pi m_p{}^*KT}{h^2}\right)^{3/2}\exp\left[\frac{-(E_F-E_V)}{KT}\right] \quad ...(2.20)$$

Now, let us define a parameter N_V, as under:

$$N_V = 2\left(\frac{2\pi m_p{}^*KT}{h^2}\right)^{3/2} \quad ...(2.21)$$

Which is called the *effective density of states function in the valence band.*

Now, The thermal-equilibrium concentration of holes in the valence band may be expressed as under:

$$p_0 = N_V\exp\left[\frac{-(E_F-E_V)}{KT}\right] \quad ...(2.22)$$

Important Point: The magnitude of N_V is also on the order of 10^{19} cm^{-3} at T = 300 K for most semiconductors.

EXAMPLE 2.3. In an N-type semiconductor, the Fermi-level lies 0.3 EV below the conduction band at room temperature. If the temperature is increased up to 330°K, determine the position of Fermi level.

Given Data: $E_C - E_F = 0.3$ eV, T′ = 330°K

The expression required: $N_C = \dfrac{N}{1+e^{\left(\frac{E_C-E_F}{KT}\right)}}$

Solution: The number of electrons in conduction band having energy E_C at room temperature T = 300°K will be

$$N_C = \frac{N}{1+e^{\left(\frac{E_C-E_F}{KT}\right)}}$$

Hence, in an appriximate manner, we can write

$$E_C - E_F = KT.\log\frac{N}{N_C}$$

or $$0.3 = K(300)\log\frac{N}{N_C} \quad ...(i)$$

Fermi-level at T′ = 330°K is expressed by

$$E_C - E_F = K(300)\log\frac{N}{N_C} \quad ...(ii)$$

Equation (ii) divided by equation (i) yields

$$\frac{E_C-E_F}{0.3} = \frac{330}{300}$$

or $$E_C - E_F = 0.33 \text{ eV}.$$

Important Point: Therefore, the Fermi level will be 0.33 eV below the conduction band.

EXAMPLE 2.4. Determine the probability that a state in the conduction band is occupied by an electron and evaluate the thermal equilibrium electron concentration of silicon at a temperature of T = 300 K.

Assume the Fermi energy is 0.25 eV below the conduction band. The value of N_C for silicon at T = 300 K is $N_C = 2.8 \times 10^{19}$ cm^{-3}.

Solution: We know that the probability that an energy state at $E = E_C$ is occupied by an electron may be expressed as under:

$$f_F(E_C) = \frac{1}{1+\exp\left(\frac{E_C - E_F}{KT}\right)} \approx \exp\left[\frac{-(E_C - E_F)}{KT}\right]$$

or

$$f_F(E_C) = \exp\left(\frac{-0.25}{0.0259}\right) = 6.43 \times 10^{-5}$$

The electron concentration may be expressed as under:

$$n_0 = N_C \exp\left[\frac{-(E_C - E_F)}{KT}\right] = (2.8 \times 10^{19}) \exp\left(\frac{-0.25}{0.0259}\right)$$

or

$$n_0 = 1.8 \times 10^{15} \text{ cm}^{-3}$$

EXAMPLE 2.5. Determine the thermal equilibrium hole concentration in silicon at T = 400 K. Assume that the Fermi energy is 0.27 eV above the valence band energy. The value of N_V for silicon at T = 300 K is $N_V = 1.04 \times 10^{19}$ cm^{-3}.

Solution: We know that the parameter values at T = 400 K may be found as under:

$$N_V = (1.04 \times 10^{19})\left(\frac{400}{300}\right)^{3/2} = 1.60 \times 10^{19} \text{ cm}^{-3}$$

and

$$KT = (0.0259)\left(\frac{400}{300}\right) = 0.03453 \text{ eV}$$

Then, the hole concentration will be

$$p_0 = N_V \exp\left[\frac{-(E_F - E_V)}{KT}\right]$$

or

$$p_0 = 1.60 \times 10^{19} \exp\left(\frac{-0.27}{0.03453}\right) = 6.43 \times 10^{15} / \text{cm}^3$$

2.15.2. Quasi-Fermi Energy Levels

(i) As discussed earlier, the thermal-equilibrium electron and hole concentrations are functions of the Fermi energy level. Now, we can write.

$$n_0 = n_i \exp\left(\frac{E_F - E_{Fi}}{KT}\right) \quad ...(2.23)$$

and

$$p_0 = n_i \exp\left(\frac{E_{Fi} - E_F}{KT}\right) \quad ...(2.24)$$

where E_F and E_{Fi} are the Fermi energy and intrinsic Fermi energy, respectively, and n_i is the intrinsic carrier concentration.

(ii) Figure 2.25(a) illustrates the energy-band diagram for an n-type semiconductor in which $E_F > E_{Fi}$. For this case, we may note from equations (2.23) and (2.24) that $n_0 > n_i$ and $p_0 < n_i$, as we would expect.

(iii) In a similar manner, figure 2.25(b) illustrates the energy-band diagram for a p-type semiconductor in which $E_F < E_{Fi}$. Again we may note from equations (2.23) and (2.24) that $n_0 < n_i$ and $p_0 > n_i$ as we would expect for the p-type material. These results are for thermal equilibrium.

(iv) If excess carriers are created in a semiconductor, we are no longer in thermal equilibrium and the Fermi energy is strictly no longer defined. However, we can define a quasi-Fermi level for electrons and a quasi-Fermi level for holes that apply from non-equilibrium. If dn and dp are the excess electron and hole concentrations, respectively. We may write:

$$n_0 + \delta n = n_i \exp\left(\frac{E_{Fn} - E_{Fi}}{KT}\right) \quad ...(2.25)$$

$$p_0 + \delta n = n_i \exp\left(\frac{E_{Fi} - E_{Fp}}{KT}\right) \quad ...(2.26)$$

where E_{Fn} and E_{Fp} are the quasi-Fermi energy levels for electrons and holes, respectively.

(v) The total electron concentration and the total hole concentration are functions of the quasi-Fermi levels.

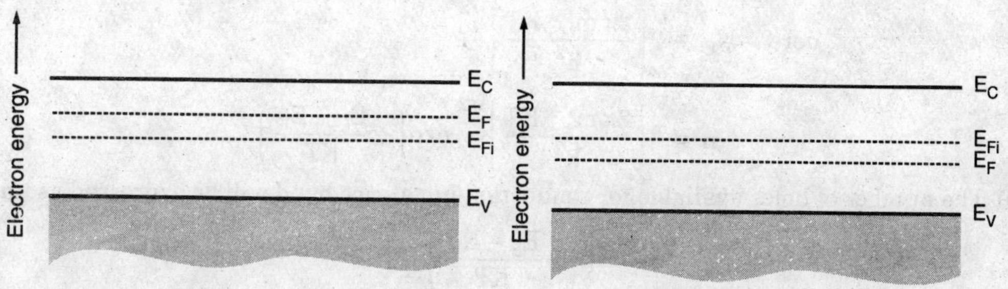

(a) n-type semiconductor

(b) p-type semiconductor

Fig. 2.25. Thermal equilibrium energy band diagrams for

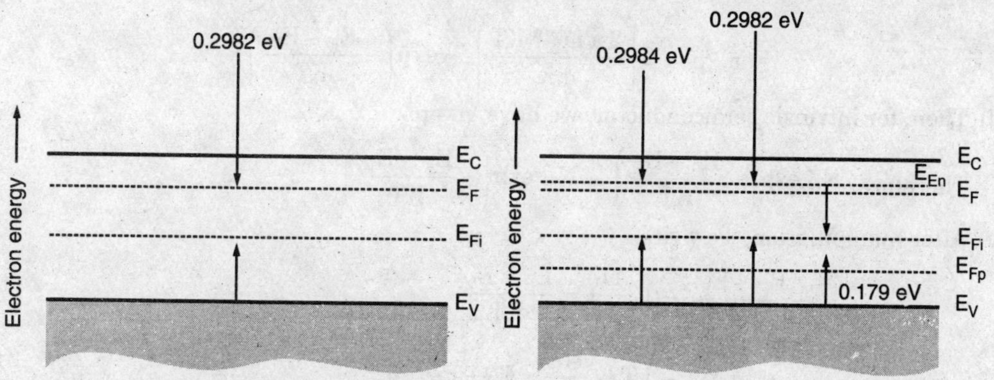

(a) Thermal equilibrium energy band diagram for $N_d = 10^{15}/cm^3$ and $n_i = 10^{10}/cm^3$

(b) Quasi-Fermi levels for electrons and holes. If $10^{13}/cm^3$ excess carriers are present

Fig. 2.26.

(vi) Figure 2.26(a) illustrates the energy-band diagram with the Fermi energy level corresponding to thermal equilibrium. Figure 2.26(b) now shows the energy-band diagram under the non-equilibrium condition.

(vii) Since, the majority carrier electron concentration does not change significantly for this low-injection condition, the quasi-Fermi level for electrons is not much different from the thermal-equilibrium Fermi-level.

(viii) The quasi-Fermi energy level for the minority carrier holes is significantly different from the Fermi level and illustrates the fact hat we have deviated from thermal equilibrium significantly.

(ix) Because the electron concentration has increased, the quasi-Fermi level for electrons has moved slightly closer to the conduction band.

(x) The hole concentration has increased significantly so that the quasi-Fermi level for holes has moved much closer to the valence band.

2.15.3. Exact Position of Fermi-Level in an Intrinsic Semiconductor

(i) We know that in an intrinsic semiconductor the free electron and hole concentrations are equal i.e., n = p. The number of electrons available for conduction in conduction band may be expressed as under:

$$n = N_C \exp\left(-\frac{E_C - E_F}{KT}\right)$$

$$\text{here,} \quad N_C = 2\left[\frac{2\pi m_e{}^* KT}{h^2}\right]^{3/2}$$

$$n = 2\left[\frac{2\pi m_e{}^* KT}{h^2}\right]^{3/2} \exp\left(-\frac{E_C - E_F}{KT}\right)$$

(ii) The number of holes available for conduction in valence band will be expressed as under:

$$p = N_v \exp\left(-\frac{E_F - E_V}{KT}\right)$$

$$\text{here} \quad N_V = 2\left[\frac{2\pi m_h{}^* KT}{h^2}\right]^{3/2}$$

$$\text{or} \quad p = 2\left[\frac{2\pi m_h{}^* KT}{h^2}\right]^{3/2} \exp\left(-\frac{E_F - E_V}{KT}\right)$$

(iii) Then, for intrinsic semiconductor, we have, n = p

$$\text{Therefore,} \quad N_C \exp\left(-\frac{E_C - E_F}{KT}\right) = N_V \exp\left(-\frac{E_F - E_V}{KT}\right)$$

(iv) After manipulation, we write

$$-\frac{E_C - E_F}{KT} = l_n \frac{N_V}{N_C} - \frac{E_F - E_V}{KT}$$

$$\text{or} \quad -E_C + E_F = kT\, l_n \frac{N_V}{N_C} - E_F + E_V$$

$$\text{or} \quad 2E_F = kT\, l_n \frac{N_V}{N_C} + (E_C + E_V)$$

$$\text{or} \quad E_F = \frac{E_C + E_V}{2} + \frac{1}{2} KT\, l_n \frac{N_V}{N_C}$$

(v) After substituting the value of N_V and N_C, we obtain

$$E_F = \frac{E_C + E_V}{2} - \frac{3}{4} KT\, l_n \frac{m_e{}^*}{m_h{}^*}$$

Let us assume $m_e{}^* = m_h{}^*$

Then, we have $E_F = \dfrac{E_C + E_V}{2}$

Again, we know that $E_C - E_V = E_g$

Therefore, $$E_F = \frac{E_C - E_V + 2\,E_V}{2} = \frac{E_g}{2} + E_V$$

Let us assume that, the top of valence band E_V is taken as zero level, then, we have

$$E_F = \frac{E_g}{2} - \frac{3}{4} KT\, l_n\left(\frac{m_e^*}{m_h^*}\right) = \frac{E_g}{2} + \frac{3}{4} KT\, l_n\left(\frac{m_h^*}{m_e^*}\right)$$

2.16. ELECTRICAL CONDUCTIVITY OF CONDUCTORS

(i) According to free electron theory of metals, a metal piece contains of a large number of atoms that closely packed. The valence electrons of these atoms are very loosely bound to individual atoms and therefore, are free to move randomly within the metal. These free electrons are known as *conduction electrons.*

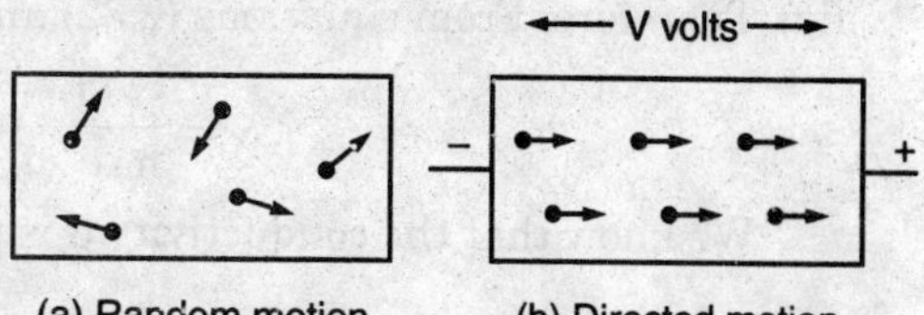

(a) Random motion (b) Directed motion

Fig. 2.27.

(ii) In absence of an external electric field, the free electrons move randomly in all directions as shown in figure 2.27(a) when an external electric field is applied, the randomly motion becomes directed as shown in figure 2.27(b).

(iii) Such directed motion is known as drift. The electron acquires a velocity v, known as the drift velocity, which depends on the strength of applied electric field E.

Therefore, $V \alpha E$

or $$V = \mu_e \, . \, E \qquad ...(2.27)$$

where μ_e is called electron mobility.*

$\mu_e = \frac{V}{E}$ and is defined as the average drift velocity acquired by the electron under unit electric field.

(iv) Let us, consider a conducting wire of length l and cross-section A as depicted figure 2.28. Let V be the voltage applied across it and E be the electric field generated. Under the effect of this electric field, the electrons acquire a drift velocity v and a current I flows through the conductor.

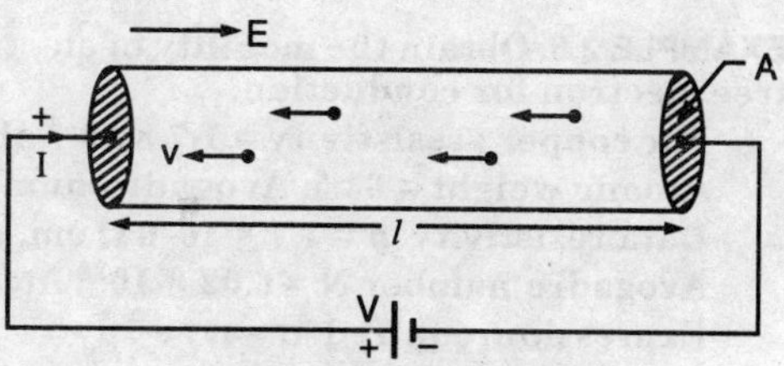

Fig. 2.28. Conductor

(v) Let e = Charge of an electron

n = electron density (i.e. the number of free electrons per unit volume of conductor)

$$E = \frac{v}{l} \qquad ...(2.28)$$

Now, the current I is the charge crossing the cross section A per unit time

Therefore, $$I = n \times (v \times A) \times e \qquad ...(2.29)$$

(vi) Substituting for v from equation (2.27), we have

$$I = n \, . \, \mu_e \, E \, . \, A \, . \, e \qquad ...(2.30)$$

Now, $E = \frac{V}{l}$ and hence, we have

$$I = n \, . \, \mu_e \, . \, \frac{V}{l} \, . \, A \, . \, e \qquad ...(2.31)$$

* Mobility may be defined as the average particle drift velocity per unit electron.

or $$\frac{V}{I} = \frac{l}{A}.\frac{1}{ne\mu_e} \qquad ...(2.31)$$

(vii) Also, by Ohms law, $\frac{V}{I}$ = R = Resistance of the wire, we have

$$R = \frac{l}{A}\cdot\frac{1}{n.e.\mu_e} \qquad ...(2.32)$$

(viii) R is also expressed as $R = \rho.\frac{l}{A}$ ...(2.33)

where $\rho \rightarrow$ Resistivity of the material of conductor

(ix) Therefore, From equations (2.32) and (2.33), we write

$$\rho = \frac{1}{n.e.\mu_e}$$

We know that the conductivity σ is the reciprocal of resistivity.

Therefore, $$\sigma = \frac{1}{\rho}$$

or $$\sigma = n.e.\mu_e \qquad ..(2.34)$$

From equation (2.30), we can write that

$$\frac{I}{A} = n.e.\mu_e \qquad ...(2.34)$$

(x) We know that the current density J is defined as the current per unit area.

Therefore, $$J = \frac{I}{A} = n.e.\mu_e.E$$

or $$J = \sigma E \qquad ...(2.35)$$

EXAMPLE 2.6. Obtain the mobility of electrons in copper assuming that each atom contributes one free electron for conduction.

For copper : resistivity = 1.7 × 10^{-6} ohm . cm. density = 8.96 gm/cc

Atomic weight = 63.5, Avogadro number = 6.02 × 10^{23} / gm. mole

Data resistivity ρ = 1.7 × 10–6 Ω cm, density d = 8.96 gm/cc,

Avogadro number N = 6.02 × 10^{23} At. wt. A = 63.5

Expressionrequired: $\sigma = n.e.\mu_e$

Solution: We know that Number of atoms of copper present per unit volume is

$$= \frac{6.02\times10^{23}}{63.5}\times 8.96 = 8.496\times10^{22}$$

Sicne, every atom contributes one free electron, the number of free electrons per unit volume, *i.e.*, the electron density is 8.496 × 10^{22} per c.c.

We know that the conductivity is given by, $\sigma = n.e.\mu_e$

Therefore, the mobility of electrons will be

$$\mu_e = \frac{\sigma}{n.e}$$

or $$\mu_e = \frac{1}{\rho.n.e}\cdot\left(\text{if } \sigma = \frac{1}{\rho}\right) = \frac{1}{(1.7\times10^{-6}\times 8.496\times10^{22}\times1.6\times10^{-19})} = 43.273 \text{ cm}^2/\text{volt.sec.}$$

Hence, σ = 43.273 cm²/volt.sec.

EXAMPLE 2.7. A copper wire 0.1 m long and 1.7 mm² cross-section has a resistance 0.1 ohm when subjected to 1 volt potential difference between its ends. Evaluate the density of free electrons in the metal and the mobility of these electrons. You may assume that

Density of copper = 8.96 gm/cc
Atomic weight of copper = 63.5
Avogadro number = 6.02×10^{23}/gm mole
Data: Length l = 0.1 m
Cross-section A = 1.7 mm^2 = 1.7×10^{-6} m^2
Resistance R = 0.1 Ω
63.5 gm of copper contain 6.02×10^{23} atoms
Density of copper = 8.96 gm/cc

Expressions required: $R = \rho . \frac{l}{A}, \sigma = n . e . \mu_e$

Solution: The number of copper atoms per unit volume are;

$$= \frac{6.02 \times 10^{23}}{63.5} \times 8.96 = 8.49 \times 10^{22} \text{ per c.c} = 8.49 \times 10^{28} \text{ per m}^3$$

Sicne, each copper atom contributes one free electron for conduction, the number of free electrons per unit volume = 8.49×10^{28}/m^3

We know that resistance of copper is given by

$$R = \rho . \frac{l}{A}$$

where $\rho \rightarrow$ resistivity of copper

Therefore, $$\rho = \frac{R \times A}{l} = \frac{0.1 \times 1.7 \times 10^{-6}}{0.1}$$

or $$\rho = 1.7 \times 10^{-6} \ \Omega.\text{m}$$

We also know that conductivity of copper is given by

$$\rho = \frac{1}{\sigma} = \frac{1}{1.7 \times 10^{-6}} \text{ mho/m}$$

Now, we have $\sigma = n . e . \mu_e$

where $n \rightarrow$ electron concentration

$\mu_e \rightarrow$ electron mobility

Therefore, Mobility $$\mu_e = \frac{\sigma}{ne} = \frac{1}{1.7 \times 10^{-6} \times 8.49 \times 10^{28} \times 1.6 \times 10^{-19}}$$

or $$\mu_e = 0.0433 \times 10^{-3} \text{ m}^2/\text{V.sec}$$

EXAMPLE 2.8. Determine the drift velocity for an electron in silver wire of radius 1 mm and carrying a current of 2A. Density of silver is 10.5 gm/cc.

Avogadro number = 6.025×10^{23}/gm mole
Data: Density of silver = 10.5 gm/cc
Atomic weight of silver = 108
Expression required : I = q . n . v . A

Solution: The number of electrons per unit volume = $n = \frac{6 \times 10^{23}}{108} \times 10.5$

Therefore, $$n = \frac{6 \times 10^{23}}{10.3} = 6 \times 10^{22} \text{ per cm}^3$$

or $$n = 6 \times 10^{28} \text{ per m}^3$$

Cross sectional area $A = \pi r^2 = \pi \times (10^{-3})^2 = 3 \times 10^{-6}$ m^2

We know that current is given by, I = q . n . V . A

Therefore, $$V = \frac{I}{nqA} = \frac{2}{(6 \times 10^{28}) \times (1.6 \times 10^{-19}) \times (3 \times 10^{-6})} = 7 \times 10^{-4} \text{ m/s.}$$

EXAMPLE 2.9. It is given that The resistance of copper wire of diameter 1.03 mm is 6.51 ohm per 300 m. The concentration of free electrons in copper is 8.4×10^{28}/m^3. If the current is 2 Amp, determine the mobility of charge carriers and the conductivity of copper.

Data: Diameter d = 1.03 mm = 1.03 × 10⁻³ m = 1.03 × 10⁻³ m

$$\textbf{Radius } r = \frac{1.03}{2} \times 10^{-3} \text{ m}$$

$$R = 6.51\ \Omega$$

$$l = 300 \text{ mm}$$

$$n = 8.4 \times 10^{28}/\text{m}^3$$

Expressions required: $R = \rho \cdot \frac{l}{A}$, $\mu_e = \frac{\sigma}{n.e}$

Solution: We know that the resistance R of a wire of length l and cross-section A is given by,

$$R = \rho . \frac{l}{A}$$

where $\rho \rightarrow$ resistivity of copper

Therefore, $$\rho = \frac{R \times A}{l} \qquad (A = \pi r^2)$$

or $$\rho = \frac{6.51 \times 3.14 \times \left(\frac{1.03}{2} \times 10^{-3}\right)^2}{300}$$

We know that the current I is due to both the current carriers i.e. electrons and holes in the semiconductor.

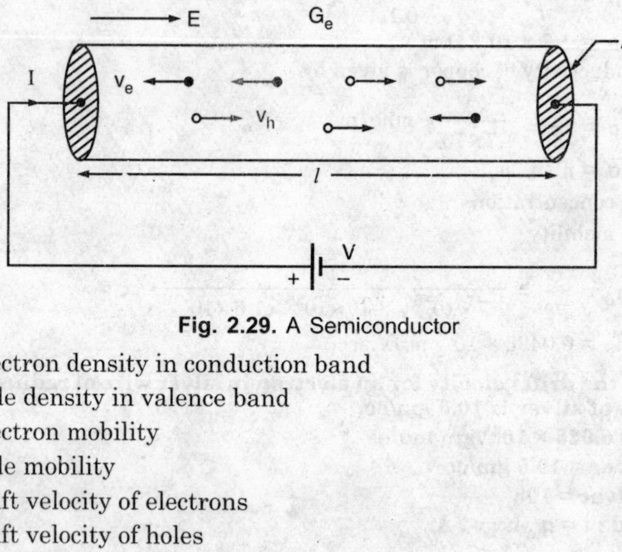

Fig. 2.29. A Semiconductor

Now, let n = electron density in conduction band

p = hole density in valence band

m_e = electron mobility

m_h = hole mobility

V_e = drift velocity of electrons

V_h = drift velocity of holes

Total current is given by $I = I_e + I_h$

or $$I = n . eV_e . A + p . e V_h . A \qquad ...(2.36)$$

or $$I = e. A(n.V_e + p.V_h) \qquad ...(2.37)$$

Now, we have $$V_e = \mu_e . E = \mu_e \cdot \frac{V}{l}$$

and $$V_h = \mu_h . E = \mu_h \cdot \frac{V}{l}$$

Equation (2.36) yields,

$$I = e.A\left[n.\mu_e . \frac{V}{l} + p . \mu_h . \frac{V}{l}\right]$$

or $$I = e.A.\frac{V}{l}[n.\mu_e + p.\mu_h]$$

Therefore, Resistance $$R = \frac{V}{I} = \frac{1}{e.A(n.\mu_e \cdot p.\mu_h)} = \rho \cdot \frac{l}{A}$$

Here, $\rho \rightarrow$ Resistivity of the semiconductor

Hence, Resistivity $$\rho = \frac{1}{e(n.\mu_e + p.\mu_h)}$$

Therefore, Conductivity of semiconductor is $\sigma = \frac{1}{\rho} = e(n.\mu_e + p.\mu_h)$

or $$\sigma_{s.c} = \sigma_e + \sigma_h \quad ...(2.38)$$

1. Conductivity of intrinsic semiconductor

For an intrinsic semiconductor, the electron concentration is equal to hole concentration.

i.e., $n = p = n_i$ = intrinsic concentration of charge carriers.

∴ Conductivity of intrinsic semiconductor is given by

$$\sigma_i = n_i \cdot e \cdot \mu_e + n_i.e.\mu_h$$

or $$\sigma_I = n_i \cdot e\,(\mu_e + \mu_h) \quad ...(2.39)$$

2. Conductivity of extrinsic semiconductor

(i) N-type semiconductor

For the N-type semiconductor, $n >> p$

Therefore, ignoring the hole concentration, the conductivity of N-type semiconductor will be written as,

or $$\sigma_n = n.e\,\mu_e \quad ...(2.40)$$

$$\sigma_n = d.e\,\mu_e$$

where d = concentration of donor atoms

(ii) P–type semiconductor

For the P-type semiconductor.

$$p >> n$$

Therefore, ignoring the electron concentration the conductivity of P-type semiconductor can be written as

$$\sigma_p = p.e\,\mu_h$$

$$\sigma_p = a.e\,\mu_h \quad ...(2.41)$$

where a = concentration of acceptor atoms.

EXAMPLE 2.10. Determine the conductivity of pure silicon at room temperature when the concentration of carriers is 1.6×10^{10} per cm^3. You can take μ_e = 1500 cm^2/volt.sec. and μ_h = 500 cm^2/volt.sec. at room temperature.

Data: μ_e = 1500 cm^2/volt.sec. μ_h = 500 cm^2/volt sec.

$n = p = n_i = 1.6 \times 10^{10}$ per cm^3.

Expression required = $\sigma_i = n_i\,(\mu_e + \mu_h) \cdot e$

Solution: We know that the conductivity of pure semiconductor is given by,

$$\sigma_i = ni\,(\mu_e + \mu_h) \cdot e$$

Therefore, $$\sigma_i = 1.6 \times 10^{10}\,(1500 + 500) \times 1.6 \times 10^{-19}$$

or $$\sigma_i = 5.12 \times 10^{-6} \text{ mho.cm}$$

or $$\sigma = 5.12 \times 10^{-6} \text{ mho/cm.}$$

EXAMPLE 2.11. An N-type semi conductor is to have a resistivity 10 Ω cm. Evaluate the number of donor atoms which must be added to achieve this.

Given that μ_d = 500 cm^2/V.S

Data: Resistivity $\rho = 10\ \Omega$ cm., $\mu_d = 500$ cm²/V.S.

Expression required: Conductivity $\sigma = \dfrac{1}{\rho}$ **and** $\sigma = n_d \cdot e \cdot \mu_d$

Solution: We know that $n_d = \dfrac{\sigma}{e.\mu_d} = \dfrac{1}{\rho.\sigma.\mu_d}$

or $$n_d = \frac{1}{10\times1.6\times10^{-19}\times500} = 1.25\times10^{15} \text{ per cm}^3$$

EXAMPLE 2.12. Evaluate the conductivity of specimen if a donor impurity is added to an extent of one part in 10^8 Ge atoms at room temperature?

Data: Avodadro number = 6.02×10^{23} atoms/gm.mole.At.wt. of Ge = 72.6,

Density of Ge = 5.32 gm/c.c, mobility μ = 3800 cm²/v.s

Expression required: $\sigma = n \cdot d \cdot \mu_e$

Solution: Concentration of Ge atoms $= \dfrac{6.02\times10^{23}}{72.6}\times5.32 = 4.41\times10^{22}$ per cm³

Since, there is one donar atom per 10^8 atoms of Ge. Therefore, we have

$$n_d = \frac{4.41\times10^{22}}{10^8} = 4.41\times10^{14} \text{ per cc}$$

We know that conductivity $\sigma = n_d \cdot \mu_e = 4.41\times10^{14}\times3800\times1.6\times10^{-19} = 0.268$ mho/cm.

EXAMPLE 2.13. A germanium crystal is doped with a pentavalent impurity of concentration 1 ppm. if the resistivity of doped germanium is 0.3623×10^{-3} ohm-m, determine the mobility of electrons in germanium. You may assume that all the impurity atoms are ionised and the density of germanium atoms is 4.42×10^{28} per m³.

Data: Resistivity $\rho = 0.3623\times10^{-3}\ \Omega$m

Ge density = 4.42×10^{28} atoms/m³

Rate of doping = 1 ppm = 1 impurity atom per 10^6 Ge atoms.

Expression required: $\sigma = \dfrac{1}{\rho}\ \mu = \dfrac{\sigma}{n_d \cdot e}$

Solution: We know that conductivity of germanium is given by:

$$\sigma = \frac{1}{\rho} = \frac{1}{0.3623\times10^{-3}} \text{ mho/m}$$

or $\sigma = 2760$ mho/m

The rate of impurity doping is 1 impurity atom per 10^6 atoms and density Ge atoms is 4.42×10^{28}/m³.

Since, the impurity atoms are ionised, therefore the concentration of impurity atom

$$n_d = \frac{4.42\times10^{28}}{10^6} = 4.42\times10^{22} \text{ atoms/m}^3$$

We know that conductivity is given by $\sigma = n_d \cdot e \cdot \mu$

Therefore, Electron mobility will be

$$\mu = \frac{\sigma}{n_d.e} = \frac{2760}{4.42\times10^{22}\times1.6\times10^{-19}} = 0.39 \text{ m}^2\text{/V.sec}$$

EXAMPLE 2.14. A germanium semiconductor consists of 10^{-6} % Boron and has a resistivity of 0.42 ohm.m. Evaluate the density and mobility of holes in the semiconductor.

Assume density of germanium = 5.36×10^3 kg/m³

Atomic weight of germanium = 72.59

Avogadro number = 6.025×10^{26} / kg – mole

Data: Resistivity = 0.42 Ω m

Boron impurity = 10^{-6} %

Density of Ge = 5.36×10^3 kg/m³

Atomic weight of Ge = 72.59

Expression required: Conductivity = $\frac{1}{\text{resistivity}}$

Solution: Avogadro number = 6.025×10^{26}/kg. Mole.

Therefore, 72.59 kg of Ge contains 6.025×10^{26} atoms

Density = Mass per unit volume

Hence, Number of Ge atoms present per unit volume will be $\frac{6.025\times10^{26}}{72.59}\times5.36\times10^{3} = 0.445 \times 10^{29}$

Rate of Boron impurity is 10^{-6} % i.e. 10^{-6}

Thus, 1 Boron atom is present per 10^{8} Ge atoms.

If the impurity atoms are completely ionised, then the number of holes per unit volume will be

$$= \frac{0.445\times10^{29}}{10^{8}} = 4.45\times10^{20}$$

Thus, Holes per unit volume = 4.45×10^{20}

Now one hole is available per acceptor Boron atom.

Therefore, Concentration of acceptor atoms is a = $4.45 \times 10^{20}/m^3$

The conductivity of P-type semiconductor is $\sigma_p = a.e.\mu_h$

and conductivity = $\frac{1}{\text{resistivity}} = \frac{1}{0.42}$ mho/m

Therefore, $\frac{1}{0.42} = (4.45 \times 10^{20}) \times (1.6 \times 10^{-19}) \times \mu_h$

Hence, Mobility of holes = $\frac{1}{0.42\times4.45\times10^{20}\times1.6\times10^{-19}} = 0.0334\ m^2/V.sec$

EXAMPLE 2.15. Determine the current produced in a small germanium plate of area 1 cm² and of thickness 0.3 mm when a potential difference of 2V is applied across the faces. Given that the concentration of free electrons of Ge is $2 \times 10^{19}/m^3$ and mobilities of electrons and holes are 0.36 m²/V.s and 0.17 m²/v.s respectively.

Data: $n_i = 2 \times 10^{19}\ /m^3$ $\mu_e = 0.36\ m^2/v.s.$

$\mu_h = 0.17\ m^2/v.s.$, Area A = $1 \times 10^{-4}\ m^2$

Voltage V = 2 volts

Length l = 0.3 mm = 3×10^{-4} m.

Expression required: J = σE

$$I = \frac{\sigma_i \times V \times A}{l}$$

Solution: We know that, the current density is given by J = σE

or $J = \sigma.\frac{V}{l}$ also $J = \frac{I}{A}$

Therefore, $\frac{I}{A} = \sigma.\frac{V}{l}$

or $I = \frac{\sigma_i \times V \times A}{l}$

Now, we have $\sigma_i = n_i\ .\ e\ .\ (\mu_e + \mu_h) = 2 \times 10^{19} \times 1.6 \times 10^{-19}\ (0.36 + 0.17) = 1.696$ mho/m

Therefore, current $I = \frac{1.696\times2\times10^{-4}}{3\times10^{-4}}$

or I = 1.13 amp.

EXAMPLE 2.16. A specimen of pure Germanium at 300° K has a density of charge carriers of 2.5 × $10^{19}/m^3$. It is doped with donor impurity atoms at the rate of one impurity atom for every 10^6

atoms of germanium. All impurity atoms may be supposed to be ionized. The density of germanium atoms is 4.2×10^{28} atoms/m^3. Determine the resistivity of doped germanium if the electron mobility is 0.36 m^2/vs.

Data: Density of charge carriers at 300°K = 2.5×10^{19}/m^3

Expression required: $\sigma_n = N_d \cdot e \cdot \mu_e \qquad \rho_n = \frac{1}{\sigma_n}$

Solution: Intrinsic carrier concentration = 2.5×10^{19}/m^3

Density of added impurity atoms i.e. $N_d = \frac{4.2 \times 10^{28}}{10^6} = 4.2 \times 10^{22}$ atoms/m^3

The donor concentration is very large as compared to intrinsic carrier concentration and hene the intrinsic concentration can be neglected.

Therefore, conductivity of doped material is given by

$$\sigma_n = N_d \cdot e \cdot \mu_e = 4.2 \times 10^{22} \times 1.6 \times 10^{-19} \times 0.36 = 2.492 \times 10^3 \text{ mho/m}$$

Hence, the resistivity will be

$$\rho_n = \frac{1}{\sigma_n} = \frac{1}{2.4192 \times 10^3} \text{ ohm m} = 0.4133 \times 10^{-3}\ \Omega\text{m}.$$

EXAMPLE 2.17. Evaluate the current produced in a germanium sample of area 2 sq.cm and thickness 0.1 mm when a potential difference of 4 Volts is applied across it.

Given: $n_i = 1 \times 10^{19}$ / m^3

$\mu_e = 0.36$ m^2/volt.sec

$\mu_h = 0.17$ m^2/volt.sec

Data: Area $A = 2$ cm^2 $= 2 \times 10^{-4}$ m^2

Thickness $t = 0.1$mm $= 10^{-4}$ m

$V = 4$ volts

Expression required: $\sigma_i = n_i \cdot e \cdot (\mu_e + \mu_h)$

Solution: We know that the conductivity is given by

$$\sigma_i = n_i \cdot e \cdot (\mu_e + \mu_h)$$
$$= 1 \times 10^{19} \times 1.6 \times 10^{-19} (0.36 + 0.17) = 1.6 \times 0.53 = 0.848 \text{ mho/m}$$

We know that the current density is given by

$$J = \sigma_i \times E = \sigma_i \times \frac{V}{t}$$

Therefore, $J = 0.848 \times \frac{4}{10^{-4}}$ Amp / m^2

Hence, Current produce will be, $I = J \times A$

Therefore, $I = 0.848 \times \frac{4}{10^{-4}} \times 10^{-4} \times 2 = 6.784$ Amp.

EXAMPLE 2.18. Evaluate conductivity of pure silicon at room temperature if concentration of carriers is 1.6×10^{10} per cm^3. Given that mobilities of electron and hole are 1500 cm^2/v.s and 500 cm^2/v.s. respectively.

Data: $\mu_h = 500$ cm^2/V.s

$\mu_e = 1500$ cm^2/V.s

$n_i = 1.6 \times 10^{10}$/cm^3

Expression required: $\sigma_l = n_i\, e\, (\mu_h + \mu_e)$

Solution: We know that intrinsic conductivity is given by

$$\sigma_i = n_i\, e(\mu_e + \mu_h)$$

or $\sigma_i = 1.6 \times 10^{10} \times 1.6 \times 10^{-19}$ (1500 + 500)

or $\sigma_i = 5.12 \times 10^{-6}$ mho/cm

2.17. TEMPERATURE DEPENDENCE OF CONDUCTIVITY

(i) For the case of metals, which are the good conductors of electricity, the resistance increase with the increase in the temperature of the conductor. The resistance at temperature t°C is expressed by $R_t = R_o (1 + \propto t)$
Here, $R_o \rightarrow$ resistance at 0°C.
and $\propto \rightarrow$ temperature coefficient of resistance

(ii) Therefore, the conductvity of conductors decreases with the increase of temperature. In case of semiconductors, the current is because of two types of carriers, the electrons and the holes. The conductivity of a semiconductor is expressed by,

$$\sigma_{s.c} = e(n \,.\, \mu_e + p \,.\, \mu_h)$$

(iii) For the intrinsic semiconductor the electrons in C.B. and holes in V.B. are equal in number at any temperature.

$$n = p = n_i = \text{intrinsic concentration of carriers}$$

(iv) The conductivity of semiconductors depends on temperature. With the increase in temeprature more number of electron-hole pairs are generated. Hence, the intrinsic carrier concentration n_i also increases.

(v) Hence, the conductivity of the intrinsic semiconductor also increases. The intrinsic carrier concentration n_i increases with temperature as per the relation.

$$n_i^2 = A_o . T^3 e^{\frac{-E_g}{KT}}$$

where A_o = Constant, independent of temperature
T = Absolute temperature of the semiconductor
E_g = Forbidden energy gap at 0° K
K = Boltzmann constant expressed in eV/°K

Important Point: Thus, the current through the semiconductor also increases with the increase in temperature.

2.18. HALL EFFECT AND HALL COEFFICIENT (U.P. Tech. Sem. Exam, 2005-06, 2008-09, 2009-10)

1. Definition

If a metal or a semiconductor strip carrying a current I is placed in transverse magnetic field B, an electric field E is induced in the direction perpendicular to both B and I. This phenomenon is known as **Hall effect**. It is used to detect whether a semiconductor is N-type or P-type.

2. Derivation of Expression

(i) Let us consider a rectangular strip of a N-type semiconducting material as shown in figure 2.30.

(ii) Under the application of some potential difference, let current I flow through it along the positive X-direction. Let the magnetic field B act along the positive Z direction (i.e., B acts perpendicular to plane of paper in the outward direction).

(iii) Hence, a force F acts on the charge carriers along the negative Y direction. As the semiconductor is N-type, the charge carriers are electrons. This force due to the magnetic field has the magnitude.

$$F_1 = B.e.v \qquad ...(2.42)$$

where $v \rightarrow$ drift velocity of electron
$e \rightarrow$ charge of electron

(iv) Thus, the electrons will be forced to move downward towards side 1 in figure 2.30 and side 1 becomes negatively charged with respect to side 2.

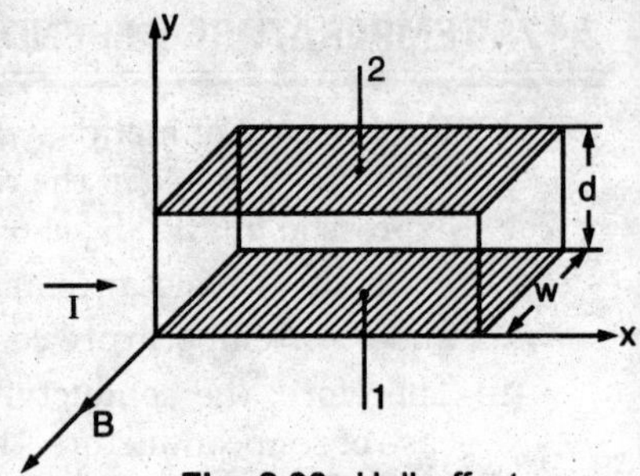

Fig. 2.30. Hall effect

(v) As a result, an electric field is produced between the surfaces 1 and 2; and is directed along the negative Y direction. This field will oppose the further piling up of the electrons on surface 1. In the equilibrium condition, the force due to the electric field balances the force due to the magnetic field and then the electrons do not move along the Y direction.

(vi) At this stage, a steady potential difference is developed across the surfaces 1 and 2. This voltage is called the Hall voltage V_H. Let E be the electric field intensity due to Hall voltage V_H, then the force due to this electric field balances the force due to magnetic field.

Therefore, $$e.E = B.e.v \quad ...(2.43)$$

or $$E = B.v$$

But, we have $$E = \frac{V_H}{d}$$

Here, d = distance between the surfaces 1 and 2

Hence, we have $$\frac{V_H}{d} = B.v \quad ...(2.44)$$

or $$V_H = B.v.d \quad ...(2.45)$$

The current $I = n.e.v.A$

We have, $$v = \frac{I}{n.e.A}$$

Therefore, $$V_H = \frac{B.I.d}{n.eA} = \frac{1}{n.e} \cdot \frac{B.I.d}{A} \quad ...(2.46)$$

Here, v = Carrier concentration

A = Cross-section of the strip

(vii) Now, we have $\frac{I}{A}$ = J = Current density

Therefore, the Hall voltage is given by,

$$V_H = \frac{1}{n.e} \cdot \frac{B.I.d}{A} = \frac{1}{n.e} \cdot BJd \quad ...(2.47)$$

The charges density ρ is the charge per unit volume of material.

Therefore, $\rho = n.e$

And area $A = d.W$

where W = Width of the specimen in the direction of magnetic field

Therefore, Hall voltage $$V_H = \frac{B.I}{\rho.W} \quad ...(2.48)$$

(viii) Hence, if the polarity of V_H is positive at surface 2 (as in the above discussion) then the carriers must be electrons. But if the surface 1 has positive polarity with respect to surface 2, then the carriers are positively charged holes and the semiconductor is P-type. Thus measurement of Hall voltage helps to determine whether the given semiconductor is P-type or N-type.

(ix) From equation (2.48), we see that if V_H, B, 1 and W are measured, the charge density ρ can be determined. Hence, ρ = n.e can give the carrier concentration n.

The Hall coefficient R_H is defined as

$$R_H = \frac{1}{n.e} = \frac{1}{\rho} \quad \text{...(2.49)}$$

Therefore, Equation (2.48) gives the Hall voltage as under:

$$V_H = R_H.\frac{B.I}{W}$$

(x) Therefore, Hall coefficient

$$R_H = \frac{V_H.W}{B.I}$$

or

$$R_H = \frac{E.d.W}{B.I} = \frac{E}{J.B} \quad \text{...(2.50)}$$

Since,

$$J = \frac{I}{d.W}$$

Important Point: *Thus, the Hall coefficient R_H of a semiconductor material may be defined as the electric field strength per unit surface current density due to unit transverse magnetic field.*

(xi) The conductivity of the material is given as $\sigma = n \, . \, e\mu$
where $\mu \to$ mobility of the carrier

Therefore,

$$\sigma = \frac{I}{R_H}.\mu \quad \text{or} \quad \text{Mobility } \mu = R_H \, . \, \sigma \quad \text{...(2.51)}$$

Important Point: Thus, measurement of conductivity and Hall coefficient determines the mobility of carrier.

2.18.1. Applications of Hall Effect

The Hall effect may be used for following applications:

(i) Determination of type of semiconductor.
(ii) Determination of carrier concentration.
(iii) Determination of mobility of carriers.

EXAMPLE 2.19. A silver wire is in the form of a ribbon 0.50 cm. wide and 0.10 mm thick. When a current of 2 ampere passes through the ribbon, perpendicular to 0.80 Tesla magnetic field, then evaluate the Hall voltage produced. The density of silver is = 10.5 gm/cc.

Data: B = 0.8 Tesla, Density = 10.5 gm/cc

Expression required: $V_H = B.v.d, \quad v = \frac{I}{nAq}$

Solution: The number of electrons in 1 cc of silver are:

$$n = 6.025 \times 10^{23} \times \frac{10.5}{108} \cong 6 \times 10^{22} \text{ per c.c.}$$

Since, each silver atom contributes one electron, the number of electrons per $m^3 = 6 \times 10^{28}$

Also, Area $A = 0.5 \times 10^{-2} \times 0.1 \times 10^{-3} = 5 \times 10^{-7}$ m^2

We know that hall voltage, $V_H = B.v.d.$

Also, the drift velocity is given by $v = \frac{I}{n\,A\,q}$

Therefore,

$$V_H = \frac{I}{n.q}.\frac{B.I.d}{A} = \frac{1}{6\times10^{28}\times1.6\times10^{-19}} \times \frac{0.8\times2\times0.1\times10^{-3}}{5\times10^{-7}}$$

or

$$V_H = 0.333 \times 10^{-7} \text{ volts}$$

EXAMPLE 2.20. A copper specimen having length 1 metre, width 1 cm and thickness 1 mm is conducting 1 amp, current along its length and is applied with a magnetic field of 1 Tesla along its thickness. It experiences Hall effect and a hall voltage of 0.074 microvolts appears along its width. Determine the Hall coefficient and the mobility of electrons in copper.
(Conductivity of copper is $\sigma = 5.8 \times 10^7\ (\Omega m)^{-1}$)

Data: $l = 1$ m, $d = 1$ cm $= 10^{-2}$m

$W = 1$ mm $= 10^{-3}$ m, $I = 1$ Amp

$B = 1$ Tesla, $V_H = 0.074 \times 10^{-6}$ Volts

$\sigma = 5.8 \times 10^7$ mho/m

Expression required:

$$V_H = \frac{1}{n.q} \cdot \frac{B.I.d}{A}$$

Solution: We know that The Hall voltage is given by,

$$V_H = \left(\frac{1}{n.q}\right) \cdot \frac{B.I.d}{A}$$

or
$$V_H = R_H \times \frac{B.I.d}{A}$$

Thus, we can write
$$R_H = \frac{V_H \times A}{B.I.d} = \frac{0.074 \times 10^{-6} \times (10^{-2} \times 10^{-3})}{1 \times 1 \times 10^{-2}}$$

or
$$R_H = 7.4 \times 10^{-11}\ m^3/c$$

Also, Mobility
$$\mu = \sigma \,.\, R_H = 5.8 \times 10^7 \times 7.4 \times 10^{-11}$$

or
$$\mu = 4.3 \times 10^{-3}\ m^2/volt.sec$$

2.19. HALL ANGLE

(U.P. Tech, Sem. Exam; 2009-10)

1. Definition

The net electric field E in the specimen is vector sum of electric field component in the x-direction because of flow of current, E_x and electric field due to Hall effect i.e. E_H depicted in figure 2.31. Therefore, resultant electric field E acts at an angle θ_H to the x-axis and this angle is known as **Hall angle**.

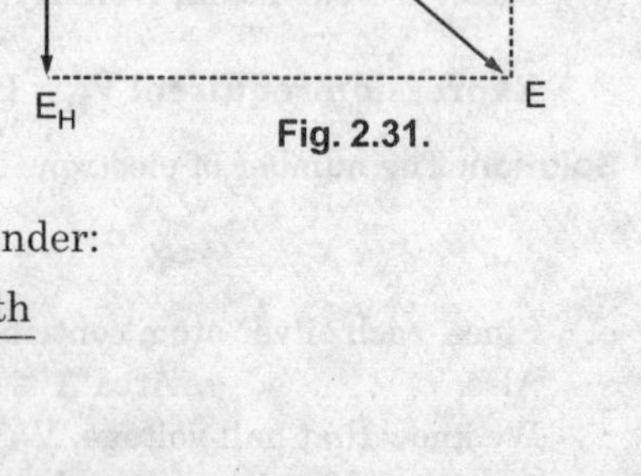

Fig. 2.31.

Therefore, Hall angle
$$\theta_H = \tan^{-1}\left(\frac{E_H}{E_x}\right) \quad ...(2.52)$$

We have,
$$E_H = \frac{V_H}{d} \text{ and}$$

We can write electric field component in the x-direction as under:

$$E_x = \frac{\text{Voltage drop along the length}}{\text{Length of specimen}}$$

$$E_x = \frac{IR}{l} = \frac{I \times \text{resistivity} \times l}{l \times A}$$

$$E_x = \frac{I}{A} \cdot \rho_0 = \frac{J}{\sigma}, \text{ where } \rho_0 = \text{resistivity.}$$

Substituting, the value of E_x and E_H in equation (2.30), we have,

$$\theta = \tan^{-1}\left(\frac{E_H}{E_x}\right) = \tan^{-1}\frac{V_H/d}{j/\sigma} = \tan^{-1}\frac{BI/\rho wd}{j/\sigma}$$

or
$$\theta_H = \frac{BJ/\rho}{j/\sigma} = \tan^{-1}\frac{B\sigma}{\rho} = \tan^{-1}(\sigma B.R_H) = \tan^{-1}(\mu B)$$

EXAMPLE 2.21. Obtain the concentration of holes in Si crystals having donor concentration of $1.4 \times 10^{24}/m^3$ when the intrinsic carrier is $1.4 \times 10^{18}/m^3$. Calculate the ratio of electron to hole concentration.

Solution: Intrinsic carrier concentration is $n_i = 1.4 \times 10^{18}/m^3$

Donor concentration is $n_D = 1.4 \times 10^{24}/m^3$

Concentration of electron is $n = N_D = 1.4 \times 10^{24}/m^3$

Concentration of holes is given by

$$p = \frac{n_i}{n} = \frac{(1.4\times10^{18})^2}{1.4\times10^{24}} = 1.4 \times 10^{12}/m^3$$

Ratio of electron to hole concentration will be

$$= \frac{n}{p} = \frac{1.4\times10^{24}}{1.4\times10^{12}} = 1 \times 10^{12}$$

EXAMPLE 2.22. The resistivity of semiconductor material is known to be 0.00912 Ω-m at room temperature. The flux density in the Hall model is 0.48 Wb/m². Evaluate the Hall angle for a Hall coefficient of 3.55×10^{-4} m³ /coulomb.

Solution: Flux density in Hall Model B = 0.48 wb/m²

Hall coefficient $R_H = 3.55 \times 10^{-4}$ m³/c

We know that, Conductivity $\sigma = \frac{1}{\text{resistivity}} = \frac{1}{9\times10^{-3}} = 111.11\ (\text{ohm-m})^{-1}$

Also, Charge density $\rho = \frac{1}{R_H} = \frac{1}{3.6\times10^{-4}} = 2{,}778$ coulomb/m³

Density of charge carriers $n = \frac{\text{charge density}}{e} = \frac{2{,}778}{1.6\times10^{-19}} = 1.73625 \times 10^{22}/m^3$

We know that, Mobility $\mu = \sigma R_H = 111.11 \times 3.6 \times 10^{-4} = 0.04\ m^2/v\text{-}s$ **Ans.**

2.20. MOBILITY, DRIFT VELOCITY AND DRIFT CURRENT

(U.P. Tech, Sem. Exam; 2005-06, 2008-09, 2009-10)

1. Drift velocity of charged particles

(i) Let us consider a material that contains a large number of free electrons. When we apply an external electric field of E volts per meter to such a material, the free electrons inside the material are accelerated in a definite direction. These accelerated electrons may collide with the immobile ions while travelling and lose their energy.

(ii) Eventually, a steady state shall be reached when the electrons will travel at a steady velocity.

(iii) This steady state velocity attained by the electrons under the influence of the externally applied electric field E is known as the **Drift Velocity** or **Drift Speed**. Drift speed is measured in meters/sec and denoted by v.

2. Mobility

(i) The drift velocity v is proportional to the externally applied electric field E. Hence, we have

$$v \propto E \qquad ...(2.53)$$

(ii) Using μ as constant of proportionality, we write,

$$v = \mu E \qquad ...(2.54)$$

(iii) This constant of proportionality μ may be defined as the *mobility* of charged particles. Rearranging equation (2.54), we write as under:

$$\mu = \frac{v}{E} \quad ...(2.55)$$

(iv) However, the units of drift velocity are m/sec and that of electric field E are volts/m. Therefore, units of mobility μ are:

$$\mu = \frac{m/sec}{v/m} = \frac{m^2}{v - sec} \quad ...(2.56)$$

Important Point: *Thus, mobility (μ) is expressed in square meters per volt-sec.*

3. Drift current

(i) Under the influence of the externally applied electric field, the electrons are accelerated in one particular direction. They travel at a speed equal to drift speed.

(ii) This movement of electrons will give rise to a current which is defined as the **Drift current.**

2.21. RELAXATION TIME, COLLISION TIME AND MEAN FREE PATH

(U.P. Tech. Sem. Exam 200-01, 2005-06, 2008-09)

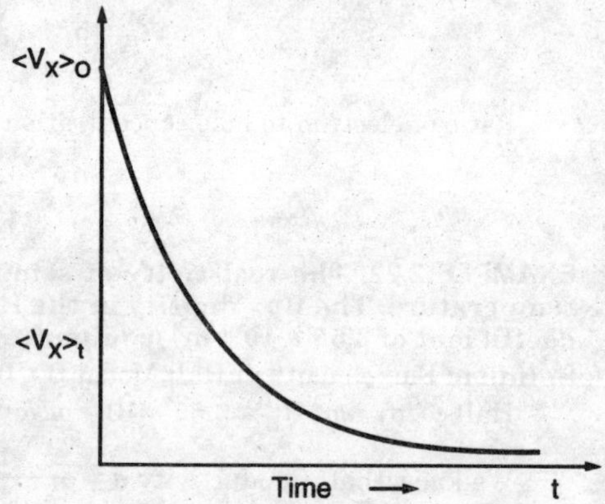

Fig. 2.32. Decay of drift velocty

1. Definition and Basic Concepts

(i) The relaxation time is closely related to the mean time of flight between collision and also to be mean free path of the conduction of electrons, which is the average distance of undisturbed motion between collision.

(ii) The collision process taking place in the free electron gas can be explained through relaxation time.

(iii) The collisions between the electrons occurring due to thermal excitation and structural imperfections in the lattice.

(iv) The relaxation time is introduced as the characteristic time governing the establishment of equilibrium, through collision, from an initial disturbed situation in which $V_x \neq 0$.

2. Derivation of Mathematical Expression

(i) The mathematical expression of relaxation time can be expressed as follows.

(ii) Let us, assume the electrons in a metal have an average velocity $<V_x>_o$ at any instant t = 0; say for t > 0 there is no external field, the electrons interact only with lattice structure and the average velocity ultimate reduces to zero. Say at any instant (t > 0), the velocity of electrons can be written as

$$<V_x>_t = <V_x>_o e^{-\tau/t} \quad ...(2.57)$$

where τ is the relaxation time of electrons.

(iii) The rate of change of velocity due to collision is given by,

$$\left[\frac{d<V_x>}{dt}\right]_{collision} = -\frac{1}{\tau}<V_x> \quad ...(2.58)$$

(iv) Considering the relaxation time, the frictional force on the electrons due collision may be written as,

$$\frac{m<V_x>}{\tau}$$

(v) Therefore, the modified equation of electron is given by,

$$m\frac{d<V_x>}{dt}+\frac{m<V_x>}{\tau} = -eE \quad ...(2.59)$$

As there is no electric field, the above equation becomes,

$$\frac{d<V_x>}{dt}+\frac{1}{\tau}\cdot<V_x> = 0 \quad ...(2.60)$$

(vi) If $<V_x>_0$ is the initial drift velocity in the non-equilibrium distribution, the solution of above equation can be written as,

$$<V_x>_t = <V_x>_0\, e^{-t/\tau} \quad ...(2.61)$$

(vii) This equation implies that disturbances die out exponentially from equilibrium, as shown in figure 2.32. Let us consider the probability for an electron to collide with an obstacle during a short time interval dt is given by,

$\frac{dt}{\tau}$, where τ' is a constant with the dimensions of time. Assume F(t) represent the probability that an electron moves for a time t without suffering a collision and F(t + dt) represents the same for a period (t + dt). Hence, we can write,

$$F(t+dt) = F(t)+\frac{dF}{dt}\cdot dt \quad ...(2.62)$$

(Taylor series expansion)

and from the theory of probability, we write,

$$F(t+dt) = F(t)\,.\,F(dt) = F(t)\left[1-\frac{dt}{\tau'}\right] \quad ...(2.63)$$

(viii) Comparing equations (2.63) and (2.62), we write,

$$\frac{dF}{dt} = -\frac{F}{\tau'}$$

or

$$F(t) = e^{-t/\tau'} \quad ...(2.64)$$

Integration constant can be determined by putting t = 0, i.e. F(t) = 1

(ix) Hence, the average time elapses between collisions can be written as

$$<t> = -\int_0^{\alpha} t\,\frac{dF}{dt}\cdot dt \quad ...(2.65)$$

(x) Since, $-\frac{dF}{dt}$ provides the probability that the electron collides between t and (t + dt). Put the value of F(t) from equation (2.64) to equation (2.65), we get

$$<t> = \tau'$$

(xi) Therefore, the constant τ' represents the average time between collisions. For isotropic materials, the mean time of collision is the same as the relaxation time i.e. $\tau' = \tau$, which is not true for non-isotropic materials. Hence, the mean free path of an electron λ, can be defined as,

$$\lambda = v\tau' \quad ...(2.66)$$

where v is the total velocity of electron.

(xii) In the case of metals, the conductivity always refers to a single total velocity, the one corresponding to Fermi energy. The velocity of the electrons with the Fermi-energy, V_F is given by

$$\frac{1}{2} m v_F^2 = E_F$$

(xiii) Therefore, the mean free path of electron can be written as

$$\lambda = V_F \cdot \tau' \qquad ...(2.67)$$

2.22. ELECTRICAL CONDUCTIVITY AND RELAXATION TIME

(i) From figure 2.28, the force F on the particle of charge q when on electric field E is applied is given by

$$F = qE$$

Therefore, force acting on an electron having charge e is given by F = eE

According Newton's Law, we write, F = ma

Therefore, F = eE = ma

Therefore, Acceleration of electrons due to applied electric field E is expressed as under:

$$a = \frac{eE}{m}$$

(ii) Because of collisions of electrons during motion, the electrons will not get accelerated indefinitely. In fact, velocity of electrons will reduce to zero. If τ is the relaxation time, then the average velocity of electrons, known as drift velocity can be written as

$$v = a \times \tau = \frac{eE}{m}\tau \qquad ...(2.68)$$

Current I is the charge crossing the cross-section A per unit time.

$$I = n \times (v \times A) \times e$$

(iii) Current density is given by

$$J = \frac{I}{A} = n\,e\,v = ne \cdot \frac{eE}{m}\tau$$

$$J = \sigma E \quad \text{where} \quad \sigma = \frac{n e^2 \tau}{m}$$

(iv) For a particular material, the quantity $\sigma = \frac{n e^2 \tau}{m}$ is constant at particular temperature and is called electrical conductivity of the material. Electrical conductivity of the material is also given as

$$\sigma = \frac{n e^2 \tau}{m} = n\,e\,\mu_e$$

where $\mu_e = \frac{e\tau}{m}$ is the mobility acquired by electrons due to presence of electric field. Assume λ be the mean free path and v′ be the root mean square velocity of electrons, then the relaxation time τ is given by,

$$\tau = \frac{\lambda}{v'}$$

$$\text{but} \quad v' = \sqrt{\frac{3 K_B T}{m}}$$

$$\text{So,}\quad \tau = \lambda\sqrt{\frac{m}{3\,K_B\,T}}$$

(v) Therefore, the electrical conductivity σ can be written as

$$\sigma = \frac{n\,e^2\tau}{m} = \frac{n\,e^2}{m}\cdot\lambda\sqrt{\frac{m}{3\,K_B\,T}}$$

or
$$\sigma = \frac{n\,e^2\lambda}{\sqrt{3\,m\,K_B T}}$$

(vi) Therefore, the electrical resistivity ρ is given by,

$$\rho = \frac{1}{\sigma} = \frac{\sqrt{3mK_BT}}{ne^2\lambda}$$

or
$$\rho \propto \sqrt{T}$$

The above equation proves that resistivity varies as $\sqrt{T}$ but it is vary actually linearly with temperature.

2.23. DEPENDENCE OF FERMI LEVEL ON TEMPERATURE AND DOPING

1. Effect of temperature on n-type material

(i) At low temperature

When the temperature in the semiconductor is low, only few donor atoms get ionized and electrons move from the donor level to the conduction band.

Let us, assume, N_d = concentration of donor atoms and

n_d = number of donor atoms not ionized

$N_d - n_d$ = number of donor atoms ionized

Also, $N_d - n_d = n_c$ = is the number of electrons in conduction band.

Therefore, we can write, $n_c = N_d - n_d = [1 - f(E_d)]$

Here, $f(E_d)$ = probability of vacant donor level

or,
$$n_c = N_d\left[1 - \frac{1}{1 + \exp\left(\frac{E_d - E_F}{KT}\right)}\right]\text{, after simplification}$$

we obtain,
$$n_c = N_d \exp\left(+\frac{E_d - E_F}{KT}\right) \qquad \text{...(2.69)}$$

However, the number of electrons in conduction band is given by (using Fermi-Dirac statistics)

$$n_c = N_c \exp\left(+\frac{E_c - E_F}{KT}\right) \qquad \text{...(2.70)}$$

Solving equation (2.69) and (2.70), we obtain

$$N_d \exp\left(\frac{E_d - E_F}{KT}\right) = N_c \exp\left(+\frac{E_c - E_F}{KT}\right)$$

On solving at T = 0° K, we obtain

$$E_F = \frac{E_c + E_d}{2}$$

Important Point: Therefore, we can conclude that fermi level for n-type semiconductor at 0° K lies midway between the bottom of the conduction band and donor level.

(ii) At moderate temperature:

At moderate temperature all donor atoms are ionized. Hence, the concentration of electrons in conduction band is equal to the concentration of donor atoms.

$$n_c = N_d \quad ...(2.71)$$

$$n_c = N_c \exp\left(+\frac{E_C - E_F}{KT}\right) \quad ...(2.72)$$

or

$$N_d = N_c \exp\left(+\frac{E_C - E_F}{KT}\right)$$

$$\frac{N_d}{N_c} = \exp\left(+\frac{E_C - E_F}{KT}\right)$$

$$+\frac{E_C - E_F}{KT} = \ln\left(\frac{N_d}{N_c}\right)$$

$$E_F = E_C - KT\ln\left(\frac{N_d}{N_c}\right)$$

Important Point: When the temperature increases upto moderate value, Fermi level shifts away from the conduction band and moves towards the centre of the forbidden gap.

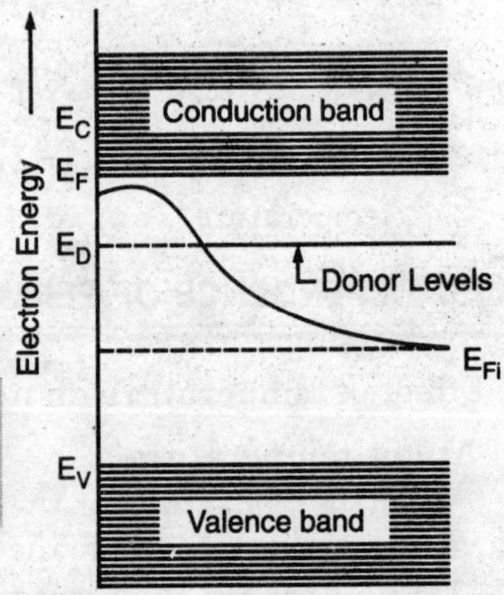

Fig. 2.33. Variation of E_F with temperature in n-type material

(iii) At higher temperature

At high temperature, the concentration of transfer of electrons from valence band to conduction band is more compared to concentration of electrons from donor atoms and fermi level is given by

$$E_F = \frac{E_C + E_V}{2} + \frac{KT}{2}\ln\left(\frac{N_c}{N_d}\right)$$

The variation of Fermi level with temperature for N-type of material has been shown in figure 2.33.

2. Effect of temperature on P-type material

(i) At low temperature

At low temperature, only few acceptor levels are occupied simultaneously holes are produced in valence band. Say N_a is the concentration of acceptor atoms and n_a is the number of acceptor levels are vacant. Hence, the number of holes present in valence band.

Therefore,

$$n_v = N_a - n_a = N_a[1 - f(E_a)] = N_a\left[1 - \frac{1}{1 + \exp\left(\frac{E_a - E_F}{KT}\right)}\right]$$

$$n_v = N_a \exp\left(+\frac{E_a - E_F}{KT}\right) \quad ...(2.73)$$

Hole concentration in valence band is given by (using Fermi-Dirac statistics)

$$n_v = N_v \exp\left(+\frac{E_v - E_F}{KT}\right) \quad ...(2.74)$$

Solving equations (2.73) and (2.74), we obtain

$$E_F = \frac{E_a + E_v}{2}, \text{ at } T = 0° \text{ K.}$$

Important Point: Therefore, Fermi-level lies in the middle of the top of the valence band and the acceptor level.

(ii) At moderate temperature

At moderate temperature, all acceptor levels are filled so, we write, $n_v = N_a$

Hence, from equation (2.74), we have

$$N_a = N_v \exp\left(\frac{E_v - E_F}{KT}\right)$$

or
$$\ln\left(\frac{N_a}{N_v}\right) = +\frac{E_v - E_F}{KT}$$

or
$$E_F = E_V + KT \ln\left(\frac{N_v}{N_a}\right)$$

Important Point: Hence, at moderate temperature, fermi-level gradually moves up i.e. moves towards the middle of the forbidden gap.

(iii) At higher temperature

(i) At very higher temperature, the contribution of conduction band for the formation of holes in the valence band is more compared to acceptor impurity.

(ii) Therefore, the position of Fermi-level is given by

$$E_F = \frac{E_C + E_V}{2} - \frac{KT}{2} \ln\left(\frac{N_a}{N_v}\right)$$

(iii) Hence, at very higher temperature, fermi-level approaches the middle of the energy gap i.e. the position of E_{Fi} for intrinsic semiconductor.

(iv) The variation of E_F with temperature in p-type material has been shown in figure 2.34.

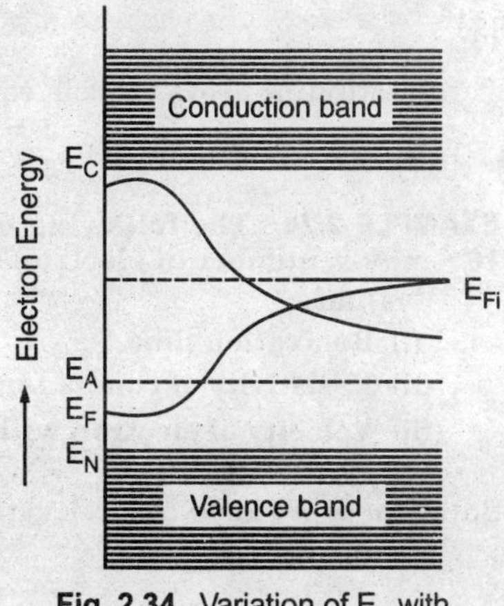

Fig. 2.34. Variation of E_F with temperature in p-type material

3. Effect of impurity concentration on Fermi-level for n-type semiconductor

(i) If impurity atoms like phosphorus, antimony, bismuth are added to intrinsic semiconductor, discrete, donor levels are introduced just below the conduction band.

(ii) When the concentration of doping is small, donor levels are represented by discrete localized energy levels. As doping concentration is increased, the interatomic distance between impurity atoms reduces, they interact each other.

(iii) Therefore, donor energy level splits into current energy levels and form an impurity band as shown in figure 2.35.

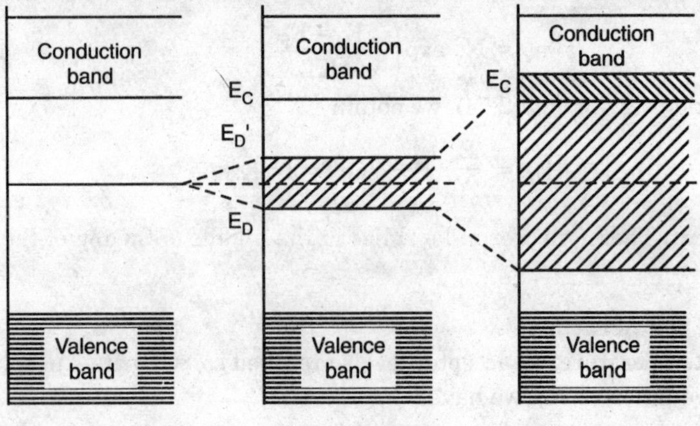

Fig. 2.35.

(iv) Higher the concentration of doping, wider the impurity band. With the increase of doping concentration, fermi-level moves closer to conduction band and finally impurity band overlaps the conduction band.

(v) In a similar manner, in P-type semiconductors, when doping concentration is increased, acceptor energy levels split into several energy levels and form the impurity energy band. As the doping concentration is increased impurity band becomes wider and ultimate it overlaps with the valence band.

> **Important Point:** Thus, Fermi-level shifts towards the valence band and finally overlaps with the valence band for very high doping concentration.

EXAMPLE 2.23. An electric field of 100 V/m is applied to a specimen of N-type semiconductor for which Hall co-efficient is 0.0145 m³/coulomb. Calculate the current density in the specimen. (Given that μ_e = 0.36 m²/v-s).

Solution: We know that the current density is given by,

$$J = \sigma E \text{ and } \sigma = n e \mu_e$$

Therefore,
$$J = n e \mu_e E \qquad ...(i)$$

We have
$$R_H = \frac{1}{ne}$$

$$n = \frac{1}{e R_H} = \frac{1}{1.6\times10^{-19}\times0.0145} = 4.31 \times 10^{20}/m^3$$

Substituting above value in equation (i), we obtain

$$J = n e \mu_e E = 4.31 \times 10^{20} \times 1.6 \times 10^{-19} \times 0.36 \times 100$$

or
$$J = 2.482 \times 10^3 \text{ A/m}^2$$

EXAMPLE 2.24. The following data is given for a conductor, Fermi energy = 5.5 eV, μ_e = 7.04 × 10⁻³ m²/v-s, number of electron/m³ = 5.8 × 10²⁸.

Evaluate:

(i) Relaxation time

(ii) Resistivity of conductor

(iii) Velocity of electron with the Fermi energy.

Solution: (i) We know that relaxation time $\tau = \frac{\mu_e}{e} \times m$

$$\tau = \frac{7.04\times10^{-3}\times9.1\times10^{-31}}{1.6\times10^{-19}} = 40.04 \times 10^{-15} \text{ sec.}$$

(ii) We know that Resistivity of conductor is given by

$$\rho = \frac{1}{\sigma} = \frac{1}{ne\,\mu_e} = \frac{1}{5.8\times10^{28}\times1.6\times10^{-19}\times7.04\times10^{-3}} = 1.531\times10^{-18}\ \Omega\ \text{cm}$$

(iii) We know that Velocity of electrons with fermi-energy is given by,

$$v_F = \sqrt{\frac{2E_F}{m}} = \sqrt{\frac{2\times5.5\times1.6\times10^{-19}}{9.1\times10^{-31}}} = 1.3907\times10^{6}\ \text{m/s}$$

EXAMPLE 2.25. To determine the temperature at which there is a 1 percent probability that an energy state is emply.

You may assume that the Fermi energy level for a particular material is 6.25 eV and that electrons in this material follow the Fermi-Dirac distribution function. Evaluate the temperature at which there is a 1 percent probability that a state 0.30 eV below the Fermi energy level will not contain an electron.

Solution: We know that the probability that a state is empty is

$$1 - f_F(E) = 1 - \frac{1}{1+\exp\left(\frac{E-E_F}{KT}\right)}$$

Then, we have

$$0.01 = 1 - \frac{1}{1+\exp\left(\frac{5.95-6.25}{KT}\right)}$$

Solving for kT, we obtain kT = 0.06529 eV, so that the temperature is T =756 K.

Important Point: The Fermi probability function is a strong function of temperature.

EXAMPLE 2.26. Calculate the thermal equilibrium hole concentration in silicon at T = 400 K. Assume that the Fermi energy is 0.27 eV above the valence band energy. The value of N_v for silicon at T = 300 K is $N_v = 1.04\times10^{19}\ \text{cm}^{-3}$.

Solution: The parameter value at T = 400 K can be found as under:

$$N_v = (1.04\times10^{19})\left(\frac{400}{300}\right)^{3/2} = 1.60\times10^{19}\ \text{cm}^{-3}$$

And

$$KT = (0.0259)\left(\frac{400}{300}\right) = 0.03453\ \text{eV}$$

$$P_0 = N_1 \exp\left[\frac{-(E_F - E_U)}{KT}\right] = (1.60\times10^{19})\exp\left(\frac{-0.27}{0.03453}\right)$$

$$P_0 = 6.43\times10^{15}\ \text{cm}^{-3}$$

Important Point: The parameter values at any temperature can easily by determined by using the 300 K values the temperature dependence.

EXAMPLE 2.27. To determine the required doping concentration in a semiconductor material. A silicon device with n-type material is to be operated at T = 550 K. At this temperature, the intrinsic carrier concentration must contribute no more than 5 percent of the total electron concentration. Determine the minimum donor concentration required to meet this specification.

Solution: We know that, at T = 550 K, the intrinsic carrier concentration is given by,

$$n_i^2 = N_c N_v \exp\left(\frac{-E_g}{KT}\right) = (2.8\times10^{19})(1.04\times10^{19})\left(\frac{550}{300}\right)^3 \exp\left[\frac{-1.12}{0.0259}\left(\frac{300}{550}\right)\right]$$

or $n_i^2 = 1.02\times10^{29}$

Thus, we have $n_i = 3.20\times10^{14}\ \text{cm}^{-3}$

For the intrinsic carrier concentration to contribute to more than 5 percent of the total electron concentration, let us set $n_0 = 1.05\ N_d$.

We have, $$n_0 = \frac{N_d}{2} + \sqrt{\left(\frac{N_d}{2}\right)^2 + n_i^2}$$

$$1.05\, N_d = \frac{N_d}{2} + \sqrt{\left(\frac{N_d}{2}\right)^2 + (3.20\times10^{14})^2}$$

$$N_d = 1.39 \times 10^{15}\ \text{cm}^{-3}$$

Important Point: If the temperature remains less than T = 550 K, then the intrinsic carrier concentration will contribute less than 5 percent of the total electron concentration for this donor impurity concentration.

EXAMPLE 2.28. To calculate the quasi-Fermi energy levels, consider an N-type semiconductor at T = 300 K with carrier concentration of $n_0 = 10^{15}$ cm^{-3}, $n_i = 10^{10}$ cm^{-3}, $n_i = 10^{10}$ cm^{-3} and $p_0 = 10^5$ cm^{-3}. In non-equilibrium, you may assume that the excess carrier concentration are $\delta n = \delta p = 10^{13}$ cm^{-3}.

Solution: The Fermi level for thermal equilibrium can be determine from equation (2.23). We have

$$E_F - E_{Fi} = KT \ln\left(\frac{n_0}{n_i}\right) = 0.2982 \text{ eV}$$

We can use equation (2.25) to determine the quasi-Fermi level for electrons in non-equilibrium. The we can write

$$E_{Fn} - E_{Fi} = KT \ln\left(\frac{n_0 + \delta n}{n_i}\right) = 0.2984 \text{ eV}$$

Equation (2.26) can be used to evaluate the quasi-Fermi level for holes in non-equilibrium. We can write

$$E_{Fi} - E_{Fp} = KT \ln\left(\frac{p_0 + \delta p}{n_i}\right) = 0.179 \text{ eV}$$

Important Point: We may note the quasi-Fermi level for electron is above E_{Fi} while the quasi-Fermi level for holes is below E_{Fi}.

SUMMARY

- Energy bands are formed in a solid due to large number of atoms very close.
- The uppermost allowed energy band occupied by valence electrons is the valence band.
- The allowed energy band immediately above the V.B. is the conduction band.
- The V.B. and C.B. are separated by energy band gap E_g. It is the energy required to break a covalent bond so as to make an electron free.
- Solids are classified on the basis of band theory as conductors, semiconductors and insulators.
- E_g is very high for insulators, small for semiconductors and zero for conductors.
- Materials used for semiconductors are generally Germanium and silicon.
- Intrinsic semiconductors are the materials in very pure form. The current carriers are of two types free electrons and holes; whose number is same.
- Doped semiconductors are called extrinsic. Their types are P and N. Pure semiconductor doped with pentavelent impurity element like Phosphorous, Antimony etc. give N type semiconductor where electrons are majority current carriers. Pure semiconductors doped with trivalent impurity element like Boron, Indium give P type semiconductor where holes are the majority current carriers.
- The conductivity expression are:
  (i) $\sigma = n \cdot e \cdot \mu$ for conductors
  (ii) $\sigma_{sc} = e\,(n \cdot \mu_e + p \cdot \mu_n)$ for semiconductor

(iii) $\sigma_i = n_i \cdot e\,(\mu_e + \mu_n)$ for intrensic semiconductor
(iv) $\sigma_N = n_d \cdot e \cdot \mu_e$ for N-type semiconductor
(v) $\sigma_p = p_a \cdot e \cdot \mu_n$ for P-type semiconductor

- Fermi level for conductors is the highest occupied allowed energy level at 0°K. For semiconductors, it is the energy that corresponds to the centre of gravity of conduction electrons and holes weighted according to their energies.
- For intrinsic semiconductors the Fermi level lies exactly at the centre of the forbidden gap i.e. $E_F = \frac{E_C + E_V}{2}$
- For N type semiconductor the fermi level is below the bottom of C.B.
- An electron is a negatively charged particle having negligible mass. Some of the important properties of an electron are

  Charge on an electron, $e = 1.602 \times 10^{-19}$ C
  Mass of an electron, $m = 9.107 \times 10^{-31}$ kg
  Radius of an electron, $r = 1.9 \times 10^{-15}$ m
  Radio of e/m of an electron $= 1.759 \times 10^{11}$ C/kg
  Total energy of an electron $= \frac{-e^2}{8\pi\varepsilon_0 r}$
- Neils Bohr put forward his model of atom in 1913. He postulated the following three fundamental laws.
  (i) The electrons continue revolving in their respective orbits without losing energy.
  (ii) Energy is emitted or absorbed by an atom only when an electron moves from one level to another.
  In a transition from one stationary state corresponding to a definite energy W_2 to another stationary state, with an associated energy W_i radiation will be emitted. The frequency of this radiant energy is given by

$$f = \frac{W_2 - W_1}{h}$$

  (iii) The angular momentum of an electron moving round the nucleus is quantized. The energy level in joules of each state is given by

$$W = \frac{-me^4}{8\epsilon_0^2 n^2 h^2}$$

- The electrons closest to the nucleus need the greatest energy application to extract them from the atom and, therefore, the electrons in the inner orbits remain bound to the nucleus and are known as *bound electrons*.
  A definite amount of energy is required to be spent in order to remove an electron from its orbit (level).
  Excited electrons are unstable, and they tend to fall back to low energy level by giving up their additional energy. The electrons that do leave their parent atom are termed free electrons.
- A semiconductor is a material that has a resistivity somewhere between that of a good conductor and that of an insulator. The resistivity of different semiconductor materials lies in a very wide range from 10^{-4} to about 0.5 Ω-m whereas the resistivities of conductors and insulators are of the order of 10^{-4} ω-m and 10^{12} Ω-m respectively. Most semiconductor materials used in electronics industry have negative temperature coefficients.
  The semiconductors have another very important property—*the resistivity (or conductivity) of semiconductor changes considerably when even minute amounts of certain other substances, called the impurities, are added to them. Semiconductor materials are non-linear resistors.*
- Within any given material there are two distinct energy bands in which electrons may exist. These two energy bands are valence energy band (or valence band) and conduction energy band (or conduction band) and are separated by an energy gap in which no electron can normally exist. This energy gap is termed as forbidden energy gap (or forbidden gap). The distinction between conductors, insulators and semiconductors is largely concerned with the relative width of the forbidden energy gaps (no forbidden gap in case of conductors, about 1 eV in case of semiconductors and more than 5 eV in case of insulators).
- The two basic charge transport mechanisms are drift, due to an applied electric field, and diffusion, due to a density gradient.

The steady flow of electrons in one direction caused by the applied electric field constitutes an electric current called the drift current. The process whereby particles flow from a region of high concentration to a region of low concentration is known as diffusion process and the current that results from the diffusion of charged particles is termed as diffusion current.

- A bonding of atoms, strengthened by sharing of electrons between neighbouring atoms, is called covalent bonding. The covalent bonding arrangement exists in silicon and germanium.
- Intrinsic materials are those semiconductors that have very low level of impurities, whereas extrinsic materials are semiconductors that have been exposed to a doping process. Silicon (Si) and germanium (Ge) are the two most widely used semiconductor materials in electronics industry.

  At absolute zero temperature E_G for silicon and germanium are 1.21 eV and 0.785 eV respectively. Hole, strictly speaking, is not a particle; but for all practical purposes it is viewed as a positive charged particle capable of conducting current. The mobility of hole is about half that of an **electron.** The resistance of a semiconductor depends largely on (a) *temperature* — decreases with the increase in temperature, (b) *illumination* — decreases in brighter surroundings and (c) *magnitude of the electric field* — the current does not increase in proportion to the applied potential, as in case of conductors obeying Ohm's law but increases by far more than the voltage i.e., *the semiconductors are non-linear.*

  In intrinsic semiconductors, the Fermi level lies between the conduction and valence bands. Conduction in a pure semiconductor consists of electron flow in the conduction band and hole flow in the valence band. A germanium device is seldom used at a temperature higher than 75°C, a silicon seldom higher than 175°C.
- Depending on the type of impurity added, the extrinsic semiconductors conductors can be divided into two classes viz. (i) N-type semiconductors and (ii) P-type semiconductors.

  An N-type material is formed by adding donor atoms (such as arsenic, antimony, bismuth or phosphorous) that have five valence electrons to establish a high level of relatively free electrons. In an N-type material, the electron is the majority carrier and the hole is the minority carrier.

  A P-type material is formed by adding acceptor atoms (such as boron, gallium, indium or aluminium) that have three valence electrons to establish a high level of holes in the material. In a P-type material, the hole is the majority charge carrier and the electron is the minority charge carrier.

SHORT QUESTIONS WITH ANSWERS

Q.1. Why copper is a good conductor ?

Ans. In case of copper, the electrons in the conduction band are very large. An electron in the conduction band experiences almost negligible nuclear attraction. In fact, an electron in the conduction band does not belong to any particular atom. But it moves randomly throughout the solid. Due to overlapping of valence and conduction bands, a slight potential difference across conductors causes the electrons in conduction band, called the free electrons, to constitute electric current. So copper is a good conductor of electricity.

Q.2. Classify materials into conductors, semiconductors and insulators.

Ans. According to the free electron theory, the total energy which is equal to the kinetic as well as the potential energy is assumed to be zero, is related to the wave number by the relation

$$E = \frac{h^2k^2}{2m} \quad \text{where } k = \frac{2\pi}{\lambda}$$

In the free electron theory, when one considers the interaction of the electron with the period potential of the lattice, one finds that band gaps occur at Brillouin zone boundaries. In case the lower valence band is half filled, the material is said to be a good conductor. There is overlapping of zones/bands. This also refers to case where the material behaves like a metal or good conductor. At higher temperature ($K_BT \geq E_G$) an appreciable number of electrons get excited from the top of the valence band into the bottom of the conduction band. The material is then said to exhibit intrinsic semiconductivity; the magnitude of conductivity depends on the density and mobility of electrons in the conduction band and holes in the valence band. If the band gap is very large so that thermal excitation of electrons from valence to conduction band is negligible even at high temperatures, the material is deemed to be as an insulator.

Q.3. What is the basis for classifying a material as a conductor, semiconductor or a dielectric ? What is the conductivity of perfect dielectric ? **[U.P.S.C.I.E.S., Examination, 2002]**

Ans. Conductors possess high conductivity whereas the characteristic property of insulating materials (or dielectrics) is poor conductivity. Semiconductors occupy an intermediate position between conductors and insulators. Though there is no rigid line separating the conductors from semiconductors and semiconductors from insulators, but still according to resistivity the materials of resistivity of the order from 10^{-8} to 10^{-3}, 10^6 to 10^{18} and 10^6 to 10^{18} ohm-metres may be classified as conductors, semiconductors and dielectrics respectively.

Another classification is based on temperature coefficient of resistivity. Metals have positive temperature coefficient of resistivity; semiconductors have small negative temperature coefficient of resistivity and insulators have large negative temperature coefficient of resistivity.

Q.4. Differentiate semiconductors, conductors and insulators on the basis of band gap.
(U.P. Tech., Sem. Exam., 2002-03, 2008-09)

Ans. The distinction between conductors, insulators and semiconductors is largely concerned with the relative width of the forbidden energy gaps in their energy band structures. There is a wide forbidden gap (more than 5 eV) for insulators, narrow forbidden gap (about 1 eV) in case of semiconductors and no forbidden gap in case of conductors.

Q.5. Why the energy levels of an atom become energy bands in a solid.

Ans. For the single isolated atom, there are discrete energy levels, Is, 2s, 2p, 3s, that will be occupied by the electrons of atom. All the atoms of a solid can have identical electronic schemes for their energy levels. If atoms are isolated from one another then the electrons fill the levels in each atom independently but when isolated atoms are brought together to form a solid, the energy levels are considerably affected because of interaction between nearer atoms. There will be a coupling between the outer-shell electrons of the atoms resulting in a band of closely spaced energy states, instead of widely separated energy levels of the isolated atom.

Q.6. What is the importance of valence shell and valence electrons ?

Ans. The outermost shell of an atom is called the valence shell and electrons in this shell are called the valence electrons. Formation of energy bands occur owing to overlapping of energy levels of these valence electrons in valence shells. With the decrease in interatomic distance" between the atoms in a crystal, the energy levels of electrons in outermost shells of atoms overlap to form energy bands.

Q.7. What is the forbidden energy gap ? How does it occur ? What is its magnitude for Ge and Si?

Ans. The energy gap between the valence band and conduction band is known as forbidden energy gap. It is a region in which no electron can stay as there is no allowed energy state. Magnitude of forbidden energy gap in germanium and silicon is 0.72 eV and 1.12 eV respectively at 300 K and 0.785 eV and 1.21 eV respectively at absolute zero temperature.

Q.8. Is hole a fundamental particle in an atom ? **(U.P. Tech., Sem. Exam., 2007-08)**

Ans. Hole is not a fundamental particle in an atom. Holes may be thought of as positive particles, and as such, they move through an electric field in a direction opposite to that of electrons.

Q.9. Define a hole in a semiconductor. **(U.P. Tech., Sem. Exam., 2007-08)**

Ans. When an energy is supplied to a semiconductor, a valence electron is lifted to a higher energy level, the departing electron leaves a vacancy in the valence band. This vacancy is called a hole. Thus, a vacancy left in the valence band because of lifting of an electron from valence band to conduction band is known as hole.

Q.10. What is hole current ?

Ans. The movement of the hole (positively charged vacancy in the valence band) from positive terminal of the supply to the negative terminal through semiconductor constitutes hole current.

Q.11. What is intrinsic semiconductor?

Ans. An intrinsic semiconductor is one which is made of the semiconductor material in the extremely pure form (impurity content not exceeding one part in 100 million parts of semiconductors).

Q.12. Why silicon and germanium are the two widely used semiconductor materials ?

Ans. Because the energy required to release an electron from their valence band *(i.e.,* to break their . covalent bonds) is very small (1.12 eV for Si and 0.72 eV for Ge).

Q.13. Which of the two semiconductor materials Si or Ge has larger conductivity at room temperature? Why ?

Ans. Since energy required in transferring electrons from valence band to conduction band is more in case of silicon than that in case of germanium, the conductivity of Ge will be more than that of Si at room temperature.

Q.14. Why does a pure semiconductor behave like an insulator at absolute zero temperature?

(U.P. Tech., Sem. Exam., 2006-07)

Ans. For a pure semiconductor at a temperature of absolute zero (- 273.15°C) the valence band is usually full and there may be no electron in the conduction band and it is difficult to provide additional energy required for lifting electron from valence band to conduction band by applying electric field. Hence the conductivity for a pure semiconductor at absolute zero temperature is zero and it behaves like an insulator.

Q.15. What is the main factor for controlling the thermal generation and recombination ?

Ans. Temperature, because with the increase in temperature, concentrations of free electrons and holes increase and the rate of recombination is proportional to the product of concentrations of free electrons and holes and also the rate of production of electron-hole pairs (thermal generation) increases with the rise in temperature.

Q.16. Define mean lifetime of a carrier.

Ans. The amount of time between the creation and disappearance of a free electron is called the lifetime. It varies from a few nanoseconds to several microseconds depending how perfect the crystal is and other factors.

Q.17. In which bands do the mpvement of electrons and holes take place ?

Ans. Free electrons move in conduction band while holes in valence band.

Q.18. What is the mechanism by which conduction takes place inside the semiconductor.

Ans. Conduction occurs in any given material when an applied electric field causes electrons to move in a desired direction within the material. This may be due to one or both of two processes, electron motion and hole transfer. In case of former process, free electrons in the conduction band move under the influence of the applied electric field. Hole transfer involves electrons which are still attached to the atoms i.e. those in valence band.

Q.19. What do you mean by mobility of a carrier ? How does it depend on temperature and doping concentrations? (U.P. Tech., Sem. Exam., 2005-06; 2009-10)

Ans. Mobility of charge carriers is essentially a measure of the ease with which the carriers can move under the influence of an electric field. The drift velocity per unit electric field is known as mobility of charge carriers.

The mobility u of charge carriers varies as T^{-m} over a temperature range of 100 K and 400 K. For silicon m = 2.5 for electrons and 2.7 for holes. For germanium m = 1.66 for electrons and 2.33 for holes.

With the increase in cocentration of dopant, the carrier mobility at a given temperature increases.

Q.20. What do you mean by drift velocity and mobility of a free electron ?

Ans. The average velocity of a free electron is known as drift velocity whereas mobility of an electron is defined as the drift velocity per unit electric field.

Q.21. Define mobility of a carrier. Show that the mobility constant of electron is larger than that of a hole.

Ans. Mobility is defined as the average particle drift velocity per unit electric field. The mobility of electrons is more than that of holes because the probability of an electron having the energy required to move to an empty state in the conduction band is much greater than the probability of an electron having the energy required to move to the empty state in valence band. The mobility of electron is about double that of an hole.

Q.22. Define diffusion current in a semiconductor. (U.P. Tech., Sem. Exam., 2004-05)

Ans. The diffusion of charge carriers is as a result of a gradient of carrier concentration (i.e., the difference of carrier concentration from one region to another). In this case concentrations of charge carriers (either electrons or holes) tend to distribute themselves uniformly throughout the semiconductor'crystal.

This movement continues until all the carriers are evenly distributed throughout the material. This type of movement of charge carriers is called the diffusion current.

Q.23. Define drift current in a semiconductor.

Ans. The steady flow of electrons in one direction caused by applied electric field constitutes an electric current, called the drift current.

Q.24. What happens to the conductivity of the semiconductor with the rise in temperature? Compare with the conductivity of metals.

Ans. With the increase in temperature, the concentration of charge carriers increases resulting in increase of conductivity of semiconductors. The conductivity of metal decreases with the increase in temperature.

Q.25. Why the temperature coefficient of resistance of a semiconductor is negative?

(U.P. Tech., Sem. Exam., 2007-08)

Ans. With the increase in temperature, the concentration of charge carriers (electrons and holes) increases. As more charge carriers are made available, the conductivity of a pure semiconductor increases i.e. resistivity of a pure semiconductor decreases with the rise in temperature i.e. semiconductors have negative temperature coefficient of resistance.

Q.26. What is meant by Fermi level in semiconductor ? Where does the Fermi level lie in an intrinsic semiconductor ?

Ans. Fermi level in a semiconductor can be defined as the maximum energy that an electron in a semiconductor have at absolute zero temperature. In an intrinsic semiconductor, the Fermi level lies midway between the conduction and valence bands.

Q.27. Differentiate between intrinsic semiconductors and extrinsic semiconductors ? Give examples. **(U.P. Tech., Sem. Exam., 2009-10)**

Ans. An intrinsic semiconductor is one which is made of the semiconductor material in its extremely pure form.

When a small amount of impurity is added to a pure semiconductor crystal during the crystal growth in order to increase its conductivity, the resulting crystal is called the extrinsic semiconductor.

Q.28. Give reasons — Doping is done in semiconductors.

Ans. Intrinsic (or pure) semiconductor by itself is of little significance as it has little current conduction capability at ordinary room temperature but if very small amount of impurity (of the order of one atom per million atoms of pure semiconductor) is added to it in the process of crystallization, the electrical conductivity is increased many times.

Q.29. Describe the difference between P-type and N-type semiconductor materials.

(U.P. Tech., Sem. Exam., 2010-11)

Ans. When a small amount of trivalent impurity (such as boron, gallium, indium or aluminium) is added to a pure semiconductor crystal during crystal growth, the resulting crystal is called the P-type semiconductor.

When a small amount of pentavalent impurity (such as arsenic, antimony, bismuth or phosphorus) is added to a pure semiconductor crystal during crystal growth, the resulting crystal is called the N-type semiconductor.

Q.30. What do you mean by donor and acceptor impurities? **(U.P. Tech., Sem. Exam., 2006-07)**

Ans. Donor impurities (such as arsenic, antimony, bismuth or phosphorus) when added to a pure semiconductor lattice, form N-type extrinsic semiconductor. The pentavalent impurities are called the donor type impurities as such impurities donate electrons to the lattice. Acceptor impurities (such as boron, gallium, indium or aluminium) when added to a pure semiconductor lattice form P-type extrinsic semiconductor. The trivalent impurities are called the acceptor type impurities because such impurities accept electrons from the lattice.

Q.31. Explain the term doping and its need.

Ans. The electrical conductivity of intrinsic semiconductor, which has little current conduction capability at ordinary room temperature and so is of little use, can be increased many times by adding very small amount of impurity (of the order of one atom per million atoms of pure semiconductor) to it in the process of crystallization. This process is called the doping.

Q.32. What are the majority and minority carriers in P-type semiconductor materials ?

Ans. Holes are majority carriers and free electrons are minority carriers in P-type semiconductor materials.

Q.33. What is the effect of temperature on extrinsic semiconductors ?

Ans. With the increase in temperature of an extrinsic semiconductor, the number of thermally generated carriers is increased resulting in increase of concentration of minority carriers. At temperature exceeding critical temperature the extrinsic semiconductor behaves like an intrinsic semiconductor but with higher conductivity.

Q.34. What are the charge carriers in N-type and P-type semiconductors ?

Ans. Free electrons in N-type semiconductors and holes in P-type semiconductors are the charge carriers.

Q.35. For the same order of doping, why does N-type semiconductor exhibit larger conductivity than P-type semiconductor?

Ans. Since the mobility of electrons is greater than that of holes, for same level of doping, N-type semiconductor exhibits larger conductivity.

Q.36. What is the ratio of majority and minority carriers in intrinsic and extrinsic semiconductors?

Ans.

Types of Semiconductor Material	Ratio of Majority and Minority Carriers
Intrinsic	Unity
Extrinsic	Very large

Q.37. State law of mass action.

Ans. Under thermal equilibrium conditions, the product of concentration *n* of free electrons and concentration *p* of holes is constant and is independent of the amount of doping by donor and acceptor impurities, i.e. np - nf. This is known as mass action law.

Q.38. Define the term intrinsic concentration. **(U.P. Tech., Sem. Exam., 2006-07, 2008-09)**

Ans. Intrinsic concentration of holes or free electrons is called the intrinsic concentration, n_{it} which varies with temperature according to the relation $n_i = A_0 T^3 {}_e{}^{-EGo/kT}$.

where A_0 is a constant and independent of temperature, k is a Boltzmann constant in eV per K and E_{Go} is the energy gap in eV at absolute zero temperature, T is the temperature in degrees absolute and n_t is the concentration in number/m³. A_0 is taken as 2.33×10^{43}.

Q.39. Define diffusion constant of an electron and hole.

Ans. Diffusion constant of any charge carrier (electron or hole) may be defind from the Einstein relationship

$$\frac{D_h}{\mu_h} = \frac{D_e}{\mu_e} = V_T$$

where D_h and D_e are diffusion constant, μ_h and μ_h are mobilities and V_T is the volt-equivalent of temperature.

Q.40. Describe Hall effect. **(U.P. Tech., Sem. Exam., 2005-06)**

Ans. When a specimen (metal or semiconductor) is placed in a transverse magnetic field and a direct current is passed through it, then an electric field is induced across its edges in the perpendicular direction of current as well as magnetic field. This phenomenon is called the Hall effect.

Q.41. "Hall effect has 'played a decisive role in revealing the mechanism of conduction in semiconductors." Explain the statement. **(U.P. Tech., Sem. Exam., 2007)**

Ans. The conductivity of a semiconductor does determine the mobility and density of charge carriers separately. But in case there is only one type of charge carriers (either electrons or holes), the density of charge carriers can be found from a measurement of the Hall coefficient of the material. Hence, the given statement is justified.

Q.42. What properties of a semiconductor are determined fi om a Hall effect ? **(U.P. Tech., Sem. Exam., 2007-08)**

Ans. The Hall effect may be used to determine *(i)* whether a semiconductor is of N-type or P-type *(ii)* the carrier concentration and (ii) measure the conductivity of a material and then compute mobility.

Q.43. What is Hall coefficient ? Where is it used ? **(U.P. Tech., Sem. Exam., 2006-07)**

Ans. Hall coefficient $R_H = \frac{1}{\rho} = \frac{\mu}{\sigma}$

where ρ is the charge density, μ is the mobility and σ is the conductivity of specimen. It is used to determine mobility of specimen.

REVIEW QUESTIONS

1. Define conductor, semiconductor and insulator.
2. Differentiate conductor, semiconductor and insulator on the basis of band gap.
3. Define forbidden energy gap.
4. Energy band gap of some unknown materials is given below:
   (*i*) 6-10 ev, (*ii*) 0.7 – 1 ev, (*iii*) overlapping
   Distinguish their respective class as metal, semiconductor and insulator.
5. Define valence band and conduction band.
6. What is the magnitude of energy gap for Ge and Si.
7. Define the concept of hole.
8. What is intrinsic and extrinsic semiconductor?
9. Why Silicon and Germanium are the two widely used semiconductor materials.
10. Which of the two semiconductor materials Si ors Ge has larger conductivity at room temperature? Why?
11. Why does a pure semiconductor behave like an insualtor at absolute zero temperature?
12. Define life time of a carrier.
13. Define mobility of a carrier.
14. What are the two conduction processes in semiconductor?
15. Define diffusion current in semiconductor.
16. Define drift current in semiconductors.
17. Why the temperature coefficient of resistance of a semiconductor in negative?
18. Define Fermi-level and Fermi-energy in semiconductor.
19. Define P and N-type semiconductors.
20. Write the name of impurities used to fabricate N and P-type semiconductors.
21. What are the majority and minority carriers in N-type semiconductor.
22. What is Hall effect? Where is it used?
23. Define Hall Angle.
24. Define relaxation time of electrons.
25. What properties of a semiconductor are determined from the Hall effect?
26. Explain the existance of the various electron energy bands in solids. Based on these bands distinguish between insulators, conductors and semiconductors. **(U.P. Tech., Sem. Exam., 2006-07)**
27. Explain, why the energy levels of an atom become energy bands in a solid. Define a hole in a semiconductor. Explain why a semiconductor acts as an insulator at 0 K and why its conductivity increases with increasing temperature?
28. Explain the various electron energy bands in solids. Based on these bands distinguish between insulators, conductors and semiconductors. **(U.P. Tech., Sem. Exam., 2005-06)**
29. Define and explain drift and diffusion currents.
30. Define the Fermi-Dirac distribution function and Fermi energy level of a material. Describe briefly the temperature dependence of the above distribution function. **(U.P. Tech., Sem. Exam., 2006-07)**
31. Define the effective density of states function in the conduction band. **(U.P. Tech., Sem. Exam., 2006-07)**
32. Define an intrinsic material, a negative temperature coefficient and covalent bonding. List three materials that have negative temperature coefficient. **(U.P. Tech., Sem. Exam., 2007-08)**
33. What is concept of hole ? What do you understand by covalent bonding ? **(U.P. Tech., Sem. Exam., 2005-06)**

34. Is hole a fundamental particle in an atom ? Between electron and hole which one has greater mobility? Give a reason for the difference.
35. Differentiate among conductor, insulator and semiconductor using energy band concept. Also differentiate between N type and P type semiconductor materials. **(U.P. Tech., Sem. Exam., 2008-09)**
36. Draw the atomic structure of a P type material and explain the effect of acceptor impurities on the energy band structure. **(U.P. Tech., Sem. Exam., 2004-05)**
37. Derive an expression for the thermal equilibrium concentration of electrons in the conduction band of a semiconductor. **(U.P. Tech., Sem. Exam., 2009-10)**
38. What is doping ? How does it affect the conductivity of the parent semiconductor material ? **(U.P. Tech., Sem. Exam., 2005-06)**
39. What are the advantages and disadvantages of impurities in semiconductor solids ? Explain them with suitable examples. **(U.P. Tech., Sem. Exam., 2010-11)**
40. Distinguish between intrinsic and extrinsic semiconductors. What is the effect of temperature on the conductivity of a semiconductor ?
41. Distinguish between 'majority' and 'minority' carriers in a semiconductor. Define mobility of charge carriers.
42. How does conduction take place in an intrinsic semiconductor ? Explain with examples. What are limitations of intrinsic semiconductors and how are these taken care of by extrinsic semiconductors ? Show that the minimum value of conductivity of a semiconductor with impurity is $\sigma_{rain} = 2n_e\, e\, \mu_e$ where n_e-concentration of conduction electrons (m^{-3}), e-electron charge (C), μ_e-mobility of electrons ($m^2\, V^{-1}\, S^{-1}$). **(U.P. Tech., Sem. Exam., 2008-09)**
43. Define the terms conductivity, intrinsic concentration and energy gap of semiconductors.
44. An N-type semiconductor has excess of electrons and P-type excess of holes. Does that mean N-type materials have a net negative charge and P-type materials have a net positive charge. Comment. **(U.P. Tech., Sem. Exam., 2006-07)**
45. What do you mean by mobility of a carrier ? How does it depend on temperature and doping concentrations. **(U.P. Tech., Sem. Exam., 2009-10)**
46. Define mobility of a charge carrier. Show that $\frac{\mu}{D} = \frac{e}{kT}$ **(U.P. Tech., Sem. Exam., 2009-10)**
47. Using the concept of diffusion and drift of carriers, establish Einstein relation and modified form of Ohms law. **(U.P. Tech., Sem. Exam., 2009-10)**
48. Obtain the relation $\frac{D_p}{\mu_p} = \frac{kT}{q}$ **(U.P. Tech., Sem. Exam., 2005-06)**
49. Given an intrinsic semiconductor specimen, state two physical processes for increasing its conductivity. Explain briefly.
    Write the equation for the net electron current in a semiconductor. What is the physical significance of each term ? How is this equation modified for a metal.
50. Explain electron and hole mobilities and drift velocities. **(U.P. Tech., Sem. Exam., 2010-11)**
51. How are the carrier mobilities related with resistivity of a semiconductor ? Does a 'hole' in a semiconductor contribute a flow of current ? If yes, how and if no, how ?
52. Starting from Fermi-Dirac energy distribution function, show how it is possible to obtain definition of Fermi level in a solid at 0 K and finite temperatures.
53. What is the meaning of energy density function ? Estimate the position of Fermi level in a conductor having N number of free electrons per cc. Deduce all steps necessary for deriving the result.
54. Explain the current conduction mechanism in a semiconductor. Explain how doping changes the conductivity of a semiconductor ? Show that the Fermi level lies half way between the valence and conduction bands of an intrinsic semiconductor. **(U.P. Tech., Sem. Exam., 2008-09)**
55. Draw the energy band diagram of an N-type semiconductor. State the factors on which the position of Fermi level depends. **(U.P. Tech., Sem. Exam., 2006-07)**
56. What is Hall effect and Hall angle ? Show that $\theta_H = \tan^{-1} \mu_n\, B_z$ where symbols have their usual meaning. **(U.P. Tech., Sem. Exam., 2009-10)**
57. Explain Hall effect in semiconductors. What is Hall angle ? **(U.P. Tech., Sem. Exam., 2009-10)**

58. What is Hall effect ? Derive the relation between Hall coefficient and carrier density. Assume the presence of only one type of charge carrier. **(U.P. Tech., Sem. Exam., 2006-07)**
59. Explain the Hall effect. Show that the Hall coefficient is inversely proportional to the number of charge carriers density. **(U.P. Tech., Sem. Exam., 2005-06)**
60. What properties of semiconductor are determined from Hall effect experiment ? Discuss the effect in detail. Derive the expression for Hall effect. **(U.P. Tech., Sem. Exam., 2005-06)**
61. Explain the Hall effect in semiconductors. Derive the expression for Hall coefficient of P-type semiconductor. **(U.P. Tech., Sem. Exam., 2006-07)**

NUMERICAL PROBLEMS

1. The mobilities of carriers in intrinsic germanium sample at room temperature are μ_e = 3600 cm²/v-s and μ_h = 1700 cm²/v-s. If the density of electrons is same as holes and is equal to 2.5×10^{13} per cm³, calculate the conductivity. **(Ans. 2.12 mho/m)**
2. Calculate the number of acceptors to be added to a germanium sample to obtain the resistivity ρ = 10 ohm-cm. Given μ_h = 1700 cm²/v.s **(Ans. 3.676×10^{14}/cm³)**
3. At room temperature the conductivity of a Si crystal is 5×10^{-4} mho/cm. If the electron and hole mobilities are 0.14 m²/v-s and 0.05 m²/v-s, determine the density of carriers. **(Ans. 1.64×10^{16}/cm³)**
4. The specific density of tungsten is 18.8 g/cm³ and its atomic wt. is 184.0. Assume that there are two free electrons per atom. Calculate the concentration of free electrons. Avogadro number = 6.025×10^{23}/g-mole. **(Ans. 2.5×10^{23}/cm³)**
5. The resistivity of copper wire of diameter 1.03 mm is 6.51 ohm per 300 m. The concentration of free electrons in copper is 8.4×10^{28}/m³. If the current is 2A, find the (a) mobility, (b) drift velocity (c) conductivity. **(Ans. 0.413 m²/v-s, 0.286×10^{-20} m/s, 55.5×10^{8} mho/m)**
6. Calculate the energy gap in Si if it is given that it is transparent to radiation of wavelength greater than 11000 Å. **(Ans. 1.13 eV)**
7. An N-type semiconductor is to have a resistivity of 10 ohm-cm. Calcualte the number of donor atoms which must be added to achieve this.
   Given μ_e = 500 cm²/v-s. **(Ans. 12.5×10^{14}/cm³)**
8. A silver wire is in the form of a ribbon 0.50 cm. wide and 0.10 mm thick. When a current of 2 Amp passes through the ribbon, perpendicular to 0.80 Tesla magnetic field, calculate the Hall voltage produced. The density of silver = 10.5 gm/cc. **(Ans. 1.7×10^{-6} volts)**
9. Determine the mobility of holes in P-type silicon bar. Assume the bar resistivity as 200 kΩ-cm, the magnetic field B_Z = 0.1 Wb/m² and d = W = 3 mm. The measured values of the current and Hall voltage are 10 μA and 50 mV respectively. **(Ans. 0.075 m²/v-s)**
10. Find the magnitude of Hall voltage V_H in an N-type germanium bar, having majority carrier concentration $N_D = 10^{17}$/cm³. Assume B_Z = 0.1 Wb/m², W = d = 4 mm, E_x = 5 V/cm, and μ_e G_e = 3800 cm²/v-s. **(Ans. 76 mv)**
11. Calculate the Hall angle for Germanium specimen which has flux density of 0.1 Wb/m². electron mobility for Ge is 3800 cm²/v-s. **(Ans. 2.18°)**
12. A sample of pure germanium at room temperature has a density of charge carriers of 2.5×10^{19}/m³. It is doped with donor impurity atoms at the rate of one impurity atom for every 10^6 atoms of germanium. The density of germanium atoms is 4.2×10^{28} atoms/m³. Determine the hole density. **(Ans. 1.42×10^{16}/cm³)**
13. The Hall coefficient of a specimen of a doped silicon is found to be 3.66×10^{-4} m³ c⁻¹ the resistivity of the specimen is 8.93 × 10–3 Ω-m. Find the mobility and density of the charge carriers, assuming single carrier concentration. **(Ans. 0.04 m²/v-s, 1.707×10^{22}/m³)**
14. In a Hall experiment a current of 25 A passes through a long silver foil of width 3 cm and 0.1 mm thick. The magnetic field of 1.4 Wb/m² is applied in the direction such that the thick side is perpendicular to B. Find the Hall voltage developed if carrier concentration in silver is 5.8×10^{28}/m³. **(Ans. 1.257×10^{-7} volts)**
15. Determine the number of electrons per m³ if the resistivity of the conductor material is 0.01531×10^{-6} Ω-m and mobility is 7.04×10^{-3} m²/V-s. Also determine the relaxation time. Given that e = 1.6×10^{-19} C and m = 9×10^{-31} kg. **(Ans.** 5.8×10^{28}/m³; 4.004×10^{-14} second)
16. A conduction wire has a resistivity of 1.54×10^{-8} H-m at room temperature. There are 5.8×10^{28} conduction electrons per m³. Calculate the mobility and relaxation time of electrons. **(Ans.** 6.997×10^{-3} m²/V-s; 3.979×10^{-14} second)

17. For silicon the intrinsic concentration is approximately 10^{16} carriers/m^3. If an impurity concentration of 10^{22} donor atoms/m^3 is doped, determine the electron and hole concentration. **(Ans.** 10^{22}; 10^{10}**)**
18. A silicon wafer (intrinsic carrier concentration 1×10^{16} m^{-3}) is doped with 2×10^{22} aluminium atoms/m^3 and 1×10^{22} arsenic atoms/m^3. Determine the minority carrier concentration. Assume complete dopant ionization. **(Ans.** 1×10^{10}/m^3**)**
19. The electron and hole concentration in a P-type semiconductor is 10^{12}/m^3 and 10^{22}/m^3 respectively. If the mobility of electrons and holes are 0,04 and 0.06 m^2/V-s respectively, determine the conductivity of the material. **(Ans.** 96 (ohm-metre^{-1})**)**
20. Determine the number of electrons per m^3 if the resistivity of the conductor material is 0.01531×10^{-6} Ω-m and mobility is 7.04×10^{-3} m^2/V-s. Also determine the relaxation time. Given that $e = 1.6 \times 10^{-19}$ C and $m = 9.1 \times 10^{-31}$ kg. **(Ans.** 5.8×10^{28}; 4.004×10^{-14} second**)**
21. Find the concentration of holes and electrons in an N-type silicon at 300 K if the conductivity is 0.1 (Ω-cm)$^-$. Given that n_i at 300 K for silicon is 1.5×10^{10}/cm^3. μ_e at 300 K for silicon is 1,300 cm^2/V-s. **(Ans.** $n = 4.8 \times 10^{14}$ electrons/cm^3, $p = 4.688 \times 10^5$ holes/cm^3**)**
22. A sample of pure germanium at room temperature has a density of charge carriers of 2.5×10^{19}/m^3. It is doped with donor impurity atoms at the rate of one impurity atom for every 10^6 atoms of germanium. The density of germanium atoms is 4.2×10^2 atoms/m^3. Determine hole density. **(Ans.** 1.42×10^{16}/m^3**)**
23. What is the donor concentration in a sample of N-type germanium having a resistivity of 0. 1 Ω-m at 300 K.

(Hint: $N_D = \dfrac{\sigma_n}{e\,\mu_n}$ and $\mu n = 3{,}800$ **)** **(Ans.** 1.645×10^{16}/m^3**)**

□□□

UNIT-II

Syllabus

Excess Carriers in Semiconductors: Optical Absorption, Luminescence, Carrier Life Time and Photo Conductivity, Diffusion of carriers.

Above syllabus has been covered in following chapter:

Chapter 3: Excess Carriers in Semiconductors

Chapter 3

Excess Carriers in Semiconductors

Syllabus

Excess carriers in Semiconductors: Optical absorption, Luminescence, Carrier life time and photo conductivity, Diffusion of carriers.

Chapter Outline

- Introduction
- Photo Electric Cell
- Optical Absorption Coefficient
- Luminescence
- Photoluminescence
- Cathodoluminescence
- Electroluminescence
- Carrier Lifetime and Photoconductivity
- Recombination Process
- Photoconductor
- Electric Conductivity Due to Drift
- Diffusion Mechanism
- Carrier Concentration
- Mass Action Law
- Einstein Relationship
- Semiconductor Materials used in Opto-electronic Devices
- Photo-Luminescent Devices
- Semiconductor Materials used in Opto-electronic Devices
- Properties of Germanium and Silicon
- The Continuity Equation
- Various Physical Constants

3.1. INTRODUCTION

(i) All of us already know about that the importance of semiconductors to modern technology.

(ii) The world-wide utilization of semiconductors in various applications such as radio, welders, solar batteries, LSI and VLSI design, computers and many more devices has started intensive research efforts to understand their properties more deeply.

(iii) Basically, semiconductor devices operate by the creation of charge carriers in excess of the thermal equilibrium values.

(iv) In fact, the excess carrie in semiconductors are produced by thermal excitation, optical absorption, electron bombardment, electric field etc.

(v) Also these excess carriers are primarily responsible for conduction process in the semiconductors devices.

(vi) This chapter discusses in details the production of excess carriers by the absorption of light that leads to the properties of photoluminescence and photoconductivity.

(vii) Basically, creation of electron-hole pair and mechanism of electron-hole recombination, effect of trapping centre.

(viii) Also, the diffusion of excess carriers due to carrier gradient, which is the basic mechanism of current conduction and the mechanism of drift of carr[illegible] in the electric field has been discussed in [illegible] chapter.

3.2. PHOTO ELECTRIC CELL

3.2.1. Definition and Basic Aspects

It is device that converts light energy into electrical energy.

1. Construction

(i) A Photo electric cell consists of an evacuated glass tube, two electrodes cathode C and anode A are fixed to it. Basically, the cathode is cylindrical in shape. It is concave surface and is coated with a photo sensitive material.

(ii) The anode A is in form of a platinum rod placed at the axis of the cylindrical concave cathode.

(iii) The glass tube is mounted on a non-conducting base through which leads from the cathode and anode are taken out and then connected to connecting pin.

(iv) Also, the cathode C is connected to negative terminal and anode A is connected to positive terminal of the dc source.

2. Working Operation

(i) When light rays of suitable frequency are incident on the concave surface of the cathode, it emits electrons. All these photoelectrons are attracted by the anode and flows into the external circuit that produce an electric current.

(ii) This current is measured with the help of milliammeter connected in series with the anode.

(iii) When we increase the intensity of the incident light, the rate of emission of photoelectrons also increase, which, in turn, increases the current in the external circuit.

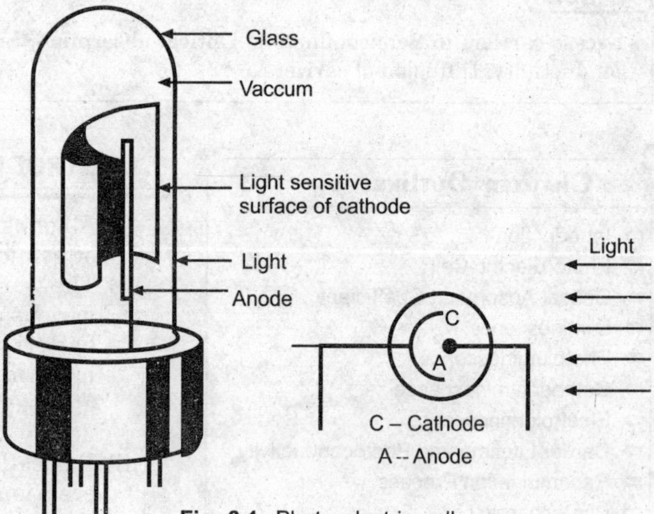

Fig. 3.1. Photo electric cell

3.2.2. Photo-electric Effect Characteristics

The photo-electric effect characteristics may be listed as under:

(i) The photoelectric emission is an instantaneous process. There is no time lag between incident light and emission of photoelectrons.

(ii) Photoelectrons are emitted only if the frequency of incident radiation is greater than a certain minimum frequency called the threshold frequency (v_0). It is different for different materials. The wavelength corresponding to the threshold frequency is called the threshold wavelength (λ_0) and is given by the relation.

$$\lambda_0 = \frac{C}{v_0}$$

(iii) The photoelectric emission is directy proportional to the intensity of the incident radiation.

(iv) The maximum kinetic energy with which photoelectrons are emitted is dependent on the frequency of incident radiation. It does not depend on the intensity of the incident radiation.

(v) The emissions of the photoelectrons is independent of temperature of the emitting metal but depends on the nature of the emitting metal.

3.2.3. Planck's Quantum Theory

Prior to Planck's quantum theory, there was the electromagnetic theory of radiation that states that bodies emit radiation continuously. *According to planck's quantum theory emission of radiation from a body is not a continuous process.* The emission takes place at certain intervals and thus, it is a discrete process. Plancks quantum theory states that,

(i) The emission or absorption of energy are not a continuous process.

(ii) The energy is either emitted or absorbed intermittently in the form of small energy packets, with velocity of light.

(iii) The energy carried by each small packet is known as quanta or photon.

(iv) The energy possessed by a photon is given by the relation/

$$E = h\nu$$

where h = Planks constant = 6.625×10^{-34} Joule-second

$$\nu = \text{Frequency of radiation} = \frac{C}{\lambda}$$

where C is the velocity of light = 3×10^8 m/s and λ is the wavelength of the radiation.

> **Important Point:** The photons are electrically neutral, as such are not affected by magnetic and electric fields.

3.2.4. Planck Einstein Equation

1. Basic Concepts

(i) According to Planck's quantum theory, the light consists of stream of photons particles and each photon is having energy equal to $h\nu$.

(ii) When light is incident on a photosensitive metal plate, photons transfer its energy to the atoms of metal through inelastic collisions.

(iii) This energy is utilized in two ways. A fraction of energy is used in detaching the electrons from the atoms and remaining energy is used to set free and increase the kinetic energy (K.E.) of the electrons.

(iv) Let us consider $\frac{1}{2}mv^2$ = K.E. possessed by the photo-electron

W_0 = Energy required to detach the electrons.

Then, we may write

$$h\nu = W_0 + \frac{1}{2}mv^2$$

or

$$\frac{1}{2}mv^2 = h\nu - W_0 \quad ...(3.1)$$

This is known as **Planck - Einstein's equation**.

2. Work Function

(i) The minimum energy required to detach or to set free the electrons from the surface of the metal is known as work function (W_0) of the given metal. It is different for different metal.

(ii) For a given metal, the photoelectric work function W_0 is constant. Hence, the K.E. of photoelectron depends on the frequency of the incident light. If the frequency of the incident light is gradually decreased, the K.E. of the emitted photoelectrons will also decrease.

(iii) At a particular frequency (ν_0), K.E. of electrons becomes zero i.e., photoelectrons just reach outside the metal plate. The frequency ν_0 is known as threshold frequency.

(iv) Then, we can write

$$\frac{1}{2}mv^2 = 0 = h\nu_0 - W_0$$

or $$W_0 = h\,\nu_0$$

(v) Substituting this value in equation (3.1), we get

$$\frac{1}{2}mv^2 = h\nu - h\nu_0$$

or $$\frac{1}{2}mv^2 = h(\nu - \nu_0)$$

(vi) We can conclude that, the photoelectrons are emitted only when $\nu > \nu_0$, where ν is the frequency of the incident light.

3. Optical Absorption

(i) Optical absorption is the absorption of optical energy as explained above. At a constant incident light intensity, the maximum kinetic energy of the photoelectrons varies linearly with frequency with a limiting frequency $\nu = \nu_0$ below which no photoelectron is produced.

(ii) Figure 3.2 shows the reflection of photoelectrons and the variations of the maximum kinetic energy with respect to the frequency.

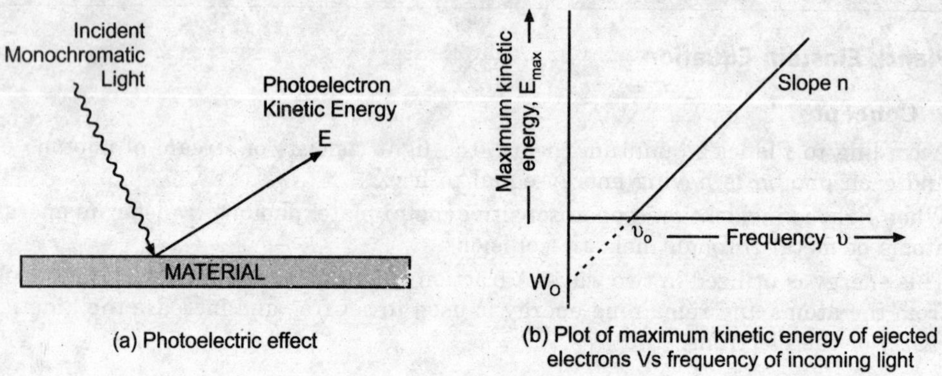

Fig. 3.2.

3.2.5. Optical Absorption: Detailed Aspects

(i) The optical band gap of a semiconductor can be measured by the absorption of photons of suitable frequency into the material.

(ii) The experiment can be carried out by directing the photons of particular wavelength at the sample and the relative absorption and transmission of the different photons may be observed.

(iii) Basically, Photons having energies greater than the band gap energy are absorbed and photons having energies less than the band gap are transmitted. Optical band gap of semiconductor can be measured precisely simply by using this technique.

(iv) Photons with energies greater than the band gap are absorbed in a semiconductor as depicted in figure 3.3.

(v) In semiconductor valence band possess many electrons and the conduction band contains many empty states in which electrons can be excited, that increases the probability of photon absorption.

(vi) When an electron excited to the conduction band by light absorption it possess more energy than commonly available electrons in conduction band (that is nearly equal to E_C).

(vii) The excited electron loses its energy to the lattice due to scattering with it until its velocity reaches to the thermal equilibrium velocity like other conduction band electrons.

(viii) Thus, electron and hole created in conduction band and valence band respectively due to this light absorption process are excess carriers as they are out of balance from their natural environment and they must be immediately recombine.

(ix) The excess carriers are free to move in their respective bands and they are ready to contribute to the conductivity of semiconductor material.

(x) Photons with energies less than band gap ($h\nu < E_g$) energy are not able to excite the electron from valence band to the conduction band, so there is negligible absorption of these photons into semiconductor material.

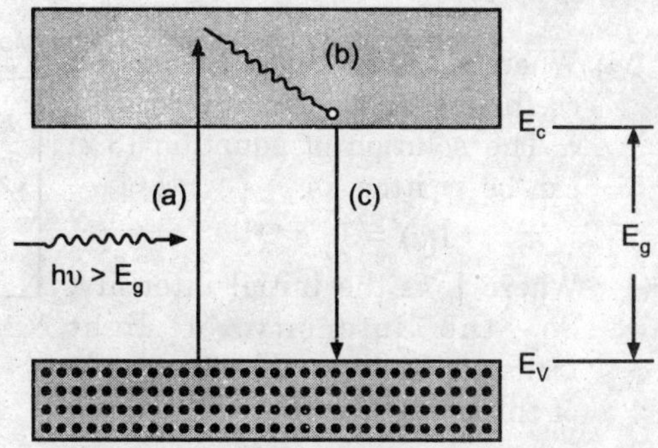

(a) An EHP is created during photon absorption
(b) The excited electron the lattice by gives up energy to scattering events
(c) The electron recombines with a hole in the valence band

Fig. 3.3. Optical absorption of a photon with $h\nu > E_g$

(xi) This fact proves that some semiconductor materials are transparent in particular wavelength range.

(xii) Insulators due to its large band gap do not possess electrons states in the material. If the band gap of the material is of the order of (1.5 – 2.00) eV only infrared spectrum is transmitted and if the band gap is of the order of (2.5 – 3.00) eV, then infrared and visible spectrum are transmitted.

(xiii) The energy of a photon is $E = h\nu$, where h is Plancks constant and ν is the frequency.

Wavelength and energy can be related by $\lambda = \frac{c}{\nu} = \frac{hc}{E} = \frac{1.24}{E}\,\mu m$

Where E is photon energy is eV and c is the velocity of light.

3.3. OPTICAL ABSORPTION COEFFICIENT

(U.P. Tech, Sem. Exam; 2008-09)

1. Basic Concepts

(i) When a stream of photons of energy more than the band gap ($h\nu > E_g$) incident on the semiconductor material, appreciable amount of light energy is absorbed by the material, whcih depends on the properties of the material thickness of the material and also on the wavelength of the incident photons.

(ii) To measure the intensity of the transmitted photons, let us consider that a beam of photon of intensity I_0 is incident at a sample of thickness d as shown in figure 3.4. The beam contains photons of particular wavelength as adjusted by the monochromator.

(iii) As the beam passes through the sample, its intensity at a distance x from the surface can be calculated by considering the probability of absorption within any increment dx.

(iv) The probability of absorption of photons at any element dx depends on the intensity of light at distance x i.e., I(x).

(v) Hence, we can write the degradation of light intensity at distance x is equal to

$$-\frac{dI(x)}{dx} \propto I(x)$$

or, $$-\frac{dI(x)}{dx} = \alpha\, I(x) \quad ...(3.2)$$

(vi) Where α is called the absorption coefficient and it has unit of cm^{-1}. The solution of equation (3.2) can be written as

$$I(x) = I_0\, e^{-\alpha x}$$

Where I_0 is the initial intensity. So, the intensity of light transmitted through the sample of thickness d is

$$I_t = I_0\, e^{-\alpha d}$$

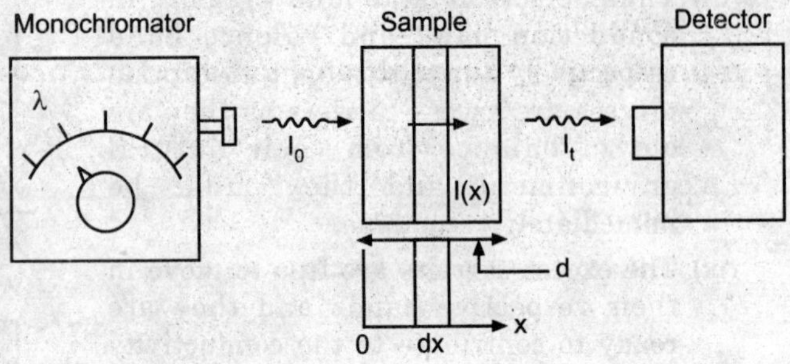

Fig. 3.4. Optical absorption in bulk semiconductor

(vii) The coefficient α depends on the wavelength of photon and nature of the material.

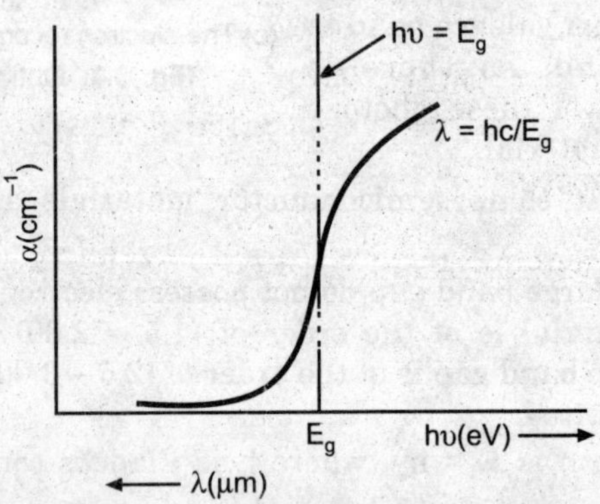

Fig. 3.5. Dependence of optical absorption coefficient α for a semiconductor on the wavelength of incident light

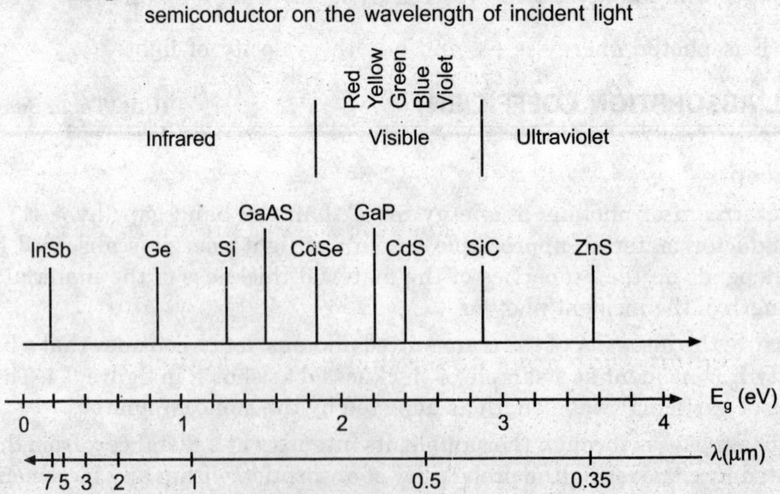

Fig. 3.6. Band gaps of semiconductors relative to optical spectrum

(viii) A graph is plotted between absorption coefficient (α) and wavelength of photons as shown in figure 3.5. The graph indicates that photons of shorter wavelength are absorbed by the material appreciably whereas photons of longer wavelength are negligibly absorbed.

(ix) The band gap energies of few semiconductors related to ultraviolet, visible and infrared range of the spectrum are shown in figure 3.6.

(x) The graph indicates that In Sb, Ge, Si and GaAs lie outside the visible region and their band gap matches with infrared range of the spectrum. The semiconductors like GaP, CdSe and CdS have large band gap such that photons of visible range are absorbed significantly. We can conclude that semiconductors absorb photons with energies equal to the band gap or larger. As example Si absorbs not only band gap light ($\lambda \cong 1$ μm) but also shorter wavelengths and Ge absorbs not only light of wavelength ($\lambda \cong 2$ μm) but also shorter wavelength lies in the visible spectrum.

3.4. LUMINESCENCE (U.P. Tech., Sem. Exam.; 2005-06, 2008-09)

When excess carriers (electron-hole pairs) are created in the semiconductor or when carriers are excited into higher impurity levels due to absorption of light or electron bombardment from which they fall to their equilibrium states, light of particular wavelength depending on the nature of material is emitted. The compound semiconductors like GaAs, GaP with direct band gap are most suitable for light emission. Light emission in this process by the semiconductors is known as luminescence. There are different types of luminescence which depends on the excitation mechanism.

(i) **Photoluminescence:** When carriers are excited by the absorption of photons of suitable frequency, then the resulting radiation due to the recombination of excited carriers (electron-hole pair) is known as photoluminescence.

(ii) **Cathodoluminescence:** If the carriers are excited by the bombardment of highly energetic electrons into the material and the resulting radiation is called cathodoluminescence.

(iii) **Electroluminescence:** If the excitation of carriers is produced due to the introduction of forward current into the sample and the resulting radiation due to the recombination is called electroluminescence.

The above mentioned three types of luminescence are mostly important for semiconductor device applications.

3.5. PHOTOLUMINESCENCE (U.P. Tech., Sem. Exam.; 2006-07, 2010-11)

1. Definition

In semiconductor excited carriers emit light due to direct and indirect transitions i.e., recombination of an electron-hole pair as shown in figure 3.7.

2. Recombination

When the recombination of electron-hole pair takes place directly rather than through a defect level, it emits light of energy same as band gap of the semiconductor. For steady state excitation, the recombination of electron-hole pair takes place at the same rate as the generation, i.e., one photon is generated per each photon absorbed. The mean life time of the electron-hole pair recombinations for direct process is of the order of 10^{-8} sec, hence it is a fast process. In this process, the emission of photons or light stops within approximately 10^{-8} sec, after the excitation source is switched off. This type of fast process of light emission is known as fluorescence.

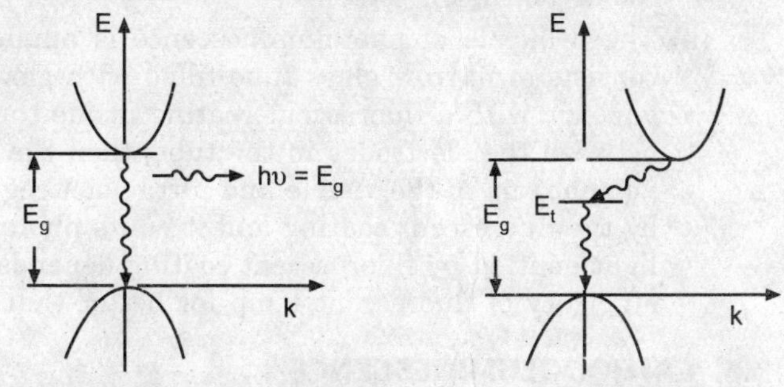

(a) Direct transition with accompanying photon emission (b) Indirect transition via a defect level

Fig. 3.7. Direct and indirect electron transitions in semiconductors

In some semiconductor materials, the process of recombination continues for the period of few seconds to few minutes after the excitation source is switched off. The slow process of light emission from the semiconductor material is known as phosphorescence, which is shown in figure 3.8.

3. Impurity or Defect Level

If the semiconductor contains impurity or defect level in the forbidden band gap, which acts as traping centre and its tendency is always capture electron from conduction band. The figure 3.8, explains that,

(i) A photon with energy greater than band gap then it is absorbed and an electron-hole pair is produced.

(ii) The excited, electrons lose its energy to the lattice by the process of scattering until it reaches its energy equal to bottom of the conduction band (E_C).

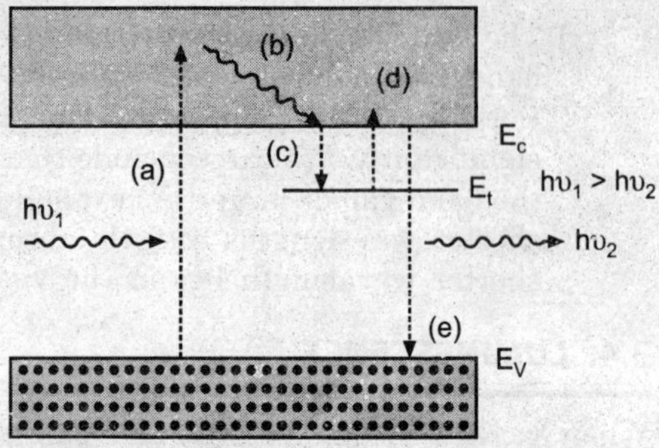

Fig. 3.8. Excitation and recombination mechanisms in photoluminescence witgh a trapping level for electrons

(iii) The electron is trapped by the impurity level (E_A) and remains in the trap centre until it is re-excited to the conduction band again.

(iv) Ultimate the electron recombine with the empty states present in the valence band directly and in this process photon of energy equal to band gap (h ν_2) is emitted.

The delay time between excitation and recombination may be comparatively large when the probability of thermal reexcitation electron from the trap centre to the conduction band is less. We may expect more longer time if the electron is retrapped several times before recombination. If the trapping probability is more than the recombination then electron moves from traping centre to conduction band and come down to trapping centre for long time, before final recombination takes place. This type of semiconductor material emits light for long time even after the excitation source is removed.

4. Applications

(i) The colour of light emitted by phosphorescence material (phosphor) like ZnS depends mainly on the impurities presents in the band gap, because radiative transition takes place between impurity levels and conduction band and finally with valence band. The selection of colours by selecting proper impurity is useful for device applications like colour television screen.

(ii) The principle of photoluminescence is applied to design fluorescent lamp. This lamp consists of narrow glass tube filled with gas at low pressure (96% Argon, 4% mercury vapour), with a fluorescent coating inside the tube. If an electric discharge is produced between the electrodes in the tube, then the gas atoms are excited and they emit light *i.e.*, photons in the visible and ultraviolet region of the spectrum . The light is absorbed by the fluorescent coating and it emits photons in the visible range. The wavelength of light emitted by fluoroescent coating depends on the material used for the coating. The efficiency of fluorescent lamp for better than the incandescent lamp.

3.6. CATHODOLUMINESCENCE **(U.P. Tech., Sem. Exam.; 2006-07; 2008-09)**

(i) The excitation of luminescent materials by electron bombardment is applied to cathode-ray-tube (CRT). The oscilloscope, television set and other visual display systems designed using the principle of light emitting tube.

(ii) The main principle of the CRT is the selective excitation of a phosphorescent screen by a beam of energetic electrons within vacuum tube as shown in figure 3.9.

(iii) Electrons are emitted from electron gun and are accelerated by the electric field applied to positively charged anode.

(iv) The fine and energetic beam of electrons move towards the deflection system *i.e.*, xx and yy plates which consists of electric fields and magnetic field respectively. Electron beam can be deflected horizontally or vertically using these plates and ultimately it can be focussed on any spot on the screen, which is made of luminescent material.

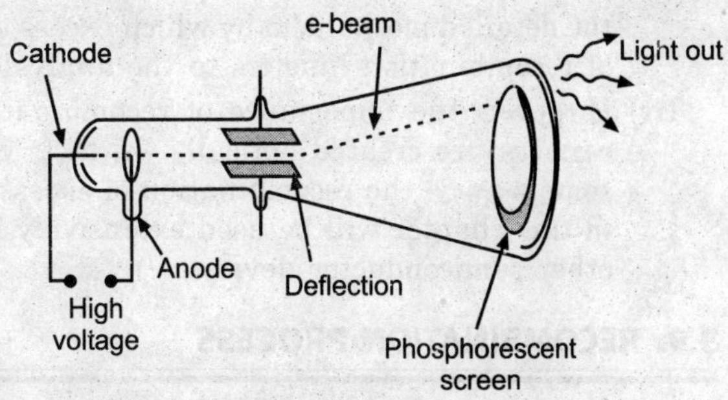

Fig. 3.9. Simplified view of a cathode ray tube

(v) As the energetic electrons strike the luminescent material, electron present in the material are excited to higher energy states and during recombination, it emits the light at particular spot on the screen. In television picture tube, the intensity and position of the beam is adjusted to produce exact replica of the image being received.

(vi) In colour television the screen is coated with a pattern of phosphorescent materials dots which are doped to emit the primary additive colours *i.e.*, red, green and blue.

(vii) To excite the appropriate dots, three electron beam with different energy are swept through the pattern. Each beam are suitable to excite one type of phosphor dots.

(viii) By adjusting the relative intensity of light emitted by the each dots, a broad range of colours can be synthesized.

3.7. ELECTROLUMINESCENCE

(U.P. Tech., Sem. Exam.; 2006-07)

(i) Electroluminescence effect was first observed by certain phosphor materials which emits light under the influence of alternating electric field. In this device, phosphorescence material like ZnS powder held in a binder material of a high dielectric constant like plastic.

(ii) If an alternating electric field is applied to this material, it starts emitting light. These cells are useful for lighting panels and its efficiency is not good enough for large scale commercial applications.

(iii) Now a days light emitting diode (LED) is designed based on the principle of electroluminescence. LED is made from p-n junction diode prepared with the material like GaAs.

(iv) When forward current passes through this diode, it causes the generation of minority carriers into the junctions where they recombine again, which causes the emission of radiation.

(v) The wavelength of the emitted radiation depends on the nature of the material and amount of doping in it. The efficiency of the LED's are better and they are used commercially in display devices.

3.8. CARRIER LIFETIME AND PHOTOCONDUCTIVITY

(U.P. Tech., Sem. Exam.; 2005-06, 2009-10)

(i) If excess electrons and holes are produced in a semiconductor, there is a corresponding increase in the conductivity of the sample is given by the equation

$$J_x = q\,(n\,\mu_n + p\,\mu_p)\,E_x = \sigma\,E_x$$

(ii) When the excess carriers are created due to the optical excitation, the corresponding increase in conductivity is known as photoconductivity.

(iii) This is an important effect, with useful applications in the analysis of semiconductor materials and in the operation of different types of semiconductor devices. In this chapter the detailed mechanism by which excess electrons and holes recombine and the application of recombination kinetics to the analysis of photoconductive devices are discussed.

(iv) However, the importance of recombination is not limited to cases in which the excess carriers are created optically. In fact, virtually every semiconductor device depends in some way on the recombination of excess electrons and holes. Hence, the ideas developed in this chapter will be used extensively in the analysis of diodes, transistors, lasers and other semiconductor devices.

3.9. RECOMBINATION PROCESS (U.P. Tech., Sem. Exam.; 2006-07, 2007-08)

Every semiconductor device depends on the recombination of excess electrons and holes. These concepts are useful in analysis of diodes, transistors and other devices. Electrons in the conduction band of a semiconductor recombine with holes in the valence band (i) directly (ii) indirectly.

3.9.1. Direct Recombination

(i) In direct recombination, excess concentration of electrons and holes, decay by electrons falling from conduction band to empty states (holes) in the valence band. It occurs spontaneously.

(ii) The energy lost by the electron during the transition is given up as a photon. The probability that an electron and hole will recombine is constant, all the time.

Let us consider an intrinsic semiconductor material with

n_i = intrinsic carrier concentration per cm^3

n_o = electron concentration at thermal equilibrium per cm^3

p_o = hole concentration at thermal equilibrium.

For an intrinsic material at thermal equilibrium, we have

$$n_o = p_o = n_i$$

Therefore, $$n_o p_o = n_i^2$$

(iii) The rate of recombination (r_i) is proportional to the equilibrium concentrations of electrons and holes.

Therefore, $$r_i \propto n_o p_o$$

$$r_i = \alpha_r n_o p_o$$

$$r_i = \alpha_r \cdot n_i^2$$

where α_r = proportionality constant

= recombination rate in thermal equilibrium the generation rate (g_i) is equal to the recombination rate (r_i)

Therefore, $$g_i = r_i = \alpha_r \cdot n_i^2$$

If $n(t)$ = Electron concentration at time t per cm^3,

$p(t)$ = Hole concentration at time t per cm^3

Then, we write net rate of change of conduction band electrons as under:

$$\frac{d[n(t)]}{dt} = \alpha_r \cdot n_i^2 - \alpha_r\, n(t)p(t)$$

Here, $\frac{d[n(t)]}{dt}$ = net rate of change (decay) of electron concentration

$\alpha_r \cdot n_i^2$ = rate of change of electron concentration due to thermal generation

$\alpha_r - n(t)\ p(t)$ =rate of change of electron concentration due to direct recombination

Let us assume n_o = equilibrium value of electron concentration per cm^3

p_o = equilibrium value of hole concentration per cm^3

$\delta_n(t)$ = excess electron concentration at time t

$\delta_p(t)$ = excess hole concentration at time t

Therefore, $n(t) = n_o + \delta_n(t)$

$p(t) = p_o + \delta_p(t)$

(iv) If excess electrons holes are created at time t = 0 and initial excess concentrations are Δn for electrons and Δp for holes, then rate of change of excess concentration. Therefore, we have

$$\frac{d[\delta_n(t)]}{dt} = \alpha_r n_i^2 - \alpha_r[n_o + \delta_n(t)] \times [p_o + \delta_p(t)] = -\alpha_r[(n_o + p_o)\delta_n(t) + \delta_n^2(t)]$$

$$= \delta_n(t) = \Delta n\, e^{-\alpha_r . p_o . t} = \Delta n\, e^{-t/\tau_n} \quad ...(3.3)$$

Here, $\tau_n = \dfrac{1}{\alpha_r p_o}$ = recombination life time of electrons in P-type semiconductor

(v) In a similar manner,

$$\tau_p = \frac{1}{\alpha_r n_o} = \text{recombination life time of holes in N-type semiconductor.}$$

Here, τ_p and τ_n are called minority carrier life time in N-type and P-type semiconductors respectively.

A more general expression for carrier life time is given by

$$\tau = \frac{1}{\alpha_r (n_o + p_o)}$$

(vi) This equation is valid for both P or N type semiconductor with low level of injection.

(vii) The generation and recombination excess carriers with time has been illustrated in figure 3.10. Let us consider an N-type silicon bar containing the thermal equilibrium concentration p_0 and n_0.

(viii) Let us assume that at any instant t = t′ the bar is illuminated as shown in figure 3.10 and excess EHP are produced uniformly throughout the crystal.

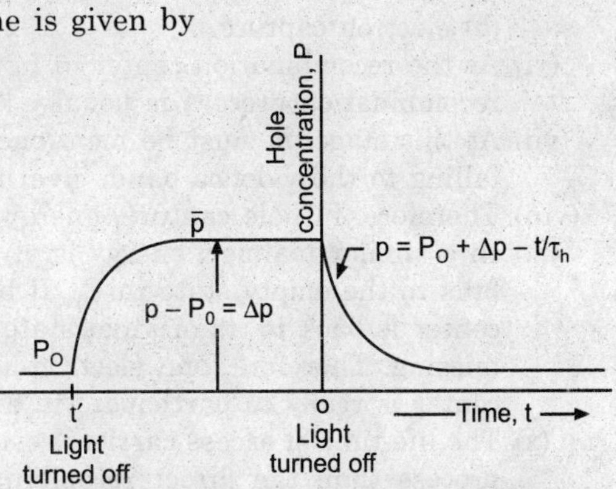

Fig. 3.10. Hole (Minority) concentration in an N-type semiconductor bar as a function of time due to generation and recombination

(ix) An equilibrium conditions is attained and the new concentrations are $\bar{p}$ and $\bar{n}$ under the influence of radiation. The photo-injected excess concentration is $\bar{p} - p_0 = \Delta p$ for holes and $\bar{n} - n_0 = \Delta n$ for electrons. As EHP's are generated due to radiation then we can write

$$\Delta n = \Delta p$$

(x) Although the increase in hole concentration and electron concentration are equal, the percentage increase for electrons in N-type semiconductor is very small while percentage increase in holes may be tremendous.

(xi) This is due to the fact that electrons are majority carriers in N-type semiconductor and holes are minority carriers. Hence, majority carrier concentration is hardly affected due to radiation. We will be limited to the behaviour of the minority carriers. When a steady-state condition is reached, at t = 0 (as shown in figure 3.10), the radiation is removed. Now, in the graph, it has been shown that excess carrier density returns to zero exponentially with time.

3.9.2. Indirect Recombination: Trapping (U.P. Tech. Sem. Exam 2005-06, 2008-09)

(i) As a matter of fact, the probability of direct electron-hole recombination is quite small for semiconductors such as Ge, Si etc. These semiconductors emit very weak radiation during recombination process and may be detected by some sensitive equipments.

(ii) The maximum recombination process in indirect materials usually takes place through recombination levels within the band gap and the resulting energy loss by recombining electrons is generally given up to the lattice as heat rather than by the emission of photons.

(iii) Any impurity or lattice defect may serve as a recombination center, if it is capable of receiving a carrier of one type and hence capturing the opposite type of carrier, there by annihilating the pair.

(iv) Figure 3.11 shows a recombination level E_r which is below fermi level (E_F) at equilibrium and hence is substantially filled with electrons.

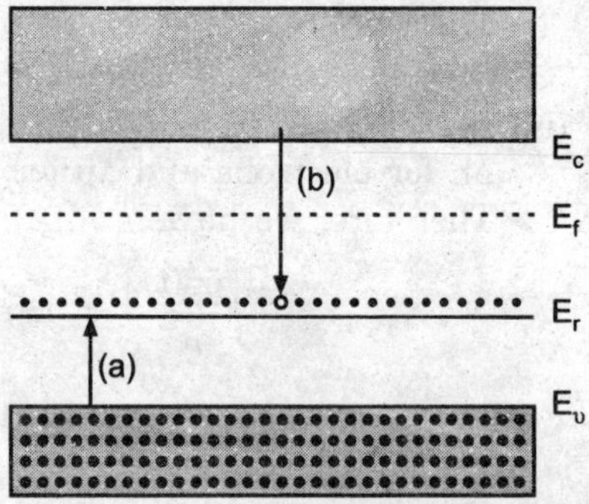

(a) Hole capture at a filled recombination center
(b) Electron capture at an empty center

Fig. 3.11. Capture processes at a recombination level

(v) When excess electrons and holes are created in this material, each electron-hole pair recombines at E_r in two steps
(a) hole capture and
(b) electron capture.

(vi) As the recombination center in figure 3.11 is filled at equilibrium, the first event in the recombination process is usually hole capture.

(vii) At this stage, it must be mentioned that this process is equivalent to an electron at E_r falling to the valence band, creating an empty states in the recombination level.

(viii) Therefore, in hole capture, energy is given up as heat to the lattice.

(ix) In a similar manner, energy is given up when a conduction band electron subsequently falls to the empty state in E_r. If both of these events have occurred, the recombination center is back to its original state (filled with an electron), but an electron-hole pair is missing. Therefore, one electron-hole pair recombination has taken place and now the center is ready to participate in another recombination event by capturing a hole.

(x) The life-time of excess carriers resulting from indirect recombination is more complicated process than the direct recombination, since carriers need unequal times to capture charges in indirect recombination.

(xi) Thus, recombination is often delayed by the tendency for a captured carrier to the thermally reexcited to its original band before capture of the opposite type of carrier can occur.

(xii) As an example, when electron capture does not follow immediately after the hole capture as shown in figure 3.11, then hole may be thermally, reexcited to the valence band. Energy is required for this process, which is equivalent to a valence band electron being raised to the empty state in the recombination level.

(xiii) This process delays the recombination, since the hole must be captured again before recombination can be completed.

(xiv) If a carrier is trapped temporarily at a center and then reexcited without recombination taking place, the process is known as temporary trapping. As per definition in general we refer to an impurity or defect center as a trapping center, if after capture of one type of carrier, the most probable next event is reexcitation.

(xv) When the next probable event is capture of the opposite type of carrier, the center is predominately a recombination center. The recombination process may be slow or fast, depending on the average time the first carrier is held before the second carrier is captured.

Usually, trapping levels situated deep in the band gap are slower in releasing trapped carriers than are the levels located near one of the bands.

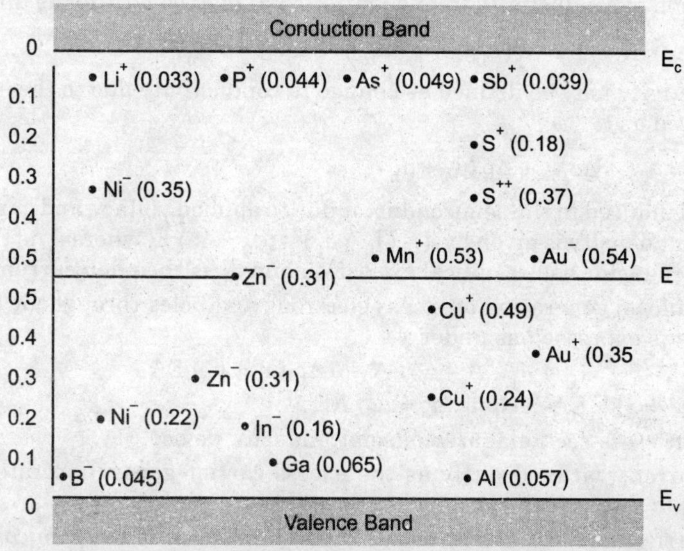

Fig. 3.12. Energy levels of impurities. The energies are measured from the nearest band edge (E_v or E_c); donor levels are designed by plus sign and acceptors by a minus sign

(xvi) This concludes that more energy is required to reexcite a trapped electron from a level near to the conduction band.

(xvii) As an example of impurity levels in semiconductors, figure 3.12 illustrates the energy level positions of various impurities in Si. In this figure, superscript represents whether the impurity is positive *i.e.*, donor type or negative *i.e.*, acceptor type when ionized.

3.10. INTRODUCTION TO PHOTOCONDUCTOR (U.P. Tech., Sem. Exam.; 2005-06, 2009-10)

(i) Let us consider a semiconductor bar with ohmic contacts at each end and a voltage applied between the terminals, as depicted in figure 3.13. The initial thermal-equilibrium conductivity is expressed by

$$\sigma_0 = e(n_0\mu_e + p_0\,\mu_h)$$

(ii) When the excess carriers are created in the semiconductor because of the incident light, the conductivity becomes

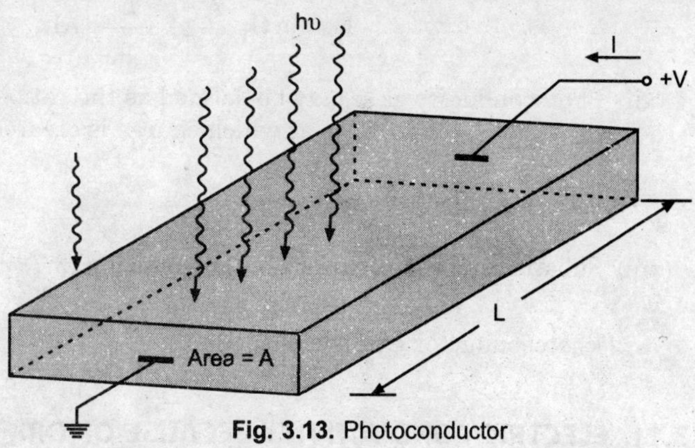

Fig. 3.13. Photoconductor

$$\sigma = e\left[(n_0 + \Delta n)\,\mu_e + (p_0 + \Delta p)\,\mu_h\right] \quad ...(3.4)$$

where Δn and Δp are the excess electron and hole concentrations, respectively. Assuming N-type semiconductor, it may be considered that $\Delta n = \Delta p$.

(iii) Considering Δp as the concentration of excess carriers, in steady state the carrier concentration will be given by

$$\Delta p = G_L\,\tau_h$$

Here, G_L is the generation rate of excess carriers and τ_h is the life time of excess minority carriers.

(iv) The expression of conductivity from equation (3.4) may be written as under:

$$\sigma = e\left(n_0\mu_e + p_0\,\mu_h\right) + e\,\Delta p\,(\mu_e + \mu_h)$$

(v) Photoconductivity may be defined as change in conductivity due to the optical excitation is then expressed as

$$\Delta\sigma = e\,\Delta p\,(\mu_e + \mu_h)$$

(vi) Electric field induced in the semiconductor due to applied voltage and current flows through it. The current density is given by $J = (J_0 + J_L) = (\sigma_0 + \Delta\sigma)\,E$, where J_0 is the current density in the semiconductor before optical excitation and J_L is the photocurrent density.

(vii) In case of uniform generation of excess electrons and holes throughout the semiconductor, photocurrent is expressed as under :

$$I_L = J_L \,.\, A = \Delta\sigma EA = e\,\Delta p\,(\mu_e + \mu_h)\,EA$$

$$I_L = eG_L\,\tau_h\,(\mu_e + \mu_h)\,AE \qquad ...(3.5)$$

Here, A is the cross-sectional area of semiconductor device.

(viii) The photocurrent varies directly as the excess carrier generation rate, which in turn is proportional to the incident photon flux.

(ix) If excess carriers are not generated uniformly throughout the semiconducutor material, then the total photocurrent may be determined by integrating the photoconductivity over the cross-sectional area.

(x) Electron drift velocity ($v_d = \mu_e E$) and the electron transit time t_n (i.e., time needed for an electron to flow through the photoconductor) is expressed as under :

$$t_n = \frac{L}{\mu_e E} \qquad ...(3.6)$$

Here, L is the length of the semiconductor device.

(xi) Then, photocurrent I_L may be written as under:

$$I_L = e\,G_L\,\frac{\tau_h}{t_n}\left(1 + \frac{\mu_h}{\mu_e}\right)AL \qquad ...(3.7)$$

(xii) Photoconductor gain may be defined as the ratio of the rate at which charge is collected by the contacts to the rate at which charge is created within the photoconductor, i.e.,

$$\text{Photoconductor gain} = \frac{I_L}{e\,G_L\,AL} \qquad ...(3.8)$$

(xiii) Substituting the value of I_L from equation (3.7) in equation (3.8), we obtain

$$\text{Photoconductor gain} = \frac{\tau_h}{t_n}\left(1 + \frac{\mu_h}{\mu_e}\right) \qquad ...(3.9)$$

3.11. ELECTRIC CONDUCTIVITY BECAUSE OF DRIFT

(i) In a semiconductor, movement of carriers under the applied electric field is known as **drift**.

Let, n = concentration of electrons

p = concentration of holes

$$J = \text{Current density} = \frac{\text{current}}{\text{cross–sectional area}} = \frac{I}{A}$$

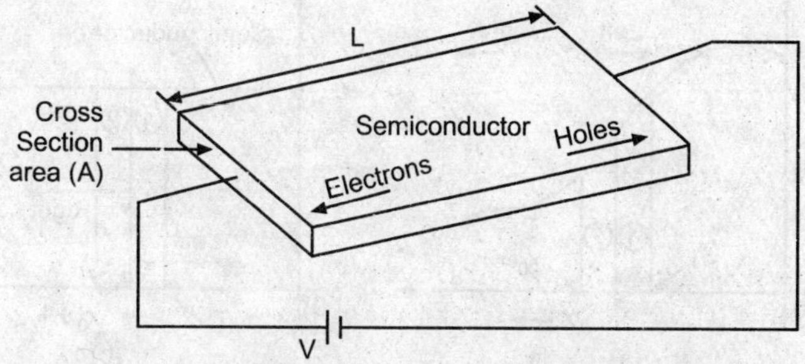

Fig. 3.14. Current due to drift in a semiconductor

$(J_p)dr$ = current density for hole current due to drift

$(J_n)dr$ = current density for electron current due to drift

μ_p = mobility of holes

μ_n = mobility of electrons

(ii) We can write, $J = (J_p)\ dr + (J_n)\ dr = \sigma E$ ...(A)

where σ = conductivity of a semiconductor

$$E = \text{electric field applied} = \frac{V}{L}$$

(iii) Each electron and hole has a charge $q = 1.6 \times 10^{-19}$ coulomb.

If $\bar{\upsilon}_n$ = average drift velocity of electrons

$\bar{\upsilon}_p$ = average drift velocity of holes

$nq\,\bar{\upsilon}_n$ = net flow of electrons across section = (J_n) dr

$pq\,\bar{\upsilon}_p$ = net flow of holes across section = (J_p) dr

$$(J_p)\ dr = qp\,\bar{\upsilon}_p = qp\,\mu_p\,E \quad ...(3.10)$$

$$(J_n)\ dr = q = qn\,\mu_n\,E \quad ...(3.11)$$

$$J = qp\mu_p E + qn\,\mu_n E = (qp\,\mu_p + qn\,\mu_n)\,E = \sigma E$$

(iv) Conductivity $\sigma = qp\,\mu_p + qn\,\mu_n = q\,(p\,\mu_p + n\,\mu_n)$

3.12. DIFFUSION MECHANISM: DETAILED ASPECTS (U.P. Tech, Sem. Exam; 2009-10)

(i) If charge carriers are non-uniformly created/present in a semiconductor then, the concentrations of electrons (n) and holes (p) vary with position inside it. Such a gradient (variation) in space gives rise to net motions of charge carriers from a region of higher concentration to a region of lower concentration of carriers. This type of motion is called as **Diffusion**. It is very important charge transport process in semiconductors.

(ii) We evaluate the rate at which the 'electrons' diffuse for a one dimensional problem with an arbitrary distribution n(x) in a semiconductor.

When $\bar{l}$ = mean free path between collisions.

We divide n(x), along X axis into segments of length $\bar{l}$. Then, n(x) is evaluated at the center of each segment.

(iii) Let us consider figures 3.15(a) and (b).

(iv) The electrons in each segment have equal chance of moving to left or right. In the mean free time ($\bar{t}$), half of the electrons move to the right, other half to the left.

A = cross-sectional area of the semiconductor bar.

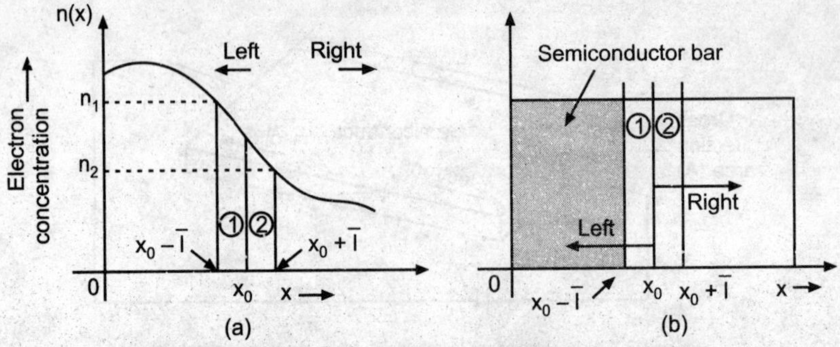

Fig. 3.15. Electron concentration in a semiconductor

At 'x_0' we have $\frac{1}{2} n_1 \bar{l} A$ electrons moving to right,

here, n_1 = concentration of electrons in segment (1)

At 'x_0' we have $\frac{1}{2} n_2 \bar{l} A$ electrons moving to left from segment (2) to segment (1),

here, n_2 = concentration of electrons in segment (2).

(v) Net flow of electrons at x_0 from left to right (+ ve 'x' direction) = $\frac{1}{2} n_1 \bar{l} A - \frac{1}{2} n_2 \bar{l} A$

These many electrons move in time equal to mean free time '$\bar{t}$'.

(vi) The rate of flow of electrons in + ve 'x' direction per unit area may be defined as electron flux density (ϕ_n).

Therefore,
$$\phi_n = \frac{\text{Net flow of electrons in + ve 'x' direction}}{\text{Mean free time}} = \frac{\bar{l}}{2\bar{t}}(n_1 - n_2) \quad ...(3.12)$$

(vii) The difference in electron concentration $(n_1 - n_2)$ is written as

$$n_1 - n_2 = \frac{n(x) - n(x + \Delta x)}{\Delta x} \bar{l} \quad ...(3.13)$$

Substituting above value of $(n_1 - n_2)$ in equation (3.12), we obtain

$$\phi_n(x) = \frac{\bar{l}^2}{2\bar{t}} \lim_{\Delta x \to 0} \frac{n(x) - n(x + \Delta x)}{\Delta x} = -\frac{\bar{l}^2}{2\bar{t}} \frac{dn(x)}{dx} \quad ...(3.14)$$

(viii) Let us, we define $\frac{\bar{l}^2}{2\bar{t}}$ as diffusion constant (D_n) with units cm^2/sec.

Therefore,
$$\phi_n(x) = -D_n \frac{d[n(x)]}{dx} \quad ...(3.15)$$

(ix) In a similar manner, for a semiconductor material with gradient in concentration of holes, we obtain movement of holes as under:

$$\phi_p(x) = -D_p \frac{d[p(x)]}{dx} \quad ...(3.16)$$

The minus sign implies that electrons in 'n' type material and holes in 'p' type material with concentration gradient move in a direction (*i.e.*, diffuse) of reducing (decreasing) concentration.

(x) Diffusion coefficients D_p and D_n are determined experimentally for a particular material.

Parameter	Si	Ge	GaAs
Diffusion constant D_n (cm^2 / sec)	36	100	290
Diffusion constant D_p (cm^2 / sec)	13	45	< 12

3.12.1. Current Density because of Diffusion

(i) The diffusion current crossing per unit area (i.e., the diffusion current density) is equal to the particle (electron/hole) flux density multiplied by charge of the carrier.

(ii) We write, for electrons as under:

$$(J_n)\text{ diff} = -(-q)\,D_n \frac{d[n(x)]}{dx} = q\,D_n \frac{d[n(x)]}{dx} \qquad ...(3.17)$$

(iii) For holes, we write as under:

$$(J_p)\text{ diff} = -(+q)\,D_p \frac{d[p(x)]}{dx} = -q\,D_p \frac{d[p(x)]}{dx} \qquad ...(3.18)$$

(iv) Substituting (J_n) dr, (J_p) dr from equations (3.13) and (3.12) and (J_p) diff and (J_n) diff from equations (3.18) and (3.17) in equation (A), we obtain

$$J_n(x) = q\,\mu_n\,n(x)\,E(x) + qD_n \frac{dn(x)}{dx} \qquad ...(3.19)$$

$$J_p(x) = q\,\mu_p\,p(x)\,E(x) - q\,D_p \frac{dp(x)}{dx} \qquad ...(3.20)$$

(v) The total current density is given by $J(x) = J_n(x) + J_p(x)$ ...(3.21)

> **Important Point:** The above equation refer for behaviour of a semiconductor material with carrier distribution and applied electric field. Minority carriers contribute quite significantly to the current through diffusion.

3.13. CARRIER CONCENTRATIONS

1. Definition

It is essential to know charge carrier concentrations i.e., number of charge carriers per cm^3 in the material. We can evaluate electrical properties of a semiconductor material from the concentrations. We can analyse the device behaviour once we know the carrier charge concentrations.

2. Expressions

(i) To obtain equations for charge carrier concentration, we must know the distribution of carriers over the available energy states in the energy band diagram.

(ii) Electrons in solids usually obey Fermi-Dirac statistical distribution. The distribution of electrons over a range of allowable energy levels at 'thermal equilibrium' is

$$f(E) = \frac{1}{1 + e^{(E - E_f)/kT}} \qquad ...(3.22)$$

Here, f(E) = Fermi-Dirac distribution function
= Probability that energy state at 'E' will be occupied by an electron at temperature T°K

E = Energy state

k = Boltzman's constant = 8.62×10^{-5} eV/°K = 1.38×10^{-5} eV/°K

E_F = Fermi level
= Energy level of a state with 50% probability that it is occupied.

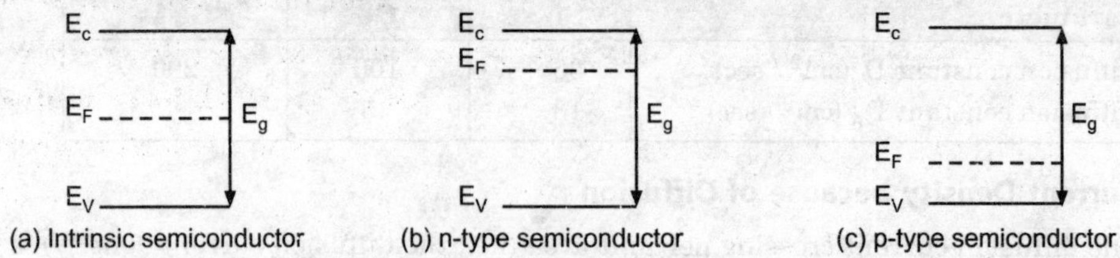

Fig. 3.16. Schematic band diagrams in different types of a semiconductor material

For an intrinsic material, E_F lies at the centre of the band gap. Fermi level for intrinsic material (E_i) is equal to average of E_c and E_v.

Therefore, we have $E_i = \dfrac{E_c + E_v}{2}$

Here, E_c = energy of conduction band bottom

E_v = energy of valence band top

(iii) The position of Fermi level in N type material is near to the conduction band in the band gap. The Fermi level depends on the doping density.

(iv) The position of Fermi level in P type material is near to the valence band in the band gap. The actual Fermi level depends upon the doping density.

(v) At equilibrium, the electron and hole copncentrations are expressed as under:,

$$n_o = n_i e^{(E_F - E_i)/kT}$$

$$p_o = n_i e^{(E_i - E_F)/kT}$$

Here, n_o = Concentration of electrons in the conduction band at thermal equilibrium

p_o = Concentration of holes in the valence band at thermal equilibrium

E_i = Intrinsic Fermi level = $\dfrac{E_c + E_v}{2} = E_v + \dfrac{1}{2}E_g$

E_F = Fermi level in doped semiconductor

n_i = intrinsic carrier concentration

= 1.5×10^{10} / cm^3 for silicon at 300° K

3.14. MASS ACTION LAW (U.P. Tech, Sem. Exam; 2010-11)

1. Basic Concept

By adding N type impurities (P, As, Sb) to an intrinsic semiconductor material, there is reduction/ decreases in number (concentration) of holes. In a similar manner, by adding P type impurities (B, Al, Ga, In) to an intrinsic material, there is reduction/decrease in the concentration of free electrons below the concentration of electrons in the intrinsic material.

2. Statement

The mass action law states that under thermal equilibrium, the product of (1) the free negative (electrons) and (2) free positive (holes) carrier concentrations is constant. This product is independent of amount or concentration of donor or acceptor impurity doping in n type or p type semiconductor.

3. Derivation of Expressions

(i) For an instrinsic material at thermal equilibrium,

Concentration of electrons = Concentration of holes

Therefore, $n = p$

(ii) We deonte this concentration of charge carriers in intrinsic semiconductor by n_i.

Therefore, $n = p = n_i$

or $np = n_i \times n_i = n_i^2$

This relationship is the **Mass Action Law**.

It is applicable to extrinsic semiconductor material as well.

(iii) For a N type semiconductor, with impurity concentration of N_D atoms/cm^3, we can say,

Concentration of electrons per cm^3 = Concentration of donor atoms per cm^3

Therefore, $n = N_D$

For silicon, we have $n_i = p_i = 1.5 \times 10^{10}$ (electrons/holes) per cm^3

Silicon has 5×10^{22} atoms/cm^3

(iv) If we dope the intrinsic silicon with 10^{17} atoms/cm^3 of N type impurity, then, we have

$$n = N_D = 10^{17} \text{ electrons/cm}^3$$

Substituting this in the expression, $n \cdot p = n_i^2$, we get, $10^{17} \times p = (1.5 \times 10^{10})\,(1.5 \times 10^{10})$

Therefore, $p = 2.25 \times 10^3$ holes/cm^3 = concentration of holes

(v) In a similar manner, for a P-type material with impurity concentration of N_A atoms/cm^3, we can say

Concentration of holes (/cm^3) = Concentration of acceptor atoms (/cm^3)

Therefore, $p = N_A$

For silicon, we have $n_i = p_i$ = intrinsic concentration of electrons and holes

= 1.5×10^{10} (electrons/holes) per cm^3

Silicon has 5×10^{22} atoms/cm^3

(vi) If we dope the intrinsic silicon with 5×10^{16} atoms/cm^3 of p type impurity then, we have

$$p = N_A = 5 \times 10^{16} \text{ holes/cm}^3$$

Substituting $n_i = 1.5 \times 10^{10}$ / cm^3 and $p = 5 \times 10^{16}$ / cm^3 in the expression

$$n \cdot p = n_i^2 \text{, we get, } n \times 5 \times 10^{16} = (1.5 \times 10^{10})\,(1.5 \times 10^{10})$$

Therefore, $$n = \frac{2.25\times10^{20}}{5\times10^{16}} = 0.45 \times 10^4 = 4.5 \times 10^3 \text{ electrons/cm}^3$$

4. Alternate Illustration of Mass Action Law

(i) Alternatively, the mass action law can be understood from the electrical neutrality of a semiconductor material. For the semiconductor to be neutral, concentration of negative charge carriers = concentration of positive charge carriers.

(ii) Concentration of negative charge carriers.

= Concentration of electrons + Concentration of ionized acceptor atoms = $n + N_A$

Concentration of positive charge carriers

= Concentration of holes + Concentration of ionized donor atoms

= $p + N_D$

Therefore, $n + N_A = p + N_D$

For a N type material, $N_A = 0$

Hence, $n_n = p_n + N_D$

where n_n = concentration of electrons in N type

p_n = concentration of holes in N type

N_D = concentration of donor atoms (/cm^3)

Since $p_n << N_D$ we get, $n_n = N_D$

(iii) Applying mass action law, concentration of holes in N type, we get

$$p_n = \frac{n_i^2}{N_D}$$

(iv) In a similar manner, for a P type material, $N_D = 0$

Therefore, $n_p + N_A = p_p$

and $n_p << N_A$

hence, $p_p = N_A$ = concentration of holes in P type

n_p = concentration of electrons in P type = $\frac{n_i^2}{N_A}$

3.15. EINSTEIN RELATIONSHIP

(U.P. Tech., Sem. Exam.; 2005-06)

(i) At equilibrium there is no net current that flows in a semiconductor. Any fluctuation will initiate/begin diffusion current. The diffusion current sets up an electric field. This electric field will redistribute the carriers by drift. These requirements for equilibrium imply that the diffusion coefficients and mobility must be related.

(ii) For diffusion current densities, we have derived following expressions:

$$J_n(x) = q\,\mu_n\, n(x)\, E(x) + q\, D_n \frac{dn\,(x)}{dx}$$

$$J_p(x) = q\,\mu_p\, p(x)\, E(x) - q\, D_p \frac{dp\,(x)}{dx}$$

(iii) Also, we know that carrier concentrations in a semiconductor at thermal equilibrium are as under :

$$n_o = n_i e^{\frac{(E_F - E_i)}{kT}}$$

$$p_o = n_i e^{\frac{(E_i - E_F)}{kT}} \qquad ...(3.23)$$

(iv) For a N-type semiconductor, holes are minority carriers. Minority carriers do not provides drift current. Minority carrier currents through diffusion process, may be as large as, drift current due to majority carriers.

(v) At equilibrium in a N type material, we have

$$J_p\,(x) = 0$$

Therefore, $$0 = q\,\mu_p p(x)\, E\,(x) - q\, D_p \frac{dp(x)}{dx}$$

or $$E(x) = \frac{D_p}{\mu_p} \cdot \frac{1}{p(x)} \cdot \frac{dp(x)}{dx}$$

We know that, $$p(x) = n_i e^{\frac{(E_i - E_F)}{kT}}$$

Therefore, $$E(x) = \frac{D_p}{\mu_p} \cdot \frac{1}{p(x)} \cdot \frac{d}{dx}\left[n_i e^{(E_i - E_F)/kT}\right] = \frac{D_p}{\mu_p} \cdot \frac{1}{kT}\left[\frac{dE_i}{dx} - \frac{dE_F}{dx}\right].$$

Fermi level does not change with x.

Hence, we have $\frac{dE_F}{dx} = 0$

$$E(x) = \frac{1}{q} \cdot \frac{dE_i}{dx}$$

We obtain, $$\frac{D_p}{\mu_p} = \frac{kT}{q}$$

(vi) In a similar manner, for a P type semiconductor, by using other two equations, we obtain,

$$\frac{D_n}{\mu_n} = \frac{KT}{q}, \text{ This indicater that } \frac{D_p}{\mu_p} = \frac{D_n}{\mu_n} = V_T$$

Hence, V_T is the volt-equivalent of temperature.

This result obtained for both electrons and holes is called **Einstein Relation.**

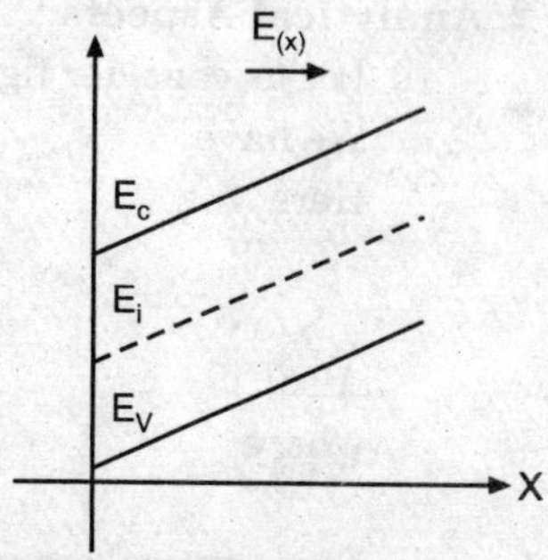

Fig. 3.17. Band diagram of semiconductor in an electric field

3.16. SEMICONDUCTOR MATERIALS USED IN OPTOELECTRONIC DEVICES

1. Relationship between photon energy and Wavelength

The relation between photon energy and wavelength may be expressed as under:

$$E = \frac{hc}{\lambda} = \frac{1.24}{\lambda}$$

here, E = Photon energy (eV)

h = Planck's constant = 4.14×10^{-15} eV-sec

c = Speed of light (2.998×10^{10} cm / sec)

λ = Wavelength of light (μm)

A semiconductor absorbs photons with energies greater than or equal to the energy band gap. We know energy band gap E_g is the property of a crystal lattice of a particular, semiconductor material. Figure 3.6 illustrates the band gap energies relative to the optical spectrum.

2. Photo-conductivity

It is the increase in conductivity due to excess carriers arising from optical excitation, this is an important effect. It is used in

(i) Analysis of semiconductor devices,

(ii) Operation of several devices like photo-diode, photo-transistor, etc.

3.16.1. Photoluminescence

The property of light emission is called luminescence. The radiation resulting from recombination of excited carriers is called photoluminescence. If the excitation occurs due to injection of current into the sample, it is called electro-luminescence. In LEDs, an electric current causes injection of minority carriers. This results in radiation due to recombination.

3.16.2. Photo-conductors and Photo-diodes

1. Defintion

We know that the bulk semiconductors behave likes photo-conductors by providing change in conductivity proportional to optical generation rate of carriers. Junction devices are used to improve speed of response and sensitivity of detector to optical radiation. A two terminal device which responds to photon absorption is known as a '**photo-diode**'. Modern electronic systems often involve both optical and electrical signals. In such systems, photo-diode serves important function of optical signal to electrical signal conversion.

2. Analytical Aspects

(i) Let us consider figure 3.4.

We have $I(x) = I_o\, e^{-\alpha x}$

Here, $I(x)$ = current at depth x in semiconductor (eV)

I_o = incident intensity of photons (eV)

α = absorption coefficient (/cm)

Also, $I_t = I_o\, e^{-\alpha l}$

where I_t = light intensity through the sample

l = thickness of the sample

(ii) Since optical signal changes intensity, conductivity of sample changes.

(iii) As conductivity changes, current i(t) through the sample changes. Also, current is converted into voltage v(t) by means of resistor R, this v(t) we can see on an oscilloscope.

(iv) The current in the diode is proportional to optical generation rate of carriers *i.e.*, optical intensity.

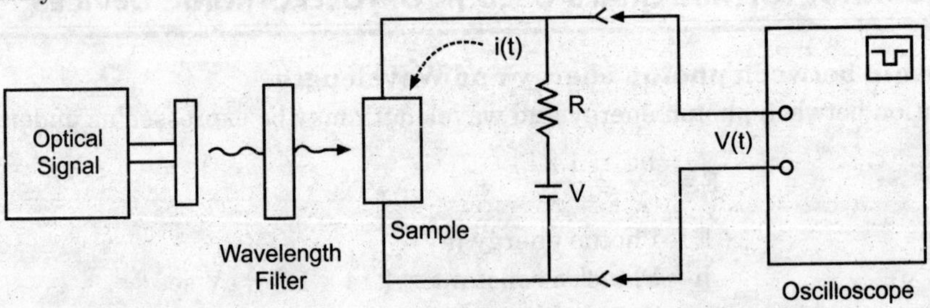

Fig. 3.18. Electrical signal from a photo-conductor

(v) The region 'W' in a p-n junction is called depletion region because it is depleted of carriers compared with rest of the crystal. The appropriate width 'W' of the depletion is chosen. For higher sensitivity, 'W' should be large. For faster response, 'W' must be small. Therefore, a compromise has to be made between sensitivity and response. The depletion layer width 'W' can be made large by doping one side of the junction lightly.

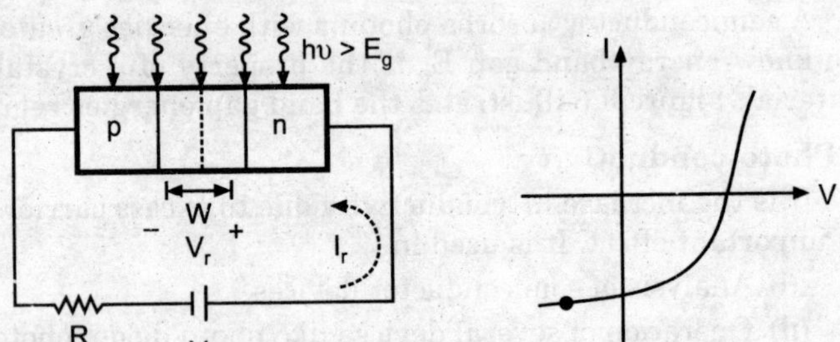

(a) Absorption of light by device (b) Operation of illuminated diode in third quadrant

Fig. 3.19. Photo-diode working

(vi) A convenient method of controlling width 'W' of depletion layer is to build a p-i-n structure. Here i means intrinsic semiconductor. This is called as **PIN photodetector**.

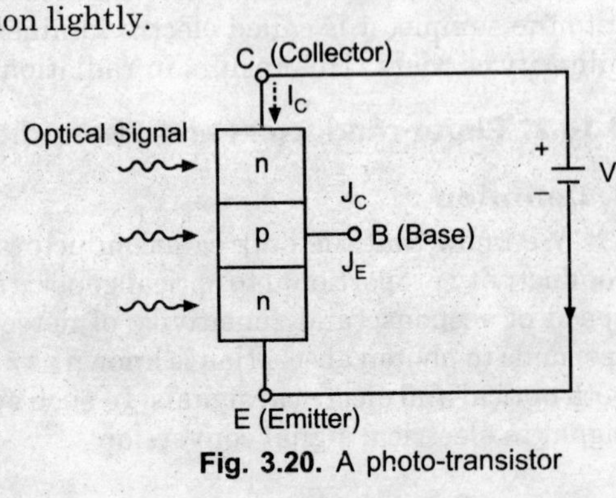

Fig. 3.20. A photo-transistor

3.16.3. Phototransistor

1. Definition

A photo-transistor is a much more photo sensitive semiconductor device than photodiode. A photo-transistor is usually connected in common emitter configuration with base open.

2. Basic Aspects

(i) When base is open in a transistor, the emitter junction (J_E) is slightly forward biased and collector junction is reverse biased. This means the transistor is in active region.

(ii) When the base is kept open, then,

Collector current $I_C = I_{CO}$ = Reverse saturation current of collector junction

(iii) When there is no optical signal, then,

Collector current, $I_C = (1 + \beta) I_{CO}$

(iv) Because of optical signal, additional carriers are photogenerated. These minority carriers contribute to the reverse saturation current of the collector junction.

Therefore, $I_C = (1 + \beta) (I_{CO} + I_L)$

where I_L = current due to light

Current I_L due to light is multiplied by $(1 + \beta)$ due to transistor action.

(v) Consideration for choosing a photo-conductor in a given application may be listed as under:

(a) Sensitive wavelength range,

(b) Time response,

(c) Optical sensitivity of the material

3.17. PHOTO-LUMINESCENT DEVICES

3.17.1. Light Emitting Diodes

(i) When carriers are injected across a forward biased junction, current flows. This current is due to recombination of charge carriers in the depletion region.

(ii) In a material characterised by direct recombination, light is given out when junction is forward biased. This phenomenon is called electro-luminescence. This provides important application of a diode as generator of light. We call them as Light Emitting Diode: (LEDs).

(iii) The light emitting diodes are used in digital displays. Other important application is Fiber Optic Communication System.

(iv) The light emission from p-n junctions (LEDs) is not coherent. It is generated by spontaneous recombination of holes and electrons injected across the junction.

3.17.2. Laser

(i) Laser is an acronym for Light Amplification by Stimulated Emission of Radiation. A LASER source is monochromatic (one wavelength), directional, coherent light. The junction LASERs are very small (0.1 × 0.1 × 0.3 mm).

(ii) They have high efficiency. The LASER output can be easily modulated by controlling the junction current.

(iii) Semiconductor LASER operate with very low power. Because of these characteristics, the semiconductor LASER provide a portable and easily controllable source of low-power coherent light radiation. These LASERs are suitable for fiber Optic Communication Systems.

(iv) A p-n junction LASER is built using a highly doped direct band gap semiconductor (n type) such as GaAs by diffusing zinc (Zn) to form p region.

(v) The cavity is by cutting these into a particular orientation. The device is mounted on a suitable header and adequate heat sink.

3.18. SEMICONDUCTOR MATERIALS USED IN OPTOELECTRONIC DEVICES

Photo-conductors

Material	Band gap (eV)	Application area
Cadmium sulphide (CdS)	2.42	Visible optical range
Germanium (Ge)	0.67	Infrared portion of optical spectrum
Indium antimonide (InSb)	0.18	Infrared portion of optical spectrum

Photo-diodes

Material	Band gap (eV)	Application area
Indium gallium arsenide (InGaAs)	0.75	Fiber optic systems
Indium phosphide (InP)	1.35	Transmission of absorbed light
Silicon (Si)	1.1	Highly sensitive low optical level avalanche photo-diode

Light emitting diodes

Material	Band gap (eV)	Application area
Zinc sulphide (ZnS)	3.6	Ultraviolet optical spectrum
Gallium arsenide (GaAs)	1.43	Infrared emitters in optical communication systems, LASERs (Infrared)
Gallium arsenic phosphide with nitrogen doping	—	Yellow and green LEDs, LASERs (visible)

3.19. PROPERTIES OF GERMANIUM AND SILICON

Table 3.1. Properties of germanium and silicon

S.No.	Name of Property	Ge	Si
1.	Atomic number	32	14
2.	Atomic weight	72.6	28.1
3.	Density, g/cm^3	5.32	2.33
4.	Dielectric constant (relative)	16	12
5.	Atoms/cm^3	4.4×10^{22}	5.0×10^{22}
6.	E_{GO}, eV, at 0°K	0.785	1.21
7.	E_G, eV, at 300°K	0.72	1.1
8.	n_i at 300 °K, cm^{-3}	2.5×10^{13}	1.5×10^{10}
9.	Intrinsic resistivity at 300°KM, Ω – cm	45	230,000
10.	μ_n, cm^2/V–s at 300 °K	3,800	1,300

S.No.	Name of Property	Ge	Si
11.	μ_p, cm^2/V–s at 300 °K	1,800	500
12.	D_n, cm^2/s $= \mu_n V_T$	99	34
13.	D_p, cm^2/s $= \mu_p V_T$	47	13

3.20. THE CONTINUITY EQUATION

1. Diffusion Equations for Electrons and Holes

(i) We must consider the effects of recombination of excess carriers to describe conduction proces in semiconductor, because recombination can produce variation in the carrier distribution.

(ii) As an example, let us consider a differential length Dx of a semiconductor sample with area A in the yz plane as shown in figure 3.21.

(iii) The hole current density leaving the volume, $J_p(x + Dx)$, can be larger or smaller than the current density entering, $J_p(x)$, depending on the generation and recombination of carriers taking place within the volume. The net increase in hole concentration per unit time, $\frac{\partial p}{\partial t}$, is the difference between the hole flux per unit volume entering and leaving minus the recombination rate.

(iv) Hole particle flux density can be obtained dividing hole current density by charge (q).

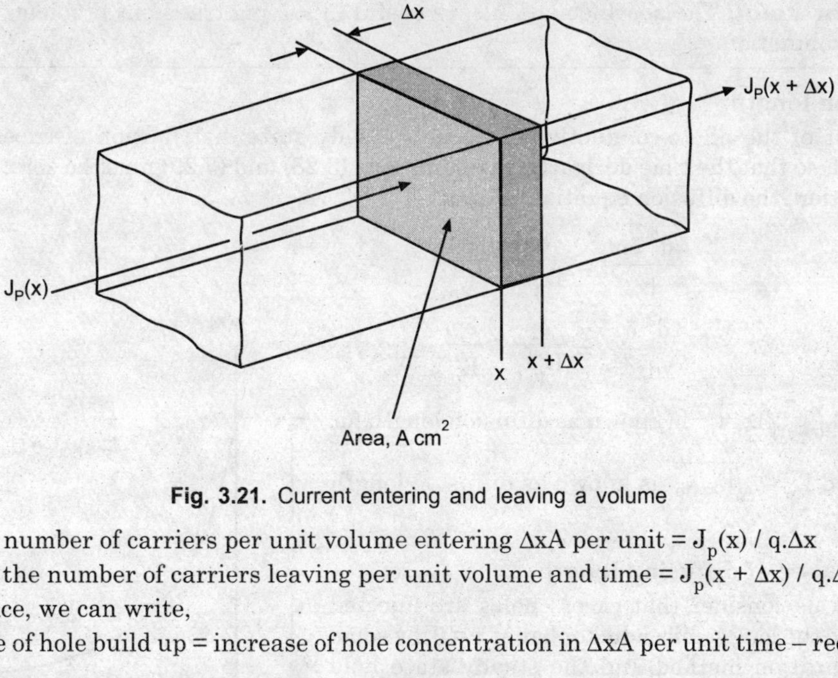

Fig. 3.21. Current entering and leaving a volume

(v) The number of carriers per unit volume entering ΔxA per unit $= J_p(x) / q.\Delta x$ and the number of carriers leaving per unit volume and time $= J_p(x + \Delta x) / q.\Delta x$.

(vi) Hence, we can write,

Rate of hole build up = increase of hole concentration in ΔxA per unit time – recombination rate

$$\left.\frac{\partial p}{\partial t}\right|_{x \to x+\Delta x} = \frac{1}{q}\frac{J_p(x) - J_p(x + \Delta x)}{\Delta x} - \frac{\delta p}{\tau_p} \quad ...(3.24)$$

As $\Delta x \to 0$, then the current change can write in derivative form,

$$\frac{\partial p(x,t)}{\partial t} = \frac{\partial\, \delta p}{\partial t} = -\frac{1}{q}\frac{\partial J_p}{\partial x} - \frac{\partial p}{\tau_p} \qquad \text{...(3.25)}$$

The above equation is known as continuity equation for holes.

(vii) In a similar manner, we can write the continuity equation for electronss

$$\frac{\partial\, \delta n}{\partial t} = \frac{1}{q}\frac{\partial J_n}{\partial x} - \frac{\delta n}{\tau_n} \qquad \text{...(3.26)}$$

When the current is carried mainly by diffusion, then we can replace the currents in equations (3.25) and (3.26) by the expressions for diffusion current. As an example, for electron diffusion, we can write,

$$J_n\,(\text{diff}) = q.D_n \frac{\partial\, \delta n}{\partial x} \qquad \text{...(3.27)}$$

(viii) By substituting equation (3.27) into (3.26), we obtain the diffusion equation for electrons as under:

$$\frac{\partial\, \delta n}{\partial t} = D_n \frac{\partial^2 \delta n}{\partial x^2} - \frac{\delta n}{\tau_n} \qquad \text{...(3.28)}$$

(ix) In a similar manner, we write diffusion equation for holes as under:

$$\frac{\partial\, \delta p}{\partial t} = D_p \frac{\partial^2 \delta p}{\partial x^2} - \frac{\delta p}{\tau_p} \qquad \text{...(3.29)}$$

Important Point: The above equations are useful in solving transient problems of diffusion with recombination.

2. Diffusion length

In most of the photo-conductive devices, a steady state distribution of excess carriers is maintained, so that the time derivatives in equations (3.28) and (3.29) must be zero. In the steady state condition, the diffusion equation becomes,

$$\frac{d^2\, \delta n}{dx^2} = \frac{\delta n}{D_n \tau_n} \cong \frac{\delta n}{L_n^2} \qquad \text{...(3.30)}$$

$$\frac{d^2\, \delta_n}{dx^2} = \frac{\delta p}{D_p \tau_p} \cong \frac{\delta p}{L_p^2} \qquad \text{...(3.31)}$$

where $L_n = \sqrt{D_n\, \tau_n}$ is known as diffusion length for electron and $L_p = \sqrt{D_p\, \tau_p}$ is known as diffusion length for holes

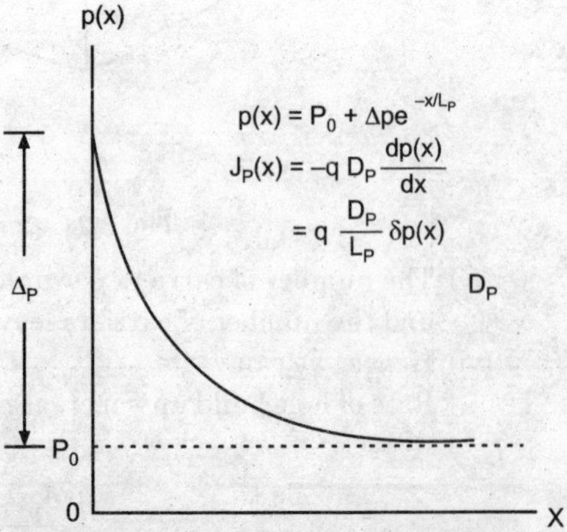

Fig. 3.22. Injection of hole at x = 0, giving to steady state hole distribution p(x) and resulting diffusion current density $J_p(x)$

3. Significance of diffusion length

(i) Let us consider that excess holes are injected into the long semiconductor bar at x = 0, by some excitation method and the steady state hole injection maintains a constant excess hole concentration at the injection point $\delta p\,(x = 0) = \Delta p$.

(ii) The injected holes diffuse along the bar, recombining with a characteristics life time τ_p.

In steady state, the distribution of excess holes to decay to zero for large values of x, due to the recombination (as shown in figure 3.22).

(iii) To solve this problem, the steady state diffusion equation (3.31) for holes can be used. The solution for this equation can be written as

$$\delta p\,(x) = A_1 e^{+x/L_p} + A_2 e^{-x/L_p}$$

The constants A_1 and A_2 can be found out from the boundary conditions. As recombination must reduce $\delta p(x)$ to zero for large values of x, hence $\delta p = 0$, for $x = \infty$. Hence, $A_1 = 0$.

(iv) In a similar manner, $\delta p = \Delta p$, at x = 0, so

$$c_2 = \Delta p$$

Therefore,, the final solution can be written as

$$\delta p(x) = \Delta p\, e^{-x/L_p}$$

(v) The injected excess hole concentration decreases exponentially in x due to recombination, and the diffusion length of holes L_p represents the distance at which the excess hole concentration is reduced to 1/e of its value at the point of injection.

Important Point: It can be proved that L_p is the average distance a hole diffuses before recombining.

3.21. VARIOUS PHYSICAL CONSTANTS

S.No.	Name of Physical Constant	Value of the Constant
1.	Avodadro's number	$N_A = 6.02 \times 10^{23}$ molecules/mole
2.	Boltzmann's constant	$K = 1.38 \times 10^{-23}$ J/K $= 8.62 \times 10^{-5}$ eV/K
3.	Electronic charge (magnitude)	$q = 1.60 \times 10^{-19}$ C
4.	Electronic rest mass	$m_0 = 9.11 \times 10^{-31}$ kg
5.	Permittivity of free space	$\varepsilon_0 = 8.85 \times 10^{-14}$ F/cm $= 8.85 \times 10^{-12}$ F/m
6.	Plank's constant	$h = 6.63 \times 10^{-34}$ J-s $= 4.14 \times 10^{-15}$ eV–s
7.	Room temperature value of KT	KT = 0.0259 eV
8.	Speed of light	$c = 2.998 \times 10^{10}$ cm/s
		Prefixes:
9.	1 A° (angstrom) = 10^{-8} cm	milli, m – = 10^{-3}
10	1 μm (micron) = 10^{-4} cm	micro-, μ = 10^{-6}
11.	1 nm = 10 A° = 10^{-7} cm	nano-, n = 10^{-9}
12.	2.54 cm = 1 inch	pico-, p- = 10^{-12}
13.	1 eV = 1.6×10^{-19} J	kilo-, k- = 10^3
		mega-, = M- = 10^6
		giga-, G-, 10^9

A wavelength λ of 1 μm corresponds to a photon energy of 1.24 eV.

SOLVED NUMERICAL PROBLEMS

EXAMPLE 3.1. **It is given that a silicon sample is doped with 10^{17} atoms/cm³ of arsenic (As). The intrinsic carrier concentration (n_i) for silicon at 300 K is 1.5×10^{10}/cm³. Determine the hole concentration at equilibrium (p_o).**

Solution: Given that, Donor concentration $N_d = 10^{17}$ atoms/cm^3
and intrinsic carrier concentration $n_i = 1.5 \times 10^{10}$/cm^3
Since, $N_d >> n_i$
Therefore, n_o = equilibrium electron concentration = $N_d = 10^{17}$/cm^3

We know that, $p_o n_o = n_i^2$

Therefore, p_o = equilibrium hole concentration

or
$$p_0 = \frac{n_i^2}{n_o} = \frac{n_i^2}{N_d} = \frac{(1.5\times10^{10})^2}{10^{17}} = 2.25 \times 10^3 \text{ holes/cm}^3$$

EXAMPLE 3.2. A semiconductor bar has electric field E, applied in +x direction and concentration gradient of electrons n(x) and concentration gradient of holes p(x) as depicted in figure 3.23. Obtain direction of electron flux (ϕ_n), hole flux (ϕ_p), drift and diffusion current densities for electrons and holes.

Solution: Diffusion:

(ϕ_n) diff

(ϕ_p) diff

The direction of flux of electrons and holes is in the direction of reducing concentration.

Drift:

⟶ (ϕ_n) dr

⟵ (ϕ_p) dr

Due to electric field applied in +ve x direction, holes move to right and electrons move to left.

Current density due to diffusion of electrons and holes:

⟶ (J_n) diff

⟵ (J_p) diff

Current density due to drift of electrons and holes:

⟶ (J_p) dr

⟵ (J_n) dr

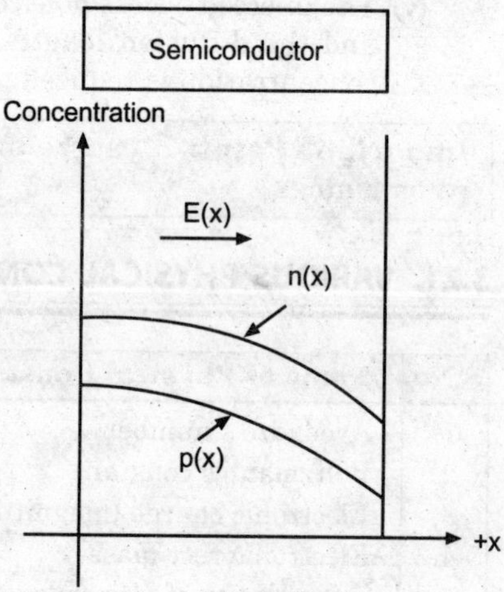

Fig. 3.23.

EXAMPLE 3.3. It is given that a silicon sample is doped with 10^{17} atoms/cm^3 of arsenic (As). Where would be the Fermi level E_f with respect to E_i?

Solution: We know that, $n_i = 1.5 \times 10^{10}$ / cm^3 for silicon

Since, N_d = doping density of donor arsenic
$= 10^{17}$ atoms / cm^3

Therefore, $N_d >> n_i = n_o = N_d = 10^{17}$ electrons / cm^3

where n_o = concentration of electrons

We know that, $n_o = n_i e^{(E_f - E_i)/KT}$

Taking natural log of both sides, we obtain

$$\ln\left(\frac{n_o}{n_i}\right) = \frac{(E_f - E_i)}{KT}$$

Hence,
$$E_f - E_i = KT/n\left(\frac{n_o}{n_i}\right) = 0.0259 \ln\left(\frac{10^{17}}{1.5\times10^{10}}\right) = 0.407 \text{ eV}$$

or
$$E_f = E_i + 0.407 \text{ eV}$$

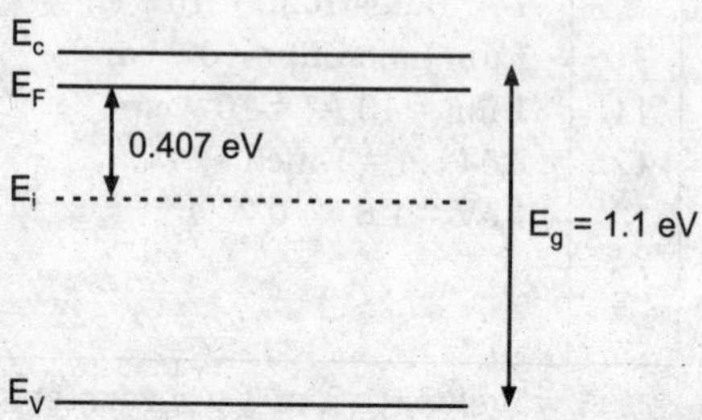

Fig. 3.24.

EXAMPLE 3.4. Determine the diffusion coefficients of electrons and holes of a silicon single crystal at 27°C, if the mobilities of electrons and holes are 0.17 and 0.025 m²/v-s, repectively at 27°C.

Solution: We know that $D = \left(\frac{KT}{e}\right)\mu$

Given that, $K = 1.38 \times 10^{-23}$ J/K, $T = 273 + 27 = 300°$ K

For electron, we have $D_n = \frac{1.38\times10^{-23}\times300}{1.6\times10^{-19}}\times0.17 = 44 \times 10^{-4}$ m²/s

For holes, we have $D_p = \frac{1.38\times10^{-23}\times300}{1.6\times10^{-19}}\times(0.025) = 6.47 \times 10^{-4}$ m²/s.

EXAMPLE 3.5. The minority carrier life time in P-type material is 10^{-7} s. The mobility of electron in silicon is 0.15 m² / v-s at 300°K. Determine

(i) What will be the diffusion length?

(ii) If 10^{20} electron/m³ are injected at x = 0, what will be the diffusion current density at x = 0?

Solution: We know that the diffusion coefficient is given by

$$D_n = \mu_n\left[\frac{KT}{e}\right] = \frac{0.15\times(1.38\times10^{-23})\times300}{1.6\times10^{-19}} = 3.88 \times 10^{3} \text{ m}^2/\text{s}$$

(i) We know that the diffusion length is given by

$$L_n = \sqrt{D_n\tau_n} = \sqrt{3.88\times10^{-3}\times10^{-7}} = 1.97 \times 10^{-5} \text{ m}$$

(ii) We know that the diffusion current density at x = 0 is given by,

$$J_n = \frac{eD_n\,(\Delta n)_0}{L_n} = \frac{1.6\times10^{-19}\times3.88\times10^{-3}\times10^{20}}{1.97\times10^{-5}} = 3.15 \times 10^{3} \text{ A/m}^2$$

EXAMPLE 3.6. Obtain the concentration of holes and electrons in n type silicon if the conductivity is 0.1 (Ω-cm)$^{-1}$

Solution: We know that the conductivity of n type material is given by,

$$\sigma_n = n_n\,\mu_n\,q$$

Thus, the concentration of electrons = $n_n = \frac{\sigma_n}{\mu_n\,q}$

$$n_n = \frac{0.1}{1300\times1.6\times10^{-19}} = 4.81 \times 10^{14} \text{ electrons/cm}^3 = 4.81 \times 10^{20} \text{ electrons/m}^3$$

Now, using the law of mass action, we have,

$$p_n \times n_n = n_i^2$$

Therefore, $p_n = \frac{n_i^2}{n_n} = \frac{(1.5\times10^{10})^2}{4.81\times10^{14}} = 4.68 \times 10^{5}$ holes/cm³ = 4.68×10^{11} holes/m³

EXAMPLE 3.7. Given an N-type silicon semiconductor with a length of 100 μm, cross sectional area A = 10^{-7} cm² and minority carrier life time of 10^{-6} s. Find out:

(i) Electron transit time and

(ii) Photoconductor gain when voltage applied to the photoconductor is 12 V.

(Given: m_e = 0.13 m² /v-s, m_h = 0.05 m² / v-s)

Solution: We know that Electron transit time $t_n = \frac{L}{\mu_e E} = \frac{L}{\mu_e \frac{V}{L}}$

Therefore, $t_n = \frac{L^2}{\mu_e V} = \frac{(100\times10^{-6})^2}{0.13\times2} = 6.4 \times 10^{-9}$ s. **Ans.**

Now, Photoconductor gain $= \frac{\tau_h}{t_n}\left(1+\frac{\mu_h}{\mu_e}\right) = \frac{10^{-6}}{6.4\times10^{-9}}\left(1+\frac{0.05}{0.13}\right) = 216.35$ **Ans.**

EXAMPLE 3.8. (i) Show that the resistivity of intrinsic germanium at 300K is 45 Ω-cm.

(ii) If a donor type impurity is added to the extent of 1 atom per 10^8 germanium atoms, prove that the resistivity drops to 3.7 Ω-cm.

Solution: (i) We know that the conductivity of a semiconductor is given by,

$$\sigma = (n\,\mu_n + p\,\mu_p)\,q$$

Since intrinsic germanium is given, n = p

Therefore, $\sigma = n_i\,(\mu_n + \mu_p)\,q = 2.5 \times 10^{13}\,(3800 + 1800) \times 1.6 \times 10^{-19} = 0.0224\,(\Omega\text{-cm})^{-1}$

Now, resistivity will be $\rho = \frac{1}{\sigma} = \frac{1}{0.0224\,(\Omega-\text{cm})^{-1}}$

$$\rho = 44.64 \approx 45\ (\Omega\text{-cm})$$

(ii) Since donor type impurity is added to the extent of 1 atom per 10^8 germanium atoms, then, we have

$$N_D = \frac{4.4\times10^{22}}{1\times10^{8}} = 4.4 \times 10^{14}\ \text{atoms/cm}^3$$

Because of donor impurity, N type of semiconductor will be formed.

Now, conductivity of n type semiconductor is expressed as under,

$$\sigma_n = N_D\,\mu_n\,q = 4.4 \times 10^{14} \times 3800 \times 1.6 \times 10^{-19}$$

$$= 0.26752\ (\Omega\text{-cm})^{-1} = \rho = \frac{1}{\sigma} = 3.73\ \Omega\ \text{cm}$$

EXAMPLE 3.9. For silicon the intrinsic concentration is approximately 10^{16} carriers/m³. If an impurity concentration of 10^{22} donor atoms/m³ is doped, find out

(a) electron concentration,

(b) hole concentration

Solution: Intrinsic carrier concentration, $n_i = 10^{16}/\text{m}^3$

Donor concentration, $N_D = 10^{22}/\text{m}^3$

(a) We have, concentration of electrons (n) = $N_D = 10^{22}$ / m³

(b) We have, concentration of holes (p) = $\frac{n_i^2}{n} = \frac{(10^{16})^2}{10^{22}} = 1 \times 10^{11}/\text{m}^3 = 10^{10}$ / m³

EXAMPLE 3.10. It is given that A donor impurity is added to intrinsic silicon and the resistivity at room temperature is observed to be 9.6 Ω-cm. Determine the ratio of donor atoms to silicon atoms per unit volume. Assume μ_n = 1300 cm² / V-s

Solution: We know that the conductivity of n type of silicon is expressed as,

$$\sigma_n = N_D\,\mu_n\,q$$

$$\sigma_n = \frac{1}{\rho} = \frac{1}{9.6\times10^{-2}} = 10.41\ (\Omega\text{m})^{-1}$$

Therefore, $N_D = \frac{\sigma_n}{\mu_n q} = \frac{10.41}{1300\times10^{-4}\times1.6\times10^{-19}}$

$$N_D = 5 \times 10^{20}\ \text{atoms/m}^3$$

From the properties of silicon, we have,

Atom density = 5×10^{22} atoms / cm³ = 5×10^{28} atoms / m³

Hence, The ratio of donor atom to silicon atom will be

$$\frac{N_D}{\text{Atom density of Si}} = \frac{5\times10^{20}}{5\times10^{28}} = 1 \times 10^{-8}$$

EXAMPLE 3.11. The intrinsic carrier density at 300 K is 1.5 × 10^{10}/cm^3 in silicon. For N type silicon doped to 2.25 × 10^{15} atoms/cm^3, evaluate the equilibrium electron and hole densities.

Solution: Given that $n_i = 1.5 \times 10^{10}$ / cm^3, $n_n = 2.25 \times 10^{15}$ atoms / cm^3

Using the law of mass action, we have

$$p_n\, n_n = n_i^2$$

where n_i = intrinsic concentration

n_n = donor impurity

p_n = concentration of holes in n type material

Therefore, $$p_n = \frac{n_i^2}{n_n} = \frac{(1.5\times10^{10}\ /\ \text{cm}^3)}{2.25\times10^{15}}$$

$$p_n = 1.0 \times 10^5/\text{cm}^3 = 1 \times 10^5/\text{cm}^3$$

$$n_n = 2.25 \times 10^{15}/\text{cm}^3$$

EXAMPLE 3.12. A silicon sample is uniformly doped with 10^{16} phosphorus atom/cm^3 and 2 × 10^{16} boron atoms/cm^3. If all the dopants are fully ionised, (i) state the type of material, (ii) calculate carrier concentration

Solution: (i) We know that boron is trivalent impurity whereas phosphorus is pentavalent impurity. Since, boron doping (2 × 10^{16} atoms/cm^3) is higher than phosphorus doping, the material will be of P type.

(ii) Carrier concentration = $N_A - N_D = 2 \times 10^{16} - 10^{16} = 1 \times 10^{16} = 10^{16}$ / cm^3

Hence, the material is P-type with carrier concentration of 10^{16} holes/cm^3.

EXAMPLE 3.13. A semiconductor is irradiated with light such that the carriers are uniformly generated throughout its volume. The semiconductor is n type with $N_D = 10^{19}$/cm^3. If the excess electron concentration in steady state is $\Delta n = 10^{15}$/cm^3 and minority carrier life time τ_p = 10 μsec, determine the generation rate due to radiation

Solution: We know that rate of generation of minority carrier is given by

$$= \frac{\text{Excess hole concentration}}{\text{Hole life time}} = \frac{\Delta n}{\tau_p} = \frac{10^{15}\ /\ \text{cm}^3}{10\times10^{-6}\ \text{sec}} = 10^{20}$$

Hence, the rate of generation of minority carrier is 10^{20} electron hole pairs/sec/cm^3.

EXAMPLE 3.14. A small concentration of minority carrier is injected into a homogeneous semiconductor crystal at certain point. When an electric field of 10 V/cm is applied across the crystal, the minority charge carriers travel a distance of 1 cm in 20 μ sec. Determine the mobility of minority charge carrier.

Solution: We know that Mobility is expressed as under:

$$\mu = \frac{v}{E}$$

where v = drift velocity which is given as $\frac{1\ \text{cm}}{20\ \mu\ \text{sec}}$

Thus, we have $$v = \frac{1\ \text{cm}}{20\times10^{-6}\ \text{sec}}$$

and $$\mu = \frac{1\ \text{cm}}{20\times10^{-6}\ \text{sec}\times10\ \text{V/cm}} = \frac{1\ \text{cm}}{200\times10^{-6}\ \text{V}-\text{sec}} = 5000\ \text{cm}^2/\text{V-sec.}$$

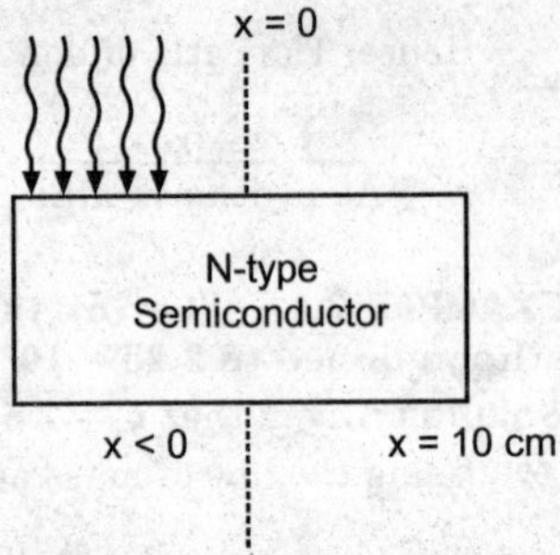

Fig. 3.25.

EXAMPLE 3.15. An N type silicon bar is doped uniformly by phosphorus atoms to a concentration of 4.5×10^{15}/cc. The bar has cross-section of 1 mm^2 and length of 10 cm. It is illuminated uniformly for region x < 0 as depicted in figure 3.25. You may assume optical generation rate 10^{21} electron-hole pairs per cm^3 per second, for this case. The hole life time and electron life time are equal and is equal to 1 μ sec.

Determine the hole and electron diffusion currents at x = 34.6 μm. Following expressions and data may be used in this evaluation:

$$J_p = -q D_p \frac{dp}{dx}; \quad J_n = q D_n \frac{dn}{dx}$$

$$D_p = 12 \text{ cm}^2/\text{sec} ; D_n = 30 \text{ cm}^2 / \text{sec}, q = 1.65 \times 10^{-19} \text{ coulomb}, \frac{KT}{q} = 26 \text{ mV}$$

Solution: Given that, Density / concentration of donor atoms (N_D) = $4.5 \times 10^{15}/cm^3$

Area of cross-section (A) = 1 mm^2 = 1×10^{-4} cm^2

Length of n type Si bar (l) = 10 cm

Hole life time (τ_p) = 1 μsec. = 1×10^{-6} cm

Electron life time (τ_n) = 1 μsec. = 1×10^{-6} sec

(i) Intrinsic concentration of holes and electrons in Si is

$$n_i = 1.5 \times 10^{10}/cm^3$$

(ii) The Si is doped with donor atoms of concentration (N_D) = 4.5×10^{15} / cm^3

Electron concentration at thermal equilibrium (n_o).

(iii) We know that hole concentration at thermal equilibrium (p_o)

$$p_o = \frac{n_i^2}{n_o} = \frac{(1.5 \times 10^{10})^2}{4.5 \times 10^{15}}$$

$$p_o = 0.5 \times 10^5 / cm^3$$

(iv) We know that diffusion length means the length to which the charge carriers diffuse in a semiconductor within one carrier life time and is given by,

$$l = \sqrt{D\tau}$$

For electrons, we have

$$\text{Diffusion length} \quad (l_n) = \sqrt{D_n \tau_n} = \sqrt{30 \times 1 \times 10^{-6}}$$

$$l_n = 5.47 \times 10^{-3} \text{ cm}$$

For holes, we have

$$\text{Diffusion length } (l_p) = \sqrt{D_p \tau_p} = \sqrt{12 \times 1 \times 10^{-6}}$$

$$l_p = 3.4 \times 10^{-3} \text{ cm}$$

(v) We know that concentration gradient of holes at distance x is given by

$$\frac{dp}{dx} = \Delta p \, e^{-x/l_p}$$

where Δp = optical generation rate at x = 0 = 10^{21} electron-hole pair/cm^3/sec

$$\frac{dp}{dx} = 10^{21} \times e^{\frac{-34.6 \times 10^{-4} \text{ cm}}{3.4 \times 10^{-3} \text{ cm}}} \qquad (\because x = 34.6 \text{ μm} = 34.6 \times 10^{-4} \text{ cm})$$

$$\frac{dp}{dx} = 1 \times 10^{22} / \text{cm}^3.\text{cm} = 1 \times 10^{22} / \text{cm}^4$$

(vi) We know that concentration gradient of electrons at distance x is given by

$$\frac{dn}{dx} = \Delta p e^{-x/l_n} = 10^{21} \times e^{\frac{-34.6 \times 10^{-4} \text{ cm}}{5.47 \times 10^{-3} \text{ cm}}}$$

$$\frac{dn}{dx} = 1.88 \times 10^{22} / cm^2 . cm$$
$$= 1.88 \times 10^{22} / cm^3$$

(vii) We know that current density of holes due to diffusion is given by

$$J_p = -q D_p \frac{dp}{dx}$$
$$= -1.6 \times 10^{-19} \times 12 \text{ cm}^2/\text{sec} \times 1 \times 10^{22} / \text{cm}^4$$
$$= -1.92 \times 10^4 \text{ A/cm}^2$$

(vii) We know that current density of electrons due to diffusion is given by

$$J_n = -q D_n \frac{dn}{dx}$$
$$= -1.6 \times 10^{-19} \text{ C} \times 30 \text{ cm}^2 / \text{sec} \times 1.88 \times 10^{22} / \text{cm}^4$$
$$= 9 \times 10^4 \text{ A/cm}^2$$

EXAMPLE 3.16. A particular green LED emits light of wavelength 5490 A°. Find out the energy band gap of the semiconductor material used. You may assume Planck's constant = 6.626 × 10^{-34} J-s

Solution: We know that the energy band gap is given by,

$$E = h\nu$$

where h = Plancks constant = 6.626 × 10^{-34} J-s = 4.137 × 10^{-15} eV

and
$$f = \frac{c}{\lambda} = \frac{3\times10^8}{5490\times10^{-10} \text{ m}}$$

Therefore,
$$E = 4.137\times10^{-15} \times \frac{3\times10^8}{5490\times10^{-10}}$$

or
$$E = 2.26 \text{ eV}$$

EXAMPLE 3.17. The electron concentration in a sample of uniformly doped n type silicon a 300 k varies from 10^{17}/cm³ at x = 0 to 6 × 10^{16}/cm³ at x = 2 μm. You may assume a situation that electrons are supplied to keep this concentration gradient constant with time. If the electronic charge is 1.6 × 10^{-19} coulomb and the diffusion constant D_n = 35 cm²/sec. Determine, the current density in the silicon if no electric field is present.

Solution: From the given data, we can have a plot as illustrated in figure 3.26.

We know that the current density in silicon is given by,

$$J_n = q D_n \frac{dn}{dx}$$

Also, concentration of gradient is given by

$$\frac{dn}{dx} = \frac{y_2 - y_1}{x_2 - x_1}$$

$$\frac{dn}{dx} = \frac{6\times10^{16} - 10^{17} / \text{cm}^3}{(2-0)\times10^{-4} \text{ cm}}$$

$$= -2 \times 10^{20}$$

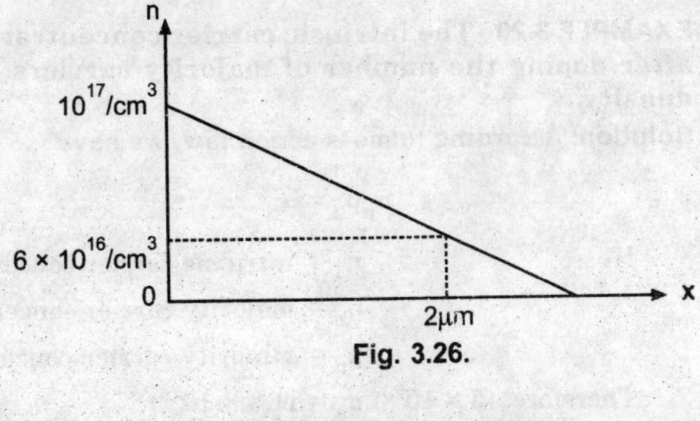

Fig. 3.26.

Given that $D_n = 35 \text{ cm}^2 / \text{sec}$

Therefore, $J_n = 1.6 \times 10^{-19} \times 35 \times -2 \times 10^{20}$

or $J_n = -1120 \text{ A/cm}^2$

EXAMPLE 3.18. An N-type silicon bar is 0.1 cm long and 100 μm² in cross-sectional area has a majority carrier concentration of 5 × 10^{20}/m³ and the carrier mobility is 0.13 m² / V-sec. at 300 K. If the charge on an electron is 1.6 × 10^{-19} coulomb, determine the resistance of the bar.

Solution: We know that the resistance of the bar is given by,

$$R = \rho \frac{l}{A}$$

where ρ = Resistivity of the bar = $\frac{1}{\sigma}$

where σ is conductivity of the bar and is given as

$$\sigma_n = q\, n_n\, \mu_n$$

Given that $q = 1.6 \times 10^{-19}$ coulomb

$$n_n = 5 \times 10^{20} / m^3 = 5 \times 10^{14} / cm^3$$

$$\mu_n = 0.13\ m^2 / V\text{-sec}$$

$$= 1300\ cm^2/V\text{-sec}$$

$$\sigma_n = 1.6 \times 10^{-19} \times 5 \times 10^{14} \times 1300$$

$$= 0.104\ \text{Ohm-cm}^{-1}$$

$$\rho = \frac{1}{\sigma_n} = 9.616$$

Using $R = \rho \frac{l}{A} = 9.616 \times \frac{0.1\ cm}{10^{-6}\ cm^2}$

$$100\ \mu m^2 = 100 \times 10^{-8}\ cm^2 = 10^{-6}\ cm^2$$

$$R = 0.9616\ M\Omega = 1\ M\Omega$$

EXAMPLE 3.19. In an abrupt P-N junction, the doping concentrations on the P side and N side are $N_A = 9 \times 10^{16} / cm^3$, $N_D = 1 \times 10^{16} / cm^3$ respectively. The p-n junction is reverse biased and the total depletion width is 3 μm. The depletion width on the P side is

(a) 2.7 μm **(b) 0.3 μm**

(c) 2.25 μm **(d) 0.75 μm**

Solution: Justification: From the given data, it is quite obvious that doping on P side is 9 times higher than on N side. The depletion layer penetrates more in lightly doped region and less in highly doped region. Therefore, the penetration on P side will be nine times less.

Therefore, Depletion width on P side = $\frac{3\ \mu m}{9} = 0.3\ \mu m$

EXAMPLE 3.20. The intrinsic carrier concentration of silicon sample at 300°K is $1.5 \times 10^{16}/m^3$. If after doping the number of majority carriers is $5 \times 10^{20}/m^3$, determine the minority carrier density.

Solution: According to mass action law, we have

$$n_n p_n = n_i^2$$

Here, n_1 = intrinsic carrier concentration = $1.5 \times 10^{16}/m^3$

n_n = majority carrier concentration = $5 \times 10^{20}/m^3$

p_n = minority carrier concentration

Therefore, $(5 \times 10^{20})\, p_n = (1.5 \times 10^{16})^2$

or
$$p_n = \frac{(1.5 \times 10^{16})^2}{5 \times 10^{20}}$$

$$= 4.50 \times 10^{11}/m^3$$

EXAMPLE 3.21. **The neutral base width of a bipolar transistor biased in the active region is 0.5 μm. The maximum electron concentration and the diffusion constant in the base are $10^{14}/cm^3$ and D_n = 25 cm^2/sec respectively. Assume negligible recombination in the base. Evaluate the collector current density. You assume q = 1.6×10^{-19} coulomb.**

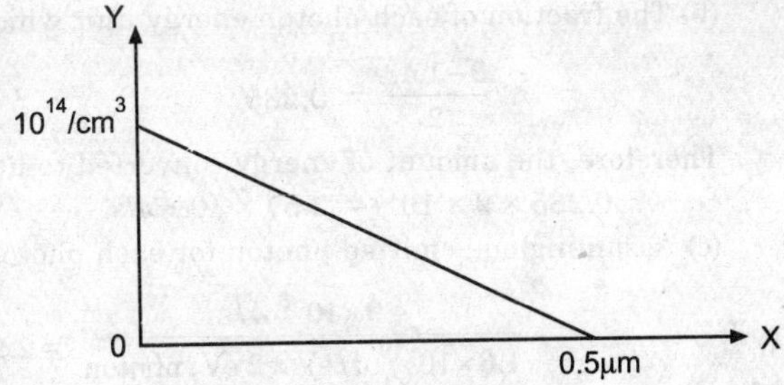

Fig. 3.27.

Solution: From the given data, we can have a plot as illustrate in figure 3.27.

From the plot, we can write,

$$\frac{dn}{dx} = \frac{y_2 - y_1}{x_2 - x_1} = \frac{(10^{14} - 0)/\text{cm}^3}{0 - 0.5\,\mu\text{m}} = \frac{10^{14}/\text{cm}^3}{0.5\times10^{-6}\times100\text{ cm}}$$

or $$\frac{dn}{dx} = 2 \times 10^{19}/\text{cm}^4$$

We know that the current density is given by

$$J_n = q D_n \frac{dn}{dx} = 1.6 \times 10^{-19}\text{ C} \times 25\text{ cm}^2/\text{sec} \times 2 \times 10^{19}/\text{cm}^4$$

or $$J_n = \frac{8\text{A}}{\text{cm}^2}$$

EXAMPLE 3.22. **The longest wavelength that can be absorbed by silicon which has the band gap of 1.12 eV is 1.1 μm. If the longest wavelength that can be absorbed by another material is 0.87 μm then the band gap of this material is**

(a) 1.416 eV
(b) 0.886 eV
(c) 0.854 eV
(d) 0.706 eV

Solution: We know that band gap $E_g = h\nu = \frac{hc}{\lambda} = \frac{6.64\times10^{-34}\times3\times10^{8}}{0.87\times10^{-6}}$ = 1.431 eV Ans. (a)

EXAMPLE 3.23. **A 0.46 - μm thick sample of GaAs is illuminated with monochromatic light of hν = 2 eV. The absorption coefficient α is $5 \times 10^4\ cm^{-1}$. The power incident on the sample is 10 mW.**

(a) Determine the total energy absrobed by the sample per second (J/s).

(b) Determine the rate of excess thermal energy given up by the electrons to the lattice befoe recombination (J/s).

(c) Determine the number of photons per second given off from recombination events, assuming perfect quantum efficiency.

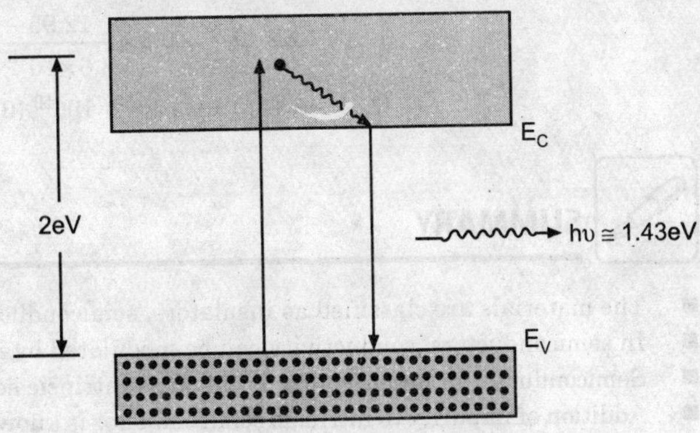

Fig. 3.28. Excitation and band-to-band recombination leading to photo-luminescence

(U.P. Tech, Sem. Exam, 2006-07) (10 Marks)

Solution: (a) We have $I_t = I_0\, e^{-\alpha l} = 10^{-2} \exp(-5 \times 10^4 \times 0.46 \times 10^{-4}) = 10^{-2}\, e^{-2.3} = 10^{-3}$ W

Hence, the absorbed power will be

$$10 - 1 = 9\text{ mW} = 9 \times 10^{-3}\text{ J/s}$$

(b) The fraction of each photon energy unit which is converted to heat is

$$\frac{2-1.43}{2} = 0.285$$

Therefore, the amount of energy converted to heat per second is

$$0.285 \times 9 \times 10^{-3} = 2.57 \times 10^{-3} \text{ J/s}$$

(c) Assuming one emitted photon for each photon absorbed (perfect quantum efficiency), we get

$$\frac{9\times10^{-3}\text{ J/s}}{1.6\times10^{-19}\text{ J/eV}\times 2\text{ eV/photon}} = 2.81 \times 10^{16} \text{ photon/s}$$

EXAMPLE 3.24. In a very long P-type Si bar with cross-sectional area = 0.5 cm² and $N_a = 10^{17}$ cm^{-3}, we inject holes such that the steady state excess hole concentration is 5×10^{16} cm^{-3} at x = 0. What will be the steady state separation between F_p and E_c at x = 1000 A°? What will be the hole current there? How much is the excess stored hole charge? You assume that μ_p = 500 cm²/V-s and $\tau_p = 10^{-10}$ s.

Solution: We know that $D_p = \frac{kT}{q}\mu_p = 0.0259 \times 500 = 12.95$ cm/s

We know that $L_p = \sqrt{D_p\tau_p} = \sqrt{12.95\times10^{-10}} = 3.6 \times 10^{-5}$ cm

We know that $p = p_0 + \Delta p e^{\frac{x}{L_p}} = 10^{17} + 5\times10^{16}\, e^{\frac{10^{-5}}{3.6\times10^{-5}}}$

$$= 1.379 \times 10^{17} = n_i\, e^{(E_i - F_p)/kT} = (1.5 \times 10^{10} \text{ cm}^{-3})\, e^{(E_i - F_p)/kT}$$

Also, $E_i - F_p = \left(\ln\frac{1.379\times10^{17}}{1.5\times10^{10}}\right)0.0259 = 0.415$ eV

or $E_c - F_p = 1.1/2$ eV + 0.415 eV = 0.965 eV

We can evaluate the hole current as under:

$$I_p = -qAD_p\frac{dp}{dx} = qA\frac{D_p}{L_p}(\Delta p)\, e^{\frac{x}{L_p}}$$

$$= 1.6\times10^{-19}\times0.5\times\frac{12.95}{3.6\times10^{-5}}\times5\times10^{16}\frac{10^{-5}}{e^{3.6\times10^{-5}}} = 1.09 \times 10^{3} \text{ A}$$

$$Q_p = qA(\Delta p)L_p = 1.6 \times 10^{-19}(0.5)(5 \times 10^{16})(3.6 \times 10^{-5}) = \mathbf{1.44 \times 10^{-7} C}$$

SUMMARY

- The materials are classified as insulators, semiconductors and conductors based on energy band gap.
- In semiconductors, conductivity can be modulated by impurity doping.
- Semiconductor in its pure form is called as intrinsic semiconductor.
- Addition of impurity to intrinsic semiconductor is known as doping and doped semiconductor is known as extrinsic semiconductor.
- n-type semiconductors are formed by adding pentavalent impurity whereas p-type semiconductors are formed by adding trivalent impurities.
- The movement of charge carriers due to influence of electric field is known as drift whereas movement of charge carriers due to concentration gradient is known as diffusion.
- Fermi level gives the energy level with 50% probability of occupation of electrons.
- The ratio of diffusion constant to mobility of charge carriers is constant as given by Einsteins relationship.

- The rate of recombination is proportional to the equilibrium concentration of electrons and holes.
- Mean carrier life time during recombination is important for minority carriers.
- The mass action law is $np = n_i^2$
- The carrier generation due to photon energy is known as photo-conduction.
- The current injection across forward bias junctions in semiconductors like GaAs and GaAsP emit light which is known as electro-luminescence.
- Miniaturization, cost reduction, high reliability are the major factors in the development of integrated circuits.
- When the surface of certain alkaline material such as sodium, potassium, cesium, or rubidium is illuminated by a beam of light or ultraviolet radiations, the electrons are emitted. This phenomenon is called the photoelectric emission.
- The use of photoelectric emission phenomenon is made in phototubes which form the basis of TV and sound films.

  Laws of Photoelectric Emission

  (i) Photoelectric emission from a metal surface takes place only when the energy content hf of the incident light exceeds the work function ϕ of the metal.

  (ii) The number of electrons emitted per second is directly proportional to the amount of incident light flux.

  (iii) Maximum energy of emitted electrons is independent of the amount of incident light flux but is linearly related with the frequency of light flux.
- When monochromatic light is incident on a clean surface of a material, then under certain conditions, electrons also called photoelectrons, aer emitted from the surface. The electrons in the metal absorb energy from the light, and some of the electrons receive sufficient energy to be ejected from the metal surface into the vacuum. This phenomenon is called the photoelectric effect.
- Wavelength and energy can be related by

$$\lambda = \frac{c}{f} = \frac{hc}{E} = \frac{1.24}{E}\ \mu m$$

  where E is the photon energy in eV and c is the velocity of light.
- When light falls on a semiconductor, the photons may be absorbed or they may propagate through the semiconductor depending on the photon energy (E) and the band gap energy E_G. Where the photon energy is less than E_G, the photons are not readily absorbed and the light is transmitted through the material and the semiconductor appears to be transparent.
- The probability of interaction between the photons and a valence electron is high whenever $E = hf > E_G$. This interaction generates an electron in the conduction band and a hole in the valence band – known as electron-hole pair (EHP). The excess energy provides the electron or hole additional kinetic energy which heats the semiconductor.
- When a system absorbs energy in some form or other, it may be partly re-emitted as radiation. This phenomenon is called luminescence. Luminescence may occur in solids (crystals or glasses) as well as in liquids and gases.
- Types of luminescence may be distinguished by the excitation mechanism. If carriers are excited by photon absorption, the radiation resulting from the recombination of the excited carriers is called photoluminescence; if the excited carriers are produced by high-energy electron bombardment of the material, the mechanism is called *cathodoluminescence*; if the excitation occurs due to introduction of current into the sample, the resulting luminescence is calld *electroluminescence*.
- The amount of time between the creation and disappearance of a free electron is called the lifetime. It varies from a few nanoseconds to several microseconds depending how perfect the crystal is and other factors.
- When the excess carriers in a semiconductor are due to optical luminescence, the resulting conductivity is called photoconductivity.
- The conductivity of a material is proportional to the concentration of charge carriers present. Radiant energy supplied to the semiconductor causes covalent bonds to be broken, and EHPs in excess of those generated thermally are produced. These increased current carriers reduce the resistance of the material and, therefore, such a device is called a *photoconductor* or *photoresistor*.

SHORT QUESTIONS WITH ANSWERS

Q.1. What is meant by photoelectric emission ?

Ans. When the surface of certain alkaline material such as sodium, potassium, cesium, or rubidium is illuminated by a beam of light or ultraviolet radiations, the electrons are emitted. This phenomenon is called the photoelectric emission.

Q.2. What is meant by the threshold frequency ?

Ans. The minimum frequency which can cause photoelectric emission is called the threshold frequency.

Q.3. What is photoelectric effect ?

Ans. Photoelectric effect is the relationship between collection of light energy and reflection of electrons from a metal.

Q.4. How wavelength and energy of a photon are related ?

Ans. Wavelength, $\lambda = \dfrac{1.24}{\text{Energy, E}}$ in micrometre.

Q.5. What is luminescence ?

Ans. When a system absorbs energy in some form or the other, it may be partly re-emitted as radiation. This phenomenon is called luminescence. Luminescence is the general property of light emission. This phenomenon may occur in solids (crystals or glasses) as well as in liquids and gases.

Q.6. What is electroluminescence ?

Ans. When excitation occurs due to introduction of current into the sample, the resulting luminescence is called the electroluminescence. There are two mechanisms for production of electroluminescence. One mechanism is photoemission in the act of electron-hole recombination. LEDs and P-N junction diodes are examples of this phenomenon. The second mechanism for electroluminescence is photoemission following electron impact excitation of an activator atom. This process has been observed in ZnS: Cu.

Q.7. What is photoluminescence ?

Ans. When carriers are excited by photon absorption, the radiation resulting from the recombination of the excited carriers is called the photoluminescence. One of the most common examples of photoluminescence is the fluorescent lamp.

Q.8. What is cathodoluminescence ?

Ans. If the excited carriers are produced by high-energy electron bombardment of the material, the mechanism is called the cathodoluminescence. This principle is used in cathode ray tube (CRT), oscilloscope and other visual display systems.

Q.9. Define mean lifetime of a carrier.

Ans. The amount of time between the creation and disappearance of a free electron (or hole) is called the lifetime. It varies from a few nanoseconds to several microseconds depending how perfect the crystal is and other factors. The lifetimes of carriers (holes and electrons) form very important factors in semiconductor devices because they indicate the time required for the hole and electron concentrations to return to their equilibrium values after being influenced by an external cause such as illumination by light.

Q.10. What is photoconductivity ?

Ans. When the excess carriers are generated in a semiconductor by optical luminescence, the resulting conductivity is called photoconductivity. This is an important effect, with usual applications in the analysis of semiconductor materials and in the operation of different types of devices.

Q.11. What is photoconductor gain ?

Ans. Photoconductor gain may be defined as the ratio of the rate at which charge is collected by the contacts to the rate at which charge is created within the photoconductor.

REVIEW QUESTIONS

1. What is photoelectric emission ? How is the electron emission affected if *(i)* the frequency, and (ii) the intensity of the incident radiations are increased ? Name a material which is suitable for photoemission.
2. What is photoelectric effect ? Explain the terms "optical absorption" and "work function". Give the expression for maximum kinetic energy of the photoelectron in terms of incident photon energy and work function.
3. What is optical absorption ? Explain the three situations which normally arise with photon energy conditions.
4. What is photo absorption coefficient ? How intensity of photoflux varies with distance through the semiconductor material. Give relation between photon energy and wavelength.
5. What is luminescence? Explain the terms 'photoluminescence', 'cathodoluminescence' and 'electroluminescence'.
6. What type of semiconductor material is suited for luminescence effect ? Explain excitation and recombination of photo-luminescence with trap level for electrons. **(U.P. Tech. Sem. Exam, 2005-06)**
7. What is electroluminescence ? Explain the two basic mechanisms of electroluminescence.
8. What is recombination ? Derive the expression for minority carrier lifetime. **(U.P. Tech. Sem. Exam, 2005-06)**
9. What is meant by mean lifetime of electron (or hole)? What is photoconductivity ? Explain photoconductive effect.
10. What is photoconductor ? Derive the expressions for photocurrent and photoconductor gain.
11. Obtain an equation for photocurrent in terms of lifetime and transit time of carriers in a sample dominated by μ_n. **(U.P. Tech. Sem. Exam, 2007-08)**
12. What is photoconductivity ? Show that photocurrent is proportional to the lifetime (τ_h) and inversely proportional to transit time (τ_t) of carrier. **(U.P. Tech. Sem. Exam, 2007-08)**
13. 'Obtain relationship between photoconductivity and mobility of carriers. **[U.P. Tech. Sem. Exam. 2009-10]**
14. Obtain equation of diffusion current in terms of lifetime of the carrier and other parameters. **(U.P. Tech. Sem. Exam, 2007-08)**
15. Obtain relationship between photoconductivity and mobility of carriers. **(U.P. Tech. Sem. Exam, 2008-09)**
16. Explain carrier diffusion phenomenon. Derive an expression for the electron current due to diffusion.
17. Show that a hole can take part in a transport of current. Define mobility of a charge carrier. Show that $\frac{\mu}{D} = \frac{e}{kT}$.
18. Obtain the relation $\frac{D_p}{\mu_p} = \frac{KT}{q}$. **(U.P. Tech. Sem. Exam, 2007-08)**
19. Derive the equation of conservation of charge.
20. What is the effect of temperature on the reverse saturation of a diode ?
21. Define the following terms with suitable examples:
    Photoluminescence, cathodoluminescence, electroluminescence and carrier lifetime. **(U.P. Tech. Sem. Exam, 2006-07)**
22. Define (i) photoconductivity, (ii) Luminescence. **(U.P. Tech. Sem. Exam, 2009-10)**
23. Write short notes on the following:
    (i) Photoelectric emission.
    (ii) Photoelectric effect.
    (iii) Luminescence.
    (iv) Electroluminescence.
    (v) Photoluminescence.
    (vi) Photoconductivity.
    (vii) Photoconductor.
    (viii) Carrier diffusion.

5. [illegible] Explain the [illegible]
6. [illegible]
7. [illegible]

9. What is [illegible]

11. Obtain an expression [illegible] (U.P. Tech. Sem. Exam. 2006-07)
12. What is photoconductivity? [illegible] (U.P. Tech. Sem. Exam. 2007-08)

16. [illegible]

19. What is the [illegible]
20. [illegible]

22. [illegible] (U.P. Tech. Sem. Exam. [illegible])

(i) [illegible] (ii) [illegible]
(iii) Luminescence (iv) [illegible]
(v) [illegible] (vi) [illegible]

UNIT-III

Syllabus

Junction Properties: Equilibrium conditions, Biased junctions, Steady state conditions, Reverse Bias Break Down, Transient and AC Conditions, Metal Semiconductor Junctions.

Above syllabus has been covered in following chapter :

Chapter 4: Junction Properties

Chapter 4

Junction Properties

Syllabus

Junction Properties: Equilibrium conditions, Biased junctions, Steady state conditions, Reverse Bias Break Down, Transient and AC Conditions, Metal Semiconductor Junctions.

Chapter Outline

4.1. INTRODUCTION

As a matter of fact, most of the semiconductor devices consists of at least one P-N junction. The P-N junction is a fundamental element to the performance of the functions like rectification, amplification, switching and other operations in the electronic circuits.

This chapter discusses the equilibrium state of the junction and the flow of electrons and holes across the junction under steady state and transient conditions. The various parameters associated with junctions, its dependence on temperature and diode models under different signal conditions are also discussed.

We shall also present a brief review of special purpose diodes such as rectifier, signal, switching and schottky diodes.

Important characteristics of a diode

(i) In diode, current flows due to majority as well as minority charge carriers, so it is a bipolar device.

(ii) In diode, current flow can take place only in one direction i.e. anode to cathode, so, it is a unidirectional device.

(iii) Diode is a current operated device.

(iv) As diode is unidirectional i.e., either it conducts heavily or does not conduct at all, so it is used for rectification.

4.2. SEMICONDUCTOR DIODE

1. Basics

(i) A device that is obtained by joining a section of P-type semiconductor material with a similar section of N-type semiconductor material is known as semiconductor diode.

(ii) The area where P and N regions meet is known as junction. The structure and symbol of a diode have been shown in figure 4.1.

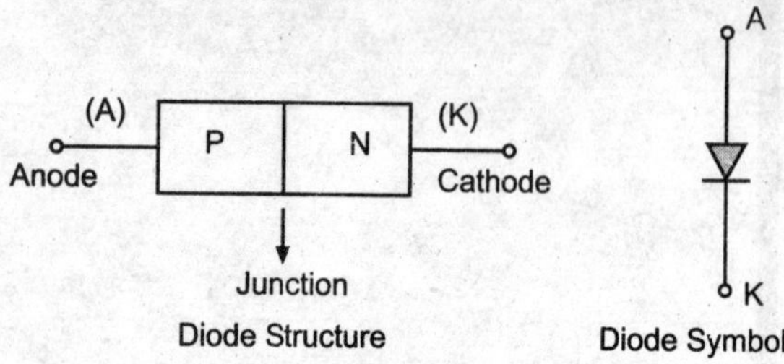

Fig. 4.1.

(iii) The device is called diode since it contains two electrodes i.e., anode (A) and cathode (k). The anode terminal is taken out of a P-type semiconductor material while cathode terminal is taken out of a N-type semiconductor material. As the current flows in the diode only in one direction, it has the rectification property, which is used to convert alternating current (ac) to direct current (dc).

(iv) In diode symbol direction of arrowhead indicates the flows of conventional current. The conventional current flows anode to cathode.

2. Diode Construction

To prepare a rectifying junction merely pressing a section of P-type semiconductor material together with a section of N-type semiconductor material is not sufficient. The semiconductor should be in one piece, with proper division of P-type impurity region and N-type impurity region, to form a P-N junction.

3. Techniques of preparation of P-N junction

(i) Grown junction, (ii) Alloy junction, (iii) Point junction

(i) Grown junction: In this type of diode, P-type and N-type impurities are mixed into a single crystal during the manufacturing process. The P-region is grown over a part of semiconductor base and N-region is grown over the other part of a semiconductor bar.

(ii) Alloy junction: In this process, an impurity of one type is being melt into a semiconductor of the opposite type impurity i.e., a pellet of acceptor impurity is placed over a wafer of an N type silicon and is heated. The acceptor impurity fuses under controlled temperature condition into the wafer to form P-region within it.

(iii) Point contact :

(i) In this process, a fine metal wire is utilized which makes the contact with a small area on the surface of an N-type semiconductor.

(ii) In this method, by P-N junction is formed by momentarily applying a high surge current to the wire and an N-type semiconductor. The heat generated by this surge current converts the material near the contact point to a P-type material. Grown junction and fused alloy junction, from which bars are cut to form a diode, are shown in figure 4.2 (a) and 4.2 (b). Point contact type of construction to form a diode has been shown in figure 4.2 (c).

(iii) Another way in point contact method is to heat a semiconductor material till melting and then to diffuse an impurity atom into the surface layer.

(iv) In all the above process, it is quite important to have a perfect bond everywhere along the union or junction between P-type and N-type material. In fact, the proper contact between the union or junction is the key for the rectifying action of a diode.

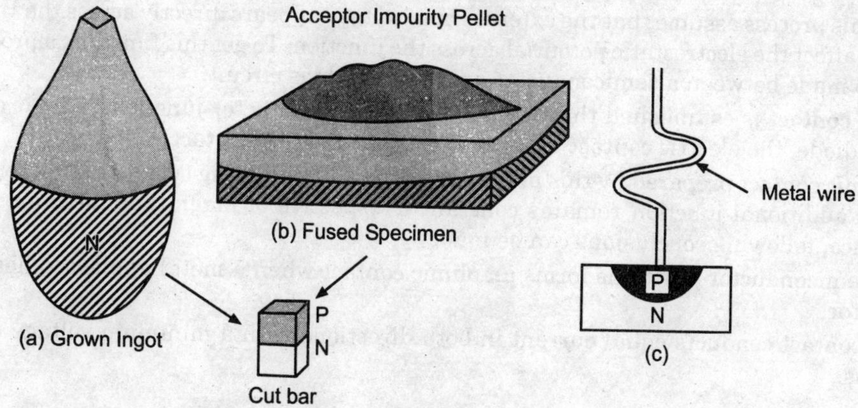

Fig. 4.2

4.3. METALLURGICAL JUNCTION AND OHMIC CONTACT

1. Basics

(i) When semiconductor junction is prepared by doping a semiconductor i.e. impurity concentration varies from N-type to P-type. The electrical junction where the transition from one doping type to another occurs is called metallurgical junction.

(ii) P-N junction and P-type and N-type concentrations are shown in figure 4.3(a) and 4.3(b), and metallurgical junction is shown in figure 4.3(c).

(iii) The interface at x = 0 is known as the metallurgical junction. A large density gradient in both hole and electron concentrations appears across the junction.

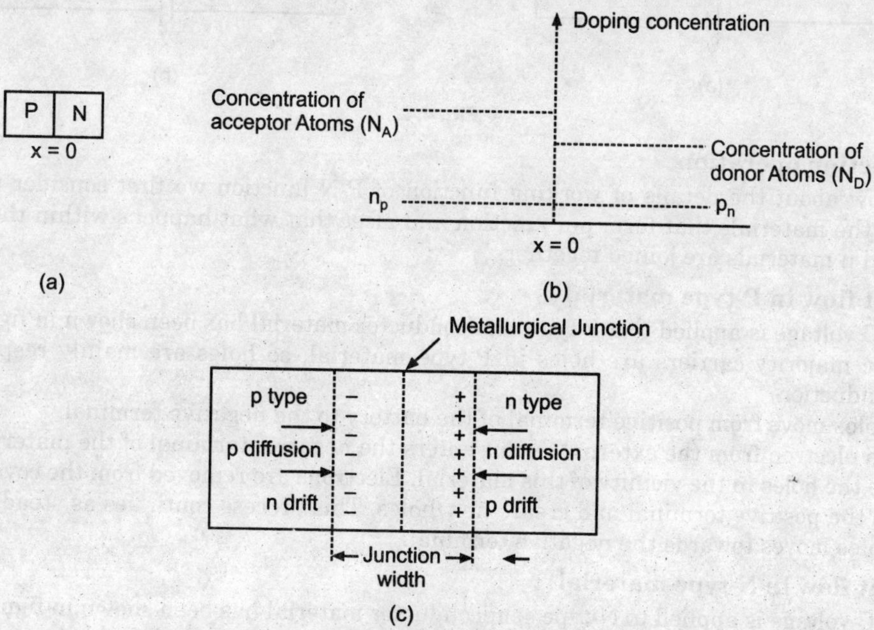

Fig. 4.3.

2. Ohmic Contact

(i) The external voltage applied to P-N junction diode is shown in figure 4.4. The external biasing is connected to a semiconductor through metallic ohmic contact.

(ii) In this process assume that the external bias voltage appears directly across the P-N junction and affect the electrostatic potential across the junction. To get this function, a proper contact is to made between a semiconductor and external bias circuit.

(iii) The contact is established through two metal semiconductor junctions, one at each end of the diode, the electric contact this nature is called ohmic contact.

(iv) Ohmic contact prepared by this process must be non-rectifying type so that potential across this additional junction remains constant irrespective of magnitude direction of current. Hence, following conclusions can be made:

Metal-semiconductor junctions forms an ohmic contact when a metal applied to heavily doped semiconductor.

Ohmic contact conducts equal current in both directions with a minimum voltage drop across the junctions.

3. Use

This type of contact is used,

(i) To connect one semiconductor device to another semiconductor device in integrated circuit.

(ii) To connect an integrated circuit to its external terminals.

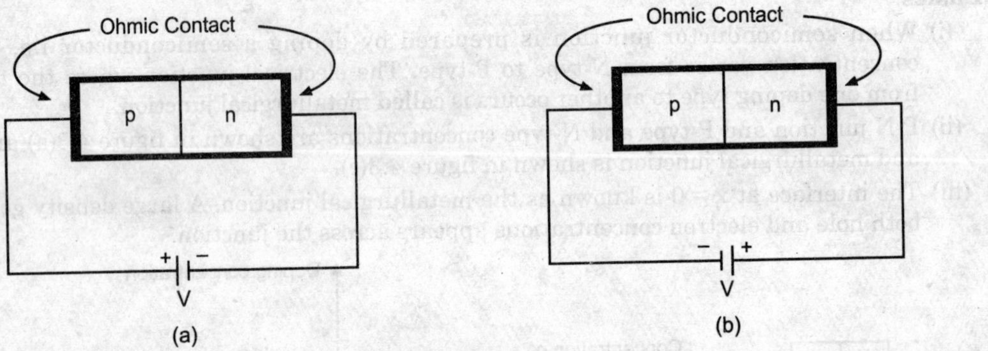

Fig. 4.4.

4. P-N junction operation

To know about the details of working function of P-N junction we first consider the flow of current in the materials that form p-n junction and after that what happens within the junctions when p and n materials are joined together.

5. Current flow in P-type material

(i) DC voltage is applied to a P-type semiconductor material has been shown in figure 4.5. As the majority carriers are holes in P-type material, so holes are mainly responsible for conduction.

(ii) Holes move from positive terminal of the battery to the negative terminal.

(iii) An electron from the external circuit enters the negative terminal of the material and fills up the holes in the vicinity of this material. Electrons are removed from the covalent bonds at the positive terminal and create new holes. This process continues as steady stream of holes moves towards the negative terminal.

6. Current flow in N-type material

(i) DC voltage is applied to N-type semiconductor material has been shown in figure 4.6. Free electrons of the material are attracted by the positive terminal of the battery.

(ii) Electrons will leave the semiconductor material and flow towards the positive terminal of the battery.

(iii) When an electron leaves the semiconductor material, an electron from battery will enter into material to complete the current flow path.

(iv) As the majority carriers in N-type material are electrons, so electrons are repelled by the negative terminal of the battery and flow through the material towards the positive terminal of the battery.

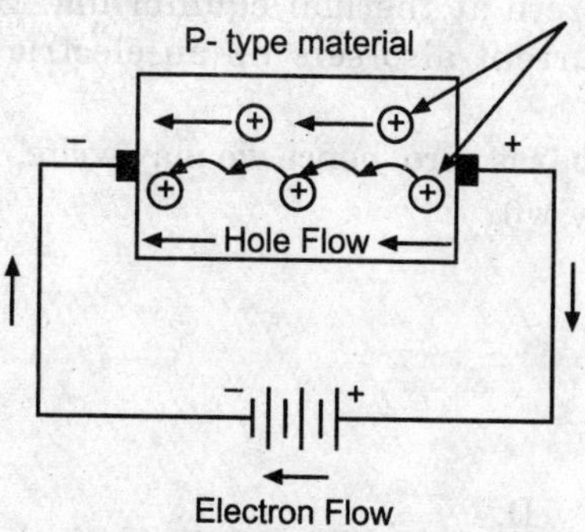

Fig. 4.5. Current flow in P-type material

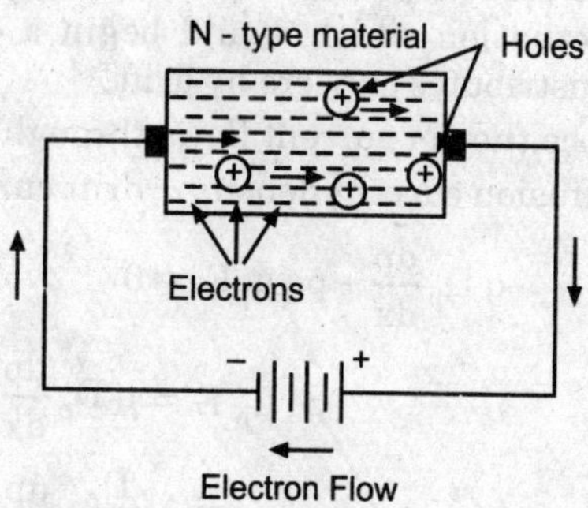

Fig. 4.6. Current flow in N-type material

7. Potential variation in graded semiconductor

(i) Non-uniformly doped semiconductor means graded semiconductor in which charge carrier concentration changes with distance has been shown in figure 4.7(a) for a non uniformly doped P-type semiconductor bar.

(ii) The variation of carrier concentration of holes in the semiconductor bar with distance x has been shown in figure 4.7(b).

(iii) The concentration of carriers (holes) is maximum at x = 0 and it decreases with increase in distance x,

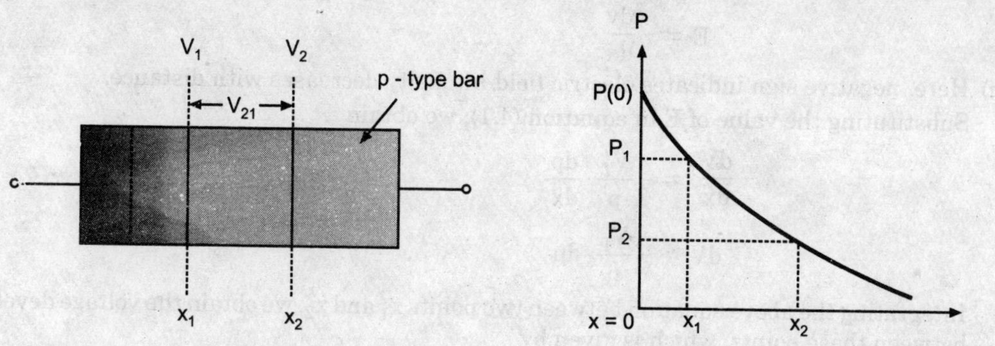

(a) Non-uniformly Doped Semiconductor

(b) Change in Concentration of Hole with x

Fig. 4.7.

(iv) If, there is no external voltage applied to P-type semiconductor bar, then net current flower through the bar is zero.

(v) In P-type semiconductor bar, there exists concentration gradient because it is non-uniformly doped. Holes from higher concentration area flows towards lower concentration area due to the presence of concentration gradient. Therefore, a diffusion current density of holes exists which may be written as under:

$$(J_p)_{diff} = q\, D_p \frac{dp}{dx}$$

(vi) Since there is no external voltage applied to semiconductor bar, the net current flows through it is zero. This shows that there is equal amount of drift current flowing in opposite direction. The drift current density may be written as under:

$$(J_p)_{dr} = pq\, \mu_p\, E$$

(vii) When external voltage is applied to semiconductor bar then only drift curent flows thorugh it. This shows the potential required for the flow of drift current is developed internally, which means that non-uniform doping creates the voltage.

(viii) Net current flows into the semiconductor is zero at thermal equilibrium. Hence, any fluctuation which would begin a diffusion current also sets up an electric field that redistributes carriers by drift.

(ix) Since the net current flows through the p-type bar is zero, hence we may write,
Diffusion current density + drift current density = 0

$$-q D_p \frac{dp}{dx} + pq \mu_p E = 0$$

$$pq \mu_p E = q D_p \frac{dp}{dx}$$

$$pE = \frac{D_p}{\mu_p} \cdot \frac{dp}{dx} \qquad \text{where } \frac{D_p}{\mu_p} = V_T \text{ (Einstein's relation)}$$

$$= V_T \frac{dp}{dx}$$

$$E = \frac{V_T}{p} \cdot \frac{dp}{dx} \qquad \text{...(4.1)}$$

Where E is the electric field induced internally in the semiconductor due to the non-uniform doping.

(x) According to the definition of electric field intensity, we can write as under:

$$E = -\frac{dv}{dx}$$

(xi) Here, negative sign indicates electric field intensity decreases with distance.
Substituting the value of E in equation (4.1), we obtain

$$-\frac{dV}{dx} = -\frac{V_T}{p} \cdot \frac{dp}{dx}$$

$$dV = -\frac{V_T}{p} \cdot dp$$

Integrating the above equation between two points x_1 and x_2, we obtain the voltage developed between these points, which is given by

$$\int_{V_1}^{V_2} dV = -V_T \int_{p_1}^{p_2} dp$$

where p_1 and p_2 are hole concentrations of x_1 and x_2 respectively.

$$[V]_{V_1}^{V_2} = V_T [\ln p]_{p_1}^{p_2}$$

$$V_2 - V_1 = -V_T \ln \frac{p_2}{p_1}$$

$$V_{21} = V_T \ln \frac{p_1}{p_2} \qquad \text{...(4.2)}$$

where V_{21} in the voltage developed between two points x_1 and x_2.

(xii) The potential difference in the semiconductor depdends on the concentration of charge carriers and not on the distance.

To determine the concentration of holes p_1 and p_2, we write from equation (4.2) as under:

$$\ln \frac{p_1}{p_2} = \frac{V_{21}}{V_T}$$

then
$$\frac{p_1}{p_2} = e^{V_{21}/V_T}$$

and
$$p_1 = p_2\, e^{-V_{21}/V_T} \text{ and}$$

$$p_2 = p_1\, e^{-V_{21}/V_T} \quad \text{...(4.3)}$$

(xiii) In a similar manner, we can determine the concentrations of electrons n_1 and n_2 at a distance x_1 and x_2 respectively for an n-type non-uniformly doping semiconductor bar, and can be written as

$$n_1 = n_2\, e^{-V_{21}/V_T} \text{ and } n_2 = n_1\, e^{V_{21}/V_T} \quad \text{...(4.4)}$$

Law of Mass Action

Using the above equations (4.3) and (4.4) we can write,

$$p_1 \times n_1 = p_2\, e^{V_{21}/V_T} \times n_2\, e^{-V_{21}/V_T} = V_T \ln \frac{p_1}{p_2} \quad \text{...(4.5)}$$

$$p_1\, n_1 = p_2\, n_2 \quad \text{...(4.6)}$$

Important Point: Equation (4.6) shows that the product of concentration of holes and electrons at x_1 is equal to their product at x_2, means concentration of holes and electrons is always constant at any point in the semiconductor bar, so, it process the law of mass action.

4.4. JUNCTION BARRIER

1. Basic Fundamentals

(i) Extrinsic semiconductor like N-type material has an excess of free electrons but still it is electrically neutral. This is due to the fact that donor atoms in the N-type semiconductor became positively charged atom when electrons are became free by breaking the covalent bonding.

(ii) Hence, for every free electron in the N-type semiconductor material, there is a corresponding positively charge atom to neutral it.

(iii) Finally, we conclude that N-type semiconductor has an overall zero charge inside it.

(iv) Similarly, by applying the same rule, the P-type semiconductor material also electrically neutral because the excess holes in the semiconductor is perfectly balanced by the number of electrons.

(v) In extrinsic semiconductor i.e. N-type or P-type semiconductor electrons and holes are still free to move inside the semiconductor because they are loosely bound with their parent atoms.

(vi) We may think that when we join the N-type and P-type semiconductor materials together by any process mentioned above, then all electrons and holes recombine together and became neutral. On the contrary this does not happen. Instead of that few electrons in the N-type semiconductor material diffuse across the junction into the P-type semiconductor material and fill some of the holes.

(vii) In a similar manner at the same time, the few holes in the P-type semiconductor material diffuse across the junction into the N-type semiconductor material and fill up some of the electrons (as shown in figure 4.8(a)). This process is known as junction recombination which reduces the numbers of free electrons and holes in the vicinity of the junction.

(viii) As there is a depletion or lack of free electrons and holes in the area near to the junction, this area is known as depletion region.

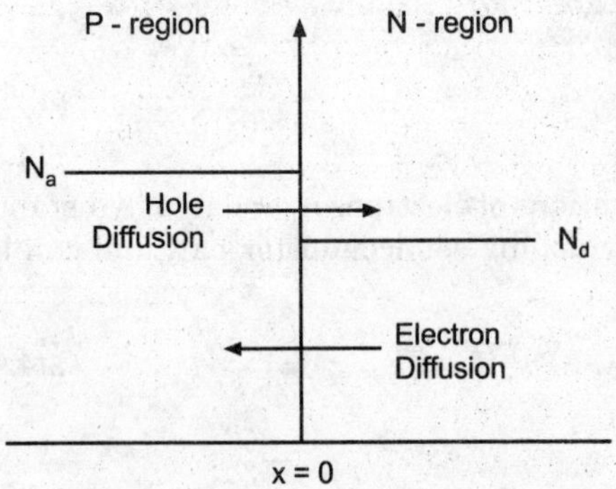

(a) Initial diffusion of electrons and holes at the metallurgical junction, establishing thermal equilibruim

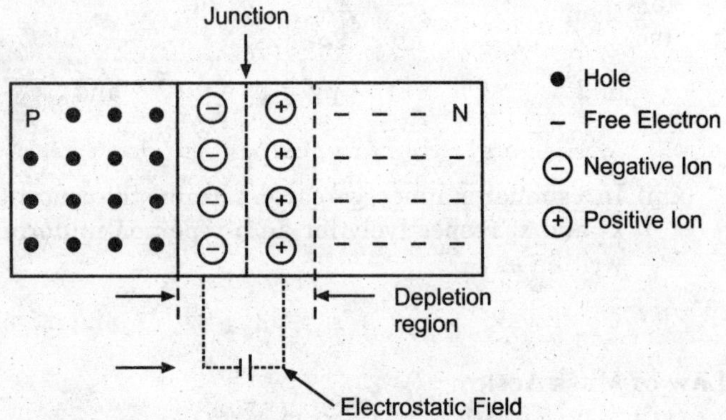

(b) p-n junction barrier formation

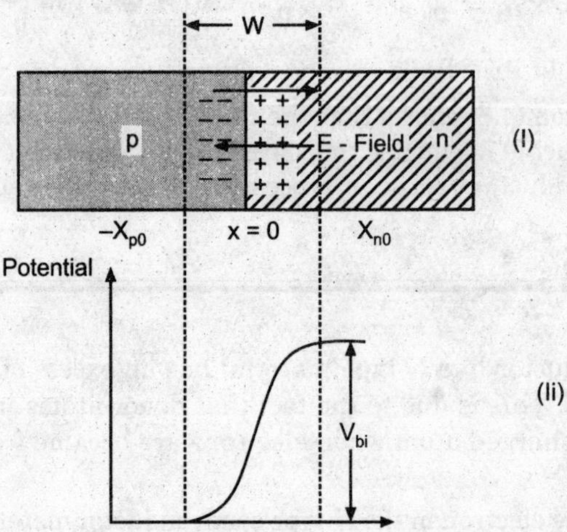

Fig. 4.8. (c) The P-N junction in thermal equilibrium, (i) the space charge region and electric field and (ii) the potential through the junction

(ix) The loss of an electron from the n-type semiconductor material created a positive ion in the N-type material. And similarly the loss of a hole from the P-type semiconductor material created a negative ion in this material. These ions are known as immobile ion because they are fixed in place in the crystal lattice structure and cannot move.

(x) The immobile ions build up a layer of fixed charges on the two sides of the junction is known of depletion layer as shown in figure 4.8(b). There is a layer of positively charged ions the N-side of the junction and there is a layer of negatively charged ions on the P-side of the junction.

(xi) An electrostatic field is developed across the junction between the opposite charged ions as shown by a small battery in the figure 4.8(b). The diffusion of electrons and holes across the junction will continue until the magnitude of the electrostatic field increased to the point where the electrons and holes no longer have enough energy overcome it, and are repelled by the negative and positive ions respectively. At this point equilibrium is established and movement of holes and electrons across the junction stop.

(xii) Hence, the electrostatic field developed by the positive and negative ions at the junction known as barrier potential (shown in figure 4.8(c)). The action just mentioned above take place instantly when P-N junction is formed, only the carriers in the immediate vicinity the junctions are affected. The carriers in the remaining part of the N-type and P-type semiconductor materials are relatively remain undisturbed and maintain the balance condition.

2. Expression for potential barrier (Built in potential barrier V_{bi} or contact potential)

(i) Recall from equation (4.2) that

$$V_{21} = V_T \ln \frac{p_1}{p_2} \qquad ...(4.7)$$

(ii) At the junction there is sudden exchange of charge carriers. Hence, p_p changes to p_n and movement from p to n is considered. Hence, substituting $p_1 = p_p$ and $p_2 = p_n$ in the above equation (4.7), we get

$$V_{21} = V_T \frac{p_p}{p_n}$$

or

$$V_j = V_T \frac{N_A}{p_n} \qquad ...(4.8)$$

where $p_p = N_A$ = concentration of acceptor atoms and V_j is known as **junction potential or barrier potential** or **height of barrier**.

(iii) According to the law of mass action, we have

$$p_n \times n_n = n_i^2$$

$$p_n = \frac{n_i^2}{n_n} = \frac{n_i^2}{N_D}$$

where $n_n = N_D$ = concentration of donor atoms.

(iv) Substituting the value of p_n in equation (4.8), we write as under:

$$V_j = V_T \ln \left[\frac{N_A \cdot N_D}{n_i^2} \right] \text{volts} = V_{bi} \qquad ...(4.9)$$

Here, n_i = intrinsic concentration

and V_T = volt equivalent of temperature

Important Point: In above equation (4.8), V_j reperesents that the height of barrier depends on doping level of N side (N_D) and P side (N_A).

3. Electron concentration

(i) A semiconductor having both donor and acceptor impurity atoms in the same regard is known as compensated semiconductor. It is prepared by diffusing donor impurities into the p-type material ($N_A > N_D$) or by diffusing acceptor impurities into N-type material ($N_D > N_A$).

(ii) We assume $N_D = N_A$, then we obtain a completely compensated semiconductor. Compensated semiconductors are generally produced during semiconductor fabrication. Energy band diagram of a compensated semiconductor is depicted in figure 4.8(d).

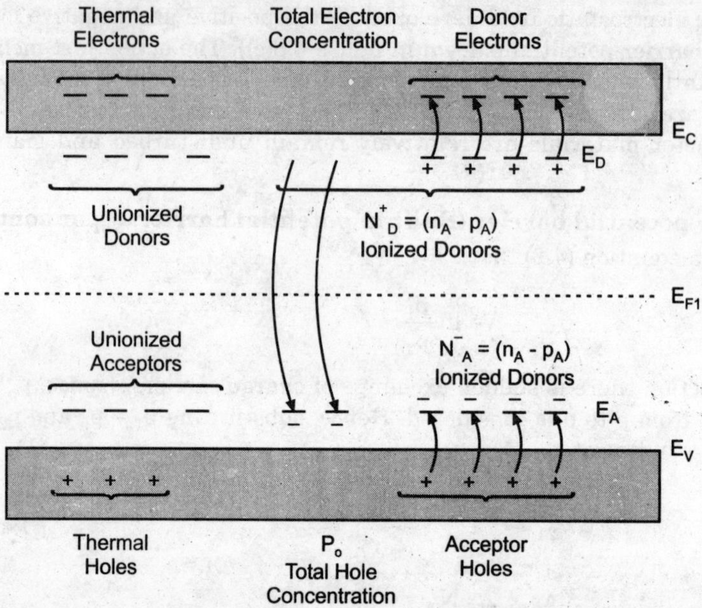

Fig. 4.8. (d) energy band diagram of compensated semiconductor showing ionized and unionized donors and acceptors

(iii) Figure 4.8 (d) represents how the electrons and holes can be distributed among the various states. The charge density of negative and positive charges can be equated for charge neutrality condition and hence, we have

$$n_0 + N_A^- = p_0 + N_D^+$$

or $$n_0 + (N_A - P_A) = p_0 + (N_D - n_D) \quad ...(4.10)$$

where p_0 and n_0 are the concentrations of holes and electrons in the equilibrium condition.

(iv) The parameter P_A in the concentration of holes in the acceptor energy states, hence $N_A^- = N_A - P_A$ is the negatively charged acceptor states concentration.

(v) In a similar manner, n_D is the concentration of electrons in the donor energy states, hence, $N_D^+ = N_D - n_D$ is the positively charged donor states concentration.

(vi) For complete ionization n_D and P_A are zero, so equation (4.10) will reduce to

$$p_0 + N_D = n_0 + N_A \quad ...(4.11)$$

p_o may be expressed as $p_0 = \frac{n_i^2}{n_0}$, then we can write equation (4.11) as

$$\frac{n_i^2}{n_0} + N_D = n_0 + N_A$$

$$n_0^2 - (N_D - N_A)\, n_0 - n_i^2 = 0$$

$$n_0 = \frac{N_D - N_A}{2} + \sqrt{\frac{(N_D - N_A)^2}{4} + n_i^2} \quad ...(4.12)$$

The positive sign in the quadratic equation is required to be used, since, in the limit of an intrinsic semiconductor when $N_A = N_D = 0$, the electron concentration must be a positive quantity or $n_o = n_i$.

Important Point: Equation (4.12) is used to evaluate the electron concentration in an N-type semiconductor or when $N_D > N_A$. Also, equation (4.12) is also valid for $N_A = 0$.

4. Electric field at the depletion region

(i) An electric field is created in the depletion region by the separation of positive and negative space charge densities.

(ii) Figure 4.8 (e) represents the volume charge density distribution in the pn junction assuming uniform doping and an abrupt junction approximation. Consider that the space charge region abruptly junction ends in the n region at $x = + x_n$ and abruptly ends in the p region at $x = - x_p$.

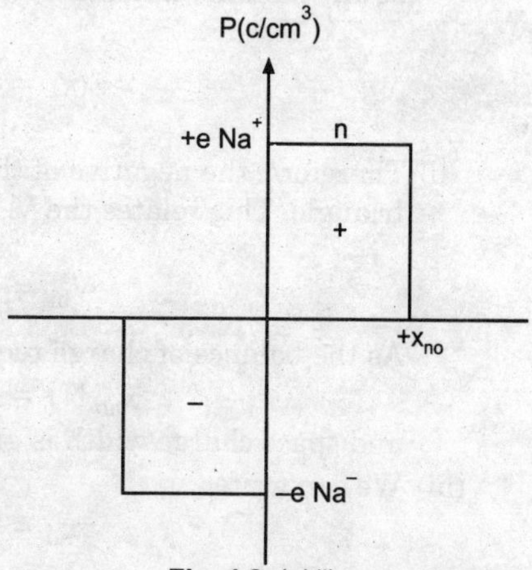

Fig. 4.8. (e)(i)

(iii) To evaluate the electric field distribution within the transition region we start with the point form of Gauss's law, which relates the gradient of the electric field to the local space charge at any point x,

$$\frac{dE(x)}{dx} = \frac{e}{\varepsilon}(p - n + N_d^+ - N_a^-)$$

The above equation is greatly simplified within the depletion region, if we negative the contribution of the carriers (P-N) to the space charge.

(iv) According to this approximation, we have two regions of constant space charge,

$$\frac{dE}{dx} = \frac{e}{\varepsilon} N_d \qquad 0 < x < x_{no}$$

$$\frac{dE}{dx} = -\frac{e}{\varepsilon} \cdot N_a \qquad -x_{po} < x < 0$$

Assuming complete ionization of the impurities $(N_d^+ = N_d)$ and $(N_a^- = N_a)$.

(v) Figure 4.8 (e)(ii) represents the plot of E(x) against x within the transition region which has two slopes, positive on the n side and negative on the p-side. There is so maximum value of the field E_0 at $x = 0$ and E(x) is everywhere negative within transition region.

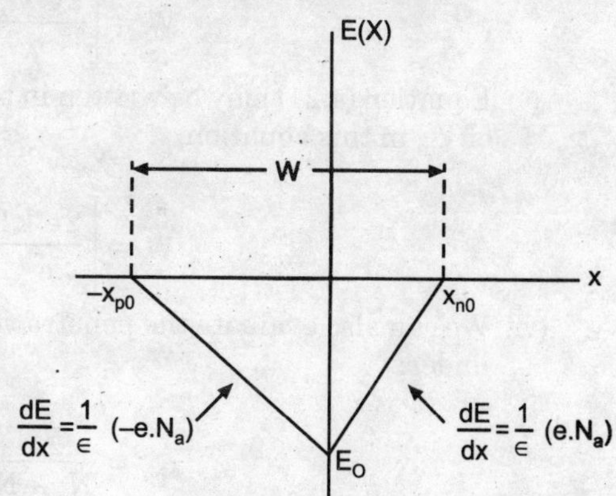

Fig. 4.8. (e) (ii)

(vi) These conclusions have been drawn from Gauss's law The value of E_0 can found by integrating equation (4.13) and (4.14) with appropriate limits.

(vii) Therefore, we can write,

$$\int_{E_0}^{0} dE = \frac{e}{\epsilon} N_d \int_{0}^{x_{n0}} dx \qquad 0 < x < x_{n0} \qquad \text{...(4.15)}$$

and

$$\int_{0}^{E_0} dE = -\frac{e}{\epsilon} N_a \int_{-x_{p0}}^{0} dx, \qquad -x_{po} < x < 0 \qquad \text{...(4.16)}$$

Hence, the maximum value of the electric field is given by

$$E_0 = -\frac{e}{\epsilon} N_d\, x_{no} = -\frac{e}{\epsilon} \cdot N_a\, x_{po} \quad ...(4.17)$$

5. Space charge width (W)

(i) We can relate the electric field to the built in potential barrier V_{bi}, since the electric field E at any distance x is the negative of the potential gradient at that point, i.e.,

$$E(x) = -\frac{dV(x)}{dx} \Rightarrow -V_{bi} = \int_{-x_{po}}^{x_{no}} E(x)\, dx \quad ...(4.18)$$

(ii) Therefore, the negative of the built in potential barrier is simply the area under E(x) Vs. x triangle. This relates the V_{bi} to the width of the depletion region, so

$$V_{bi} = -\frac{1}{2} E_0 W = \frac{1}{2} \cdot \frac{e}{\epsilon} \cdot N_d\, x_{no} \cdot W \quad ...(4.19)$$

As the balance of charge requirement is

$$x_{no} N_d = x_{po} N_a$$

and space charge width is given by $W = x_{po} + x_{no}$,

(iii) We can write

$$x_{no} = W N_a / (N_a + N_d) \text{ in equation (4.19),}$$

We get $$V_{bi} = \frac{1}{2} \cdot \frac{e}{\epsilon} \cdot \frac{N_a \cdot N_d}{N_a + N_d} \cdot W^2 \quad ...(4.20)$$

(iv) After solving, we can write the expression for space-charge width (W) as under:

$$W = \left[\frac{2\epsilon\, V_{bi}}{e} \cdot \left(\frac{N_a + N_d}{N_a N_d}\right)\right]^{\frac{1}{2}} = \left[\frac{2\epsilon\, V_{bi}}{e} \cdot \left(\frac{1}{N_a} + \frac{1}{N_d}\right)\right]^{\frac{1}{2}} \quad ...(4.21)$$

(v) Equation (4.21) may be written in terms of the doping concentrations, by substituting value of V_{bi} in this equation,

$$W = \left[\frac{2\epsilon\, kT}{e^2}\left(\ln \frac{N_a N_d}{n_i^2}\right)\left(\frac{1}{N_a} + \frac{1}{N_d}\right)\right]^{\frac{1}{2}} \quad ...(4.22)$$

(vi) We can also evaluate the penetration of the transition region into the N and P materials as under:

$$x_{po} = \frac{WN_d}{N_a + N_d} = \frac{W}{1 + \frac{N_a}{N_d}} = \left\{\frac{2\epsilon\, V_{bi}}{e}\left[\frac{N_a}{N_a\,(N_a + N_d)}\right]\right\}^{\frac{1}{2}} \quad ...(4.23)$$

$$x_{no} = \frac{WN_a}{N_a + N_d} = \frac{W}{1 + \frac{N_d}{N_a}} = \left\{\frac{2\epsilon\, V_{bi}}{e}\left[\frac{N_a}{N_d\,(N_a + N_d)}\right]\right\}^{\frac{1}{2}} \quad ...(4.24)$$

(vii) Equations (4.23) and (4.24) predict that the transition region extends further into the lighter doping.

(viii) As an example, if $N_a << N_d$, x_{po} is large compared with x_{no}. This matches with qualitative argument that a deep penetration is necessary in lightly doped material uncover the same amount of space charge compared to heavily doped material in which short penetration is required, for the same.

Important Point: To derive the space charge width (W), we consider only the equilibrium condition. However, when we apply external voltage across the pn junction, then this applied voltage can increase or decrease the built in potential across the transition region by aiding or opposing the equilibrium electric field. Thus, equation (4.21) concludes that an applied voltage will increase or decrease the width of the transition region i.e. depletion layer as well.

6. Forward Bias

(i) When an external voltage through a battery is applied to a P-N junction is known as biasing.

(ii) As an example, when battery is connected to a P-N junction for biasing in such a way that its voltage opposes the junction field and reduce the barrier height and helps current to flow across the junction.

(iii) This type of biasing is called as forward biasing and causes the junction to offer minimum resistance to the flow of current.

(iv) The positive terminal of the battery is connected to the P-type material and the negative terminal of the battery is connected to the N-type material is shown in figure 4.9 (a) and steady state minority carrier concentration in P-N junction under forward bias is shown in figure 4.9 (b).

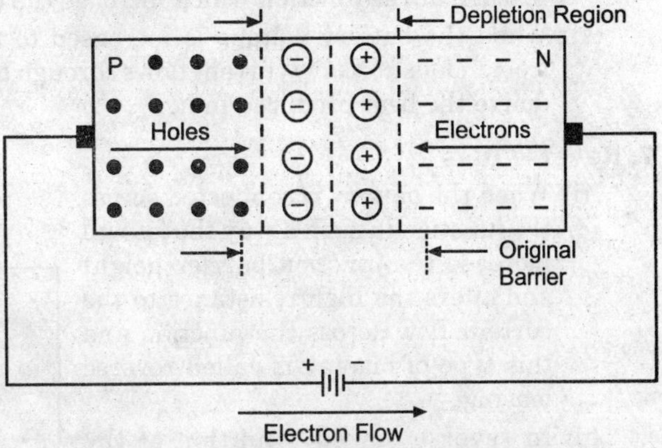

Fig. 4.9. (a) Forward biased p-n junction

(v) The positive potential of the battery repels holes toward the junction where some of the holes neutralize by the negative ions. simultaneously, the negative potential of the battery repels electrons towards the junction where some of the electrons neutralize by the positive ions. As ions on both sides of the barrier are being neutralized, the width of the barrier decreases.

(vi) Hence, the effect of battery voltage in forward bias condition reduces the barrier potential across the junction and allow majority carriers to cross the junction. Current flows in the forward biased P-N junction is maximum and also mechanism is relatively simple.

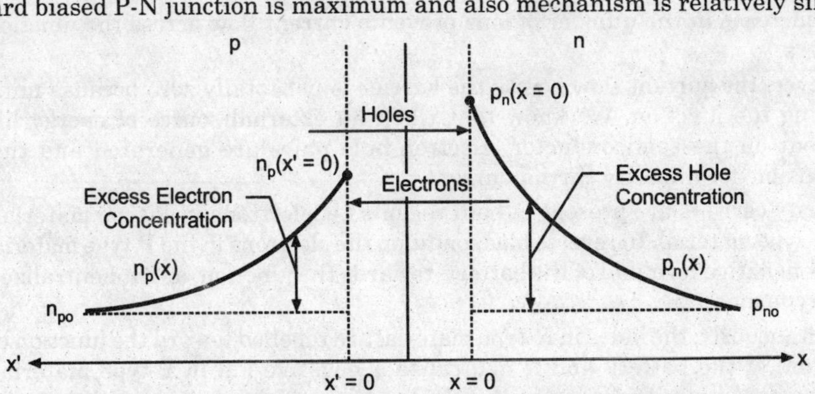

Fig. 4.9. (b) Steady-state minority carrier concentration in a P-N junction under forward bias

(vii) An electron leaves the negative terminal of the battery and moves to the terminal of the N-type material. It enters the N material, where it is the majority carrier and flows to the edge of the junction barrier.

(viii) The junction barrier offers less resistance to the electron at it is forward biased, and electrons pass through the depletion region and reaches into the P-type material. To overcome the resistance of the junction barrier, electron loses its energy and ultimate reaches to the P-type material where it combines with a hole.

(ix) The hole was produced when an electron was extracted from the P-type material by the positive potential of the battery. The created hole moves through the P material toward the junction where it combines with an electron.

(x) Hence, we can conclude that in forward biased condition conduction takes place by majority carriers means holes are mainly responsible for conduction in P-type material and electrons in N-type material. If the battery voltage is increased, then large number of majority carriers reached at the junction which increase the current flow.

(xi) When the battery voltage is increased to the point where the barrier height is reduced tremendously, heavy current flows through the junction and which may damage the junction due to the heat produced in it.

7. Reverse Bias

(i) When the battery is connected across the junction in such a way that it will increase the junction barrier height and offers the high resistance to the current flow across the junction and this type of biasing is called reverse biasing.

(ii) In reverse biased condition of the junction diode, the negative terminal of the battery is connected to the p-type material and positive teminal of the battery is connected to n-type material as shown in figure 4.10(a).

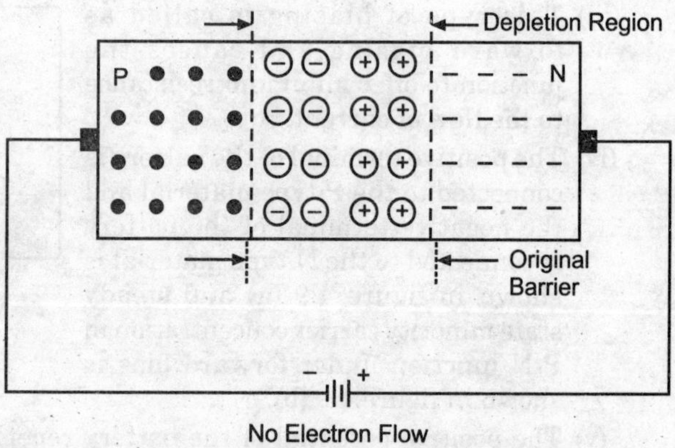

Fig. 4.10 (a) Reverse bias p-n junction

(iii) The negative potential attracts the holes away from the edge of the junction barrier on the p-side, while the positive terminal attracts the electrons away from the edge of the barrier on the N-side. This action increases the barrier width because there are more negative ions on the P-side of the junction and more positive ions on the N-side of the junction as shown in figure 4.10(a).

(iv) This increase in the number of ions prevents current flow across the junction by majority carriers.

(v) However, the current flow across the barrier is not totally zero because minority carriers crossing the junction. We know that when an external source of energy like heat, light incident on the semiconductor, electron-hole pairs are generated and these pairs are responsible for minority carrier current.

(vi) Minority carriers are present in both regions i.e. electrons in P-type material and holes in the N-type material. In reverse bias condition the electrons in the P-type material are repelled by the negative terminal of the battery towards the junction, and it neutralize a positive ion in N-type material.

(vii) Simultaneously, the holes in N-type material are repelled toward the junction by the positive terminal of the battery and it neutralize a negative ion in P-type material. Due to the movement of minority carriers across the junction, which is called as minority carrier current.

(viii) Minority carriers i.e. electron and hole involved in this process come from electron hole pairs which are generated in the crystal lattice structure due to external light energy or heat energy incident on the crystal but not due to the addition of impurity atoms.

(ix) Hence, we can conclude that in reverse biased condition of P-N junction, there is no current flow due to majority carriers but due to minority carriers a small current flow across the junction. This small amount of current may be neglected at normal operating temperature. The point is to be noted in biased P-N junction that in forward bias it offers minimum resistance to current flow and in reverse bias it offers maximum resistance to current flow. Figure 4.10(b) illustrates the energy band diagram of PN junction in reverse bias. As the positive potential is downward, the Fermi level on the N side is below the Fermi level on the P side. The difference between the two is equal to the applied voltage in units of energy. The total potential barrier, indicated by V_{total} has increased. The total potential barrier is now given by

$$V_t = |V_{Fn}| + V_{Fp}| + V_R$$

where V_R is the magnitude of the applied reverse bias voltage. The above equation may be rewritten as under:

$$V_t = V_{bi} + V_R$$

where V_{bi} is the built in potential barrier in thermal equilibrium.

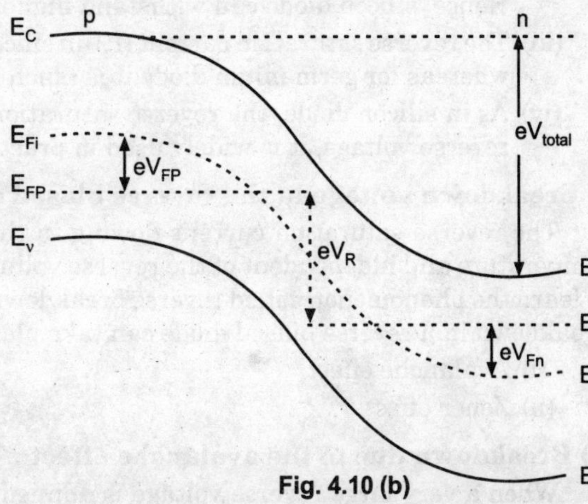

Fig. 4.10 (b)

(x) The total space charge width may be written as in reverse bias condition as under:

$$W = \left\{\frac{2\in(V_{bi}+V_R)}{e}\left[\frac{N_a+N_d}{N_a \cdot N_d}\right]\right\}^{\frac{1}{2}}$$

4.5 VOLTAGE AMPERE (V-I) CHARACTERISTICS OF DIODE

1. V-I Characteristics

The plot of voltage across diode versus diode curent gives V-I characteristics as shown in figure 4.11 for silicon and germanium diodes.

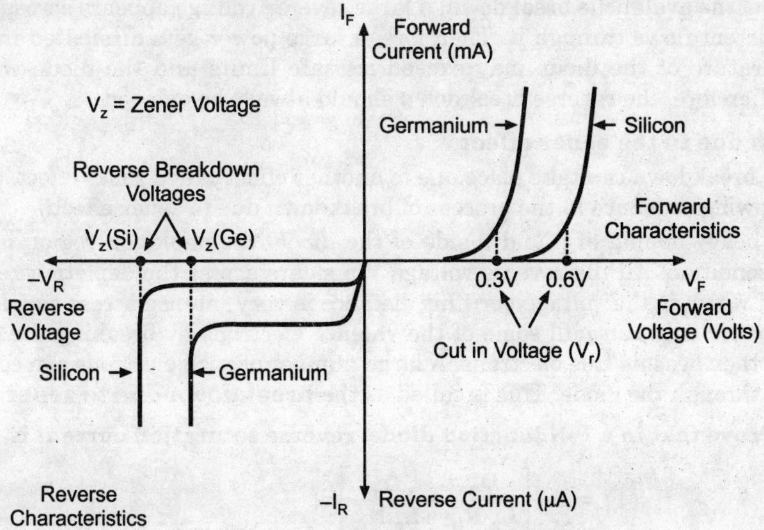

Fig. 4.11

2. Performance comparison between silicon and germanium diode characteristics

(i) The cut in voltage for silicon (Si) diode is 0.7 volts whereas for germanium (Ge) diode it is 0.3 volts.

(ii) The reverse breakdown voltage of silicon diode is greater than that of germanium diode. Hence, silicon diode can withstand higher reverse voltage compared to germanium diode.

(iii) The reverse saturation current (I_o) in silicon diode is very small (of the order of nanoamperes) whereas for germanium diode it is much higher (of the order of microamperes).

(iv) As in silicon diode, the reverse saturation current is very small and it can withstand high reverse voltage, it is widely used in practice.

3. Breakdown voltage in the reverse biased diode

The reverse saturation current flowing in the reverse biased diode is dependent only on the temperature and independent of the reverse voltage applied externally. However, now we are going to learn the phenomenon called reverse breakdown which takes place at a large reverse voltage. The breakdown in a reverse biased diode can take place due to the following effects:

(i) Avalanche effect

(ii) Zener effect

(i) Breakdown due to the avalanche effect

When a very large reverse voltage is applied to a diode the events take place in the following sequence:

Due to large reverse voltage the velocity of the minority carriers will increase to a great extent. Therefore the kinetic energy associated with them will also increase. While travelling these minority carriers will collide with the stationary atoms and impart some of the kinetic energy to the valence electrons present in the covalent bonds. Due to this additionally acquired energy, these valence electrons will break the covalent bonds and jump into the conduction band to become free for conduction. Now these free electrons will be accelerated and they knock out some more valence electrons by means of collisions. This "chain reaction" is called as "Avalanche effect". In a very short time, a large number of free minority electrons will be available for conduction and a large reverse current will flow through the reverse biased diode. The avalanche breakdown has thus taken place.

Why should we avoid breakdown?

At the time of the avalanche breakdown, a large reverse voltage appears across the diode and a large reverse current flows through it. Therefore, a large power gets dissipated in the diode. The junction temperature of the diode may exceed its safe limits and the diode will be damaged permanently. Therefore, the reverse breakdown should always be avoided.

(ii) Breakdown due to the zener effect

The reverse breakdown can take place due to another effect called zener effect. The events take place in the following sequence in the process of breakdown due to zener effect:

Due to the heavy doping of p and n-side of the diode, the depletion region is narrow in the reverse biased condition. All the reverse voltage V appears across the depletion region. Therefore the electric field which is the voltage per unit distance is very intense across the depletion region. This intense electric field can pull some of the valence electrons by breaking the covalent bonds. These electrons then become free electrons. A large number of such electrons can constitute a large reverse current through the diode. This is called as the **breakdown due to zener effect.**

EXAMPLE 4.1. Prove that in a P-N junction diode, reverse saturation current is given by,

$$I_o = Aq\left[\frac{D_p}{L_p N_D} + \frac{D_n}{L_n N_A}\right] n_i^2$$

Solution: As discussed earlier, the total diode current is expressed as under:

$$I = I_{pn}(0) + I_{np}(0)$$

Substituting the values of $I_{pn}(0)$ and $I_{np}(0)$, we obtain,

$$I = \left[\frac{q\,A\,D_p\,p_{no}}{L_p} + \frac{q\,A\,D_n\,n_{po}}{L_n}\right](e^{V/V_T} - 1)$$

But, we have $I = I_o\,[e^{V/V_T} - 1]$

Therefore, we write $$I_o = \left[\frac{q\,AD_p\,p_{no}}{L_p} + \frac{q\,AD_n\,n_{po}}{L_n}\right] = qA\left[\frac{D_p\,p_{no}}{L_p} + \frac{D_n\,n_{po}}{L_n}\right] \quad ...(i)$$

Using the law of mass action, we can write as under:

$$P_{no} = \frac{n_i^2}{N_D} \text{ and } n_{po} = \frac{n_i^2}{N_A}$$

Substituting these values into equation (i), we obtain

$$I_o = qA\left[\frac{D_p}{L_p\,N_D} + \frac{D_n}{L_n\,N_A}\right]n_i^2$$

This is the required result.

4.6. DIODE CHARACTERISTICS EQUATION

As a matter of fact, the V-I characteristics of semiconductor diode can be mathematically represented as under:

$$I = I_o\left(\frac{V}{e^{\eta V_T} - 1}\right)$$

Here, I = forward current through the diode

I_o = reverse saturation current

V = applied voltage

V_T = voltage equivalent of temperature

η = ideality factor or emission coefficient

($\eta = 1$ for Ge and $\eta = 2$ for Si).

In fact, η represents recombination effect which takes place in depletion region.

$$V_T = KT$$

Here, k = Boltzmann's constant = 8.62×10^{-5} eV/°K

T = Temperature in degree kelvin (°K)

At room temperature (27°C or 273 + 27 = 300 °K), we have

$$V_T = 8.62 \times 10^{-5} \times 300$$

$$V_T = 26 \text{ mV}$$

Reverse saturation current (I_o) and V_T are temperature dependent, therefore, the entire characteristics of diode is temperature dependent. The applied voltage (V) is positive for forward bias and is negative for reverse bias. For large forward bias, $\frac{V}{e^{\eta V_T}} >> 1$, therefore, we have

$$I = I_o\,e^{\frac{V}{nV_T}}$$

This shows that forward current (I) increases exponentially.

For reverse bias, $\frac{V}{e^{\eta V_T}} << 1$, therefore, we have

$$I = -I_o$$

Important Point: Negative sign implies that the current is in opposite direction to that of conventional current.

4.7. EXPRESSION FOR THE DYNAMIC RESISTANCE (r)

Basically the dynamic resistance of a diode is the reciprocal of the slope of the volt-ampere characteristics of a diode i.e.,

$$r = \frac{1}{\text{slope of graph}} = \frac{1}{\left(\frac{dI}{dV}\right)}$$

We know that the characteristics of a diode is expressed as under:

$$I = I_o\,(e^{V/\eta V_T} - 1)$$

Differentiating w.r.t. V, we obtain

$$\frac{dI}{dV} = I_o\left[\frac{1}{\eta V_T}\,e^{V/\eta V_T}\right]$$

or

$$\frac{dI}{dV} = \frac{I_o\,e^{V/\eta V_T}}{\eta V_T}$$

According to the definition of dynamic resistance, we have

$$r = \frac{1}{\left(\frac{dI}{dV}\right)} = \frac{\eta V_T}{I_o\,e^{V/\eta V_T}} = \frac{\eta V_T}{I_o\,e^{V/\eta V_T}}$$

Also from the equation of current, we obtain

$$I_o\,(e^{V/\eta V_{T-1}}) = I_o\,e^{V/\eta V_T} - I_0$$

$$I_o\,e^{V/\eta V_T} = I + I_o$$

Substituting in equation of 'r', we obtain

$$r = \frac{\eta V_T}{I + I_o}$$

Important Point: If diodes is forward biased then V must be positive and if diodes is reverse biased then V must be negative.

4.8. TEMPERATURE DEPENDENCE OF V-I CHARACTERISTICS OF DIODE

1. Basic Concepts

We know that the diode characteristics are given by

$$I = I_o(e^{V/\eta V_{T-1}})$$

where I_o = reverse saturation current and depends on temperature and V_T is volt equivalent of temperature is also dependent on temperature.

Because of these two parameters, whole diode characteristics are temperature dependent.

If temperature increases, the reverse saturation current (I_o) increases which further increases the diode current. Just to keep diode current constant, it is needed to reduce applied forward voltage, i.e., $\frac{dI}{dT} = 0$.

The required rate of reduction in applied voltage to maintain constant diode current is expressed as

$$\frac{dV}{dT} = \frac{V - [V_{GO} - m\eta V_T]}{T}$$

Here, V_{GO} = Forbidden energy gap
= 0.785 V for Ge and 1.21 V for Si
m = 2 for Ge and 1.5 for Si

2. Illustrative examples for silicon and Germanium Diodes

(i) For silicon diode

$$V = V_r = 0.6\text{ V}, V_{GO} = 1.21\text{ V}, m = 1.5, V_T = 26\text{ mv}$$
$$\eta = 2 \text{ and } T = 300°K$$

$$\frac{dV}{dT} = \frac{0.6 - (1.21 + 1.5 \times 2 \times 26\text{ mV})}{300}$$

$$\frac{dV}{dT} = -2.3\text{ mV/°C}$$

This shows that for silicon diode, the voltage must be reduced at a rate of 2.3 mV/°C in order to maintain constant diode current.

(ii) For germanium diode

$V = V_r = 0.2$ V, $V_{GO} = 0.785$ V, m = 2, $V_T = 26$ mV
η = 1 and T = 300 °K

$$\frac{dV}{dT} = \frac{0.2 - (0.785 \times 2 \times 1 \times 26\text{ mV})}{300}$$

$$\frac{dV}{dT} = -2.12\text{ mV/ °C}$$

i.e. for germanium diode, the voltage must be reduced at a rate of 2.12 mV/°C in order to maintain constant diode current.

Important Point: Although dV/dT values are slightly different for Si and Ge, in practice the voltage being reduced at a rate of 2.5 mV/°C for both silicon and germanium diodes.

4.8.1. Effect of Temperature on Reverse Saturation Current

(i) As temperature increases, the generation of electron-hole pairs in semiconductor material increases. It also increases the conductivity. Because of this, the current through the diode increases with temperature.

(ii) The reverse saturation current (I_o) of p-n junction diode increases approximate 7% per °C for both silicon and germanium i.e. if at certain temperature T°C, I_o is 1 μA that at 1°C rise (T + 1)°C it becomes 1.07 μA. Then for 10°C rise we get $(1.07)^{10} = 1.96 \cong 2$ μA. This shows that I_o becomes double for every 10°C rise temperature.

(iii) The mathemaical expression for this is given by

$$I_{o2} = I_{o1}\, 2^{\frac{T_2 - T_1}{10}}$$

where I_{o1} = saturation current at temperature T_1

I_{o2} = saturation current at temperature T_2

(iv) Figure 4.12 illustrates the effect of increased temperature on the characteristics curve of a diode.

(v) Figure 4.12 shows the rise in temperature results in decrease in forward voltage across the diode and increase in forward current during forward bias whereas the reverse saturation current increases during reverse bias.

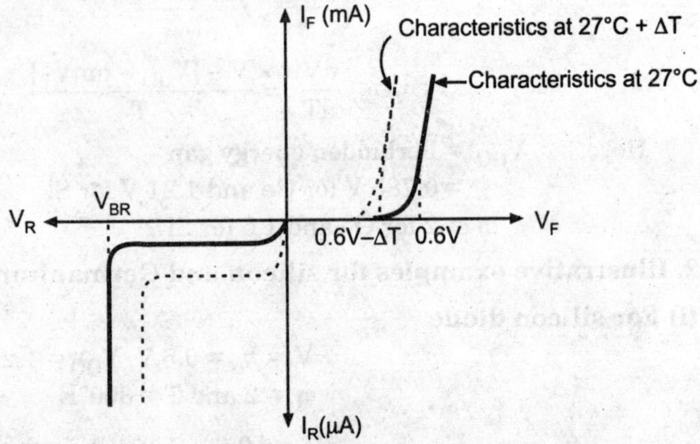

Fig. 4.12.

4.8.2. Zener Diode

1. Basics

(i) Zener diode is also known as the breakdown diode, is a P-N junction diode specially designed for the operation in the breakdown region in the reverse biased condition.

(ii) The diode may use either avalanche breakdown or zener breakdown mechanism. Despite the fact that avalanche and zener constitute two types of breakdown diodes, the name zener is commonly applied to both types of diodes.

(iii) The zener diode may be prepared from Si or Ge but Si is preferred because of its higher operating temperature and current capability. The knee point is also more sharp in case of Si diode.

(iv) The symbol of a zener diode has been shown in figure 4.13.

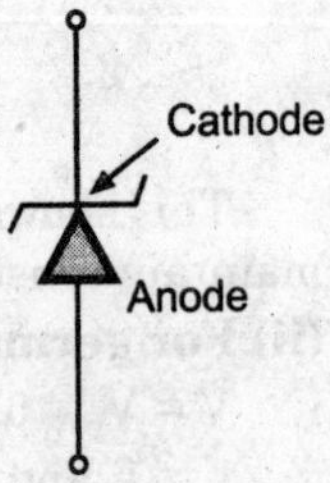

Fig. 4.13.

2. Zener Diode Construction

(i) Zener diode is like an ordinary P-N junction diode except that it is properly doped so as to have sharp breakdown voltage.

(ii) There are different methods available to manufacture the zener diodes, such as diffused structure, diffused and passivated structure and alloy diffused structure.

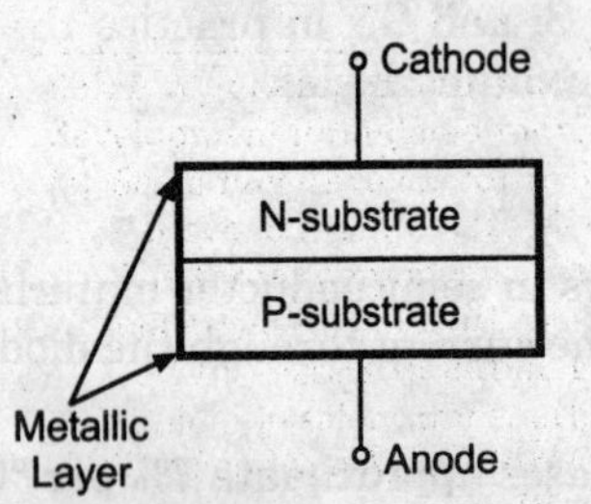

Fig. 4.14. Diffused structure of a zener diode

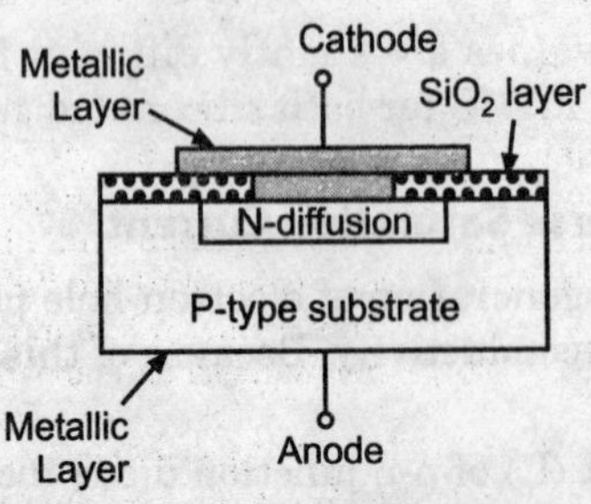

Fig. 4.15. Diffused and passivated structure of a zener diode

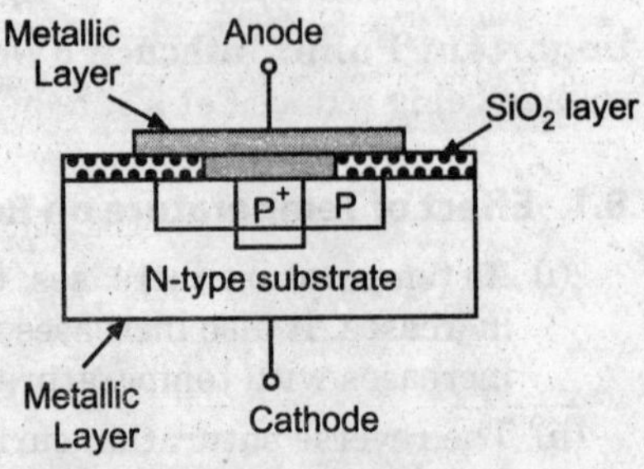

Fig. 4.16. Alloy diffused structure of a zener diode

(iii) Diffused structure zener diode consists of two N and P substrates diffused together and has metallic layers deposited on both sides to connect anode and cathode terminals as shown in figure 4.14. The drawback of such a construction is that the edges of the junction are exposed to contamination.

(iv) In passivated structure of a zener diode, the edges of the junction are covered by layer of SiO_2 as shown in figure 4.15.

(v) In a similar manner, all junctions are protected in alloy diffused structures as shown in figure 4.16. In general, alloy diffused structure provides better zener characteristics at lower zener voltages. On the contrary diffused and passivated zener diode is used at higher zener voltages.

4.8.3. Zener Diode Applications

1. Zener diode voltage regulation

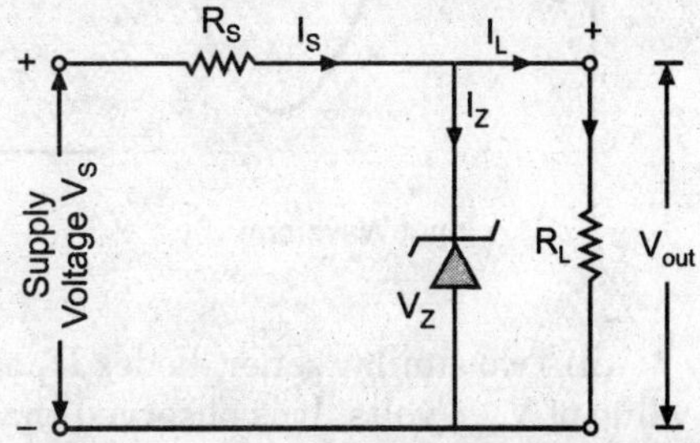

Fig. 4.17

(i) The circuit diagram shown in figure 4.17 represents how a zener diode can be used as a voltage regulator to provide a constant voltage from a source whose voltage may vary appreciably.

(ii) A resistance R_s is connected to limit the reverse current through the diode to protect the same. The source voltage V_s and the resistance R_s are selected in such a way that the diode operates in the breakdown region.

(iii) The diode voltage in this region is known as zener voltage which is also the voltage across the load (V_z), and at this stage the diode current is known as zener current I_Z.

(iv) The resistance connnected in the series R_s absorbs the output voltage fluctuations so it is possible to maintain voltage across the load constant. Zener diode is connected in reverse bias across the input voltage whose variations are to be regulated.

(v) The zener diode does not conduct as long as the voltage across the load resistance R_L is less than the breakdown voltage V_z, so the resistors R_s and R_L constitute a potential divider across V_s. At an increased supply voltage V_s, the voltage drop across load resistor becomes greater than the zener brakdown voltage.

(vi) The diode then operates in its breakdown region. The series resistance R_s controls the zener current I_z so it must not exceed its rated maximum $(I_z)_{max}$.

(vii) Current flows through the resistance R_s is given by

$$I_s = \frac{V_s - V_z}{R_s}$$

(viii) The current from the source divides at the junction of zener diode and the load resistor R_L. Hence,

$$I_s = I_z + I_L$$

(ix) When the zener diode operates in its breakdown region, the voltage across it (V_z) remains constant even though the current I_z flowing through it may vary appreciably.

(x) If the supply voltage V_s increases, the current through both the zener diode and load resistance R_L increases. At the same time, however, the zener diode resistance decreases and the current through the diode increases more than proportionately.

(xi) Therefore, a greater voltage drop takes place across the series resistance R_s and the output voltage become very close to the original value and vice versa.

(xii) Hence, a zener diode can maintain the output voltage V_{out} within a fraction of a volt when the supply voltage V_s may vary over a range of several volts. The zener diode maintains a constant voltage across the load as long as the supply voltage is more than the zener voltage.

2. Zener diode as a peak clipper

(i) Zener diode may be used clipping off the input waveform in wave shaping circuits as depicted in figure 4.18.

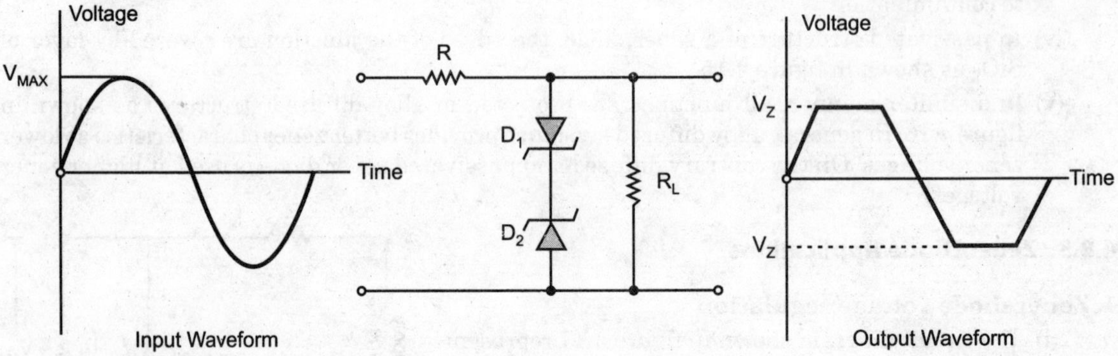

Fig. 4.18.

(ii) Two similar zener diodes D_1 and D_2 are shunted across the input sinusoidal voltage of peak value of V_{max} volts. It is observed that a zener diode acts like a short i.e. very low resistance when forward biased and open i.e. very high resistance when reverse biased till it goes into breakdown at V_z (zener voltage).

(iii) For positive input half cycle diode D_1 is forward biased and acts like a short while diode D_2 being reverse biased acts like an open upto zener voltage V_z. Hence, diode D_2 operates in breakdown and maintain the output voltage constant at V_z till input voltage falls below V_z in the latter part of the positive half cycle. This is due to the fact that during this duration of time both of the diodes acts as short.

(iv) If the supply voltage falls below V_z, diode D_2 will be out of the breakdown and again acts as an open across the load circuit. During the negative input half cycle, the operations of diodes D_1 and D_2 are reversed. As a result, the output voltage waveform is clipped off on both peaks as depicted in figure 4.18.

4.9. JUNCTION CAPACITANCES **(Important)**

As a matter of fact, there are following two types of capacitances associated with P-N junction

(i) Transition capacitance

(ii) Diffusion capacitance

Both these capacitances are important and must be taken into consideration while designing a P-N junction device for time varying signals.

The transition capacitance is dominant under reverse bias condition of the junction. The diffusion capacitance is dominant when the junction is forward biased.

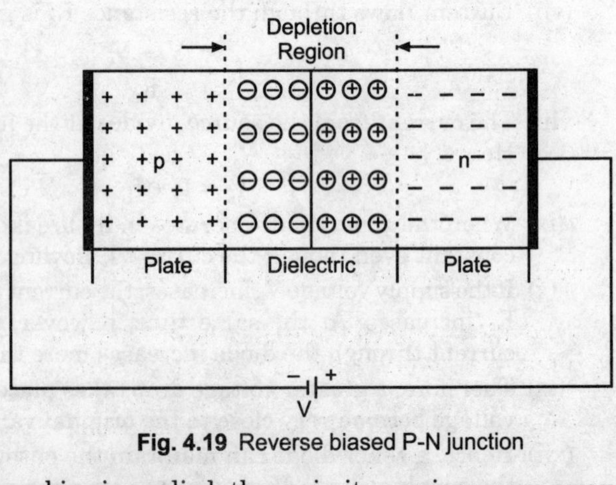

Fig. 4.19 Reverse biased P-N junction

4.9.1. Transition Capacitance

1. Basics

Let us consider a reverse biased P-N junction as depicted in figure 4.19. When reverse bias is applied, the majority carriers move away from the junction and width of the depletion region increases.

2. Representation

The movement of majority carriers gives change in charge at the junction due to applied reverse voltage. This change in charge because of applied reverse voltage is known as transition capacitance, barrier capacitance, junction capacitance or space charge capacitance. It is represented by C_T.

$$C_T = \frac{\text{Rate of change of charge at the junction}}{\text{Rate of change of applied reverse voltage}}$$

$$C_T = \frac{dQ}{dV}$$

Important Point: The depletion region acts as a dielectric while the P and N regions act as plates of capacitor.

3. Expression

It is given by,

$$C_T = \frac{\epsilon A}{W}$$

Here, A = area of cross-section of junction
W = width of depletion region
ϵ = permittivity of semiconductor
$\epsilon = \epsilon_0 \epsilon_r$
ϵ_r = relative permittivity of semiconductor
= 16 for Ge, 12 for Si

$$\epsilon_0 = \frac{1}{36\pi \times 10^9} = 8.84 \times 10^{-12} \text{ F/m}$$

The relationship between width of depletion region and barrier potential may be expressed as under:

$$V_B = \frac{qNW^2}{2\epsilon} \quad \text{when } N = N_A \text{ or } N_D$$

$$V_B = V_J - V$$

where V_J = junction potential
V = applied voltage

Important Point: The transition capacitance is inversely proportional to width of depletion region i.e. if reverse voltage increases, the value of transition capacitance decreases.

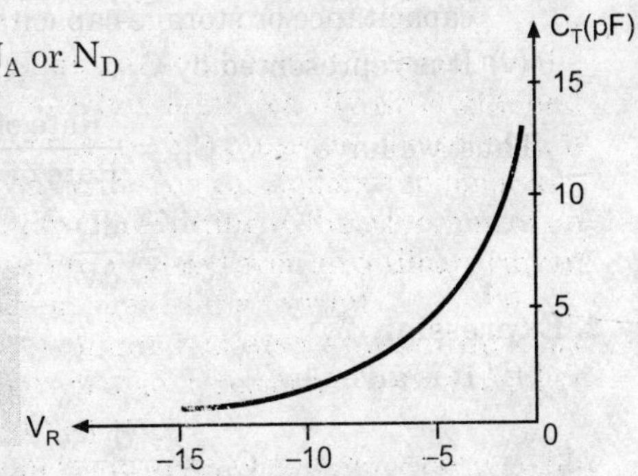

Fig. 4.20.

Figure 4.20 illustrates the transition capacitance versus applied reverse bias for a typical silicon diode.

4.9.2. Application of Transition Capacitance

(i) As discussed earlier, the transition capacitance is dominant under reverse bias condition and it may be varied simply by changing applied reverse voltage.

(ii) This property of the P-N junction is used to manufacture special purpose diode, in which the capacitance is varied by applied reverse voltage. Such a diode is knowns as a varactor diode or varicap or voltacap. Figure 4.21 illustrates the circuit symbol of a varactor diode.

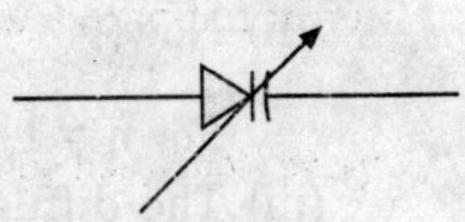
Fig. 4.21.

(iii) In varactor diode, the capacitance parameter is controlled by the technique of doping near the P-N and size and geometry of the diode construction. The varactor capacitance varies from few picofarads to several hundred picofarads.

(iv) Because of variable capacitance versus voltage characteristics of P-N junction, it is useful for electrically tunable resonant circuits.

(v) Figure 4.22 shows the use of a varactor diode in an LC resonant circuit.

(vi) The resonant frequency is given by,

$$f_r = \frac{1}{2\pi\sqrt{L(C + C_T)}}$$

(vii) From the above expression, it is obvious that by changing the C_T by varying applied voltage, the resonant frequency may be changed.

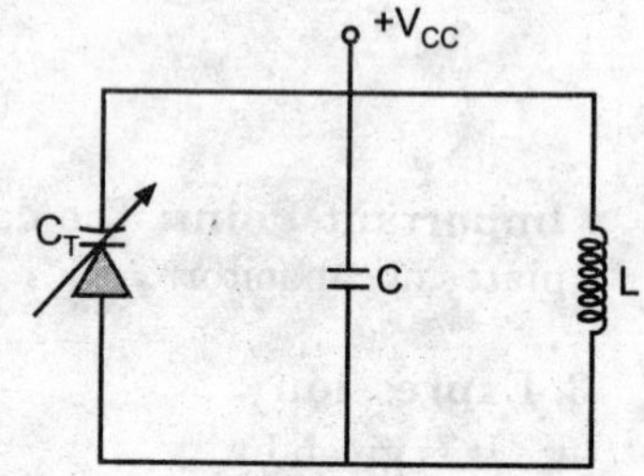

Fig. 4.22.

4.9.3. Diffusion Capacitance

1. Basics

(i) Let us consider a forward biased P-N junction as depicted in figure 4.23.

(ii) When forward bias is applied, holes get diffuse from the P side to the N side and electrons get diffuse from the N side to P side. The amount of excess charge is established by the degree of forward bias.

(iii) The movement of majority carriers towards the junction give change in charge at the junction due to applied forward voltage.

(iv) This change in charge due to applied forward voltage is called as diffusion capacitance or storage capacitance.

(v) It is represented by C_D.

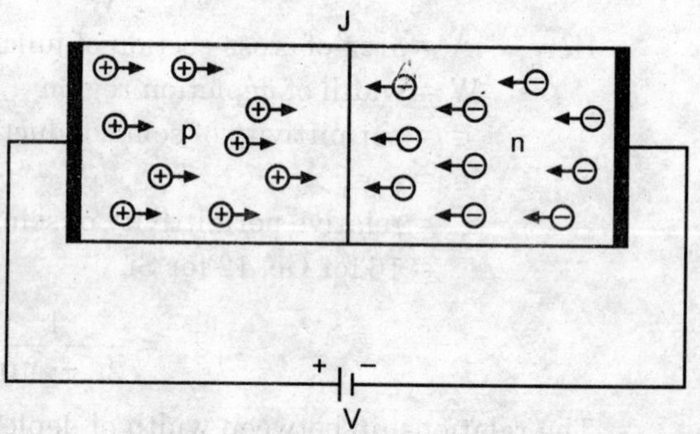

Fig. 4.23

Thus, we have $$C_D = \frac{\text{Rate of change of charge at the junction}}{\text{Rate of change of applied forward voltage}}$$

$$C_D = \frac{dQ}{dV}$$

2. Expression

(i) It is given by,

$$C_D = \frac{\tau I}{\eta V_T}$$

Here, τ = mean carrier life time

τ is measure of the recombination time for excess minority carriers.

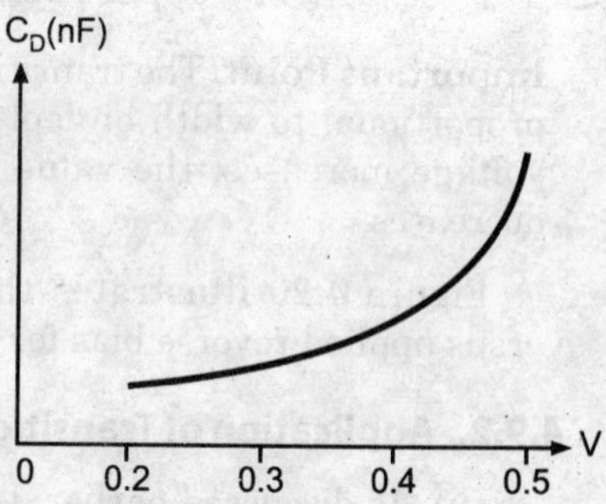

Fig. 4.24.

(ii) Now, $$C_D = \frac{\tau}{\eta r_f}$$

For $\eta = 1$, $\tau = r_f C_D$, carrier life time can be taken as diffusion time constant.

(iii) The diffusion capacitance C_D appears in parallel with forward resistance (r_f). The time constant which is a function of C_D and r_f is very small. However, at operating frequencies, its effect must be considered.

3. Diffusion capacitance versus applied forward voltage

Figure 4.24 illustrates the diffusion capacitance versus applied forward voltage.

4.10. DIODE SWITCHING TIME

1. Basics

(i) Electronic devices are built with P-N junctions. A diode as we have seen is a single junction structure. The junction properties lead to many fundamental applications.

(ii) The junction properties are
- (a) rectification
- (b) variable capacitance
- (c) tunnelling and
- (d) light emission and detection.

The diodes are designed to exploit specific properties of the junction.

(iii) For a diode, it is important that reverse-bias current and forward bias power should be minimum. The time response is also important in some applications.

(iv) Special considerations must be given to the charge control properties of a junction diode so that it can be switched rapidly between conducting state and non-conducting state.

(v) The equations governing turn-on time and reverse recovery time of a P-N junction are

$$i(t) = \frac{Q_p(t)}{\tau_p} + \frac{d\,Q_p(t)}{dt} \qquad ...(4.25)$$

Here, $i(t)$ = Total diode current
= Hole current injected across P-N junction

$\frac{Q_p}{\tau_p}$ = Recombination current

$\frac{dQ(t)}{dt}$ = Rate of charge build up (storage of charge) across junction
= 0 for steady state condition

τ_p = Mean life time of holes in 'n' type material

(vi) We have

$$t_{sd} = \tau_p \left[\text{erf}^{-1} \left(\frac{I_f}{I_f + I_r} \right) \right]^2 \qquad ...(4.26)$$

Here, erf = Error function (a tabulated function)

τ_p = Mean life time of holes in 'n' type material

I_f = Steady state forward current through the diode

I_r = Initial value of reverse current when the diode enters non-conducting state

t_{sd} = Storage delay time
= Time required for stored charge to become zero.

(vii) From above two expressions, it is obvious that a diode with fast switching properties must
- (i) Store very little charge in the neutral regions under steady forward current
- (ii) Have a very short carrier life time τ_p

2. Time variation of stored charge

(i) This can lead to an important implication that even if the forward current through the diode is suddenly terminated, voltage across the P-N junction still persists until injected hole charge Q_p disappears.

(ii) The voltage across the P-N junction cannot be changed instantaneously. The stored charge presents a problem in a diode intended for switching application.

(iii) The junction voltage varies according to the following expression:

$$v(t) = \frac{kT}{q} \ln\left[\frac{I\tau_p}{qA\, L_p p_n} e^{-t/\tau p} + 1\right] \qquad ...(4.27)$$

Here, $v(t)$ = voltage across junction
I = forward current through the junction just before switch off
τ_p = hold carrier life-time
L_p = diffusion length of holes
A = cross-section area
P_n = concentration of holes in 'n' region

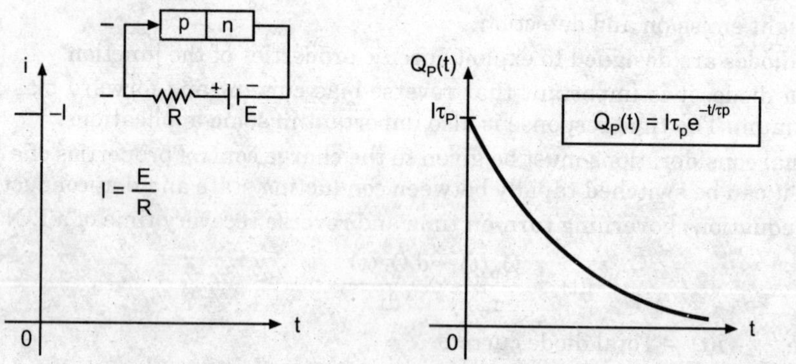

(a) Diode forward current (b) Decay of stored charge in 'n' region

Fig. 4.25. Effect of turn-off transient

(iv) Problems of stored charge are reduced by designing a diode with very narrow 'N' region. If the 'N' region is shorter than the hole diffusion length (L_p) then very little charge is stored. Therefore, the time required to switch the diode ON and OFF is very small. This type of structure is called '**Narrow Base' diode**.

(v) The switching process can be made faster by adding recombination centres. The recombination centres can be created by adding impurities such as Gold (Au) to the semiconductor material (Si).

Important Point: The recombination centres act like metals provide additional electrons for recombination of holes injected. This gives rise to higher recombination rate. Hence, switching time is reduced.

4.10.1. Reverse Recovery Transient

(i) The diode is switched from forward conduction state to a reverse biased state in most of the applications. Figure 4.26 illustrates a diode circuit, its input and output waveforms.

(ii) Equation (4.28) comes from the analsis of figure 4.26.

$$t_{sd} = \tau_p \left[\operatorname{erf}^{-1}\left(\frac{I_f}{I_f + I_r}\right)\right]^2 \qquad ...(4.28)$$

Important Point: It is desirable that t_{sd} is small as compared to switching times needed in the application. Therefore, t_{sd} is an important parameter in evaluating diodes for switching applications.

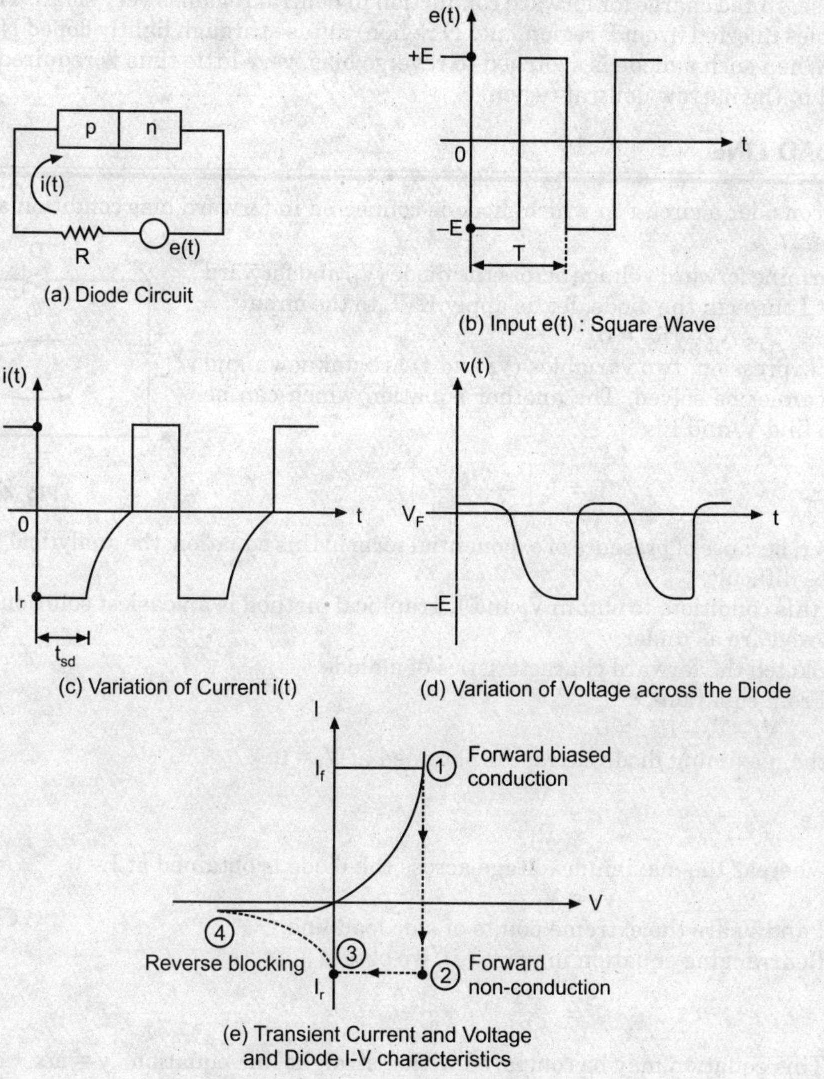

Fig. 4.26. Storage delay time in a p-n diode

4.10.2. Techniques to Improve Switching Speed of a Diode

(i) As discussed earlier, a common technique is to add recombination centres. For silicon diodes, gold (Au) doping is done for this purpose. The carrier life time varies inversely proportional to the concentration of recombination centres.

Illustrative Example

For a silicon, diode has τ_p = 1 μsec and reverse recovery time (t_{sd}) of 0.1 μsec before gold (Au) doping.

The addition of 10^{14} gold atoms/cm^3 reduces τ_p to μsec and t_{sd} to 0.01 μsec. The addition of 10^{15} gold atoms/cm^3 can reduce τ_p to 0.01 n sec and τ_{sd} to 1 nsec. However, this process cannot be continued as reverse current due to gold starts increasing appreciably.

(ii) The second approach as discussed earlier is make the 'n' region (lightly doped) shorter than the minority carrier (hole) diffusion length (L_p). This is a narrow base diode.

In this case, stored charge for forward conduction in neutral region is very small. This is because most of the holes injected (from P region, into N region) diffuse through lightly doped N region to the end contact. When such a diode is switched to reverse bias, very little time is required to eliminate charge stored in the narrow neutral region.

4.11. DC LOAD LINE

1. Let us consider a circuit in which diode is connected in forward bias condition as depicted in figure 4.27.
2. To determine forward voltage across the diode (V_f) and forward current I through the diode, let us apply KVL to the circuit.
   $$V_i - V_f - IR_L = 0$$
3. In this expression, two variables (V_f and I) are unknown and hence cannot be solved. The another equation which can be used to find V_f and I is
   $$I = I_o\left(\frac{V_f}{e^{\eta V_T} - 1}\right)$$

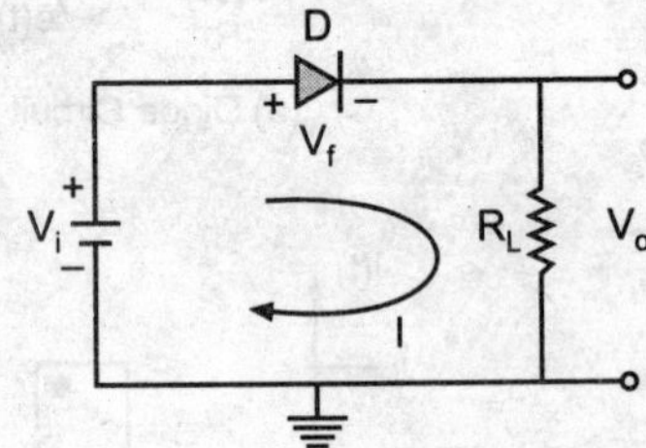

Fig. 4.27.

4. However, because of presence of exponential term in this equation, the analytical simplification becomes difficult.
   Under this condition, to obtain V_f and I, graphical method is an easiest solution. The steps to be followed are as under:
   (i) Sketch the forward characteristics of a diode.
   (ii) From equation
   $$V_i - V_f - IR_L = 0$$
   the maximum diode current is obtained at $V_f = 0$.
   i.e. $$I = \frac{V_i}{R_L}$$
   whereas the maximum voltage across the diode is obtained at I = 0
   i.e., $$V_f = V_i$$
   I and V_f are the extreme points of a dc load line.
   (iii) Rearranging equation in point (ii), we obtain
   $$I = -\frac{1}{R_L} V_f + \frac{V_i}{R_L}$$
   This equation may be compared with a straight line equation $y = mx + c$
   On comparing, we obtain
   Y axis parameter as current I, X parameter as voltage V_f,
   Slope of the line (m) $= -\frac{1}{R_L}$
   It is desirable that t_{sd} is small as compared to switching times required in the application. Therefore, t_{sd} is an important parameter in evaluating diodes for switching applications.
   Y intercept at $\frac{V_i}{R_L}$
   (iv) Figure 4.28 illustrates the forward characteristics and dc load line of diode.

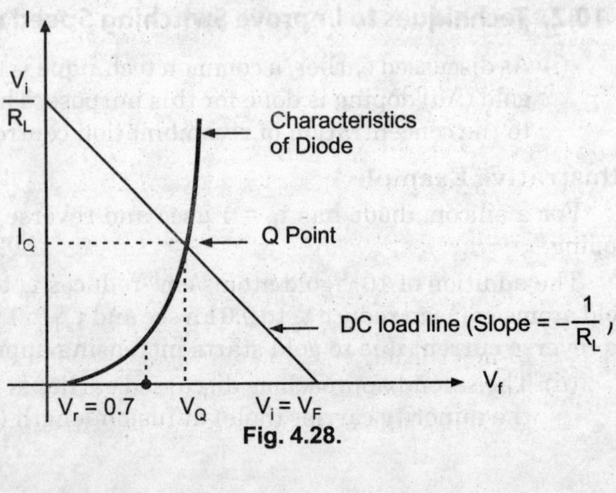

Fig. 4.28.

(v) The point of intersection of dc load line and V-I characteristics of a diode is known as quiescent (Q) or operating or dc bias point.

(vi) Q point indicates the dc voltage across the diode and dc forward current through the diode.

(vii) The slope of dc load line is reciprocal of load resistance and can be adjusted appropriately by varying load resistance (R_L).

4.12. JUNCTION PROPERTIES

(U.P. Tech, Sem. Exam; 2008-09, 2010-11)

4.12.1. Metal-metal junctions

1. Basic Concepts

(i) Figure 4.29 (a) shows the energy-band models of two dissimilar metals at absolute. The work that must be done to remove an electron that is inside the crystal to a distance far removed from the crystal is known as the **work function Φ of that crystal**.

(ii) At absolute zero, this can be represented in the band model by the energy required to raise an electron from the Fermi energy to the so-called vacuum level. When these two metals are brought in intimate contact with each other, some of the electrons in metal 1, shown to the left in figure 4.29(a), occupy quantum states that have larger energies than those of unoccupied states in metal 2. The electrons near the Fermi level in metal 1, therefore flow into metal 2 until at equilibrium the Fermi energy in both metals is at the same energy level.

(iii) In the process, the surface of metal 2 becomes negatively charged while the surface of metal 1 becomes positively charged, as indicated in figure 4.29(b).

(iv) This produces a potential difference between the two metals called the contact potential V determined by

$$eV = \Phi_2 - \Phi_1 \quad \text{...(4.29)}$$

(v) Because a local electric field cannot exist inside a metal because of its high conductivity, the change in the free-electron distribution in the two metals that produced the potential in equation (4.29) takes place on the contact surface between them.

(vi) It may also be noted that it is possible to measure the difference in the work function of the two metals according to equation (4.29) by measuring the contact potential.

(vii) When a potential is applied across the metal-metal junction so that a current may flow through it, heat is generated (or absorbed) at the junction in addition to the so-called Joule heat (I^2R) normally produced. The generated heat is proportional to the current, and the phenomenon is called the Peltier effect. If current flow from metal 1 to metal 2 generates heat, the corresponding Peltier coefficient is said to be positive.

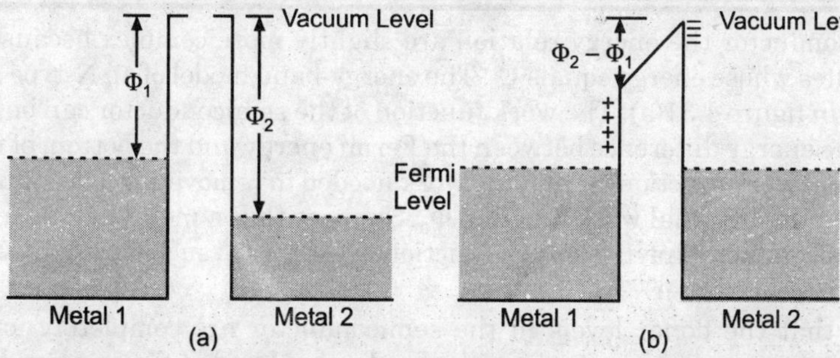

Fig. 4.29.

(viii) When the current between these two metals is reversed, the Peltier coefficient changes its sign and the reverse current absorbs heat.

(ix) It is possible to take advantage of this effect in still another way. Let us consider that the two sides of the junction are maintained at different temperatures. The electrons in the hot metal gain thermal energy so that they occupy higher quantum states.

(x) At equilibrium, the Fermi energy is at the same level in both metals so that some of the electrons must flow from the hot metal to the cold one. This produces a thermoelectric emf at the contact, and this phenomenon is known as the **Seeback**

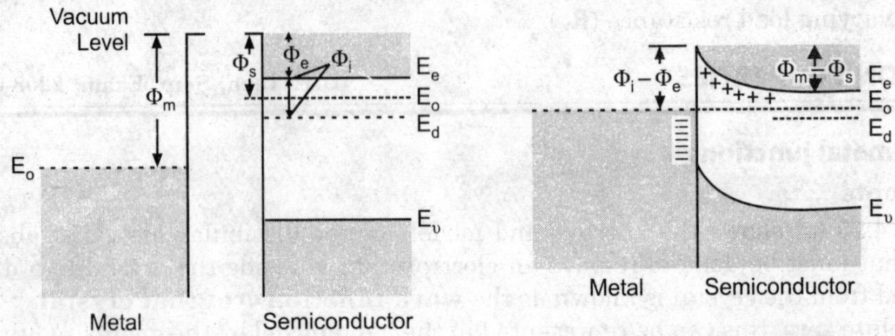

(a) Before contact is made (b) After contact is made

Fig. 4.30. Energy-level diagrams of metal n-type semiconductor contact with $\Phi_m > \Phi_s$

2. Effect

Thus, it is obvious that the thermoelectric emf produced at the metal-metal contact may be used to measure temperature; this is, of course, the explanation of how a thermocouple operates. In a similar manner, the Peltier effect may be used to heat or cool the metal-metal junction by passing an electric current through it.

Important Point: This is the principle on which the so-called electronic refrigerator is based.

Schottky Barriers

(i) We have already discussed the work function $q\Phi_m$ of a metal in a vacuum. An energy of $q\Phi_m$ is required to remove an electron at the Fermi level to the vacuum outside the metal. Typical values of Φ_m for very clean surfaces are 4.3 V for Al and 4.8 V for Au.

(ii) When negative charges are brought near the metal surface, positive (image) charges are induced in the metal. When this image force is combined with an applied electric field, the effective work function is somewhat reduced. Such barrier lowering is called the **Schottky effect**. Although the **Schottky effect** is only a part of the explanation of metal-semiconductor contacts, rectifying contacts are generally referred to as Schottky barrier diodes.

4.13. METAL-SEMICONDUCTOR JUNCTIONS

For the case of semiconductor the energy relation are slightly more complex because there are no occupied quantum states whose energy equals E_o. The energy-band model of an N-type semiconductor is shown on the right in figure 4.31(a). The work function of the semiconductor can be constructed to consist of two parts the energy difference between the Fermi energy and the bottom of the conduction band, called the internal work function Φ_i, and the work needed to remove a free (conduction) electron from the crystal, called the external work function Φ_e. Suppose that a metal having a work function Φ_m and an N-type semiconductor having a work function $\Phi_s = \Phi_e + \Phi_i$, such that $\Phi_m + \Phi_s$, are brought in intimate contact with each other.

If it is assumed that the donor levels in the semiconductor are completely ionized at room temperature, the electrons occupying s states in the conduction band of the semiconductor will flow into the metal until at the equilibrium the Fermi energy in both the metal and the semiconductor is at the same energy level. Since, the Fermi energy in both crystals has the same energy, the energy levels in the bulk of the semiconductor are lowered by an amount $\Phi_m - \Phi_s$, as shown in figure 4.30(b). Thus the region that has been depleted of electrons represents a potential barrier to further flow of electrons from the semiconductor. This barrier can be expressed in terms of diffusion potential V_D by

$$ev_D = \Phi_m - \Phi_s \qquad ...(4.30)$$

Here, the height of the barrier on the metal side is $\Phi_m - \Phi_e$.

If a positive potential $+V_o$ is applied to the semiconductor, the height of the potential barrier at the contact, for electrons flowing from the semiconductor to the metal, is increased from eV_D to $e(V_D + V_o)$, as shown in figure 4.31(a). If a negative potential $-V_D$ is applied then the height of the barrier decreases to $e(V_D - V_o)$ as shown in figure 4.31(c). The height of the potential barrier for electrons flowing from the metal to the semiconductor is not affected by the applied potential and remains $\Phi_m - \Phi_e$, as can be seen in figure 4.31(b) for the equilibrium situation. This means that the current flowing from the metal to the semiconductor remains the same whereas the current flowing from the semiconductor to the metal depends on the polarity of the applied field. Suppose that the voltage applied to the semiconductor is positive, the current flow from semiconductor to metal is impeded, and it is said that the semiconductor is biased in the direction of difficult current flow or that a reverse bias has been applied. Conversely, if the applied voltage aids the current flow, and it is said that a forward bias has been applied.

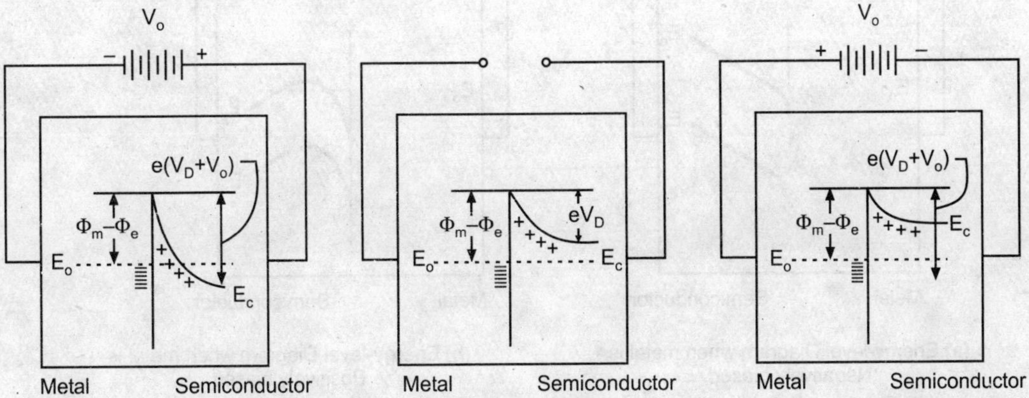

(a) Energy level diagram for reverse bias (b) Energy level diagram for equilibrium (c) Energy level diagram for forward bias

Fig. 4.31. Rectifying, metal N-type semiconductor contact

This metal-semiconductor contact, therefore acts like a rectifier, and it is called a rectifying contact.

The contact behaves differently if the work function of the metal is less than that of the n-type semiconductor. The energy-band models for the case $\phi_m < \phi_s$ are shown before contact is made in figure 4.32(a) and after contact in figure 4.32(b).

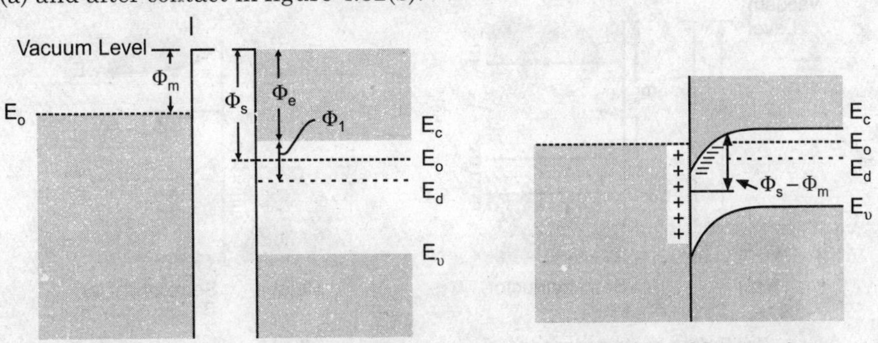

(a) Before contact is made (b) After contact is made

Fig. 4.32. Energy-level diagrams of metal-N-type semiconductor contact with $\phi_m < \phi_s$

These electrons collect on the surface of the semiconductor and form a negative surface charge. After equilibrium has been established, the negative surface charge on the semiconductor has the effect of depressing the bottom of the conduction band at the contact, as shown in figure 4.32(b) when the potential is applied.

Therefore, the energy of the quantum states in the bulk of the semiconductor is shifted, as shown in figure 4.33. Let us consider that the semiconductor is made positive with respect to the metal; then the energy levels shift as shown in figure 4.33(a) and the current carried by the electrons form the semiconductor to the metal flows 'downhill' with encountering as appreciable barrier. when the semiconductor is made negative with respect to the metal (figure 4.33(b)), then the current carried by electrons flowing from the metal to the semiconductor similarly flows *downhill*. The current flow in these cases is directly proportional to the applied voltage in accordance with Ohm's law, and the contact is said to be an ohmic contact not unlike a metal-metal contact.

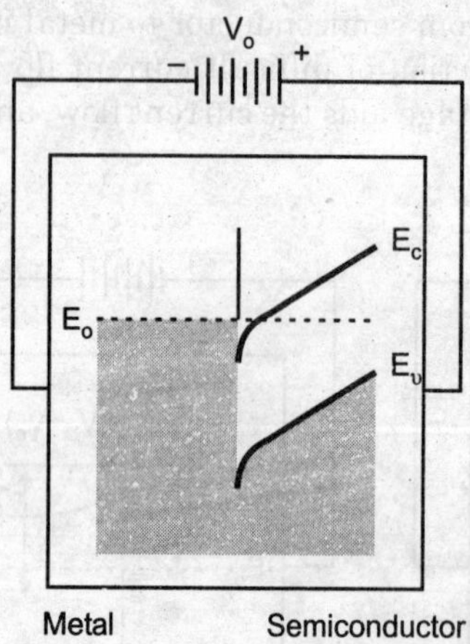

(a) Energy-level Diagram when metal is Negatively biased

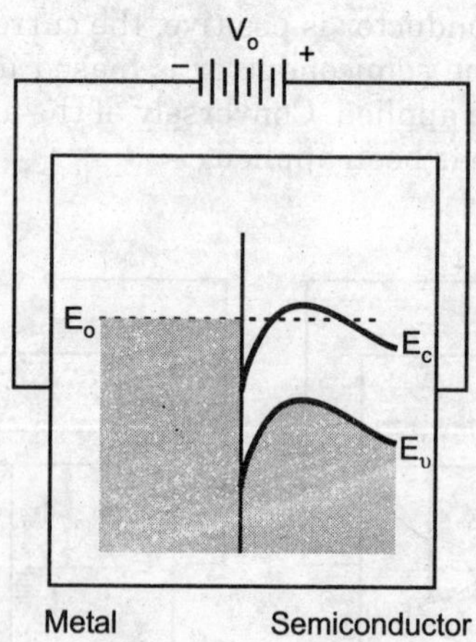

(b) Energy-level Diagram when metal is Positively biased

Fig. 4.33. Ohmic, metal n-type semiconductor contact

The situation is exactly reversed when a p-type semiconductor forms a contact with a metal. The energy-band models of a metal and p-type semiconductor, having $\Phi_m > \Phi_s$, are shown before contact in figure 4.34(a) and after contact in figure 4.34(b). Because the semiconductor is p-type, the electrons that occupied acceptor states near the crystals surface move to the metal and leave a positive surface charge on the semiconductor while creating a negative charge on the metal surface. Suppose that a potential is then applied in either direction. The holes in the semiconductor can flow

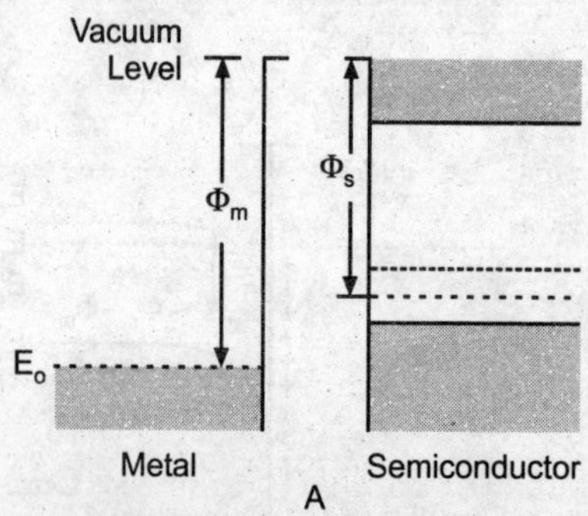

(a) Before contact is made

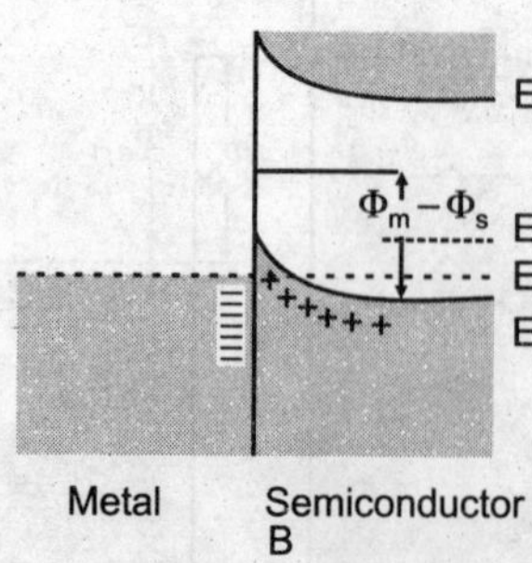

(b) After contact is made

Fig. 4.34. Energy-level diagrams of metal-p-type semiconductor contact with $\Phi_m > \Phi_s$

into the metal, or, if the field is reversed, the thermally created holes in the metal can flow *uphill* into the semiconductor, without any difficulty. The contact, therefore, is ohmic. On the other hand, $\Phi_m < \Phi_s$, a barrier layer is formed and the contact acts as a rectifier. It is left to the reader to work out the energy-band model for this case.

4.14. P-N JUNCTIONS

The junction between the N-type region and a P-type region in the same semiconductor crystal is called a P-N junction. Such a junction has already been encountered in the section describing the photovoltaic effect. There are several ways that such junctions can be prepared. If a crystal is grown from the melt by slowly withdrawing the solidified crystal, the change from P to N type is readily accomplished by suddenly adding an excess of donor impurities to the melt. Alternatively, an alloy junction can be formed by placing a suitable metal in contact with a semiconductor, say indium with N-type germanium, and heating them in a furnace until the metal melts. The molten indium dissolves the germanium in contact with it until the melting point of the solution equals the furnace temperature. Upon slow cooling, the molten region of the germanium crystal solidifies and retains the indium in solid solution. This region now in P-type whereas the host crystal was N-type. Alternatively, a semiconductor can be heated in the presence of a donor or acceptor gas so that these atoms can diffuse into part of the crystal to form a diffusion junction.

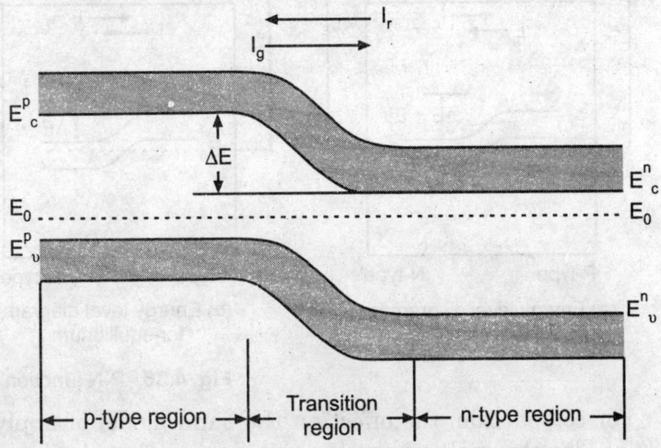

Fig. 4.35. Energy-level diagram for a p-n junction

Since, the density of electrons is larger in the N-type region than in the P-type region, elect: ons on the N side of the contact diffuse into the P-type region where they recombine with free holes. As this process continues, the P side of the junction becomes negatively charged and the N side becomes positively charged until, at equilibrium, a potential barrier eV_D deters further diffusion. The region on both sides of the contact is said to contain a space charge and is usually called the space charge region or the transition region.

A thermally generated current I_g which is directly proportional to the number of thermally excited electrons in the P-type region.

$$n_p = e^{-(E_c^p - E_0)/KT} \qquad ...(4.31)$$

where the superscript P denotes the energy values in the P-type region. At the same time, the number of electrons occupying quantum states in the conduction band of the N-type region is $n_n = e^{-(E_c^n - E_0)/KT}$.

This current is proportional to the number of electrons that can surmount the energy barrier, that is, it is proportional to

$$e^{-[\Delta E + (E_c^n - E_0)]/KT} = e^{-(E_c^p - E_0)/T} \qquad ...(4.32)$$

But, the right side of equation (4.32) is the same as that of equation (4.31). Consequently, at equilibrium,the current flowing to the right, given by equation (4.32). The equilibrium situation is indicated by the two arrows drawn at the top of figure 4.36.

Let us consider that an external voltage is applied to the junction so that the P-type region is made even more negative with respect to the N-type region. The effect on the band model of applying this so-called reverse bias is shown in figure 4.36(a). If the applied voltage is V, the energy barrier is increased from ΔE to $\Delta E + eV$ and virtually no electrons can cross over from the N-type region to the P-type region and the recombination current I_r, is very small. Thus, I_r, for holes is reduced and I_g remains unaffected, since applied voltage does not alter the number of thermally generated holes appreciably.

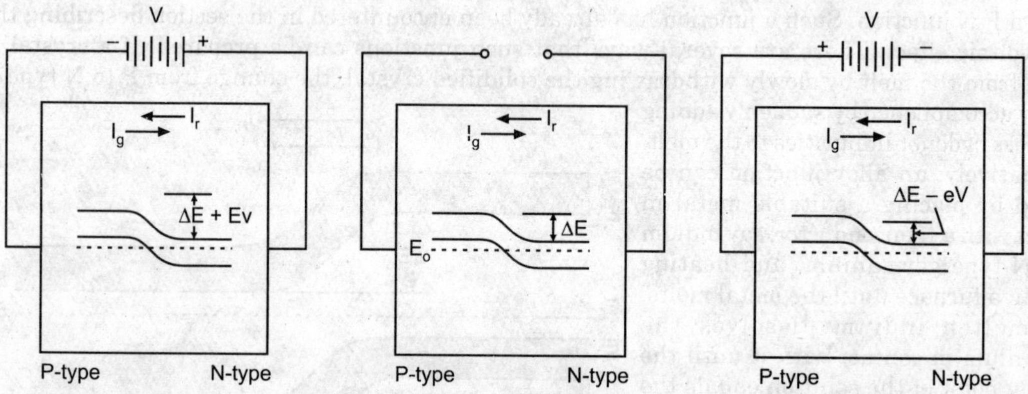

(a) Energy level diagram for a reverse bias
(b) Energy level diagram for equilibrium
(c) Energy level diagram for a forward bias

Fig. 4.36. P-N junction

For comparison, the effect on the band model of applying a forward bias is shown in figure 4.36(c). The thermal current I_g again is not affected. Since the energy barrier is decreased by the applied voltage, I_r increases according to the Boltzmann distribution law by a factor $e^{\frac{eV}{KT}}$. Since, at equilibrium, I_r equaled I_g, and since the value of I_g is not changed, the increased current may be expressed by,

$$I_r = I_g e^{eV/KT} \qquad ...(4.33)$$

This relation applies to both the electron and hole currents and is represented by a longer arrow in the direction of I_r in figure 4.36(c).

The net current I is equal to the difference between the two currents, $I_r - I_g$ which, according to equation (4.33), can be written

$$I_r - I_g = I_g e^{eV/KT} - I_g$$

$$= I_g(e^{eV/KT} - 1) \qquad ...(4.34)$$

If the combined current due to holes and electrons is represented by the so-called dark current I_o, then equation (4.34) can be written in its more familiar form

$$I = I_o(e^{eV/KT} - 1) \qquad ...(4.35)$$

It should be noted that equations (4.34) applies equally well to the case of a reverse bias. In such a case, the magnitude of eV is much larger, but since V is negative, $e^{eV/KT} << 1$ when eV ~ from 4KT. This means that, when a reverse bias is applied, an increase in the applied voltage produces a saturation effect, at which point the current has the constant value $- I_o$.

Overview of a diode

The diodes are semiconductors devices having two electrodes, namely anode (A) and cathode

(K). The diodes are basically a p-n junction. The various types of diodes are obtained based upon the amount of doping and type of material used. The various diodes are:

(i) Rectifier diode
(ii) Signal diode
(iii) Switching diode
(iv) Power diode
(v) Tunnel diode
(vi) Schottky diode
(vii) Gunn diode
(viii) PIN diode

4.14.1. Rectifier Diode

1. Basics

The semiconductor diodes are mainly used for rectification operation. In such operation, the diode being unidirectional conducts only during positive half cycle of input AC signal applied from AC mains whereas during negative half cycle of AC input, it does not conduct and block the entire voltage applied.

Thus, in such application, the diodes acts as a switch. During positive half cycle, it is ON and conducts currents heavily whereas during negative half cycle, it is OFF and blocks the entire voltage applied. Hence, it is necessary for a diode to have a good forward current rating and good reverse voltage blocking capacity. The forward current ratings of such diodes are in amperes whereas the PIV rating is hundred of volts.

For safe operation of the diode, the maximum forward current flowing in the circuit should be well below its maximum current rating whereas its PIV rating should be much above the maximum reverse voltage being applied.

2. Salient features of rectifier diodes

(i) High forward current rating
(ii) High peak inverse voltage rating
(iii) High reverse recovery time
(iv) Operates at low frequency
(v) High breakdown voltage

3. Advantage of rectifier diode

The low doping of p and n region results in wider depletion region, which increases the reverse voltage at which the device breakdown occurs.

4. Disadvantages of rectifier diode

As the reverse recovery (t_{rr}) is large, these diodes take longer time to turn OFF. Hence, such diodes can operate only for low frequencies upto 100 Hz.

5. Applications of rectifier diode

(i) Half wave and full wave rectification
(ii) Voltage multipliers
(iii) Clipping
(iv) Clamping

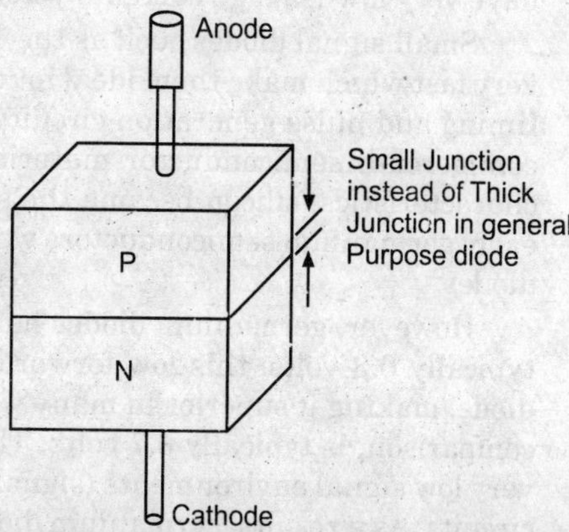

Fig. 4.37. Structure of a signal diode

4.14.2. Signal Diodes

1. Construction of signal diodes

Figure 4.37 shows the structure of signal diodes.

Signal diodes are physically small devices usually used where small currents, high voltages and high frequencies are involved. A small junction offers less resistance to high frequencies that does a wide thick junction. The name comes from the fact that these diodes are suitable for use in radio detectors to isolate the radio 'signal'. Signal diodes are very small and often glass encapsulated with a red or black band on one end. The glass is sometimes painted over to reduce the unwanted photo-voltaic effects.

2. Symbol of diode

Figure 4.38 shows the symbol of a signal diode.

Fig. 4.38. Symbol of a signal diode

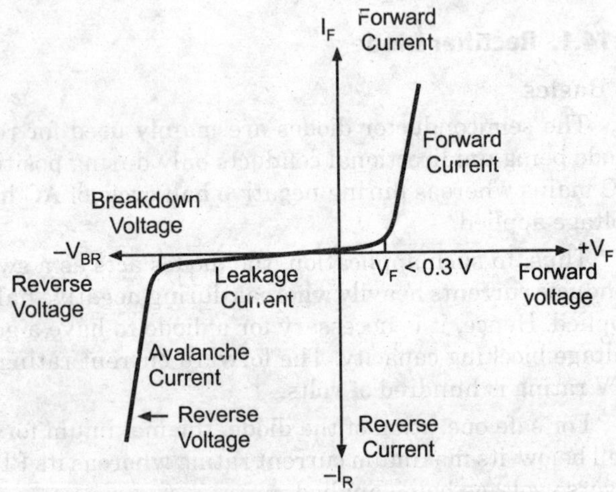

Fig. 4.39. V-I characteristics of a signal diode

3. Signal diode (small current diodes) characteristics and operation

Figure 4.39 shows the V-I characteristics of signal diode.

Signal diodes are used to process information (electrical signals) in circuits, so they are only required to pass small currents of upto 100 mA. General purpose signal diodes such as the 1N4148 are made from silicon and have a forward voltage drop of 0.7 V. Germanium diodes such as the 0A90 have a lower forward voltage drop of 0.2 V and this makes them suitable to use in radio circuits as detectors, which extract the audio signal from the weak radio signal. For general use,where the size of the forward voltage drop is less important, silicon diodes are better because they are less easily damaged by heat when soldering they have a lower resistance when conducting and they have very low leakage currents when a reverse voltage is applied.

Small signal diodes such as the 1N914 and 1N4148 are low voltage and current diodes but are very fast which make them ideal for use with very high frequency circuits. These diodes are used in timing and pulse generation circuits. Early semiconductor developments used germanium as the commercial, semiconductor material. Due to ease of processing and more stable temperature characteristics, silicon became the semiconductors of choice and as a consequence of that, mostt early germanium semiconductors were replaced with silicon. These were primarily transistors and diodes.

However, germanium diodes have the advantage of an intrinsically low forward voltage drop, typically 0.3 volts; this low forward voltage drop results in a low power loss and more efficient diode, making it superior in many ways to the silicon diode a silicon diode forward voltage drop by comparison, is typically 0.7 volts. This lower voltage drop with germanium becomes important in very low signal environments (signal detection from audio to FM frequencies) and in low-level logic circuits. As a results, germanium diodes are finding increase application in low level digital circuits with this increased interest, certain general germanium characteristics should be understood. First and most important is that of an increased leakage current for germanium at a reverse voltage. This is mitigated to some degree by the fact in low-level circuits, the reverse voltage applied to a germanium diode is also usually very low, resulting, in a low reverse leakage current (leakage current is directly proportional to reverse voltage). However, the leakage current is still larger than with silicon. A properly designed circuit will reduce this factor. Now-a-days manufacturers supply an improved, low leakage current, germanium diode for small signal applications.

4. Typical specification sheet for low signal diodes

				Fwd voltage drop		Reverse current	
	Peak Inv. voltage PIV	Max. aver. rectified current Io At 25°C	Junction temperature e T_j	V_F	@ I_F	I_R	@ V_R
Part No.	Volts	mA	Max. °C	Max. V	mA	Max. μA	Volt
1N34A	60	50	90	1.0	50	30	10
1N60	50	30	100	1.0	50	40	20
1N100N	80	70	90	1.0	40	50	50
1N270	100	200	90	1.0	200	100	50
1N277	120	100	90	1.0	100	250	50

5. Application of signal diodes

As protection diodes for relays

Figure 4.40 shows application of signal diode.

Signal diodes are used with relays to protect transistors and integrated circuits from the brief high voltage produced when the relay coil is switched OFF. The diagram shows how a protection diode is connected across the relay coil, note that the diode is connected 'backwards'so that it will normally NOT conduct. Conduction only occurs when the relay coil is switched OFF, at this moment, current tries to continue flowing through the coil and it is harmlessly diverted through the diode. without the diode, no current could flow and the coil would produce a damaging high voltage 'spike' in its attempt to keep the current flowing.

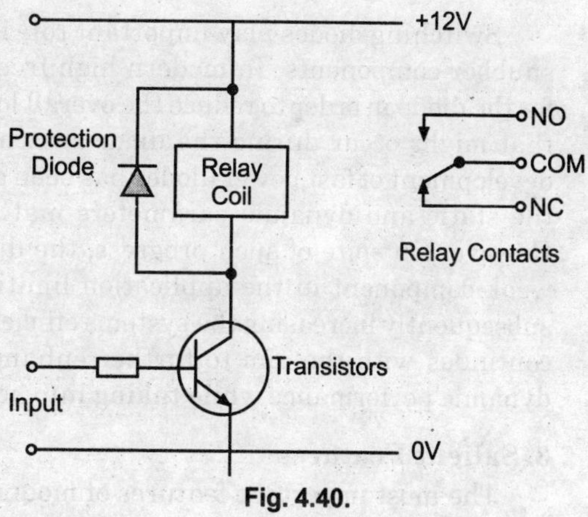

Fig. 4.40.

6. Advantages and application of signal diodes

Thus, due to low forward voltage drop, low leakage and low cost, this type of diode can be used in radio communications, radio signal detection in amplitude modulated receivers and frequency modulated receivers.

4.14.3. Switching Diode

1. Construction of switching diode

Figure 4.41 shows the structure of a switching diode.

2. Symbol

Figure 4.42 shows the symbol of a switching diode.

The rectifier diodes are required to provide improved switching performance, and especially reverse recovery characteristics. The chip structure of a switching diode employs a structure that is similar to that of conventional diodes with fast recovery. The structure normally consists of diffusion of heavy metals such as gold or platinum. However, shortening the ramp time causes a decrease in the number of carriers resulting in an increase in forward voltage drop. Although expensive, switching diodes employ epitaxial wafers having a highly precise thickness to make it possible to create a chip structure that coincides with design values of parameters, which are required in design of diodes.

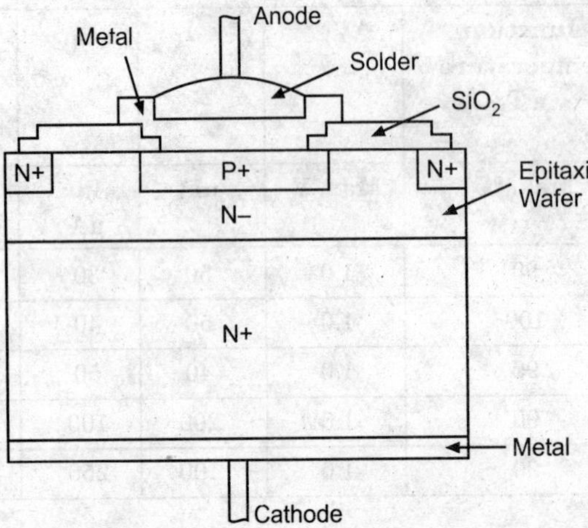

Fig. 4.41. Structure of a switching diode

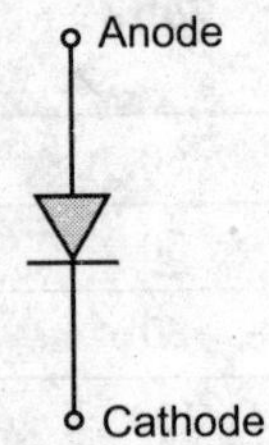

Fig. 4.42. Symbol of switching diode

Switching diodes play important role in most power electronics circuits as freewheeling and/or snubber components, In modern high frequency applications, many design features are required for the diode in order to reduce the overall losses of the circuit, and to prevent any failure mechanisms that might occur during the diode-switching transients. As a result, significant progress in the development of fast power diodes has been achieved in recent years with optimum trade-offs between the static and dynamic parameters matched with the requirements of the specific application. However, in spite of such progress, the diode is still considered by many circuit designers as the weak component in the application limiting the ability to increase the operating frequency and subsequently increasing the systems efficiency. Therefore, the development of new design techniques continues with the aim to further enhance the diode performance especially with regard to the dynamic performance while taking into account the manufacturing costs and process capabilities.

3. Salient Features

The most important features of modern switching diodes are–

(i) Low on-state voltage and positive temperature coefficient for optimum parallel diode operation.

(ii) Stable reverse blocking characteristics with low leakage current at elevated temperature.

(iii) Low reverse recovery losses, soft recovery and rugged against dynamic avalanching.

4. Switching diode characteristics and operation

The diagram shows the voltage and current waveforms for a power diode driven by currents with a specified rate of rise during turn-on and a specified rate of fall during turn-off. During turn-on and turn-off, specific physical processes occur in sequence as shown in figure 4.43.

(i) During turn-on

t_1 : During t_1 the space charge stored in the depletion region is, because of the large bias voltage, removed by the growth of the forward current.

t_2 : During this phase, the excess carriers distribution in the drift region grows towards the steady state value that can be supported by the forward current.

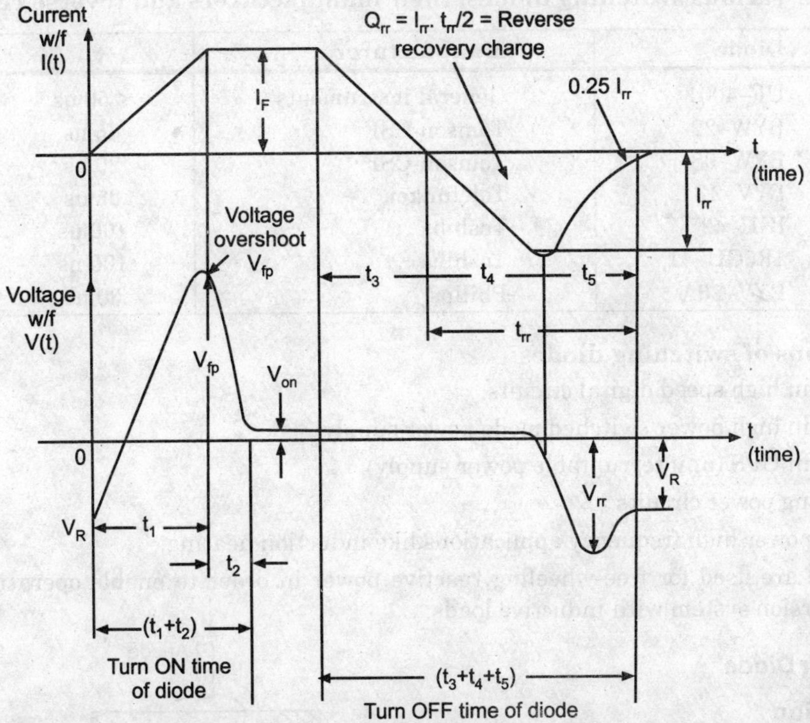

Fig. 4.43.

(ii) During turn-off

t_3 : The excess carriers stored in the drift region must be removed before the metallurgical junctions can become reverse biased.

t_4 : The depletion layer acquires a substantial amount of space charge from the reverse bias voltage. As long as there are excess carriers at the ends of drift region, the junction must be forward biased. When the current goes negative and the carrier sweep-out has proceeded for a sufficient amount of time to reduce the excess carrier density at one or both of the junctions to zero, or junctions become reverse biased.

t_5 : The diode voltage goes negative and the depletion regions from the two junctions expand towards each other into the drift region. At this time the negative diode current demanded by the stray inductance of the external circuit cannot be supported by the excess carriers. The diode current ceases its growth in the negative direction and quickly falls and becomes zero. The parameter t_{rr} is defined as the sum of t_4 and t_5. The amount of charge that has to be removed before the diode can be reverse biased in defined by,

$$Q_{rr} = \frac{I_{rr}\, t_{rr}}{2}$$

5. Advantage of switching diodes

Thus switching diodes has very short reverse recovery time, short forward recovery time in the range of few nano seconds. Also the diode capacitances very small to operate the switching diodes at very high frequencies.

Table 4.1. Various switching diodes, their manufacturers and reverse recovery time

Diode	Manufacturer	t_{rr}
UF–4001	General instruments	< 50 ns
BYW–22	Tomson-CSF	35 ns
BYW–98	Tomson-CSF	20 ns
BYV–27	Telefunken	25 ns
IGU–42	Toshiba	100 ns
1R5GU–41	Toshiba	100 ns
BYV–26A	Philips	30 ns

6. Applications of switching diodes

(i) Used in high speed digital circuits.

(ii) Used in high power switched mode power supply.

(iii) Used in UPS (uninterruptable power supply).

(iv) Welding power circuits.

(v) High power/high frequency applications like induction heating.

(vi) These are used for free-wheeling reactive power in order to enable operation of DC-AC conversion system with inductive loads.

4.14.4. Power Diode

1. Construction

The basic structure of high power diode is vertical in nature to increase the surface area for the forward current. Junction area is also kept larger to reduce the forward resistance of power diode and to reduce the power dissipation when diode is in ON state. In general purpose p-n junction diode, only p and n type of material is used to form p-n junction. In power diode lightly doped n^- layer known as 'drift' layer is introduced between heavily doped P^+ anode and heavily doped n^+ cathode. Material used for manufacturing of this power diode is silicon (Si) to have large current handling capacity at higher voltage rating. This power diode also have PIV rating and capability to work at very high temperature. Thus, high power diodes are equipped with heat sinks for heat dissipation.

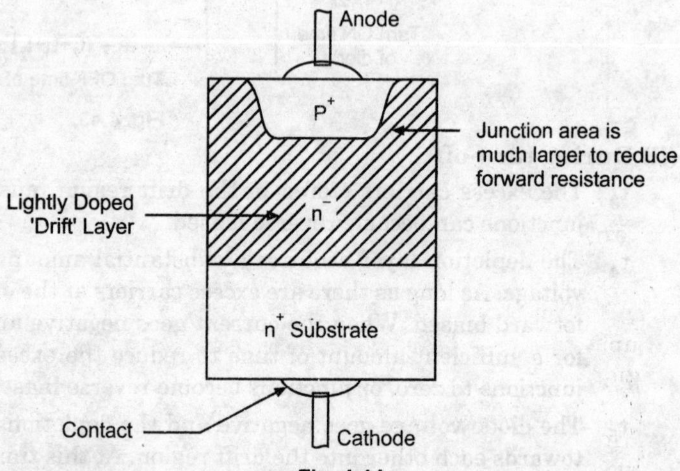

Fig. 4.44.

2. Symbol of power diode

Figure 4.45 shows the symbol of a power diode.

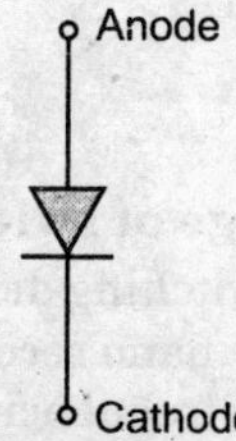

Fig. 4.45. Symbol of a power diode

3. Power diode characteristics and operation

Figure 4.46 shows the V-I characteristics of power diode.

The forward biased characteristics of the power diode is same as compared to the normal p-n junction diode except cut-in voltage in power diode is high i.e. 1 V to 1.1 V due to added drift layer between p^+ anode and n^+ cathode to increase the junction area. This causes reduction in forward resistance and hence ON state dissipation in diode causes to work this at very large current and hence higher power handling capacity. But increasing junction area causes to increase the junction capacitance which is directly proportional to the junction area and hence switching speed of diode degrades automatically.

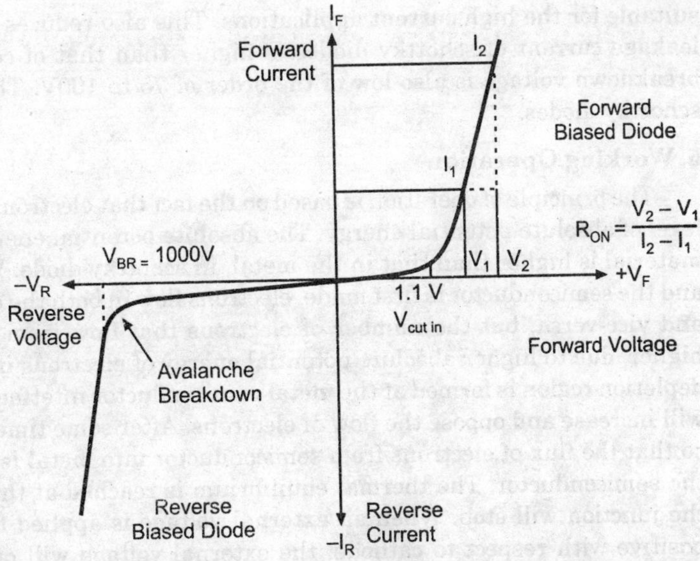

Fig. 4.46.

In reverse biased mode of operation, very small leakage current flows through the power diode. This leakage current is generated due to the thermal energy received by charge carriers in reverse biased diode. When the reverse voltage across the diode reaches to the breakdown voltage V_{BR}, the current due to avalanche breakdown increases heavily which must be limited by external load resistor. The reverse current increases rapidly with a small change in reverse voltage beyond V_{BR}. The operation in the breakdown region will not be destructive provided that the power dissipation is within a "safety level" that is specified in the manufacturer's data sheet.

4. Types of power diodes

Ideally, any diode should have zero reverse recovery time (t_{rr}) or $t_{rr} = 0$. However,the manufacturing cost of such diode will increase due to special manufacturing process. Thus, depending upon the reverse recovery characteristics and manufacturing techniques, the power diodes can be classified into three categories:

(i) Standard and general purpose diodes

(ii) Fast recovery diodes

(iii) Schottky diodes

(i) Standard and general purpose diodes: General purpose rectifier diodes have relatively high reverse recovery time typically 25 μs to 30 μs and are used in low speed application (for example, low frequency converters and rectifiers). These diodes current rating is from 0.5 A to 1200 A and voltage ratings from 60 V to 6 kV.

(ii) Fast recovery diodes: The fast recovery diodes are generally made by platinum or gold diffusion technique. This technique causes to have low recovery time, normally less than 6μs. They are used in inverters and choppers circuit. Current ratings of these diodes go upto 300 A and voltage ratings upto 3 kV.

(iii) Schottky diodes: The structure for a schottky diode is as shown in figure 4.47. This diode is formed by depositing a thin metal film directly on a semiconductor. The metal film acts as anode of the diode and the semiconductor acts as cathode. The schottky diodes have I-V characteristics very similar to that of the conventional diodes. The on state voltage is however much less as compared to the conventional diodes. (0.3 to 0.4 volts only). This makes the schottky diodes

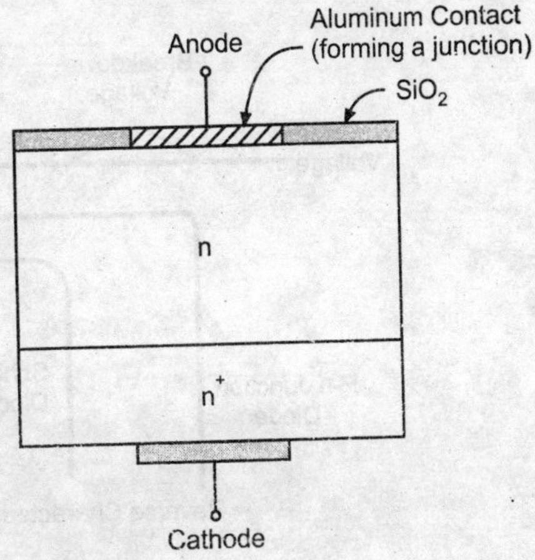

Fig. 4.47. Structure of a schottky diode

suitable for the high current applications. This also reduces the on state power loss. The reverse leakage current of schottky diodes is higher than that of conventional pn junction diodes. The breakdown voltage is also low of the order of 75 to 100V. These are some of the limitations of a schottky diodes.

5. Working Operation

The principle of operation is based on the fact that electrons in different materials have different level of absolute potential energy. The absolute potential energy of electrons in the semiconductor material is higher than that in the metal, in schottky diode. When the contact between the metal and the semiconductor is first made, electrons flow in both the directions i.e. metal to semiconductor and vice-versa, but the number of electrons that flow from the semiconductor to metal will be higher, due to higher absolute potential energy of electrons in the semiconductor. Due to this the depletion region is formed at the metal-semiconductor interface. The electrostatic potential barrier will increase and oppose the flow of electrons. After some time this potential will grow sufficiently so that the flux of electrons from semiconductor into metal is just equal to the flux from metal to the semiconductor. The thermal equilibrium is reached at this point and the current flow across the junction will stop. When an external voltage is applied to the schottky diode making anode positive with respect to cathode, the external voltage will oppose the built in potential (due to depletion region) which will make it easier for the current to flow. If the device is reverse biased, then it will increase the barrier potential and oppose the flow of current. In this way the schottky diode has the rectifying properties similar to the pn junction.

It is important to note that in the formation of depletion region and in establishing the thermal equilibrium, the carriers involved are only the majority carriers, no minority carriers are involved. Therefore, the schottky diodes are known as the "majority carrier devices". The conventional pn junction diodes are known as "minority carrier devices" or bipolar devices as both the majority and minority carriers are involved in their basic operation. The schottky power diodes are suitable for low voltage high current applications such as power supplies.

6. I-V characteristics of schottky diode

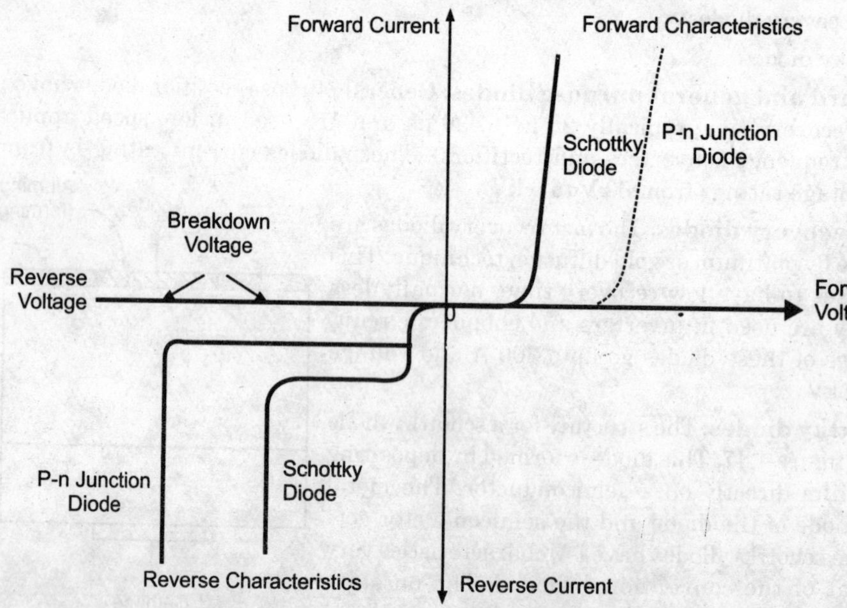

Fig. 4.48. I-V characteristics of a schottky diode

The I-V characteristics of a schottky diode is shown in figure 4.48. From this figure we can compare the characteristics of schottky and conventional p-n junction diodes.

(i) The characteristics of a schottky diode is very similar to that of a p-n junction diode. But it has a very low cut-in voltage (of the order of 0.2 Volts).

(ii) The reverse breakdown voltage of a schottky diode is less than that of a p-n junction diode. Typically it is about 50 V as compared to 150 V for a p-n junction diode.

(iii) The leakage current in the reverse biased region is higher than that of a p-n junction diode. The reverse leakage current flows due to the electrons in metal passing into the semiconductor material.

7. Equivalent circuits and circuit symbol

The approximate equivalent circuit of a schottky diode is as shown in figure 4.49(a) and its circuit symbol is shown in figure 4.49(b). C_j is the junction capacitance which appears in parallel with an ideal diode.

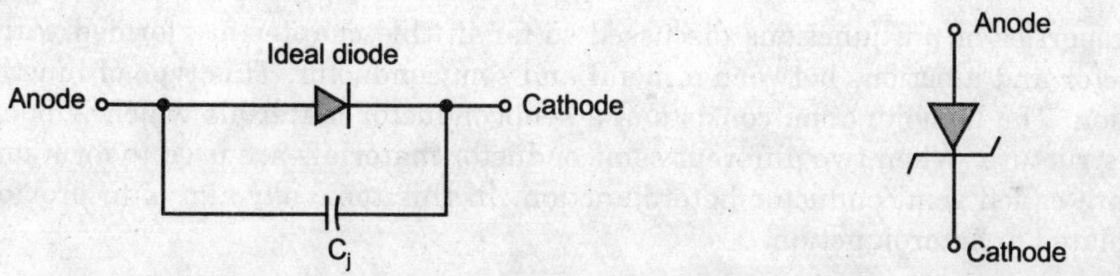

(a) Approximate equivalent circuit of a schottky diode (b) Circuit symbol of a schottky diode

Fig. 4.49.

8. Advantages of Schottky diode

The main advantages of schottky diode are:

(i) Very low on state voltage typically 0.3 to 0.4 volts, hence reduced on state power loss as compared to general purpose p-n junction diode.

(ii) Fast switching capability due to the majority carrier characteristics.

(iii) Low power dissipation

9. Disadvantages

The major limitations of schottky diodes are as follows:

(i) Low reverse breakdown voltage typically 75 to 100 V only.

(ii) Large reverse leakage current.

10. Applications

(i) Switched model power supplies.

(ii) AC to DC converters.

(iii) Radar systems

(iv) Mixers and detectors in communication equipments.

11. Performance Comparison of Schottky barrier diode and P-N junction diode

S. No.	Parameter	P-N junction diode	Schottky diode
1.	Circuit symbol		
2.	Forward voltage drop	Typically 0.7 V to 1V for silicon diodes	Between 0 to 0.4 V
3.	Peak inverse voltage	High	Low

4.	Leakage current	Low	High
5.	Carriers	Conduction takes place due to majority and minority carriers	Conduction takes place only due to majority carriers
6.	Junction is formed between switching speed	P and N type semiconductors very low. Therefore suitable only for low frequency operation	A metal and n-type semiconductor Very high. Therefore can operate at high frequencies
7.	Applications	Rectifiers, voltage multipliers	SMPS, AC-to-DC converters radar

4.15. HETEROJUNCTIONS

1. Basics

The properties of p-n junctions discussed so far in this chapter are formed within a single semiconductor and junctions between a metal and semiconductor. This type of junction is called homojunction. The homojunction consists of a semiconductor materials which is homogenous in the entire structure. When two different semiconductor materials are used to form junctions, the junctions are called semiconductor heterojunction. In this topic our aim is to provide the basic concepts related to heterojunction.

2. Heterojunction materials

As the two materials used to form a heterojunction will have different energy band gaps the energy band will have a discontinuity at the junction surface. We may have an abrupt junction in which the semiconductor changes abruptly from a narrow bandgap material to a wide-band gap material, on the other hand, if we get a material like GaAs – $Al_x\ Ga_{1-x}$ as an example, the value of x may continuously vary over a distance of several nanometers to form a graded heterojunction. Changing the value of x in the $Al_x\ Ga_{1-x}$ As material helps the engineer and scientists to design materials of different bandgap energy.

To prepare the most useful heterojunction, the lattice constants of the two semiconductor materials must be well matched. The lattice parameter match is useful because any lattice mismatch can introduce dislocations resulting in interface states. As an example, germanium and gallium arsenide have lattice parameters matched to within approximately 0.13 percent. Germanium-gallium arsenide heterojunctions have been investigated quite extensively to check its feasibility for different semiconductor device applications. Recently, gallium arsenide-aluminium gallium arsenide (GaAs-Al (GaAs) heterojunctions have been studied throughly, as the lattice constants of GaAs and the AlGaAs system vary by not more than 0.14 percent. So, these two materials are semiconductors well matched for heterojunction.

3. Energy-band diagrams

In forming a heterojunction with a wideband gap material and a narrowband gap material, the alignment on the band gap energies is important in determination of the junction characteristics. The alignment of band gap energies can be demonstrated in three modes, shown in figure 4.50.

In figure 4.50(a) we observe the case when the forbidden bandgap of the wide-gap material completely overlaps the bandgap of the narrow-gap material. This type of combination is known as straddling, and used mostly to form heterojunctions. In this chapter we will consider only this case. There are also two possibilities is known as staggered and broken gap as depicted in figure 4.50(b) and 4.50(c).

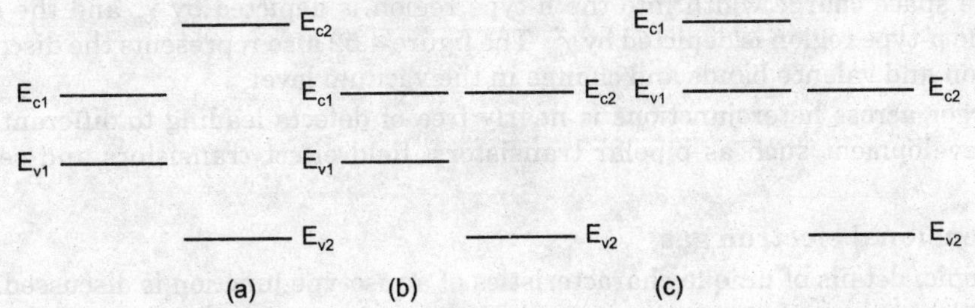

Fig. 4.50. Relation between narrowband gap and wideband gap energies

Four types of heterojunction can be formed. In anisotype hetrojunctions, the dopant type changes at the junction. nP and NP junctions can be formed where the capital leter indicates the larger-band gap material and small letter indicates the narrow band gap material. Another type of hetero-junction is isotype which has same dopant type on either side of the junction.

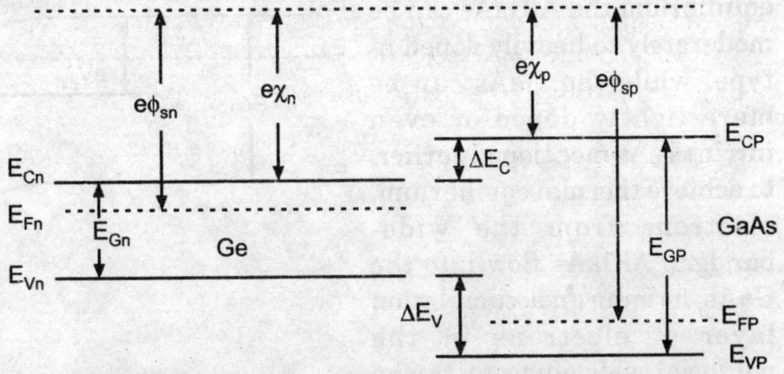

Fig. 4.51. Energy band diagrams of a narrowband gap and a wideband gap material before contact

Figure 4.51 shows the energy-band diagrams of isolated n-type and p-type materials, with the vacuum level used as a reference. The electron affinity of the narrow band gap material is greater than that of wideband gap material. ΔE_c represents the difference between the two conduction band energies whereas ΔE_v denotes the difference between the two valence band energies. According to figure 4.51, we can write

$$\Delta E_c = e\,(\chi_n - \chi_p)$$

and
$$\Delta E_c + \Delta E_v = E_{Gp} - E_{Gn} = \Delta E_G$$

In the ideal abrupt heterojunction making use of non-degenerately doped semiconductors, the vacuum level is parallel to both conduction bands and valence bands. If the vacuum level is continuous, then the same ΔE_c and ΔE_v discontinuities will exist at the heterojunction interface. This is an ideal situation and is known as the electron affinity rule. There are still some uncertainty about the applicability of this rule, but it gives good starting point for the discussion of heterojunctions.

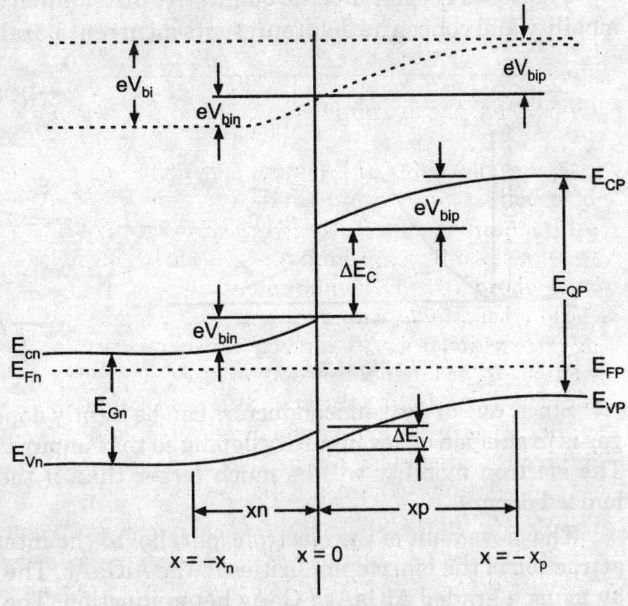

Fig. 4.52. Ideal energy band diagram of an np heterojunction in thermal equilibrium

Figure 4.52 represents a general ideal nP heterojunction in thermal equilibrium. In order for the fermi levels in the two materials to become aligned, electrons from the narrow gap n region and holes from the widegap P region must flow across the junction. As in the case of homojunction, the charge flow results in a space charge region in the vicinity of the metallurgical junction. The space charge width into the n-type region is depicted by χ_n and the space charge width into the p-type region is depicted by χ_p. The figure 4.52 also represents the discontinuities in the conduction and valence bands and change in the vacuum level.

The merger across heterojunctions is nearly free of defects leading to different possibilities for device development such as bipolar transistors, field effect transistors and semiconductor lasers.

4. Two dimensional electron gas

In this topic, details of unique characteristics of an isotype junction is discussed. Figure 4.53 represents the energy-band diagram of an nN GaAs-AlGaAs heterojunction in thermal equilibrium the AlGaAs can be moderately to heavily doped n-type, while the GaAs can be more lightly doped or even intrinsic. As mentioned earlier, to achieve thermal equilibrium, electrons from the wide-bandgap AlGaAs flow into the GaAs, forming an accumulation layer of electrons in the potential well adjacent to the interface. One basic quantum mechanical result that we know that the energy of an electron contained in a potential well is quantized.

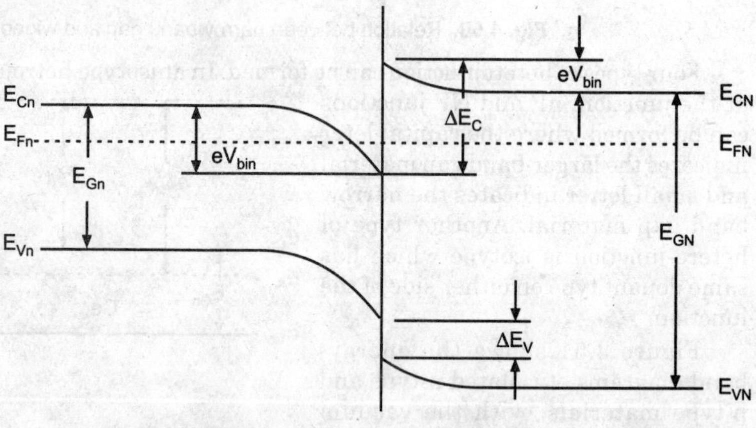

Fig. 4.53. Ideal energy band diagram of an ηN heterojunction in thermal equilibrium

The two dimensional electron gas refers to the condition in which the electrons have quantized energy levels in one spatial direction perpendicular to the interface, but are freee to move in the other two spatial directions.

Figure 4.54 represents the qualitative distribution of electrons in the potential well. The electron mobility and concentration represents a current parallel to the interface.

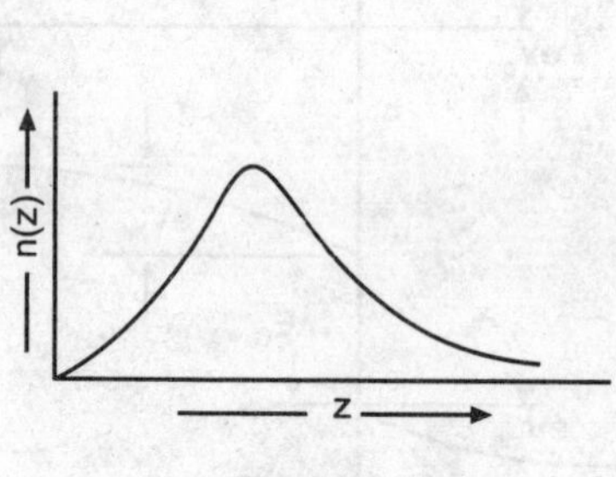

Fig. 4.54.

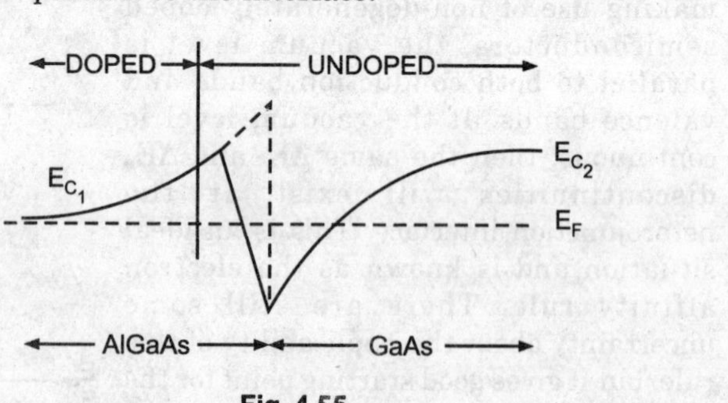

Fig. 4.55.

Since, one of the semiconductors can be lightly doped or intrinsic, the two dimensional electron gas is in a region of low impurity doping so that impurity scattering effects are reduced to minimum. The electron mobility will be much larger than if the electrons were in the same region as the ionized donors.

The movement of the electrons parallel to the interface will still be influenced by the coulomb attraction of the ionized impurities in the AlGaAs. The effect of these forces can be further reduced by using a graded AlGaAs / GaAs heterojunction. The graded layer is $Al_x\ Ga_{1-x}$ As in which the mole fraction x varies with distance. In this case, an intrinsic layer of graded AlGaAs can be sandwiched between the N-type AlGaAs and the intrinsic GaAs. Figure 4.55 represents the conduction-band edges across a graded AlGaAs/GaAs heterojunction in thermal equilibrium. The electrons in the potential well are further separated from the ionized impurities so that the electron mobility is increased above that an abrupt heterojunction.

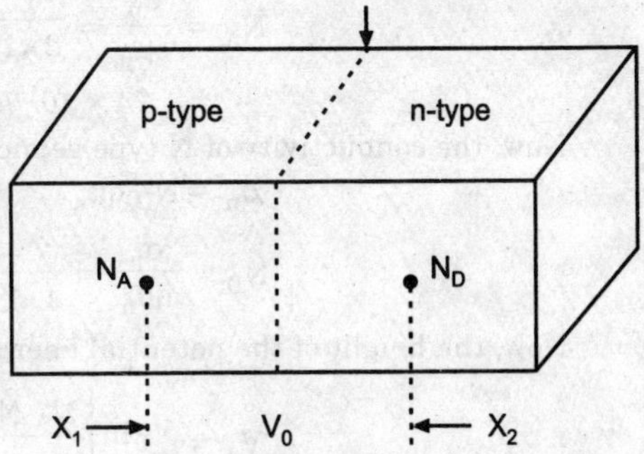

Fig. 4.56.

SOLVED NUMERICAL PROBLEMS

EXAMPLE 4.2. (a) Given the step graded germanium semiconductor as depicted in figure 4.56 with $N_D = 10^3\ N_A$ and with N_A corresponding to 1 acceptor atom per 10^3 germanium atoms, determine the contact difference of potential V_o at room temperature.

(b) Repeat part (a) for a silicon P-N junction.

Solution: We know that the concentration of acceptor atoms is given by

$$N_A = \frac{\text{Total number of Ge atoms/cm}^3}{\text{Number of impurity atoms}} = \frac{4.4\times10^{22}}{1\times10^8}$$

$$N_A = 4.4 \times 10^{14}\ \text{atoms/cm}^3 = 4.4 \times 10^{20}\ \text{atoms/m}^3$$

Now, we have $N_D = 10^3\ N_A = 10^3 \times 4.4\ 10^{20} = 4.4 \times 10^{23}\ \text{atoms/m}^3$

The intrinsic concentration for Ge is given by

$$n_i = 2.5 \times 10^{13}\ \text{atoms/cm}^3 = 2.5 \times 10^{19}\ \text{per m}^3$$

$$V_T = 26\ \text{mA at room temperature}$$

The contact potential may be expressed as under:

$$V_J = V_T \ln\left[\frac{N_A \, . \, N_D}{n_i^2}\right]$$

$$V_J = 26\ \text{mV} \ln\left[\frac{4.4\times10^{20}\times4.4\times10^{23}}{(2.5\times10^{19})^2}\right] = 0.328\ \text{V} \quad \textbf{Ans.}$$

(b) The concentration of acceptor atoms will be

$$N_A = \frac{\text{Total number of Si atoms/cm}^3}{\text{Number of impurity atoms}} = \frac{5\times10^{22}}{1\times10^8} = 5 \times 10^{14}\ \text{atoms/cm}^3$$

$$N_A = 5 \times 10^{20}\ \text{atoms/m}^3$$

Now, we have $N_D = 10^3\ N_A = 10^3 \times 5 \times 10^{20} = 5 \times 10^{23}\ \text{atoms/m}^3$

The intrinsic concentration for Si,

$$n_i = 1.5 \times 10^{10}\ \text{per cm}^3$$

$$V_T = 26\ \text{mV at room temperature}$$

Therefore, we have $V_J = V_T \ln\left[\frac{N_A\ N_D}{n_i^2}\right] = 26\ \text{mV} \ln\left[\frac{5\times10^{23}\times5\times10^{20}}{(1.5\times10^{10})^2}\right] = \mathbf{0.721\ V}$ **Ans.**

EXAMPLE 4.3. (a) The resistivities of two sides of a step-graded germanium diode are 2 Ω-cm (P side) and 1 Ω-cm (N side). Determine the height of the potential energy barrier.

(b) Repeat part (a) for silicon P-N junction.

Solution: (a) We know that the conductivity of P type semiconductor is given by

$$\therefore \quad N_A = \frac{\sigma_p}{q\mu_p} = \frac{1}{2 \times 1.6 \times 10^{-19} \times 1800} \quad \left(\text{Since } \sigma = \frac{1}{\rho}\right)$$

$$= 1.74 \times 10^{15}/\text{cm}^3$$

Now, the conductivity of N type semiconductor is given by

$$\sigma_n = N_D q\mu_n$$

$$N_D = \frac{\sigma_n}{q\mu_n} = \frac{1}{1 \times 1.6 \times 10^{-19} \times 3800} = 1.65 \times 10^{15}/\text{cm}^3$$

Now, the height of the potential energy barrier is expressed as

$$V_J = V_T \ln\left[\frac{N_A N_D}{n_i^2}\right] = 26 \text{ mV} \ln\left[\frac{1.74 \times 10^{15} \times 1.65 \times 10^{15}}{(2.5 \times 10^{13})^2}\right] = 0.22 \text{ V} \quad \textbf{Ans.}$$

(b) Similar to part (a), we can write

$$N_A = \frac{1}{2 \times 1.6 \times 10^{-19} \times 500} = 6.25 \times 10^{15} \text{ per cm}^3$$

or

$$N_D = \frac{1}{1 \times 1.6 \times 10^{-19} \times 1300} = 4.8 \times 10^{15} \text{ per cm}^3$$

Now, we have

$$V_J = V_T \ln\left[\frac{N_A N_D}{n_i^2}\right] = 26 \text{ mV} \ln\left[\frac{6.25 \times 10^{15} \times 4.8 \times 10^{15}}{(1.5 \times 10^{10})^2}\right]$$

or $V_o = 0.667$ V **Ans.**

EXAMPLE 4.4. (i) For what voltage will the reverse current in a p-n junction germanium diode reach 90 percent of its saturation value at room temperature?

(ii) What will be the ratio of the current for a forward bias of 0.05 V to the current for the same magnitude of reverse bias?

(iii) If reverse saturation current is 10 μA, calculate the forward current for voltages of 0.1 V, 0.2 V and 0.3 V respectively.

Solution: (i) We know that the equation of diode current is given by

$$I = I_o (e^{V/\eta V_T} - 1)$$

Given that $I = -90\% \ I_o = -0.9 \, I_o$

Here, negative sign is used since the reverse saturation current is in opposite direction.

$$= -0.9 \, I_o = I_o (e^{V/1 \times 0.026} - 1)$$

$$e^{V/0.026} = 0.1$$

Therefore, V = (0.026) × (– 2.3) = – 0.060 V **Ans.**

(ii) We have

$$\frac{e^{0.05/0.026} - 1}{e^{-0.05/0.026} - 1} = \frac{e^{1.92} - 1}{e^{-1.92} - 1} = \frac{6.82 - 1}{0.147 - 1} = -6.83 \quad \textbf{Ans.}$$

(iii) For V = 0.1 V, we have

$$I = I_o(e^{V/\eta V_T} - 1) = 10 \, \mu\text{A} \, (e^{3.84} - 1) = 0.455 \text{ mA}$$

For V = 0.2 V, we have

$$I = 10 \, \mu\text{A} \, (e^{0.2/0.026} - 1) = 10 \, \mu\text{A} \, (e^{7.69} - 1) = 21.7 \text{ mA}$$

For V = 0.3 V, we have

$$I = 10 \, \mu\text{A} \, (e^{0.3/0.026} - 1) = 10 \, \mu\text{A}(e^{11.54} - 1) = 1.01 \text{ A}$$

Here, it may be noted that from three answers of I, for small rise in forward voltage, the diode current increases rapidly.

EXAMPLE 4.5. (a) Determine the anticipated factor by whichs the reverse saturation current of a germanium diode is multiplied when the temperature is increased from 25 to 80°C.

(b) Repeat part (a) for a silicon diode over the range of 25 to 150°C.

Solution: (a) We know that

$$I_{o_2} = I_{o_1} 2^{\frac{T_2 - T_1}{10}}$$

$$I_{o(80°)} = I_{o(25°)} 2^{\frac{80-25}{10}} = I_{o(25°)} 2^{5.5}$$

$$I_{o(80°)} = 45\, I_{o(25°)} \quad \textbf{Ans.}$$

(b) In a similar manner, we have

$$I_{o(150°)} = I_{o(25°)} 2^{\frac{150-25}{10}}$$

$$= I_{o(25°)} 2^{12.5} = I_{o(150°)} = 5700\, I_{o(25°)} \quad \textbf{Ans.}$$

EXAMPLE 4.6. (a) It is predicted that, for germanium, the reverse sauration current should increase by 0.11°C^{-1}. It is found experimentally in a particular diode that at a reverse voltage of 10 V, the reverse current is 5 μA and the temperature dependence is only 0.07° C^{-1}. What will be the leakage resistance shunting the diode?

Solution: From the given data, the circuit may be drawn as depicted in figure 4.57.

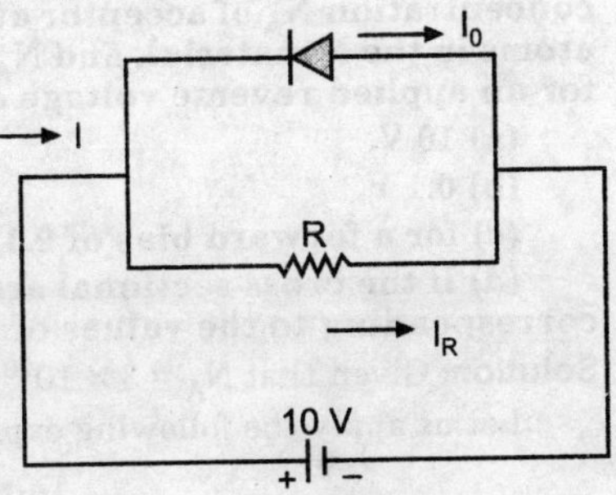

Fig. 4.57.

From the circuit, we have $I = I_o + I_R$ ...(i)

I_r is independent of T, hence, we have

$$\frac{dI}{dT} = \frac{dI_o}{dT} \quad ...(ii)$$

It is given that $\frac{1}{I_o}\frac{dI_o}{dT} = 0.11$

and $\frac{1}{I}\frac{dI}{dT} = 0.07$ ...(iii)

From equation (iii), we have

$$\frac{dI_o}{dT} = 0.11\, I_o = 0.71 = \frac{dI}{dT} \quad ...(iv)$$

$$0.07\, I = 0.11\, I_o$$

Multiplying (i) by 0.11 and subtracting (iv) from (i), we obtain

$$0.041 = 0.11\, I_R \text{ or } I_R = \frac{0.04}{0.11} I$$

$$I_R = \frac{0.04}{0.11} \times 5\,\mu A = \frac{20}{11}\,\mu A$$

Hence, we have $R = \frac{10 \times 11}{20} = 5.5\ M\Omega$ **Ans.**

EXAMPLE 4.7. An ideal germanium P-N junction diode has at a temperature of 125°C a reverse saturation current of 30 μA. At a temperature of 125°C, determine the dynamic resistance for a 0.2 V bias in

(a) the forward direction,

(b) the reverse direction.

Solution: (a) For dynamic resistance in the forward direction, V = + 0.2 V

Also, we have $V_T = kT$

Here, k = Boltzmann's constant = 8.62 × 10^{-5}

T = 125°C + 273 = 398°K

$$V_T = 8.62 \times 10^{-5} \times 398 = 34.30 \text{ mV}$$

Now, we have

$$r_f = \frac{\eta V_T}{I_o e^{V/\eta V_T}} = \frac{1 \times 34.30 \text{ mV}}{30\,\mu A \times e^{-0.2\,V/1 \times 34.30\,mV}} = 3.36\ \Omega \quad \textbf{Ans.}$$

(b) For dynamic resistance in the reverse direction, we have

$$V = -0.2\text{ V}$$

Therefore, $$r_r = \frac{\eta V_T}{I_o\, e^{V/\eta V_T}} = \frac{1\times 34.3\text{ mV}}{30\,\mu\text{A}\times e^{-0.2\text{V}/1\times 34.3\text{ mV}}} = 389.49\text{ k}\Omega \quad \textbf{Ans.}$$

EXAMPLE 4.8. Find out the barrier capacitance of a germanium P-N junction whose area is 1 mm by 1 mm and whose space-charge thickness is 2 × 10^{-4} cm. The dielectric constant of germanium (relative to free space) is 16.

Solution: We know that the barrier capacitance is given by,

$$C_T = \frac{\epsilon A}{W}$$

For Ge, we have $$\epsilon = \frac{16}{36\pi\times 10^{11}}\text{ F/cm} = 1.41\times 10^{-12}$$

Therefore, $$C_T = \frac{1.41\times 10^{-12}\times 1\times 10^{-2}}{2\times 10^{-4}} = 70.73\text{ pF}$$

EXAMPLE 4.9. The zero-voltage barrier height at an alloy-germanium P-N junction is 0.2 V. the concentration N_A of acceptor atoms in the P side much smaller than the concentration of donor atoms in the N material, and N_A = 3 × 10^{20} atoms/m^3. Determine the width of the depletion layer for an applied reverse voltage of

(a) 10 V,

(b) 0.1 V,

(c) for a forward bias of 0.1 V,

(d) If the cross-sectional area of the diode is 1 mm^2, evaluate the space-charge capacitance corresponding to the values of applied voltage in (a) and (b).

Solution: Given that $N_A = 3\times 10^{20}$ atoms/m^3, $A = 1\text{ mm}^2 = 1\times 10^{-6}\text{ m}^2$ $V_o = 0.2$ V and $\epsilon_r = 16$ for Ge

Let us apply the following expression:

$$V_o - V = \frac{1}{2}\frac{qN_A}{\epsilon}W^2$$

where $$\epsilon = \epsilon_o\,\epsilon_r = 8.854\times 10^{-12}\times 16 = 1.41\times 10^{-10}\text{ F/m}$$

(a) For V = – 10 V, we have

$$0.2-(-10) = \frac{1}{2}\frac{1.6\times 10^{-19}\times 3\times 10^{20}}{1.41\times 10^{-10}}\times W^2$$

$$W = 7.74\,\mu\text{m} \quad \textbf{Ans.}$$

(b) For V = – 0.1 V, we have

$$0.2-(-0.1\text{ V}) = \frac{1}{2}\frac{1.6\times 10^{-19}\times 3\times 10^{20}}{1.41\times 10^{-10}}W^2$$

$$W = 1.32\,\mu\text{m} \quad \textbf{Ans.}$$

(c) For V = + 0.1 V, we have

$$0.2\text{ V}-0.1\text{ V} = \frac{1}{2}\frac{1.6\times 10^{-19}\times 3\times 10^{20}}{1.41\times 10^{-10}}W^2$$

$$W = 0.766\,\mu\text{m} \quad \textbf{Ans.}$$

(d) $A = 1\text{ mm}^2 = 1\times 10^{-6}\text{ m}^2$, we have

For (a), we get $$C_T = \frac{\epsilon A}{W} = \frac{1.41\times 10^{-10}\times 1\times 10^{-6}}{7.74\,\mu\text{m}} = 18.21\text{ pF} \quad \textbf{Ans.}$$

For (b), we get $$C_T = \frac{\epsilon A}{W} = \frac{1.41\times 10^{-10}\times 1\times 10^{-6}}{1.32\,\mu\text{m}} = 106.81\text{ pF} \quad \textbf{Ans.}$$

EXAMPLE 4.10. A silicon P-N junction has a reverse saturation current of 1.8×10^{-9} A. Assuming that $\eta = 2$, determine the current in the junction when the forward biasing voltage is 0.6 V and the temperature is 27°C.

Solution: The forward current of diode is given by

$$I = I_o(e^{V/\eta V_T - 1})$$

$$= 1.8 \times 10^{-9}(e^{0.6/2 \times 26\text{ mV}} - 1) = 0.184\text{ mA} \quad \textbf{Ans.}$$

EXAMPLE 4.11. The forward current in a p-n junction is 1.5 mA at 27°C. If $I_o = 2.4 \times 10^{-14}$ and η = 1, what will be the forward biasing voltage across the junction?

Solution: We know that the diode current is given by

$$I = I_o(e^{V/\eta V_T} - 1)$$

$$1.5\text{ mA} = 2.4 \times 10^{-14}(e^{V/26\text{ mV}} - 1)$$

$$e^{V/26\text{ mV}} = 6.25 \times 10^{10}$$

$$V/26\text{ mV} = \ln(6.25 \times 10^{10})$$

$$V = 0.6463\text{ V} \quad \textbf{Ans.}$$

EXAMPLE 4.12. A certain P-N junction diode is reverse biased by 10 V source connected directly across the diode. If the reverse saturation current of the diode is 10 nA, what will be the theoretical diode current? Assume $\eta = 2$.

Solution: We know that the diode forward current is given by,

$$I = I_o(e^{V/\eta V_T - 1})$$

Because, the diode is reverse biased by large voltage, $e^{V/\eta V_T} << 1$

Therefore, $I = I_o(-1) = -I_o$

Diode current $I = -10$ nA **Ans.**

EXAMPLE 4.13. A germanium diode has static resistance of 4.5 Ω at l = 44.4 mA. Determine the diode dynamic resistance with a forward bias of 0.1 V. Assume ideal diode.

Solution: We know that the static resistance is given by

$$R = \frac{V}{I}$$

Therefore, $V = R \times I = 4.5 \times 44.4\text{ mA} = 0.199 \cong 0.2\text{ V}$

Now, let us find reverse saturation current I_o at V = 0.2 V.

$$I = I_o(e^{V/\eta V_T - 1})$$

$$44.4\text{ mA} = I_o(e^{0.2\text{ V}/26\text{ mV}} - 1)$$

$$I_o = 20.27\ \mu\text{A}$$

Therefore, the dynamic resistance at 0.1 V forward bias will be

$$r_f = \frac{\eta V_T}{I_o\, e^{V/\eta V_T - 1}}$$

or

$$r_f = \frac{1 \times 26\text{ mV}}{20.27\ \mu\text{A}\ e^{0.1\text{ V}/26\text{ mV}}} = 27.40\ \Omega \quad \textbf{Ans.}$$

EXAMPLE 4.14. A diode is connected in series with a load resistance of 250 Ω across a dc supply voltage of 10 V. How can you draw dc load line and find the quiescent point on the characteristics depicted in figure 4.58.

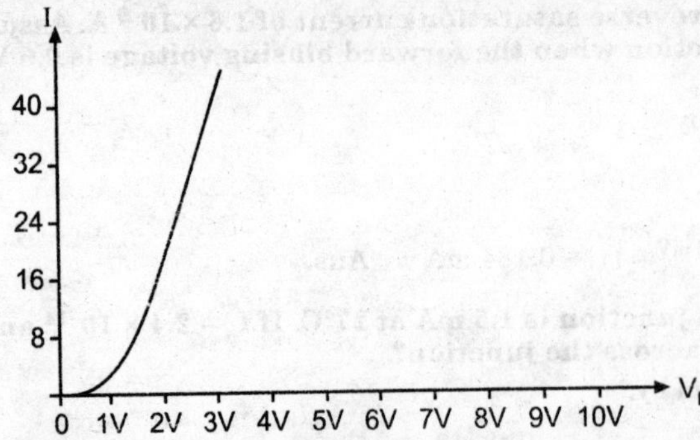

Fig. 4.58.

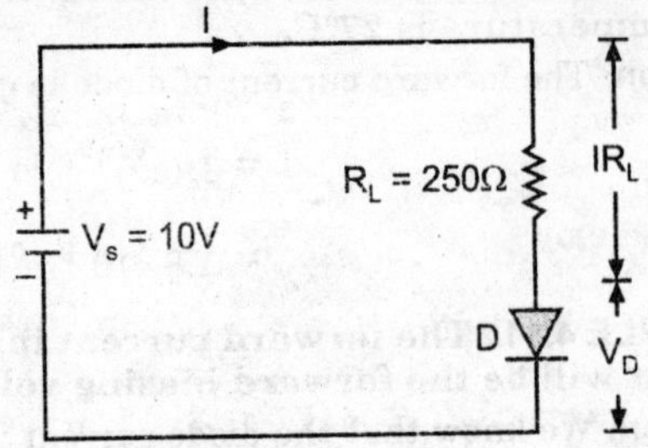

Fig. 4.59.

Solution: From the problem statement, let us draw the circuit as illustrated in figure 4.59.

From the circuit in figure 4.59. we can write

$$V_s = IR_L + V_D$$

When diode is off, no current can flow through the circuit

Therefore, $V_S = V_D$ (Since $IR_L = 0$)

$$10\text{ V} = V_D$$

When diode is ON, assuming ideal diode, we can say

$V_D = 0$

Therefore, $V_S = IR_L$

$$10\text{ V} = IR_L$$

or $$I = \frac{10\text{ V}}{250\ \Omega} = 40\text{ mA}$$

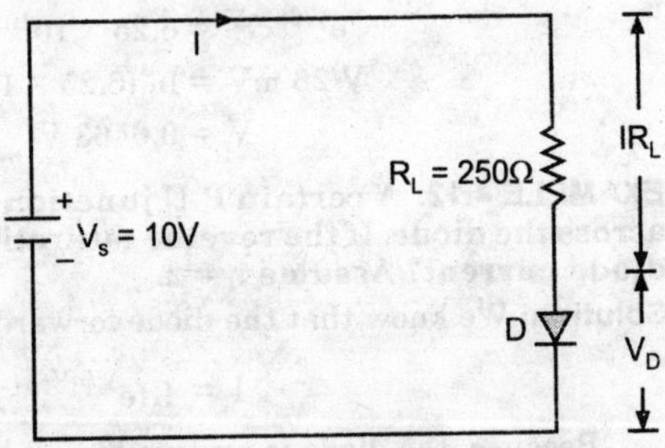

Fig. 4.60.

Hence, we get coordinates on X axis as 10 V and on Y axis as 40 mA. On joining these coordinates on the characteristics given, we obtain a line known as **dc load line** as shown in figure 4.60.

Important Point: The point at which the dc load line intersects the diode characteristics gives a quiescent point. From graph, quiescent current I = 32 mA and quiescent voltage = 2V. Thus, the diode current I is 32 mA and diode voltage V_D = 2 V.

EXAMPLE 4.15. Determine ac resistance of a Ge diode with a forward bias of 0.25 V. The reverse saturation current at room temperature is 1.2 μA.

Solution: V = 0.25 V

Given $I_o = 1.2\ \mu A = 1.2 \times 10^{-6}$A

At room temperature (300°K), V_T = 26 mV = 0.026 V

We know that the ac resistance (r) of the diode is given by

$$r = \frac{nV_T}{I_o\, e^{V/\eta V_T}} = \frac{0.026}{1.2\times10^{-6}\, e^{0.25/0.026}} \qquad (\because \eta = 1 \text{ for Ge})$$

$$r = 1.445\ \Omega \quad \textbf{Ans.}$$

EXAMPLE 4.16. Given a step graded germanium p-n junction. It has $N_D = 10^3\ N_A$ and N_A corresponds to 1 atom per 10^8 germanium atoms. Assume $n_i = 2.5 \times 10^{19}/\text{cm}^3$ and atom density of Ge as 4.4×10^{22} atoms/cm³. Determine the junction potential.

Solution: Atom density of Ge = 4.4×10^{22} atoms/cm³ and N_A corresponds to 1 atom per 10^8 Ge atoms.

Therefore, $$N_A = \frac{4.4\times10^{22}}{1\times10^{8}} = 4.4 \times 10^{14}\text{ atoms/cm}^3 = 4.4 \times 10^{20}\text{ atoms/m}^3$$

Now, we have $$N_D = 10^3 \times N_A = 10^3 \times 4.4 \times 10^{20}\text{ atoms/m}^3 = 4.4 \times 10^{23}\text{ atoms/m}^3$$

We know that the junction potential is given by,

$$V_J = V_T \ln\left[\frac{N_A N_D}{n_i^2}\right]$$

At room temperature, $V_T = 26$ mV = 0.026 .

Therefore, $$V_J = 0.026 \ln\left[\frac{4.4\times10^{20}\times4.4\times10^{23}}{(2.5\times10^{19})^2}\right] = 0.328 \text{ volt} \quad \textbf{Ans.}$$

EXAMPLE 4.17. A P-N junction germanium diode has at a temperature of 125°C, a reverse saturation current of 30 μA. Determine the dynamic resistance for 0.2 V bias in forward and reverse direction at a temperature of 125°C.

Solution: We know that the dynamic resistance r is expressed as

$$r = \frac{\eta V_T}{I_o e^{V/\eta V_T}}$$

Here, $\eta = 1$ for Ge

I_o = reverse saturation current

V = applied voltage

$$V_T = \frac{kT}{q} = \frac{1.381\times10^{-23}\ \text{J/°K}\times(125+173)}{1.6\times10^{-19}\text{C}}$$

$$V_T = 8.620\times10^{-5}\times398 = 0.03435 \text{ V} = 34.35 \text{ mV}$$

Forward dynamic resistance is given by

$$r_f = \frac{1\times0.03435}{30\times10^{-6}\times e^{0.2/0.03435}} = 3.391\Omega \quad \textbf{Ans.}$$

Reverse dynamic resistance is given by

$$r_f = \frac{1\times0.03435}{30\times10^{-6}\times e^{-0.2/0.03435}} = 386.76 \text{ k}\Omega \quad \textbf{Ans.}$$

EXAMPLE 4.18. The zero voltage barrier height at an alloy germanium P-N junction is 0.25 V. The concentration N_A of acceptor atoms in P side is much smaller than the concentration of donor atoms in the n material and $N_A = 3\times10^{20}/m^3$.

Determine the width of depletion layer for an applied reverse voltage of 10 V. Also obtain the space charge capacitance if cross-sectional area of diode is 1 μm². Assume ϵ_o = 8.854 pF/m and ϵ_r = 16.

Solution: Given that $N_A = 3\times10^{20}/m^3$; $A = 1\ \mu m^2 = 1\times10^{-6}\ m^2$; $V = -10V$

Therefore, $V_B = V_J - V = 0.25 + 10\ V = 10.25\ V$

We know that the equation for V_B is given by

$$V_B = \frac{9N_A}{2\epsilon}W^2$$

$$W = \sqrt{\frac{V_B\times2\epsilon}{9N_A}}$$

Substituting $\epsilon = \epsilon_o\,\epsilon_r$, we obtain

$$W = \frac{\sqrt{10.25\times2\times8.854\times10^{-12}\times16}}{1.6\times10^{-19}\times3\times10^{20}} = 7.77\times10^{-6}\text{ m} = 7.77\ \mu\text{m}$$

To evaluate space charge capacitance, we get

$$C_T = \frac{\epsilon A}{W} = \frac{8.854\times10^{-12}\times16\times1\times10^{-6}}{7.77\times10^{-6}} = 18.2126 \text{ pF} \quad \textbf{Ans.}$$

EXAMPLE 4.19. Determine the barrier capacitance of a germanium P-N junction whose area of cross-section is 1 mm × 1 mm and space charge thickness is 2 × 10⁻⁴ cm. The dielectric constant of germanium relative to free space is 16.

Solution: We know that the barrier capacitance (C_T) is given by,

$$C_T = \frac{\epsilon A}{W}$$

Given that $W = 2 \times 10^{-4}$ cm $= 2 \times 10^{-6}$ m

$A = 1$ mm $\times$ 1 mm $= 1$ mm$^2 = 1 \times 10^{-6}$ m^2

$\epsilon_r = 16$

We know that, $\epsilon_o = 8.854 \times 10^{-12}$

Hence, we have $C_T = \frac{\epsilon_o \epsilon_r A}{W} = \frac{8.854 \times 10^{-12} \times 16 \times 1 \times 10^{-6}}{2 \times 10^{-6}} = 70.83 \times 10^{-12}\text{F} = 70.832 \text{ pF}$ **Ans.**

EXAMPLE 4.20. A silicon diode has internal contact potential of 0.5 A reverse bias of 4.5 V is applied externally. The diode is doped such that $N_A << N_D$. The resistivity of P side is 5 Ω-cm. The diode has a circular cross-section with diameter D. The junction capacitance measured is 100 pF. Determine the diameter D.

Given: ϵ_r for Si = 12 and μ_p = 500 cm²/V-sec.

Solution: Given that

$\rho_p = 5$ Ω-cm $= 0.05$ Ω-m ; $V_j = 0.5$ V

$V = -4.5$ V ; $\mu_p = 00$ cm$^2 = 500 \times 10^{-4}$ m^2

$$\sigma_p = \frac{1}{P_p} = 20/\Omega\text{-m}$$

Also, we have $\sigma_p = qN_A\mu_p$

$$qN_A = \frac{\sigma_p}{p_p} = \frac{20}{500 \times 10^{-4}}$$

$$qN_A = 400$$

Also, we have $V_B = V_J - V = 0.5 - (-4.5) = 5$ V

We know that, $V_B = \frac{qN_A}{2\epsilon} W^2$

or $W^2 = \frac{2 V_B}{qN_A} \epsilon = \frac{5 \times 2 \times 8.854 \times 10^{-12} \times 12}{400} = 2.65 \times 10^{-12}$

or $W = 1.62 \times 10^{-6}$ m $= 1.629$ μm

From the expression $C_T = \frac{\epsilon A}{W}$, we write

$$A = \frac{C_T W}{\epsilon} = \frac{100 \times 10^{-12} \times 1.629 \times 10^{-6}}{8.854 \times 10^{-12} \times 12}$$

or $A = 1.533 \times 10^{-6}$ m $\left(\text{Since } A = \frac{\pi}{4} D^2\right)$

$$\frac{\pi}{4} D^2 = 1.533 \times 10^{-6} \text{ m}$$

$$D^2 = 1.953 \times 10^{-6}$$

$$D = 1.39 \times 10^{-3} \text{ m} = 1.39 \text{ mm}$$ **Ans.**

EXAMPLE 4.21. The intrinsic concentration is 1.5 × 10¹⁰/cm³. If the contact potential developed across the junction is 0.56 V, determine the temperature of junction in degree Celsius.

Given, μ_n = 1500 cm²/V-sec, μ_p = 500 cm²/V-sec, σ_n = 10 Ω-cm, σ_p = 3.5 Ω-cm.

Solution: We know that,

$$V_J = V_T \ln \frac{N_A N_D}{n_i^2}$$

Let us use the following expression:

$$\sigma_p = N_A \mu_p q$$

$$N_A = \frac{\sigma_p}{\mu_p q} = \frac{1}{\rho_p \mu_p q} = \frac{1}{3.5 \times 500 \times 1.6 \times 10^{-19}} = 3.57142 \times 10^{15}/\text{cm}^3$$

From expression, we have

$$\sigma_n = N_D \mu_n q$$

$$N_D = \frac{\sigma_n}{\mu_n q} = \frac{1}{\rho_n \mu_n q} = \frac{1}{10 \times 1500 \times 1.6 \times 10^{-19}} = 4.16666 \times 10^{14}/\text{cm}^3$$

Substituting N_A and N_D values in the expression of V_J, we obtain

$$0.56 = V_T \ln \frac{3.572 \times 10^{15} \times 4.167 \times 10^{14}}{(1.5 \times 10^{10})^2}$$

$$0.56 = V_T \times 22.613$$

$$V_T = \frac{0.56}{22.613} = 0.02476 \text{ V} = 24.77 \text{ mV}$$

Also, we get $$V_T = \frac{T}{11600}$$

$$T = 24.77 \text{ mV} \times 11600 = 287.27°\text{K} \qquad (1°\text{K} = 19.78°\text{C})$$

$$T = 14.516°\text{C} \qquad \textbf{Ans.}$$

EXAMPLE 4.22. For what voltage will the reverse current in a P-N junction germanium diode reach 90% of its saturation value at room temperature?

Solution: We know that the diode current is given by,

$$I = I_o [e^{V/\eta V_T} - 1]$$

At room temperature, we have

$$V_T = 26 \text{ mV, for Ge, } \eta = 1$$

$$I = -90\% \text{ of } I_o$$

$$-0.9 I_o = I_o [e^{V/26 \times 10^{-3}} - 1]$$

$$-0.9 I_o = I_o e^{V/26 \times 10^{-3}} - I_o$$

$$0.1 I_o = I_o e^{V/26 \times 10^{-3}}$$

Taking natural log of both sides, we obtain

$$\ln(0.1) = \frac{V}{26 \times 10^{-3}}$$

$$V = -0.05987 \text{ V} \qquad \textbf{Ans.}$$

The negative sign indicates a reverse bias voltage.

EXAMPLE 4.23. A static resistance of 5 Ω is observed in an ideal germanium diode at room temperature. The current flowing through the diode is l = 50 mA. If the forward biased voltage is 0.2 V, determine

(i) Reverse saturation current

(ii) Dynamic resistance of the diode.

Solution: We know that the static resistance (R) is given as

$$R = \frac{V}{I}$$

$$5\,\Omega = \frac{V}{50 \text{ mA}}$$

$$V = 250 \text{ mV}$$

Also, the diode current I is given by,

$$I = I_o[e^{V/\eta V_T} - 1]$$

or $$I_o = \frac{I}{[e^{V/\eta V_T} - 1]} = \frac{50 \text{ mA}}{[e^{250 \text{ mV}/1 \times 26 \text{ mV}} - 1]} = 3.33 \times 10^{-6} \text{ A} = 3.33 \text{ μA} \qquad \textbf{Ans.}$$

$$r = \frac{\eta V_T}{I_o \, e^{V/\eta V_T}} = \frac{1 \times 26 \times 10^{-3}}{3.33 \times 10^{-6} \times e^{0.2/1 \times 26 \times 10^{-3}}} = 3.563 \; \Omega \qquad \textbf{Ans.}$$

SUMMARY

- The semiconductor diode is formed by joining section of p-type material with a similar section of n-type material.
- The area where p and n regions meet is known as p-n junction.
- The various ways of p-n junction formation are
  (i) Grown junction (ii) Alloy junction (iii) Point contact junction
- An electrical junction where transition from one doping type to other occurs is known as metallurgical junction.
- An electric contact established through metal semiconductor junction between a semiconductor and external bias circuit is known as ohmic contact.
- The ohmic contact is non-rectifying is nature and allows, current in either direction with minimum voltage drop across it.
- Graded semiconductor means non-uniformly doped semiconductor. The non-uniform doping induces the voltage within the material.
- When p-n junction is formed, there is sudden exchange of charge carriers like a step at the junction.
- The sudden exchange of charge carriers across unbiased p-n junction forms a depletion layer or depletion region.
- The electric field developed across the junction due to diffusion of charge carriers is known as electrostatic field.
- The depletion layer does not allow free movement of charge carriers across the junction.
- The voltage which will vanish the depletion layer is known as cut-in voltage (V_r).
- The cut-in voltage is 0.3 V for Ge and 0.7 V for Si diodes.
- The reverse saturation current (I_o) in Ge is of the order of μA while in Si it is in nA.
- The reverse breakdown voltage of silicon diode is greater than that germanium diode.
- Diode conducts only when biased forward and does not conduct when biased reverse and hence can be used as a rectifying device.
- The resistance offered by diode in forward direction is classified as static and dynamic resistance whereas in reverse bias diode offers infinite resistance.
- The diode characteristics is exponential in nature during forward bias.
- The diode voltage decreases at a rate of 2.5 mV/°C for both silicon and germanium with rise in temperature.
- The reverse saturation current becomes double for every 10°C for rise in temperature.
- The dc load line is used to obtain an operating point (Q point) on forward characteristics of a diode.
- Q point indicates dc voltage across a diode and dc current through a diode.
- The representation of a semiconductor diode by an equivalent electronic function is known as diode model. The diode models are classified as small signal model and large signal model.
- The diode characteristics is approximated as a straight line segments to draw its piecewise linear model.

- The capacitances associated with p-n junction are
  - (i) Transition capacitance (C_T)
  - (ii) Diffusion capacitance (C_D)
- C_T is dominant under reverse bias condition of the junction whereas C_D is dominant when the junction is forward biased.
- The transition capacitance (C_T) is inversely proportional to width of the depletion layer.
- The diffusion capacitance (C_D) is proportional to the applied forward voltage.
- The addition of storage time (t_s) and transition time (t_r) gives reverse recovery time (t_{rr}).

SHORT QUESTIONS WITH ANSWERS

Q.1. What is a P-N junction ?

Ans. The contact surface between the layers of P-type and N-type semiconductor pieces placed together so as to form a junction is called the P-N junction.

Q.2. Which type of charges are present on the two opposite faces of the junction ?

Ans. Positive charge on N-side and negative charge on P-side of the junction.

Q.3. What types of carriers are present in space charge region ?

Ans. No mobile carrier is present in the space charge region.

Q.4. Why is space region called the depletion region ?

Ans. The region around the junction is completely ionised on formation of P-N junction. As a result, there are no free electrons on the N-side or the holes on the P-side. Since the region around the junction is depleted of mobile charges, it is called the depletion region.

Q.5. What is a compensated semiconductor ?

Ans. A semiconductor containing both donor and acceptor impurity atoms in the same region is called a compensated conductor.

Q.6. How is compensated semiconductor formed ?

Ans. Compensated conductor is formed by diffusing donor impurities in a P-type material or by diffusing acceptor impurities in an N-type material.

Q.7. What is forward and reverse biasing of a P-N junction ?

Ans. When an external field, with P-region connected to the positive terminal and N-region connected to the negative terminal of the battery, is applied across the junction, the junction is said to be forward biased.

When an external bias voltage is applied with positive terminal to N-side and the negative terminal to the P-side of a P-N junction, the junction is said to be reverse biased.

Q.8. What is the effect of reverse bias on the width of a P-N junction ?

Ans. When the P-N junction is reverse biased, the width of depletion region is increased.

Q.9. How do the transition region width and contact potential across a P-N junction vary with the applied bias voltage ?

Ans. When the P-N junction is forward biased, the transition region width is reduced and the contact potential is also reduced with the increase in applied bias voltage.

When the P-N junction is reverse biased, the transition region is widened, and the contact potential is increased with the increase in applied bias voltage.

Q.10. Why an electric field is produced in a depletion region of a P-N junction ?

Ans. The separation of positive and negative space charge densities in a P-N junction results in an electric field.

Q.11. What is space charge width ?

Ans. The space charge region extends into the N and P-regions from the metallurgical junction. The distance is known as the space charge width.

Q.12. The electric field in the space charge region decreases with forward bias and increases with reverse bias. Why ?

Ans. Because applied electric field opposes built-in field.

Q.13. Define cut-in voltage of a P-N junction diode ?

Ans. The forward voltage, at which the current through the P-N junction starts increasing rapidly, is called the cut-in voltage.

Q.14. What do you understand by reverse saturation current of a diode ?

Ans. Reverse saturation current of a diode is due to minority carriers and is caused when the diode is reverse biased. Only a very small voltage is required to direct all minority carriers across the junction, and when all minority carriers are flowing across further increase in bias voltage will not cause increase in current. This current is referred to as a reverse saturation current.

Q.15. What is the effect of temperature on the reverse current of a P-N junction ?

Ans. Reverse current of a P-N junction increases with the increase in junction temperature.

Q.16. Define peak inverse voltage ?

Ans. Peak inverse voltage is the maximum voltage that can be applied to the P-N junction without damage to the junction. If the reverse voltage across the junction exceeds its peak inverse voltage (PIV), the junction may get destroyed owing to excessive heat.

Q.17. Define the limitations in the operating conditions of P-N junction.

Ans. Every P-N junction has limiting values of *(i)* maximum forward current *(ii)* peak inverse voltage (PIV) and (iii) maximum power rating. The P-N junction provides satisfactory performance when operated within these limiting values. The P-N junction diode may get destroyed due to excessive heat if any of these values is exceeded.

Q.18. What are the two mechanisms of breakdown in a P-N junction ?

Ans. Avalanche and zener breakdown.

Q.19. Name the breakdown mechanism in a lightly doped P-N junction under reverse biased condition.

Ans. Avalanche breakdown.

Q.20. Name the breakdown mechanism in a heavily doped P-N junction under reverse biased condition.

Ans. Zener breakdown.

Q.21. What is an ideal diode ?

Ans. A two terminal device, called the diode, if acts as a perfect conductor (with zero resistance or zero voltage drop across it) when forward biased and as perfect insulator (with an infinite resistance or with no current through it) when reverse biased, is an ideal diode.

Q.22. Is reverse saturation current of a junction diode independent of reverse bias voltage ?

Ans. Yes.

Q.23. Enumerate the properties that the junction diodes designed for use as rectifiers should possess.

Ans. Junction diodes designed for use in rectifiers should have the properties : (i) V-I characteristics as close as possible to that of an ideal diode (ii) Negligible forward resistance (iii) Negligible reverse current i.e. large reverse resistance (iv) Large reverse breakdown voltage and (v) Small offset voltage V_0 in the forward direction.

Q.24. Define breakdown voltage.

Ans. The voltage where the magnitude of diode current takes a sharp turn and increases rapidly on a small increase of reverse voltage magnitude, is known as breakdown voltage

Q.25. What is a zener diode ?

Ans. Zener diode is a P-N junction diode specially designed for operation in its breakdown region.

Q.26. What is zener voltage ?

Ans. The voltage at which the zener diode breaks down is called the zener voltage.

Q.27. What is meant by the temperature coefficient ?

Ans. The effect of temperature on zener voltage is given in terms of temperature coefficient which is defined as the percentage change in nominal zener voltage for each degree centigrade of change in junction temperature.

Mathematically it is given as $\frac{\Delta V_Z}{V_Z(t - t_0)} \times 100$ where ΔV_Z is the change in zener voltage due to temperature variation of $(t - t_0)$.

Q.28. What happens to the series current, load current and zener current when the dc input voltage of a zener regulator increases ?

Ans. Zener current and series current increase while the load current remains unchanged.

Q.29. Draw the equivalent circuit of an ideal zener diode in breakdown region.

Ans. In breakdown region ideal zener diode appears like a dc battery of V_z volts.

Q.30. Why is zener diode used as voltage regulator ?

Ans. Zener diode has the property of behaving like a dc battery in 'on' state (i.e. when the voltage across zener diode exceeds its zener voltage rating V_Z). In 'on state', the voltage across zener diode remains constant until the voltage across it drops less than V_Z. This property of zener diode makes its use as voltage regulator.

REVIEW QUESTIONS

1. Explain P-N junction operation in forward and reverse biased condition.
2. Explain and compare the V-I characteristics of P-N junction.
3. Explain why a P-N diode acts as a rectifier.
4. Write the volt-ampere equation for a P-N junction diode and explain the meaning of each symbol.
5. How does the reverse saturation current of a P-N diode vary with temperature?
6. How does the diode voltage vary with temperature?
7. Define and explain static and dynamic resistance of a diode.
8. Which are the capacitances associated with junction. How the diode performance is affected due to junction capacitances for time varying signals?
9. What is d.c. load line and operating point for a diode? Explain with proper equations and graph.
10. How does the transition capacitance (C_T) vary with the depletion layer width?
11. How does the diffusion capacitance C_D vary with d.c. diode current?
12. Derive the expression of dynamic resistance of a diode.
13. Give features of signal diodes and switching diodes.
14. What is power diode? Give its important features and applications.
15. What is meant by the term reverse-recovery time? Is it due to the majority or minority carriers?
16. Define diffusion capacitance of P-N diode. Obtain an expression for the same. Why is the diffusion capacitance negligible for a reverse-biased diode.
17. Define cut-in voltage and P-N junction diode. Draw the reverse-bias characteristics of the P-N junction.
18. What is heterojunction? How does it differ from a homojunction? Why is it necessary that the lattice constants of the two semiconductor materials used to form a heterojunction must be well matched.Give the three different modes of alignment of the band gap energies.
19. How does an anisotype differ from an isotype heterojunction? Name the three possible modes of alignment of the band gap energies. Which one applies to most heterojunctions.
20. What is a varactor diode? How does the junction capacitance of a varactor diode vary with the variation in reverse bias voltage.

21. Write short notes on the following:
    (i) Switching diodes (ii) Varactor diode
    (iii) Transition and diffusion capacitance of a diode
    (iv) Schottky diode (v) Heterojunction

NUMERICAL PROBLEMS

1. Determine the thermal equilibrium hole and electron concentrations in a germanium sample at T = 300 K in which $N_A = 1 \times 10^{15}$ cm^{-3} and $N_D = 1 \times 10^{12}$ cm^{-3} and $n_i = 2.1 \times 10^{10}$ cm^{-3}. (**Ans.** 0.999×10^{15} cm^{-3}; 4.4144×10^5 cm^{-3})
2. Determine the built-in potential of a silicon P-N junction at 300 K with doping density of $N_A = 1 \times 10^{16}$ cm^{-3} and $N_D = 1 \times 10^{15}$ cm^{-3} for intrinsic concentration of 1.5×10^{10} cm^{-3}. (**Ans.** 0.635 V)
3. Determine the current flowing in the circuit shown in figure 4.61. (**Ans.** 1.94 mA)

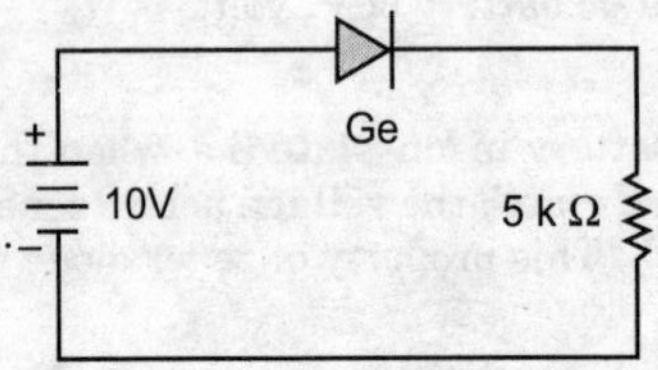

Fig. 4.61.

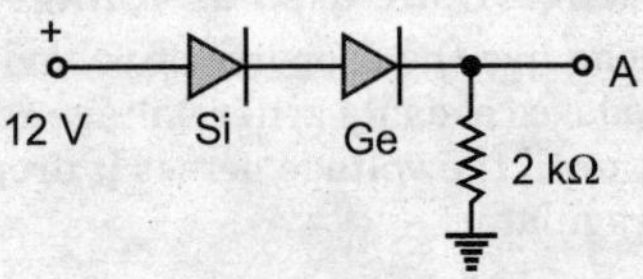

Fig. 4.62.

4. Determine the current flowing in the circuit shown in figure 4.62. Also, determine the potential of point A. (**Ans.** 5.5 mA, 11 V)
5. A zener diode shunt regulator is shown in figure 4.63. Determine
   (i) load voltage,
   (ii) voltage, drop across series resistance,
   (iii) current through the diode (**Ans.** (i) 12 V (ii) 18 V (iii) 51 mA)

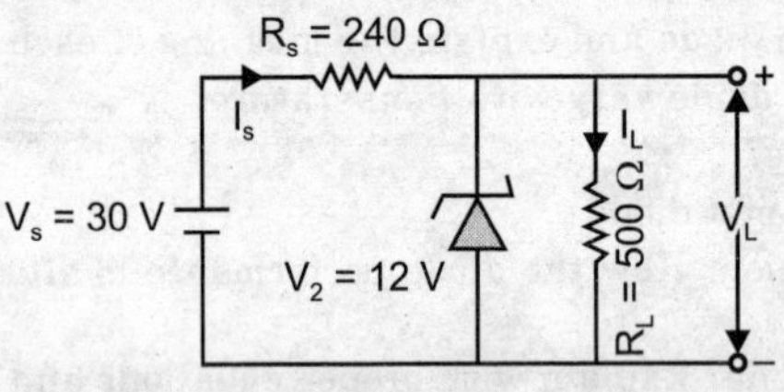

Fig. 4.63.

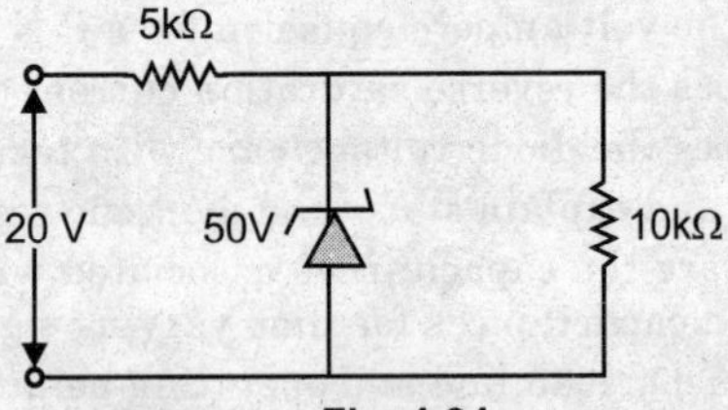

Fig. 4.64.

6. For the circuit shown in figure 4.64, find
   (a) the output voltage,
   (b) the voltage drop across series resistance,
   (c) the current through zener diode. (**Ans.** (a) 50 V (b) 70 V (c) 9 mA)
7. Determine the maximum and minimum values of zener current if in the circuit shown in figure 4.45 the load resistance $R_L = 5$ kΩ, series resistance is 2 kΩ, input voltage varies between 80 V and 100 V and the output voltage is 50 V. (**Ans.** 15 mA and 5 mA)
8. In the circuit shown in figure 4.45 the series resistance is 2 kΩ, load resistance is 5 kΩ, input voltage is 60 V and output voltage is 35 V. Determine (i) voltage drop across series resistance (ii) load current (iii) current through series resistance and (iv) zener diode current.
   (**Ans.** (i) 25 V (ii) 7.0 mA (iii) 12.5 (iv) 5.5 mA)

❑❑❑

UNIT-IV

Syllabus

Transistors: Metal-Semicondutor-Field-Effect-Transistors (MESFET), Metal-Insulator-Semiconductor-Field-Effect-Transistor (MISFET), Metal Oxide Semiconductor Field Effect Transistor (MOSFET), Construction, Operation and Characteristics of above Devices.

Bipolar Junction Transistors: Fundamentals of BJT operation, amplification with BJTs.

Above syllabus has been covered in following chapters :

Chapter 5

Bipolar Junction Transistors (BJTs)

Syllabus

Bipolar Junction Transistors: Fundamentals of BJT operation, amplification with BJTs.

Chapter Outline

5.1. INTRODUCTION

The transistor was invented By Dr. William Shockley and Dr. John Bardeen at Bell Laboratory in America in 1951. First time, in 1952 transistor was used in telephone switching circuits. Since then, it has revolutionized the field of Electronics.

The transistor has replaced the bulky vacuum tubes in most of the electronic circuits. The transistor is a basic building block of all modern electronic systems. It is a three terminal device. The output voltage, current or power are controlled by the input current in a transistor. Therefore, it is also called a current-controlled device.

In short, transistor is also called as BJT. BJT stands for Bipolar junction transistor because the transistor operation is carried out by two types of charge carriers : majority carriers and minority carriers.

A transistor has a very important property that it can raise the strength of an input weak signal. This property is called amplification. Due to this quality, the transistor is one of the most widely used semiconductor devices. At present, transistor is used in almost every field of electronics.

Few applications may be listed as under :

(i) Transistor is used in control systems.

(ii) Transistor is used in satellites and mobile phones.

(iii) In digital computer electronics, the transistor is used as a high-speed electronic switch.

DO YOU KNOW?

During the afternoon of December 23, 1947, Walter H. Brattain and John Bardeen demonstrated the amplifying action of the first transistor at the Bell Telephone Laboratories. The first transistor was called a point-contact transistor, which was the predecessor to the junction transistor invented by Schockley

(iv) In communication systems, it is widely used as the primary component in the amplifier.

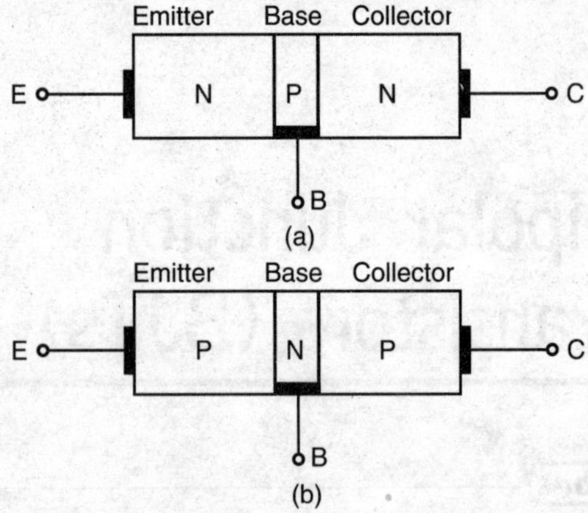

Fig. 5.1. (a) NPN Transistor (b) PNP Transistor.

5.1.1. Salient Features of a Transistor

The transistors are more desirable than the vacuum tubes because of their following salient features:

(i) Small size and ruggedness.

(ii) Do not require any filament power.

(iii) Operate at a low voltage.

(iv) Higher efficiency.

5.1.2. Why is it called as a Transistor ?

The term *transistor* was derived from the words **TRANSFER** and **RESISTOR**. This term was adopted because it best describes the operation of a transistor, which is the transfer of an input signal current from a low resistance circuit to a high resistance circuit.

5.2. JUNCTION TRANSISTOR CONSTRUCTION

A transistor is a single crystal of silicon (Si) or Germanium (Ge). A transistor can be of two types : NPN or PNP. A transistor consists of two PN junctions. The junctions are formed by sandwitching either P-type or N-type semiconductor layers between a pair of opposite types as shown in figure 5.1(a) and (b).

A transistor has three regions. They are emitter, base and collector. A brief introduction and functioning of these three junctions may be discussed as ahead:

1. Emitter : It is an outer region situated in one side of transistor. The function of emitter is to inject charge carriers (electrons in case of NPN transistor and holes in case of PNP transistors) into the base. Since emitter has to supply a large number of charge carriers, so it is heavily doped.

"More is the doping, more will be the available charge carriers."

2. Base : It is the middle region of the transistor. The base is a very thin and lightly doped region. The function of this region is to pass all the charge carriers (electrons or holes) onto the collector.

3. Collector : It is the other outer region situated in the other side of transistor. The doping of collector is between the heavy doping of emitter and light doping of Base. The function of collector is to collect charge carriers (electrons or holes). The collector region is physically larger than the emitter region. The reason for this is that the collector has to dissipate more heat. Hence, it is clear that although a BJT has two same type of outer regions, their function cannot be interchanged.

A transistor has two PN-junctions. One junction is formed between the emitter and the base. This is known as emitter-base junction. Another junction is formed between the base and the collector. This is known as collector-base junction.

Pioneers in Electronics :
George simon ohm (1787-1854)

The unit of measure for resistance, the ohm, was named for German physicist *George Simon Ohm (1787-1854).* Ohm is also known for his development of Ohm's law: I =

5.3. TRANSISTOR SYMBOLS

There are two types of transistors known as NPN or PNP. When the transistor is used as a circuit element in any electronic circuit, it is always represented by its circuit symbol as shown in figures 5.2 (a) and (b).

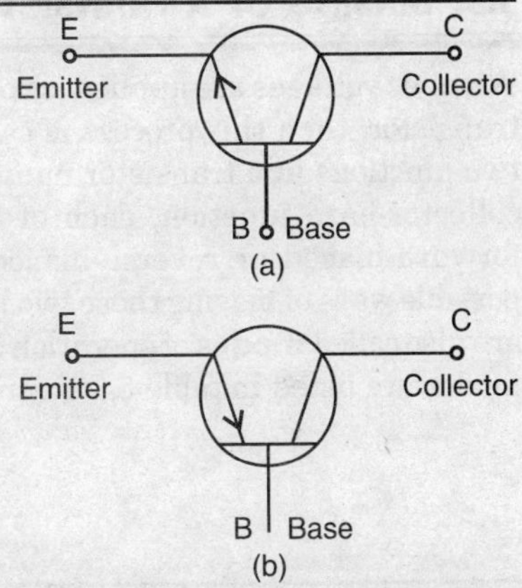

Fig. 5.2. (a) Symbol for NPN Transistor (b) Symbol for PNP Transistor

Figure 5.2(a) represents the symbol for NPN transistor. The emitter has an arrowhead. The arrowhead indicates the direction of conventional current flow in a transistor. The arrowhead points from the P-region towards the N-region. Therefore, in a NPN transistor the conventional current will flow from the base to the emitter.

Figure 5.2(b) shows the symbol for PNP transistor. The arrowhead in the emitter points from the P-region towards the N-region. Therefore in a PNP transistor the conventional current will flow from the emitter to the base.

The PNP transistor is a complement of the NPN transistor. Hence, in NPN transistor, the majority carriers are free electrons but in PNP transistor majority carriers are the holes. The currents and voltages involved in the action of PNP transistor are opposite to those of NPN transistor.

Out of these two types of transistor, the NPN transistor is mostly used. The reason for this is that in NPN transistors, the current conduction is mainly by electrons whereas in PNP transistors the current conduction is mainly by holes. Since the electrons are more mobile than holes, the conduction is higher in NPN transistors than PNP transistors.

DO YOU KNOW?

In a transistor, the emitter base depletion layer is narrower than the collector-base depletion layer. The reason can be attributed to the different doping levels of the emitter and collector regions. With much heavier doping in the emitter region, the penetration into the *n* material is minimal because of the availability of many more free electrons. However, on the collector side, fewer free electrons are available and the depletion layer must penetrate more deeply in order to set up the barrier potential.

5.4. UNBIASED TRANSISTOR

When no battery is connected between the different terminals of a transistor then the transistor is said to be in an unbiased state or in an open-circuit state. In this condition, the diffusion of charge carriers across the junction produces two depletion regions. The barrier potential for each of these two depletion layers is **0.7 V for Si** and **0.3 V for Ge.** But the three regions have different doping levels, so the depletion widths are not the same for two junctions. Since emitter is heavily doped than base and collector, the emitter-base depletion region width is smaller than that of collector-base depletion region width.

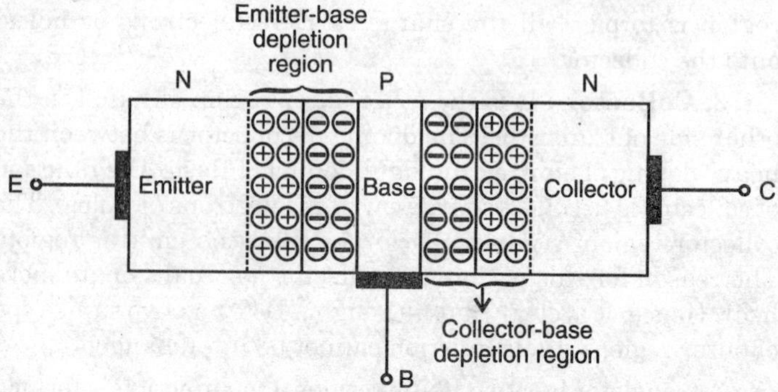

Fig. 5.3. Unbiased Transistor

5.5. BIASING OF A TRANSISTOR

When dc voltages are applied across the different terminals of a transistor, then this process is called biasing. Since, there are two junctions in a transistor namely emitter-base junction and collector-base junction, each of these two junctions may be forward-biased or reverse-biased. Therefore, there are four possible ways of biasing these two junctions. These possible ways are also called **modes** of operation of a transistor. These different modes are listed in table 5.1 given below:

DO YOU KNOW?

A transistor has three doped regions: an emitter, a base, and a collector. A *pn*-junction exists between the base and the emitter; this part of the transistor is called the emitter diode. Another *pn* junction exists between the base and the collector; this part of the transistor is called the collector diode.

Table 5.1.

Cases	Emitter-base junction	Collector-base junction	Region of operation
I	Forward-biased	Reverse-biased	Active
II	Forward-biased	Forward-biased	Saturation
III	Reverse-biased	Reverse-biased	Cut-off
IV	Reverse-biased	Forward-biased	Inverted

Case I - Active Region : In this mode the emitter-base junction is forward-biased and collector-base junction is reverse-biased.

Battery V_{EB} is connected between emitter and base so as to make emitter-base junction forward-biased. Similarly battery V_{CB} makes collector-base junction reverse-biased. In this region, transistor is used for amplification. In this region, the collector current depends upon the base current (input current).

Case II - Saturation Region : In this region, both the junctions are forward-biased. The transistor works in saturation. In this region, the collector current becomes independent of the base current. Therefore, the transistor acts like a closed switch.

Case III - Cut-off Region : In this mode, both the junctions are reverse-biased. In this region, the transistor has practically zero current because the emitter does not emit charge carriers into the base and no charge carriers are collected by the collector except a few thermally generated minority carriers. The transistor is open in this mode when it is used as an open switch.

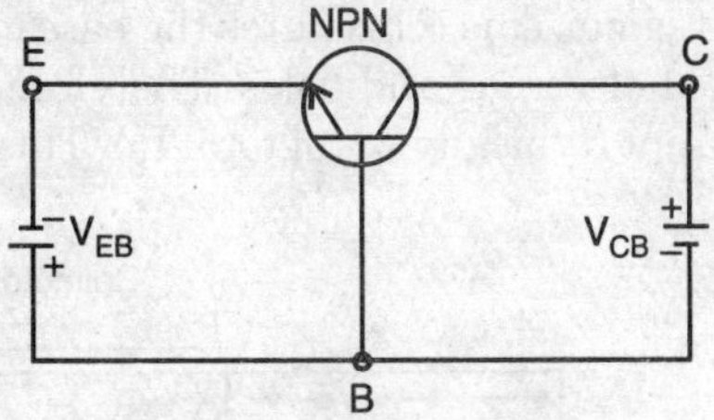

Fig. 5.4 (a). Active region

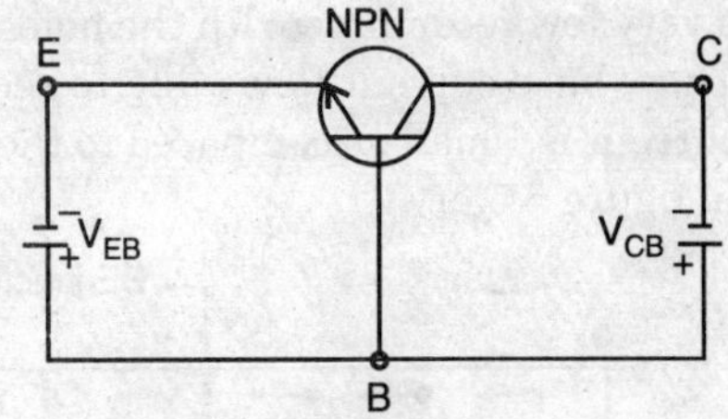

Fig. 5.4 (b). Saturation region

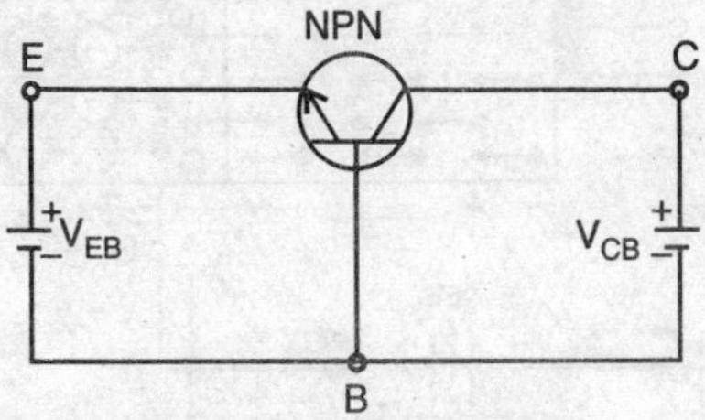

Fig. 5.4 (c). Cut-off region

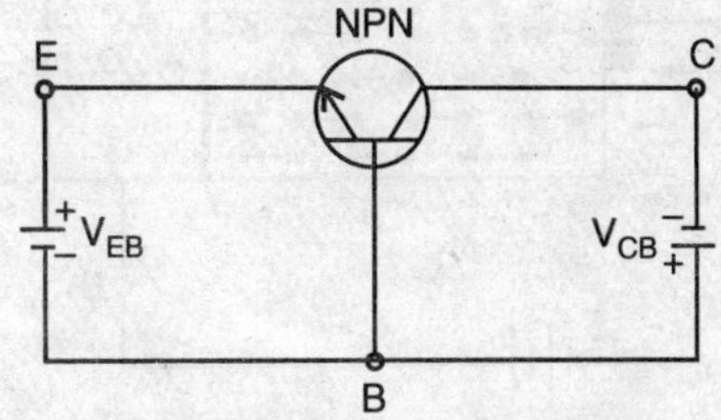

Fig. 5.4. (d). Inverted region

Case IV - Inverted Region : In this condition, the emitter-base junction is reverse-biased and the collector-base junction is forward-biased. Since, the doping level of emitter and collector is not the same, therefore, the collector cannot inject the majority charge carriers into the base. In this region, the action of transistor is poor. This region is of little importance.

5.6. TRANSISTOR OPERATION IN ACTIVE REGION

5.6.1 Operation of n-p-n-Transistor

The positive supply V_{EE} forward biases the base-emitter junction and the voltage V_{CC} reverse biases the collector to base junction as shown in figure 5.5 (a). Hence, the width of depletion region for base-emitter junction is very small, but that at the collector-base junction is large.

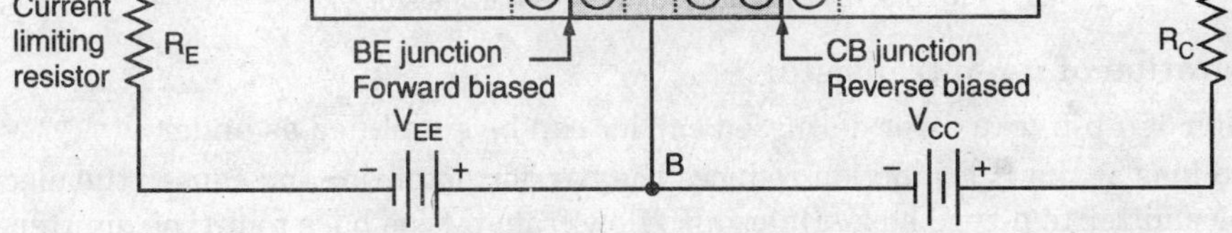

Fig. 5.5. (a) Transistor (n-p-n) biased to operate in the active region

Operation: The sequence of operation for an n-p-n transistor is as follows :

1. The electrons which are the majority carriers in the n-type emitter start flowing towards the p-type base as shown in figure 5.5 (b). This constitutes the emitter current I_E.

2. Electrons moving from the emitter to base have three options as under :

(i) They recombine with the holes present in the base. As the base region is thin and lightly doped, the number of holes is few. Hence, out of the total injected electrons from the emitter a very few recombine with the holes in the base region. This constitutes the base current I_B. Thus, base current flows due to recombination of electrons and holes. Therefore, the base current is, small as compared to the emitter current (typically 2% of total I_E). This is shown in figure 5.5(c).

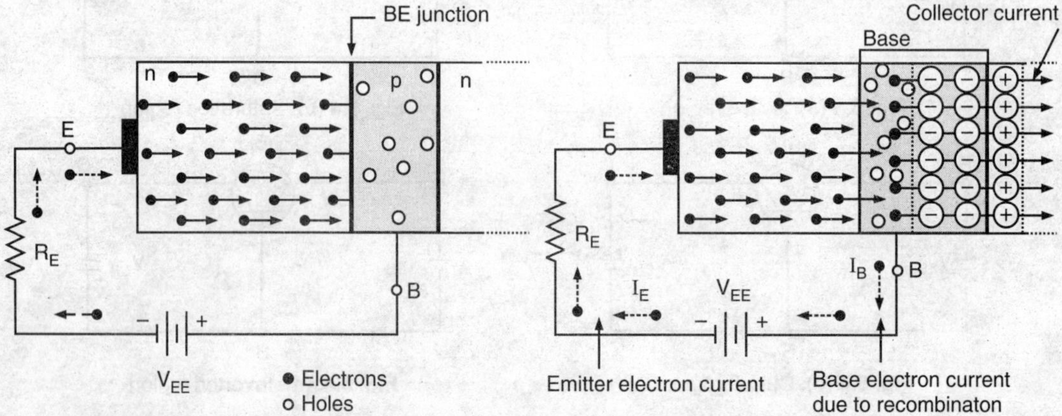

(b) Constitution of emitter current (c) Constitution of base and collector currents

Fig. 5.5 (b) and (c).

(ii) Some of the electrons diffuse through the base and out of the base connection.

(iii) The remaining large number of electrons pass through the depletion region of CB junction and pass through the collector region to the positive end of the external power supply V_{CC} as shown in figure 5.5 (d). The collector current I_C is much larger than the base current I_B.

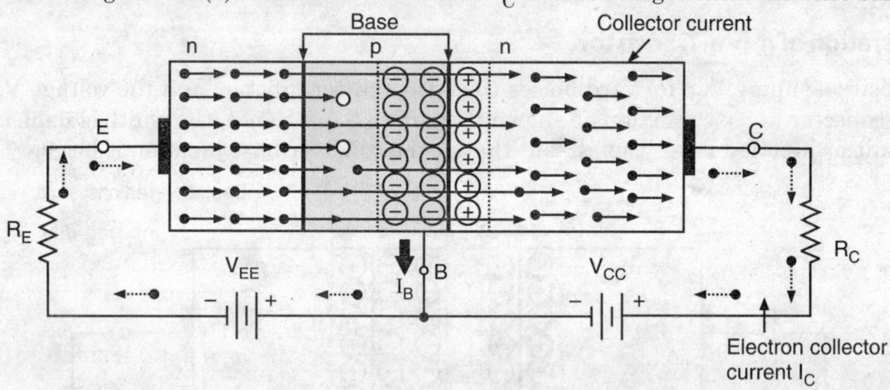

Fig. 5.5. (d) Operation of the n-p-n transistor

Simplified operation of n-p-n transistor

The operation of n-p-n transistor discussed earlier can be simplified as under :

The forward bias at the B-E junction reduces the barrier potential and causes the electrons to flow from n-type emitter to p-type base. Holes also flow from p-type base to n-type emitter. But, as the base is more lightly doped than the emitter, almost all the current flowing across the B-E junction consists of electrons entering the base from the emitter. Hence, electrons are the majority carriers in an n-p-n transistor. Some of the electrons entering into the base region do not reach the collector region. Instead they flow out of the base terminal via the base connection as shown in figure 5.6, due to recombination. As the base region is very thin and lightly doped, there are very few holes available in the base region for recombination. Hence, about 2% electrons flow out of base due to recombination. The remaining 98% electrons cross the reverse biased collector junction to constitute the collector current. They cross the collector region and are collected by the supply V_{CC}.

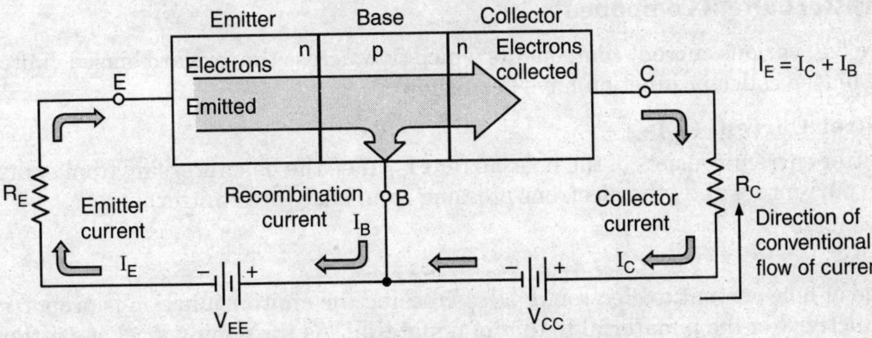

Fig. 5.6. Operation of an n-p-n transistor

The emitter current is equal to the sum of collector and base currents.

Therefore, $I_E = I_C + I_B$

5.6.2. Operation of p-n-p Transistor

The p-n-p transistor behaves exactly in the same way as the n-p-n device. The only difference is the majority charge carriers are holes instead of electrons. As shown in figure 5.7, holes are emitted from the p-type emitter across the forward biased EB junction, into the base.

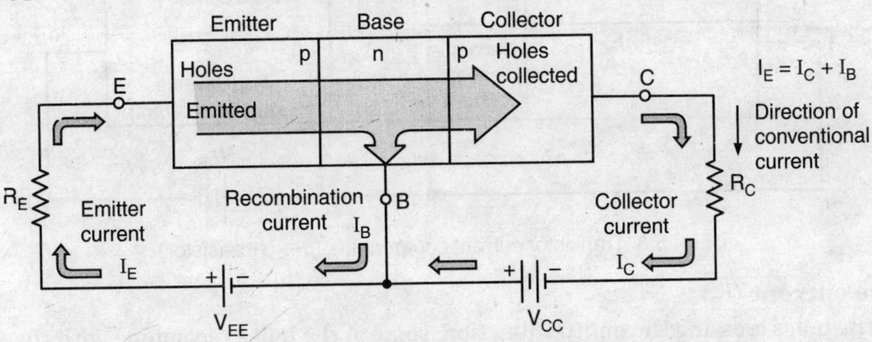

Fig. 5.7. Operation of a p-n-p transistor

In the lightly doped base there are very few number of electrons available for recombination. Therefore, about 2% of total emitted holes flow out via the base terminal and the remaining are drawn across the collector by the electric field at the reverse biased collector junction. As in case of n-p-n transistor, the forward bias at the EB junction controls the collector and emitter currents.

5.7. RELATION BETWEEN TRANSISTOR CURRENTS

As discussed earlier, the electrons injected from emitter into the base constitute the emitter current (I_E). Out of these electrons, very few combine with the holes in the thin base region to constitute the base current (I_B). The remaining electrons pass through to the collector region and then to the positive end of V_{CC} to constitute the collector current (I_C). Therefore we can write that,

$$I_E = I_C + I_B \qquad ...(5.1)$$

This means that emitter current is always equal to the sum of collector current and base current. As I_B is very small compared to I_E, we can assume the collector current to be nearly equal to the emitter current.

Hence, $I_C \cong I_E \qquad ...(5.2)$

5.7.1. Transistor Current Components

In figure 5.8, various current components which flow across the forward biased emitter junction and reverse biased collector junction have been shown.

1. The Emitter Current (I_E)

The emitter current consists of the hole current I_{pE} (i.e., the holes crossing from emitter to base) and electron current I_{nE} (i.e., the electrons crossing from base into emitter).

Thus, we have

$$I_E = I_{pE} + I_{nE} \quad ...(5.3)$$

The ratio of hole current to electron, I_{pE}/I_{nE} crossing the emitter junction is proportional to the ratio of conductivity of the p material to that of n material. As the doping level of emitter (p) side is much higher than that of the base (n) layer, $I_{pE} >> I_{nE}$ for a p-n-p transistor. Therefore, the emitter current is approximately equal to the hole current I_{pE}.

Therefore, $$I_E \cong I_{pE} \quad ...(5.4)$$

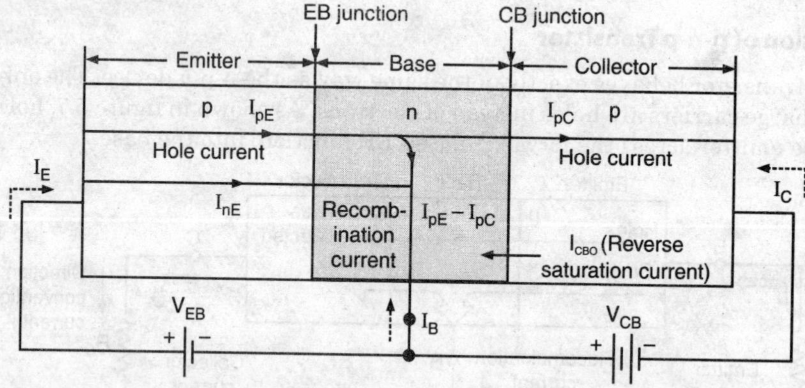

Fig. 5.8. Transistor current components (p-n-p transistor)

2. The base current (I_B)

Out of the holes crossing the emitter junction, some of the holes recombine with the electrons in the n-type base region, to constitute the base current. This is shown as recombination current ($I_{pE} - I_{pC}$) in figure 5.8.

3. The collector current (I_C)

The holes which do not recombine in the base region pass through the collector junction to constitute the collector current. It has been indicated as I_{pC} in figure 5.8. There is one more current component flowing inside the transistor which is called as the Reverse saturation current (I_{CBO}). I_{CBO} is the reverse saturation current that flows across reverse biased collector junction when emitter is open circuited as shown in figure 5.8.

Therefore, $$I_C = I_{CBO} \text{ when emitter is open circuited} \quad ...(5.5)$$

The reverse saturation current I_{CBO} flows due to the minority carriers and it increases with increase in temperature. I_{CBO} is also called as collector to base leakage current. It is made up of minority carriers. The minority carriers in case of p-n-p transistors are electrons moving from the n-type base to p-type collector. The collector current (I_C) when emitter junction is forward biased will be the sum of I_{pC} and I_{CBO}.

Therefore, $$I_C = I_{pC} + I_{CBO} \quad ...(5.6)$$

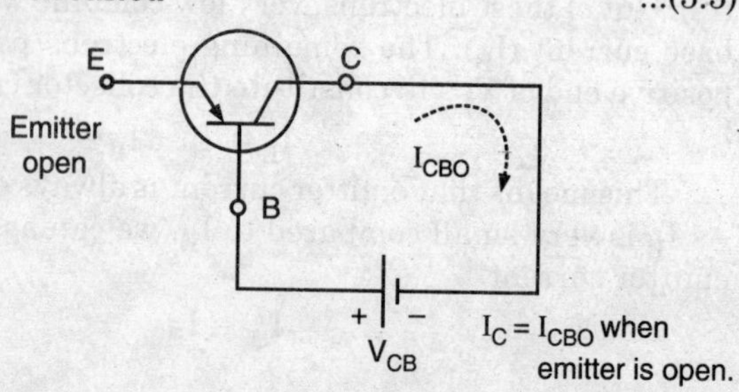

Fig. 5.9. $I_C = I_{CBO}$ when emitter is open

5.8. CIRCUIT SYMBOLS AND CONVENTIONS

The block diagram and conventional circuit symbols of n-p-n and p-n-p bipolar transistors have been shown in figure 5.10.

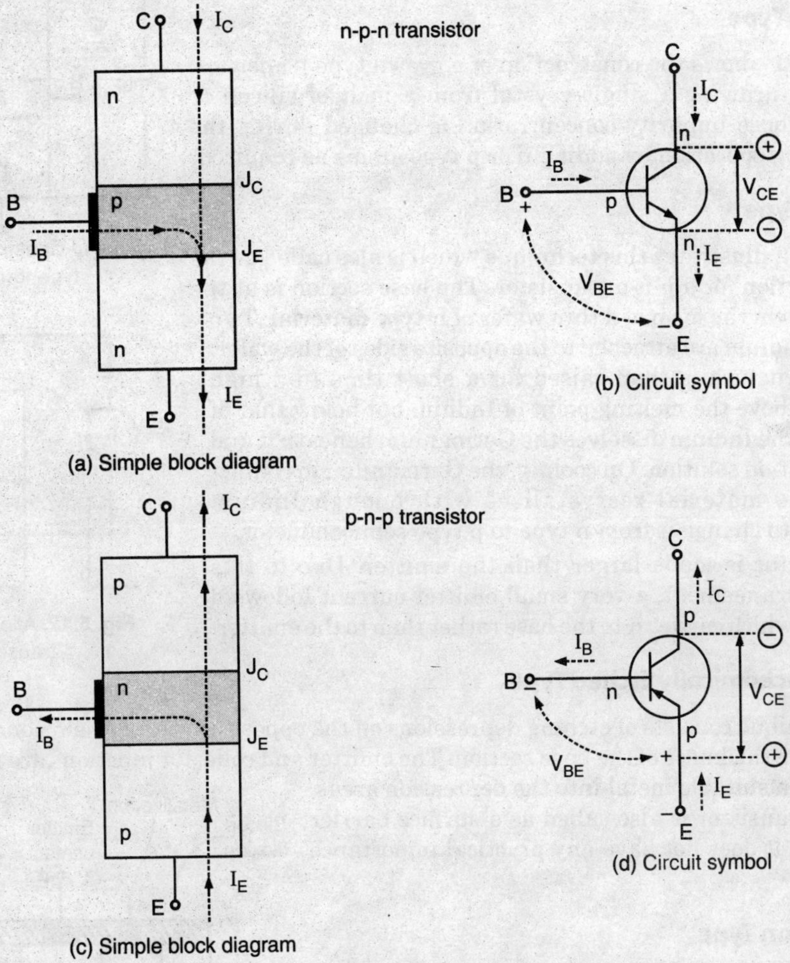

Fig. 5.10. Circuit symbols and conventions of a bipolar junction transistor (BJT)

The arrowhead is always placed on the emitter terminal and it indicates the direction of flow of the conventional emitter current. For the N-P-N transistor, the emitter current flows out of the emitter terminal whereas for the P-N-P transistor, the emitter flows into the emitter terminal. The current direction for all the other current components have been shown. Here, it may be noted that all these are the conventional currents.

5.9. TRANSISTOR CONSTRUCTION

There are five basic techniques developed for the manufactured diodes, transistors and other semiconductor devices. Hence, such devices may be classified into one of the following types:

1. Grown type
2. Alloy type
3. Electrochemically etched type

4. Diffusion type
5. Epitaxial type

Let us discuss them one by one

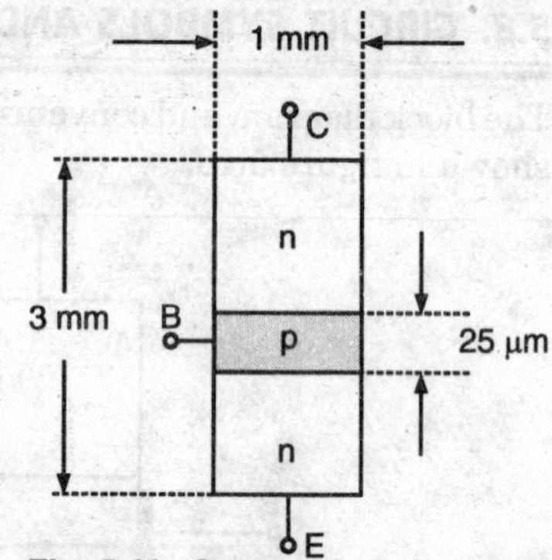

Fig. 5.11. Construction of grown type transistor

5.9.1. Grown Type

Figure 5.11 shows the construction of a grown type transistor. It is made by drawing a single crystal from a melt of silicon or germanium whose impurity concentration is changed during the crystal drawing operation by adding n or p type atoms as required.

5.9.2. Alloy Type

Figure 5.12 illustrates this technique which is also called as the fused construction for a p-n-p transistor. The base section is at the center and it is in the form of a thin wafer of n-type material. Two small dots of Indium are attached to the opposite sides of the wafer. The whole structure is then raised for a short time to a high temperature above the melting point of Indium but below that of Germanium. The Indium dissolves the Germanium beneath it and forms a saturation solution. On cooling, the Germanium in contact with the base material recrystalizes with enough Indium concentration to change it from n type to p type semiconductor.

The collector is made larger than the emitter. Due to this geometrical arrangement, a very small emitter current follows a diffusion path which carries it to the base rather than to the emitter.

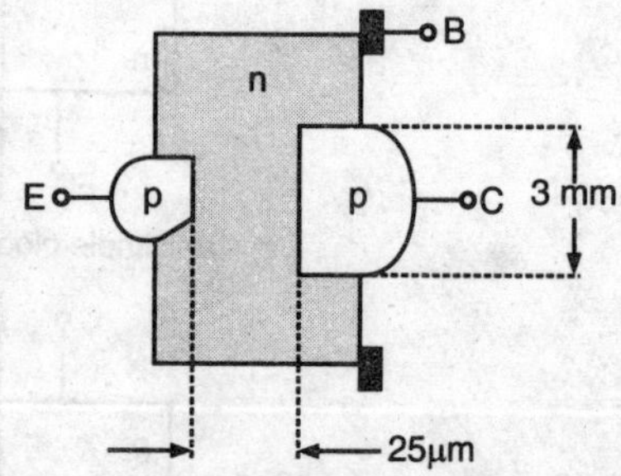

Fig. 5.12. Alloy type p-n-p transistor

5.9.3. Electrochemically Etched Type

This technique consists of etching depressions on the opposite sides of a semiconductor wafer so as to reduce the thickness of the base section. The emitter and collector junctions are then formed by electroplating a suitable metal into the depression areas. This type of transistor is also called as a surface barrier transistor and it does not have any practical importance now.

5.9.4. Diffusion Type

In this technique, a semiconductor wafer is subjected to gaseous diffusions on both n and p type impurities to from the emitter and collector junctions. Figure 5.13 shows the construction of a planer silicon transistor of the diffusion type. In this process, the base-collector junction, area is determined by a diffusion mask, which is photoetched just prior to the base diffusion. The emitter is then diffused on the base, and a final layer of silicon dioxide is thermally grown over the entire surface. Due to this layer, most surface problems are avoided and a very low leakage current results. There is also an improvement in the current gain at low currents and in the noise figure.

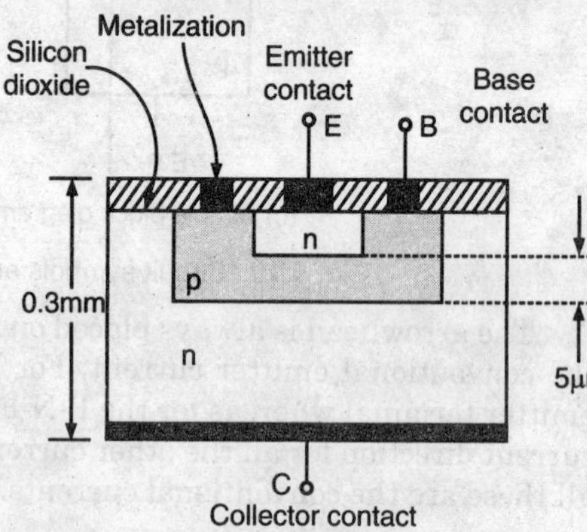

Fig. 5.13. Diffused or epitaxial planer type construction

5.9.5. Epitaxial Type

In this technique, a very thin, high purity, single crystal layer of silicon or germanium is grown on a heavily doped substrate of the same material. This augmented crystal forms the collector. The base and emitter may be diffused on it.

5.9.6. Combinations of Tecniques

The techniques discussed earlier can be combined to form a large number of methods for constructing the transistors. The examples of such combinations are as under :

(i) Diffused alloy type.

(ii) Grown diffused devices.

(iii) Alloy emitter epitaxial base transistor etc.

The V-I characteristics at low frequencies of all types of junction transistors are the same.

5.10. TRANSISTOR CONFIGURATIONS

(Important)

We know that a transistor may be represented as a two port network. A transistor has three terminals. Any of these three terminals can be made common to input and output. This common terminal is generally grounded or connected to the basis. Hence, depending upon the common terminal, there are three transistor circuit configurations namely common-base configuration, common-emitter configuration and common collector configuration. One thing should be noted that in all three configurations the transistor works in active region. This means that the emitter-base junction is always forward-biased and the collector-base junction is always reverse biased.

1. Common-base Configuration : In common-base configuration, the base of the transistor is made common. The input is applied between the emitter and base and the output is taken between the collector and base.

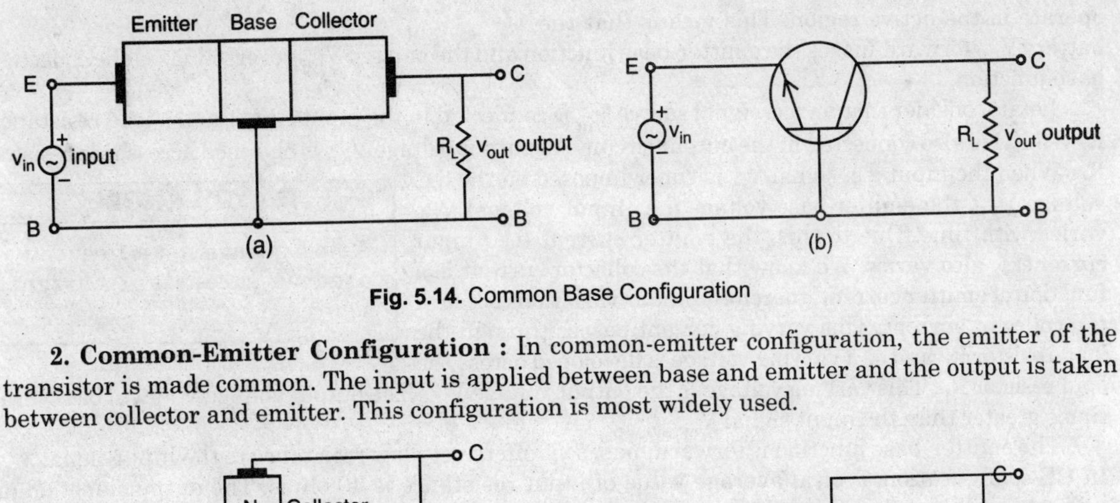

Fig. 5.14. Common Base Configuration

2. Common-Emitter Configuration : In common-emitter configuration, the emitter of the transistor is made common. The input is applied between base and emitter and the output is taken between collector and emitter. This configuration is most widely used.

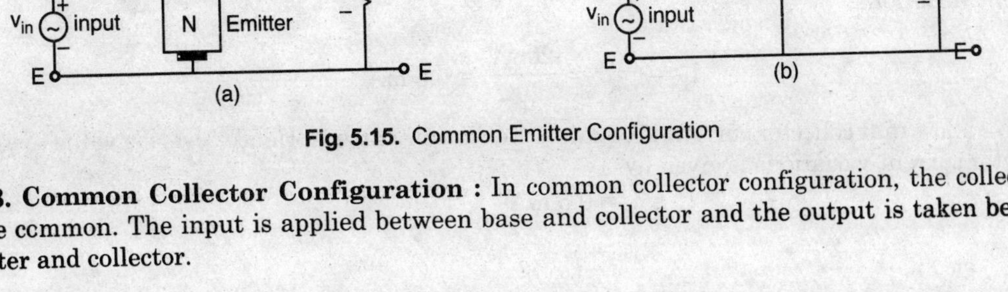

Fig. 5.15. Common Emitter Configuration

3. Common Collector Configuration : In common collector configuration, the collector is made common. The input is applied between base and collector and the output is taken between emitter and collector.

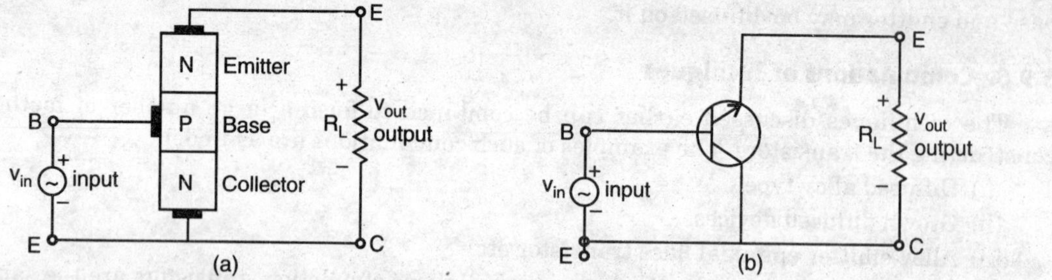

Fig. 5.16. Common Collector Configuration.

5.11. TRANSISTOR AS AN AMPLIFIER

A transistor may be used to perform a number of functions. But mainly transistor is used in amplifying electrical signals. Figure 5.17 shows the basic circuit of a transistor amplifier in which an NPN transistor is connected in common base configuration. The emitter is the input terminal and the collector is the output terminal. The weak input signal is applied between emitter-base junction and output is taken across the load resistance R_L which is connected in the collector circuit. In order to get faithful amplification, the transistor is biased to operate in the active region. This means that the battery V_{EE} forward-biases the emitter-base junction and the battery V_{CC} reverse-biases the collector-base junction.

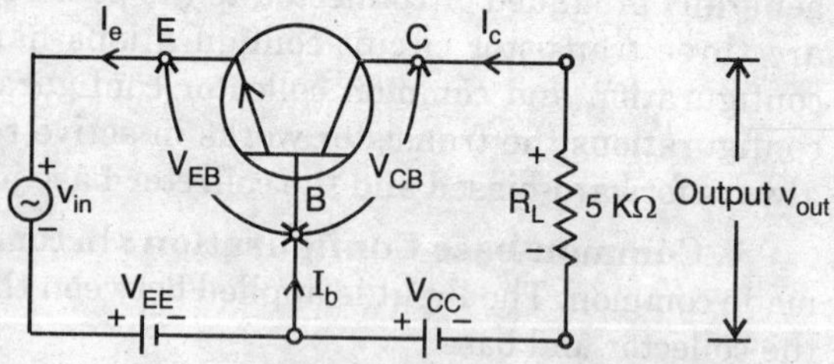

Fig. 5.17.

Let us consider that an a.c. signal source V_{in} is connected in the input circuit and a load resistance $R_L = 5\ K\Omega$ is also connected in the output circuit. An output voltage V_{out} is obtained across this resistor R_L. When the input a.c. signal V_{in} is super imposed on the d.c. voltage V_{EE}, the emitter-base voltage (i.e., input voltage) V_{EB} varies with time. Due to this, the emitter current (i.e., input current) I_e also varies. We know that the collector current is a function of emitter current, therefore similar variations occur in the collector current. This varying current passes through the load resistor R_L and so a varying voltage is developed across the load resistor R_L. This varying voltage is the output voltage V_{out}. The output voltage V_{out} will be several times greater than the input signal V_{in}.

> **DO YOU KNOW?**
>
> A bipolar transistor is frequently used as a constant current source.

The emitter-base junction is forward-bias, so it offers very low resistance to the input signal V_{in}. In CB configuration, typical average value of input resistance is 40 ohms. The output junction or collector-base junction is reverse-bias and so output resistance is very high. Its typical value is 500 KΩ.

Let the rms value of input signal be V_{in} = 20 mV, then rms value of input current or emitter current variations.

$$I_E = \frac{V_{in}}{r_i} = \frac{20\text{mV}}{40\Omega} = 0.5\text{ mA}$$

We know that collector current is equal to the emitter current. Hence, the rms value of input or collector current variations is given by

$$I_c \cong I_e = 0.5\text{ mA}$$

Since, the output resistance of transistor is very high than load resistance, therefore, almost all the current I_C passes through the load resistance R_L.

Hence, the rms value of output voltage will be

$$V_{out} = I_c.R_L = (0.5 \times 10^{-3}) \times (5 \times 10^3) = 2.5 \text{ volt}$$

Now, voltage gain or voltage amplification A of amplifier is given as ratio of output voltage V_0 to the input voltage V_{in}.

Therefore, $$A = \frac{V_{out}}{V_{in}} = \frac{2.5}{20 \times 10^{-3}} = 125$$

Hence, amplitude of input voltages increases by 125 times.

5.12. CURRENT GAIN OF A TRANSISTOR IN COMMON-BASE CONFIGURATION

1. Definition

Figure 5.18 shows the circuit of common-base configuration using NPN transistor and the actual directions of different currents in the circuit. In this configuration, emitter current is the input current and collector current is the output current. The current gain of a transistor is defined as the ratio of the transistor output current to the input current. Current gain may be of two types namely d.c. current gain and a.c. current gain depending upon whether the input and output currents are d.c. currents or a.c. currents.

2. D.C. current gain in common-base configuration

d.c. current gain may be defined as the ratio of collector current I_C to emitter current I_E. It is denoted by α or α_{dc} or h_{FB}.

$$\alpha = \frac{I_C}{I_E} = h_{FB} \quad \text{...(5.7)}$$

But we know that the value of collector current is always less than the value of emitter current. Hence d.c. current gain in CB configuration is always less than unity.

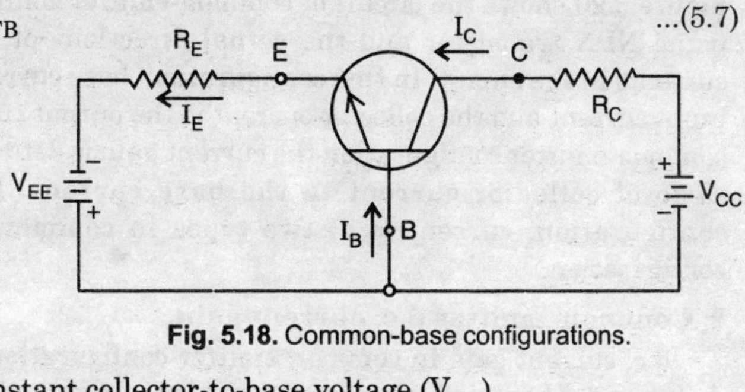

Fig. 5.18. Common-base configurations.

3. A.C. current gain in common-base configuration

A.C. current gain may be defined as the ratio of change in collector current (ΔI_C) to the change in emitter current (ΔI_E) for a constant collector-to-base voltage (V_{CB}).

It is represented by α_0, α_{ac} or h_{fb}

$$\alpha_0 = \alpha_{ac} = \left.\frac{\Delta I_C}{\Delta I_E}\right| h_{fb} \text{ constant} \quad \text{...(5.8)}$$

α_0 is also called **common-base short-circuit current gain.** The value of α_0 is also less than unity. α_0 is approximately same as α.

Therefore, for all practical purposes, we can write

$$\alpha = \alpha_0$$

$$\text{d.c. current gain} = \text{a.c. current gain} \quad \text{...(5.9)}$$

EXAMPLE 5.1. In a common base configuration, the value of collector current is 5.10 mA and the value of emitter current is 5.18 mA. Calculate the common-base d.c. current gain.

Solution : Common-base d.c. current gain is given as

$$\alpha_{dc} = \alpha = \frac{I_C}{I_E} = \frac{5.10}{5.18} \quad \textbf{Ans.}$$

EXAMPLE 5.2. Find the value of base current if the common-base d.c. current gain of a transistor is 0.987 and emitter current is 10 mA.

Solution : Common-base d.c. current gain is given as

$$\alpha = \frac{I_C}{I_E} \Rightarrow I_C = \alpha I_E$$

$$I_C = 0.987 \times 10 = 9.87 \text{ mA}$$

Emitter current is expressed as

$$I_E = I_B + I_C$$

$$I_B = I_E - I_C = 10 - 9.87 = 0.13 \text{ mA} \quad \textbf{Ans.}$$

EXAMPLE 5.3. Figure 5.19 shows a NPN transistor for which I_E = 10 mA, d.c. current gain α = 0.967. Find the values of I_C and I_B.

Solution : Common-base d.c. current gain is

$$\alpha = \frac{I_C}{I_E}$$

$$I_C = \alpha I_E = 0.967 \times 10$$

or $$I_C = 9.67 \text{ mA}$$

But $$I_E = I_B + I_C$$

$$I_B = I_E - I_C = 10 - 9.67 = 0.33 \text{ mA} \quad \textbf{Ans.}$$

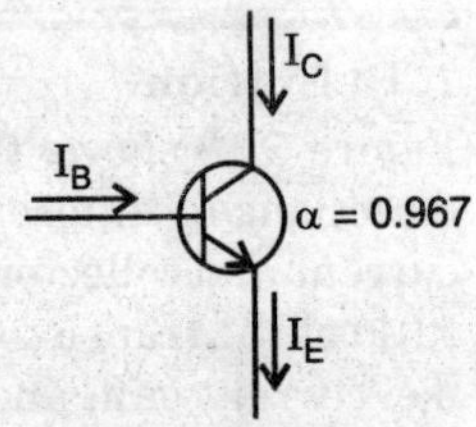

Fig. 5.19.

5.13. CURRENT GAIN OF A TRANSISTOR IN COMMON-EMITTER CONFIGURATION.

1. Definition

Figure 5.20 shows the circuit of common-emitter configuration using NPN transistor and the actual directions of different currents in the circuit. In this configuration, base current is the input current and the collector current is the output current. In common-emitter configuration the current gain is defined as the ratio of collector current to the base current. Like CB configuration, current is of two types in common-emitter configuration.

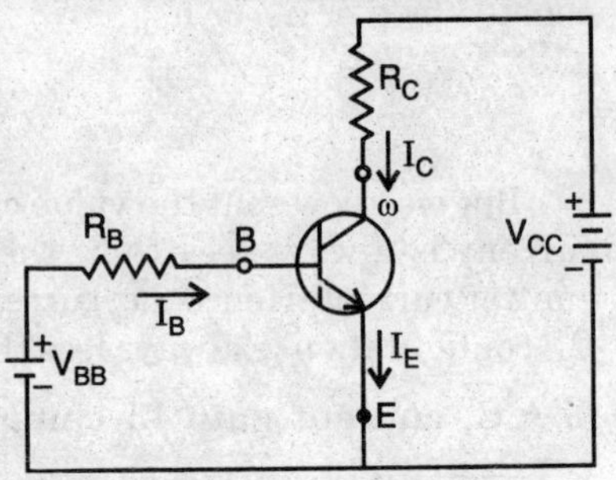

Fig. 5.20. Common emitter configuration

2. Common-Emitter d.c. current gain.

d.c. current gain in common-emitter configuration may be defined as the ratio of collector current I_C to base current I_B. It is denoted by β or β_{dc} or h_{FE}.

$$\beta_{dc} = \beta = \frac{I_C}{I_B} \quad ...(5.10)$$

Since, the collector current is much larger than the base current, the value of β is much greater than unity. Typical range of β may be from 20 to 300.

> **DO YOU KNOW?**
>
> The ratio of the collector current to the base current is called the current gain, symbolized as β_{dc} or h_{FE}. For low-power transistors, this is typically 100 to 300. The emitter current is the largest of the three currents, the collector current is almost as large, and the base current is much smaller.

3. a.c. current gain in common emitter configuration

a.c. current gain in common emitter configuration may be defined as the ratio of change in collector current ΔI_C to the change in base current ΔI_B keeping collector to emitter voltage V_{CE} constant. It is denoted by β_0, β_{ac} or h_{fe}.

$$\beta_0 = \frac{\Delta I_C}{\Delta I_B} \quad ...(5.11)$$

β_o is also known as **common-emitter short circuit current gain.** The values of β_0 are approximately same as that of β.

For all practical purposes, we can write

$$\beta = \beta_0 \qquad ...(5.12)$$

d.c. current gain = a.c. current gain

5.14. RELATION BETWEEN CURRENT GAIN α AND β.

(U.P. Tech, First Sem. Exam. 2000-2001)

We know that the emitter current I_E of a transistor is given as

$$I_E = I_B + I_C \qquad ...(5.13)$$

Here I_B = Base Current

I_C = Collector Current

Dividing above equation by I_C on both sides, we have

$$\frac{I_E}{I_C} = \frac{I_B}{I_C} + 1 \qquad ...(5.14)$$

But $$\frac{I_C}{I_E} = \alpha$$

and $$\frac{I_C}{I_B} = \beta$$

Putting these values in the last expression, we get

$$\frac{1}{\alpha} = \frac{1}{\beta} + 1 = \frac{1+\beta}{\beta}$$

or $$\alpha = \frac{\beta}{1+\beta}$$

or $$\alpha(1+\beta) = \beta$$

$$\alpha + \alpha\beta = \beta$$

$$\alpha = \beta - \alpha\beta = \beta(1-\alpha)$$

or $$\beta = \frac{\alpha}{1-\alpha} \qquad ...(5.15)$$

This is desired expression between current gain α and β.

EXAMPLE 5.4. The value of β for a transistor is 100. If the value of emitter current is 10 mA, then determine the values of collector and base current.

Solution : Given that $\beta = 100$ and $I_E = 10$ mA

value of β is given as

$$\beta = \frac{\alpha}{1-\alpha}$$

so that $$\alpha = \frac{\beta}{\beta+1} = \frac{100}{100+1}$$

$$\alpha = \frac{100}{101} = 0.98 \quad \textbf{Ans.}$$

common-base current gain α is given as

$$\alpha = \frac{I_C}{I_E}$$

$$I_C = \alpha I_E = 0.98 \times 10 \times 10^{-3}$$
$$I_C = 0.98 \times 10^{-4} = 9.8 \text{ mA} \quad \textbf{Ans.}$$

value of emitter current is given as

$$I_E = I_B + I_C$$
$$I_B = I_E - I_C = 10 - 9.8 = 0.2 \text{ mA} \quad \textbf{Ans.}$$

EXAMPLE 5.5. The value of α for a transistor is 0.950. Find the value of β. Also find the value of α if β changes to 100.

Solution : The relation between α and β is given as

$$\beta = \frac{\alpha}{1-\alpha} = \frac{0.950}{1-0.950} = \frac{0.950}{0.050} = 19$$

For second case

$$\because \quad \beta = \frac{\alpha}{1-\alpha} \quad \text{or} \quad \alpha = \frac{\beta}{\beta+1}$$

$$\alpha = \frac{100}{100+1} = \frac{100}{101} = 0.99 \quad \textbf{Ans.}$$

EXAMPLE 5.6. Figure 5.21 shows a NPN transistor for which I_E = 10 mA and β = 100. Determine the values of I_C and I_B.

Solution :

$$\text{We know that } \alpha = \frac{\beta}{1+\beta}$$

But
$$\alpha = \frac{I_C}{I_E}$$

$$\beta = \frac{I_C}{I_B}$$

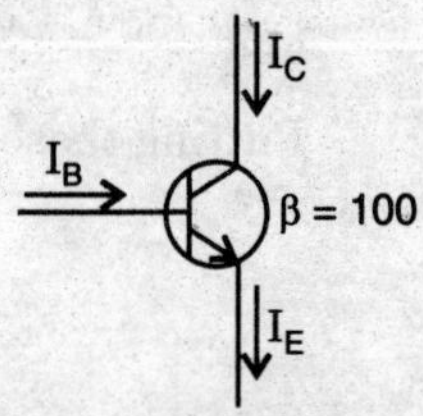

Fig. 5.21.

Putting the value of α and β, we get

$$\frac{I_C}{I_E} = \frac{I_C}{I_B} \cdot \left(\frac{1}{1+\beta}\right)$$

or
$$I_B = \frac{I_E}{1+\beta}$$

Given that $\beta = 100$, $I_E = 10$ mA

$$I_B = \frac{10}{1+100} = \frac{10}{101} = 0.098 \text{ mA}$$

Emitter current is given as

$$I_E = I_C + I_B$$
$$I_C = I_E - I_B = 10 - 0.098 = 9.902 \text{ mA} \quad \textbf{Ans.}$$

5.15. LEAKAGE CURRENT IN A COMMON-BASE TRANSISTOR

Figure 5.22 shows the circuit of a common base NPN transistor. The polarity of battery V_{EE} is such that it makes the emitter-base junction forward biased. The battery V_{CC} sets the collector-base junction in reverse biased. Now, we consider two cases

(i) Switch S is in closed position : In this case, electrons are injected into base region due to forward-bias. The most of these electrons are attracted by collector due to reverse-bias.

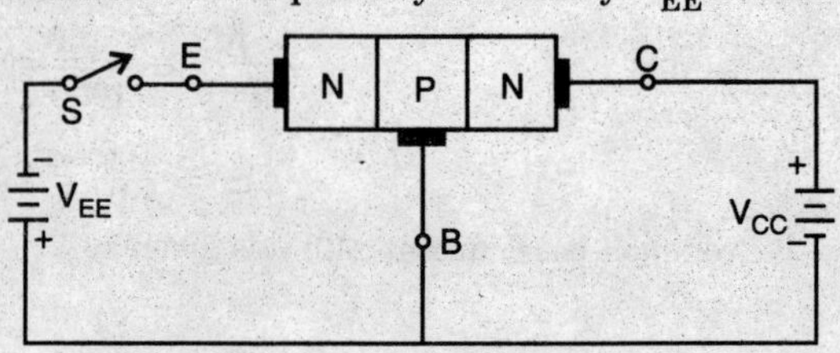

Fig. 5.22. Common-base transistor circuit

This accounts for collector current. Few electrons are combined with holes in base region and constitute base current. Therefore, total emitter current breaks up into two parts namely the collector current I_C and base current I_B.

$$I_E = I_B + I_C \qquad ...(5.16)$$

(ii) Switch S is opened : Let us assume that the switch is opened.

In this case, emitter-base junction of transistor is open-circuited. Therefore there is no emitter current which means that there is no base current and collector current. But since the collector-base junction is reverse-biased, minority carriers diffuse across the collector base junction and this produce a small current. This small current is called leakage current and is denoted by I_{CBO} or I_{CO}. I_{CBO} means the leakage current between collector and base with emitter open. It is also known as reverse saturation current and it is like reverse saturation current in a PN-junction.

> **DO YOU KNOW?**
>
> The graph of base current versus base-emitter voltage looks like the graph of an ordinary diode. Because of this, we can use any of the three diode approximations to calculate the base current. Most of the time, the ideal and the second approximation are all that is necessary.

Hence, now it is clear that total collector current for a transistor contains following parts:

(a) The injected current due to majority carriers or a fraction of emitter current is equal to $\alpha \cdot I_E$.

(b) The leakage current I_{CBO} or I_{CO} produced by thermally generated minority carriers.

Therefore,
$$I_C = \alpha I_E + I_{CBO} \qquad ...(5.17)$$

or
$$\alpha I_E = I_C - I_{CBO}$$

$$\alpha = \frac{I_C - I_{CBO}}{I_E} = \frac{I_C - I_{CBO}}{I_B + I_C} \qquad \because I_E = I_B + I_C$$

or
$$\alpha I_B + \alpha . I_C + I_{CBO} = I_C$$

$$\text{or } I_C(1-\alpha) = \alpha I_B + I_{CBO}$$

$$I_C = \frac{\alpha . I_B}{1-\alpha} + \frac{I_{CBO}}{1-\alpha} \qquad ...(5.18)$$

For silicon transistor, at room temperature, the value of I_{CBO} is of the order of nanoamperes and for germanium transistor, it is of the order of microamperes. For both Si and Ge, the value of I_{CBO} strongly depends on temperature. It becomes doubled for every 10° C rise in temperature. Since the value of I_{CBO} is less for Si than Ge, the silicon transistors can be used upto 200° C whereas Ge transistors can only be used upto 100° C. Therefore silicon transistors are preferred over germanium transistors.

5.16. LEAKAGE CURRENT IN COMMON-EMITTER CONFIGURATION

Figure 5.23 shows a circuit of common-emitter configuration with NPN transistor. If the switch S is closed, then, the usual currents I_E, I_C and I_B flow. It means that emitter current is the sum of collector and base currents.

When the switch is open, the emitter-base junction is an open circuit and so the value of input or base current is zero. But a leakage current I_{CEO} flows between collector and emitter. This leakage current I_{CEO} is not only due to the thermally generated minority carriers (I_{CBO}) across the collector-base junction but also due to the movement of holes which flow across the base-emitter junction. Due to flow of holes, the emitter base junction is slightly forward biased.

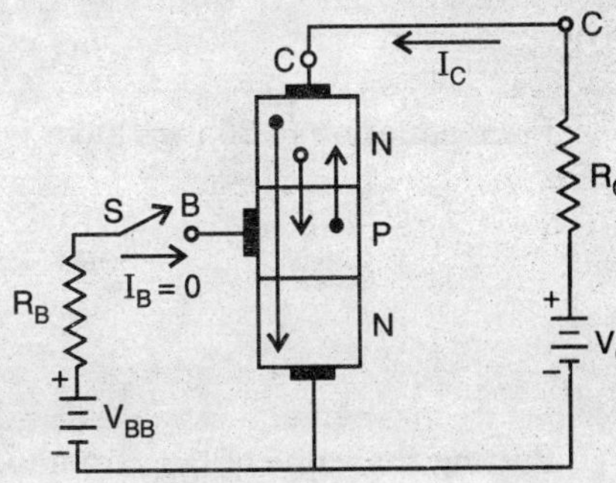

Fig. 5.23. Common emitter transistor circuit

Due to this reason, it can be assumed that a small base current I_{CBO} is supplied to the transistor. This small base current I_{CBO} is equal to additional collector current $\beta.I_{CBO}$.

Therefore, with emitter-base junction open, the total leakage current flowing through the transistor

$$I_{CEO} = I_{CBO} + \beta.I_{CBO} \quad ...(5.19)$$

$$I_{CEO} = I_{CBO}(1 + \beta) \quad ...(5.20)$$

Hence, the total collector current in transistor is the sum of injected current $\beta.I_B$ and I_{CEO}

$$I_C = \beta.I_B + I_{CBO}(1 + \beta)$$

$$I_C = \beta.I_B + I_{CBO}(1 + \beta) \quad ...(5.21)$$

The above relation can also be proved in the following way :

We know that $I_E = I_C + I_B$ ...(5.22)

For CB configuration, we have

$$I_C = \alpha I_E + I_{CBO} \quad ...(5.23)$$

In CE configuration, I_B is the input current and I_C is the output current. Therefore, we have to find the relation between I_B and I_C.

Putting the value of I_E from equation (5.22) in equation (5.23), we get

$$I_C = \alpha(I_C + I_B) + I_{CBO} = \alpha I_C + \alpha I_B + I_{CBO}$$

$$I_C(1 - \alpha) = \alpha I_B + I_{CBO}$$

or

$$I_C = \frac{\alpha}{1-\alpha}.I_B + \frac{1}{1-\alpha}.I_{CBO}$$

To simplify the above equation, defining

$$\frac{\alpha}{1-\alpha} = \beta \quad ...(5.24)$$

and

$$\frac{I_{CBO}}{1-\alpha} = I_{CEO} \quad ...(5.25)$$

$$I_C = \beta I_B + I_{CEO} \quad ...(5.26)$$

where β = common-emitter d.c. current gain

I_{CEO} = Leakage current in common emitter configuration.

Using equation (5.24), we have

$$\alpha = \frac{\beta}{1+\beta}$$

From equation (5.25), we have

$$I_{CEO} = \frac{I_{CBO}}{1-\alpha} = \frac{I_{CBO}}{1-\left(\frac{\beta}{1+\beta}\right)} = \frac{I_{CBO}(1+\beta)}{1+\beta-\beta}$$

$$I_{CEO} = (1 + \beta)\, I_{CBO} \quad ...(5.27)$$

Putting the value of I_{CEO} from (5.27) in (5.26) we get

$$I_C = \beta I_B + (1+\beta) I_{CBO} \quad ...(5.28)$$

5.17. COMMON COLLECTOR (CC) CONFIGURATION

(U.P. Tech, Sem. Exam; 2006-07) (05 marks)

Figure 5.24 shows the common collector (CC) configuration for NPN transistor. In this configuration, the input is applied between base and collector whereas the output signal is taken across the emitter and collector. This means that the collector of the transistor is common to both input and output circuits and hence the name common-collector configuration.

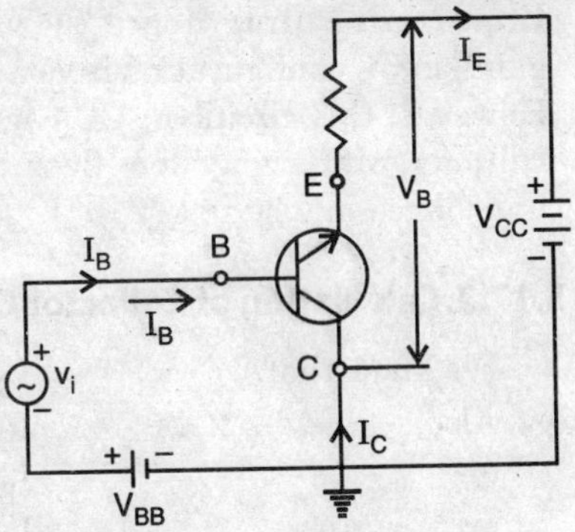

Fig. 5.24. Common-collector configuration for NPN transistor

5.17.1. Current Gain or Current Amplification Factor in CC Configuration

In CC configuration, the ratio of change in emitter current to the change in base current is called as current-gain or current amplification factor. It is denoted by letter γ (Gamma).

Mathematically,

$$\gamma = \frac{\Delta I_E}{\Delta I_B}$$

Relation between current gain γ and α

We know that

$$\alpha = \frac{\Delta I_C}{\Delta I_E} \quad \text{and} \quad \gamma = \frac{\Delta I_E}{\Delta I_B}$$

$$I_E = I_C + I_B$$

Therefore $\quad \Delta I_E = \Delta I_C + \Delta I_B = \Delta I_E - \Delta I_C$

Putting this value of ΔI_B in equation, $\gamma = \frac{\Delta I_E}{\Delta I_B}$, we get

$$\gamma = \frac{\Delta I_E}{\Delta I_E - \Delta I_C}$$

$$= \frac{\Delta I_E / \Delta I_E}{\Delta I_E / \Delta I_E - \Delta I_C / \Delta I_E} = \frac{1}{1 - \frac{\Delta I_C}{\Delta I_E}}$$

But $\quad \frac{\Delta I_C}{\Delta I_E} = \alpha$

Therefore, $\quad \gamma = \frac{1}{1-\alpha}$

Also, we know that $\beta = \frac{\alpha}{1-\alpha}$

or $\quad \beta + 1 = \frac{\alpha}{1-\alpha} + 1 = \frac{1}{1-\alpha}$

Therefore, $\quad \gamma = \beta + 1$

DO YOU KNOW?

The four distinct operating regions of a transistor are the active region, the sturation region, the cutoff region, and the breakdown region. When it is used as an amplifier, the transistor operates in the active region. When it is used in digital circuits, the transistor usually operates in the saturation and cutoff regions. The breakdown region is avoided because the risk of transistor destruction is too high.

* Common collector (CC) configuration is also popularly known as emitter follower. It is so called because here output follows the input (gain is almost one). Although, it is not used in amplification, yet it is used in so many other applications. We shall discuss emitter follower in detail in chapter 6.

Important Point: Hence, the value of γ is almost equal to β. This means that the value of current gain γ in CC configuration is very high and is comparable to the current gain β in CE configuration. However, this circuit in CC configuration is rarely used for amplification purpose because the voltage gain is very low (less than 1)*. Common-Collector configuration is mainly used for impedance-matching.

5.17.2. Calculation of Collector Current

We know that $$I_C = \alpha I_E + I_{CBO}$$

Also, $$I_E = I_C + I_B$$

or $$I_E = (\alpha I_E + I_{CBO}) + I_B$$

or $$I_E - \alpha I_E = I_B + I_{CBO}$$

$$I_E(1-\alpha) = I_B + I_{CBO}$$

or $$I_E = \frac{I_B}{1-\alpha} + \frac{I_{CBO}}{1-\alpha}$$

But $$\frac{1}{1-\alpha} = \beta + 1 = \gamma$$

Therefore, $$I_E = I_B(\beta+1) + I_{CBO}\,(\beta+1)$$

5.18. COMPARISON BETWEEN THREE TRANSISTOR CONFIGURATIONS (VERY IMPORTANT)

It may be observed that a transistor may be connected in any one of the three configurations namely CB, CE and CC configuration. The transistor behaves differently in different configuration. The configuration in which a transistor has to be connected depends upon the requirement or application. A configuration may be suitable for a particular application, but it may not be suitable for the other.

Some important parameters which govern the suitability of a configuration are input dynamic resistance, output dynamic resistance, dc current gain, ac current gain, ac voltage gain and leakage current of the transistor.

For example, out of the three configuration, the common collector configuration has very high input resistance but a small value of voltage gain. Hence it is best suited for impedance matching. Therefore, for the application of a particular configuration, we must know its characteristics.

Table 5.3. shows the comparison of the characteristics of three configurations.

Table 5.3. Comparison between the three configurations

S.No.	Characteristics	CB configuration	CE configuration	CC configuration
1.	Input resistance	Very low (40Ω)	Low (50 KΩ)	Very high (750 KΩ)
2.	Output resistance	Very high (1MΩ)	High (10 KΩ)	Low (50Ω)
3.	Current Gain	Less than unity (0.98)	High (100)	High (100)
4.	Voltage Gain	Small (150)	High (500)	Less than unity
5.	Leakage Current	Very small (5 μA for Ge and 1μA for Si)	Very large (500 μA for Ge and 20 μA for Si)	Very large (500 μA for Ge and 20 μA for Si)
6.	Applications	For high-frequency applications	For audio freq. applications	For impedance matching

5.19. CE CONFIGURATION IS MOST WIDELY USED IN AMPLIFIER CIRCUITS

For an ideal amplifier circuit, the input impedance should be as high as possible and the output impedance should be as low as possible. Although the input impedance for CC configuration is highest but gain for CC configuration is less than unity. So generally this is not preferred for amplifier circuits.

The input impedance of CE configuration is much higher than CB configuration since

$$r_i \text{ for CB} = \left.\frac{\Delta v_{EB}}{\Delta i_E}\right|_{V_{CB}=\text{Const.}} \quad ...(5.29)$$

$$r_i \text{ for CE} = \left.\frac{\Delta v_{BE}}{\Delta i_B}\right|_{V_{CE}=\text{Const.}} \quad ...(5.30)$$

The numerators are same in above two equations. But the denominator of the first is of the order of few μA, whereas that of the second is of the order of a few μA. Hence r_i is much higher for CE than CB.

Similarly output impedance is much lower for CE than CB since the output characteristic curves are more horizontal for CB than CE. Except this, the current gain of CB configuration is less than unity. It is generally 0.98. The current gain of CE configuration is quite high. It is generally 100 and may be as high as 250. Such high current gain in the CE configuration makes it possible to have quite high voltage gain as well as high power gain.

Hence, for the above reasons, CE configuration is most popular in amplifier circuits.

5.20. COMPLETE STATIC TRANSISTOR CHARACTERISTICS (Expected)

We know about some parameters of transistor such as α and β and leakage current. They give a small idea of transistor behaviour. The complete behaviour of transistor may be observed with the help of some types of curves known as characteristics curves. These characteristics curves relate the transistor currents and voltages. There are two types of characteristics curve:

(a) Input characteristics : Input characteristics curves relate the input current and input voltage for a given value of output voltage.

(b) Output characteristics : Output characteristics curves relate the output current and output voltage for a given value of input current.

With the help of these two types of characteristics curves the complete static operation of transistor can be explained. Hence, these curves are also known as static characteristics curves.

5.21. TRANSISTOR CHARACTERISTICS IN COMMON-BASE CONFIGURATION

In common-base configuration, two types of characteristics are :

(a) Input characteristics : Input characteristics curves relate the input or emitter current I_E and input or emitter-to-base voltage V_{EB} keeping output or collector-to-base voltage V_{CB} constant.

(b) Output Characteristics : Output characteristics curves relate the output or collector current I_C and output or collector-to-base voltage V_{CB} keeping the input or emitter current I_E constant.

Both these characteristics curves can be drawn with the same circuit arrangement shown in figure 5.25.

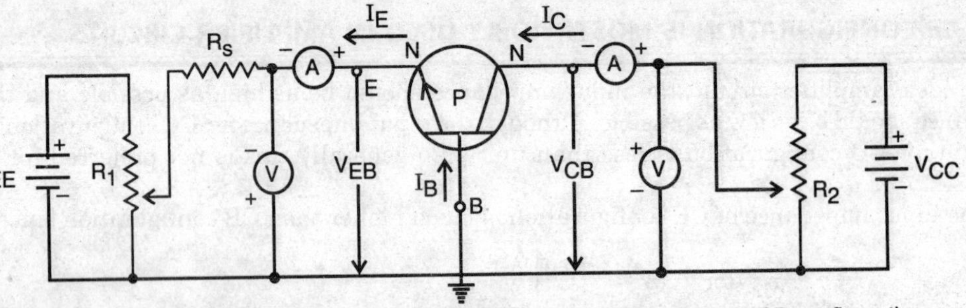

Fig. 5.25. Circuit arrangement for input and output characteristics curve in CB configuration

5.21.1. Transistor Input Characteristics in CB Configuration

Input characteristics curves are obtained with the help of circuit arrangement shown in fig. 5.25. In this circuit NPN transistor is used. The series resistor R_s is connected to limit the emitter current. The value of voltage V_{EB} may be varied with the help of a potentiometer R_1. The collector voltage V_{CB} is varied by adjusting the potentiometer R_2.

The d.c. millimetres and d.c. voltmeters have been connected in the emitter and collector circuits to measure the voltages and currents.

The input characteristics in CB configuration are plotted between emitter current I_E and the emitter-base voltage V_{EB}, for different values of collector-base voltage V_{CB}. Input characteristics are shown in figure 5.26.

Few Points

(i) Below the cut-in voltage or knee voltage the emitter current is negligibly small. This value of cut-in-voltage is approximately 0.7 V for silicon and 0.3 V for germanium transistors.

(ii) After the knee voltage, the emitter current I_E increases rapidly with a small increase in emitter-to-base voltage V_{EB}. This indicates that input resistance in CB configuration is very small.

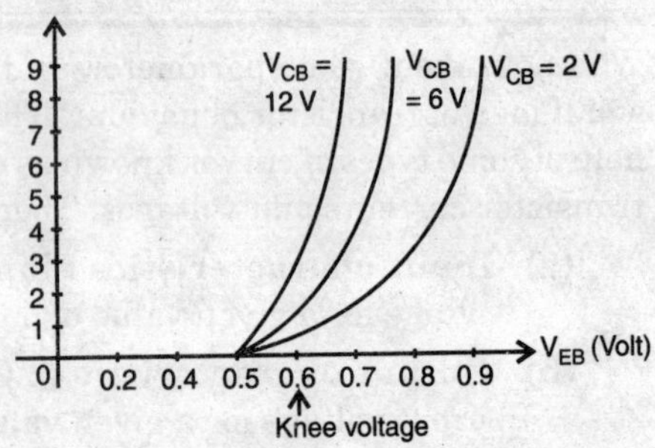

Fig. 5.26. Input characteristics in CB configuration

(iii) With the increase in V_{CB}, the curves shift upwards or becomes vertical. Hence r_i decreases.

(iv) At any point on the curve, the value of a.c. or dynamic input resistance is given as

$$r_i = \frac{\Delta V_{EB}}{\Delta I_E}\Bigg|\ V_{CB} \text{ constant} \qquad \ldots(5.31)$$

where, $\Delta V_{EB} \rightarrow$ Small change in emitter-to-base voltage V_{EB}.

$\Delta I_E \rightarrow$ Small change in emitter current I_E.

The dynamic input resistance r_i is very small nearly 25 to 100 Ω.

5.21.2. The Early Effect or Base Width Modulation

Referring to figure 5.27 and the width of the region, the total base width is equal to the sum of the widths of the depletion regions extended into the base region from the collector and emitter side and the width of the region occupied by the free charge particles. But as the EB junction is forward biased, the width of its depletion region is very narrow. As the CB junction is reverse biased, the width of its depletion region is much larger. Thus neglecting the width of depletion region at the EB junction we can write :

Total base width = Width of depletion region at the CB junction inside the base region + width of the region containing free charge particles

(i) Now as V_{CB} is increased, the reverse voltage applied to the CB junction increases. This widens the depletion region at the collector junction. Due to this, effective width of the base region decreases as shown in figure 5.27.

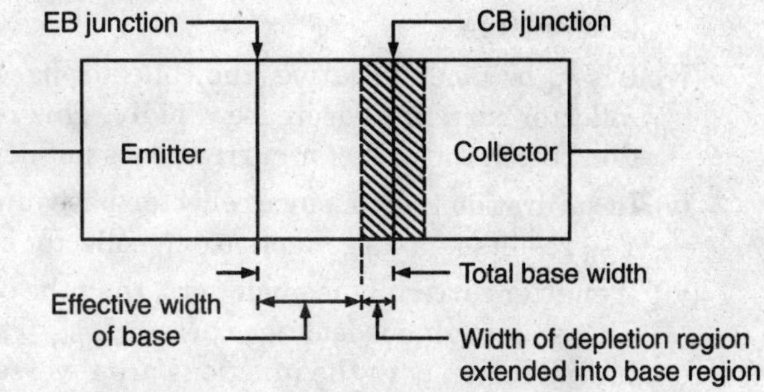

Fig. 5.27. Width of the base region

(ii) This will increase the charge concentration gradient in the base region. Due to increased charge carrier concentration, more number of electrons diffuse from the emitter to base i.e., emitter current increases.

(iii) Thus with increase in V_{CB}, the input current I_E increases slightly.

The reduction in effective base width due to increase in V_{CB} is called as **Early effect** or **"Base width Modulation"**.

Other effects of base width modulation :

There are two more effects of the base width modulation as under:

(i) As the base width decreases due to early effect, the possibility of recombination in the base region will reduce. This will increase the values of both, the amplification factor β and the common base current amplification factor α_{dc}, with increase in the voltage V_{CB}.

(ii) Due to increase in the charge gradient within the base, more number of "minority" charge carriers are injected across the junction. This will increase the current of minority carriers.

5.21.3. Output Characteristics in CB Configuration

The output characteristics curves in CB configuration can be obtained with the help of same circuit arrangement shown in figure 5.25. First, a suitable value of input current (emitter current) is fixed. Now keeping emitter current I_E constant. Collector-to-base voltage V_{CB} is increased and corresponding changes in collector current are noted and a graph is plotted. Figure 5.28 shows the output characteristics curve for CB configuration.

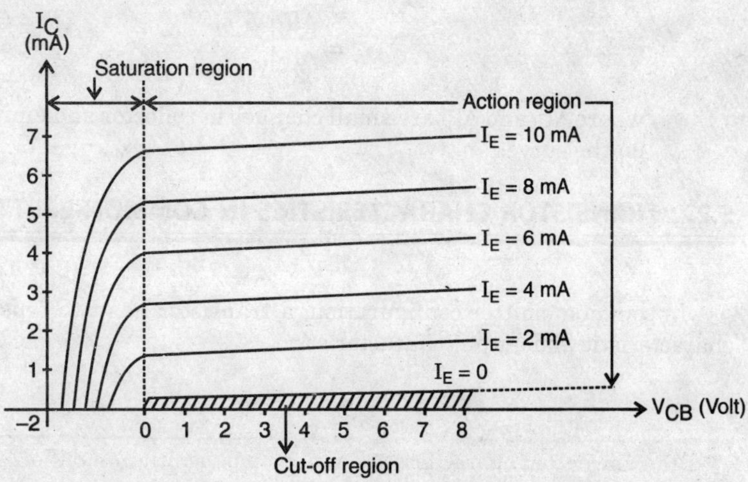

Fig. 5.28. Output characteristics in a CB configuration

5.21.4. Few Points

(i) The characteristics curves show the three regions namely active region, cut-off region and saturation region.

(ii) In active region, the collector current is approximately equal to the emitter current I_E.

(iii) Emitter current increases slowly for increase in V_{CB} in active region. Hence, in active region, the curves are nearly flat. This also means that the output resistance in this configuration is very large.

(iv) If V_{CB} becomes negative, the collector-base junction is set in forward bias which causes collector current to decrease rapidly. This region is known as saturation region. In saturation region the collector current does not depend much on emitter current.

(v) In saturation region, since collector-base junction is also forward biased, a small increase in V_{CB} results in a large increase in collector current.

(vi) If emitter current I_E is made zero, then the collector current is not zero. It has a small value. This small value is leakage current I_{CBO}. The curve for $I_E = 0$ corresponds to cut-off region. In this region both the junctions are reverse-biased.

(vii) With the help of output characteristics, the dynamic output resistance r_0 can be found by the expression

$$r_0 = \left.\frac{\Delta V_{CB}}{\Delta I_C}\right|_{I_E=\text{Constant}} \quad ...(5.32)$$

Here ΔV_{CB} and ΔI_C are small changes in collector-to-emitter voltage and collector current for a given point.

(viii) With the help of output characteristics curve d.c. and a.c. current gains in CB configuration can also be determined.

d.c. current gain

$$\alpha_{dc} = \alpha = \frac{I_C}{I_E} \quad ...(5.33)$$

where I_C and I_E are the values of collector and emitter currents at any point on the curve.

a.c. current gain is

$$\alpha_0 = \left.\frac{\Delta I_C}{\Delta I_E}\right|_{V_{CB}=\text{Constant}} \quad ...(5.34)$$

where ΔI_C and ΔI_E are small changes in collector and emitter currents around a given point on the curve.

5.22. TRANSISTOR CHARACTERISTICS IN COMMON-EMITTER CONFIGURATION*

(U.P. Tech, Sem. Exam; 2006-07) (05 marks)

In common-emitter configuration, a transistor has two types of characteristics namely input characteristic and output characteristic.

* Draw the output characteristics of a common emitter amplifier. Also draw the load line.
(U.P. Tech, Sem. Exam; 2006-07) (05 marks)

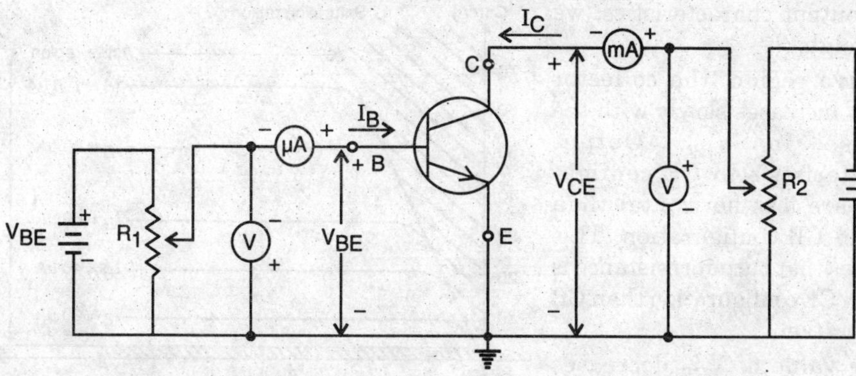

Fig. 5.29. Circuit arrangement for input and output characteristics

(a) Input characteristics : Input characteristics curves relate the base current I_B and the base-to-emitter voltage V_{BE} keeping collector-to-emitter voltage V_{CE} constant.

(b) Output characteristics : Output characteristics curves relate the collector current I_C and collector-to-emitter voltage V_{CE} keeping base current I_B constant.

Both the characteristics may be obtained with the help of single circuit arrangement shown in above figure. Potentiometer R_2 is used to vary the collector-to-emitter voltage V_{CE} and potentiometer R_1 is used to vary base-to-emitter voltage V_{BE}. To measure different voltages and currents, d.c. millimetres and d.c. voltmeters are connected in base and collector circuit of transistor.

5.22.1. Input Characteristics in Common-Emitter (CE) Configuration

Input characteristics curves may be obtained with the help of circuit arrangement shown in figure 5.29 by increasing the V_{BE} in steps and recording the corresponding values of I_B while keeping V_{CE} constant. In figure 5.30, the typical input characteristics curves which relate I_B and V_{BE} for various values of V_{CE}.

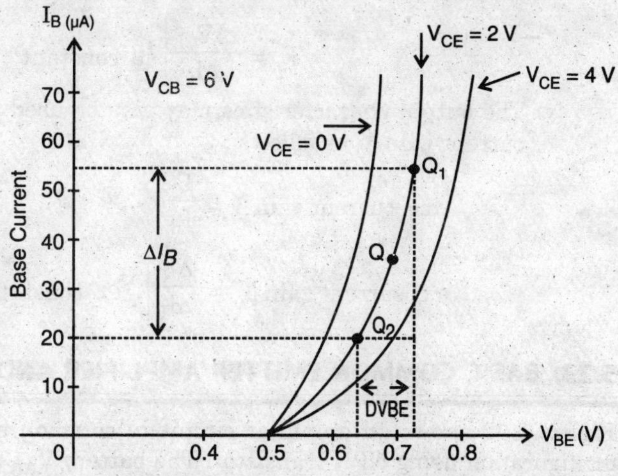

Fig. 5.30. Input characteristics in CE configuration

With the help of input characteristics, we can find the value expressed dynamic input resistance at a given point. At a given point, it is expressed as the ratio of small change in V_{BE} to the resulting small change in I_B keeping V_{CE} constant.

$$r_i = \left.\frac{\Delta V_{BE}}{\Delta I_B}\right| V_{CE} \text{ constant} \qquad ...(5.35)$$

5.22.2. Ouput Characteristics in Common-Emitter Configuration

Output characteristics in CE configuration can be plotted with the help of circuit arrangement shown in figure 5.29. The related output current I_C to the collector-to-emitter voltage for different values of input current I_B.

Figure 5.31 shows the typical output characteristics.

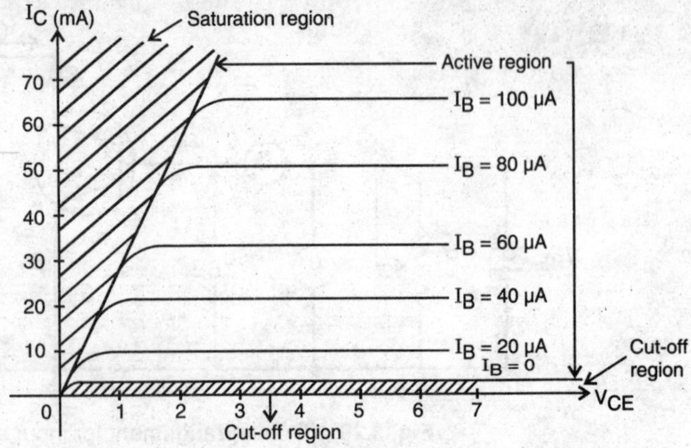

Fig. 5.31. Output characteristics in CE configuration

From the output characteristics, we can draw few points :

(i) In active region, the collector current increases slowly with the increase in V_{CE}. Output characteristics in CE configuration are less horizontal than those in CB configuration. This indicates that output resistance is small in CE configuration than CB configuration.

(ii) As the value of V_{CE} decreases below zero, I_C decreases rapidly. This occurs due to reason that when V_{CE} decreases below zero, collector-base junction becomes forward-biased. This means that now both the junctions are forward-biased. This is called saturation region. In saturation region, the collector current I_C does not depend upon the input current I_B.

(iii) If base current is made zero, then the collector current is not zero. It has a value known as reverse leakage current I_{CEO}. This condition is called the cut-off region of transistor.

(iv) The output characteristics may be used to find the dynamic output resistance at any given point. Output resistance is given

$$r_0 = \left.\frac{\Delta V_{CE}}{\Delta I_C}\right|_{I_B \text{ constant}} \quad \text{...(5.36)}$$

(v) The output characteristics may also be used to calculate the d.c. current gain β and a.c. current gain β_0 as follows :

$$\text{d.c. current gain } \beta = \frac{I_C}{I_B} \quad \text{...(5.37)}$$

$$\text{a.c. current gain } \beta_0 = \left.\frac{\Delta I_C}{\Delta I_B}\right|_{V_{CE} \text{ constant}} \quad \text{...(5.38)}$$

5.23. BASIC COMMON EMITTER AMPLIFIER AND D.C. LOAD LINE

Figure 5.32 shows an amplifier circuit in common emitter configuration using NPN transistor. The battery V_{BB} is used to forward bias the emitter-base junction and battery V_{CC} is used to reverse bias the collector-base junction. The resistor R_B is connected to limit the input or base current. The resistor R_C acts as a load resistance. The amplified a.c. output voltage appears across load resistor R_C.

The input a.c. signal to be amplified is V_{in}. This is applied to the base through capacitor C_1. The capacitors C_1 and C_2 are called coupling capacitors. The coupling capacitor C_1 blocks d.c. Hence d.c. current flows only through R_B and is prevented from reaching the input signal source. The coupling capacitor C_2 also blocks d.c. from reaching the output terminals. Therefore, at the output terminals. Only amplified output a.c. signal voltage V_{out} appears, D.C. load line. With the help of

DO YOU KNOW?

The emitter is grounded or common in a CE circuit. The base-emitter part of a transistor acts approximately like an ordinary diode. The base-collector part acts like a current source that is equal to β_{dc} times the base current. The transistor has an active region, a cutoff region, and a breakdown region. The active region is used in linear amplifiers. Saturation and cutoff are used in digital circuits.

d.c. load line, we can easily analyse the performance of an amplifier circuit. In this section, we will learn how to draw a d.c. load line.

For this, first we will find the d.c. equivalent circuit of the amplifier circuit shown in figure 5.32. This condition of having no input a.c. signal is also known as quiescent condition.

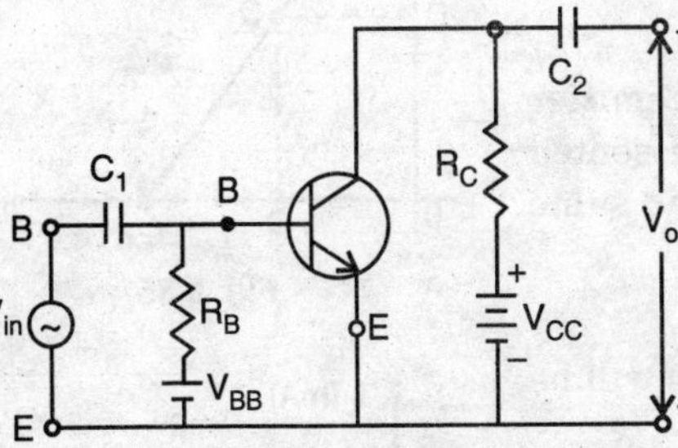

Fig. 5.32. Basic common emitter amplifier for NPN transistor

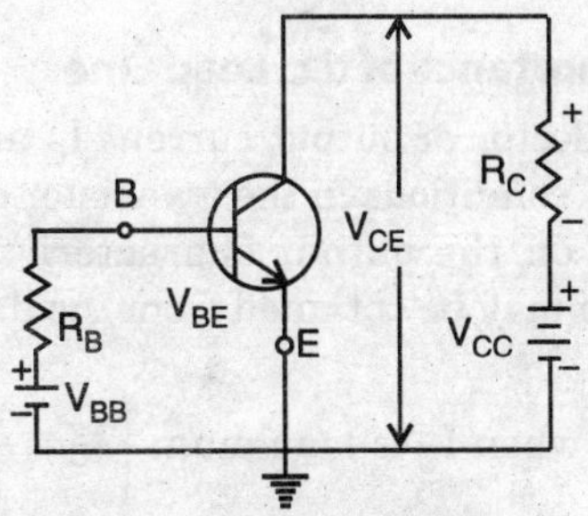

Fig. 5.33. Basic common emitter (d.c. equivalent circuit)

Figure 5.33 shows the d.c. equivalent circuit of one shown in figure 5.32. This circuit is obtained by opening the two capacitors and make the value of signal source equal to zero. For d.c. the capacitor works as an open circuit.

> **DO YOU KNOW?**
>
> Centering the Q-point on a transistor load line is important because it allows for the maximum ac output voltage from the amplifier. Centering the Q-point on the dc load line is sometimes referred to as mid-point bias.

Applying KVL to the collector portion of the circuit, we get

$$I_C R_C + V_{CE} = V_{CC}$$

or
$$I_C = \left(-\frac{1}{R_C}\right)V_{CE} + \frac{V_{CC}}{R_C} \quad \text{...(5.39)}$$

The above equation is in the form of

$$y = mx + c \quad \text{...(5.40)}$$

The equation (5.40) is an equation of straight line.

Comparing equations (5.39) and (5.40), the slope of straight line will be

$$m = -\frac{1}{R_C}$$

Therefore, if equation (5.39) is plotted on the output characteristics curves, we will get a straight line having slope equal to $\left(-\frac{1}{R_C}\right)$ and intercept on vertical axis equal to $\left(\frac{V_{CC}}{R_C}\right)$.

This line plotted on the output characteristics is called d.c. load line. To plot the d.c. load line, we use the concept that on V_{CE} axis, the I_C is zero and on I_C axis, the V_{CE} is zero.

Substituting these conditions in the equation (5.39), i.e.

$$V_{CC} = V_{CE} + I_C R_C \text{, we get}$$

when $I_C = 0 \Rightarrow V_{CE} = V_{CC}$, Point A = $(V_{CC}, 0)$

when $V_{CE} = 0 \Rightarrow V_{CC} = I_C R_C$

$$\text{or } I_C = \frac{V_{CC}}{R_C}\text{, Point B} \equiv \left(0, \frac{V_{CC}}{R_C}\right)$$

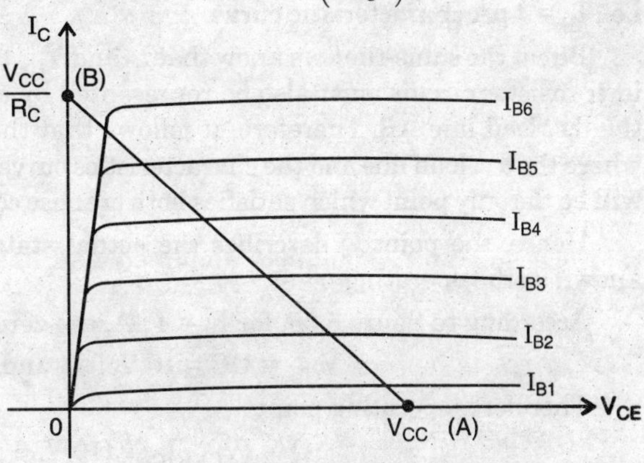

Fig. 5.34. The dc load line

If two points are located on the output characteristics and joined by a straight line, we get a d.c. load line. This line is called d.c. load line because its slope depends upon the value of resistor R_C which is the d.c. load of the circuit.

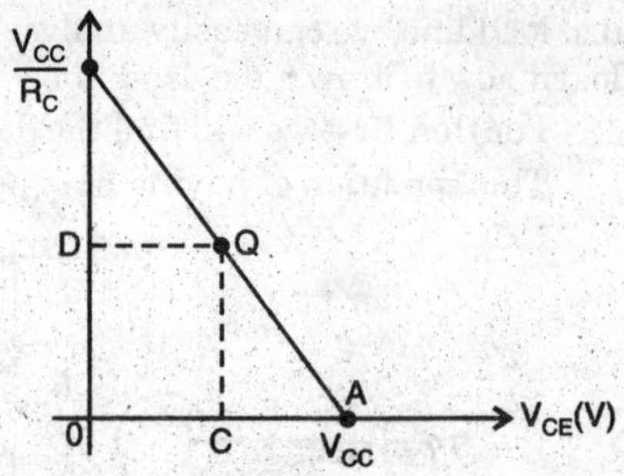

Fig. 5.35.

5.23.1. Importance of d.c. Load Line

The collector or output current I_C and collector to emitter voltage V_{CE} conditions in the transistor circuit are represented by a point on the output characteristics curve. This same information may be obtained from the d.c. load line.

Hence, when I_C is maximum $\left(I_C = \frac{V_{CC}}{R_C}\right)$ then V_{CE} will be zero ($V_{CE} = 0$) and when $I_C = 0$, then V_{CE} is maximum ($V_{CE} = V_{CC}$) as shown in figure 5.35.

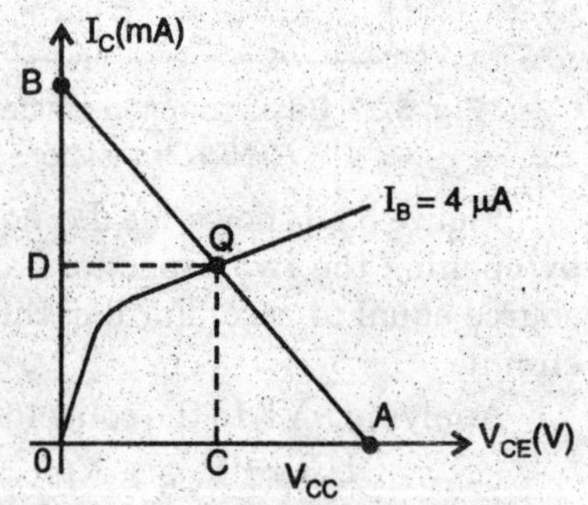

Fig. 5.36.

For some other value of output or collector current I_C (say OD), the collector-to-emitter voltage V_{CE} will be equal to OD. It means that d.c. load line provides a more convenient and direct solution to the problem.

From above, it also follows that if we plot the d.c. load line on the output characteristics curve of the transistor, we may investigate the behaviour of transistor amplifier.

5.23.2. Operating Point

The zero signal values (no a.c. signal is applied) of collector current I_C and collector-to-emitter voltage V_{CE} are called as the **operating point.**

Since the variations of I_C and V_{CE} take place about this point, therefore this is known as the operating point.

This point is also called as quiescent (stable) point or simply Q-point since it is a point on output characteristics curve when the transistor is in a silent condition (i.e., the absence of a.c. input signal).

> **DO YOU KNOW?**
>
> The operating point of the transistor is on the dc load line. The exact location of this point is determined by the collector current and the collector-emitter voltage. With base bias, the *Q* point moves whenever any of the circuit values change.

Let us consider that in the absence of a.c. input signal, the base current is 4 μA. Then, in transistor circuit, the I_C and V_{CE} conditions must be represented by a point on I_B curve i.e., $I_B = 4\ \mu A$ characteristic curve.

But at the same-time we know that I_C and V_{CE} conditions in transistor circuit must also be represented by a point on the d.c. load line AB. Therefore, it follows that the point Q where the d.c. load line and the characteristics curve intersect will be the only point which satisfies both of these conditions.

Hence, the point Q describes the actual state of circuit in the zero signal conditions and is known as the operating point.

According to figure 5.36, for $I_B = 4\ \mu A$, the zero signal values are

$$V_{CE} = OC = 10 \text{ Volts} \quad \text{and} \quad I_C = OD = 6 \text{ mA}$$

Therefore, operating point

$$Q \equiv (V_{CE}, I_C) \equiv (10 \text{ V}, 6 \text{ mA})$$

EXAMPLE 5.7. For the circuit shown in figure 5.37, draw the d.c. load line.

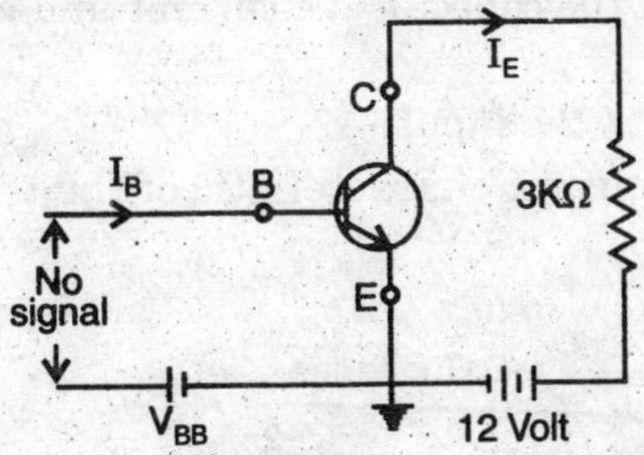

Fig. 5.37.

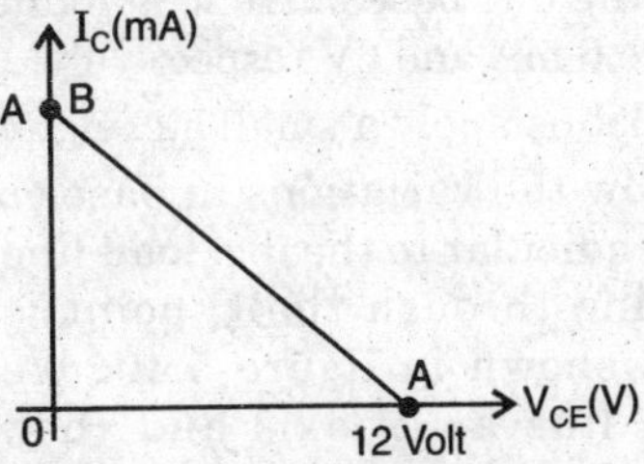

Fig. 5.38.

Solution : Applying KVL in the output portion, we get

$$V_{CC} = I_C R_C + V_{CE}$$

or $$V_{CE} = V_{CC} - I_C.R_C \quad \text{...(i)}$$

Now, when $I_C = 0$

from equation (i) $$V_{CE} = V_{CC} - 0 = V_{CC} = 12 \text{ Volt}$$

This locates the point A of the d.c. load line on the collector-emitter voltage (V_{CE}) axis.

Similarly, when $V_{CE} = 0$

from equation (i) $$0 = V_{CC} - I_C.R_C$$

or $$I_C R_C = V_{CC}$$

or $$I_C = \frac{V_{CC}}{R_C} = \frac{12\text{V}}{3\text{K}\Omega} = 4 \text{ mA}$$

This located the point B of the d.c. load line on the collector current (I_C) axis.

Now, by joining these two points A and B, we get the d.c. load line AB as shown in figure 5.38.

5.23.3. Analysis of Amplifier Using D.C. Load Line

As discussed earlier, a transistor may be used to amplify a.c. signals after fixing its operating point suitably.

In last article, we learnt how to fix the Q-point on the output characteristics of a transistor. In fact, the Q-point must lie in the middle portion on the output characteristics of a transistor. This helps the transistor to amplify a.c signals faithfully i.e., without producing distortion in waveshape of input signal.

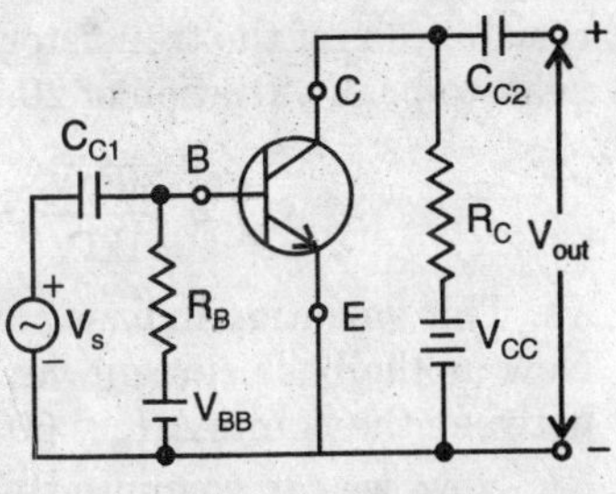

Fig. 5.39. A basic amplifier circuit

In quiescent condition, the base current is fixed at a constant d.c value. It is found from the Q-point. After this, when we apply an a.c signal at the input of the amplifier circuit, the base voltage varies as per the signal voltage V_s. Due to this, the base current will also vary. Now, as the base current varies, the instantaneous operating point of the transistor moves along the d.c. load line. This means that the instantaneous values of the collector current and voltage also vary according to the input signal. This variation in collector voltage is several times larger than the variation of the input signal.

Also, this collector-voltage variation reaches the output terminals through coupling capacitor. Thus, the output signal is several times larger than the input signal. Let us consider an example. Figure 5.39 shows an amplifier circuit. Let us assume that

> **DO YOU KNOW?**
>
> The dc load line contains all the possible dc operating points of a transistor circuit. The upper end of the load line is called saturation, and the lower end is called cutoff. The key step in finding the saturation current is to visualize a short between the collector and the emitter. The key step to finding the cutoff voltage is to visualize an open between the collector and emitter.

V_{CC} = 12 V, R_C = 1 kΩ, V_{BB} = 10.7 V and R_B = 200 kΩ.

Here the d.c. base current is found to be 50μA and the collector d.c current and collector d.c voltage are 6 mA and 6V respectively.

Now, let us apply a small a.c. signal voltage of 7mV at the input.

To show the variations in base current I_B on the output characteristic of the transistor, a line perpendicular to the d.c. load line and passing through the Q-point is drawn as shown in figure 5.40. We take this line as ωt-axis and then variations in I_B are plotted.

As the instantaneous operating point moves along the d.c. load line between the points A and B, both the collector-current and collector voltage would vary.

The current I_C varies between the points A_1 (I_C = 7.3 mA) and B_1 (I_C = 4.8 mA). These variations are shown on the left side of the output characteristics. The voltage V_{CE} varies between points A_2 (V_{CE} = 4.9 V) and B_2 (V_{CE} = 7.1 V).

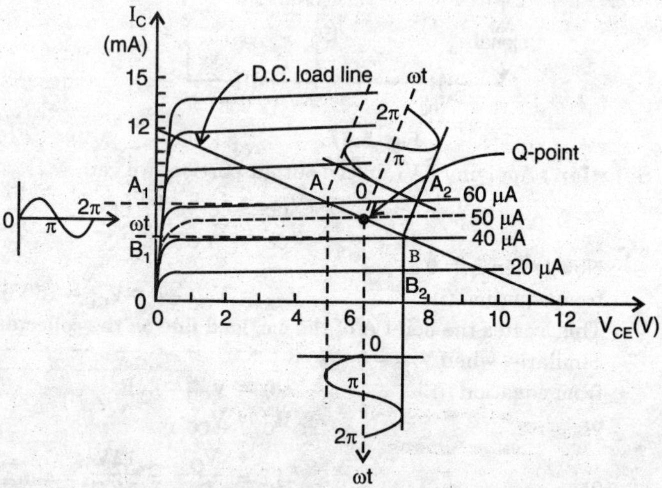

Fig. 5.40. Illustration of variation in base current which produces variation in collector current voltage in a CE amplifier

Also the collector-voltage variations are shown at the bottom of the output characteristics.

This voltage would have about 20 mV peak-to-peak variations. Now, if the input dynamic resistance r_i of the transistor is taken to be 1 kΩ, then the input voltage variation would produce a peak-to-peak variation of 20 μA in base current, since

$$\Delta I_B = \frac{20\text{mV}}{1\text{k}\Omega} = 20\ \mu\text{A}$$

This variation in base current exists around its quiescent value (i.e., operating point) of 50 μA. Now as the base current varies, the instantaneous operating point moves along the d.c. load line between the points A(I_B = 60 μA) and B(I_B = 40 μA).

Now, we can compute the current gain, voltage gain and the power gain of the amplifier on the basis of peak-to-peak variation as under :

(i) Current gain

$$A_i = \frac{\Delta I_C}{\Delta I_B}$$

Substituting all the values, we get

or $$A_i = \frac{(7.3-4.8)\times 10^{-3}}{(60-40)\times 10^{-6}} = 125$$

(ii) Voltage gain $$A_v = \frac{\Delta V_{CE}}{\Delta V_{BE}} = \frac{7.1-4.9}{20\times 10^{-3}} = 110$$

(iii) Power gain $$= \frac{\text{Output ac power}}{\text{Input ac power}} = \frac{i_c.v_{ce}}{i_b.v_{be}} = A_i \times A_V = 125 \times 110 = 13750$$

> **Important Point:** *From above, it is obvious that a common emitter amplifier has sufficiently large values of current gain, voltage gain and the power gain.*

5.24. NUMBERING SYSTEM USED FOR NAMING THE SEMICONDUCTOR DEVICES

After the fabrication of Semiconductor devices, a numbering is required. Several numbering systems have been used from time to time by different countries. But, the numbering system invented by proelectron standardization Authority in Belgium, is accepted and adopted internationally. According to this numbering system following points may be noted :

(i) Different semiconductor devices are numbered by fine alpha-numeric symbols like BF 194, BFX 65 etc. If the device number contains three numeral digits (i.e., BF 194), the device is usually used for consumer equipment (i.e., for entertainment etc.) But if the device number contains two numeral digits (i.e., BF X 15), the device is usually used for industrial or professional equipment.

(ii) In this numbering system, the first letter indicates the nature of Semiconductor material as

A – Germanium B– Silicon

C – Gallium Arsenide R – Compound (e.g., cadmium sulphide)

Thus, AC 125 is a germanium transistor whereas BF 194 is a silicon transistor.

(iii) The second letter indicates the device and its function in the circuit as :

A – Diode
B – Varactor
C – A.F. low power transistor
D – A.F. Power transistor
E – Tunnel diode
F – H.F. low power transistor
G – Multiple device
H – Magnetic Sensitive diode
K – Hall-effect modulator
L – H.F. power transistor
M – Hall-effect modulator
P – Radiation Sensitive diode
Q – Radiation generating diode
R – Thyristor (SCR or TRIAC)
S – Low power switching transistor
T – Thyristor, high power
U – switching transistor, power
X – diode, multiplier
Y – Power multiplier
Z – Zener diode

However, it may be noted that only this numbering system is not applied in electronics. Some other numbering systems are also used. As an example, IN 4007 is a silicon diode, 2N 3055 is a power transistor, BHT F 25 is a silicon controlled rectifier etc.

Thus, in this numbering system IN represents one junction device i.e., diode, 2N represents 2-junctions device like transistor and 3N represents two gates device like FET.

Also, in Japan, different numbering system is used for transistors and thyristors. According to Japanese Industries standards 2 A 777 is a transistor where 2 indicates number of junctions. S means semi-conductor device and letters A to M are used to indicate polarity and application, whereas the last digits 777 are for specific characteristics. Here letter A to M stand for:

A – pnp transistor, high frequency
C – npn transistor, high frequency
E – p-gate thyristor
H – n-base unijunction transistor (UJT)
K – n-channel. FET
B – pnp transistor, low frequency
D – npn transistor, low frequency
G – n-gate thyristor
J – p-channel. FET
M – triac (bidirectional)

5.25. TRANSISTOR AS SWITCH

(U.P. Tech, Sem. Exam; 2006-07) (05 marks)

The transistor can be used for two types of applications viz. amplification and switching. For the amplification applications, a transistor is biased in its active region. Whereas for switching applications, it is biased to operate in the saturation (fully on) or cut off (fully off) regions.

(i) Transistor in cutoff region (Transistor as an open switch)

In the cut off region, both the junctions of a transistor are reverse biased and a very small reverse current flows through the transistor. The voltage drop across the transistor (V_{CE}) is high. Thus, in the cut off region, the transistor is equivalent to an open switch.

The input voltage is zero so $I_B = 0$ and $I_C = 0$.

Applying KVL to the collector circuit of figure 5.41, we obtain

$$V_{CC} = I_C R_C + V_{CE}$$

But $$I_C = 0$$

Therefore, $$V_{CE} = V_{CC}$$

The voltage across the transistor is $V_{CE} = V_{CC}$ i.e., high and the current flowing through it $I_C = 0$. This shows that the resistance offered by the transistor is infinite and the transistor operates as an open switch. Thus, in the cutoff region, the transistor acts as an open switch as shown in figure 5.41.

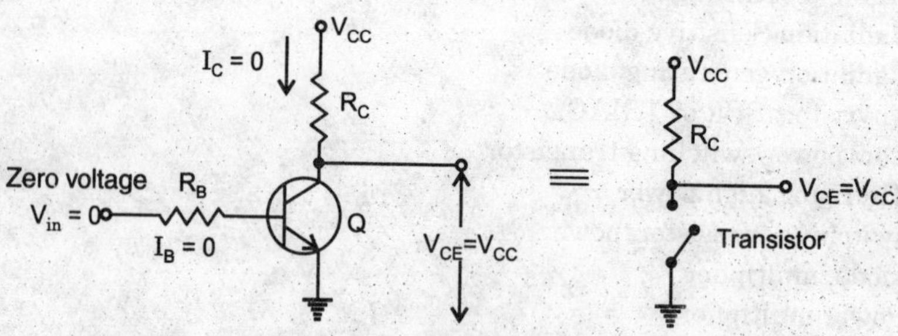

Fig. 5.41. Transistor in cut off is equivalent to an open switch

(ii) Transistor in the saturation region (Transistor as closed switch)

Figure 5.42 shows the biasing of transistor in the saturation region. A high input voltage V_{in} is applied at the base. The value of V_{in} and resistance R_B are adjusted such that a large current I_B flows. This will saturate the transistor. The value of I_B should be such that the following condition should be satisfied.

$$I_B \geq \frac{I_{C(sat)}}{\beta}$$

In the saturation region, both the junctions of a transistor are forward biased. The voltage drop across the transistor (V_{CE}) is very small of the order of 0.2 V to 1V depending on the type of transistor and collector current is very large. A low (almost 0) voltage drop across the transistor and a large current through it indicates that the transistor is equivalent to a closed switch.

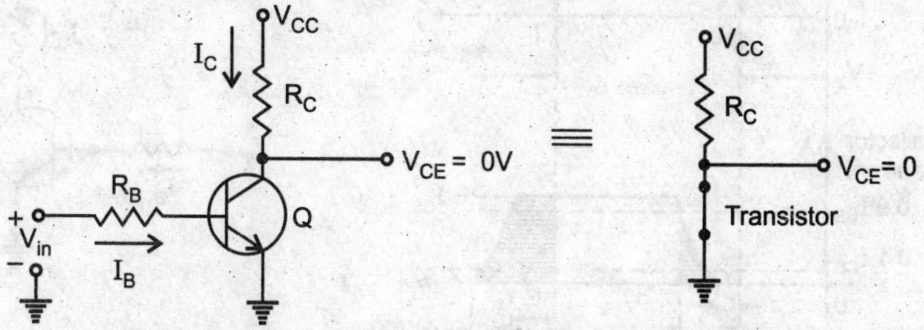

Fig. 5.42. Transistor in saturation is equivalent to an closed switch

5.25.1. Switching Waveforms of an Ideal Transistor Switch

Let us consider the circuit diagram of figure 5.43 which shows the biasing arrangement for a transistor to operate as a switch. A rectangular waveform is applied at the base and the transistor is assumed to be an ideal one. Hence, the transistor will change its state from ON to OFF or vice versa instantaneously.

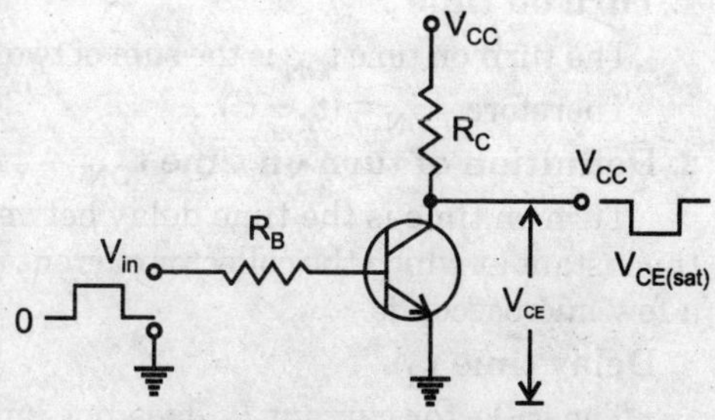

Fig. 5.43. Circuit diagram showing transistor biased to operate as a switch

When the input is low (0), the transistor is off, it acts as an open switch and the output voltage is $V_{CE} = V_{CC}$. When the input voltage goes high, the transistor is driven into saturation. It acts as a closed switch and the output voltage $V_{CE} = V_{CE(sat)} \cong 0$ V. The input and output voltage waveforms are shown in figure 5.44 which show that the transistor state change takes place instantly.

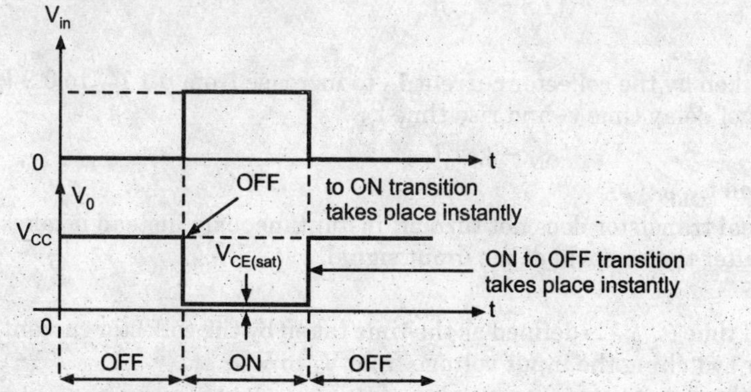

Fig. 5.44. Input output waveforms for transistor switching of an ideal transistor

5.25.2. Transistor Switching Times

The practical transistor does not switch its state instantaneously. It needs a finite time to turn on and to turn off completely. Hence, let us consider a practical transistor being driven by a pulse waveform as shown in figure 5.45. The input voltage and collector current waveforms for a non ideal transistor is shown in figure 5.45. Various delays and time intervals are involved in the switching waveforms of the transistor as shown in figure 5.45.

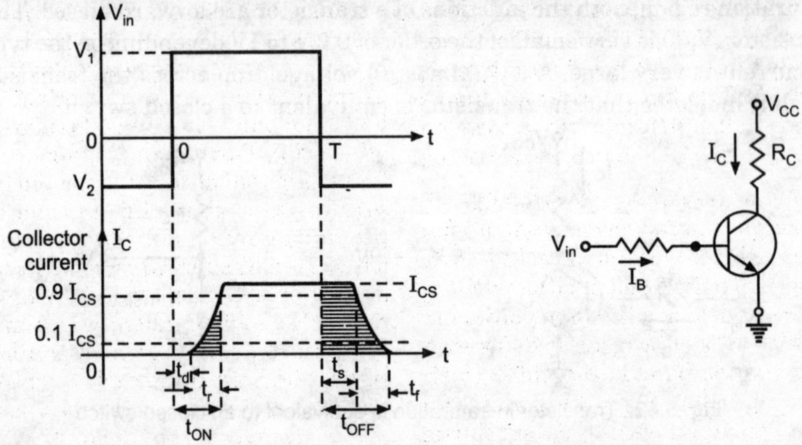

Fig. 5.45. Switching waveforms for the non ideal transistor

1. Turn on time

The turn on time t_{ON} is the sum of two time intervals namely, the delay time t_d and the rise time t_r.

Therefore, $t_{ON} = (t_d + t_r)$

2. Definition of turn on time t_{ON}

Turn on time is the time delay between the instant at which a high base voltage is applied and the instant at which the collector current reaches 90% of its maximum value. Turn on time is typically a few microseconds.

3. Delay time t_d

The collector current I_C does not immediately respond to the input signal. Instead, there is a delay. The delay time t_d is the time required for the collector current to attain 10% of the maximum value of I_C i.e., I_{CS}, after the application of input signal. I_{CS} is the saturation value of I_C and it is expressed as

$$I_{CS} = V_{CC}/R_C$$

4. Rise time t_r

It is time taken by the collector current I_C to increase from 0.1 I_{CS} to 0.9 I_{CS}. The total turn on time is the sum of delay time t_d and rise time t_r.

Therefore, $$t_{ON} = t_d + t_r \qquad ...(5.41)$$

5. Turn off time t_{OFF}

The non-ideal transistor does not turn off instantaneously instead it takes a finite time to turn off completely after the removal of the input signal.

Definition

The turn off time (t_{OFF}) is defined as the time taken by the collector current to reduce from I_{CS} to 10% of I_{CS} after switching the input voltage from V_1 to V_2.

The turn off time is the sum of two components namely the storage time (t_s) and the fall time (t_f).

Therefore, $$t_{OFF} = t_s + t_f$$

6. Storage time (t_s)

It is defined as the time interval which elapses between the transition of input waveform (from V_1 to V_2) and the time when I_C reduces to 90% of I_{CS}.

7. Fall time (t_f)

The storage interval is followed by the fall time interval t_f. It is defined as the time required for I_C to fall from 90% to 10% of the maximum value of I_C i.e., I_{CS}.

The turn off time is the sum of t_s and t_f.

5.25.3. Typical Values of Various Time Delays

The typical values of delay time, rise time, storage time and time for the n-p-n transistor 2 N 3830 are as under:

S. No.	Time	Value
1.	Delay time t_d	10 n sec
2.	Rise time t_r	50 n sec
3.	Storage time t_s	40 n sec
4.	Fall time t_f	30 n sec

5.25.4. Importance of Turn ON and Turn OFF Time

The turn on and turn off times are important in deciding the maximum switching frequency of the transistor. In order to operate the transistor at a high frequency, the turn on and turn off times should be as short as possible.

5.26. TRANSISTOR BIASING (i.e., BIASING THE BJT FOR DISCRETE-CIRCUIT DESIGN)

The transistor is used in a wide number of applications. The most basic application of a transistor is amplification. Amplification means to raise the strength or amplitude of a weak signal without any change in its original shape. For faithful amplification two things are necessary. Firstly, the transistor must operate in active region. It means that the emitter-base junction is forward-bias and collector-base junction is reverse bias. Active region for amplification is must because amplification is a linear process and transistor operates linearly only in active region. Secondly, we should get a fixed dc collector current at a fixed d.c. collector voltage. These fixed values of d.c. collector current and d.c. collector voltage are expressed by a term called operating point or quiescent point or Q-point.

Therefore, the primary aim of biasing is to obtain a fixed collector current at a fixed collector voltage. The circuits used for transistor biasing are simply called **biasing circuits.**

5.27. SELECTION OF OPERATING POINT

For faithful amplification of an input signal, a proper or judicious selection of operating point is of utmost importance.

Figure 5.46(a) shows the circuit of common emitter configuration and corresponding output characteristics of transistor.

To study the effects of biasing conditions on the performance of a transistor, it is necessary to draw a d.c. load line on the output characteristics of a transistor. Figure 5.46(b) shows that a d.c. load line is drawn on the characteristics. The operating point will essentially lie somewhere on this load line. Figure 5.46(b) shows three operating points. These three operating points can be set depending upon the values of base current. We will consider three cases to find which one of these three operating points is most suitable.

Case (i) If point Q_1 is the operating point, then Q_1 will lie near the saturation region. Therefore, the positive cycle of the signal will be clipped off. Hence distortion will occur. Thus, the point Q_1 is not the suitable operating point.

Case (ii) If Q_2 is considered the operating point, then this point will lie near the cut-off region. Therefore, the negative cycle will be clipped off. Hence, again distortion will occur. So, the point Q_2 is also not the suitable operating point.

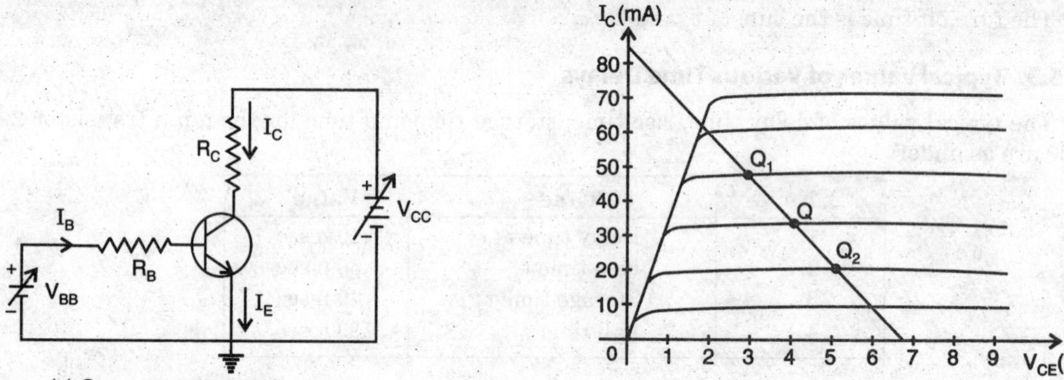

(a) Common emitter configuration (b) Output characteristics

Fig. 5.46.

Case (iii) If Q is chosen the operating point, then this will lie in the active region. Also this point is in the middle of active region. Therefore the signal will not be clipped off. Hence, there will be no distortion at all. Therefore, point Q is the suitable operating point.

Hence, we can conclude that for faithful amplification, the operating point must lie in the middle of the active region.

5.28. BIAS STABILIZATION

We know that for proper operation of transistor, the operating point should be properly fixed. However, only the fixing of operating point is not sufficient. The operating point must remain stable throughout the operation where it was fixed originally. This type of shift of operating point may drive the transistor into an unwanted region. Therefore bias stabilization is of utmost importance. Following are the factors which are responsible for the shift of operating point :

(i) Temperature dependence of transistor parameters

We know that the collector current is given as

$$I_C = \beta I_B + (1 + \beta)I_{CBO} \quad \text{...(5.42)}$$

Here β = Common emitter current gain

I_B = Base current

I_{CBO} = leakage current

Important Point: The above three variables are strongly temperature dependent. Hence, with the increase in temperature, if any of these parameters changes, then collector current changes and so operating is shifted to any other region.

(ii) Variation in parameters between transistors of same type : Even after the tremendous advances in semiconductor technology, the transistor parameters such as β vary widely between transistors are of same type. Change in β will shift the operating point in any undesired region.

(iii) Thermal runway : The collector current is given as

$$I_C = \beta I_B + (1 + \beta)I_{CBO} \quad \text{...(5.43)}$$

If temperature changes then I_{CBO} changes which in turn change I_C and operating point. Flow of collector current in the collector circuit produces heat at the collector junction. This

DO YOU KNOW?

The current gain of a transistor is an unpredictable quantity. Because of manufacturing tolerances, the current gain of a transistor may vary over as much as a 3 : 1 range when you change from one transistor to another of the same type. Changes in the temperature and the collector current produce additional variations in the dc gain.

raises the temperature. Hence I_{CBO} increases which in turn increases collector current I_C. This again increases the temperature of collector junction and the whole process repeats again. Such a successive increase in I_C, will drive the operating point into saturation region. This process is also called thermal runway. This is very dangerous process because due to excessive heat the transistor may burn out.

5.29. NEED OF STABILIZATION

In order to avoid thermal runway (i.e., self-destruction of transistor due to rise in temperature) and to nullify the effect of inherent variations of transistor parameters, it is very essential to stabilize the operating point i.e., to fix the operating point at all conditions. In practice, it is achieved by selecting a suitable biasing circuit which decreases I_B automatically with the rise in temperature (or rise in β while replacing the transistor). Then decrease in βI_B will compensate for the increase in $(\beta + 1)\, I_{CBO}$, keeping I_C almost constant. Hence, the associated circuitry can be applied to stabilise the operating point in the transistors.

Hence, **stabilization** may be defined as under :

The process of making operating point in dependent of temperature changes or inherent variations in transistor parameters is known as **stabilization**.

5.30. REQUIREMENT OF BIASING CIRCUITS **(Important)**

Following are the requirements of a biasing circuit :

(a) A biasing circuit must set the operating point in the middle of the active region of the transistor characteristics.

(b) A biasing circuit must stabilize the collector current against temperature variations.

(c) A biasing circuit must ensure that the operating point is independent of the transistor parameters such as β so that operating point is not shifted if the transistor is replaced by another transistor of same type.

5.31. THE STABILITY FACTOR

1. Definition

The stability factor is expressed as the rate of change of collector current with respect to the reverse saturation current while keeping the common-emitter current gain β and base current I_B as constant.

Mathematically, the stability factor may be stated as

$$S = \frac{dI_C}{dI_{CO}} \quad ...(5.44)$$

Important Point: The stability factor S is a measure of bias stability of a transistor circuit. It may be observed that higher the value of stability factor, poor is the stability whereas lower is the value of stability factor, good is the stability.

2. Mathematical expression for stability factor

Since, the value of collector current in a transistor is given as

$$I_C = \beta . I_B + (1+\beta) I_{CO} \quad ...(5.45)$$

Here, I_C = Collector current

I_B = Base current

I_{CO} = Reverse saturation current

β = common-emitter current gain

Differentiating equation (5.45) with respect to I_C

$$1 = \frac{d}{dI_C}(\beta.I_B) + \frac{d}{dI_C}\left[(1+\beta)I_{CO}\right]$$

Assuming β as constant, we have

$$1 = \beta \times \frac{dI_B}{dI_C} + (1+\beta).\frac{dI_{CO}}{dI_C} = \beta.\frac{dI_B}{dI_C} + (1+\beta).\frac{1}{S} \qquad \left[\because S = \frac{dI_C}{dI_{CO}}\right]$$

or
$$S = S.\beta\left(\frac{dI_B}{dI_C}\right) + (1+\beta)$$

or
$$S\left[1-\beta\left(\frac{dI_B}{dI_C}\right)\right] = 1+\beta$$

or
$$S = \frac{1+\beta}{1-\beta\left(\frac{dI_B}{dI_C}\right)} \qquad ...(5.46)$$

The above expression for stability factor is a general expression for any biasing circuit.

3. Stability factor for common base configuration

For common base configuration, the expression for collector current is

$$I_C = \alpha.\, I_E + I_{CO} \qquad ...(5.47)$$

where I_E = emitter current

I_{CO} = Reverse saturation current

Differentiating equation (5.47) with respect to I_{CO}

$$\frac{dI_C}{dI_{CO}} = 0 + 1 = 1$$

$$\text{or} \quad S = 1 \qquad ...(5.48)$$

The value of stability factor S is unity for common base circuits which means that these circuits are highly stable. Hence, there is no need of bias stabilization in these types of circuits.

4. Stability factor for common emitter configuration

For common emitter configuration, the expression for collector current is

$$I_C = \beta I_B + (1+\beta)\, I_{CO} \qquad ...(5.49)$$

Differentiating the equation (5.49) with respect to I_{CO}

$$\frac{dI_C}{dI_{CO}} = 0 + (1+\beta) = 1+\beta$$

or
$$S = 1+\beta \qquad ...(5.50)$$

Hence, according to above equation if β = 50 then value of S = 1 + 50 = 51.

This indicates that the collector current changes 51 times the change in reverse saturation current. Hence, in a CE configuration the collector current highly depends on I_{CO} which in turn depends on the temperature. Therefore, for CE configuration bias stabilization is of utmost importance to improve the stability factor.

5.32. METHODS OF TRANSISTOR BIASING

There are following methods of transistor biasing which are most widely used:

1. Fixed bias
2. Modified fixed bias or emitter resistor biasing
3. Collector to base bias
4. Self bias or voltage divider bias
5. Emitter bias

5.33. FIXED BIAS CIRCUIT

1. Mathematical Analysis

Figure 5.47 shows a fixed bias circuit. In this circuit, two d.c. batteries V_{BB} and V_{CC} are used. The battery V_{BB} is a low value battery and V_{CC} is a high value battery.

The fixed bias circuit may be replaced with more practical circuit in which a single supply V_{CC} is used. In this circuit, both the base and collector resistors are connected to the positive side of the battery V_{CC} as shown in figure 5.48.

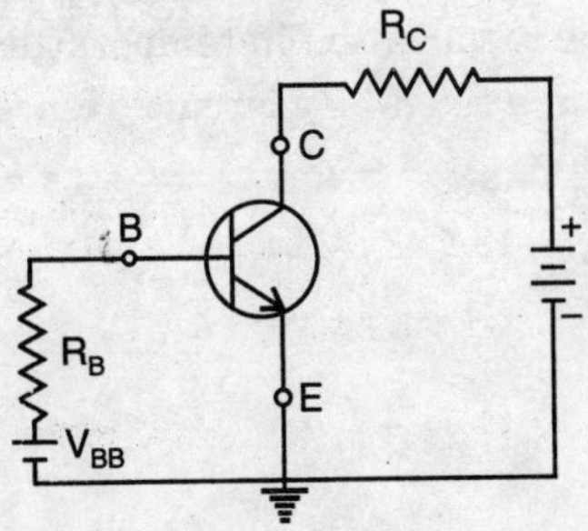

Fig. 5.47. Fixed bias circuit with two batteries.

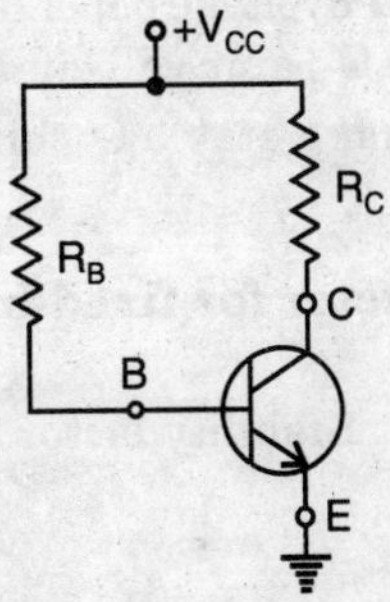

Fig. 5.48. Fixed bias circuit with single battery.

Now, we will find the value of d.c. bias voltages and currents in the base and collector portions of transistor.

Case (i) Considering the input portion of the fixed bias circuit (figure 5.49)

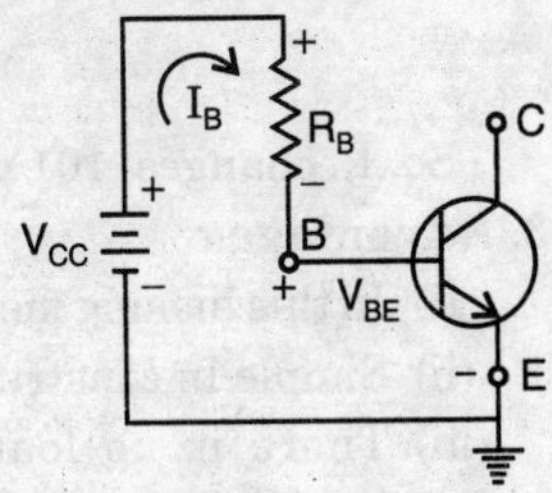

Fig. 5.49. Input portion of fixed bias

Applying Kirchoff's voltage law for the input side

$$V_{CC} = I_B R_B + V_{BE} \quad \text{...(5.51)}$$

or $$I_B R_B = V_{CC} - V_{BE}$$

or Base Current $$I_B = \frac{V_{CC} - V_{BE}}{R_B} \quad \text{...(5.52)}$$

The values of voltages V_{CC} and V_{BE} are fixed, therefore the value of base current depends on the base resistor R_B. For simplification, we may neglect the value of V_{BE} because it is small.

Hence,

base current $$I_B \simeq \frac{V_{CC}}{R_B} \quad \text{...(5.53)}$$

Case (ii) Let us consider the output portion of the fixed bias circuit (figure. 5.50)

Applying KVL for output portion, we get

$$R_C I_C + V_{CE} = V_{CC} \quad \text{...(5.54)}$$

or $$V_{CE} = V_{CC} - I_C . R_C$$

The collector current I_C is given by the relation

$$I_C = \beta I_B + I_{CO} \quad ...(5.55)$$

As an approximation, neglecting I_{CO}, we have

$$I_C \cong \beta I_B$$

Substituting the value of I_B, we get

$$I_C = \beta . \frac{V_{CC}}{R_B} = \frac{V_{CC}}{R_B/\beta} \quad ...(5.56)$$

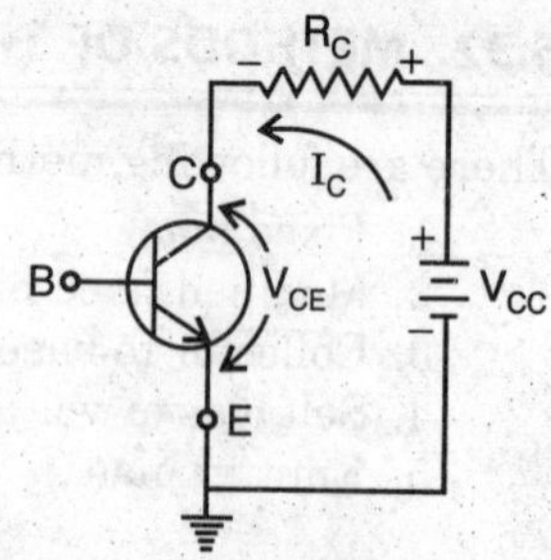

Fig. 5.50. Output portion of fixed bias circuit

The collector current is β times greater than the base current and does not depend upon the collector resistor.

Hence, Q-Point for fixed-bias circuit is

$$Q \equiv (V_{CE}, I_C) \quad ...(5.57)$$

where $\quad V_{CE} = V_{CC} - I_C . R_C$ and $I_C = \beta I_B = \dfrac{V_{CC}}{R_B/\beta}$

Design Tip: From Q-point, it may be observed that the values of collector current I_C and collector-emitter voltage depend upon the value of β which in turn depends strongly on temperature. It means that Q-point will change with change in β due to variation in temperature. This shows that a stable Q-point can not be achieved in a fixed-bias circuit. Hence, it is seldom used.

2. Stability Factor for fixed-bias circuit

$$\text{Stability factor } S = \frac{\beta + 1}{1 - \beta\left(\dfrac{dI_B}{dI_C}\right)} \quad ...(5.58)$$

In fixed bias circuit, $\dfrac{dI_B}{dI_C} = 0$, because I_B is independent of I_C.

$$\text{Hence, } S = \frac{\beta + 1}{1 - \beta . 0} = \beta + 1 \quad ...(5.59)$$

If $\beta = 100$, then $S = 101$

So, I_C changes 101 times for any change in I_{CO}. Due to large value of S, stability is poor.

3. Advantages

(i) In this biasing method, calculations are simple and it is easy to obtain the required conditions.

(ii) Simple in construction because only one resistor R_B is required to setup the conditions.

(iii) There is no loading of the source as no resistor is employed across base-emitter junction.

4. Disadvantages

(i) It provides poor thermal stability (i.e., high stability factor) and so there are changes of thermal runway.

(ii) This method provides poor stability against inherent variations of transistor parameters. For instant, if β increases due to transistor replacement, then I_C also increases by the same factor as $I_C = \beta I_B$ where I_B is constant.

EXAMPLE 5.8. Figure 5.51 shows a fixed bias circuit. Determine with β =100
(i) the operating point, (ii) the stability factor.

Solution :

(i) Given that, $\quad V_{CC} = 6V$

$$R_B = 530 \text{ k}\Omega, R_C = 2\text{k}\Omega$$

Since transistor is silicon transistor

Therefore, $V_{BE} = 0.7 \text{ V}$

Applying KVL for input side, we have

$$V_{CC} = I_B R_B + V_{BE}$$

or
$$I_B = \frac{V_{CC} - V_{BE}}{R_B} = \frac{6 - 0.7}{530 \text{ k}\Omega}$$

or
$$I_B = \frac{5.3}{530 \times 10^3} = \frac{53 \times 10^5}{53} = 10 \text{ }\mu\text{A}$$

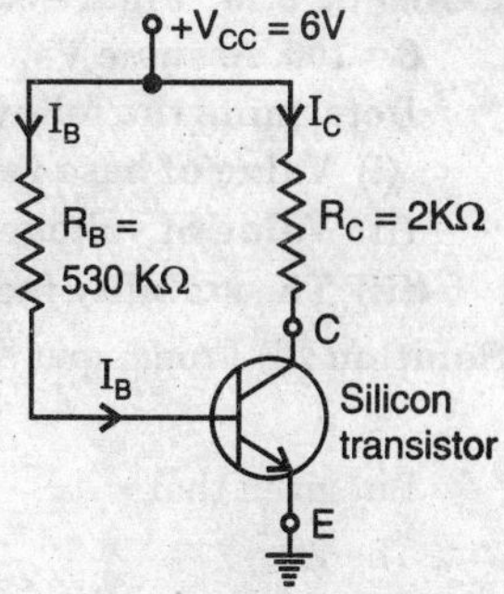

Fig. 5.51. Circuit for example 5.8.

Collector current is given by

$$I_C = \beta I_B = \beta I_B = 100 \times 10 \text{ }\mu\text{A}$$
$$I_C = 1000 \text{ }\mu\text{A} = 1 \text{ mA}$$

Applying KVL to the output side, we get

$$V_{CE} = V_{CC} - I_C R_C$$
$$V_{CE} = 6 - 1 \times 10^{-3} \times 2 \times 10^3$$
$$V_{CE} = 6 - 2 = 4\text{V}$$

Thus, operating point $Q \equiv (V_{CE}, I_C) \equiv (4 \text{ V}, 1 \text{ mA})$

(ii)For fixed bias circuit, stability factor is given by

$$S = \beta + 1 = 100 + 1 = 101 \quad \textbf{Ans.}$$

EXAMPLE 5.9. Figure 5.52 shows a fixed bias circuit with β = 75. Determine the following:

(i) Value of base current,

(ii) Value of collector current,

(iii) Value of collector-to-emitter voltage,

(iv) Stability factor (neglect V_{BE}).

Solution : (i) Base current

From input section, we write

$$V_{CC} = I_B R_B + V_{BE} \qquad \text{But } V_{BE} = 0$$

Thus,
$$V_{CC} = I_B R_B$$

or
$$I_B = \frac{V_{CC}}{R_B} = \frac{20}{200 \times 10^3}$$

or
$$I_B = \frac{1}{10 \times 10^3} = 10^{-4} = 100 \text{ }\mu\text{A} \quad \textbf{Ans.}$$

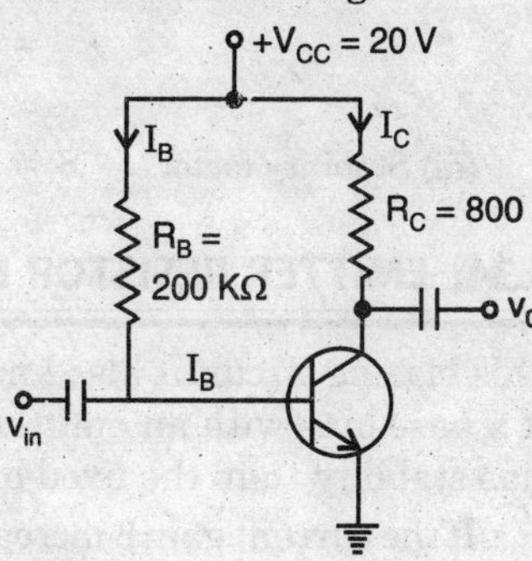

Fig. 5.52.

(ii) Value of collector current is given as

$$\because I_C = \beta I_B = 75 \times 100 \text{ }\mu\text{A}$$
$$= 7500 \text{ }\mu\text{A} = 7.5 \text{ mA} \quad \textbf{Ans.}$$

(iii) Collector to emitter voltage

From output section, we have

$$V_{CE} = V_{CC} - I_C R_C$$
$$= 20 - 7.5 \times 10^{-3} \times 800$$
$$V_{CE} = 20 - 60.0 \times 10^{-1}$$
$$= 20 - 6 = 14 \text{ V} \quad \textbf{Ans.}$$

(iv) Stability factor for fixed bias circuit is expressed as

$$S = \beta + 1 = 75 + 1 = 76 \quad \textbf{Ans.}$$

EXAMPLE 5.10. Figure 5.53 shows the circuit of fixed bias circuit with $\beta = 100$. Assume $V_{BE} \approx 0$
Determine the following.
(i) Value of base resistor R_B,
(ii) Value of voltage beτween the collector and ground.
(iii) The stability factor.

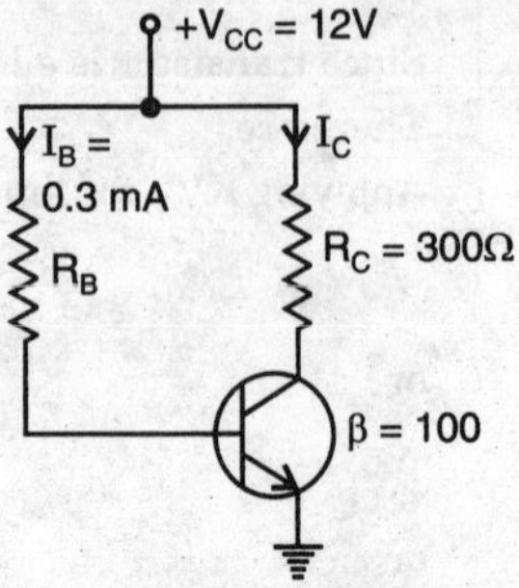

Fig. 5.53.

Solution : (i) From input section, we have

$$V_{CC} = I_B R_B + V_{BE}$$

But given that $V_{BE} \cong 0$

Therefore, $V_{CC} = I_B R_B$

or $I_B = \frac{V_{CC}}{R_B}$

or $R_B = \frac{V_{CC}}{I_B}$

Given $I_B = 0.3 \text{ mA}$

Thus, $R_B = \frac{12}{0.3 \text{ mA}} = \frac{12}{0.3 \times 10^{-3}}$

$R_B = 40 \times 10^3 = 40 \text{ k}\Omega$ **Ans.**

(ii) $\because \; I_C = \beta I_B = 100 \times 0.3 \times 10^{-1}$

$I_C = 3 \times 10^{-2} \text{ A} = 30 \text{ mA}$

Because there is no emitter resistor, the voltage between collector and ground

= Voltage between collector and emitter

$= V_{CE} = V_{CC} - I_C R_C$

$= 12 - 30 \times 10^{-3} \times 300 = 12 - 9 = 3 \text{ V}$ **Ans.**

(iii) Stability factor $S = \beta + 1 = 100 + 1 = 101$ **Ans.**

5.34. EMITTER RESISTOR BIASING

This biasing circuit is also known as base bias with emitter feedback. Figure 5.54 shows the circuit of a base bias with an emitter resistor. In this biasing circuit, the emitter resistor provides better bias stability than the fixed bias circuit discussed in last article

If the current gain β increases (say due to temperature), then the collector current in the circuit also increases. It produces more voltage across the emitter resistor R_E which reduces the voltage drop across the base resistor. This causes the base current to decreases. Thus, because of this, the collector current also decreases. Hence, the increase in collector current, due to increases in the value of current gain, will be compensated.

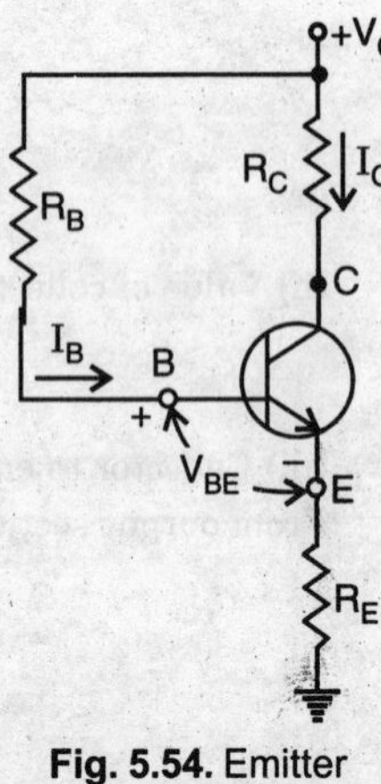

Fig. 5.54. Emitter resistor Biasing

Thus, from the above discussion, it is evident that the emitter resistor provides compensation for the variation in the value of current gain β by reducing the base current. Hence the emitter resistor is known as emitter feedback resistor. The emitter feedback circuit is effective only if the emitter resistor is as large as possible. However, in actual practice, it is not possible because the emitter resistor is kept relatively small to avoid the transistor to operate in saturation.

Now, let us determine the d.c. bias currents and voltages in the base, collector and emitter of the transistor. Firstly, let us consider the base-emitter circuit loop in the d.c. bias circuit.

Thus, applying KVL to the base-emitter circuit loop we have

$$V_{BE} + I_E R_E - V_{CC} + I_B . R_B = 0$$

Now, substituting $I_E = (\beta + 1)I_B$ in the above equation and solving it for the base current we have

$$I_B = \frac{V_{CC} - V_{BE}}{R_B + (\beta+1)R_E}$$

Assuming $\beta + 1 \cong \beta$, we get

$$I_B = \frac{V_{CC} - V_{BE}}{R_B + \beta.R_E}$$

Therefore, collector current will be

$$I_C = \beta.I_B = \frac{V_{CC} - V_{BE}}{R_E + R_B/\beta}$$

Assuming $V_{CC} >> V_{BE}$, we have

$$I_C = \frac{V_{CC}}{R_E + R_B/\beta}$$

Thus, from the above equation, it is evident that if the value of emitter resistor R_E is kept very large as compared to the value of R_B/β, the biasing circuit is independent of variations in current gain β. However, in actual paractice, it is not possible to make the emitter resistor large enough to compensate for the effects of current gain β without saturating the transistor. Hence, some compromise will have to be made between the bias stability and the performance of the circuit.

Now, let us consider the collector-emitter circuit loop in the d.c. bias circuit.

Thus, applying KVL to this loop, we have

$$V_{CE} + I_E.R_E - V_{CC} + I_C.R_C = 0$$

Replacing I_E with I_C in the last equation and solving it for the voltage across collector-emitter of a transistor, we have

$$V_{CE} = V_{CC} - I_C.(R_C + R_E)$$

The voltage measured from emitter to ground will be

$$V_E = I_E.R_E = I_C.R_E \qquad (\because I_E \cong I_C)$$

The voltage measured from collector to ground will be

$$V_C = V_C - I_C.R_C$$

Note:

(i) The voltage at which the transistor is biased, can also calculated by the expression

$$V_{CE} = V_C - V_E$$

(ii) The saturation value of collector current can be obtained by using the equation and its value is given as

$$I_C\,(\text{sat}) = \frac{V_{CC} - V_{CE}}{R_C + R_E} = \frac{V_{CC} - V_{CE}}{R_C + R_E}$$

EXAMPLE 5.11. Determine all the d.c. bias voltages and currents in the circuit shown in figure 5.55 if $R_B = 400\ k\Omega$, $R_C = 2k\Omega$, $R_E = 1\ k\Omega$, $V_{CC} = 20V$ and $\beta = 100$.

Also neglect V_{BE} of the transistor.

Solution : Given that

$$R_B = 400\ k\Omega = 400 \times 10^3\Omega$$
$$R_C = 2k\Omega = 2 \times 10^3\Omega$$
$$R_E = 1k\Omega = 1 \times 10^3\ \Omega$$
$$V_{CC} = 20\ \text{Volt}$$

and

$$\beta = 100$$

(i) **Base current**

The base current is given by the expression

$$I_B = \frac{V_{CC}}{R_B + \beta.R_E}$$

Substituting all the values, we get

$$I_B = \frac{20}{(400\times10^3)+(100\times1\times10^3)} = 0.04 \times 10^{-3}\ A$$

or $I_B = 0.04$ mA **Ans.**

(ii) Collector Current:

The collector current is expressed as

$$I_C = \beta.I_B = 100 \times 0.04 = 4 \text{ mA} \quad \textbf{Ans.}$$

(iii) Collector-to-emitter voltage :

The collector-to-emitter voltage is given as

$$V_{CE} = V_{CC} - I_C.(R_C + R_E)$$
$$= 20 - (4 \times 10^{-3})\ [(2 \times 10^3) + (1 \times 10^3)]$$
$$= 20 - 12 = 8 \text{ Volt} \quad \textbf{Ans.}$$

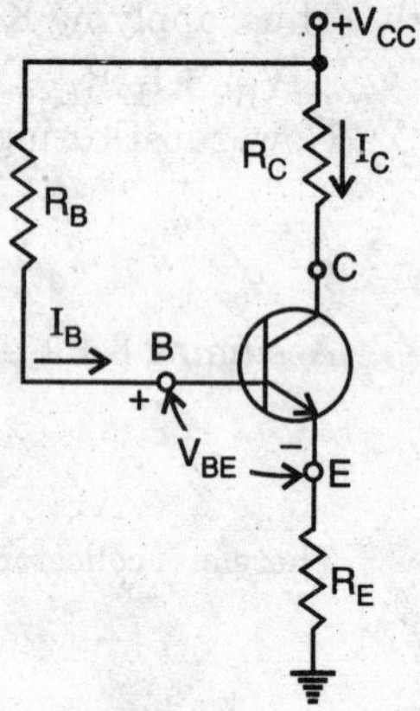

Fig. 5.55.

5.35. STABILITY FACTOR FOR EMITTER RESISTOR BIASING

1. Mathematical Expression

As discussed already in last article that if we apply KVL to the base-emitter circuit loop shown in figure 5.54, we get

$$V_{BE} + I_E.R_E - V_{CC} + I_B.R_B = 0$$

Now, replacing I_E with $I_B + I_C$ in the last equation and solving it for the base current, we have

$$I_B = \frac{V_{CC} - V_{BE} - I_C.R_E}{R_E + R_B}$$

Differentiating the last expression with respect to I_C, we get

$$\frac{dI_B}{dI_C} = \frac{0-0-1.R_E}{R_E+R_B} = -\frac{R_E}{R_E+R_B}$$

Substituting the value of dI_B/dI_C in the general expression for stability factor, we get

$$S = \frac{1+\beta}{1-\beta\left(\frac{dI_B}{dI_C}\right)} = \frac{1+\beta}{1-\beta\left(-\frac{R_E}{R_E+R_B}\right)} = \frac{1+\beta}{1+\beta.\frac{R_E}{R_E+R_B}}$$

2. Advantages

(i) Although three resistors R_C, R_B and R_E are employed in the circuit, still it is quite simple to explain.

(ii) It provides far better stabilization.

3. Disadvantages

(i) The stability factor of this biasing circuit is quite high. Hence, the operating point does change, although to lower extent, due to temperature rise or variations in transistor parameters.

(ii) The d.c. feedback helps in the stabilisation of operating point but at the same time a.c. feedback reduces the voltage gain of the amplifier. This is an undersirable feature. However, this drawback can be removed by simply putting a capacitor C_E across the resistor R_E. The capacitor C_E offers very low impedance to the a.c. current (signal) and does not allow it to pass through R_E and hence a.c. feedback is restricted. Thus, the process of amplification remains unaffected.

This biasing circuit is also not used practically because of the following reasons:

We know that

$$I_C = \frac{V_{CC}}{R_E + R_B/\beta} \qquad ...(5.60)$$

For stability, the denominator should be independent of β. It is only possible if $R_E >> \frac{R_B}{\beta}$.

This condition can only be obtained either using R_E of very large value or by using R_B of very small value. Now, a large value of R_E will cause a large voltage drop across it and to obtain required Q point we have to apply a voltage source V_{CC} of high value. On the other hand, if R_B is of very low value we have to use a separate voltage source of low value for base circuit. Both these alternatives are quite impractical. Hence, this biasing circuit is not used practically.

EXAMPLE 5.12. Figure 5.56 shows the fixed-biased circuit with emitter feedback resistor. The transistor has a current gain β equal to 80. Find the value of collector current, collector-to-emitter voltage and the stability factor.

Assume that V_{BE} = 0.7 Volt.

Solution: Given that :

$$\beta = 80,$$
$$V_{BE} = 0.7 \text{ Volt}$$
$$V_{CC} = 25 \text{ Volt},$$
$$R_C = 820\ \Omega$$
$$R_B = 180\ k\Omega$$
$$R_B = 180 \times 10^3\ \Omega$$

and $R_E = 200\ \Omega$

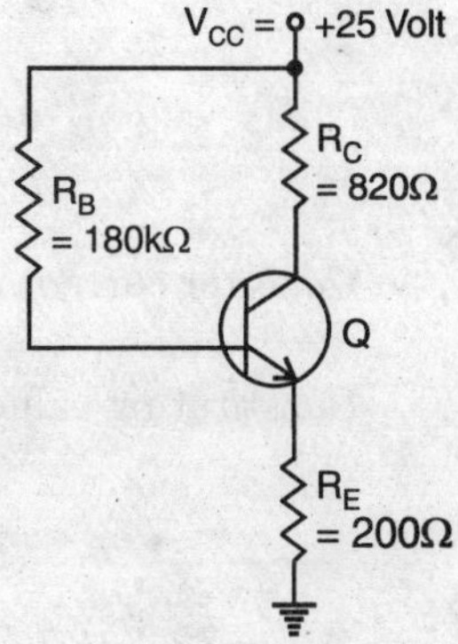

Fig. 5.56. Circuit for example 5.12.

(i) Collector Current

The value of collector current is given by

$$I_C = \frac{V_{CC} - V_{BE}}{R_E + \frac{R_B}{\beta}}$$

Substituting all the values, we get

$$I_C = \frac{25 - 0.7}{200 + \frac{180 \times 10^3}{80}} = 9.9 \times 10^{-3}\ A = 9.9\ mA \qquad \textbf{Ans.}$$

(ii) Collector-to-emitter voltage

The value of collector-to-emitter voltage is given as

$$V_{CE} = V_{CC} - I_C \cdot (R_C + R_E)$$
$$= 25 - (9.9 \times 10^{-3}) \times (820 + 200)$$

or V_{CE} = 15.1 Volt **Ans.**

5.36. COLLECTOR TO BASE BIAS CIRCUIT

(U.P. Tech. Sem. Exam., 2001-02)(5 marks)

1. Mathematical Analysis

Figure 5.57 shows the collector to base bias circuit. In this circuit, the base resistor is connected to the collector terminal of the transistor. This resistor R_B works as a feedback resistor. Figure 5.57 also shows the directions of all currents.

Now, we shall find the values of d.c. currents and voltages in the base and collector portions of the transistor.

First, let us consider the input or base-emitter loop.

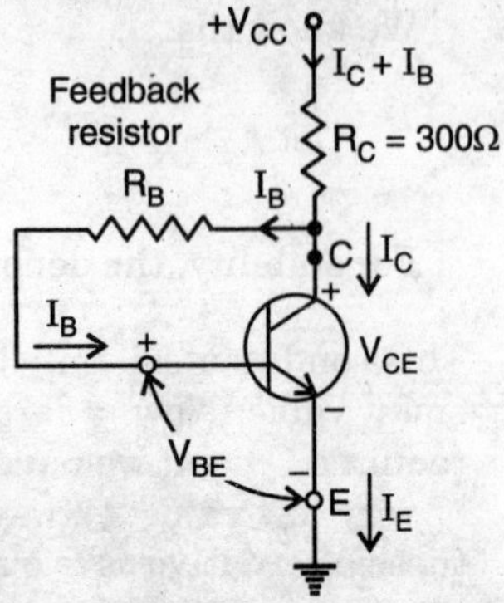

Fig. 5.57. Collector-to-base bias circuit

Applying Kirchhoff's voltage law to this loop, we get

$$-V_{CC} + (I_B + I_C) R_C + I_B R_B + V_{BE} = 0 \quad ...(5.61)$$

or $$(I_B + I_C)R_C + I_B R_B = V_{CC} - V_{BE}$$

or $$I_B(R_B + R_C) + I_C R_C = V_{CC} - V_{BE}$$

or $$I_B(R_B + R_C) + \beta I_B R_C = V_{CC} - V_{BE}$$

or $$I_B[R_B + R_C(1+\beta)] = V_{CC} - V_{BE}$$

or $$I_B = \frac{V_{CC} - V_{BE}}{R_B + R_C(1+\beta)} \quad ...(5.62)$$

Since β is quite large, therefore $1 + \beta \cong \beta$

$$I_B = \frac{V_{CC} - V_{BE}}{R_B + R_C(\beta)}$$

$$= \frac{V_{CC} - V_{BE}}{R_B + \beta R_C} \quad ...(5.63)$$

Collector current is given as

$$I_C = \beta I_B$$

Substituting value of I_B, we get

$$I_C = \beta\left(\frac{V_{CC} - V_{BE}}{R_B + \beta R_C}\right) = \frac{V_{CC} - V_{BE}}{R_C + R_B/\beta} \quad ...(5.64)$$

$\because V_{CC} >> V_{BE}$

Therefore, $$I_C = \frac{V_{CC}}{R_C + R_B/\beta} \quad ...(5.65)$$

Let us consider the output or collector-emitter loop. Applying KVL, we get

$$V_{CC} = V_{CE} + (I_C + I_B) R_C \quad ...(5.66)$$

Since $I_C >> I_B$, therefore $I_C + I_B \cong I_C$

$$V_{CE} = V_{CC} - I_C R_C \quad ...(5.67)$$

2. Stability factor for collector to base bias circuit

Stability factor is given by the expression

$$S = \frac{1+\beta}{1-\beta\left(\frac{dI_B}{dI_C}\right)} \quad ...(5.68)$$

Applying KVL to the emitter-base portion of collector-to-base bias circuit of figure 5.57, we have

$$V_{CC} = V_{BE} + (I_B + I_C)R_C + I_B R_B \quad ...(5.69)$$

or $$I_B(R_B + R_C) + I_C R_C = V_{CC} - V_{BE}$$

or $$I_B(R_B + R_C) = V_{CC} - V_{BE} - I_C R_C \quad ...(5.70)$$

or $$I_B = \frac{V_{CC} - V_{BE} - I_C R_C}{R_B + R_C}$$

Differentiating the above equation with respect to I_C, we have

$$\frac{dI_B}{dI_C} = \frac{0-0-1.R_C}{R_B+R_C} = -\frac{R_C}{R_B+R_C} \quad ...(5.71)$$

Substituting the value of $\frac{dI_B}{dI_C}$ from equation (5.71) to equation (5.68), we get

$$S = \frac{1+\beta}{1+\beta\left(\frac{R_C}{R_B+R_C}\right)} \quad ...(5.72)$$

Important Point: From the above expression, it is clear that stability factor of collector-base bias circuit is smaller than that of fixed bias circuit. Therefore, this biasing circuit has greater stability than fixed bias circuit.

3. Advantages

(i) It is a simple biasing arrangement since only one resistor R_B is required.

(ii) This biasing circuit provides some stabilization of the operating point as explained below: Suppose temperature increases, this causes increase in leakage current and at the same time there is an increase in β due to inherent variation in transistor parameters. This increases the collector current since $I_C = \beta I_B + (\beta + 1)I_{CBO}$. As the collector current increases, the voltage V_{CE} decreases as $V_{CE} = V_{CC} - I_C R_C$. The decrease in V_{CE} causes decrease in base current since $I_B = \frac{V_{CE} - V_{BE}}{R_B}$. This decrease in base current, in turn, decreases the collector current to bring it to original value. Hence, there is a tendency of this circuit to stabilise the operating point.

4. Disadvantages

(i) The stability factor of this biasing method is quite high. Hence, the operating point does change, although to lower extent, due to temperature rise or inherent variations in transistor parameters.

(ii) The resistor R_B not only provides d.c. feedback for stabilisation of operating point, but it also causes a.c. feedback. This reduces the voltage gain of the amplifier which is not desirable. Because of this drawback, this circuit is **seldom used in amplifiers.**

EXAMPLE 5.13. Figure 5.58 shows the circuit of collector-to-base bias using N-P-N transistor. Assuming V_{BE} = 0.7 V, calculate the following:

(i) The collector current,

(ii) The collector-to-emitter voltage,

(iii) Stability factor.

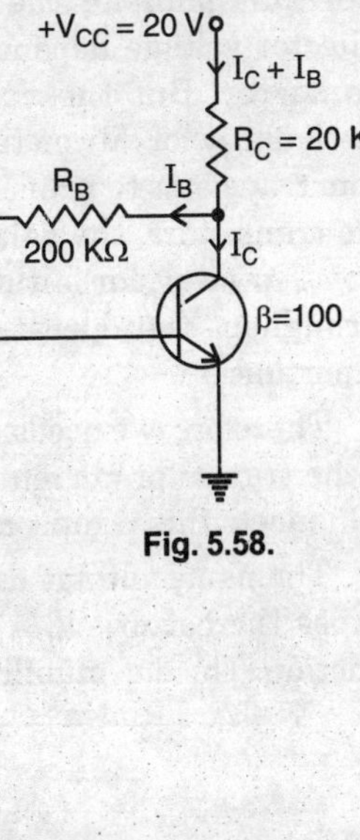

Fig. 5.58.

Solution : We know that, collector current is given by

(i) $$I_C = \frac{V_{CC} - V_{BE}}{R_C + \frac{R_B}{\beta}} = \frac{20-0.7}{\left(20+\frac{200}{100}\right)\times 10^3}$$

$$I_C = \frac{19.3}{(20+2)10^3} = \frac{19.3}{(22)10^3} = 0.877 \times 10^{-3} = 0.877 \text{ mA} \quad \textbf{Ans.}$$

(ii) The collector-to-emitter voltage is expressed as under:

$$V_{CE} = V_{CC} - I_C.R_C = 20 - 0.877 \times 10^{-3} \times 20 \times 10^3$$

$$V_{CE} = 20 - 17.540 = 20 - 17.54 = 2.46 \text{ Volt} \quad \textbf{Ans.}$$

(iii) Stability factor, is expressed as

$$S = \frac{1+\beta}{1+\beta\left(\frac{R_C}{R_C+R_B}\right)} = \frac{1+100}{1+100\left(\frac{20}{20+200}\right)}$$

$$S = \frac{101}{1+\frac{100\times 20}{220}} = \frac{101\times 22}{22+200} = \frac{2222}{222} = 10.009 \quad \textbf{Ans.}$$

EXAMPLE 5.14. Figure 5.59 shows the collector-to-base bias circuit with β = 100. Assuming $V_{BE} \cong 0$, determine the following :

(i) the value of I_B, (ii) the value of I_C, (iii) the value of V_{CE}, (iv) the stability factor. (U.P. Tech Tutorial Question Bank)

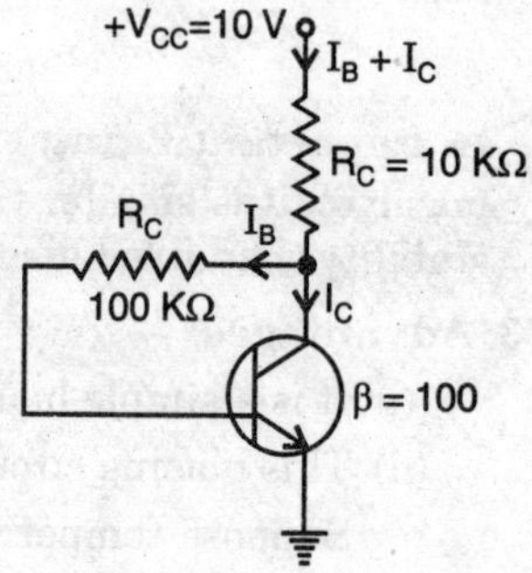

Fig. 5.59.

Solution : (i) Base current I_B is given by

$$I_B = \frac{V_{CC} - V_{BE}}{R_B + \beta.R_C} = \frac{10-0}{(100+100\times 10)10^3}$$

$$I_B = \frac{10\times 10^{-3}}{100+1000} = \frac{10^{-2}}{1100} = 0.09 \times 10^{-4} = 9\mu A$$

(ii) $I_C = \beta I_B = 100 \times 9\mu A$

$= 900 \times 10^{-6} A = 0.9$ mA **Ans.**

(iii) $V_{CE} = V_{CC} - I_C.R_C = 10 - 0.9 - 10^{-3} \times 10 \times 10^3$

$V_{CE} = 10 - 9 = 1$ Volt **Ans.**

(iv) Stability factor $S = \frac{1+\beta}{1+\beta\left(\frac{R_C}{R_C+R_B}\right)} = \frac{1+100}{1+100\left(\frac{10}{10+100}\right)}$

$$= \frac{101}{1+0.09} = \frac{101}{1.09} = 92.6 \quad \textbf{Ans.}$$

5.37. SELF BIAS OR VOLTAGE DIVIDER BIAS

(U.P.Tech – Tutorial Question Bank)

1. Mathematical Analysis

> **DO YOU KNOW?**
>
> The most famous circuit based on the emitter-bias prototype is called voltage-divider bias. You can recognize it by the voltage divider in the base circuit.

For fixed bias circuit and collector-to-base bias circuit, the operating point or the values of d.c. collector current and collector voltage depend upon the value of current gain β of transistor. But the value of β is temperature dependent particularly for silicon transistors. Also, the value of β changes from transistor to transistor. This means that for same type of two transistors, the values of β are different. Therefore, due to variation in β, operating point may vary. However, for faithful amplification, we need that the operating point should remain at a point throughout the operation where it was fixed. Hence, stability of operating point is of utmost importance.

Therefore, we need a particular type of d.c. bias circuit in which the operating point is independent of the transistor current gain β. The voltage divider bias circuit or self bias circuit shown in figure 5.60 meets this requirement. Hence, this is most widely used d.c. bias circuit.

The name voltage divider comes from the fact that resistors R_1 and R_2 form a potential divider across the battery V_{CC}. The voltage drop across the resistor R_2 forward biases the emitter-base junction. The d.c. stability is provided by emitter resistor R_E.

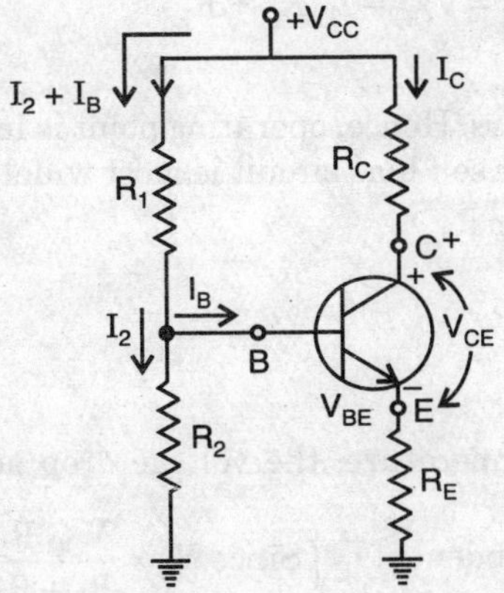

Fig. 5.60. Voltage divider or self-bias circuit

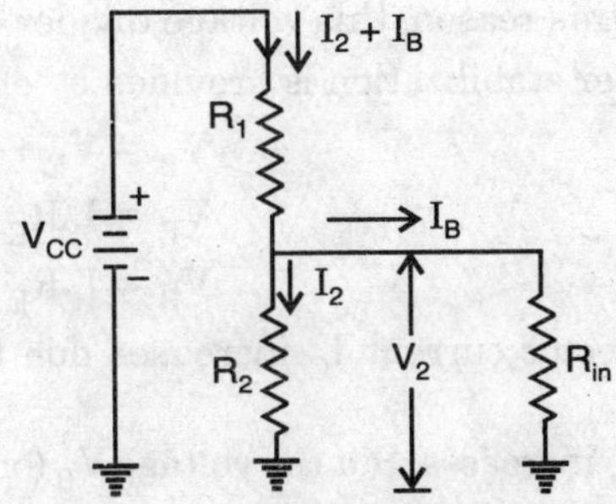

Fig. 5.61. Input section of the voltage divider bias circuit

Let us consider the emitter-base loop or input loop of voltage divider bias circuit.

Here, we take help of basic assumption that the resistance looking into the base or input resistance is much larger than the resistance R_2.

According to above assumption, the base current flowing into the input resistance is negligible small and the current through resistor R_1 flows almost completely into resistor R_2 because the current always takes low resistance path. Therefore two resistors R_1 and R_2 effectively comes in series. Let the voltage across resistor R_2 is V_2 which is also the base voltage V_B.

This voltage V_2 or V_B may be calculated with the help of voltage divider network of R_1 & R_2 and the supply voltage.

> **DO YOU KNOW?**
>
> The key idea is for the base current to be much smaller than the current through the voltage divider. When this condition is satisfied, the voltage divider holds the base voltage almost constant and equal to the unloaded voltage out of the voltage divider. This produces a solid Q point under all operating conditions.

Therefore,
$$V_B = V_{CC} \times \frac{R_2}{R_1 + R_2} \qquad ...(5.73)$$

The voltage across the emitter resistor R_E is given by
$$V_E = V_B - V_{BE} \qquad ...(5.74)$$

The emitter current is then
$$I_E = \frac{V_E}{R_E} = \frac{V_B - V_{BE}}{R_E} \qquad ...(5.75)$$

Hence, the value of collector current
$$I_C \cong I_E = \frac{V_B - V_{BE}}{R_E} \qquad ...(5.76)$$

The voltage at the collector
$$V_C = V_{CC} - I_C R_C \qquad ...(5.77)$$

The collector-to-emitter voltage is given by
$$V_{CE} = V_C - V_E = V_{CC} - I_C R_C - I_E R_E$$

Since $I_E = I_C$

Therefore, $$V_{CE} = V_{CC} - I_C R_C - I_C R_E = V_{CC} - I_C(R_C + R_E) \quad ...(5.78)$$

2. Stabilization

From the above, it is clear that in no expression β appears. Hence, operating point is independent of β. Due to this reason, this voltage divider bias circuit or self bias circuit is most widely used.

Further stabilization is provided by emitter resistor R_E.

Since $$V_B = V_E + V_{BE} \quad ...(5.79)$$

$$V_B = I_E R_E + V_{BE}$$

$$V_B = I_C R_E + V_{BE} \quad ...(5.80)$$

If collector current I_C increases due to rise in temperature, the voltage drop across emitter resistor R_E increases. But the voltage V_B (or V_2) is independent of I_C $\left(\text{Since } V_B = \frac{V_{CC}.R_2}{R_1 + R_2}\right)$, therefore, V_{BE} decreases. This will decrease the value of I_B which in turn reduce the value of I_C to its original level.

Hence, due to above two reasons, the voltage divider bias circuit or self bias circuit is widely used.*

3. Stability factor of Voltage Divider Bias or Self Bias

To obtain stability factor, first we will find thevenin's equivalent voltage and thevenin's equivalent resistance for voltage divider bias circuit.

Assuming two terminals a and b in the network. Thevenizing the circuit on the left of terminals a and b.

Thevenin's voltage is

$$V_{TH} = \frac{R_2}{R_1 + R_2}.V_{CC} \quad ...(5.81)$$

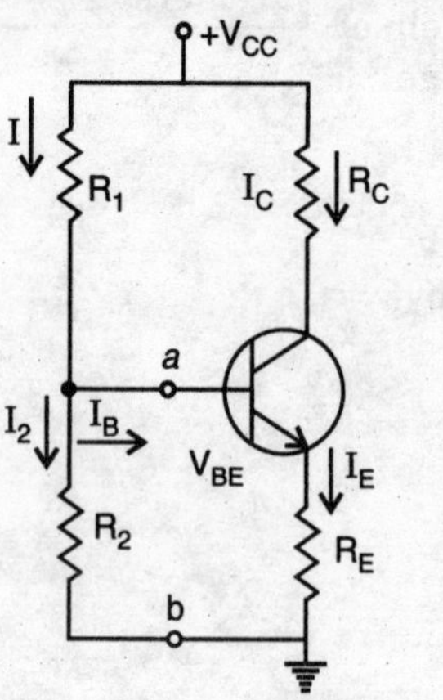

Fig. 5.62. Voltage divider or self bias circuit

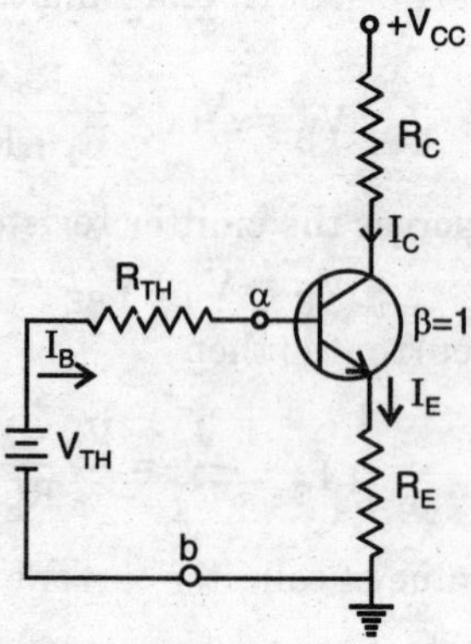

Fig. 5.63. Thevenin's equivalent circuit

and Thevenin's equivalent resistance

$$R_{TH} = \frac{R_1 R_2}{R_1 + R_2} \quad ...(5.82)$$

Now, applying Kirchhoff's voltage law to the emitter base loop of the Thevenin's equivalent

* Why is a voltage divider bias circuit most widely used? **(U.P.Tech., Sem. Exam. 2004) (5 marks)**

circuit.

$$V_{TH} = I_B \cdot R_{TH} + V_{BE} + I_E \cdot R_E \quad ...(5.83)$$

$$I_B \cdot R_{TH} + (I_B + I_C)R_E = V_{TH} - V_{BE} \qquad [\because I_E = I_B + I_C]$$

or $$I_B(R_{TH} + R_E) = V_{TH} - V_{BE} - I_C R_E$$

or $$I_B = \frac{V_{TH} - V_{BE} - I_C R_E}{R_{TH} + R_E} \quad ...(5.84)$$

> **DO YOU KNOW?**
>
> Because the values of I_C and V_{CE} are dependent on the values of beta in a base-biased circuit, the circuit is said to be beta-dependent.

Differentiating the above equation with respect to I_C, we get

$$\frac{dI_B}{dI_C} = \frac{0 - 0 - 1.R_E}{R_{TH} + R_E} = -\frac{R_E}{R_{TH} + R_E}$$

Since, Stability factor is given by the expression

$$S = \frac{1+\beta}{1-\beta\left(\frac{dI_B}{dI_C}\right)} = \frac{1+\beta}{1-\beta\left(-\frac{R_E}{R_{TH}+R_E}\right)} = \frac{1+\beta}{1+\beta\left(\frac{R_E}{R_{TH}+R_E}\right)} \quad ...(5.85)$$

or $$S = \frac{1+\beta(R_{TH}+R_E)}{R_{TH}+R_E+\beta R_E} = \frac{1+\beta\left(1+\frac{R_{TH}}{R_E}\right)}{1+\beta+\frac{R_{TH}}{R_E}}$$

5.38. EMITTER BIAS CIRCUIT

Figure 5.64 shows the emitter bias circuit. In this circuit, both the positive (+ V_{CC}) and negative ($-V_{EE}$) supply voltages are used. The name emitter bias comes from the fact that the negative supply V_{EE} is used to forward-bias the emitter-base junction through resistor R_E. The supply V_{CC} reverse-biases the collector-base junction. This circuit provides almost same stability of operating point as a voltage divider circuit.

If the base resistance (R_B) is small enough in the emitter bias circuit, then the base voltage (V_B) is approximately zero. The base-emitter junction of a transistor is forward biased by the $-V_{EE}$ supply. The d.c. bias currents and voltages may be obtained as under:

The base voltage (V_B) is approximately zero, because the value of resistance R_B is usually very high.

Hence, $V_B = 0$, and the voltage at the emitter (measured with respect to ground) will be,

$$V_E = -V_{BE} \quad ...(5.86)$$

Therefore, value of emitter current will be

$$I_E = \frac{V_E - V_{EE}}{R_E} = \frac{-V_{BE} - V_{EE}}{R_E} \quad ...(5.87)$$

since V_{EE} supply is negative, therefore, the emitter current, will be

$$I_E = \frac{-V_{BE} - (-V_{EE})}{R_E}$$

or $$I_E = \frac{V_{EE} - V_{BE}}{R_E} \quad ...(5.88)$$

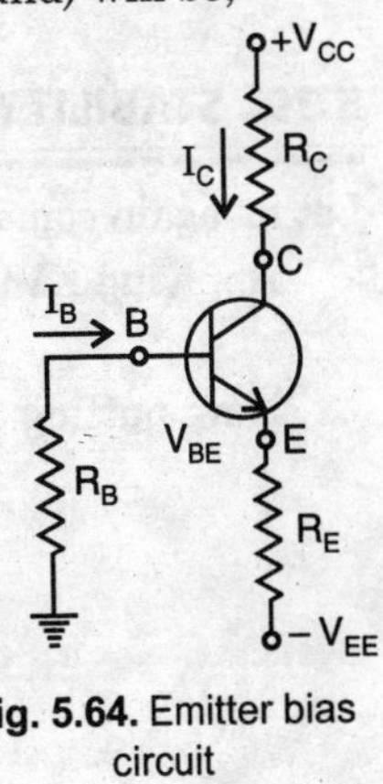

Fig. 5.64. Emitter bias circuit

We know that the value of collector current is approximately the same as that of emitter current, i.e., $I_C = I_E$

∴ Collector voltage (measured with respect to ground), will be

$$V_C = V_{CC} - I_C \cdot R_C$$

and the collector-to-emitter voltage will be

$$V_{CE} = V_C - V_E \qquad ...(5.89)$$

> **DO YOU KNOW?**
>
> This design uses two power supplies: one positive and the other negative. The idea is to set up a constant value of emitter current. The circuit is a variation of the emitter-bias prototype.

EXAMPLE 5.15. Determine the values of emitter current, collector current and collector-to-emitter voltage for the emitter bias circuit shown in figure 5.65. Assume V_{BE} = 0.7 Volt.

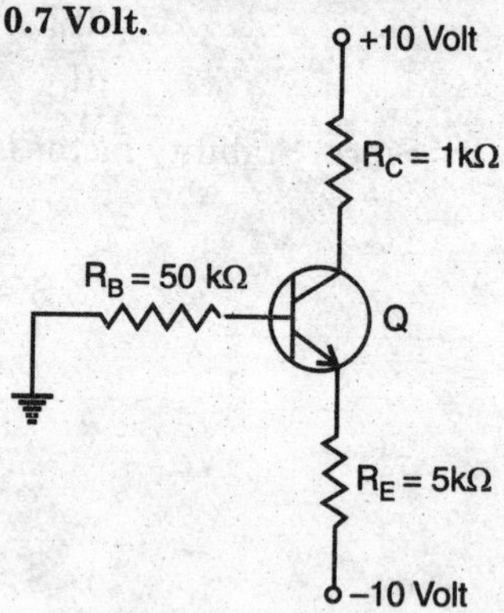

Fig. 5.65.

Solution : Given that

$$V_{CC} = 10 \text{ Volt}$$
$$V_{EE} = 10 \text{ Volt}$$
$$R_C = 1 \text{ k}\Omega = 1 \times 10^3 \ \Omega$$
$$R_E = 5 \text{ k}\Omega = 5 \times 10^3 \ \Omega$$

and
$$R_B = 50 \text{ k}\Omega = 50 \times 10^3 \ \Omega$$

(i) Emitter Current

The emitter voltage is given as

$$V_E = -V_{BE} = -0.7 \text{ Volt}$$

Also, the value of emitter current is given as

$$I_E = \frac{V_{EE} - V_{BE}}{R_E}$$

Substituting all the values, we get

$$I_E = \frac{10 - 0.7}{5 \text{ Ÿ} 10^3} = 1.86 \times 10^{-3} \text{ A} = 1.86 \text{ mA} \quad \textbf{Ans.}$$

(ii) Collector Current

The value of collector current is given by

$$I_C \cong I_E \cong 1.86 \times 10^{-3} = 1.86 \text{ mA} \quad \textbf{Ans.}$$

(iii) Collector-to-emitter voltage :

The value of collector-to-emitter voltage is given as

$$V_{CE} = V_{CC} - I_C \cdot R_C = 10 - (1.86 \times 10^{-3}) \times (1 \times 10^3) = 8.14 \text{ Volts}$$

Thus, the collector-to-emitter voltages is given by

$$V_{CE} = V_C - V_E = 8.14 - (-0.7) = 8.84 \text{ Volts} \quad \textbf{Ans.}$$

5.39. STABILITY OF EMITTER BIAS CIRCUIT

Let us again consider the emitter bias circuit shown in figure 5.64.

Applying KVL to the base-emitter loop of this circuit, we have

$$I_B \cdot R_B + V_{BE} + I_E \cdot R_E - V_{EE} = 0$$

Now, putting the value of base current I_B (equal to $I_C/\beta = I_E/\beta$) in the last equations, we get

$$I_E \left(\frac{R_B}{\beta} + R_E \right) = V_{EE} - V_{BE}$$

$$I_E = \frac{V_{EE} - V_{BE}}{\frac{R_B}{\beta} + R_E}$$

or $$I_E = \frac{V_{EE} - V_{BE}}{R_E} \qquad (\because R_E >> R_B/\beta)$$

or $$I_E = \frac{V_{EE}}{R_E} \qquad (\because V_{EE} >> V_{BE})$$

Important Point: This expression reveals the fact that the emitter current is independent of the current gain β and base-to-emitter voltage V_{BE} as long as the value of resistance R_E and V_{EE} supply are very large. However, in actual practice, the value of resistor R_E is selected such that it is 100 times greater than R_B/β.

Mathematically, it is equivalent to

$$R_E \geq 100,\ R_B/\beta \leq 0.01\ \beta.R_E{}^*$$

5.40. STABILITY FACTOR OF EMITTER BIAS CIRCUIT

As already discussed in the last article that if we apply KVL to the base-emitter loop of the emitter bias circuit shown in figure 5.64, we get

$$I_B.R_B + V_{BE} + I_E.R_E - V_{EE} = 0$$

Now, Substituting the value of I_E (equal to $I_B + I_C$) in the last equation and solving it for base current, we have

$$I_B = \frac{V_{EE} - V_{BE} + I_C.R_C}{R_E + R_B}$$

Differentiating this expression with respect to I_C, we get

$$\frac{dI_B}{dI_C} = \frac{0 - 0 - 1.R_E}{R_E + R_B} = -\frac{R_E}{R_e + R_B}$$

Lastly, substituting the above value of $\frac{dI_B}{dI_C}$ in the general expression for stability factor, we get

$$S = \frac{1+\beta}{1-\beta\left(\frac{dI_B}{dI_C}\right)} = \frac{1+\beta}{1-\beta\left(\frac{-R_E}{R_E+R_B}\right)} = \frac{1+\beta}{1+\beta\left(\frac{R_E}{R_E+R_B}\right)}$$

EXAMPLE 5.16. Figure 5.66 shows the emitter bias circuit using an NPN transistor. Find how much the Q-point will change over a temperature range where β increases from 50 to 100 and V_{BE} decreases from 0.7 Volt to 0.6 Volt.

Solution : Given that :

V_{CC} = 20 Volts
V_{EE} = 20 Volts
$R_C = 5\ k\Omega = 5 \times 10^3\ \Omega$
$R_E = 10\ k\Omega = 10 \times 10^3\ \Omega$
$R_B = 10\ k\Omega = 10 \times 10^3\ \Omega$
β = 50 and 100
V_{BE} = 0.7 Volt and 0.6 Volt

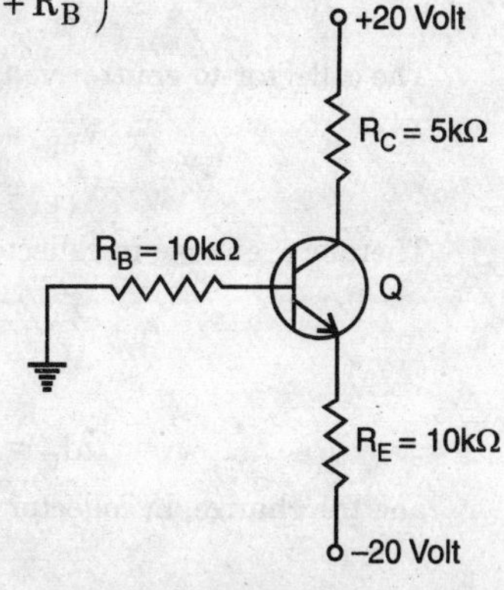

Fig. 5.66.

* Since the emitter current is independent of β and V_{BE}, therefore, the Q-point is not affected appreciably by the variations in these parameters. Hence, the emitter bias circuit provides a reasonably stable bias point.

We know that the Q-point (i.e., operating point) is defined by the values of collector current I_C and collector-to-emitter voltage V_{CE}. Thus, in order to determine the change in Q-point, we shall find the variation in collector current and the collector-to-emitter voltage in both the cases.

when $\beta = 50$ and $V_{BE} = 0.7$ volt, the value of emitter current will be

$$I_E = \frac{V_{EE} - V_{BE}}{R_E + \frac{R_B}{\beta}}$$

Substituting all the values, we get

$$I_E = \frac{20 - 0.7}{10 \times 10^3 + \frac{10 \times 10^3}{50}} = 1.892 \times 10^{-3} \text{ A} = 1.892 \text{ mA}$$

Also, the collector voltage is

$$V_C = V_{CC} - I_C.R_C$$
$$V_C = 20 - (1.892 \times 10^{-3}) \times (5 \times 10^3) = 20 - 9.46 = 10.54 \text{ Volt}$$

and the collector-to-emitter voltage V_{CE} will be

$$V_{CE} = V_C - V_E = 10.54 - (-0.7) = 11.24 \text{ Volt}$$

Similarly the value of the emitter current, when $\beta = 100$ and $V_{BE} = 0.6$ Volt, will be

$$I_E = \frac{V_{EE} - V_{BE}}{R_E + \frac{R_B}{\beta}} = \frac{20 - 0.6}{10 \times 10^3 + \frac{10 \times 10^3}{100}}$$

or $$I_E = 1.921 \times 10^{-3} \text{ A} = 1.921 \text{ mA}$$

and the collector current will be

$$I_C \cong I_E = 1.921 \times 10^{-3} \text{ A}$$
$$I_E = 1.921 \text{ mA}$$

We know that the collector voltage will be

$$V_C = V_{CC} - I_C.R_C$$
$$V_C = 20 - (1.921 \times 10^{-3}) \times (5 \times 10^3)$$
$$V_C = 20 - 9.605 = 10.395 \text{ Volt}$$

The collector-to-emitter voltage will be

$$V_{CE} = V_C - V_E$$
$$V_{CE} = 10.395 - (-0.6) = 10.995 \text{ Volt}$$

Therefore, change in collector current

$$\Delta I_C = \frac{1.921 - 1.892}{1.892}$$

$$\Delta I_C = 0.0153 = 1.53\% \quad \textbf{Ans.}$$

and the change, in collector to emitter voltage will be

$$\Delta V_{CE} = \frac{11.24 - 10.995}{11.24}$$

$$\Delta V_{CE} = 0.0218 = 2.18\% \quad \textbf{Ans.}$$

Table 5.3. Comparison of different biasing circuits

S. No.	Parameter of comparison	Fixed bias circuit	Collector to base bias circuit	Self bias circuit
1.	Emitter resistance	Not used	Not used	Used
2.	Negative feedback	Not used	Included	Included
3.	Stability factor	$S = (1+\beta)$	$S = \dfrac{1+\beta}{1+\beta\left[\dfrac{R_C}{R_C+R_B}\right]}$	$S = 1+\beta\dfrac{1+(R_B/R_E)}{1+\beta+(R_B/R_E)}$
4.	Q point stability	Poor	Moderate	Good
5.	Configuration	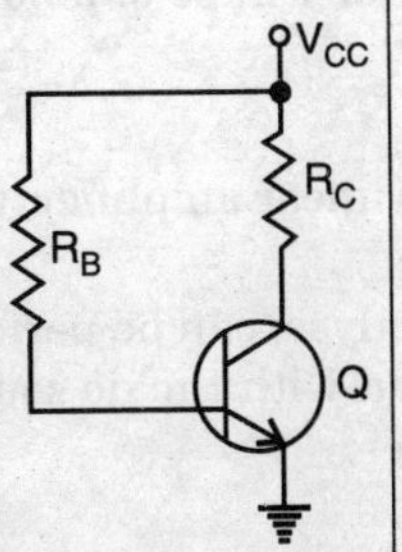	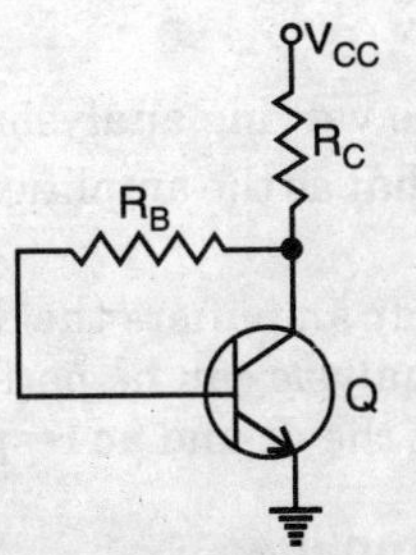	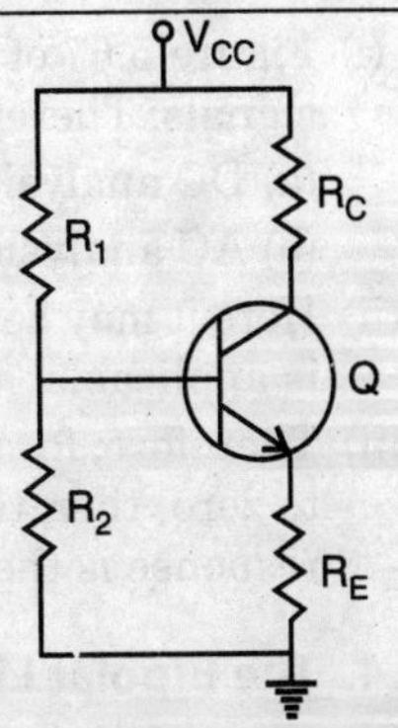

5.41. TRANSISTOR SMALL SIGNAL AMPLIFIERS

5.41.1. Basic Concepts

1. We have already discussed the operating principle and biasing circuits for the BJT. Now, let us understand how to use it as a linear amplifier.
2. The **linear amplifiers** are meant to amplify the analog signals. The magnitude of analog signal can take any value and the signal can vary continuously with time.

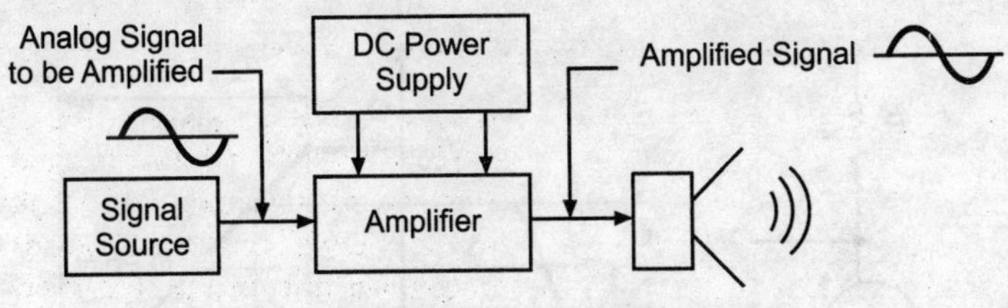

Fig. 5.67. Block diagram of an amplifier

3. A linear amplifier is then suppose to multiply the input signal by a constant to produce the output. This multiplying factor is greater than 1 and called as *gain* of the amplifier.

5.41.2. Analog Signals and Linear Amplifiers

1. Let us consider the block diagram of figure 5.67. The analog signal applied at the amplifier input is an information signal. It may be derived from a CD player, a microphone or from some kind of a transducer.
2. The magnitude of this analog signal can take on any value and may vary continuously with time. Electronic circuits which process the analog signals are called as **analog circuits.**

3. **A Linear Amplifier** magnifies the input signal and produces an output signal which is larger in size and has same shape as that of the input.
4. The dc power supply in figure 5.67 supplies power to the amplifier which contains a transistor acting as an amplifying device.

Why linear amplifier is required ?

We want the output signal to be linearly proportional to the input signal because only then the amplified signal will exactly have the same shape as that of the input. If the amplifier is non-linear, then the shape of the output signal will be different from that of the input and the output signal is said to be distorted.

Hence, the amplifier must be linear.

5.41.3. Amplifier Analysis

(i) Figure 5.67 of amplifier block diagram shows that it consists of DC voltages as well as AC signals. Therefore, the analysis of the amplifier will be of following two types :
   (a) DC analysis
   (b) AC analysis or time varying analysis

(ii) Here it may be noted that as the amplifier is a linear amplifier, the principle of superposition is applicable.

(iii) Therefore, for the linear amplifiers the dc analysis can be performed with the ac source set to zero, then the ac analysis can be performed with the dc source set to zero and the total response is the sum of the dc and ac responses.

5.41.4. The Bipolar Linear Amplifier

(i) The transistor is heart of an amplifier. Let us consider the transistor circuit shown in figure 5.68 (a). The voltage transfer characteristics of this circuit has been shown in figure 5.68(b).

(ii) To operate the transistor as an amplifier, we have to bias it at the Q point shown in figure 5.68(b) so that it is biased in the forward active region.

Working Operation

The dc voltage at the base is say V_{BB}. If a time varying signal say a sine wave is superimposed on the dc voltage V_{BB}, then the output voltage will change along the transfer curve. This will produce a time varying output voltage. If the time varying output voltage is proportional to and larger than the time varying input signal, then the circuit is a linear amplifier.

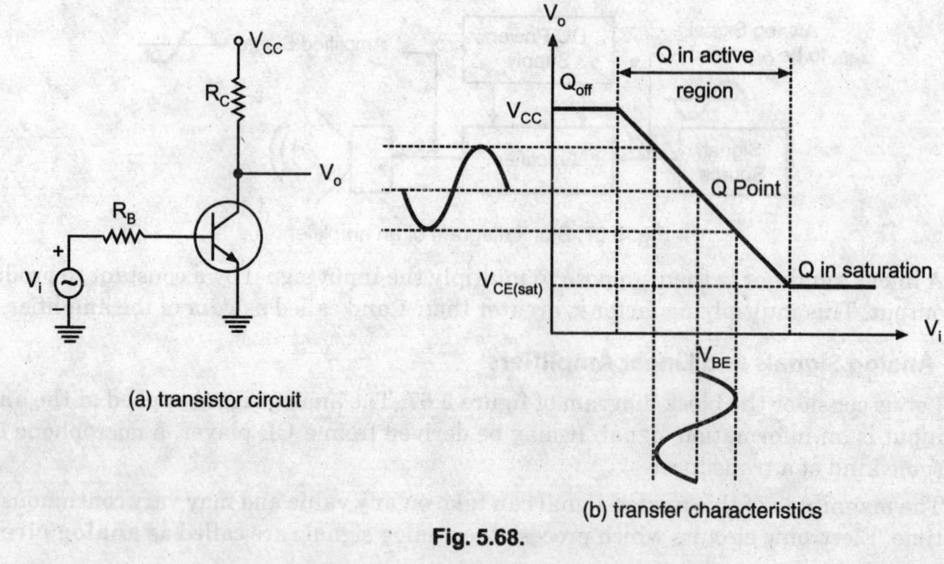

Fig. 5.68.

If the transistor is not biased in its forward active mode, then its output will not vary proportional to the input. Hence the circuit will no more act as a linear amplifier.

5.41.5 AC Analysis

Now, let us perform the ac analysis of bipolar transistor amplifiers. For AC analysis we will use two techniques namely :

1. Graphical analysis and
2. Mathematical analysis using a small signal equivalent model.

Notations used

The notations used here may be summarised in the following table.

Table 5.4. Summary of notations

Variable	Meaning
1. i_B, V_{BE}, i_C etc.	Instantaneous values
2. I_B, I_C, V_{BE}, V_{CE} etc.	DC values
3. I_b, Ic, V_{be}, V_{ce} etc.	AC values

5.41.5.1. Graphical Analysis and AC Equivalent Circuit

1. Circuit Diagram

Let us consider a practical transistor amplifier circuit, discuss its operation and then analyze the behaviour of this circuit using the dc load line. A single stage RC coupled amplifier using transistor as an active device has been shown in figure 5.69.

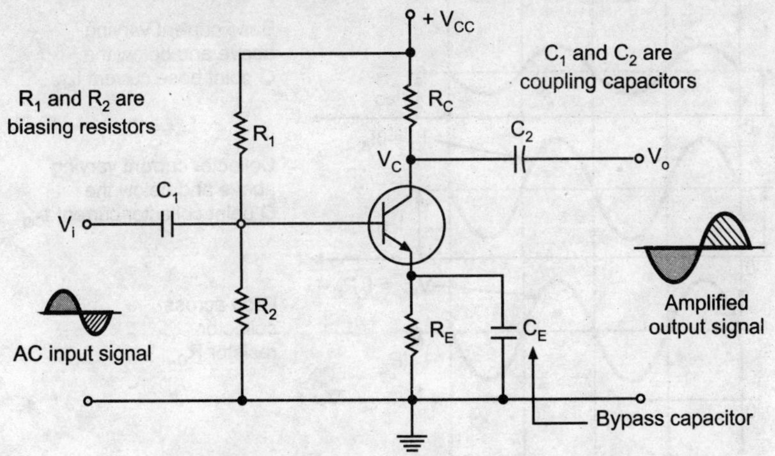

Fig. 5.69. Single stage RC coupled CE amplifier

2. Description

The capacitors C_1 and C_2 are known as the coupling capacitors. As the load resistor R_L is coupled to the amplifier through the coupling capacitor, this amplifier is called as RC coupled amplifier. The transistor is connected in the common emitter (CE) configuration. Therefore this amplifier is called CE amplifier.

3. Circuit Components and their Functions

Now, let us know about the function of each component in the RC coupled amplifier.

(i) Resistors R_1, R_2 and R_E are used for biasing the transistor in the active region, because for operating the transistor as an amplifier it is necessary to bias it in the active region. The type of biasing circuit used here is voltage divider bias or self bias.

(ii) R_C is the collector resistor used for controlling the collector current.

(iii) Input coupling capacitor C_1 : The input coupling capacitor C_1, is used for coupling the ac input voltage V_i to the base of the transistor. As capacitors block dc, this capacitor helps to block any dc component present in V_i and couples only the ac component of the input signal. This capacitor also ensures that the dc biasing conditions of the transistor remain unchanged even after application of the input signal.

(iv) Bypass capacitor C_E · The capacitor connected in parallel with the emitter resistor R_E is called as the emitter bypass capacitor. This capacitor offers a low reactance to the amplified ac signal. Therefore the emitter resistor R_E gets bypassed through C_E for only the ac signals. This will increase the voltage gain of the amplifier, Moreover as C_E acts as an open circuit for dc voltages, it does not bypass R_E for dc conditions. Thus presence of C_E does not alter the dc biasing conditions.

(v) Output coupling capacitor C_2 : This capacitor couples the amplifier output to the load resistance or to the next stage of the amplifier. It is used for blocking the dc part and passing only the ac part of the amplified signal to the load.

4. Working Operation of the RC Coupled Amplifier

1. When the ac input signal is absent$_3$ the value of dc base current of the transistor is I_{BQ} which is called as the quiescent point base current. Corresponding to $I_{BQ;}$ a quiescent dc collector current I_{CQ} also flows through the transistor ($I_{CQ} = \beta_{dc} I_{BQ}$) and the collector emitter voltage is V_{CEQ}. All this happens since we have biased the transistor in its active region using the biasing components.

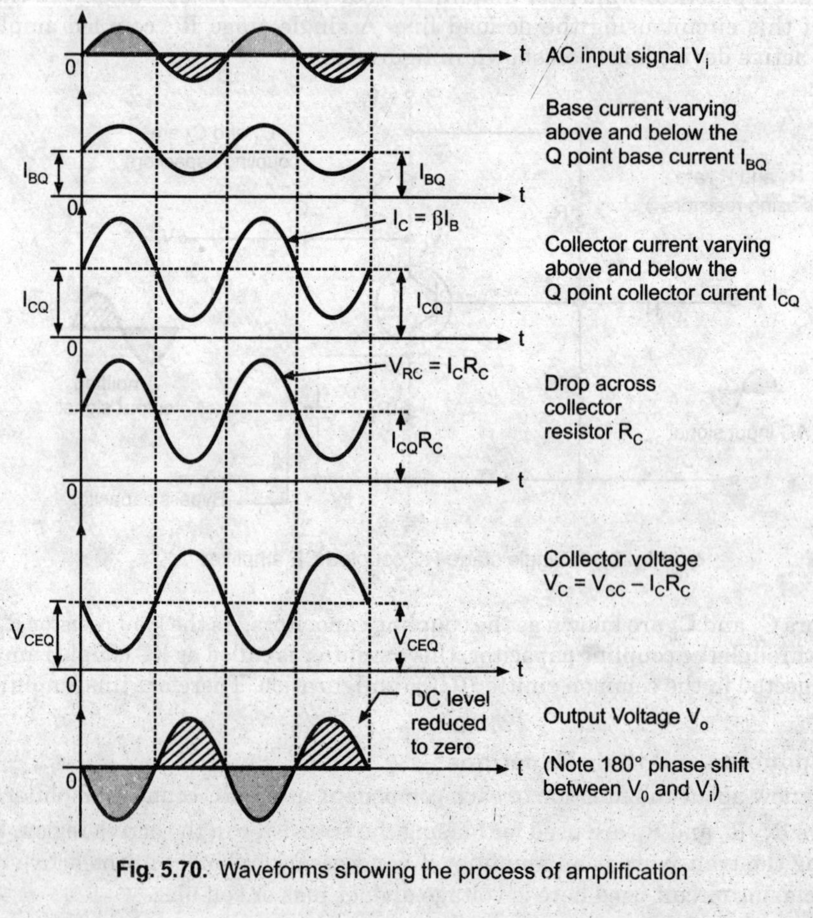

Fig. 5.70. Waveforms showing the process of amplification

2. When small ac sinusoidal signal is applied at the input of the amplifier, an alternating base current, starts flowing in the circuit. This base current varies above and below the Q point value of the base current (I_{BQ}) as shown in figure 5.70. Thus we can say that the ac signal is superimposed on the DC current I_{BQ}.
3. Due to these variations in the base current, proportional variations take place in the collector current, because $I_c = \beta I_B$. As I_B increases I_c also increases and with decrease in I_B, I_C will also decrease as shown in figure 5.70. Thus I_B and I_C are in phase but I_C is a magnified version of I_B. I_C varies above and below its Q point value I_{CQ}.
4. This varying I_C passes through the collector resistor R_C to produce a varying voltage drop $I_C R_C$ across it. This voltage drop $I_C R_C$ is in phase with the collector and base currents as shown in figure 5.70.
5. The collector voltage is given by, $V_C = V_{CC} - I_C R_C$ ...(5.86)
Therefore with changes in the voltage drop $I_C R_C$ the collector voltage also will vary as shown in figure 5.70. However V_C and $I_C R_C$ will vary in exactly opposite manner with respect to each other, because as $I_C R_C$ increases V_C has to decrease according to Equation (5.86).
6. This collector voltage is then coupled to the load through the coupling capacitor C_2. It will block the dc part of V_c and allow only ac part to pass through as shown in figure 5.70
7. See the ac signal amplitude obtained after C_2. This is the output voltage. Its magnitude **is** much higher than that of the input signal and its shape is exactly same as that of the input signal. **Thus the input ac signal has been successfully amplified.**

5. Summary of operation of RC coupled CE amplifier

(*i*) In the absence of ac input signal $I_B = I_{BQ}$, $I_C = I_{CQ}$ and $V_{CE} = V_{CEQ}$. The Q point is selected to be in the active region of transistor

↓

(*ii*) As V_i is applied, the base current varies above and below I_{BQ}.

↓

(*iii*) Hence, $I_C = \beta I_B$ varies above and below I_{CQ}. Variation in I_C is large.

↓

(*iv*) Therefore, voltage across R_C varies, because $V_{RC} = I_C \times R_C$.

↓

(*v*) Hence, collector voltage V_C varies above and below V_{CEQ} as $V_C = V_{CC} - I_C R_C$.

(*vi*) Through C_2 only the ac part of V_C is coupled to the load. V_o is of same shape as V_i, but of larger size. Thus amplification has taken place.

6. Phase relationship between input and output:

As observed from there is a 180° phase shift between the output and input or the output is said to be an "inverted" version of input. Hence, for a CE amplifier there is a phase shift of 180° between V_i and V_o.

7. Role of coupling capacitor C_2

The coupling capacitor C_2 couples on the AC part of the signal to the output and blocks the DC part as shown in figure 5.70. Therefore the signal obtained after the coupling capacitor C_2 has a zero DC value.

8. Does the frequency change in the amplification process ?

Comparing the waveforms of V_i and V_o, we can conclude that the frequency of the amplified output signal V_o is same as that of the input voltage. Thus in the amplification process, the frequency of the signal remains unchanged.

5.41.5.2. Working of the Amplifier with the Help of DC Load Line

1. Graphical Representation

The amplifier operation discussed earlier can also be explained with the help of DC load line. Let us consider figure 5.71 which illustrates the amplification process graphically.

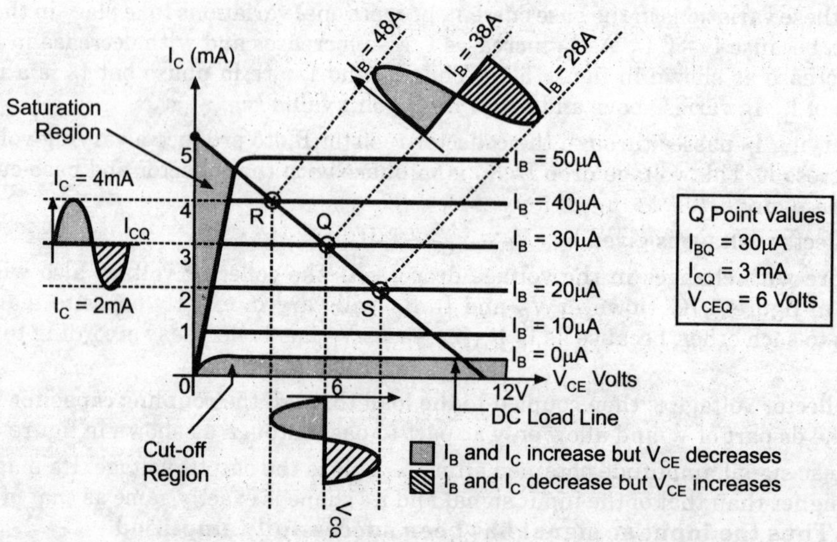

Fig 5.71. Graphical representation of amplification process

2. Selection of Q-Point

The criteria for selecting quiescent point to operate the transistor as an amplifier is as follows:

(i) The Q point should not be too close to the cutoff region.

(ii) The Q point should not be too close to the saturation region.

The conditions (i) and (ii) must be satisfied in order to avoid any waveform distortion in the amplified output signal,

(iii) The Q point must be located at the centre of the dc load line so that the variation in the amplified voltage is equal corresponding to the positive and negative half cycles of the input signal. This will ensure that the amplified signal will be an exact replica of the input signal,

(iv) In figure 5.71, the component values are so adjusted that the Q point is situated exactly at the center of the dc load line. The co-ordinates of the Q point are :

$$Q \text{ point} = (V_{CEQ}, I_{CQ}) = (6 \text{ V}, 3 \text{ mA})$$

The value of quiescent base current i.e. $I_{BQ} = 30\ \mu A$

3. Amplification Process

1. As shown in figure 5.71, due to application of the input signal, the base current varies sinusoidally above and below I_{BQ}. The change in base current i.e.

$$\Delta I_B = I_B(\max) - I_{B(\min)}$$

Hence, $\Delta I_B = 40\ \mu A - 20\ \mu A = 20\ \mu A$

2. The operating point will move along the load line from Q to R and then from R to S corresponding to change in I_B.

3. Because of this, the collector current I_C will change above and below I_{CQ} as shown in figure 5.71. Note that $\Delta I_C = \beta\, \Delta I_B = 2$ mA and I_c and I_B are in phase.

4. The collector to emitter voltage V_{CE} (which is same as V_o) will vary above and below its Q point value V_{CEQ}. This is the amplified output voltage. Note that V_{CE} is 180° out of phase with I_c and I_B.

Important Point: Here, it may be noted that the shape of the output voltage is exactly same as that of the input. This means that no waveform distortion has taken place. Thus, when Q point is at the centre of the load line, no waveform distortion will take place.

5.41.5.3. Effects of Q-Point Position

We have already observed how distortionless amplification takes place if Q point is at the center of load line.

Now, let us observe the effect of adjusting the Q point near the saturation or cutoff region of the transistor.

1. Effect of Q point close to cutoff

Figure 5.72(a) shows that the Q point is adjusted closer to the cutoff region and an input signal is applied which changes the base current by 20 μA. Because of this, the positive half cycle of the output voltage gets distorted as shown in figure 5.72(a).

2. Effect of Q point close to saturation

Figure 5.72(b) shows the effect of Q point adjusted close to the saturation region. In the positive half cycle of I_B, the transistor goes into saturation. Hence, the collector current remains constant at $I_{c\,(sat)}$. Hence, the negative half cycle of the output voltage and positive half cycle of I_c gets distorted as shown in figure 5.72(b).

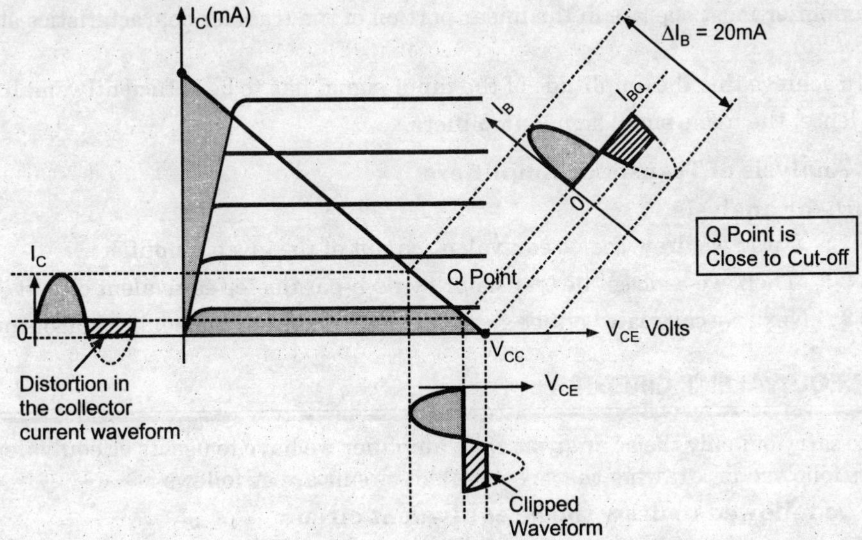

(a) Effect of Q point close to cutoff region

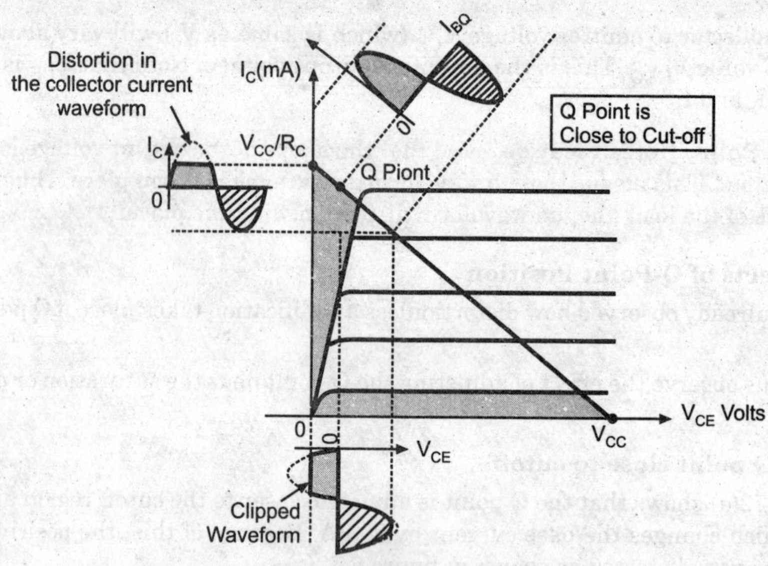

(b) Effect of Q point close to saturation region

Fig. 5.72.

5.41.5.4. Meaning of Small Signal Amplifier

- The *linear amplifier* is also called as a *small signal amplifier*. The reason for this is as follows.
- In order to amplify the input signal faithfully i.e. without introducing any distortion, the amplifier must operate in the linear portion of the transfer characteristics shown in figure 5.72(b).
- To achieve this the amplitude of the input signal has to be sufficiently small.
- Hence, the name small signal amplifiers.

5.41.5.5 . Analysis of Transistor Amplifiers

Procedure for analysis

Step 1 : First, we draw the ac equivalent circuit of the given amplifier.

Step 2 : Then, we replace the transistor by its h-parameter equivalent circuit.

Step 3 : Next, we calculate voltage gain, current gain, input impedance and output impedance.

5.42. AC EQUIVALENT CIRCUIT

In order to carry out only the ac analysis of an amplifier we have to use its ac equivalent circuit. The rules to be followed for drawing the ac equivalent circuit are as follows :

Rules to be followed to draw the ac equivalent circuit

(i) First, we replace all the dc sources by a short circuit. The dc voltages are important only in deciding the Q point of the amplifier.

(ii) Then, the coupling capacitors C_1 and C_2 and the bypass capacitor C_E are to be replaced by short circuit. This is because the reactance of these capacitors is very small at the frequency of application.

EXAMPLE. 5.17. Draw the ac equivalent circuit of the amplifier shown in figure 5.73.

Solution: The ac equivalent circuit shown in figure 5.74 is drawn by following the procedure given below:

(i) As the bypass capacitor C_E is assumed to be short circuited, the resistor R_E is bypassed i.e. shorted out, thus emitter is directly connected to ground point,

(ii) As the dc supplies are to be replaced by short circuits, the + V_{cc} point is connected to ground. Therefore R_C appears between collector and ground and R_1 appears in parallel with R_2 as shown in figure 5.74(b).

The ac equivalent circuit has been shown in figure 5.74(a).

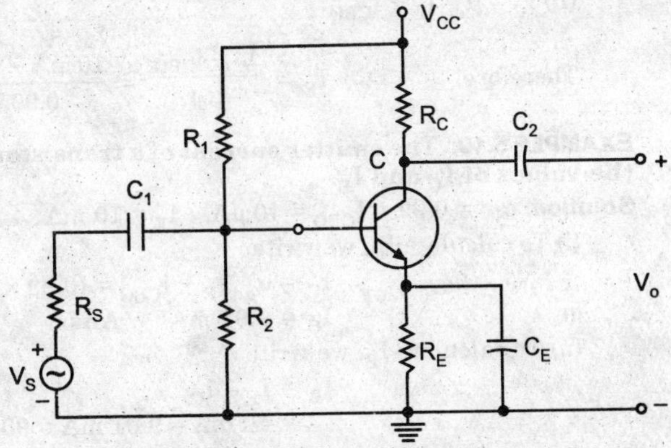

Fig. 5.73. CE single stage transistor amplifier

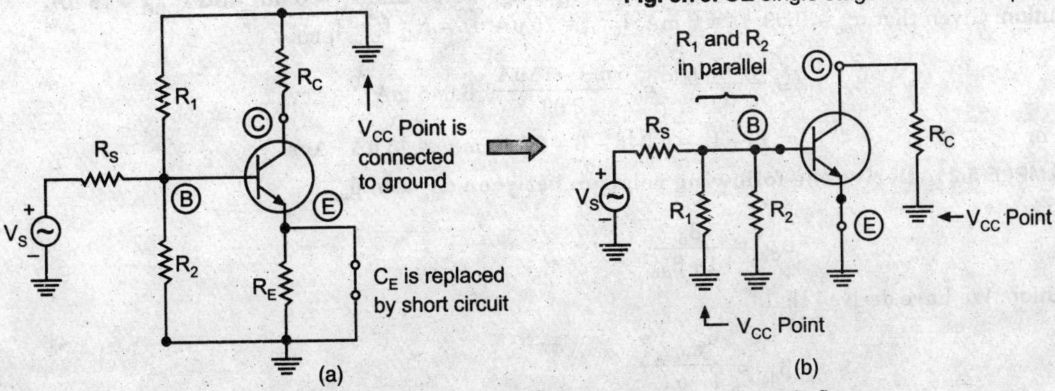

Fig. 5.74. AC equivalent circuit of the amplifier shown in figure

The final ac equivalent circuit has been shown in figure 5.75.

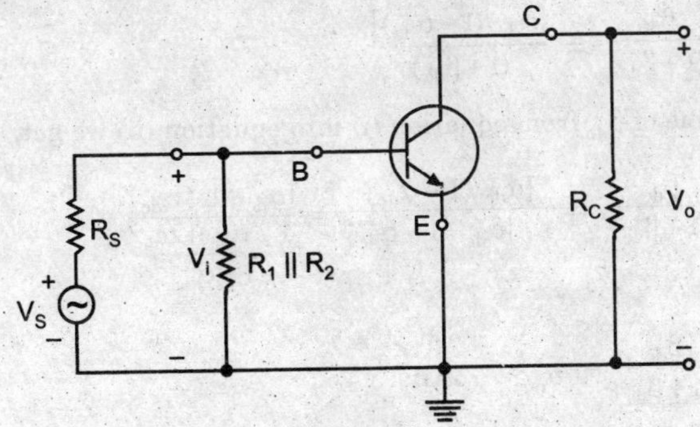

Fig. 5.75.

SOLVED NUMERICAL PROBLEMS

EXAMPLE 5.18. In a transistor 0.5% of the carriers injected into the base region constitute the base current and 99.5% cross the base region to enter into the collector region. The leakage current is 3 μA and collector current is measured to be equal to 15 mA. Determine α_{dc} and I_E.

Solution: Given that $I_B = 0.005\ I_E$, and $I_C = 0.995\ I_E$, $I_{CBO} = 3\ \mu A$, $I_C = 15$ mA

(i) $\alpha_{dc} = \dfrac{I_C}{I_E} = 0.995$

(ii) $I_C = \alpha_{dc} I_E + I_{CBO}$

Therefore, $$I_E = \frac{I_C - I_{CBO}}{\alpha_{dc}} = \frac{15\text{ mA} - 3\ \mu\text{A}}{0.995} = 15.07\text{ mA.}\quad \textbf{Ans.}$$

EXAMPLE 5.19. The emitter current of a transistor is 10 mA. If α_{dc} = 0.99 and I_{CBO} = 10 μA, determine the values of I_C and I_B.

Solution: α_{dc} = 0.99, I_{CBO} = 10 μA, I_E = 10 mA

(i) To calculate I_C, we write

$$I_C = \alpha_{dc} I_E + I_{CBO} = (0.99 \times 10\text{ mA} + 10\ \mu\text{A})$$

or $$I_C = 9.91\text{ mA}\quad \textbf{Ans.}$$

(ii) To calculate I_B, we write

$$I_B = I_E - I_C$$
$$= 10\text{ mA} - 9.91\text{ mA} = 90\ \mu\text{A}\quad \textbf{Ans.}$$

EXAMPLE 5.20. For a transistor, determine I_B if α_{dc} = 0.99 and I_C = 6 mA and I_{CBO} = 15 μA.

Solution: Given that α_{dc} = 0.99, I_C = 6 mA, I_{CBO} = 15 μA, $I_C = \alpha_{dc} I_E + I_{CBO}$

$$I_E = \frac{I_C - I_{CBO}}{\alpha_{dc}} = \frac{6\text{ mA} - 15\ \mu\text{A}}{0.99}\ 6.045\text{ mA}$$

or $$I_B = I_E - I_C = 6.045 - 6 = 0.045\text{ mA} = 45\ \mu\text{A}\quad \textbf{Ans.}$$

EXAMPLE 5.21. Derive the following relation between α_{dc} and β_{dc}.

$$\alpha_{dc} = \frac{\beta_{dc}}{1+\beta_{dc}}$$

Solution: We have derived that,

$$\beta_{dc} = \frac{\alpha_{dc}}{1-\alpha_{dc}} \quad ...(i)$$

Hence dividing both the sides by $(1 + \beta_{dc})$, we get

$$\frac{\beta_{dc}}{(1+\beta_{dc})} = \frac{[\alpha_{dc}/(1-\alpha_{dc})]}{(1+\beta_{dc})} \quad ...(ii)$$

Substituting the value of β_{dc} from equation (i) into equation (ii) we get,

$$\frac{\beta_{dc}}{(1+\beta_{dc})} = \frac{[\alpha_{dc}/1-\alpha_{dc}]}{1+[\alpha_{dc}/1-\alpha_{dc}]} = \frac{[\alpha_{dc}/1-\alpha_{dc}]}{\frac{1-\alpha_{dc}+\alpha_{dc}}{1-\alpha_{dc}}}$$

Therefore, $$\frac{\beta_{dc}}{(1+\beta_{dc})} = \alpha_{dc}\quad \textbf{Ans.}$$

EXAMPLE 5.22. Determine the value of I_E if α_{dc} = 0.98, I_{CBO} = 12 μA and I_B = 120 μA

Solution: The expression for I_E is expressed as under:

$$I_E = (1 + \beta_{dc})\, I_B + (1 + \beta_{dc})\, I_{CBO}$$

As we do not know β_{dc}, let us calculate it first

We have $$\beta_{dc} = \frac{\alpha_{dc}}{1-\alpha_{dc}} = \frac{0.98}{1-0.98} = 49$$

$$I_E = (1 + 49)\ 120 \times 10^{-6} + (1 + 49)\ 12 \times 10^{-6}$$
$$= 6.6\text{ mA}\quad \textbf{Ans.}$$

EXAMPLE 5.23. Determine the region of operation of the silicon transistor shown in figure 5.76. You may assume β = 100, $V_{BE\ sat}$ = 0.8 V and $V_{BE\ (act)}$ = 0.7 V.

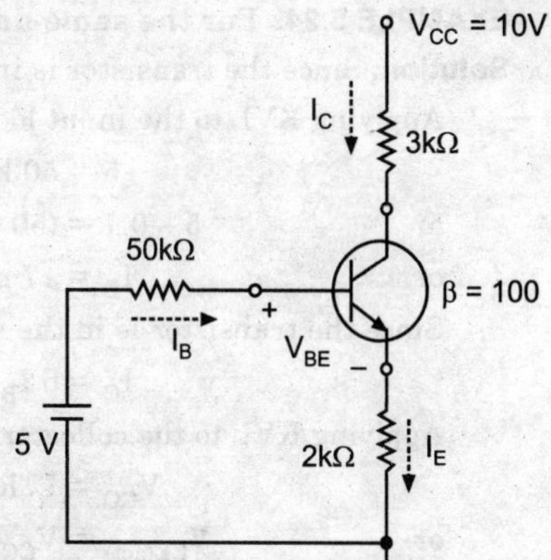

Fig. 5.76.

Solution: (i) First, we assume the transistor to be in saturation. Applying KVL to collector loop and base loop obtain the value of $I_{C\ (sat)}$ and I_B.

(ii) Then, we calculate the minimum base current $I_{B\ (min)} = I_{C\ (sat)} / \beta$, required to saturate the transistor.

(iii) Next, we compare the values of $I_{B\ (min)}$ and I_B.

Let us assume the transistor to be in saturation. Applying KVL to collector loop. In figure 5.77 (b), we have

$$V_{CC} = I_{C(sat)} R_C + V_{CE(sat)} + I_E R_E$$
$$= I_{C(sat)} R_C + 0.2 + I_{C(sat)} R_E + I_B R_E$$
(Since $V_{CE\ (sat)}$ = 0.2 V, and $I_E = I_C + I_B$)

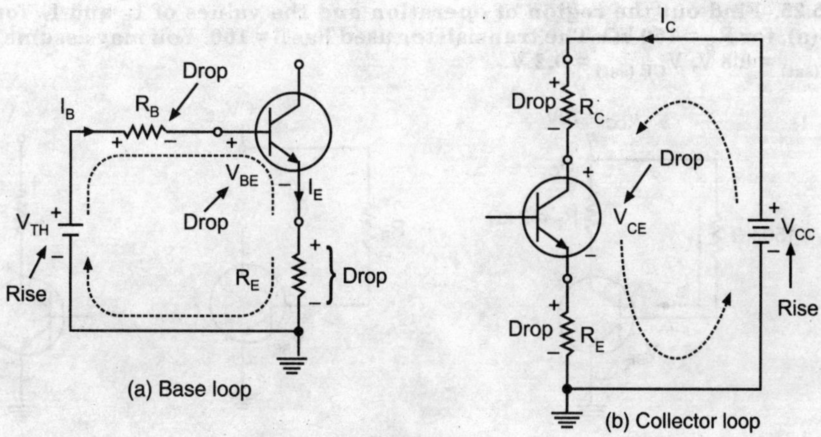

Fig. 5.77.

$$10 = I_{C\ (sat)} (R_C + R_E) + 0.2 + I_B R_E \quad ...(i)$$

Therefore, $\quad 9.8 = 5\ I_{C(sat)} + 2\ I_B$

Applying KVL to the base loop in figure 5.77 (a), We have

$$5 - 50\ I_B - V_{BE\ (sat)} - I_E R_E = 0$$

or $\quad 5 - 0.8 = 50\ I_B + R_E (I_{C\ (sat)} + I_B)$

or $\quad 4.2 = 50\ I_B + 2\ I_{C(sat)} + 2\ I_B \quad ...(ii)$

or $\quad 4.2 = 2\ I_{C(sat)} + 52\ I_B$

Solving equation (i) and (ii), we obtain

$$I_{C(sat} = 1.96 \text{ mA}$$

[Here we have not taken value of resistors in kΩ, hence, the currents obtained will directly be in mA]

Hence, $\quad I_B = 5.46\ \mu A$

Minimum base current required to saturate the transistor is given by,

$$I_{B\ (min)} = \frac{I_{C\ (sat)}}{\beta} = \frac{1.96}{\beta} = \frac{1.96 \text{ mA}}{100} = 19.6\ \mu A$$

Comparing the values of I_B and $I_{B(min)}$, we can conclude that $I_B < I_{B\ (min)}$.

Therefore, the transistor is not in the saturation region. However, it is conducting. Thus, it can not be in cutoff. Therefore, the transistor is in the active region.

EXAMPLE 5.24. For the same data of example 5.23, determine I_B, I_C and V_{CE}.

Solution: Since the transistor is in the active region, $V_{BE(act)} = 0.7$ V

Applying KVL to the input loop, we get

$$5 = 50\text{ k } I_B + V_{BE\,(act)} + (101) \times I_B \times 2\text{k}$$

or $$5 - 0.7 = (50 + 202)\text{ k } I_B$$

or $$I_B = 17\ \mu\text{A} \quad \textbf{Ans.}$$

Since the transistor is in the active region, we have

$$I_C = \beta I_B = 100 \times 17 \times 10^{-6} = 1.7\text{ mA}$$

Applying KVL to the collector circuit we get,

$$V_{CC} = I_C R_C + V_{CE(act)} + (I_C + I_B) R_E$$

or $$V_{CE(act)} = V_{CC} - I_B R_E - I_C (R_C + R_E)$$

or $$V_{CE(act)} = 10 - 17\ \mu\text{A} \times 2\text{k} - 1.7\text{ mA } (3 + 2)\text{ k} = 1.466\text{ V} \quad \textbf{Ans.}$$

EXAMPLE 5.25. Find out the region of operation and the values of I_B and I_C for the circuit in figure 5.78(a), for $R_B = 150\text{ k}\Omega$. The transisistor used has $\beta = 100$. You may assume $V_{BE\,(active)} = 0.7$ Volts, $V_{BE\,(sat)} = 0.8$ V, $V_{CE\,(sat)} = 0.2$ V.

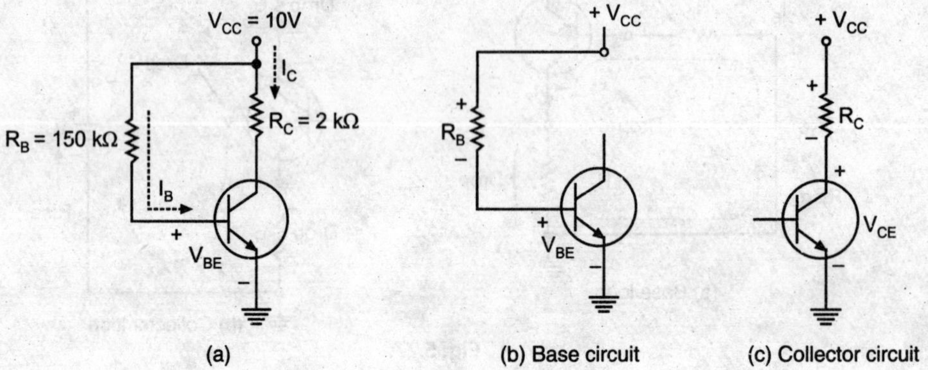

Fig. 5.78.

Solution:

(i) First, we assume that the transistor is in saturation.

(ii) Then, we apply KVL to the base circuit and calculate I_B.

(iii) Next, we apply KVL to the collector circuit and calculate I_C.

(iv) Then, we calculate $I_{B\,min} = I_C/\beta$ required for the transistor to saturate.

(v) Lastly, if I_B calculated in step 2 is greater than $I_{B\,(min)}$, then the transistor is in saturation. Otherwise it is in the active region.

Assume the transistor to be in saturation. Hence, 0.8 V. Applying KVL to the base circuit to write, in figure 5.78 (b), we get

$$V_{CC} = I_B R_B + V_{BE} \qquad \text{...(i)}$$

or $$I_B = \frac{V_{CC} - V_{BE}}{R_B} = \frac{10 - 0.8}{150\text{ k}\Omega} = 61.33\ \mu\text{A}$$

Applying KVL to the collector circuit to write, in figure 5.78 (c), we get

$$V_{CC} - I_C R_C - V_{CE\,(sat)} = 0$$

or $$I_C = \frac{V_{CC} - V_{CE\,(sat)}}{R_C} = \frac{(10 - 0.2)}{2\text{ k}\Omega} = 4.9\text{ mA} \qquad \text{...(ii)}$$

Now, we have

$$I_B\,(\text{min}) = \frac{I_C}{\beta}$$

$$= \frac{4.9\text{ mA}}{100} = 49\ \mu\text{A} \qquad \text{...(iii)}$$

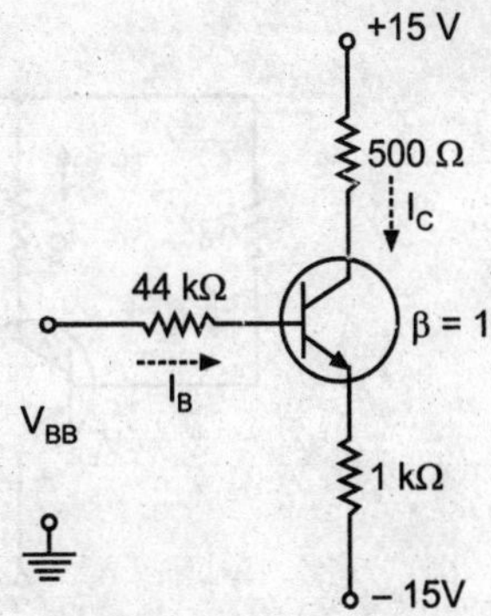

Fig. 5.79. Given circuit

Comparing the values of I_B and $I_{B(min)}$ gives in equation (i) and (iii) we can conclude that,

$$I_B > I_{B(min)}$$

Therefore the transistor is in saturation mode.

EXAMPLE 5.26. Find out the value of which just barely saturates the transistor. Consider figure 5.79(a).

Solution:

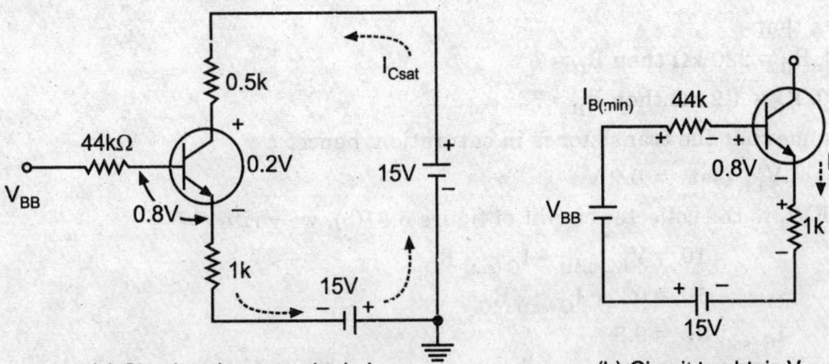

(a) Circuit redrawn to obtain $I_{B\,min}$ (b) Circuit to obtain V_{BB}

Fig. 5.80.

(i) First, we calculate the value of minimum base current required to saturate the transistor (I_B min), assuming the transistor in barely saturatin.

(ii) Then, we obtain the value of V_{BB} to barely saturate the transistor.

Let us assume the transistor to be barely in saturation. Hence, we assume that,

$$V_{CE\,(sat)} = 0.2\text{ V}$$

and

$$V_{BE\,(sat)} = 0.8\text{ V}$$

The circuit has been redrawn as shown in figure 5.80(a). Applying KVL to the collector circuitm we have

$$30 - I_{C(sat)}\,(0.5\text{ k}) - 0.2 - I_E\,(1\text{k}) = 0 \qquad \text{...(i)}$$

Substituting $I_{C\,sat} = \beta\, I_{B\,min}$, and $I_E = (1 + \beta)$ in equation (i), we get

$$30 - (0.5\text{ k} \times \beta \times I_{B\,min}) - 0.2 - [1\text{ k} \times (1 + \beta)\, I_{B\,min}] = 0$$

$$\text{or } 29.8 - I_{Bmin}\,[50\text{ k} + 101\text{ k}] = 0$$

$$\text{or} \quad I_{B\,min} = 0.197\text{ mA} \qquad \text{...(ii)}$$

Now, we consider the circuit shown in figure 5.80(b). Applying KVL to the base emitter circuit, we obtain

$$V_{BB} - (I_{B\,min} \times 44\text{ k}) - V_{BE\,sat} - (I_E \times 1\text{k}) + 15 = 0 \qquad \text{...(iii)}$$

But $\quad I_E = (1 + \beta)\, I_{B\,min} = 19.897\text{ mA}$

or $\quad V_{BB} = (0.197 \times 44) + 0.8 + (19.897 \times 1) - 15 = 14.4\text{ Volt} \quad$ **Ans.**

EXAMPLE 5.27. Determine the minimum value of R_C required to produce saturation of silicon transistor of figure 5.81(a), if R_B = 220 kΩ. If R_C = 1.2 kΩ find the largest value of R_B that will saturate the transistor. You may assume β = 100.

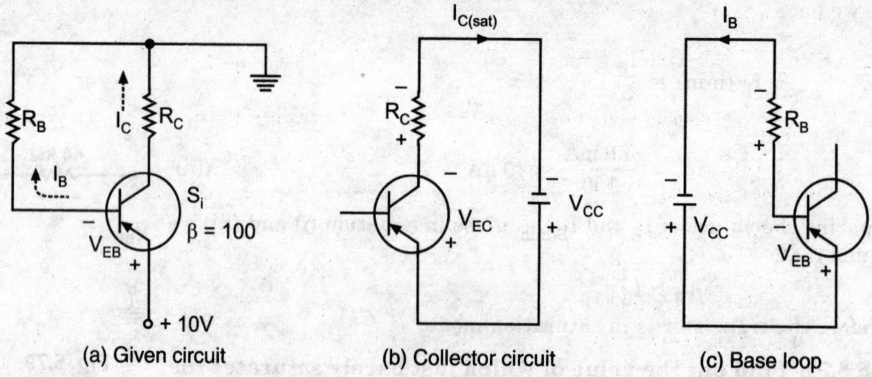

Fig. 5.81.

Solution: Given that

(i) $\beta = 100$, $R_B = 220$ kΩ then R_C = ?

(ii) $\beta = 100$, $R_C = 1.2$ kΩ then R_B = ?

Let us assume that the transistor is in saturation, hence,

$$V_{EC}\text{ (sat)} = 0.2\text{ V}$$

Applying KVL to the collector circuit of figure 5.81(b), we write

$$10 = V_{EC\,(sat)} + I_{C\,(sat)}\,R_C \quad \text{...(i)}$$

or $$10 = 0.2 + I_{C\,(sat)}\,R_C \quad \text{...(ii)}$$

or $$I_{C\,(sat)}\,R_C = 9.8$$

Now, applying KVL to the input loop of figure 5.81(c), we have

$$10 = V_{EB(sat)} + I_B \,.\, R_B \quad \text{...(iii)}$$

Let $V_{EB(sat)} = 0.8$ V and $R_B = 220$ kΩ given,

or $$10 = 0.8 + (I_B \times 220\text{ k}\Omega) \quad \text{...(iv)}$$

or $$10 = 0.0418\text{ mA}$$

For saturation, we have $I_B > \dfrac{I_C}{\beta}$

Just at saturation, we have $I_B = \dfrac{I_C}{\beta}$, ...(v)

Therefore, $I_C = \beta\, I_B = 100 \times 0.0418\text{ mA} = 4.18\text{ mA}$

Substituting this value in equation (ii), we get

$$R_{C\,(min)} = \frac{9.8}{4.18\times10^{-3}}$$

or $R_{C\,(min)} = 2.343$ kΩ **Ans.** ...(vi)

If $R_C > R_{C\ min}$, then I_C will decrease and the existing value of I_B will be sufficient to ensure the saturation of the transistor.

Given that $R_C = 1.2$ kΩ

Substituting this value of R_C into equation (ii) we get the new value of $I_{C\,(sat)}$ as under:

$$I_{C\,(sat)} = \frac{9.8}{1.2\times10^{-3}} = 8.16\text{ mA} \quad \text{...(vii)}$$

Just for saturation, we have $I_B = \dfrac{I_C\text{ (sat)}}{\beta} = \dfrac{8.16\text{ mA}}{100} = 0.0816\text{ mA}$ ...(viii)

Substituting this new value of I_B into equation (iii), we obtain

$$10 = V_{BE(sat)} + I_B R_B$$

Therefore, $\frac{8.16 \text{ mA}}{100} = R_B$

or $R_{B\,max} = 112.75 \text{ k}\Omega$ **Ans.**

EXAMPLE 5.28. The silicon transistor used in the circuit of figure 5.82 has $V_{CE\,(sat)} = 0.2$ V, $V_{BE\,(sat)} = 0.8$ V, $V_{BE\,(active)} = 0.7$ V, $V_{BE(cut\,in)} = 0.5$ V and $\beta = 100$

(a) Show that the transistor is in saturation.

(b) Evaluate the value of R_E for which the transistor just come out of saturation.

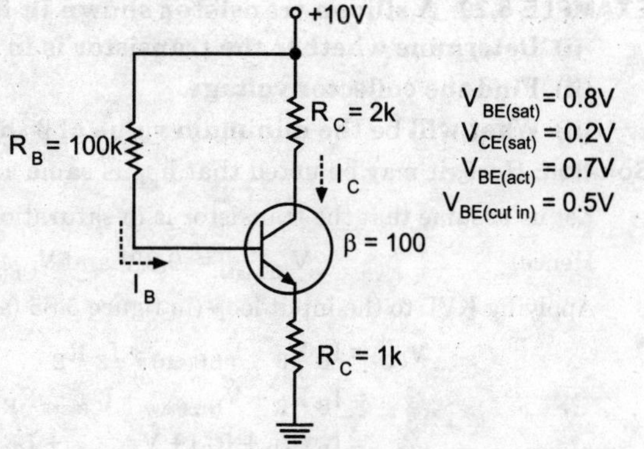

Fig. 5.82.

Solution: Part (a) to shows that the transistor is in saturation.

(i) First, we calculate $I_{C(sat)}$ by assuming the transistor to be in saturation. From that, we evaluate $I_{B(min)}$.

(ii) Then, we calculate the actual value of I_B and show that $I_B > I_{B\,min}$.

Let us assume that the transistor is in saturation.

Therefore, $V_{CE\,(sat)} = 0.2$ and $V_{BE\,(sat)} = 0.8$ V

Applying KVL to the collector circuit of figure 5.82, we obtain

$$10 = [2\text{k} \times I_{C\,(sat)}] + V_{CE\,(sat)} + (1\text{ k} \times I_E) \quad ...(i)$$

But, we have $I_E = \alpha I_C = \frac{\beta}{1+\beta} I_C = \frac{100}{101} I_C = 0.99\, I_C$

or $10 = 2\text{k}\, I_{C(sat)} + 0.2 + (1\text{ k} \times 0.99\, I_{C(sat)})$

or $9.8 = I_{C(sat)}\,[2\text{ k} + 0.99\text{ k}]$

or $I_{C(sat)} = \frac{9.8}{2.99\text{ k}} = 3.28$ mA ...(ii)

We have $I_{B\,(min)} = \frac{I_{C\,(sat)}}{\beta} = \frac{3.28 \times 10^{-3}}{100} = 32.8\ \mu\text{A}$ ...(iii)

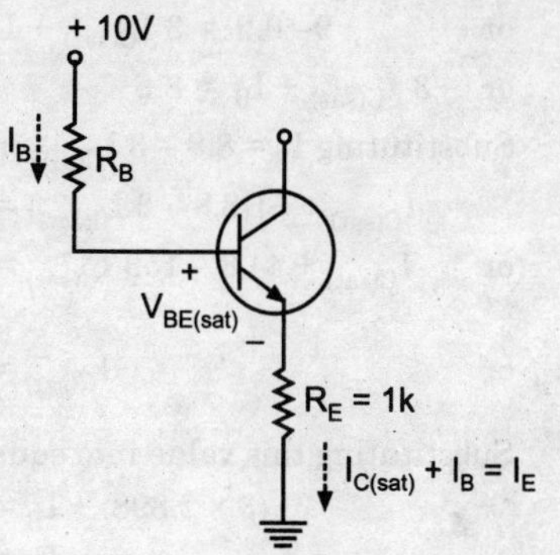

Fig. 5.83.

Applying KVL to the base loop shown in figure 5.83, we write,

$$10 = R_B I_B + V_{BE\,(sat)} + I_E R_E \quad ...(iii)$$

But, $I_E = I_{C\,(sat)} + I_B$

Therefore, $10 = R_B I_B + 0.8 + I_{C(sat)} R_E + I_B R_E$

or $10 = 0.8 + (3.28 \times 1) + I_B (R_B + R_E)$

or $I_B = \frac{10 - 4.08}{101\text{ k}\Omega} = 58.61\ \mu\text{A}$ ...(iv)

Comparing equations (iii) and (iv), we conclude that $I_B > I_{B\,min}$

Therefore, the transistor is in saturation

Part (b) Calculate the value of R_E

In order to bring the transistor just out of saturation, we have to reduce I_B below $I_{B\,(min)}$, by increasing R_E

i.e., $I_B << 32.8\ \mu\text{A}$

Hence, equation (iiia) gets modified to

$$10 = 100\text{ k}\, I_{B\,(min)} + V_{BE\,(act)} + I_E R_E$$

Substituting $I_E = (1 + \beta)\, I_{B\,min}$, we obtain

$$10 = 100\text{ k}\, I_{B\,(min)} + 0.7 + (1 + \beta)\, I_{B\,(min)} R_E$$

or $10 = (100 \times 10^3 \times 32.8 \times 10^{-6}) + 0.7 + (101 \times 32.8 \times 10^{-6})\, R_E$

or $R_E = \dfrac{10 - 0.7 - 3.28}{3.31 \times 10^{-3}} = 1.818\ k\Omega$

Hence, R_E should be greater than 1.818 kΩ in order to bring the transistor just out of saturation.

EXAMPLE 5.29. A silicon transistor shown in figure 5.84 has $h_{FE} = 70$

(i) Determine whether the transistor is in saturation or active region.

(ii) Find the collector voltage.

(iii) What will be the minimum value of β that will change the state of the transistor?

Solution: Here, it may be noted that h_{FE} is same as β. Hence = 70

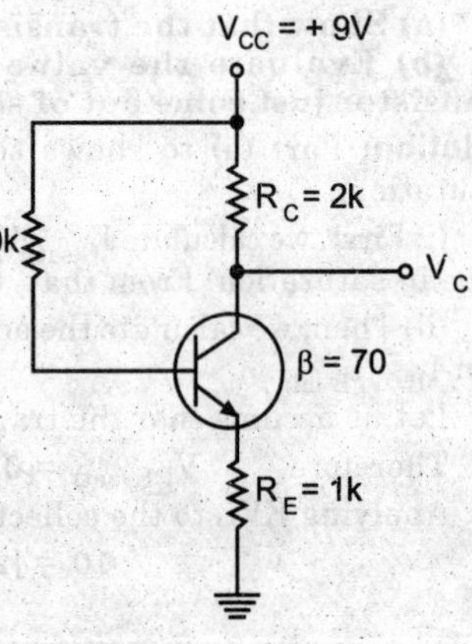

Fig. 5.84.

Let us assume that the transistor is in saturation.

Hence, $V_{BE(sat)} = 0.8$ V and $V_{CE(sat)} = 0.2$ V.

Applying KVL to the input loop (in figure 5.85 (a)), we get

$$V_{CC} = I_B R_B + V_{BE(sat)} + I_E R_E$$
$$= I_B R_B + V_{BE\,(sat)} + I_{C(sat)} R_E + I_B R_E$$
$$= I_B (R_B + R_E) + V_{BE(sat)} + I_{C(sat)} R_E$$

or $9 = 51\ I_B + 0.8 + 1\ I_{C(sat)}$

or $I_{C(sat)} = 8.2$ ...(i)

Now, applying KVL to the output loop in figure 5.85 (b), we get

$$V_{CC} = R_C I_{C(sat)} + V_{CE(sat)} + I_{C(sat)} R_E + I_B R_E$$

or $V_{CC} - V_{CE(sat)} = I_{C(sat)} (R_C + R_E) + I_B R_E$

or $9 - 0.2 = 3\ I_{C(sat)} + I_B$

or $3\ I_{C(sat)} + I_B = 8.8$ ...(ii)

Substituting $I_B = 8.8 - 3\ I_{C(sat)}$ into equation (i), we obtain

$$I_{C(sat)} + 51[8.8 - 3\ I_{C(sat)}] = 8.2$$

or $I_{C(sat)} + 44.8 - 153\ I_{C(sat)} = 8.2$

or $I_{C(sat)} = \dfrac{440.6}{152} = 2.898$ mA ....(iii)

Substituting this value into equation (ii), we obtain

$$(3 \times 2.898) + I_B = 8.8$$
$$I_B = 0.106 \text{ mA} \quad ...(iv)$$

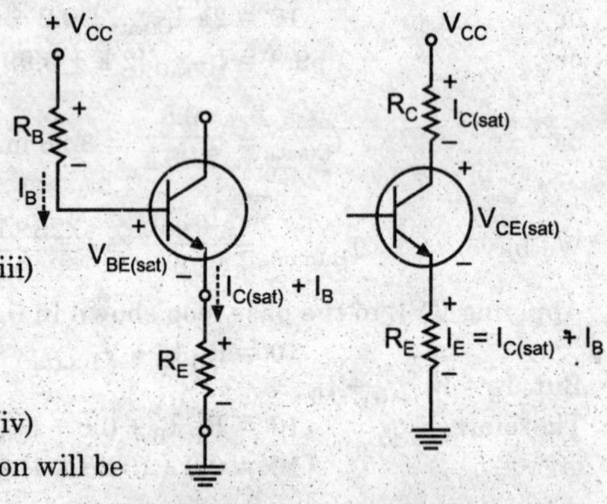

(a) Input loop (b) Output loop

Fig. 5.85.

The minimum base current required to ensure the saturation will be

$$I_{B(min)} = \frac{I_{C(sat)}}{h_{FE}} = \frac{2.898}{70} = 0.0414 \text{ mA}$$

Since, $I_B > I_{B(min)}$ the transistor is in saturation.

Now, we have $V_C = V_{CC} - I_{C(sat)} R_C$

$= 9 - (2.898 \times 2) = 3.204$ V **Ans.**

We have $I_{B(min)} = 0.0414$ mA,

$I_B = 0.106$ mA

To bring the transistor in active region from saturation, h_{FE} or β should be reduced to such a level that I_B is equal to $I_{B(min)}$.

Therefore, $h_{FE} = \beta = \dfrac{I_{C\,(sat)}}{I_B} = \dfrac{2.898}{0.106}$

or β = 27.339 **Ans.**

Hence, if β is less than 27.339, then the transistor will change its state from saturation to active.

EXAMPLE 5.30. A silicon transistor with β = 50 is used in a self bias circuit using CE configuration. The operating point is V_{CEQ} = 11.5V and I_{CQ} = 1.5 mA. For S ≤ 2, determine the values of R_1, R_2 and R_E.

Solution : Given that

$$\text{Q point} \equiv (V_{CE}, I_C) = (11.5\text{ V}, 1.5\text{ mA})$$

i.e., $V_{CE} = 11.5$ V,

and $I_C = 1.5$ mA,

also, $\beta = 50$

We know that $I_B = \frac{I_C}{\beta} = \frac{1.5\text{mA}}{50} = 30\ \mu\text{A}$

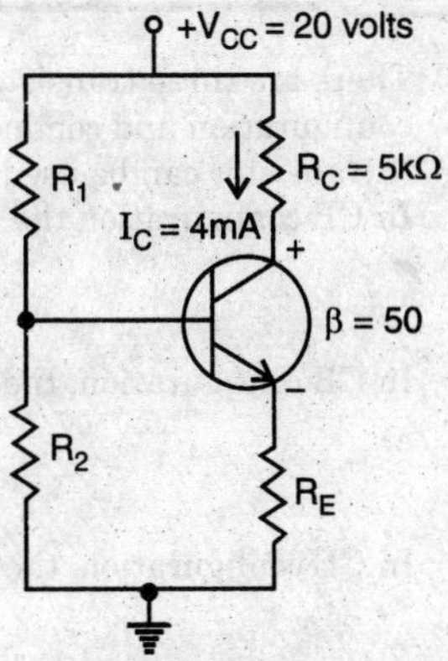

Fig. 5.86.

Applying KVL to the collector loop, we have

$$V_{CC} = I_C R_C + V_{CE} + (I_C + I_B) R_E$$

or $$R_E = \frac{V_{CC} - V_{CE} - I_C R_C}{(I_C + I_B)}$$

Substituting all the values, we get

$$R_E = \frac{20 - 11.5 - (1.5 \times 5)}{(1.53\text{ mA})} = 653.6\ \Omega \quad \textbf{Ans.}$$

For the value of R_B, we take help of the following expression

$$S = \frac{(1+\beta)[1+(R_B/R_E)]}{1+\beta+(R_B/R_E)} \leq 2$$

For S = 2, we have

$$\frac{51[1+(R_B/653.6)]}{1+50+(R_B/653.6)} = 2$$

or $$51 + \frac{51R_B}{653.6} = 102 + \frac{2R_B}{653.6}$$

or $$\frac{49R_B}{653.6} = 51$$

or $$R_B = 680.2\ \Omega \quad \text{...(ii)}$$

or $$R_B = \frac{R_1 R_2}{(R_1 + R_2)} = 680.2\ \Omega \quad \text{...(iii)}$$

Again, we write $$V_{R2} = V_{BE} + I_E R_E$$

$$V_{R2} = 0.8 + (1.53\text{ mA} \times 653.6) = 1.8\text{V}.$$

But, $$V_{R2} = \frac{R_2 \times V_{CC}}{R_1 + R_2} = 1.8 \quad \text{...(iv)}$$

Comparing equations (iii) and (iv), we get

$$\frac{680.2}{R_1} = \frac{1.8}{V_{CC}}$$

or $$R_1 = 7.55\text{ k}\Omega \quad \textbf{Ans.}$$

and $$\frac{R_2}{7.55 + R_2} = 0.09$$

or $$R_2 = 746.7\ \Omega \quad \textbf{Ans.}$$

SUMMARY

- There are three transistor configurations namely common-base (CB) configuration, common-emitter (CE) configuration and common-collector (CC) configuration.
- A transistor can be used as an amplifier.
- In CB configuration the d.c. current gain is given as

$$\alpha = \frac{I_C}{I_E}$$

- In CB configuration, the a.c. current gain is given as

$$\alpha_0 = \left.\frac{\Delta I_C}{\Delta I_E}\right|_{h_{fb}\text{ constant}}$$

- In CE configuration, the d.c. current gain is given as

$$\beta_{dc} = \beta = \frac{I_C}{I_B}$$

- In CE configuration, the a.c. current gain is given as

$$\beta_0 = \frac{\Delta I_C}{\Delta I_B}$$

- The relationship between current gain α and β is

$$\beta = \frac{\alpha}{1-\alpha}$$

- The collector current, in terms of leakage current in CE configuration is given as

$$I_C = \beta I_B + I_{CEO}$$

where

$$I_{CEO} = (1+\beta)\, I_{CBO}$$

or

$$I_C = \beta I_B + (1+\beta)\, I_{CBO}$$

- There are two types of transistor characteristics
  (i) input characteristics (ii) output characteristics
- In CB-configuration, the a.c. or dynamic input resistance is given as

$$r_i = \left.\frac{\Delta V_{EB}}{\Delta I_E}\right|_{V_{CB}\text{ constant}}$$

- In CB-configuration, the dynamic output resistance is given as

$$r_0 = \left.\frac{\Delta V_{CB}}{\Delta I_C}\right|_{I_E\text{ constant}}$$

- In CE-configuration, the dynamic input resistance is given as

$$r_i = \left.\frac{\Delta V_{BE}}{\Delta I_B}\right|_{V_{CE}\text{ constant}}$$

- In CE-configuration, the dynamic output resistance is given as

$$r_0 = \left.\frac{\Delta V_{CE}}{\Delta I_C}\right|_{I_B\text{ constant}}$$

- CE configuration is most widely used configuration in amplifier circuits.
- A transistor is a three terminal device. The output voltage, current or power are controlled by the input current in a transistor. Therefore, it is also called a current-controlled device.
- A transistor is also called a BJT. BJT stands for Bipolar junction transistor.
- A transistor has a very important property that it can raise the strength of an input weak signal.
- A transistor has three regions namely emitter, base and collector.

- When no battery is connected across the different terminals of a transistor then the transistor is said to be in an unbiased state or in an open-circuit state.
- When dc, voltages are applied across the different terminals of a transistor, then this process is called biasing.
- There are two junctions in a transistor namely emitter-base junction and collector-base junction.
- There are four possible ways of biasing a transistor. They are as under :
  (i) active region
  (ii) saturation region
  (iii) cut-off region
  (iv) inverted region
- A transistor can be treated as a two-port network.

SHORT QUESTIONS WITH ANSWERS

Q.1. Explain why an ordinary transistor is called bipolar?

Ans. Because the transistor operation is carried out by two types of charge carriers (majority and minority carriers), an ordinary transistor is called bipolar.

Q.2. What is done to the base region of a transistor to improve its operation?

Ans. Base is made thin and very lightly doped in comparison to either emitter or collector so that it may pass most of the injected charge carriers to the collector.

Q.3. Why transistor is called current controlled device?

Ans. The output voltage, current or power is controlled by the input current in a transistor so it is called the current controlled device.

Q.4. What are 'emitter injection efficiency' and 'base transport factor' and how do they influence the transistor operation?

Ans. The ratio of current of injected carriers at emitter junction to the total emitter current is called the emitter injection efficiency. The ratio of collector current to base current is known as transport factor

i.e., $$\beta = \frac{I_C}{I_B}$$

The larger the value of emitter injection efficiency, the larger the injected carriers at emitter junction and this increases the collector current. The larger the β^* value, larger the injected carriers across collector junction and hence collector current increases.

Q.5. Why silicon type transistors are more often used than germanium type?

Ans. Because silicon has smaller cut-off current I_{CBO}, small variations in I_{CBO} due to variations in temperature and high operating temperature as compared to those in case of germanium.

Q.6. Why collector is made larger than emitter and base?

Ans. Collector is made physically larger than emitter and base because collector is to dissipate much power.

Q.7. Why the width of the base region of a transistor is kept very small compared to other regions?

Ans. Base region of a transistor is kept very small and very lightly doped so as to pass most of the injected charge carriers to the collector.

Q.8. Why emitter is always forward biased?

Ans. Emitter is always forward biased w.r.t base so as to supply majority charge carriers to the base.

Q.9. Why collector is always reverse-biased w.r.t base?

Ans. Collector is always reverse-biased w.r.t. base so as to remove the charge carriers away from the base-collector junction.

Q.10. What is α_{dc}?

Ans. α_{dc} is the dc current gain and is equal to the ratio of collector current to the emitter current.

Q.11. What is β ?

Ans. The beta factor (β) is the current gain factor of a common-emitter circuit and is defined as the ratio of collector current and base current.

Q.12. How α and β are related with each other?

Ans. α and β are related as below :

$$\alpha = \frac{\beta}{1+\beta} \quad \text{or} \quad \beta = \frac{\alpha}{1-\alpha}$$

Q.13. Define beta of a transistor. What will be the maximum value of beta at high frequencies?

Ans. The β factor of a transistor is the common emitter current gain of that transistor and is defined as the ratio of collector current to base current.

i.e.,
$$\beta = \frac{I_C}{I_B}$$

Q.14. What are the three configurations of a transistor amplifier?

Ans. The three configurations of a transistor amplifier are common base, common-emitter and common-collector configurations.

Q.15. Why is there a maximum limit of collector supply voltage for a transistor?

Ans. Although collector current is practically independent of collector supply voltage over the transistor operating range, but if V_{CB} is increased beyond a certain value, collector current I_C eventually increases rapidly and possibly destroys the device.

Q.16. For a PNP transistor in the active region what is the sign (positive or negative) of I_E, I_C, I_B, V_{CB} and V_{EB}.

Ans. Signs of I_E, I_C, I_B, V_{CB} and V_{EB} for a PNP transistor ar as follows:

Quantity	Sign
I_E	Positive
I_C	Negative
I_B	Negative
V_{CB}	Negative
V_{EB}	Positive

Q.17. Why CE configuration is most popular in amplifier circuits?

Ans. CE configuration is mainly used because its current, voltage and power gains are quite high and the ratio of output impedance and input impedance are quite moderate.

Q.18. Why is CC configuration seldom used?

Ans. CC configuration is seldom used because its voltage gain is always less than unity.

Q.19. Which configuration among CE, CB, CC gives highest input impedance and no voltage gain?

Ans. Common collector configuration has the highest input impedance and voltage gain less than unity.

Q.20. Which of the following configurations (CB, CE, CC) has the (i) highest R_i (ii) lowest R_i (iii) highest R_0 (iv) lowest R_0 (v) lowest A_i?

Ans. Characteristic Configuration

(i) Highest input resistance, R_i CC
(ii) Lowest input resistance, R_i CB
(iii) Highest output resistance, R_0 CB
(iv) Lowest output resistance, R_0 CC
(v) Lowest current gain, A_i CB

Q.21. What is Early effect?

Ans. The modulation of the effective base width by the collector voltage is known as Early effect. An increase in collector voltage increases the space charge width at the output junction diode and thus the effective base width W_b is reduced.

Q.22. What is punch through effect ?

Ans. When on increasing collector voltage effective base width is reduced to zero the emitter barrier becomes smaller than normal value $V_{out} - |V_{EB}|$ because the collector voltage has reached through the base region. This lowering of emitter-junction voltage causes an excessively large emitter current, thus, placing an upper limit on the magnitude of collector voltage. This phenomenon is known as punch through.

Q.23. What do you understand by collector reverse saturation? In which configuration does it have a greater value?

Ans. When input current (I_E in case of CB configuration and I_B in case of CE configuration) is zero, collector current I_C is not zero although it is very small. In fact, this is the reverse leakage current or collector reverse saturation current (I_{CBO} or simply I_{CO} in CB configuration and I_{CEO} in CE configuration). In case of CE confuguration, it is much more than that in case of CB configuration.

Q.24. What is quiescent point?

Ans. Quiescent point is a point on the dc load line which represents V_{CE} and I_C in the absence of ac signal and variations in V_{CE} and I_C take place around this point when a.c. signal is applied.

Q. 25. Explain why an ordinary junction transistor is called bipolar ?

Ans. Because the transistor operation is carried out by two types of charge carriers (majority and minority carriers), an ordinary transistor is called bipolar.

Q. 26. What is done to the base region of a transistor to improve its operation ?

Ans. Base is made thin and very lightly doped in comparison to either emitter or collector so that it may pass most of the injected charge carriers to the collector.

Q. 27. Why transistor is called current controlled device ?

Ans. The output voltage, current or power is controlled by the input current in a transistor so it is called the current controlled device.

Q. 28. What is the significance of arrowhead in the transistor symbol ?

Ans. Arrowhead is always marked on the emitter. The direction indicates the conventional direction of current flow (from emitter-to-base in case of P-N-P transistor and from base-to-emitter in case of N-P-N transistor). Generally, no arrowhead is marked for collector since its reverse leakage current is always opposite to the direction of emitter current.

Q. 29. Discuss the need for biasing the transistor.

Ans. For normal operation, base-emitter junction should be forward-biased and the collector-base junction reverse-biased. The amount of bias required is significant for establishment of the operating or Q-point which is dictated by the mode of operation desired. In case the transistor is not biased properly, it would (i) work inefficiently, (ii) produce distortion in the output signal and (iii) with the change in transistor parameters or temperature rise, the operating point may shift and the amplifier output will be unstable.

Q. 30. What are 'emitter injection efficiency' and 'base transport factor' and how do they influence the transistor operation ?

Ans. The ratio of current of injected carriers at emitter junction to the total emitter current is called the emitter injection efficiency. The ratio of collector current to base current is known as transport factor

i.e.,
$$\beta^* = \frac{I_C}{I_B}$$

The larger the value of emitter injection efficiency, the larger the injected carriers at emitter junction and this increases the collector current. The larger the β* value larger the injected carriers across collector junction and hence collector current increases.

Q. 31. Which of the transistor currents is always the largest ? Which is always the smallest ? Which two currents are relatively close in magnitude ? **(U.P. Technical Univ. 2008-09)**

Ans. The emitter current I_E is always the largest one. The base current I_B is always the smallest. The collector current I_c and emitter current I_E are relatively close in magnitude *i.e.* $I_C \sim I_E$.

Q. 32. Can a transistor be obtained by connecting two semiconductor diodes back-to-back ?

Ans. No. Because in case of two discrete back-to-back connected diodes there are four doped regions instead of three and there is nothing that resembles a thin base region between an emitter and a collector.

Q. 33. Explain base-width modulation (Early effect).

Ans. The modulation of the effective base-width by the collector voltage is known as Early effect. An increase in collector voltage increases the space charge width at the ouput junction diode and thus the effective base width W_b is reduced.

Q. 34. Why point contact transistors are not commercially used ?

Ans. Because of poor reliability and reproducibility.

Q. 35. What is the major difference between the epitaxial mesa transistor and the diffused mesa transistor ?

Ans. The major difference between the epitaxial mesa transistor and diffused mesa transistor is that an epitaxial layer is added on the original collector substrate in epitaxial mesa transistor.

Q. 36. Explain how a BJT can be used as an amplifier.

Ans. A transistor operates as an amplifier by transfer of the current from low impedance loop to high impedance loop.

Q. 37. What limits the maximum current a transistor is capable of handling ?

Ans. The junction area of the collector region limits the maximum current a transistor can handle. Large power ratings are obtained from larger junction areas. More is the surface area, more is the effective heat dissipation and more is the magnitude of current, a transistor can handle.

Q. 38. What is an amplifier ?

Ans. The device that amplifies the amplitude of the input signal is called the amplifier. An amplifier may be defined as a device that increases the current, voltage or power of an input signal with the help of a transistor by furnishing the additional power from a separate source of supply.

Q. 39. What is meant by small signal amplifier ?

Ans. When the input signal is quite weak and produces small fluctuations in the output current in comparison to its quiescent value, the amplifier is called the small signal or voltage amplifier.

Q. 40. What is meant by phase reversal ?

Ans. In a CE configuration, the output voltage increases in the negative direction when the input signal voltage increases in the positive direction and vice-versa. This is called the phase reversal and causes a phase difference of 180° between the input signal voltage and output voltage.

REVIEW QUESTIONS

1. Draw a fixed bias circuit and obtain the value of d.c. voltage and currents in the circuit.
2. Define stability factor. Derive general expression for stability factor of a biasing circuit.
3. Draw a self-bias circuit and derive an expression for stability factor.
4. State various methods of improving stability. **(U.P.Tech – Tutorial Question Bank)**
5. Define Stability factor w.r.t. transistor biasing. **(U.P.Tech – Tutorial Question Bank)**
6. Explain the function of emitter in the operation of a junction transistor. **(U.P.Tech – Tutorial Question Bank)**
7. What is a tuned voltage amplifier ? Explain it briefly.
8. Derive an expression for the resonance frequency of a parallel circuit.
9. Explain the meaning of a resonance curve. Plot a resonance curve for a parallel tuned circuit.
10. What is a Bipolar junction transistor ?
11. Explain the different configurations of BJT.
12. Explain the input and output characteristics of a transistor in CB configuration.
13. Define α and β for a transistor.

14. Derive the relationship between α and β. **(U.P.Tech – First Semester Examination, 2000-2001)**
15. What is meant by Q-Point?
16. What is the need for biasing a transistor?
17. Explain the requirements of a biasing circuit.
18. What are the factors responsible for the instability of operating point.

(U.P.Tech – Tutorial Question Bank)

19. Explain the difference between ac and dc load lines. Derive the load line equation of a BJT in CE configuration.
20. Explain the manufacturing process for diffused planner and annular transistors. Give their merits and demerits.
21. Explain how the nature (PNP or NPN) of a transistor is determined experimentally. How three terminals (emitter, base and collector) of a transistor can be identified experimentally, explain?
22. Explain heat dissipation curve and its importance, when selecting heat sink.
23. Write short notes on the following:
    (i) Base spreading resistance.
    (ii) Use of CE configuration of transistors for amplifiers.
    (iii) Comparison of CB, CE and CC configurations.
    (iv) Fabrication of transistor by planar technology.
    (v) Transistor as an amplifier.
    (vi) Transistor data sheets.

NUMERICAL PROBLEMS

1. For a fixed bias circuit using an NPN transistor, the value of β is 80. If R_B = 390 kΩ, R_C = 1.5kΩ and V_{CC} = 30V. Find the coordinates of the Q-point. **[Ans.** 21V, 6mA]
2. For a fixed bias circuit using a germanium NPN transistor, the value of β is 100. If V_{BE} = 0.2V, R_1 = 10kΩ, R_2 = 5kΩ, R_C = 1kΩ and R_B = 790 kΩ, then, determine its operating point.
3. The various parameter of a CE amplifier which uses the self-bias method are V_{CC} = 12V, R_1 = 10kΩ, R_2 = 5kΩ, R_C = 1kΩ, R_E = 2kΩ and β = 100. Determine the following
   (i) operating point (ii) the stability factor. **[Ans.** V_{CE} = 7.05V, I_C = 1.65mA, S = 2.62]
4. Find the value of operating point for the circuit of a potential divider bias arrangement with R_2 = R_C = 5 kΩ, R_E = 1KΩ and R_1 = 40 kΩ. **[Ans.** V_{CE} = 6V, I_C = 1 mA]
5. Determine emitter current I_E, collector current I_C for a transistor with α_{dc} = 0.97 and collector to base leakage current 10 μA. Base current is 50 μA. **[Ans.** 2.00 mA, 1.95 mA]
6. For a certain transistor, collector current is 6.25 mA, base current is 100 μA and collector-to-base leakage current 5 μA. Determine emitter current and α_{dc}. **[Ans.** 6.35 mA, 0.983]
7. In a particular transistor, the collector current is 5.6 mA and the emitter current is 5.75 mA. Determine α_{dc}. **[Ans.** 0.974]
8. In a transistor, determine base current if emitter current is 1.00 mA and collector current is 0.92 mA. **[Ans.** 80 μA]
9. In a transistor, determine α if collector current is 4.9 mA and emitter current is 5 mA. **[Ans.** 0.98]
10. A transistor has α = 0.98. If emitter current of the transistor is 1 mA, determine the base current and gain factor β. **[Ans.** 20 μA ; 49]
11. Determine β if α = 0.99 **[Ans.** 100]
12. A bipolar junction transistor has a collector current of 4 mA and base current of 20 μA. Determine its β. **[Ans.** 200]
13. A BJT has a base curent of 200 μA and emitter current of 20 mA. Determine collector current and β. **[Ans.** 19.8 mA; 99]

14. A BJT has a base current of 25 μA and its $\beta = 200$. What is collector current? **[Ans.** 5 mA]

15. In a transistor, $I_C = 0.95$ mA, $I_E = 1$ mA. Determine base current, α and β in CB configuration. **[Ans.** 50 μA, 0.95; 19]

16. In a transistor CE configuration, the voltage across 5 kΩ load resistance is 5 V. Determine collector current, base current and emitter current if $\beta = 50$.
[Ans. 1.0 mA; 20 μA; 1.02 mA.]

17. In a grounded base configuration, the voltage drop across load resistance of 4 kΩ is 3.2 V. Determine collector current, emitter current and base current if $\alpha = 0.96$. **[Ans.** 0.8 mA; 0.8333 mA; 0.0333 mA]

18. A CE transistor amplifier has input resistance of 100 Ω and output resistance of 6 kΩ. Determine the voltage gain and power gain if current gain factor is 90. **[Ans.** 5,400 ; 4.86,000]

19. Increase in collector-emitter voltage from 4 V to 8 V causes increase in collector current from 5 mA to 6 mA. Determine the output resistance. **[Ans.** 4 kΩ]

20. A change of 200 mV in base-emitter voltage causes a change of 60 μA in the base current. Determine its input resistance. **[Ans.** 3.33 kΩ]

21. A transistor is connected in CB configuration. When the emitter voltage is changed by 200 mV, the emitter current changes by 5 mA. During this variation, the collector to base voltage is kept fixed. Calculate the dynamic input resistance of transistor. **(Ans.** 40 Ω)

❑❑❑

Chapter 6

Field-Effect Transistors

Syllabus

Transistors: Metal-Semiconductor-Field-Effect-Transistors (MESFET), Metal-Insulator-Semiconductor-Field-Effect-Transistor (MISFET), Metal Oxide Semiconductor Field Effect Transistor (MOSFET), Construction, Operation and Characteristics of above Devices.

Chapter Outline

6.1. INTRODUCTION

We have already discussed in chapter 5 that a bipolar junction transistor (BJT) is a current controlled device. Due to this reason, the output characteristics of BJT are controlled by the base current and not the base voltage. Another type of transistor, known as the field-effect transistor (FET), is a voltage-controlled device. This means that the output characteristics of FET are controlled by the applied input voltage and not by the input current. In addition to this, FET is an unipolar device. This means that the current flowing through it is only due to one type of charge particles, i.e., holes or electrons. BJT, on the other hand, is a bipolar device because holes and electrons both contribute to the flow of current.

In a broad sense, there are two main types of field-effect transistors as under:

(i) Junction field-effect transistors (JFET).

(ii) Metal oxide semiconductor field-effect transistor (MOSFET) or insulated gate field-effect transistor (IGFET).

Both of these types of FETs may be further sub-divided as under:

Now, let us discuss the construction, operation and

characteristics of all the above-mentioned FET in the sections to follow.

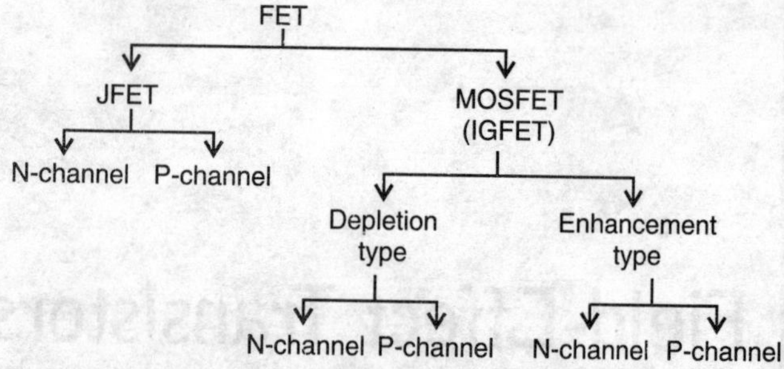

Fig. 6.1. Different types of FETs.

6.2. JUNCTION FIELD-EFFECT TRANSISTOR (JFET)

As shown in figure 6.1, the junction field-effect transistor (JFET) may be divided, depending on their structure, into the following two categories:

(i) N-channel JFET.

(ii) P-channel JFET.

Figure 6.2 shows the basic construction of an N-channel JFET. It simply consists of an N-type semiconductor bar with two P-type heavily doped regions diffused on opposite sides of its middle part. The P-type regions form two PN-junctions. The space between the junctions (i.e., N-type regions) is known as a **channel.** Both the P-type regions are connected internally and a single wire is taken out in the form of a terminal known as the gate (G). The electrical connections (known as ohmic contacts) are made to both ends of the N-type semiconductor and are taken out in the form of two terminals known as drain (D) and source (S). The drain (D) is a terminal through which electrons leave the semiconductor bar and source (S) is a terminal through which the electrons enter the semiconductor.

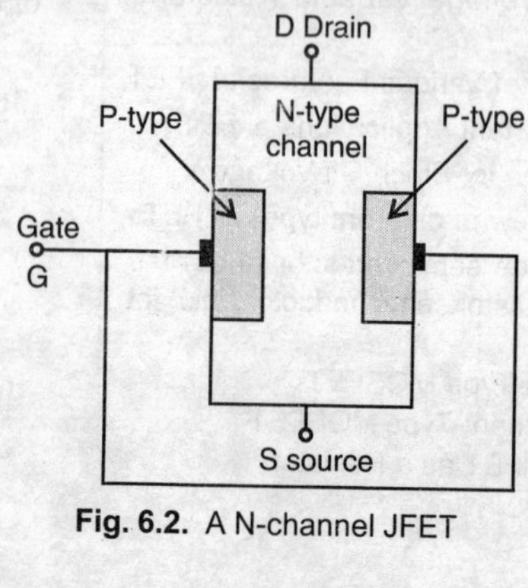

Fig. 6.2. A N-channel JFET

Whenever a voltage is applied across the drain and source terminals, a current flows through the N-channel. The current consists of only one type of carriers (i.e., electrons). Therefore, the field-effect transistor (FET) is known as an unipolar device. In fact, this quality distinguishes an FET from a BJT (i.e., a bipolar Junction transistor) where the current consists of the flow of both the electrons and holes.

Figure 6.3 shows a P-channel JFET. Its construction is similar to that of N-channel JFET, except that it consists of a P-channel and N-type junctions. The current carriers in JFET are the holes which flow through the P-type channel.

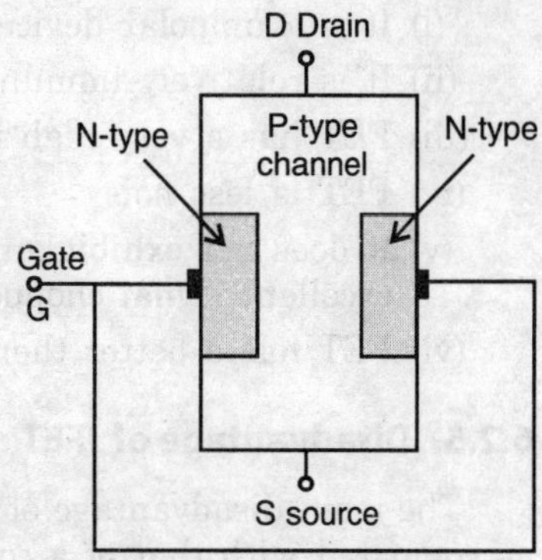

Fig. 6.3. A P-channel JFET

6.2.1. Symbol for JFET

Figure 6.4(a) shows the schematic symbol for an N-channel JFET. Similarly, figure 6.4(b) shows the schematic symbol for a P-channel JFET.

It may be noted that in an N-channel JFET, the arrow points towards the vertical line. The vertical line represents the N-channel. On the other hand, in a P-channel symbol, the arrow points away from the vertical line. In this case, the vertical line represents the P-channel.

> **DO YOU KNOW?**
>
> Because of the extremely high input impedance of a JFET, the input current is generally assumed to be 0 μA, and the current gain of a JFET amplifier is an undefined quantity.

Now let us discuss the operation and characteristics of JFET with reference to N-channel JFET. In fact, the operation of a P-channel JFET is similar to N-channel except that all voltages and currents are reversed.

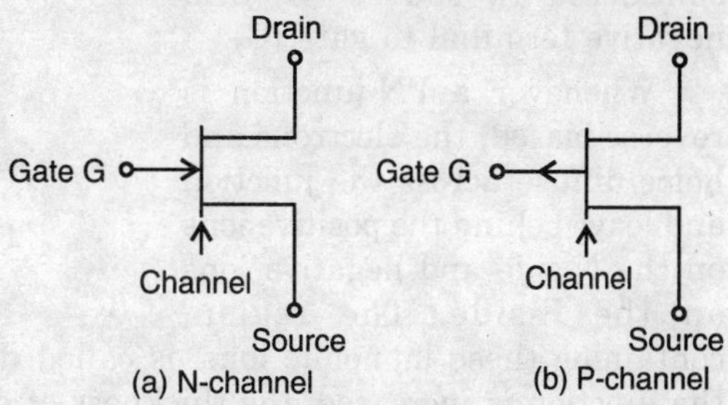

Fig. 6.4. Symbols for JFET

6.2.2. Voltage Controlled or Current Controlled?

The voltage applied between gate and source (V_{GS}) controls the drain current I_D. Therefore, FET is a voltage controlled device. The transistor on the contrary is a current controlled device because the output current I_C is controlled by varying the base current I_B. The name field effect is derived from the fact that the current flow in FET is controlled by an electric field set by an external voltage.

6.2.3. Unipolar or Bipolar?

FET is a unipolar device. This means that the current flowing through it is only due to one type of charge particles, holes or electrons. Transistor, on the other hand, is a bipolar device as holes and electrons both contribute to the flow of current. Just as there are NPN and PNP transistors, there are n-channel and p-channel field effect transistors. One of the most important characteristics of the FET is its very high input impedance. Typically, it is in the range of 1 to several megaohms, which is much higher than the input resistance of a BJT. FETs are more temperature stable as compared to the BJT and it requires less space than that for a BJT. Therefore FETs are preferred in integrated circuits.

6.2.4. Advantages of JFET Over BJT

The FET enjoys several advantages over the conventional bipolar junction transistor (BJT). Some of them may be listed as under :

(i) It is a unipolar device so its operation depends on the flow of majority carriers only.

(ii) It is relatively immune to radiation.

(iii) FET has a very high input resistance (typically few megaohms)

(iv) FET is less noisy

(v) It does not exhibit any offset voltage at zero drain current hence it can be used as an excellent signal chopper.

(vi) FET has a better thermal stability.

6.2.5. Disadvantage of JFET

The main disadvantage of the JFET is that it has a relatively small gain-bandwidth product as compared with that of a conventional transistor (i.e., BJT).

6.3. FORMATION OF DEPLETION REGION IN JFET

Let us consider an N-channel JFET as shown in figure 6.5 (a). Here the P-type gate and N-type channel constitute a PN-junction. This PN-junction is always reverse-biased in JFET operation. The reverse bias is applied by a voltage V_{GG} connected between the gate and source terminals as shown in figure 6.5 (b). Here, it can be noted that the positive terminal of the voltage V_{GG} is connected to source (S) and negative terminal to gate (G).

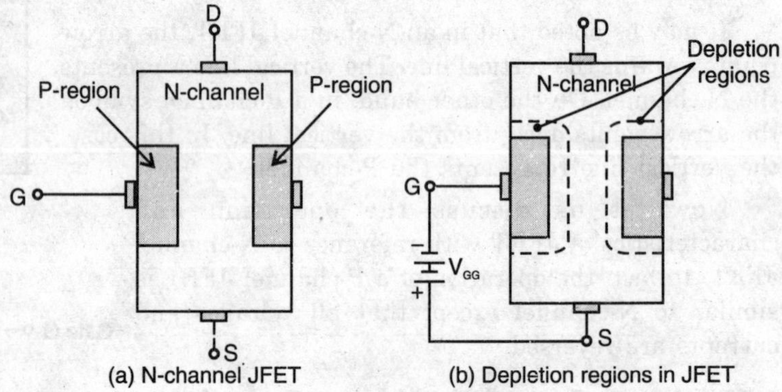

(a) N-channel JFET (b) Depletion regions in JFET

Fig. 6.5.

Whenever a PN-junction is reverse-biased, the electrons and holes diffuse across the junction and leave behind the positive ions on the N-side and negative ions on the P-side. The region, containing these immobile ions, is called **depletion region.** As the reverse bias voltage across the junction is increased, the thickness of depletion region is also increased. If both the P-side and N-side of the junctions are equally doped, the depletion region will extend equally in both the regions.

However, if one side of the junction is heavily doped as compared to the other side, the depletion region extends more into the region of lower doping.

A similar action takes place in JFET. Because the P-region of an N-channel JFET is heavily doped as compared to the N-channel, the depletion region extends less into the P-region and deeper into the N-channel as depicted in figure 6.5 (b). Incidentally, when there is no applied voltage between the drain and the source, the depletion region is symmetrical around the junction. The conductivity of depletion region is zero since there are no mobile charge carriers in this region. Therefore, the effective width of the N-channel is reduced. It further reduces with the increased reverse bias voltage applied across the gate and source terminals of the JFET.

> **DO YOU KNOW?**
>
> In a transistor, the emitter-base depletion layer is narrower than the collector-base depletion layer. The reason can be attributed to the different doping levels of the emitter and collector regions. With much heavier doping in the emitter region, the penetration into the n material is minimal because of the availability of many more free electrons. However, on the collector side, fewer free electrons are available and the depletion layer must penetrate more deeply in order to set-up the barrier potential.

The reverse-bias across the gate-source junction of a JFET can also be achieved simply by applying a voltage across the drain and source terminals as shown in figure 6.6 (a)

It may be noted that the drain (D) is connected to the positive terminal of the d.c. supply (V_{DD}) and source (S) is connected to the negative terminal.

Now, let us discuss as to how the drain-to-source voltage produces a reverse-bias across the gate-source junction in the same way as the gate-to-source voltage. In order to do so, we represent the channel resistances r_a and r_b as shown in figure 6.6 (a). These resistances have been shown variable since their values depend on the magnitude of the drain-to-source voltage (V_{DS}) and gate-to-source voltage (V_{GS}). In the presence of positive supply voltage V_{DD} (with gate open) the electrons flow from source to drain through the N-channel and constitutes a current known as **drain current (I_D).** The conventional direction of the drain current is indicated from drain-to-source through the device. The drain current causes a voltage drop across the resistance r_b, which has the effect of reverse-biasing the gate-to-source junction. Hence, even if the gate is open, the gate-to-source junction is reverse biased by the drain-to-source voltage. It creates the depletion region within the channel as shown in figure 6.6 (b).

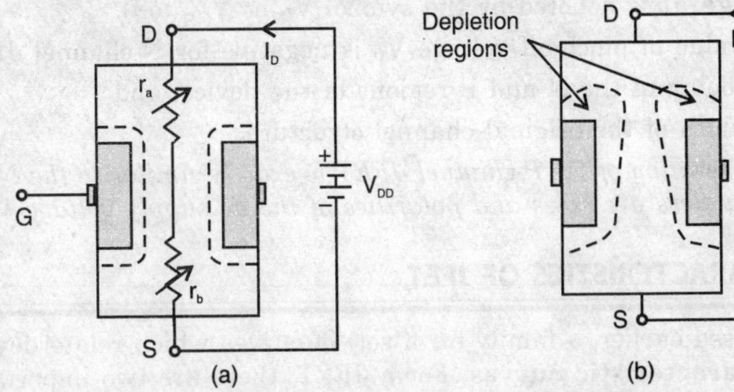

Fig. 6.6. Effect of drain-to-source supply

It may be noted from figure 6.6 (b) that the depletion region is not symmetrical around the gate-to-source junction. As a matter of fact, it is extended more deeper into the channel near the drain terminal and less on the source terminal. It is due to the fact that voltage drop across r_a is greater than that across r_b. Thus, because of this reverse bias voltage is higher near the drain end of the channel as compared to the source end.

6.4. OPERATION OF JFET* (V. Important)

In last article we have discussed the application of negative gate voltage or positive drain voltage, with respect to source, reverse biases the gate-source junction of an N-channel JFET. The effect of reverse-bias voltage is to form the depletion regions within the channel. When a voltage is applied between the drain and source with a dc supply voltage (V_{DD}), the electrons tend to flow from source to drain side through the narrow channel existing between the depletion regions. This constitutes the drain current (I_D) and its conventional direction

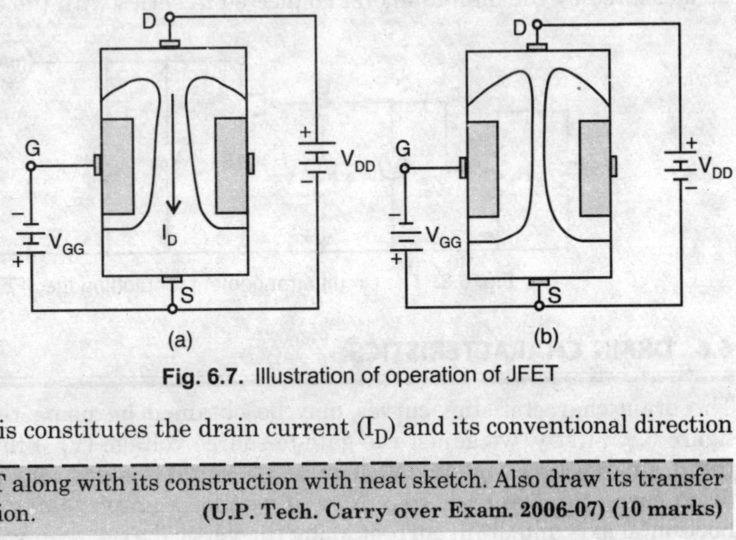

Fig. 6.7. Illustration of operation of JFET

* Explain the working of a JFET along with its construction with neat sketch. Also draw its transfer characteristic curve with explanation. **(U.P. Tech. Carry over Exam. 2006-07) (10 marks)**

is indicated from drain-to-source. The value of this drain current is maximum when no external voltage is applied between the gate and the source and is designated by the symbol I_{DSS}.

When the gate-to-source voltage (V_{GS}), is applied by a dc supply (V_{GG}) and increased above zero as depicted in figure 6.7 (a), the reverse-bias voltage across the gate-source junction is increased. Because of this, the depletion regions are widened. This reduces the effective width of the channel and hence controls the flow of drain current through the channel.

When the gate-to-source voltage (V_{GS}) is further increased, a stage is reached at which two depletion regions touch each other as shown in figure 6.7 (b). At this gate-to-source voltage, the channel is completely blocked or pinched off and drain current is reduced to zero. The gate-to-source voltage (V_{GS}) at which the drain current is zero (or completely cut-off) is known as **pinch off voltage**. It is denoted by the symbol V_P or V_{GS} (off).

The value of pinch-off voltage V_P is negative for N-channel JFET. It depends upon

(i) doping of the N and P regions of the device, and

(ii) width of the original channel structure.

The operation of the P-channel JFET is exactly similar to the N-channel JFET except that the current carriers are holes and polarities of the dc supply voltage V_{GG} and V_{DD} are reversed.

6.5. CHARACTERISTICS OF JFET

As discussed earlier, a family (or a set) of curves which relate device currents and voltages are called **characteristic curves**. For a JFET, there are two important characteristics as under:

(i) V-I or drain characteristics

These curves give relationship between the drain current (I_D) and drain-to-source voltage (V_{DS}) for different values of gate-to-source voltage (V_{GS}).

(ii) Transfer characteristics

These curves give relationship between drain current (I_D) and gate-to-source voltage (V_{GS}) for different values of drain-to-source voltage (V_{DS}).

The drain and transfer characteristics of JFET can be obtained by using an N-channel JFET connected in the common source mode as shown in figure 6.8. Here, the potentiometers R_1 and R_2 are used to vary the voltages V_{GS} and V_{DS} respectively. The voltages V_{DS} and V_{GS} can be measured by the voltmeters connected across the JFET terminals. The drain current (I_D) can be measured by the milli-ammeter connected in series with the JFET and the supply voltage V_{DD}.

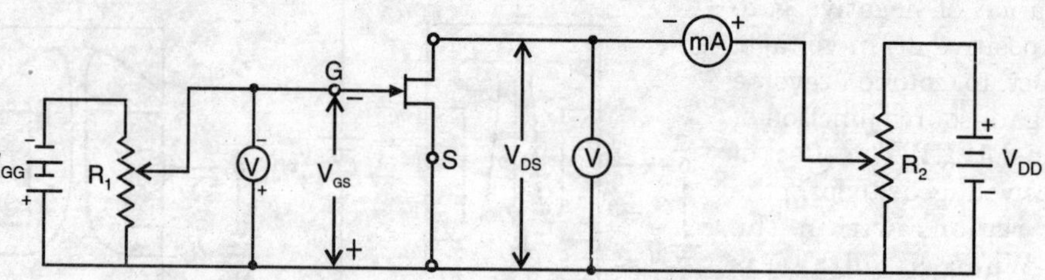

Fig. 6.8. The circuit arrangement for plotting the JFET characteristics

6.6. DRAIN CHARACTERISTICS

The drain characteristics curves may be obtained by using the circuit arrangement shown in figure 6.8. Firstly, we adjust the gate-to-source voltage (V_{GS}) to zero volt. After this, we increase the drain-to-source voltage (V_{DS}) in small suitable steps and note the corresponding values of drain current (I_D) at each step. Now, if we plot a graph taking drain-to-source voltage along the horizontal axis and drain current along the vertical axis, then we shall obtain a curve marked V_{GS}

= 0 as shown in figure 6.9. A similar procedure can be followed to obtain curves for different values of gate-to source voltage i.e., V_{GS} = 1, 2, 3 and 4 volt etc.

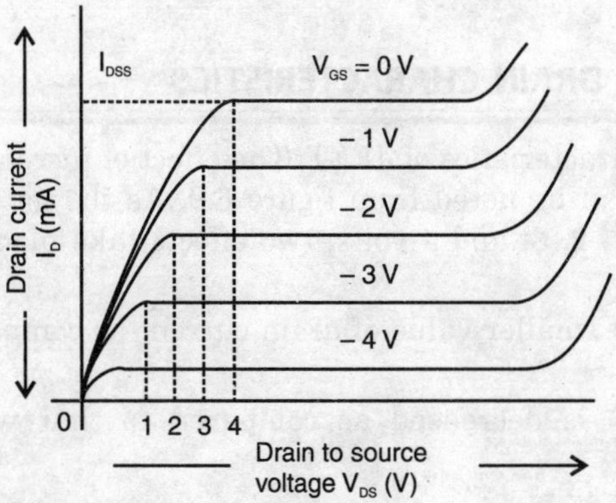

Fig. 6.9. The drain characteristics of JFET

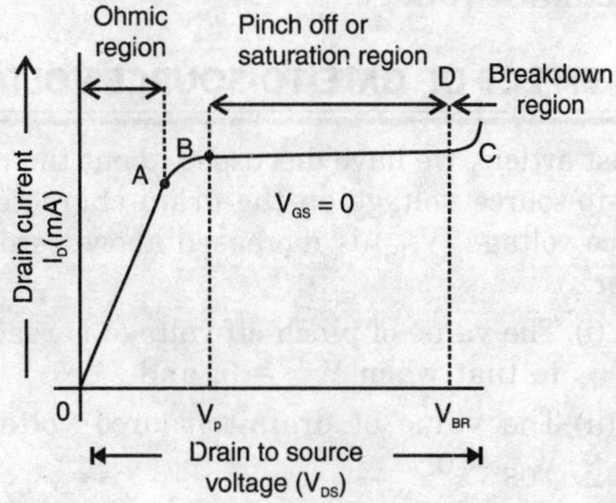

Fig. 6.10. The drain characteristics with V_{GS} = 0 Volt

Now, in order to explain the typical shape of drain characteristics, let us select the curve with V_{GS} = 0 volt as shown in figure 6.10. The curve can be sub-divided into the following regions:

(i) Ohmic region

Ohmic region has been shown as a curve OA in the figure 6.10. In this region, the drain current increases linearly with the increase in drain-to-source voltage obeying Ohm's law. The linear increase in drain current is due to the fact that N-type semiconductor bar acts like a simple resistance.

(ii) Curve AB

In this region, the drain current increases at the reverse square law rate with the increase in drain-to-source voltage. It means that drain current increases slowly as compared to that in the ohmic region. It is due to the fact that with the increase in drain-to-source voltage, the drain current increases. This, in turn, increases the reverse-bias voltage across the gate-source junction. As a result of this, the depletion region grows in size thereby reducing the effective width of the channel. At the drain-to-source voltage, corresponding to point B, the channel width is reduced to a minimum value and is called **pinch off**. The drain-to-source voltage at which the channel pinch-off occurs, is called **pinch off voltage (V_P)**.

(iii) Pinch-off region

This region has been shown by the curve BC. It is also known as the **saturation region** or **constant current region**. In this region, the drain current remains constant at its maximum value (i.e., I_{DSS}). The drain current in the pinch off region depends on the gate-to-source voltage and is given by the following relation:

$$I_D = I_{DSS}\left(1 - \frac{V_{GS}}{V_P}\right)^2$$

This relation is known as ***Shockley's equation****. The pinch-off region is the normal operating region of JFET, when used as an amplifier.*

> **DO YOU KNOW?**
>
> The pinch off voltage. V_p is the point at which further increases in V_{DS} are offset by a proportional increase in the channel's resistance. This means that if the channel resistance is increasing in direct proportion to V_{DS} above V_P, I_D must remain the same above V_P.

(iv) Breakdown region

This region has been shown by the curve CD. In this region the drain current increases rapidly as the drain-to-source voltage is also increased. It happens because of the breakdown of gate-to-source junction due to the avalanche effect. The drain-to-source voltage corresponding to point C is known as the **breakdown voltage.**

6.7. EFFECT OF GATE-TO-SOURCE VOLTAGE ON DRAIN CHARACTERISTICS

In last article, we have discussed about the drain characteristics of JFET. The effect of increasing gate-to-source voltage on the drain characteristics can be noted from figure 6.9. As the gate-to-source voltage (V_{GS}) is increased above zero to the 1, 2, 3, and 4 volts, two effects take place as under:

(i) The value of pinch-off voltage is reached at a smaller value of drain current as compared to that when $V_{GS} = 0$, and

(ii) The value of drain-to-source voltage (V_{DS}) is decreased as compared to that when $V_{GS} = 0$.

Now, the first effect can be understood from the fact that reverse bias voltage across the gate-to-source junction of JFET is the sum of gate-to-source voltage (V_{GS}) and drain-to-source voltage (V_{DS}). Hence, for fixed value of V_{DS}, the increased value of voltage V_{GS} would increase the reverse-bias voltage across the junction. For example, if V_{GS} of –1 volt is applied between the gate and the source of JFET, then the gate-to-source junction becomes reverse-biased even when the drain current is zero. This causes the depletion regions to be formed within the channel. As the voltage V_{DS} is increased above zero, the drain current increases. This produces voltage drop across channel resistance, which in turn, produces reverse-bias voltage across the gate-source junction. The amount of reverse-bias voltage required to be produced by drain current will obviously be decreased by one volt. In other words, a smaller voltage drop along the channel (i.e., smaller than that when $V_{GS} = 0$) will increase the depletion regions to the point where they will pinch off the drain current. As a result of this the pinch-off voltage (V_P) is reached at a lower value of drain current than that when $V_{GS} = 0$.

Also the second effect is caused due to the fact that the increased value of gate-to-source voltage (V_{GS}) keeps adding to the reverse-bias voltage at the junction produced by the drain current. Hence, its lower value is required to reach the breakdown voltage.

It can be noted that with zero gate-to-source voltage the drain current saturates at I_{DSS} value and the characteristic indicates the pinch-off voltage (V_P) equal to 4 volt. However, if V_{GS} of – 1 volt is applied, the channel still requires –4 volt to achieve pinch-off condition. It means that now the channel requires only 3 volt drop (instead of previously 4 volt) from the drain-to-source voltage. This 3 volt drop can be achieved with the lower value of drain current. Similarly when V_{GS} is –2 and –3 volts, pinch-off is achieved with 2 volt and 1 volt drops respectively along the channel. These voltage drops are obtained with further reduced values of the drain current. It can be noted that when V_{GS} is –4 volt (i.e., equal to pinch-off voltage), no channel drop is required and thus the drain current is zero.

6.8. TRANSFER CHARACTERISTICS (V. Important)

Transfer characteristics are also known as transconductance curves which gives us the relationship between drain current (I_D) and gate-to-source voltage (V_{GS}) for a constant value of drain-to-source voltage (V_{DS}). The transfer characteristics can be obtained by using the circuit arrangement shown in figure 6.8. Firstly, we adjust the drain-to-source voltage to some suitable value and increase the gate-to-source voltage in small suitable steps. After this, note the corresponding values of

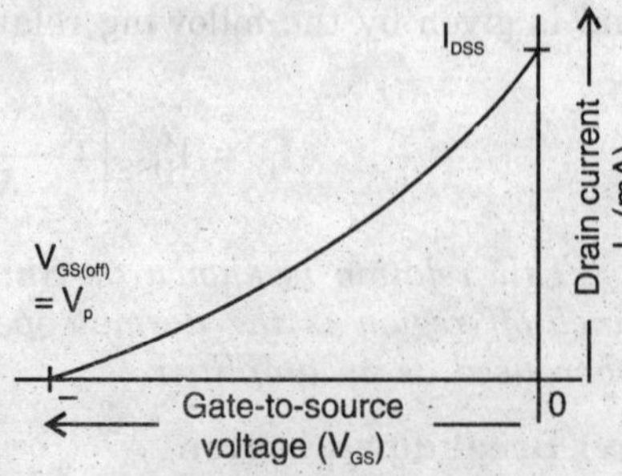

Fig. 6.11. The transfer characteristics of N-channel.

drain current at each step. If we plot a graph with gate-to-source voltage (V_{GS}) along the horizontal axis and the drain current (I_D) along the vertical axis, we shall obtain a curve as shown in figure 6.11. A similar procedure can be used to obtain curves at different values of the gate-to-source voltage (V_{GS}).

It may be observed that the upper end of the curve has been shown by the drain current value equal to I_{DSS}, while the lower end is indicated by a voltage equal to $V_{GS(off)}$ or V_P. It can be noted that the curve is part of a parabola and hence can be expressed by the following expression

> **DO YOU KNOW?**
>
> The transconductance curve of a JFET is unaffected by the circuit or configuration in which the JFET is used.

$$I_D = I_{DSS}\left[1-\frac{V_{GS}}{V_{GS(off)}}\right]^2 = I_{DSS}\left(1-\frac{V_{GS}}{V_P}\right)^2$$

There is one important point which must be mentioned at this stage. Most JFET specifications sheet will list more than one value of V_{GS} (off) and I_{DSS}.

For example, the specifications sheet for the JFET 2N 5457 lists the following:

$$V_{GS\,(off)}\ (\text{minimum}) = -\ 0.5 \text{ volt}$$

and $$I_{DSS}\ (\text{min}) = 1 \text{ mA}$$

$$V_{GS(off)}\ (\text{maximum}) = -\ 6 \text{ volt}$$

and $$I_{DSS}\ (\text{max}) = 5 \text{ mA}$$

In such a situation, we must use the two minimum values to plot one curve and the two maximum values to plot another curve on the same graph.

EXAMPLE 6.1 The data sheet of a certain JFET indicates that I_{DSS} is equal to 15 mA and $V_{GS(off)}$ is equal to –5 volt. Find the drain current for V_{GS} equal to 0 volt, –1 volt and – 4 volt.

Solution: Given that I_{DSS} = 15 mA and, $V_{GS(off)}$ = –5 volt

We know that the drain current is given by

$$I_D = I_{DSS}\left[1-\frac{V_{GS}}{V_{GS(off)}}\right]^2$$

Therefore, drain current when (V_{GS} = 0), will be

$$I_D = 15\left(1-\frac{0}{-5}\right)^2 = 15 \text{ mA} \qquad \textbf{Ans.}$$

Similarly, drain current when V_{GS} – 1 V, will be

$$I_D = 15\left(1-\frac{-1}{-5}\right)^2 = 9.6 \text{ mA} \qquad \textbf{Ans.}$$

and the drain current when V_{GS} = –4 V, will be

$$I_D = 15\left(1-\frac{-4}{-5}\right)^2 = 0.6 \text{ mA} \qquad \textbf{Ans.}$$

EXAMPLE 6.2 A JFET has parameters of $V_{GS(off)}$ equal to –20 volt and I_{DSS} equal to 12 mA. Plot the transconductance curve for the device using V_{GS} values of 0 V, –5 V, –10 V, –15 V, and –20 V.

Solution: Firstly, let us find the value of drain currents for the given values of V_{GS} (0 V , –5 V, –10 V, –15 V and –20 V) as discussed below:

Given that $V_{GS(off)}$ = –20 volt and, I_{DSS} = 12 mA

We know that the drain current is given by

$$I_D = I_{DSS}\left[1-\frac{V_{GS}}{V_{GS(off)}}\right]^2$$

Therefore, the drain current when $V_{GS} = -5$ V, will be

$$I_D = 12\left(1 - \frac{-5}{-20}\right)^2 = 6.75 \text{ mA}$$

Similarly, the drain current when $V_{GS} = -10$ V, will be

$$I_D = 12\left(1 - \frac{-10}{-20}\right)^2 = 3 \text{ mA}$$

and the drain current when $V_{GS} = -15$ V, will be

$$I_D = 12\left(1 - \frac{-15}{-20}\right)^2 = 0.75 \text{ mA}$$

The value of V_{GS} and the corresponding values of I_D obtained above have been shown in Table 6.1.

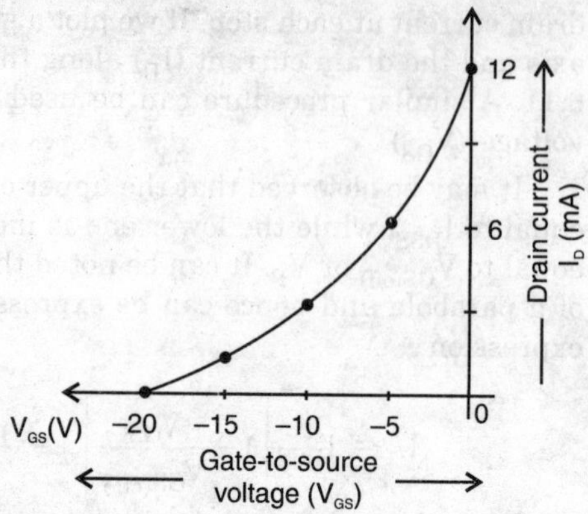

Fig. 6.12.

Table 6.1.

V_{GS} (V)	I_D (mA)
−20	0
−15	0.75
−10	0.3
−5	6.75
−0	12

These points are now plotted and joined together to form a curve shown in figure 6.12.

EXAMPLE 6.3 A 2N 5486 JFET has values of $V_{GS(off)}$ equal to −2 volt to −6 volt and I_{DSS} is equal to 8 mA to 20 mA. Plot the minimum and maximum transconductance curves for the device.

Solution: Given that

$$V_{GS\,(off)} = -2 \text{ volt to } -6 \text{ volt} \quad \text{and} \quad I_{DSS} = 8 \text{ mA to } 20 \text{ mA}$$

Firstly, let us find the values of drain currents for the given values of V_{GS}.

We know that the drain current is given by

$$I_D = I_{DSS}\left[1 - \frac{V_{GS}}{V_{GS(off)}}\right]^2 \qquad \text{...(i)}$$

In this equation, substituting the values of $V_{GS} = -0.5$, -1.0 and -1.5 with $V_{GS(off)} = -2.0$ volt, we get the corresponding values of $I_D = 4.5$ mA, 2 mA and 0.5 mA.

With these values, we have the set of V_{GS} versus I_D value as shown in Table 6.2.

Table 6.2. For Minimum Transconductance Curve

V_{GS} (V)	I_D (mA)
− 2.0	0
− 1.5	0.5
− 1.0	2.0
− 0.5	4.5
0	8

Similarly, using the maximum values of $V_{GS\,(off)} = -6$ Volt and $I_{DSS} = 20$ mA, we can use equation (i) again for $V_{GS} = -4$ V and -2 V to obtain the values of I_D in between the end points. The end points of the transcondutance curve are (−6 V, 0 mA) and (0 V, 20 mA). The values of I_D between points are 2.2 mA and 9.9 mA. The values obtained have been shown in Table 6.3.

Table 6.3. For Maximum Transconductance Curve

V_{GS}(V)	I_D (mA)
–6	0
–4	2.2
–2	3.9
0	20

Now, if we plot the values corresponding to the sets shown in Tables 6.2 and 6.3 and joining them, we get the maximum and minimum transconductance curve as shown in figure 6.13.

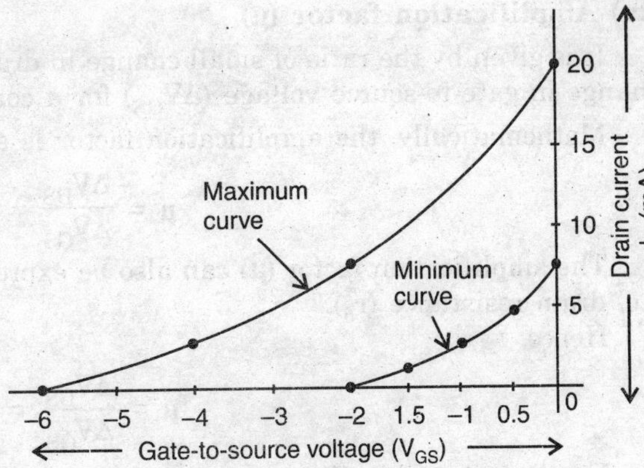

Fig. 6.13.

6.9. THE JFET PARAMETERS

The electrical behaviour of the JFET can be described in terms of certain parameters known as JFET parameters. Such parameters can be obtained from the characteristic curves. The important parameters of JFET have been discussed below:

(i) D.C. drain resistance (R_{DS})

It is also known as the static or ohmic resistance of the channel and is given by the ratio of voltage (V_{DS}) to the drain current (I_D). Mathematically, the d.c. drain resistance is expressed as

$$R_{DS} = \frac{V_{DS}}{I_D}$$

(ii) A.C. drain resistance (r_d)

It is also known as the dynamic drain resistance and is the a.c. resistance between the drain and the source terminal when the JFET is operating in the pinch-off or saturation region. It is given by the ratio of small change in drain-to-source voltage (ΔV_{DS}) to the corresponding change in drain current (ΔI_D) for a constant gate-to-source voltage (V_{GS}).

Mathematically, the a.c. drain resistance is expressed as

$$r_d = \frac{\Delta V_{DS}}{\Delta I_D}$$

It is, sometimes, written as r_{ds}, which indicates that it is the resistance from drain-to-source terminals. Since r_d is the output resistance of JFET, it can also be expressed as an output admittance y_{DS} (i.e., equal to $1/r_d$).

(iii) Transconductance (g_m) **(U.P. Tech. Sem. Exam. 2005-06) (03 marks)**

It is also known as the forward transconductance or forward transmittance. It is given by the ratio of small change in drain current (ΔI_D) to the corresponding change in gate-to-source voltage (ΔV_{GS}) for a constant drain-to-source voltage (V_{DS}).

Mathematically, the transconductance is expressed as

$$g_m = \frac{\Delta I_D}{\Delta V_{GS}}$$

The value of transconductance (g_m) is expressed in siemens (S) or mhos. We know that transfer characteristic of a JFET is a part of a parabola i.e., it is non-linear. Therefore, the value of g_m varies depending upon the location on the curve as set by the gate-to-source voltage (V_{GS}). It has a greater value near the top of the curve than near the bottom.

(iv) Amplification factor (μ)

It is given by the ratio of small change in drain-to-source voltage (ΔV_{DS}) to the corresponding change in gate-to-source voltage (ΔV_{GS}) for a constant drain current (I_D).

Mathematically, the amplification factor is expressed as

$$\mu = \frac{\Delta V_{DS}}{\Delta V_{GS}}$$

The amplification factor (μ) can also be expressed as a product of transconductance (g_m) and a.c. drain resistance (r_d).

Hence,

$$\mu = \frac{\Delta V_{DS}}{\Delta V_{GS}} = \frac{\Delta V_{DS}}{\Delta I_D} \times \frac{\Delta I_D}{\Delta V_{GS}} = r_d \times g_m$$

(v) Input resistance (R_i)

We know that a JFET operation with its gate-to-source junction reverse biased. As the reverse current of a junction is, usually, very small, therefore, gate-reverse current of a JFET is also very small.

Because the gate-source junction of JFET is used as an input, therefore, the input resistance of a JFET is very high. This high input resistance is an advantage of JFET over the bipolar transistor.

Mathematically the input resistance is given by

$$R_i = \frac{V_{GS}}{I_{GSS}}$$

where V_{GS} = Gate-to-source voltage

and I_{GSS} = Gate reverse current

It is of the order of nano-amperes.

6.10. THE MATHEMATICAL EXPRESSION FOR TRANSCONDUCTANCE (V. Important)

As discussed earlier, the drain current is expressed by the relation

$$I_D = I_{DSS}\left(1 - \frac{V_{GS}}{V_P}\right)^2$$

Differentiating the above equation on both sides with respect to V_{GS}, we have

$$\frac{dI_D}{dV_{GS}} = 2I_{DSS}\left(1 - \frac{V_{GS}}{V_P}\right) \times \left(\frac{0-1}{V_P}\right)$$

or

$$g_m = -\frac{2I_{DSS}}{V_P}\left(1 - \frac{V_{GS}}{V_P}\right) \quad \text{...(i)}$$

Substituting $V_{GS} = 0$ in the above expression, we get

$$g_m = -\frac{2I_{DSS}}{V_P} = g_{mo}$$

The value of g_m at $V_{GS} = 0$ is the maximum and is equal to $-\frac{2I_{DSS}}{V_P}$. This value of g_m is designat[illegible]s g_{mo}. The equation (i) can now be rewritten as

$$g_m = g_{mo}\left(1-\frac{V_{GS}}{V_P}\right)$$

EXAMPLE 6.4 The following information is included on the data sheet for an N-channel JFET:

I_{DSS} = 20 mA V_P = – 8 Volt and, g_{mo} = 5000 µs

Find the values of the drain current and transconductance at V_{GS} = –4 Volts.

Solution: Given that: I_{DSS} = 20 mA; V_P = –8 Volts; g_{mo}= 5000 µs; and V_{GS} = – 4 Volts

Now, we know that the drain current is given by

$$I_D = I_{DSS}\left(1-\frac{V_{GS}}{V_P}\right)^2$$

Substituting values, we get

or $$I_D = 20\left(1-\frac{-4}{-8}\right)^2 = 5 \text{ mA} \quad \textbf{Ans.}$$

Also the transconductance is given by

$$g_m = g_{mo}\left(1-\frac{V_{GS}}{V_P}\right) = 5000\left(1-\frac{-4}{-8}\right) = 2500 \text{ µs} \quad \textbf{Ans.}$$

EXAMPLE 6.5 The data sheet for a certain enhancement-type MOSFET reveals that $I_{D(on)}$ = 10 mA at V_{GS} = – 12 V and $V_{GS(th)}$ = – 3 V. Is this device P-channel or N-channel? Find the value of I_D, when V_{GS} = – 6 V. **(M.D. University, 2005)**

Solution: Given that: $I_{D(on)}$ = 10 mA; V_{GS} = – 12 volts and $V_{GS(th)}$ = – 3 Volts.

Since the value of V_{GS} is negative for the enhancement-type MOSFET, this indicated that the device is P-channel.

We know that the drain current is given by

$$I_D = K\,(V_{GS} - V_{GS(th)})^2 \qquad \text{...(i)}$$

or
$$I_{D(on)} = K\,[(V_{GS} - V_{GS(th)}]^2$$
$$10 = K\,[-12-(-3)]^2 = 81\,K$$
$$K = 10/81 = 0.12 \text{ mA/V}$$

Substituting this value of K and V_{GS} (equal to – 6) in the equation (i),

$$I_D = 0.12\,[-6-(-3)]^2 = 1.08 \text{ mA} \quad \textbf{Ans.}$$

EXAMPLE 6.6 Show that if $|V_{GS}| << |V_P|$, then drain current can be approximated as

$$\mathbf{I_{DS} = I_{DSS} - g_{mo}\,V_{GS}}$$

Solution: We know that

$$I_{DS} = I_{DSS}\left[1-\frac{V_{GS}}{V_P}\right]^2$$

Simplifying, we get
$$I_{DS} = I_{DSS}\left[1-2\frac{V_{GS}}{V_P}+\left(\frac{V_{GS}}{V_P}\right)^2\right]$$

Neglecting the square term, we get

$$I_{DS} = I_{DSS} - \frac{2I_{DSS}V_{GS}}{V_P}$$

Putting $$g_{mo} = -\frac{2I_{DSS}}{V_P}, \text{ we get}$$

$$I_{DS} = I_{DSS} + g_{mo} V_{GS} \quad \textbf{Proved.} \quad \textbf{Ans.}$$

EXAMPLE 6.7 An N-channel JFET has I_{DSS} = 8 mA and V_P = – 5 V. Find the minimum value of V_{DS} for pinch-off region and the drain current I_{DS}, for V_{GS} = – 2 V in the pinch-off region.

Solution: The minimum value of V_{DS} for pinch-off to occur for V_{GS} = – 2 V is

$$V_{DS(min)} = V_{GS} - V_P = -2 - (-5) = 3 \text{ V}$$

$$I_{DS} = I_{DSS}\left[1 - \frac{V_{GS}}{V_P}\right]^2$$

Substituting values, we get

$$I_{DS} = 8 \times 10^{-3} [1 - (-2)/(-5)]^2$$

Solving, we get $$I_{DS} = 2.88 \text{ mA} \quad \textbf{Ans.}$$

EXAMPLE 6.8 V_P = – 2 V, I_{DSS} = 1.55 mA for the circuit in figure 6.14. It is desired to bias the circuit at I_D = 0.8 mA, V_{DD} = 24 V. Find: (i) V_{GS} (ii) g_m.

Solution: (i) We know that

$$V_{GS} = V_P\left[1 - \sqrt{\frac{I_D}{I_{DSS}}}\right]$$

Substituting values, we get

$$V_{GS} = -2\left[1 - \sqrt{\frac{0.8 \times 10^{-3}}{1.65 \times 10^{-3}}}\right]$$

Simplifying, we get $$V_{GS} = -0.6074 \text{ V}$$

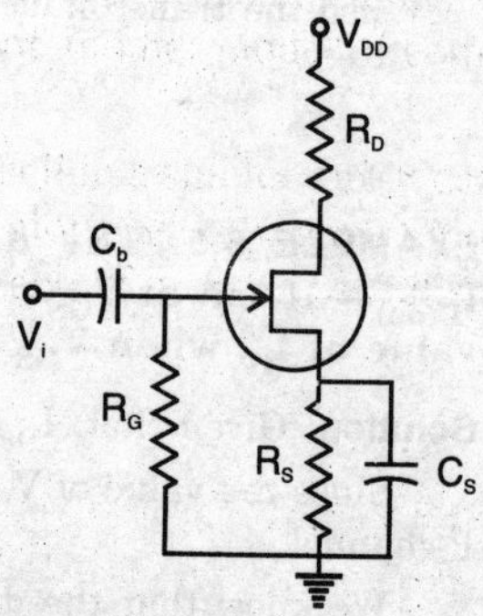

Fig. 6.14.

(ii) $$g_{mo} = -\frac{2I_{DSS}}{V_P} = -\frac{2 \times 1.65}{-2} = 1.65 \text{ mS} = g_{mo}\left(1 - \frac{V_{GS}}{V_P}\right)$$

or $$g_m = 1.65\left(1 - \frac{0.6074}{2}\right) \times 10^{-3} = 1.65 \times 10^{-3} \times 0.6963 = 1.15 \text{ mS} \quad \textbf{Ans.}$$

EXAMPLE 6.9 The pinch-down voltage of a P-channel junction FET is V_P = 5 V and the drain-to-source saturation current I_{DSS} = – 40 mA. The value of drain-source voltage V_{DS} is such that the transistor is operating in the saturated region. The drain current is given as I_D = – 15 mA. Determine the gate-source voltage V_{GS}. (GATE Examination, 2001)

Solution: It is found experimentally that a square-law characteristic closely approximate the drain current in saturation:

$$I_{D(sat)} \approx I_{DSS}\left(1 + \frac{V_{GS}}{V_P}\right)^2$$

Givan that : $V_P = 5$ V; $I_{DSS} = -40$ mA; $I_D = -15$ mA

$$-15 \text{ mA} = -40 \text{ mA}\left(1 + \frac{V_{GS}}{5}\right)^2$$

or $$\sqrt{\frac{15}{40}} = 1 + \frac{V_{GS}}{5}$$

$$\therefore \quad 1+\frac{V_{GS}}{5} = 0.612$$

$$\frac{V_{GS}}{5} = 0.612 - 1 = -0.3876$$

$$V_{GS} = 1.938 \text{ volt.} \quad \textbf{Ans.}$$

EXAMPLE 6.10 Determine the value of transconductance of a FET, when the drain current changes from 1 mA to 1.9 mA with a change in gate-source voltage from –3.3 V to –3 V.

Solution: Change in drain current,

$$\Delta I_D = 1.9 - 1.0 = 0.9 \text{ mA}$$

Change in gate-source voltage,

$$\Delta V_{GS} = 3.3 - 3.0 = 0.3 \text{ V}$$

Now, transconductance is given by

$$g_m = \frac{\Delta I_D}{\Delta V_{GS}} = \frac{0.9 \text{ mA}}{0.3 \text{ V}} = 3 \text{ mA/V}$$

$$= 3000 \ \mu \text{ mhos} \quad \textbf{Ans.}$$

EXAMPLE 6.11 The following readings were obtained experimentally from a FET:

V_{GS}	**– 0.1 V**	**– 0.1 V**	**– 0.4 V**
V_{DS}	**5 V**	**14 V**	**14 V**
I_D	**3 mA**	**3.3 mA**	**0.71 mA**

Obtain (i) a.c. drain resistance, (ii) transconductance and (iii) amplification factor.

Solution: (i) At constant V_{GS} (i.e., at $V_{GS} = -0.1$ V), when V_{DS} is increased from 5 V to 14 V the drain current I_D increases from 3 mA to 3.3 mA.

Therefore, change in drain-source voltage, $\Delta V_{DS} = 14 - 5 = 9$ V

Thus, change in drain current, $\Delta I_D = 3.3 - 3 = 0.3$ mA

Hence, a.c. drain resistance will be

$$r_d = \frac{\Delta V_{DS}}{\Delta I_D} \text{ at constant } V_{GS}$$

$$r_d = \frac{9 \text{ V}}{0.3 \text{ mA}} = \mathbf{30 \ K \ ohms}$$

(ii) At constant V_{DS} (i.e., at $V_{DS} = 14$ V), when V_{GS} is changed from – 0.1 V to – 0.4 V, the drain current I_D changes from 3.3 mA to 0.71 mA.

The change in gate-source voltage will be

$$\Delta V_{GS} = 0.4 - 0.1 = 0.3 \text{ V}$$

The change in drain current will be

$$\Delta I_D = 3.3 - 0.71 = 2.59 \text{ mA}$$

Therefore, transconductance will be

$$g_m = \frac{\Delta I_D}{\Delta V_{GS}} \text{ at constant } V_{DS}$$

or $$g_m = \frac{2.59 \text{ mA}}{0.3 \text{ V}} = 8.63 \text{ mA/V} = 8630 \ \mu\text{mhos}$$

(iii) Amplification factor is given by

$$\mu = r_d \times g_m = (30 \times 10^3) \times (8630 \times 10^{-6}) = 258.9 \quad \textbf{Ans.}$$

6.11. COMPARISON BETWEEN FET AND BJT

Table 6.4 shows some of the important points of comparison between the field effect transistor (FET) and bipolar junction transistor (BJT).

Table 6.4. Comparison Between Field-effect Transistor (FET) and Bipolar Junction Transistor (BJT) **(Important)**

S.No.	Field-Effect Transistor (FET)	Bipolar junction transistor (BJT)
1.	It is an unipolar device i.e., current in the device is carried either by electrons or holes.	It is bipolar device, i.e., current in the device is carried by both electrons and holes.
2.	It is a voltage-controlled device, i.e., voltage at the gate (or drain) terminal controls the amount of current flowing through the device.	It is a current-controlled device, i.e., the base current controls the amount of collector current.
3.	Its input resistance is very high and is of the order of several megaohms.	Its input resistance is very low as compared to FET and is of the order of few kilo-ohms.
4.	It has a negative temperature coefficient at high current levels. It means that current decreases as the temperature increases. This characteristic prevents the FET from thermal breakdown.	It has a positive temperature coefficient at high current levels. It means that collector current increases with the increase in temperature. This characteristic leads the BJT to thermal breakdown.
5.	It does not suffer from minority-carrier storage effects and therefore, has higher switching speeds and cut-off frequencies.	It suffers from minority carrier storage effects and therefore, has lower switching speed and cut-off frequencies than that of FET's.
6.	It is less noisy than a BJT or vacuum tube and is thus more suitable as an input amplifier for low-level signals. It is used extensively in high fidelity frequency modulated receivers.	It is comparatively more noisy than a field-effect transistor.
7.	It is much simpler to fabricate as an integrated circuit (IC) and occupies a less space on IC chip than that of BJT.	It is comparatively difficult to fabricate as an integrated circuit (IC) and occupies more space on IC chip than that of FET.

6.12. FET AS VOLTAGE VARIABLE RESISTANCE (VVR)

In almost all the linear applications of FET, it is operated in the constant current portion of its output characteristics (i.e., in the saturation region).

Now, let us observe the operation of JFET in the region just before pinch off. In this region, V_{DS} is small but I_D is large. This region is marked as ohmic region in the output characteristics of a JFET. In this region, the FET can be used as a voltage variable resistor (VVR) or voltage dependent resistor (VDR). The drain to source resistance r_{DS} in this region is dependent on the gate to source voltage. As V_{GS} becomes less negative, more drain current flows and r_{DS} decreases. For higher negative gate voltage, I_D decreases and r_{DS} will increase. Thus, the JFET acts as a voltage variable resistor (VVR).

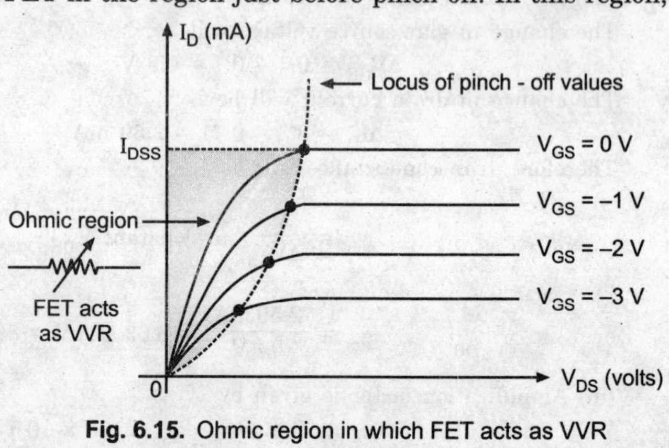

Fig. 6.15. Ohmic region in which FET acts as VVR

6.12.1. Application of VVR

(i) The FET used as VVR finds applications in many areas where this property can be used.

(ii) The VVR, for example, can be used to vary the voltage gain of a multistage amplifier A as the signal level is increased. This is called as **automatic gain control**.

(iii) A typical arrangement for controlling the gain of an amplifier automatically is shown in figure 6.16.

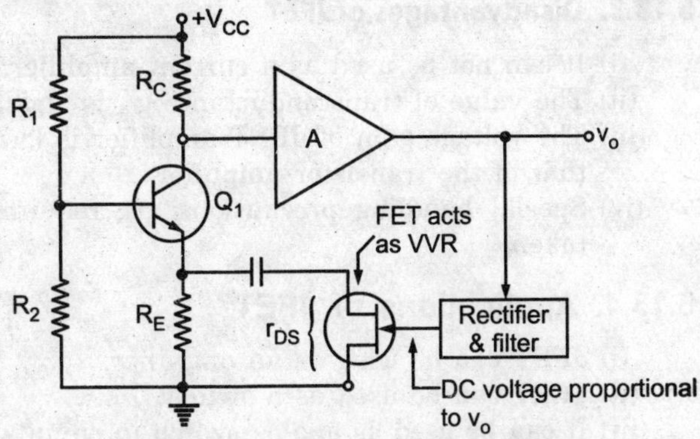

Fig. 6.16. AGC amplifier using FET as VVR

Working Operation

The gain of the common emitter amplifier Q_1 is controlled. The output signal v_0 is rectified and filtered and the dc voltage proportional to v_0 is applied to the gate of FET operating in the ohmic region. Depending on the dc gate voltage, the resistance r_{DS} of FET will change, this will change the emitter resistance of Q_1 since R_E is in parallel with r_{DS}.

The gain of CE amplifier is approximately equal to $-\dfrac{R_C}{R_E}$. Hence, change in R_E will change the gain automatically. If v_0 increases, then dc voltage at the rectifier filter output will increase. Hence, V_{GS} becomes more negative. This decreases I_D or increases r_{DS} of FET. Hence, the parallel combination (R_E || r_{DS}) will increase and the gain of Q_1 will increase. Exactly opposite action takes place to increase gain of Q_1 if v_0 decreases. This is how the automatic gain control is achieved.

6.13. SUMMARY OF VARIOUS FEATURES OF JFET

6.13.1. Peculiarities

(i) It is a voltage controlled device. (V_{GS} controls I_D).
(ii) It is equivalent to a voltage controlled current source.
(iii) The gate-source junction is always reverse biased.
(iv) Very small gate current ($I_G = 0$).
(v) High input resistance and input capacitance.
(vi) It can be used as switch (operated in ohmic region) or as an amplifier (operated in saturation region).
(vii) It can be used as voltage variable resistance VVR.

6.13.2. Advantages of FET

The advantages of FET over BJT are as under :

(i) Very high input resistance.
(ii) Better thermal stability.
(iii) Very small input current, hence, to loading on the input source.
(iv) Produces less noise.
(v) Long life and small size.
(vi) Higher power gain.
(vii) It is relatively immune to radiations.

6.13.3. Disadvantages of JFET

(i) It can not be used as a current amplifier.
(ii) The value of transconductance g_m is small.
(iii) The voltage gain of JFET amplifier is lower than that of the transistor amplifier.
(iv) Special handling precautions are required to be taken.

DO YOU KNOW?

In general, JFETs are more temperature stable than bipolar transistors. Furthermore, JFETs are typically much smaller than bipolar transistors. This size difference makes them particularly suitable for use in ICs, where the size of each component is very critical.

6.13.4. Applications of JFET

(i) JFET can be used as an amplifier.
(ii) JFET can be used as a switch.
(iii) It can be used as analog switch in circuits like sample and hold, amplitude modulation, ADC/DAC (analog to digital or digital to analog) converters.
(iv) As a voltage variable resistor (VVR).
(v) In digital circuits.

6.14. FEW IMPORTANT APPLICATIONS OF FET **(Expected)**

Since a FET has high input impedance, low output impedance and low noise level, therefore, FET is considered far superior to a bipolar transistor. Some of the important applications of FET can be listed as under:

(i) As a buffer amplifier: A buffer amplifier stage is shown in figure 6.17. It is a stage of amplification which isolates the preceding stage from the following stage. Since, FET has a high input impedance (well into the megaohms at low frequencies) and low output impedance (a few hundred ohms), it can act as an excellent buffer amplifier. The high input impedance means high load on the preceding stage. This permits almost the entire output from stage A to appear at the buffer input. The low output impedance means that the buffer (FET) can drive heavy loads. This ensures that the output from buffer reaches at the input of stage B.

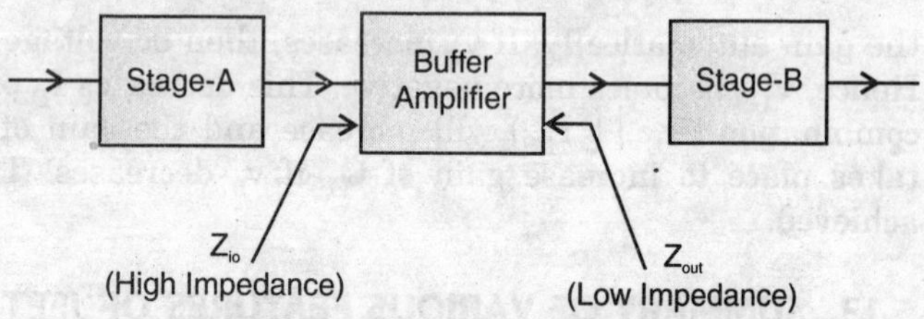

Fig. 6.17. The buffer amplifier stages

(ii) As a low-noise amplifier: Noise is an unwanted disturbance superimposed upon the useful signal. The useful signal contains the required information. As the noise level increases the information becomes inferior.

For instance, the noise in radio receivers produces hissing; severe noise may completely mask the voice or music. Similarly, the noise in television receivers produces white spots (snow) on the picture; a heavy noise may wipe off the picture. Noise is independent of the signal strength because it exists even when the signal is off.

Since FET is an outstanding low-noise device, therefore, it is preferred to apply it near the front end of the receivers and other electronic equipment. Hence, when we use FET amplifier at the front end, we get less amplified noise (disturbance) at the final output.

(iii) As phase-shift oscillator: Although all the oscillators can also work with FETs, however, the high input impedance of FET is especially valuable

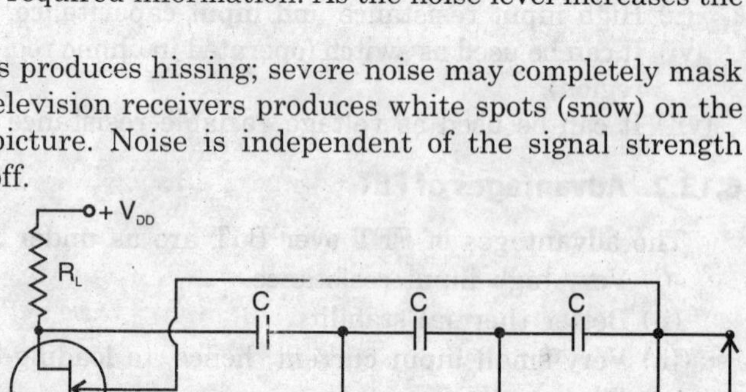

Fig. 6.18. Phase-shift oscillator using N-channel JFET

in a phase-shift oscillator to minimise loading. The circuit diagram of a phase-shift oscillator using N-channel JFET is shown in figure 6.18.

6.15. EXPRESSION FOR PINCH OFF VOLTAGE V_P

Let us now derive the expression for the pinch off voltage V_P or the gate reverse voltage, which removes all the free charges from the channel. This analysis was first made by shockley using the structure of figure 6.19 (a).

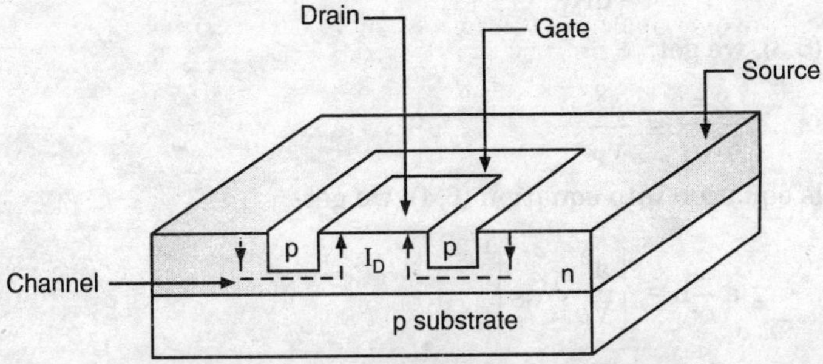

Fig. 6.19. (a) Single-ended geometry JFET

In the structure of figure. 6.19 (a), a slab of n-type semiconductor has been sandwiched between two layers of p type material forming two p-n junctions. Assume that the p-type region is doped with N_A acceptor impurity atoms per m^3 and that the n-type region is doped with N_D donor impurity atoms per m^3 The p-n junctions so formed are **abrupt** junctions. It is known that if $N_A >> N_D$ then $W_P << W_n$ i.e., the width of depletion region is more in the lightly doped n side.

The space-charge width W_n (x) = W(x) at a distance x along the channel in figure 6.19 (b).

$$W(x) = a - b(x) \left\{\frac{2\epsilon}{qN_D}[V_0 - V(x)]\right\}^{1/2} \quad ...(6.1)$$

Where ϵ = dielectric constant of channel material

q = charge on an electron.

V_0 = Junction contact potential at x.

V(x) = Applied potential across the space charge region at x and is a negative number for an applied reverse bias.

a – b (x) = Penetration W(x) of depletion region into channel at a point x along channel of figure 6.19 (b)

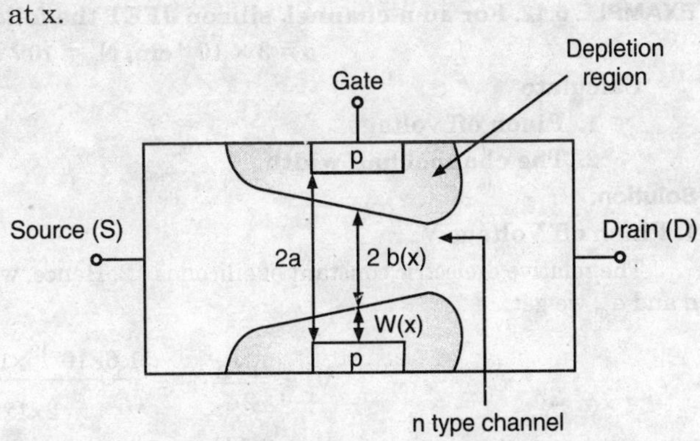

Fig. 6.19. (b) Internal structure of n-channel JFET

If the drain current is zero, then b(x) and V(x) will be independent of x and b(x) = b.

If we substitute b(x) = b = 0 in Equation (6.1) for W(x) to get,

$$W(x) = a = \left\{\frac{2\epsilon}{qN_D}[V_0 - V(x)]\right\}^{1/2} \quad ...(6.2)$$

Now, solve for V assuming $|V_0| << |V|$. To obtain the expression for the pinch off voltage V_P, the diode reverse voltage which removes all the free charges from the channel as follows.

$$|V_P| = \frac{qN_D}{2\epsilon} a^2 \quad \text{...(6.3)}$$

If we substitute V_{GS} for $V_0 - V(x)$ and $(a - b)$ for x in equation (6.1), then, we get,

$$a - b = \left\{\frac{2\epsilon}{qN_D}.V_{GS}\right\}^{1/2} \quad \text{...(6.4)}$$

From equation (6.3), we get,

$$\frac{2\epsilon}{qN_D} = \frac{a^2}{V_P} \quad \text{...(6.5)}$$

Substituting this equation into equation (6.4), we get,

$$a - b = \left\{\frac{a^2}{V_P}.V_{GS}\right\}^{1/2}$$

or $$a - b = a\left[\frac{V_{GS}}{V_P}\right]^{1/2}$$

or $$\frac{a-b}{a} = \left[\frac{V_{GS}}{V_P}\right]^{1/2}$$

or $$\left[1 - \frac{b}{a}\right]^2 = \frac{V_{GS}}{V_P}$$

or $$V_{GS} = V_P\left[1 - \frac{b}{a}\right]^2 \quad \text{...(6.6)}$$

The equation represents the reverse bias across the gate junction (V_{GS}) and is independent of distance along the channel if $I_D = 0$.

EXAMPLE 6.12. For an n-channel, silicon JFET the following information is available.

$a = 3 \times 10^{-4}$ cm, $N_D = 10^{15}$ electrons/cm^3

Calculate

1. **Pinch off voltage**
2. **The channel half width**

Solution:

1. Pinch off Voltage V_P

The relative dielectric constant of silicon is 12. Hence, we have to substitute $\epsilon = 12\,\epsilon_0$. Using the values of q and ϵ_0, we get,

$$V_P = \frac{qN_D}{2\epsilon} a^2 = \frac{1.6\times10^{-19}\times10^{21}\times(3\times10^{-4}\times10^{-2})^2}{2\times12\times(36\pi\times10^9)^{-1}}$$

$$V_P = 6.8\ \text{V}$$

2. The channel half width

$$V_{GS} = V_P\left[1 - \frac{b}{a}\right]^2$$

or $$\left[1 - \frac{b}{a}\right]^2 = \frac{V_{GS}}{V_P} = \frac{1}{2}$$

or $$1-\frac{b}{a} = 0.707$$

or $$b = (1 - 0.707)\, a = 0.293 \times 3 \times 10^{-6}$$

or $$b = 0.879 \text{ m.}$$

Hence, the channel width has been reduce to about one third of its value for $V_{GS} = 0$

EXAMPLE 6.13. An n-channel, JFET has I_{DSS} = 8 mA and V_P = – 4 Volts.
a = 3 × 10^{-4} cm, ND = 10^{15} electrons/cm

1. **If I_D = 3 mA calculate the value of V_{GS}.**
2. **Calculate $V_{DS\,(sat)}$ for I_D = 3 mA.**

Solution: Given: I_{DSS} = 8 mA, V_P = – 4 Volts and I_D = 3 mA.

(a) To calculate V_{GS}

We know that, $$I_D = I_{DSS}\left[1-\frac{V_{GS}}{V_P}\right]^2$$

or $$V_{GS} = V_P\left[1-\sqrt{\frac{I_D}{I_{DSS}}}\right]$$

Substituting the values, we get,

$$V_{GS} = -4\left[1-\sqrt{\frac{3}{8}}\right] = -1.55 \text{ Volts } \textbf{Ans.}$$

(b) To calculate V_{DS} (sat)

$$V_{DS(sat)} = V_{GS} - V_P$$

Substituting the values, we get,

$$V_{DS\,(sat)} = -1.55 + 4 = 2.45 \text{ Volts } \textbf{Ans.}$$

EXAMPLE 6.14. A JFET has a pinch off voltage of – 4 Volts and the saturation current of 9 mA. Calculate the drain current if V_{GS} = – 2 volts.

Solution: It has been given that

$$V_P = -4 \text{ Volts, } I_{DSS} = 9 \text{ mA and } V_{GS} = -2 \text{ Volts.}$$

Therefore, $$I_D = I_{DSS}\left[1-\frac{V_{GS}}{V_P}\right]^2 \qquad ...(i)$$

Substituting the values, we get,

$$I_D = 9 \times 10^{-3}\,[1 - (-2/-4)]^2 = 2.25 \text{ mA. } \textbf{Ans.}$$

EXAMPLE 6.15. A certain JFET has I_{DSS} = 12 mA and pinch off voltage V_P = – 6 Volts. Calculate the value of transconductance for V_{GS} = – 1V

Solution: It has been given that I_{DSS} = 12 ,mA and V_P = – 6 Volts.

We know that, $$g_m = g_{mo}\left[1-\frac{V_{GS}}{V_P}\right] \qquad ...(i)$$

where, $$g_{m0} = \frac{-2I_{DSS}}{V_P} \qquad ...(ii)$$

Substituting the values, we get,

$$g_{m0} = \frac{-2\times12\times10^{-3}}{-6} = 4 \times 10^{-3} = 4 \text{ mA/V} \qquad ...(iii)$$

Substituting this value into equation (i), we get,

$$\text{Transconductance, } g_m = 4 \times 10^{-3}\left[1-\left(\frac{-1}{-6}\right)\right] = 3.33 \times 10^{-3} \text{ S or } 3.33 \text{ mS } \textbf{Ans.}$$

Example 6.16. For a JFET the data is as follows:

$$I_{DSS} = 10 \text{ mA}, V_P = -5\text{V}, g_{m0} = 2 \text{ mS}.$$

Determine the transconductance and drain current of the JFET for $V_{GS} = -2.5$ Volt.

Solution: (i) To calculate transconductance g_m

$$g_m = g_{mo}\left[1-\frac{V_{GS}}{V_P}\right] = \frac{-2I_{DSS}}{V_P}\left[1-\frac{V_{GS}}{V_P}\right]$$

Substituting the values we get,

$$g_{m0} = \frac{-2\times12\times10^{-3}}{-5}\left[1-\left(\frac{-2.5}{-5}\right)\right]$$

Therefore, $g_m = 2 \times 10^{-3}$ S or 2 mS **Ans.**

(ii) To calculate the drain current I_D:

$$I_D = I_{DSS}\left[1-\frac{V_{GS}}{V_P}\right]^2 = 10\,[1-(-2.5/-5)]^2 = 2.5 \text{ mA } \textbf{Ans.}$$

6.16. AN OVERVIEW OF DIFFERENT TYPES OF FETs

The different types of FET may be listed as under:

(i) Junction FET

(ii) Metal oxide semiconductor (MOS) FET

(iii) MESFET

In this article, let us discuss the peculiarities and applications of different types of FETs.

6.16.1. MESFET

1. Definition

The MESFET is nothing but a junction field effect transistor fabricated in Gallium Arsenide (GaAs) which employs a metal semiconductor gate region (Schottky diode). The principle of operation and characteristics of the MESFET are similar to those for the silicon JFET discussed in earlier sections.

2. Construction

The simplified cross section of MESFET is shown in figure. 6.20.

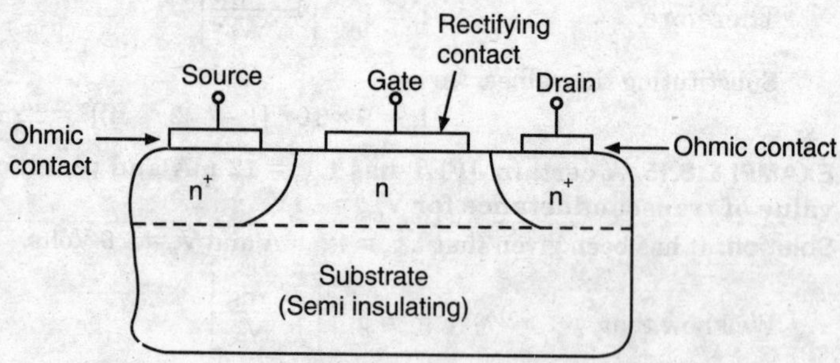

Fig 6.20. Simplified cross section of an N channel MESFET

A thin epitaxial layer of GaAs is used for the active region. The substrate is a very high resistivity GaAs material. It is referred to as semi-insulating substrate. Electron mobility of GaAs is 5 to 10 times higher than that in silicon. Therefore MESFET can be operated at frequency much higher than that obtainable using silicon JFET. This is due to the reduced transit time and fast response. The parasitic capacitance is also reduced.

Since the hole-mobility in GaAs is less than that in silicon the n-channel MESFETs are preferred to p-channel MESFET. The frequency range of operation of MESFETs was 1 to 10 GHz initially. MESFETs can therefore be used in microwave circuits which operate at a frequency higher than 1 GHz. From 1984 onwards, high-speed logic circuits using MESFET's are being produced commercially. These high speed circuits are designed to be compatible with the high speed bipolar logic family called emitter- coupled logic (ECL). In the MESFET of figure. 6.20, a reverse bias gate to source voltage needs to be applied. It induces the depletion region under the metal gate. The voltage applied to the gate modulates the channel transconductance as in case of JFET. The MESFET shown in figure 6.20 is a depletion type MESFET.

3. Enhancement Type MESFET

This is an another type of MESFET The channel width of this device is zero even at $V_{GS} = 0$ V as shown in figure. 6.21 (a). In order to open the channel, we have to reduce the depletion regions and for that a forward biased voltage should be applied to the gate semiconductor junction.

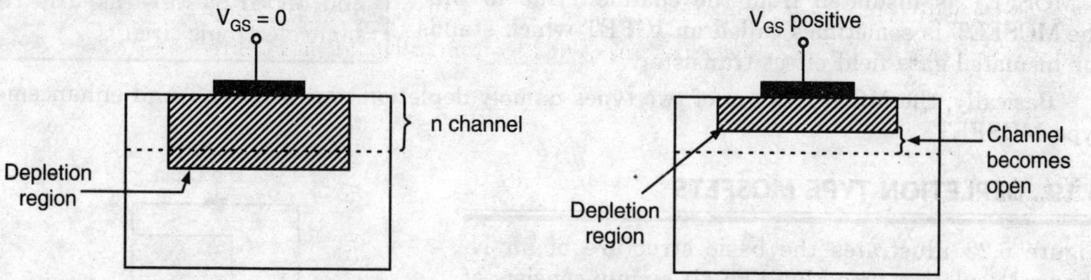

Fig. 6.21. Effect of V_{GS} on enhancement MESFET

The effect of applying a positive V_{GS} is shown in figure 6.21 (b). The width of depletion region is reduced and the channel is open for conduction. This is an n-channel enhancement MESFET but it is possible to fabricate the p-channel enhancement MESFET also. The output voltage swing of these devices is very small.

4. Peculiarities of MESFET

The material used for construction is Gallium Arsenide GaAs. Construction and characteristics same as those of silicon JFET. The gate junction is metal semiconductor type. Operating frequency range, 1 to 10 GHz, due to use of GaAs. Voltage controlled device. Reduced parasitic capacitance.

5. Applications of MESFET

(i) In microwave circuits operating in the range of 1 to 10 GHz.

(ii) Compatible with high speed bipolar logic family called ECL.

6.17. METAL OXIDE SEMICONDUCTOR (MOS) FET/METAL INSULATOR SEMICONDUCTOR (MIS) FET

The metal-insulator-semiconductor field effect transistor (MISFET) is one of the most widely used electronic devices, particularly in digital circuits, due to its relatively small size, thousand of devices can be fabricated in a single integrated circuits (IC). As it constructed with the gate terminal insulated from the channel, it may be referred to generally as an insulated gate field effect transistor (IGFET). However since most of such devices are fabricated using silicon for the semiconductor, SiO_2 for the insulator and metal or heavily doped polysilicon for the gate electrode, the term metal-oxide semiconductor field-effect- transistor (MOSFET) or simply MOS is commonly used.

MISFET is the more general term, where the insulator is not necessarily SiO_2 and the semiconductor is not necessarily silicon (Si). In this section, we will discuss about MOSFET although the same physics applies to the MISFETs. Similar to JFET, a MOSFET is also a three terminal (source, gate and drain) device and drain current in it is also controlled by gate bias. The operation of MOSFET is similar to that of JFET. It can be employed in any of the circuits covered for the JFET and so, all the equations apply equally well to the MOSFET and JFET in amplifier connections. However MOSFET has lower capacitance and input impedance much more than that of a JFET owing to small leakage current. In case of a MOSFET the positive voltage may be applied to the gate and still the gate current remains zero.

6.18. MOSFETS

The MOSFET is an abbreviation for metal-oxide semiconductor field-effect transistor Like JFET, it has a source, gate and drain. However, unlike JFET, the gate of a MOSFET is insulated from the channel. Due to this, the MOSFET is sometimes called an IGFET which stands for insulated-gate field effect transistor.

> **DO YOU KNOW?**
> In many cases, bipolar transistors and MOSFETs are used in the same electronic circuit.

Basically, the MOSFETs are of two types namely depletion type MOSFET and enhancement type MOSFET.

6.19. DEPLETION-TYPE MOSFETS*

Figure 6.22 illustrates the basic structure of an N-channel depletion type MOSFET. It simply consists of a conducting bar of N-type material with an insulated gate on the left and P-region on the right. Free electrons can flow from source to drain through the N-type material. The P-region is called substrate (or body). It physically reduces the conducting path a narrow channel. A thin layer of silicon dioxide SiO_2 is deposited on the left side of the channel. This layer insulates the gate from the channel. Because of this, a negligible gate current flows even when this gate voltage is positive. It will be interesting to know that a PN junction, which exists in a JFET, has been eliminated in the MOSFET.

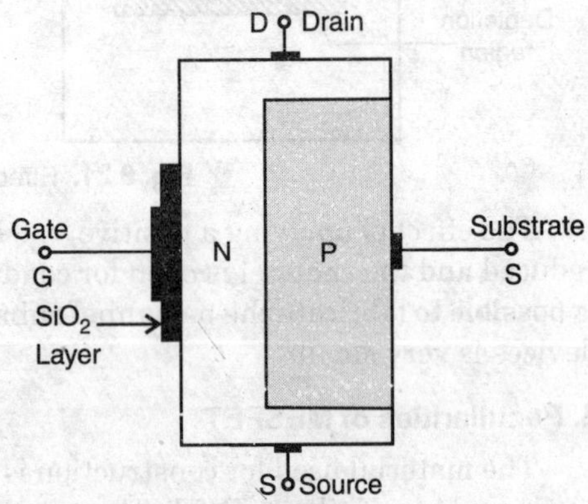

Fig. 6.22. Basic structure of an N-channel depletion-type MOSFET

It may also be observed that the basic construction of a depletion-type P-channel MOSFET is similar to that of N-channel except that the conducting bar is of P-type material and the substrate is of N-type material.

> **DO YOU KNOW?**
> Like a JFET, a depletion-mode MOSFET is considered a normally on device. This is because both devices have drain current when V_{GS} = 0 V. Recall that for a JFET, I_{DSS} is the maximum possible drain current. With a depletion-mode MOSFET, the drain current can exceed I_{DSS} if the gate voltage is of the correct polarity to increase the number of charge carriers in the channel. For an n-channel D-MOSFET, I_D is greater than I_{DSS} when V_{GS} is positive.

6.19.1. Working Operation of a Depletion-Type MOSFET

The depletion-type MOSFET can be operated in two different modes as given below:

(i) *Depletion mode.* The device operates in this mode, when the gate voltage is negative.

(ii) *Enhancement mode.* The device operates in this mode, when the gate voltage is positive.

Since the depletion-type MOSFET can be operated in either depletion or enhancement mode, therefore, this device is commonly known as depletion-enhancement (DE) type MOSFET. The working of a MOSFET may be explained easily, if we visualize the entire structure of the device as a parallel plate capacitor. One of the plates is formed by the gate and the other by the semiconductor channel. The plates are separated by a dielectric (SiO_2) layer. We know that

* Write notes on the following:
(i) depletion type MOSFET (ii) pinch-off voltage (iii) transfer characteristics of JFET (iv) Transconductance

if one plate of a capacitor is made negative, it induces a positive charge on the opposite plate and vice versa. This principle is used below in explaining the working of MOSFETs in the depletion and enhancement modes.

(i) Depletion mode: Figure 6.23 (a) shows a MOSFET with a negative gate-to-source voltage. The negative voltage, on the gate, induces a positive charge in the channel. Because of this, free electrons in the vicinity of positive charge are repelled away in the channel. As a result of this, the channel is depleted of free electrons. This reduces the number of free electrons (which constitute the drain current) passing through the channel. Thus, as the value of negative gate-to-source voltage is increased, the value of drain voltage, called $V_{GS\,(off)}$, the channel is totally depleted of free electrons and, therefore, the drain current reduces to zero. Thus, with the negative gate voltage, the operation of MOSFET is similar to that of a JFET.

It is evident from the above discussion that negative gate voltage depletes the channel of free electrons. It is due to this fact that the working of a MOSFET, with a negative gate voltage, is called depletion mode.

(ii) Enhancement mode: Figure 6.23 (*b*) shows a MOSFET with a positive gate-to-source voltage. The positive gate voltage increases the number of free electrons passing through the channel. The greater the gate voltage, greater is the number of free electrons passing through the channel. This increases *i.e.*, enhances the conducting of the channel. Because of this fact, positive gate operation is called **enhancement mode**.

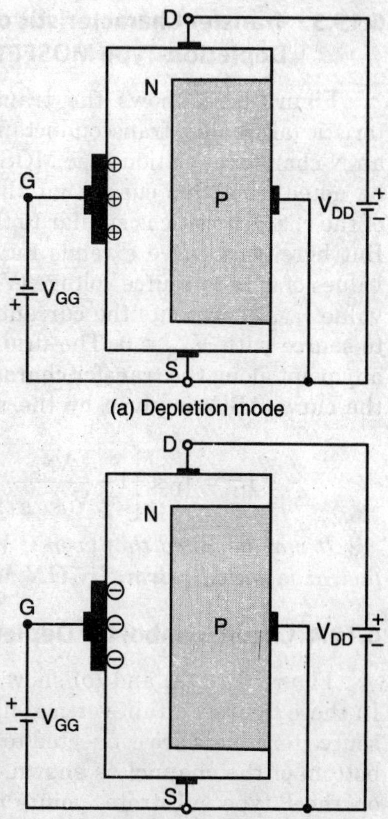

(b) Enhancement mode

Fig. 6.23. Working of a depletion-type MOSFET

*It will be interesting to know that depletion-type MOSFET can conduct even if the gate-to-source voltage (V_{GS}) is zero. Because of this, it is commonly known as **Normally-ON MOSFET**.*

6.19.2. Drain Characteristic of Depletion-Type MOSFET

Figure 6.24 shows the drain characteristic for the N-channel depletion-type MOSFET in the common source configuration. These curves are plotted for both negative and positive values of gate-to-source voltage (V_{GS}). The curves shown above the curve for $V_{GS} = 0$ have a positive zero whereas those below it have a negative value of V_{GS}. When V_{GS} is zero and negative, the MOSFET operates in the depletion-mode. On the other hand, if V_{GS} is zero and positive, the MOSFET operates in the enhancement-mode.

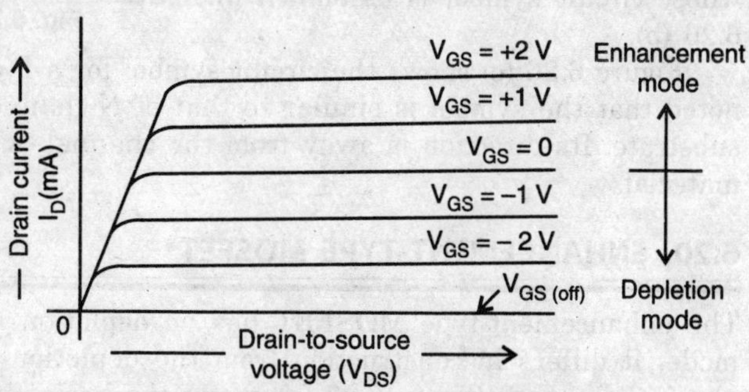

Fig. 6.24. Drain characteristic of N-channel depletion-type MOSFET

It may be noted that the drain characteristics of depletion-type MOSFETs are similar to that of JFET. The only difference is that JFET does not operate for positive values of gate-to-source voltage (V_{GS}).

6.19.3. Transfer Characteristic of Depletion-Type MOSFET

Figure 6.25 shows the transfer characteristic (also called transconductance curve) for an N-channel depletion-type MOSFET. It may be noted from this curve that the region AB of the characteristic is similar to that of JFET. But here, this curve extends for the positive values of gate-to-source voltage (V_{GS}) also. The value I_{DSS} represents the current from drain-to-source with $V_{GS} = 0$. The drain current at any point along the transfer characteristic (i.e., the curve ABC) is given by the relation,

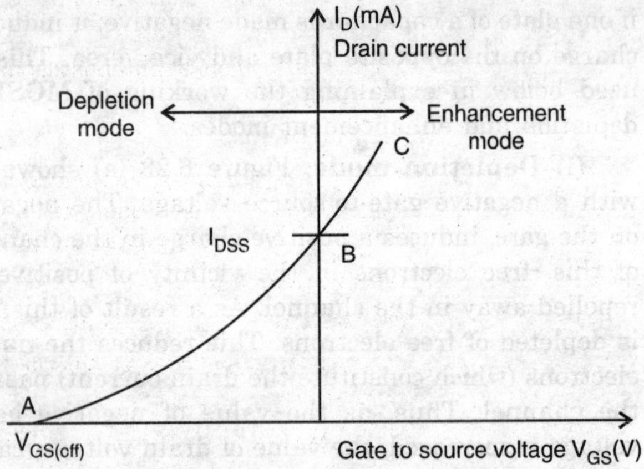

Fig. 6.25. Transfer characteristics of a N-channel depletion-type MOSFET

$$I_D = I_{DSS}\left[1 - \frac{V_{GS}}{V_{GS(off)}}\right]^2$$

*It may be noted that even if $V_{GS} = 0$, the device has a drain current equal to I_{DSS}. Due to this fact, it is called **normally-ON MOSFET.***

6.19.4. Circuit Symbol for Depletion-Type MOSFET

Figure 6.26 (a) and (b) show the circuit symbols for the N-channel depletion-type MOSFET. In these figures, a thin vertical line (just right to the gate) represents the channel. The drain and source terminals are connected to the top and bottom of the channel as shown. The arrow, on the P-type substrate, points towards the channel. This indicates that the channel is N-type.

In some MOSFETs, a connection from the substrate is also taken out. Such MOSFETs have 4-terminals as indicated in the figure 6.26 (a). However, in most of the MOSFETs, the substrate is internally connected to the source. This results in a three terminal device, whose circuit symbol is as shown in figure 6.26 (b).

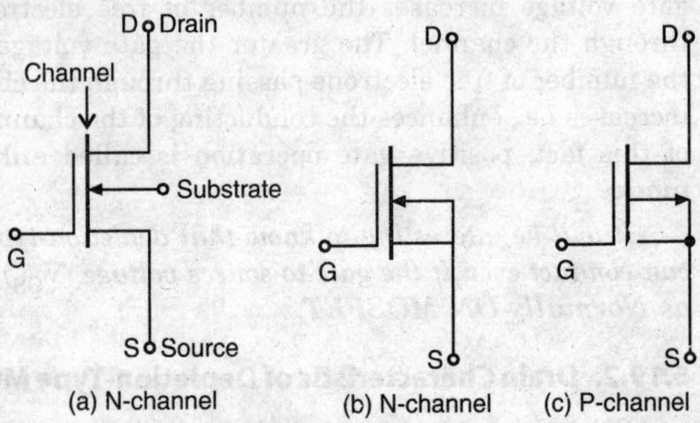

Fig. 6.26. Circuit symbols for depletion type MOSFETs

Figure 6.26 (c) shows the circuit symbol for a P-channel depletion type MOSFET. It may be noted that the symbol is similar to that of N-channel, except the direction of the arrow on the substrate. Its direction is away from the channel, which indicates that the channel is of P-type material.

6.20. ENHANCEMENT-TYPE MOSFET*

The enhancement-type MOSFET has no depletion mode and it operates only in enhancement mode. It differs in construction from the depletion-type MOSFET in the sense that it has no physical channel. Figure 6.27 (a) shows the basic structure of the N-channel enhancement-type MOSFET. It may be noted that the P-type substrate extends the silicon dioxide (SiO_2) layer completely.

* What is the significant difference between the construction of an enhancement type MOSFET and depletion type MOSFET ? Explain with suitable diagram ?

(U.P. Tech. Second Sem. Exam. 2005-06) (10 marks)

> **DO YOU KNOW?**
> With the E-MOSFET, V_{GS} has to be greater than $V_{GS(th)}$ to get any drain current at all.

Figure 6.27 (b) shows the normal biasing polarities for the N-channel enhancement-type MOSFET. It must be remembered that this MOSFET is always operated with the positive gate-to-source voltage (V_{GS}). When the gate-to-source voltage is zero, the V_{DD} supply tries to force free electrons from source-to-drain. But the presence of P-region does not permit the electrons to pass through it. Thus, there is no drain current for $V_{GS} = 0$. Because of this fact, the enhancement type MOSFET is also called **normally-OFF MOSFET**.

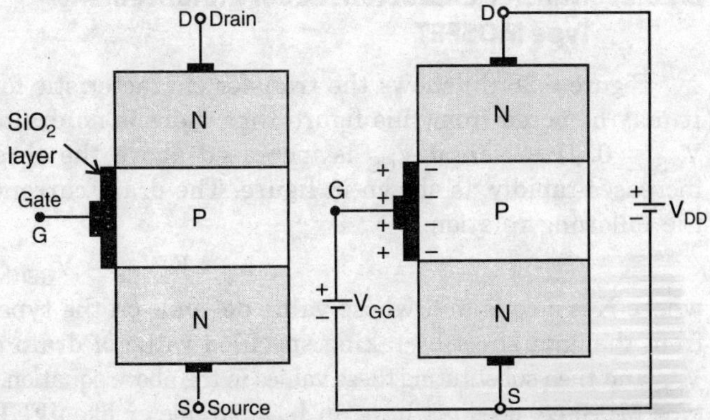

(a) Basic construction, (b) Formation of inversion layer.

Fig. 6.27. Enhancement type MOSFET

Now, if some positive voltage is applied at the gate, it induces a negative charge in the P-type substrate just adjacent to the silicon dioxide (SiO_2) layer. The induced negative charge is produced by attracting the free electrons from the source. When the gate is positive enough, it can attract a number of free electrons. This forms a thin layer of electrons, which stretches from source to drain. This effect is equivalent to producing a thin layer of N-type channel in the P-type substrate. This layer of free electrons is called **N-type inversion layer.**

The minimum gate-to-source voltage (V_{GS}), which produces inversion layer, is called **threshold voltage** and is designated by the symbol $V_{GS(th)}$. When the voltage V_{GS} is less than $V_{GS(th)}$, no current flows from drain to source. However, when the voltage V_{GS} is greater than $V_{GS(th)}$, the inversion layer connects the drain and source and we get a significant value of current.

6.20.1. Drain Characteristics for Enhancement-Type MOSFET

Figure 6.28 (a) shows the drain characteristics for N-channel enhancement-type MOSFET. It may be noted from this figure, that the gate-to-source voltage (V_{GS}) is less than threshold voltage, $V_{GS(th)}$, there is no drain current. However, in actual practice, an extremely small value of drain current does flow through the MOSFET. This current flow is due to the pressure of thermally generated electrons in the P-type substrate. When the value of V_{GS} is kept above $V_{GS(th)}$, a significant amount of drain current flows.

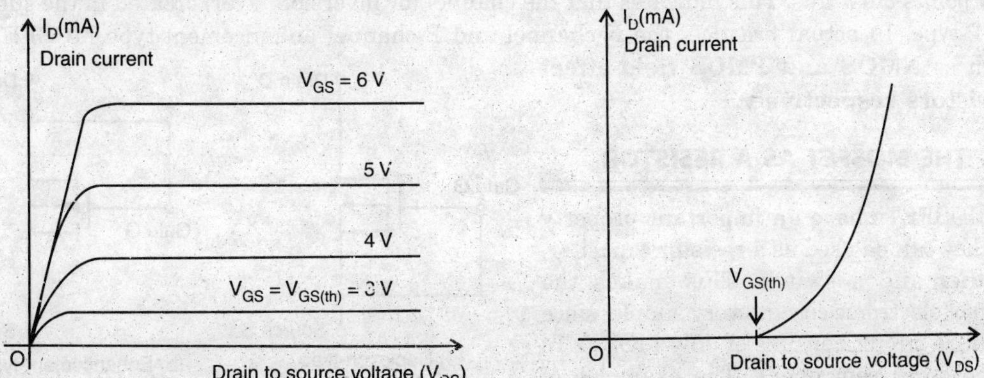

(a) Drain characteristic (b) Transfer characteristic

Fig. 6.28. Characteristics for N-channel enhancement-type MOSFET

The values of drain current increases with the increase in gate-to-source voltage. It is because of the fact that the width of inversion layer widens for increased values of V_{GS} and, therefore, allows more number of free electrons to pass through it. The drain current reaches its saturation value above a certain value of drain-to-source voltage (V_{DS}).

DO YOU KNOW?

The depletion-mode MOSFET was part of the evolution toward the enhancement-mode MOSFET, abbreviated E-MOSFET. Without the E-MOSFET, the personal computers that are now so widespread would not exist.

6.20.2. Transfer Characteristic for Enhancement-Type MOSFET

Figure 6.28 (b) shows the transfer characteristic for N-channel enhancement type MOSFET. It may be noted from this figure that there is no drain current when the gate-to-source voltage, V_{GS} = 0. However, if V_{GS} is increased above the threshold voltage, $V_{GS(th)}$, the drain current increases rapidly as shown in figure. The drain current at any point along the curve is given by the following relation:

$$I_D = K[V_{GS} - V_{GS(th)}]^2$$

where K is a constant, whose value depends on the type of MOSFET. Its value can be determined from the data sheet by taking specified value of drain current called $I_{D(ON)}$ at the given value of V_{GS} and then substituting these values in the above equation. Incidently, it may be noted that enhancement-type MOSFET does not have an I_{DSS} parameter like JFET and depletion-type MOSFET.

6.20.3. Circuit Symbol for Enhancement-Type MOSFET

Figure 6.29 (a) shows the circuit symbol for N-channel enhancement-type MOSFET. In this figure, the broken line indicates that there is no conducting channel between drain and source, when V_{GS} = 0. Because of this fact, this device is also known as **"Normally-OFF MOSFET"**. The drain and source terminals are shown at the top and bottom end of the broken line. The substrate is internally connected to the source as shown. The arrow points in the direction of channel (or inversion layer), which is created when V_{GS} is increased above $V_{GS(th)}$.

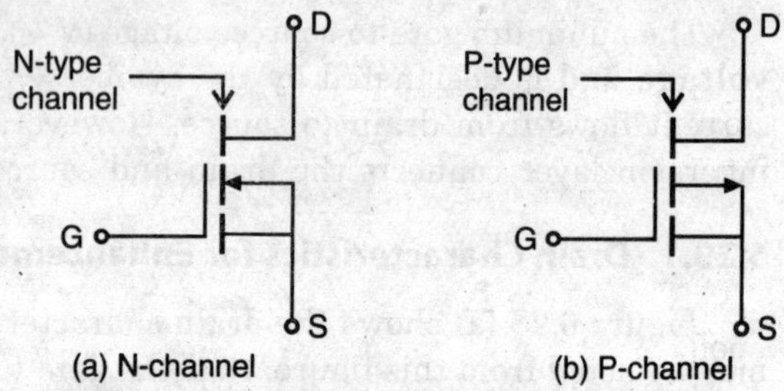

Fig. 6.29. Circuit symbols for enhancement-type MOSFET

Figure 6.29 (b) shows the circuit symbol for the P-channel enhancement-type MOSFET. This symbol is similar to N-channel except that the arrow points outwards. This indicates that the channel (or inversion layer) created in the substrate is of P-type. In actual practice, the N-channel and P-channel enhancement-type MOSFETs are known as **NMOS and PMOS field-effect transistors respectively.**

6.21. THE MOSFET AS A RESISTOR

The MOSFETs have an important property that they can be used as a resistor capacitor, amplifier and a switch. This makes the design of electronic circuits very simple, since the entire circuit consists of only MOSFETs and no other component. The examples of such circuits are microprocessors and m circuits.

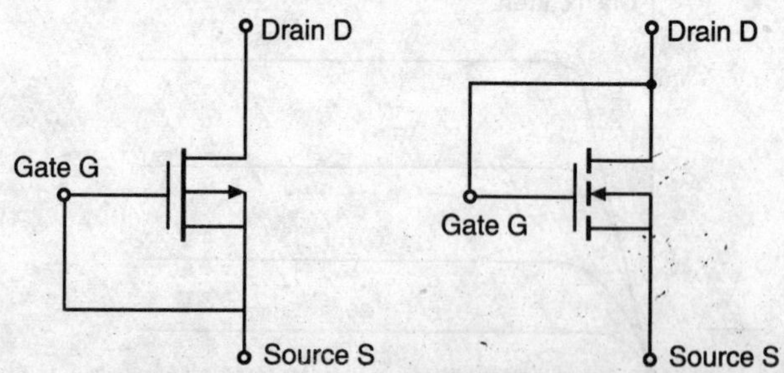

Fig. 6.30 MOSFET as a resistor

The depletion type MOSFET may be connected as a resistor, if its gate is connected to the source so that the gate-to-source voltage (V_{GS}) is zero, as shown in figure 6.30 (a).

The enhancement-type MOSFET may be connected as a resistor, if its gate is connected to drain so that $V_{GS} = V_{DS}$ as shown in figure 6.30 (b).

6.22. ADVANTAGES OF N-CHANNEL MOSFETS OVER P-CHANNEL

These days N-channel MOSFETs have become much popular than P-channel for fabricating MOS circuits and systems. The P-channel MOSFETs find their use only in complementary metal-oxide semiconductor (CMOS) integrated circuits. The main advantage of N-channel MOSFET, over P-channel, is due to the fact that charge carriers in N-channel devices are the electrons, which have a mobility of about 1300 cm^2/V.S. On the other hand, the charge carriers in P-channel devices are the holes which have a mobility of about 500 cm^2/V.S. Since the current, in a semiconductor, is directly proportional to the mobility, therefore, the current in the N-channel MOSFET is more than two times that of P-channel for the same dimension.

The ON resistance of the N-channel MOSFET is one-third of that for the P-channel. It means that in order to achieve the same value of current and ON resistance, the P-channel MOSFET requires three times the area of an equivalent N-channel MOSFET. Thus, electronic currents using N-channel MOSFETs are much smaller in size than those of P-channel. This results in higher packing density.

6.23. COMPARISON OF JFET AND MOSFET

Table 6.5. Comparison of JFET and MOSFET

S.No.	JFET	MOSFET
1.	JFET can be of two types, P-channel and N-channel JFETs.	MOSFETs can be of depletion type or enhancement type.
2.	The symbols of JFETs are as under : 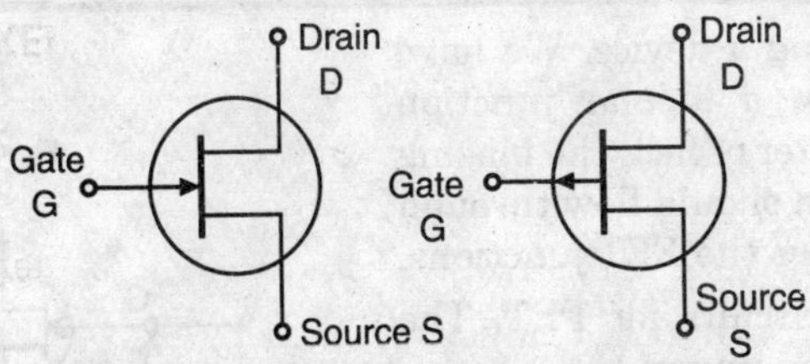 N-channel JFET P-channel JFET	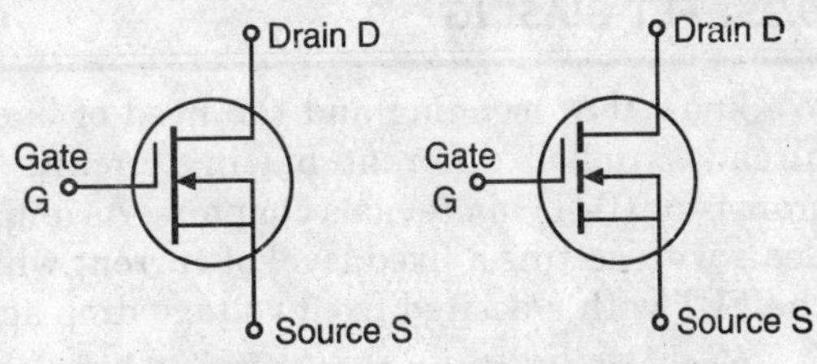 N-channel depletion MOSFET N-channel enhancement MOSFET
3.	JFETs do not have the insulated gate.	MOSFETs have the insulated gate structure.
4.	Input impedance is lower than that of the MOSFETs.	Input impedance is higher than that of JFET due to the insulated gate structure.
5.	Drain resistance is lower than that of a MOSFET.	Drain resistance is higher than that of a JFET.
6.	The input resistance r_{gs} and the feedback resistance r_{gd} are very much smaller than that of MOSFET.	r_{gs} and r_{gd} are much larger than that of a JFET.
7.	Biasing methods used are self bias, fixed bias and voltage divider bias.	Biasing circuits of depletion MOSFET are same as those used for JFET but biasing circuits for enhancement MOSFET are, feedback biasing and voltage divider biasing.

6.24. COMPARISON OF JFET AND DMOSFET

In this section, let us compare JFET and DMOSFET in the form of a Table 6.6

Table 6.6

S.No.	JFET	DMOSFET
1.	Symbol of N-channel JFET. 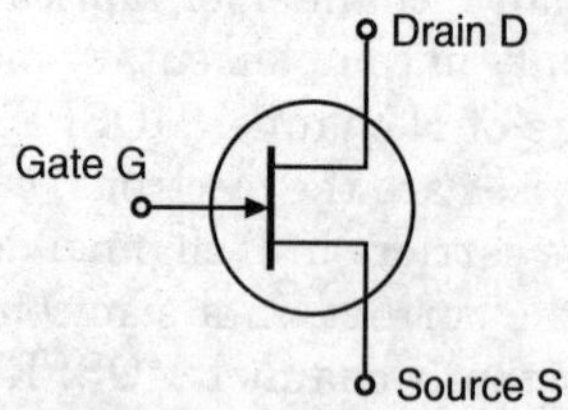	Symbol of N-channel depletion-type MOSFET. 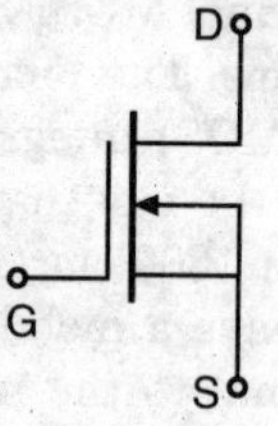
2.	The oxide layer is absent hence gate is not isolated from the channel.	The insulating layer of SiO_2 is present between the gate and channel.
3.	Drain current decreases with increase in the negative V_{GS} voltage.	Drain current decreases as negative V_{GS} is increased.
4.	Reduction in drain current is due to the narrowing of the channel width with increase in $-V_{GS}$.	Reduction in I_D is due to the recombination process taking place under the influence of negative V_{GS}.
5.	Drain characteristics are drawn only for $V_{GS} \leq 0$ volts.	Drain characteristics are same as those of JFET except for the part corresponding to positive V_{GS}.
6.	Input impedance is high	Input impedance is still higher due to presence of SiO_2 layer.

6.25. FET BIASING

We know the meaning and the need of biasing a device. We have already studied different biasing circuits for a bipolar junction transistor (BJT) in previous chapters. As a matter of fact, the biasing deals with setting a fixed level of current which should flow through the FET with a desired fixed voltage drop across the FET junctions.

Now, let us study the different biasing circuits for FET. The commonly used biasing circuits for FET are as under:

(i) Fixed-bias circuit.

(ii) Self-bias circuit.

(iii) Voltage divider bias circuit.

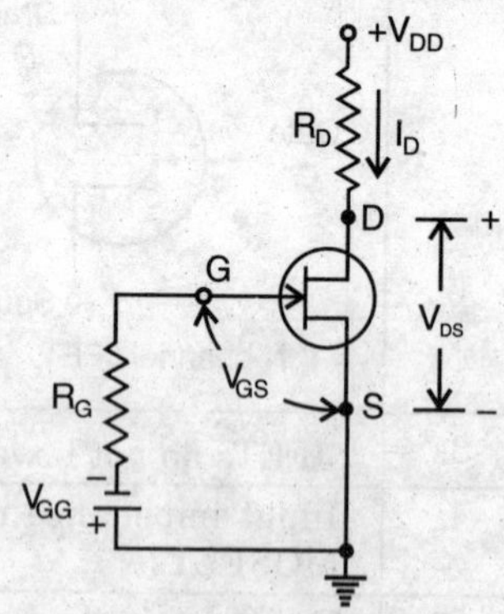

Fig. 6.31. Fixed-bias circuit for JFET

Whatever be the biasing method, the gate-source junction of a JFET will always be reverse-biased. Even when the gate and the source terminals are shorted, the drain current (I_D) will cause the junction to be reverse biased. As a matter of fact, the gate-source junction of a JFET must never be allowed to become forward-biased. It is because of the reason that the gate material is not constructed to handle any significant amount of current. If the junction is allowed to become forward-biased, current will start to flow in the gate circuit. Under most circumstances, this current will destroy the JFET.

> **DO YOU KNOW?**
>
> When biased in the active region, the JFET is equivalent to a current source.

The general relationships which can be applied to the d.c. analysis of all FET amplifiers are:

$$I_G \cong 0 \quad ...(6.7)$$

and $$I_D = I_S \quad ...(6.8)$$

6.26. FIXED-BIAS CIRCUIT

Fixed-bias circuit for JFET is also known as gate-bias circuit. Figure 6.32 shows the fixed-bias circuit or gate-bias circuit for the N-channel JFET. In this circuit, the gate supply voltage ($-V_{GG}$) is used to ensure that the gate-source junction is reverse-biased. Since, there is no gate current, therefore, there is no voltage drop across resistor R_G.

It will be interesting to know that the fixed bias circuit cannot provide a stable Q-point from one JFET to another. The reason for this is that for a given JFET, there is a range of values for $V_{GS\,(off)}$ and I_{DSS} which can lead to a major problem with the fixed bias circuit.

It can also be noted that even though there is no gate current or any voltage drop across R_G, but still the circuit contains R_G. Actually, R_G is used for ac operation purpose. Thus R_G has little to do with the d.c. operation of the amplifier, but it is a vital part of the ac operation of the circuit.

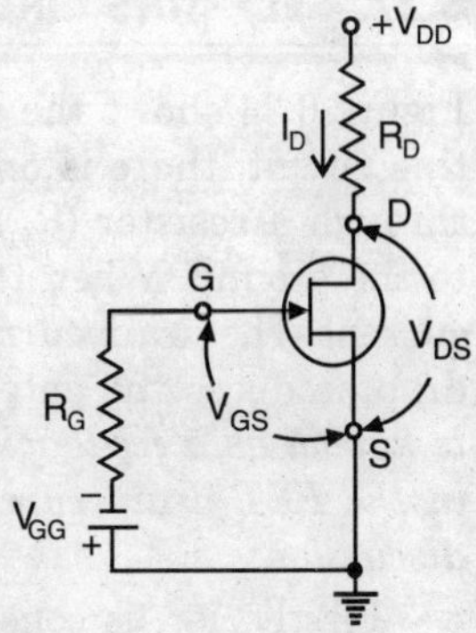

Fig. 6.32.

We know that for dc analysis,

$$I_G = 0$$

Applying KVL to the input circuit or gate circuit, we have

$$V_{GG} + V_{GS} = 0$$

or $$V_{GS} = -V_{GG} \quad ..(6.9)$$

The expression for the drain current I_D is as under:

$$I_D = I_{DSS}\left[1 - \frac{V_{GS}}{V_P}\right]^2 \quad ...(6.10)$$

This is nothing else but the Shockley's equation as discussed earlier.

The drain-to-source voltage can be obtained by applying the KVL to the output circuit as under :

$$V_{DD} - I_D R_D - V_{DS} = 0$$

or $$V_{DS} = V_{DD} - I_D R_D \quad ...(6.11)$$

EXAMPLE 6.17 Figure 6.33 shows the fixed-bias circuit for JFET. Find the values of V_{GSQ}, I_{DQ} and V_{DS}. Assume I_{DSS} = 8 mA and V_P = – 8 Volts.

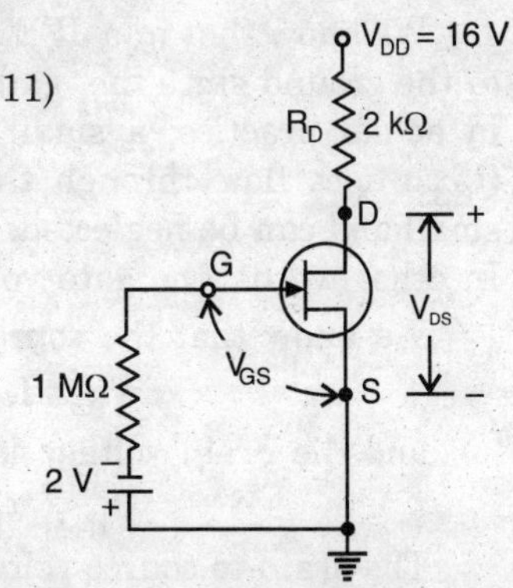

Fig. 6.33.

Solution:

Applying KVL to the gate circuit, we have

$$V_{GG} + I_G R_G + V_{GS} = 0$$

Substituting all the given values, we get

$$2 + (I_G \times 1 \times 10^6) + V_{GS} = 0$$

But for dc analysis, we have

$$I_G = 0$$

Therefore, $$2 + V_{GS} = 0$$

$$V_{GS} = -2 \text{ Volts} \quad \textbf{Ans.}$$

Now, using Shockley's equation, we get

$$I_{DQ} = I_{DSS}\left[1-\frac{V_{GS}}{V_P}\right]^2$$

Substituting values, we get

$$I_{DQ} = 8\times10^{-3}\left[1-\frac{-2}{-8}\right]^2$$

Solving, we get $\qquad I_{DQ} = 4.5 \text{ mA}$ **Ans.**

For drain-to-source voltage V_{DS}, applying KVL to the drain circuit, we have

$$V_{DD} = I_D R_D + V_{DS}$$

or $\qquad V_{DS} = V_{DD} - I_D R_D$

Substituting the values we get

$$V_{DS} = 16 - (4.5 \times 2) = 16 - 9 = 7 \text{ Volt}$$ **Ans.**

6.27. SELF-BIAS CIRCUIT

Figure 6.34 shows the self-bias circuit for an N-channel junction field-effect transistor (JFET). In this circuit, there is only one drain supply and no gate supply. The gate terminal is connected through a resistor (R_G) to the ground. The source terminal is connected through a resistor (R_S) to the ground. When the drain voltage is applied, a drain current flows even if there is no gate current. The drain current produces a voltage drop across resistor R_S (equal to $I_D.R_S$). The voltage drop produces the gate-to-source reverse voltage required for an FET operation. The resistor R_S *is known as a feedback resistor.* Its function is to prevent any *variation in the FET drain current, which can be understood from the following discussion:*

Firstly, let us consider an increase in the drain current, which will increase the voltage drop across the resistor (R_S). The increased voltage drop increases the reverse gate-to-source voltage, which decreases the effective width of the channel. It reduces the value of drain current. Now, if the drain current decreases, then the reverse action takes place, i.e., the reduced drain current decreases the gate-to-source voltage, which in turn increases the effective width of the channel and thereby increasing the value of drain current.

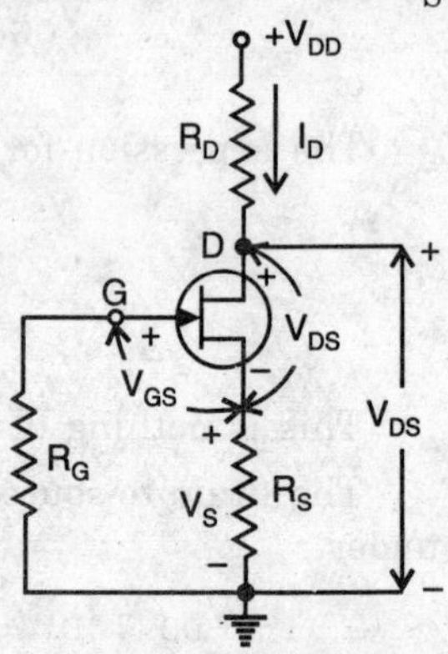

Fig. 6.34. The self-bias circuit

We know that in a JFET, no current flows through the gate terminal to the ground since the gate-source junction is reverse-biased. However, in actual practice, a small current known as reverse leakage current (I_{GSS}) does flow through the gate terminal. Usually, its value is very small and can be neglected. Therefore, we shall assume that the gate-terminal is at zero voltage. In other words, the gate voltage with respect to ground, i.e., $V_G = 0$.

We know that the source voltage with respect to ground is given by

$$V_S = I_D.R_S \qquad ...(6.12)$$

and the drain voltage is given by

$$V_D = V_{DD} - I_D.R_D \qquad ...(6.13)$$

The drain-to-source voltage (V_{DS}) is equal to the difference between the drain voltage (V_D) and the source voltage (V_S)

i.e., $\qquad V_{DS} = V_D - V_S \qquad ...(6.14)$

or $\qquad V_{DS} = (V_{DD} - I_D.R_D) - I_D.R_S = V_{DD} - I_D\ (R_D + R_S) \qquad ...(6.15)$

The gate-to-source voltage (V_{GS}) is equal to the difference between the gate voltage (V_G) and the source voltage (V_S)

i.e., $V_{GS} = V_G - V_S = 0 - I_D.R_S = -I_D.R_S$ ...(6.16)

It is evident from this expression that the gate-to-source voltage is equal to the negative of the voltage across the source resistor. Hence, it is obvious that greater the value of drain current, more negative will be the gate-to-source voltage.

Rearranging equation (6.16), we get

$$I_D = -\frac{V_{GS}}{R_S} \quad ...(6.17)$$

Now, if we plot a graph using the value of drain current obtained from equation (6.17) for a constant value resistor (R_S) against the gate-to-source voltage, we get a straight line as shown in figure 6.35. This straight line is known as a **self-bias line.**

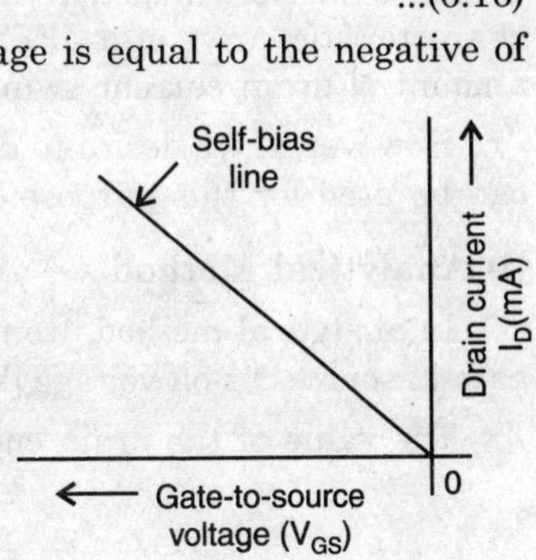

Fig. 6.35.

EXAMPLE 6.18 **Determine the values of drain-to-source voltage (V_{DS}) and the gate-to-source voltage (V_{GS}) for the circuit shown in figure 6.36. Given I_D = 5 mA.**

Solution: Given that $I_D = 5 \text{ mA} = 5 \times 10^{-3}$ A ; V_{DD} = 10 Volts

$R_D = 1 \text{ k}\Omega = 1 \times 10^3 \, \Omega$ and $R_S = 500 \, \Omega$

We know that the source voltage is given by

$$V_S = I_D.R_S$$

or $$V_S = 5 \times 10^{-3} \times 500 = 2.5 \text{ Volt}$$

and the drain voltage V_D is given by

$$V_{DD} = V_D + I_D.R_D$$

or $$V_D = V_{DD} - I_DR_D$$

or $$V_D = 10 - (5 \times 10^{-3}) \times (1 \times 10^3)$$

or $$V_D = 10 - 5 = 5 \text{ Volt}$$

Therefore, the drain-to-source voltage will be

$$V_{DS} = V_D - V_S = 5 - 2.5 = 2.5 \text{ Volt}$$

Also, the gate-to-source voltage is given by

$$V_{GS} = -V_S = -2.5 \text{ Volt} \quad \textbf{Ans.}$$

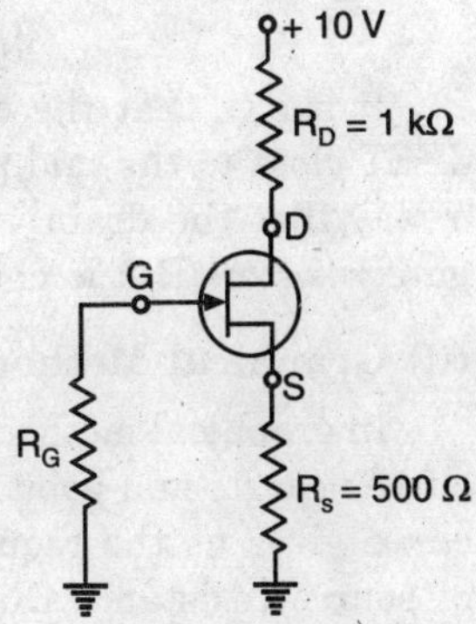

Fig. 6.36. Circuit for Example 6.18

6.28. SETTING OF A Q-POINT

The Q-point (i.e., quiescent or operating point or bias point) for a self-biased JFET is established by determining the value of drain current (I_D) for a desired value of gate-to-source voltage (V_{GS}) or vice-versa. However, if the data sheet of JFET includes a transfer characteristics curve (also known as transconductance curve), then we may determine the Q-point by using the procedure discussed as under:

(i) Firstly, select a convenient value of drain current. Generally, its value is taken half of the maximum possible value of drain current (i.e., I_{DSS}). After this, find the voltage drop across source resistor (R_S) by the expression

$$V_S = I_D.R_S \quad ...(6.18)$$

and the gate-to-source voltage from the expression

$$V_{GS} = -V_S \quad ...(6.19)$$

(ii) Now, plot the assumed value of drain current (I_D) and the corresponding gate-to-source voltage (V_{GS}) on the transfer characteristics curve.

(iii) Draw a line through the plotted point and the origin. The point of intersection of the line and the curve gives the desired Q-point. Now, read the co-ordinates of Q-point.

As a matter of fact, it is required to set the Q-point near the mid-point of the transfer characteristic curve of a JFET. Under signal conditions, the mid-point bias allows a maximum amount of drain current swing between the values of I_{DSS} and O.

However, if we desire to design a self-bias circuit, then anyone of the following two methods can be used for this purpose:

(i) Analytical Method

In analytical method, we note down the values of the maximum drain current (I_{DSS}) and the gate-to-source cut-off voltage [$V_{GS(off)}$] from the data sheets of JFET supplied by the manufacturer.

The value of the drain current is found by the expression

$$I_D = I_{DSS}\left[1 - \frac{V_{GS}}{V_{GS(off)}}\right]^2 \quad ...(6.20)$$

For example, if we select the gate-to-source voltage (V_{GS}) equal to $V_{GS(off)}/4$, then the value of the drain current would be

$$I_D = I_{DSS}\ [\ 1 - 0.25\ ]^2 = I_{DSS}\ (0.75)^2 = 0.56\ I_{DSS} \quad ...(6.21)$$

It shows that the drain current is slightly more than one-half of I_{DSS}. But it will bias the JFET close to the mid-point of the curve. The value of the drain resistor (R_D) is selected in such a way that the drain voltage (V_D) is equal to half the drain supply voltage (V_{DD}). The value of gate resistor (R_G) is chosen arbitrarily large, so that it prevents loading on the driving stages.

(ii) Graphical Method

In graphical method, we draw a self-bias line such that it intersects the transfer characteristic curve near its mid-point. The point of intersection of the self-bias line and the transfer characteristic curve gives us the required Q-point. After this, read the co-ordinates of the Q-point. The value of source resistance (R_S) is expressed by the ratio of gate-to-source voltage (V_{GS}) to the drain current (I_D).

Mathematically, the source resistance is given by

$$R_S = \frac{V_{GS}}{I_D} \quad ...(6.22)$$

However, a more accurate method is to draw a self-bias line through the co-ordinates of I_{DSS} and $V_{GS(off)}$ as shown in figure 6.37. Then the point of intersection of self-bias line and the transfer characteristic curve locates the Q-point. The value of the source resistor is expressed by the relation

$$R_S = \frac{V_{GS(off)}}{I_{Dss}} \quad ...(6.23)$$

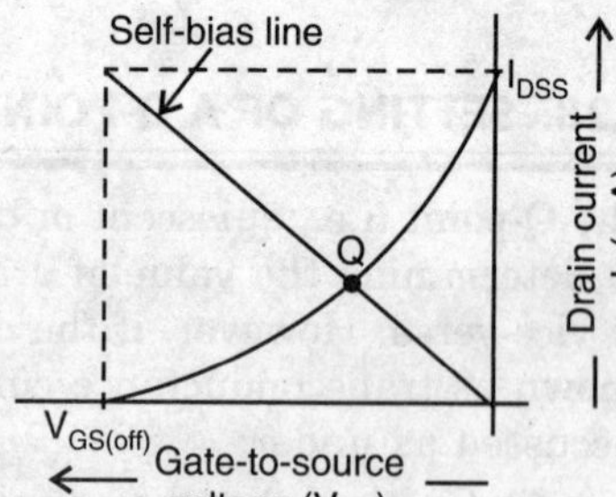

Fig. 6.37. Setting of a Q-Point in self-bias

The value of drain resistor (R_D) and the gate resistor (R_G) are selected in the same way as discussed above for the analytical method.

EXAMPLE 6.19 Figure 6.38(a) shows that circuit of a self-biased JFET amplifier and figure 6.38(b) indicates the transfer characteristic curve of the JFET.

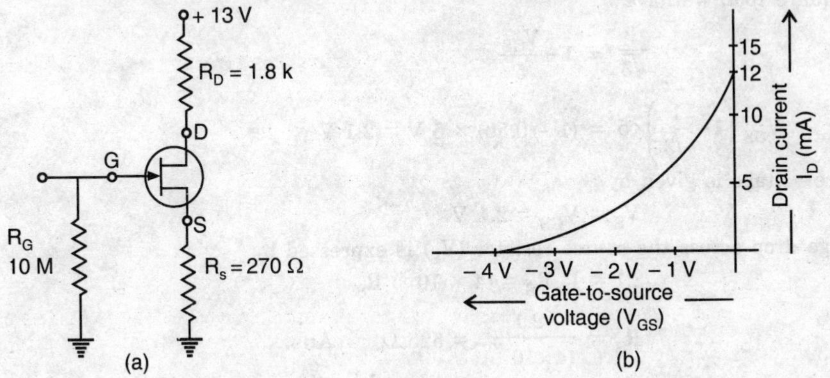

Fig. 6.38.

Determine the quiescent values of I_D and V_{GS}. Compute the values of d.c. voltage between the drain and the ground.

Solution: Given that

$$R_D = 1.8\text{ k}\Omega = 1.8 \times 10^3\ \Omega$$
$$R_S = 270\ \Omega;\quad R_G = 10\text{ M}\Omega \text{ and}$$
$$I_{DSS} \text{ (from transfer characteristic curve)} = 12 \text{ mA}$$

(i) Firstly, let us assume the drain current (I_D) equal to the half of I_{DSS} (i.e., 6 mA or 6×10^{-3} A). Using this value of drain current, the voltage drop across resistor (R_S) is given by

$$V_S = I_D.R_S = (6 \times 10^{-3}) \times 270 = 1.62 \text{ V}$$

and the gate-to-source voltage is given by

$$V_{GS} = -V_S = -1.62 \text{ V}$$

(ii) Now, let us plot the values of drain current (I_D = 6 mA) and the corresponding value of voltage V_{GS} (i.e., – 1.62 V) on the graph of transfer characteristic curve as shown in figure 6.39

(iii) Now, let us draw a line through the plotted point (P) and the origin as shown in the figure and read the co-ordinates of Q-point.

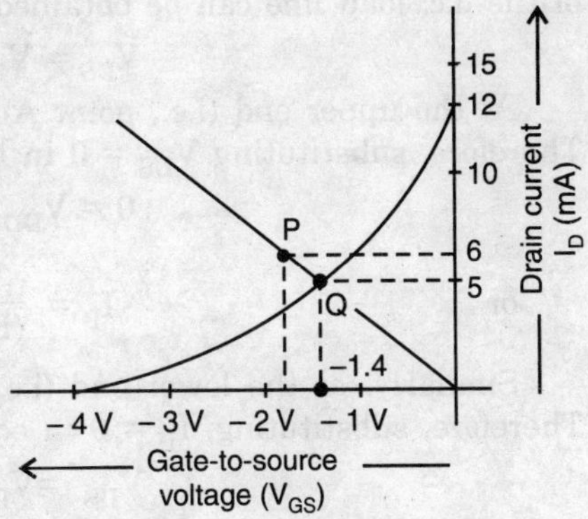

Fig. 6.39.

It is clear from this graph that the coordinates of Q-point are as under :

$$I_D = 5 \text{ mA}$$
$$V_{GS} = -1.4 \text{ V} \quad \textbf{Ans.}$$

Further, we know that the d.c. voltage between drain and ground is given by

$$V_D = V_{DD} - I_D.R_D = 15 - (5 \times 10^{-3}) \times (1.8 \times 10^3) \text{ V} = 15 - 9 = 6 \text{ V} \quad \textbf{Ans.}$$

EXAMPLE 6.20 A self-biased P-channel JFET has a pinch off voltage (V_P) = 5 V and I_{DSS} = 12 mA. The supply voltage available is equal to 12 V. Find the values of resistors R_D and R_S, so that I = 5 mA and V_{DS} = 6 V.

Solution: Given that : V_P, i.e.,

$$V_{GS\,(off)} = 5 \text{ Volt};\quad I_{DSS} = 12 \times 10^{-3} \text{ A};\quad V_{DD} = 12 \text{ volt};\quad V_{DS} = 6 \text{ volt}$$

We know that the drain current (I_D) is expressed as

$$I_D = I_{DSS}\left[1 - \frac{V_{GS}}{V_{GS(off)}}\right]^2$$

or

$$4 \times 10^{-3} = I_{DSS}\left[1 - \frac{V_{GS}}{V_{GS(off)}}\right]^2 = 12 \times 10^{-3}\left[1 - \frac{V_{GS}}{5}\right]^2 = \left(1 - \frac{V_{GS}}{5}\right)^2$$

Taking square root, we have

$$\frac{1}{\sqrt{3}} = 1 - \frac{V_{GS}}{5}$$

or $$V_{GS}\left(1-\frac{1}{\sqrt{3}}\right)\times 5 = (1 - 0.58) \times 5 \text{ V} = 2.1 \text{ V}$$

The source voltage is given by

$$V_S = V_{GS} = 2.1 \text{ V}$$

and voltage drop across the source resistor (V_S) is expressed by

$$2.1 = I_D.R_S = (4 \times 10^{-3})\ R_S$$

or $$R_S = \frac{2.1}{(4\times10^{-3})} = 525\ \Omega \qquad \textbf{Ans.}$$

and drain-to-source voltage (V_{DS}) is given as under:

$$6 = V_{DS} - I_D.R_D = 12 - (4 \times 10^{-3}) \times R_D$$

or $$R_D = 1500\ \Omega = 1.5\ \text{k}\Omega \qquad \textbf{Ans.}$$

6.29. SETTING OF A Q-POINT USING D.C. LOAD LINE

The Q-point for the self-biased JFET circuit can also be obtained graphically by using d.c. load line. The d.c. load line can be drawn easily on the drain characteristics as shown in figure 6.40(a). The values of drain current (I_D) and drain-to-source voltage (V_{DS}) at the upper and lower ends of the d.c. load line can be obtained by using the expression

$$V_{DS} = V_{DD} - I_D.\ (R_D + R_S) \qquad ...(6.24)$$

At the upper end (i.e., point A) the value of drain-to-source voltage (V_{DS}) is equal to zero. Therefore, substituting $V_{DS} = 0$ in last equation (6.24), we have

$$0 = V_{DD} - I_D.\ (R_D + R_S) \qquad ..(6.25)$$

or $$I_D = \frac{V_{DD}}{(R_D + R_S)} \qquad ...(6.26)$$

Similarly, at the lower end (i.e., point B) the value of drain current (I_D) is equal to zero. Therefore, substituting, $I_D = 0$ in equation (6.24), we have

$$V_{DS} = V_{DD} - 0(R_D + R_S) = V_{DD} \qquad ...(6.27)$$

Hence the upper and lower ends of the d.c load line are given by the points,

$$I_D = V_{DD}/(R_D + R_S) \text{ and } V_{DS} = 0 \quad \text{(Upper end)} \qquad ...(6.28)$$

$$I_D = 0 \quad \text{and} \quad V_{DS} = V_{DD} \quad \text{(Lower end)} \qquad ...(6.29)$$

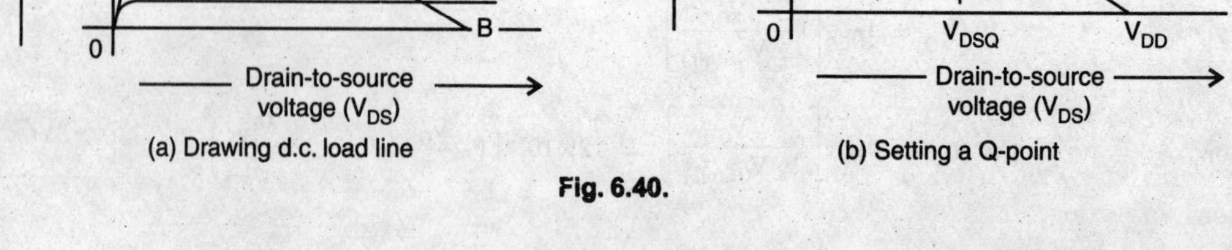

(a) Drawing d.c. load line (b) Setting a Q-point

Fig. 6.40.

The Q-point is set at mid-point of the d.c. load as shown in figure 6.40(b). In this case, the value of drain current at Q point will be given by

$$I_{DQ} = \frac{V_{DD}}{2(R_D + R_S)} \quad ...(6.30)$$

and the drain-to-source voltage at the Q-point is given by

$$V_{DSQ} = \frac{V_{DD}}{2} \quad ...(6.31)$$

EXAMPLE 6.21 **Calculate the self-bias operating point for the FET circuit shown in figure 6.41. Also find the values of resistors R_D and R_S to obtain this bias condition. Given the maximum value of drain current as 10 mA and V_{GS} = – 2.2 volt at I_D = 5 mA.**

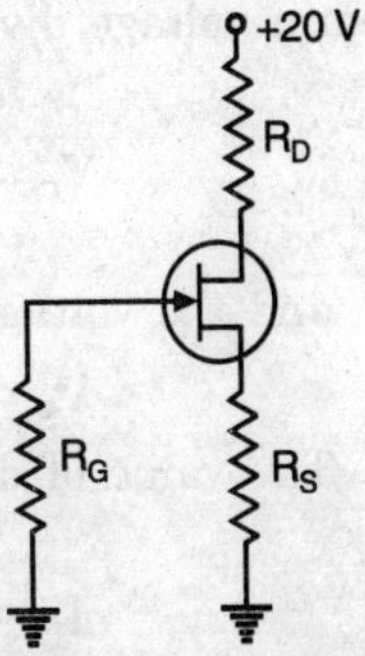

Fig. 6.41. Circuit for Example 6.21

Solution: Given that I_{DSS} = 10 mA and V_{DD} = 20 volt.

We know that the values of drain current at Q-point is given by

$$I_{DQ} = \frac{I_{DSS}}{2} = \frac{10}{2} = 5 \text{ mA}$$

and the value of drain-to-source voltage is expressed as

$$V_{DSQ} = \frac{V_{DD}}{2} = \frac{20}{2} = 10 \text{ V}$$

Therefore, operating point will be given as

$$I_D = 5 \text{ mA and } V_{DS} = 10 \text{ V} \qquad \textbf{Ans.}$$

Now, V_{GS} = – 2.2 volt and $I_D = 5 \text{ mA} = 5 \times 10^{-3}$ A

The drain-to-source voltage (V_{DS}) is given by

$$10 = V_{DD} - I_D.R_D = 20 - (5 \times 10^{-3})R_D$$

or $$R_D = \frac{20 - 10}{5 \times 10^{-3}} = 2 \times 10^3 \ \Omega = 2 \text{ k}\Omega \qquad \textbf{Ans.}$$

The source voltage is given by

$$V_S = -V_{GS} = -(-2.2) = 2.2 \text{ V}$$

and the value of voltage drop across the source resistor (R_S) is expressed by

$$2.2 = I_D.R_S = (5 \times 10^{-3}) \times R_S$$

or $$R = 2.2/(5 \times 10^{-3}) = 440 \ \Omega \qquad \textbf{Ans.}$$

6.30. BIASING AGAINST DEVICE PARAMETER VARIATION

As already discussed that for self-biased JFET, the Q-point is fixed at mid-point of the transfer characteristic curve. However, it has been observed that the JFET parameters like maximum drain current (I_{DSS}) and the gate-to-source cut-off [$V_{GS(off)}$] vary with the temperature. In this case, the Q-point will also vary with the temperature and thus will not remain stable. This degrades the performance of the FET amplifier circuit.

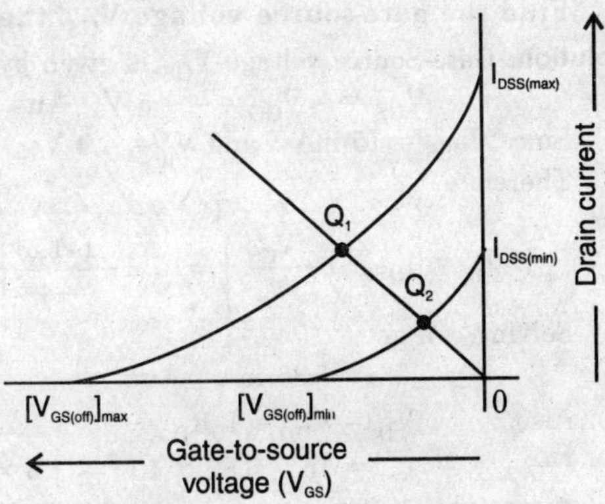

Fig. 6.42. Use of two transfer characteristic curves for setting a Q-point

In order to provide compensation for the device parameter variation the manufacturers provide two transistor characteristic curves as shown in figure 6.42. The first current corresponds to the maximum values of I_{DSS} and $V_{GS\ (off)}$. In such a case, we draw a bias line through both the curves which is near the mid-point of either curve. Then we read the values of gate-to-source voltage (V_{GS}) and the drain current (I_D) for either Q-point. The optimum value of source resistor R_S is given by the ratio of gate-to-source voltage (V_{GS}) and drain current (I_D).

Sometimes the manufacturers do not supply the transfer characteristic curves, but provide the maximum and minimum values of the I_{DSS} and $V_{GS(off)}$. In this case, we calculate the maximum and minimum values of source resistor (R_S) and then take the average of these two values. This average value will provide a Q-point near the mid-point of the either transfer characteristic curve.

6.31. VOLTAGE DIVIDER BIAS

Figure 6.43 shows the voltage divider bias circuit. The name '*voltage divider*' is derived from the fact that resistor R_1 and R_2 are connected on the gate side (between the V_{DD} supply and ground) from a voltage divider. Assuming the current to be zero, the voltage is given by the expression

$$V_G = \left(\frac{R_2}{R_1 + R_2}\right) V_{DD} \qquad ...(6.32)$$

and the voltage from source to ground is given by,

$$V_S = V_G - V_{GS} \qquad ...(6.33)$$

The value of drain current will be

$$I_D = \frac{V_S}{R_S} = \frac{V_G - V_{GS}}{R_S} \qquad ...(6.34)$$

and the d.c. voltage from drain to ground is given as

$$V_D = V_{DD} - I_D.R_D \qquad ...(6.35)$$

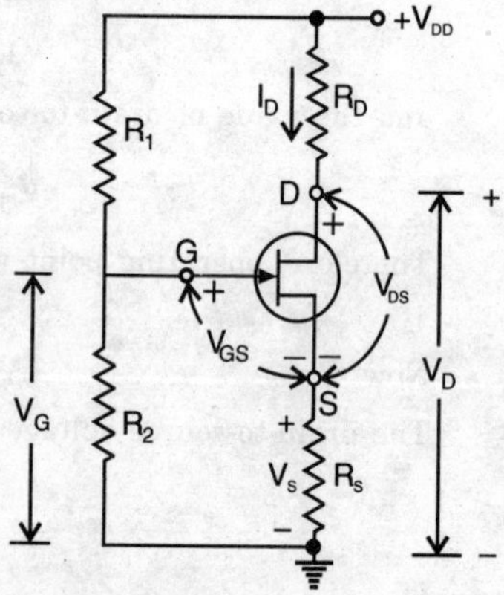

Fig. 6.43. Voltage divider bias circuit

Now, if the gate voltage is very large as compared to gate-to-source voltage, then the drain current is approximately constant for any JFET. It has been found that in actual practice the voltage divider bias is less effective with JFET than bipolar transistors. It is because of the fact that in bipolar transistors the base-to-emitter voltage (V_{BE}) is approximately 0.7 V with only minor variations from one transistor to next. However, in a JFET, the gate-to-source voltage (V_{GS}) can vary several volts from one JFET to another. Because of this, it is difficult to make gate voltage (V_G) large enough than the gate-to-source voltage (V_{GS}).

EXAMPLE 6.22 **In the circuit shown in figure 6.44, the following data are given: V_{GG} = 1.5 V, V_{DD} = 15 V, R_D = 1.5 kΩ, R_G = 2 MΩ, I_{DSS} = 15 mA and V_p = – 4 V. Find the gate-source voltage V_{GS}, the drain current I_D and drain-to-source voltage V_{DS}.**

Solution: Gate-Source voltage V_{GS} is given by

$$V_{GS} = -V_{GG} = -1.5 \text{ V} \quad \textbf{Ans.}$$

Since, I_{DSS} = 15 mA and V_p = – 4 V

Therefore,

$$I_D = I_{DSS}\left(1 - \frac{V_{GS}}{V_p}\right)^2 = 15\left[1 - \frac{(-1.5)}{-4}\right]^2$$

Solving, we get

$$I_D = 5.86 \text{ mA} \quad \textbf{Ans.}$$

Also, $V_D = V_{DD} - I_D R_D$

$$= 15 - 5.86 \times 10^{-3} \times 1.5 \times 10^3 = 6.2 \text{ V}$$

Lastly, the drain-to-source voltage V_{DS} is given by

$$V_{DS} = V_D - V_S = 6.2 - 0 = 6.2 \text{ V} \quad \textbf{Ans.}$$

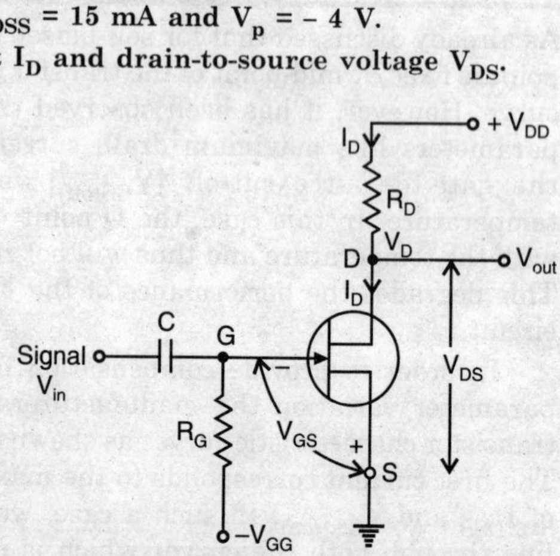

Fig. 6.44. Circuit for Example 6.22

EXAMPLE 6.23 **In the JFET circuit shown in figure 6.45, the following data are given:**
V_{DD} = 25 V, R_D = 3 kΩ, R_S = 400 Ω, I_D = 2 mA.
Find: (i) V_{DS} (ii) V_{GS}.

Solution: Given that

Drain supply voltage, V_{DD} = 25 V
Drain resistance, R_D = 3 kΩ
Source resistance, R_S = 400 Ω = 0.4 kΩ
Also, drain current, I_D = 2 mA
Now,
(i) Drain-to-source voltage is given by

$$V_{DS} = V_{DD} - I_D (R_D + R_S) = 25 - (2 \text{ mA}) [3 + 0.4) \text{ k}\Omega]$$

or V_{DS} = 18.2 V **Ans.**

(ii) Gate-to-source voltage, V_{GS}

$$= - I_D R_S = - 2.0 \text{ mA} \times 0.4 \text{ k}\Omega = - 0.8 \text{ V}$$ **Ans.**

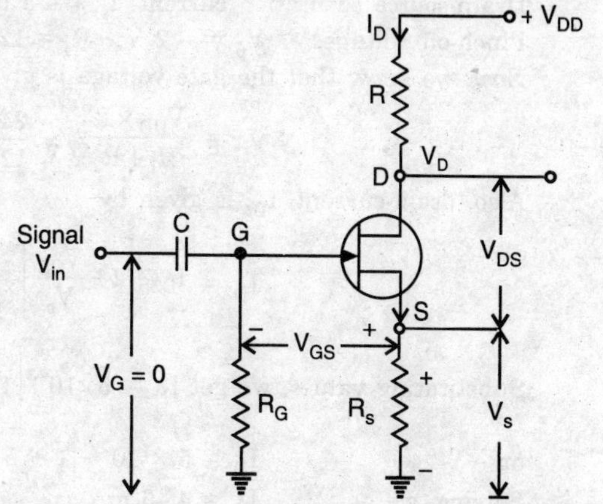

Fig. 6.45.

EXAMPLE 6.24 **In an N-channel JFET biased by the voltage-divider method, determine the value of resistance R_S so that the operating point values are,**

I_D = 4 mA and V_{DS} = 8 V.
Given that, V_{DD} = 25 V, R_{G1} = 1.2 MΩ, R_{G2} = 0.6 MΩ.
JFET parameters are I_{DSS} = 12 mA and V_p = – 4 V.

Solution: Given that V_{DD} = 25 V; R_{G1} = 1.2 MΩ or 1200 kΩ
R_{G2} = 0.6 MΩ or 600 kΩ; I_D = 4 mA; I_{DSS} = 12 mA
V_{DS} = 8 V; V_p = – 4 V

We know that
$$I_D = I_{DSS}\left[1 - \frac{V_{GS}}{V_p}\right]^2$$

Therefore, we have
$$1 - \frac{V_{GS}}{V_p} = \sqrt{\frac{I_D}{I_{DSS}}} = \sqrt{\frac{4}{12}} = 0.577$$

or
$$\frac{-V_{GS}}{-4} = 0.577 - 1 = 4(0.577 - 1) = - 1.69 \text{ V}$$

Also, since,
$$V_G = \frac{V_{DD}}{R_{G1} + R_{G2}} \times R_{G2}$$

Therefore,
$$V_G = \frac{25 \times 600}{1200 + 600} = 8.33 \text{ Volt}$$

Further,
$$R_S = \frac{V_G - V_{GS}}{I_D} = \frac{8.33 - (-1.69)}{4 \text{ mA}} = 2.5 \text{ k}\Omega$$ **Ans.**

EXAMPLE 6.25. **A JFET amplifier with stabilized biasing circuit shown in figure 6.46, has following parameters:**

V_p = – 2 V, V_{DSS} = 5 mA, R_L = 910 Ω, R_F = 2.29 kΩ, R_1 = 12 MΩ, R_2 = 8.57 MΩ and V_{DD} = 24 V.

Determine the value of drain current I_D at the operating point. Also, verify that FET will operate in pinch-off region. (A.M.I.E. Examination)

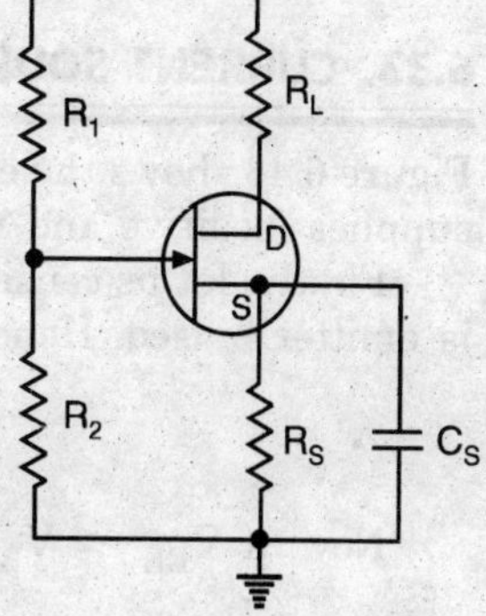

Fig. 6.46.

Solution: Given that

Drain supply voltage, V_{DD} = 24 Volt
Drain circuit or load resistance, R_L = 910 Ω

Source resistance, $R_S = 2.29$ kΩ

Drain-source saturation current, $I_{DSS} = 5$ mA $= 5 \times 10^{-3}$ A.

Pinch-off voltage, $V_p = -2$ V; $R_1 = 12$ MΩ; $R_2 = 8.57$ MΩ

Now, we know that the gate voltage is given by

$$V_G = \frac{V_{DD} \times R_2}{R_1 + R_2} = \frac{24 \times 8.57}{12 + 8.57} = 10 \text{ V}$$

Also, drain-current, I_D, is given by

$$I_D = I_{DSS}\left[1 - \frac{V_{GS}}{V_p}\right]^2$$

Substituting values, we get $I_D = 5 \times 10^{-3}\left[1 - \frac{10 - 2.29 \times 10^3 I_D}{-2}\right]^2$

or $\quad I_D = 5 \times 10^{-3}\ [1 + 5 - 1, 145\ I_D]^2 = 0.005\ [6 - 1, 145\ I_D]^2$

Solving, we get $\quad I_D = 4.46$ mA

Gate-source voltage, V_{GS} is given by

$$V_{GS} = V_G - I_D R_S = 10 - 2.29 \times 10^3 \times 4.46 \times 10^{-3} = -0.2134 \text{ Volt}$$

Current at operating point is given by

$$I_D = \frac{V_G - V_{GS}}{R_S} = \frac{10 - 0.2134}{2.29 \times 10^3} = 4.274 \text{ mA} \quad \textbf{Ans.}$$

It may be noted that the value of I_D at operating point is almost equal to previously computed value of I_D. Therefore, we can say that the FET is operating in pinch-off region.

6.32. SOURCE BIAS

Figure 6.47 shows the circuit of a source bias for JFET. It can be noted that this circuit is similar to the emitter bias circuit used for biasing bipolar transistors.

The value of drain current is given by the expression,

$$I_D = \frac{V_{SS} - V_{GS}}{R_S} \quad \text{...(6.36)}$$

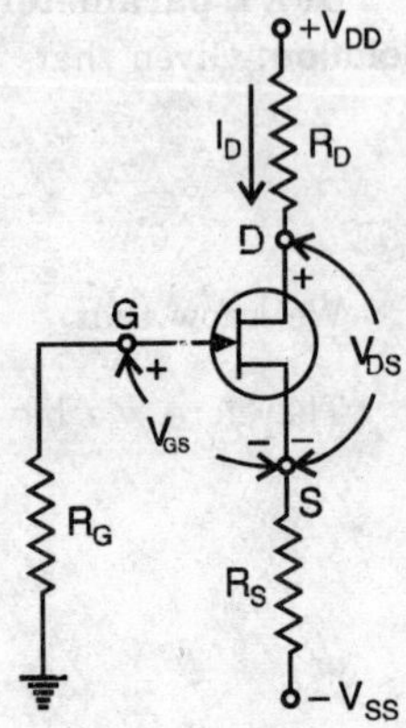

Fig. 6.47. Source bias

Because $V_{SS} >> V_{GS}$, therefore, drain current will be

$$I_D \cong \frac{V_{SS}}{R_S} \quad \text{...(6.37)}$$

6.33. CURRENT SOURCE BIAS

Figure 6.48 shows the current-source bias circuit for a JFET. The circuit of figure 6.48(a) uses two supplies (positive and negative), whereas figure 6.48(b) employs only one (i.e., positive) supply.

Firstly, let us consider the two supply current-source bias circuit. Here the bipolar transistor is emitter biased. Hence, its collector current is given by the expression

$$I_C = \frac{V_{EE} - V_{BE}}{R_E} \quad \text{...(6.38)}$$

Now, if $V_{EE} >> V_{BE}$

then $\quad I_C = \frac{V_{EE}}{R_E}$

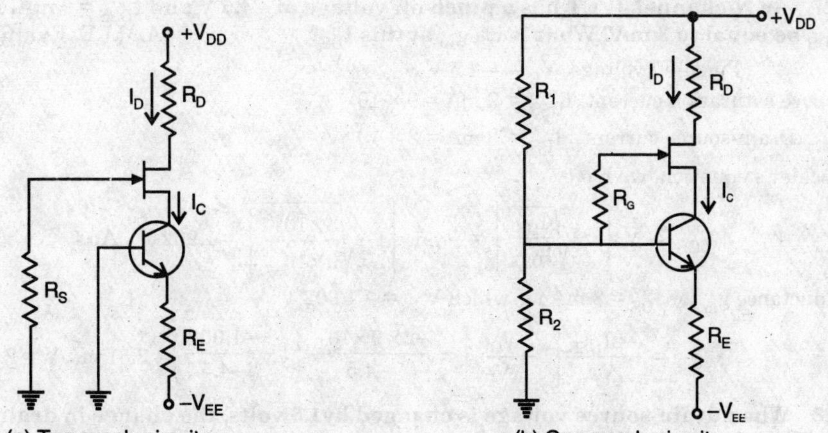

(a) Two supply circuits (b) One supply circuits

Fig. 6.48. Current-source bias circuit.

The drain current of the JFET is equal to the collector current because the bipolar transistor acts like a d.c. current-source, i.e.,

$$I_D = I_C \qquad ...(6.39)$$

Now, let us consider the one supply current-source bias circuit. Here, the bipolar transistor is voltage divider biased. Its collector current is given by the expression,

$$I_C = \frac{V_B - V_{BE}}{R_E} \qquad ...(6.40)$$

and the drain current will be

$$I_D = I_C \qquad ...(6.41)$$

The current-source bias circuit provides a solid Q-point with JFET. Thus, this circuit is very useful in actual practice.

EXAMPLE 6.26 A FET follows the following relation :

$$I_D = I_{DSS}\left[1 - \frac{V_{GS}}{V_P}\right]^2$$

I_{DSS} = 8.4 mA, V_P = – 3 V. What is the value of I_D for V_{GS} = – 1.5 V? Find g_m at this point.

(A.M.I.E. Examination, 1993)

Solution: Drain-source saturation current,

I_{DSS} = 8.4 mA

Pinch-off voltage, V_P = – 3 V

Gate-source voltage, V_{GS} = – 1.5 V

We know that the drain current is given by

$$I_D = I_{DSS}\left[1 - \frac{V_{GS}}{V_P}\right]^2$$

Substituting all the values,

$$I_D = 8.4\left(1 - \frac{-1.5}{-3}\right)^2 = 2.1 \text{ mA} \quad \textbf{Ans.}$$

Transconductance for $V_{GS} = 0$ is given by

$$g_{mo} = \frac{-2I_{DSS}}{V_P} = \frac{-2 \times 8.4}{-3} = 5.6 \text{ mA/V or } 5.6 \text{ mS}$$

Transconductance,

$$g_m = g_{mo}\left(1 - \frac{V_{GS}}{V_P}\right) = 5.6\left(1 - \frac{-1.5}{-3}\right) = 2.8 \text{ mS} \quad \textbf{Ans.}$$

EXMAPLE 6.27 **An N-channel JFET has a pinch off voltage of – 4.5 V and I_{DSS} = 9 mA. At what value of V_{GS} will I_{DS} be equal to 3 mA? What is its g_m at this I_{DS}? (A.M.I.E. Examination, 1994)**

Solution: Pinch-off voltage, $V_P = -4.5$ V

Drain-source saturation current, $I_{DSS} = 9 \text{ mA} = 9 \times 10^{-3}$ A

Drain-source current, $I_{DS} = 3 \text{ mA} = 3 \times 10^{-3}$ A

From Shockley's equation, we have

$$V_{GS} = V_P\left[1-\sqrt{\frac{I_{DS}}{I_{DSS}}}\right] = -4.5\times\left[1-\sqrt{\frac{3\times10^{-3}}{9\times10^{-3}}}\right] = 1.902 \text{ V} \quad \textbf{Ans.}$$

Transconductance g_m for I_{DS} = 3 mA for which V_{GS} = – 1.902 V

$$g_m = \frac{-2I_{DSS}}{V_P}\left[1-\frac{V_{GS}}{V_P}\right] = \frac{-2\times9\times10^{-3}}{-4.5}\left(1-\frac{-1.902}{-4.5}\right) = 2.31 \text{ mA/V} = 2.31 \text{ mS} \quad \textbf{Ans.}$$

EXAMPLE 6.28 **When drain-source voltage is changed by1.5 volts, the change in drain current is of 120 μA, the gate-source voltage remaining unchanged. Determine the a.c. drain resistance of the JFET.**

Solution: Change in drain-source voltage,

$$\Delta V_{DS} = 1.5 \text{ V}$$

Change in drain-current, $\Delta I_o = 120 \times 10^{-6}$ A

$$\text{AC drain resistance of the JFET, } r_d = \frac{\Delta V_{DS}}{\Delta I_D} = \frac{1.5}{120\times10^{-6}} = 12.5 \text{ k}\Omega \quad \textbf{Ans.}$$

EXAMPLE 6.29 **The readings obtained experimentally from a JFET are as follows:**

Drain-source voltage, V_{DS} in volts	**5**	**12**	**12**
Gate-source voltage V_{GS} in volts	**0**	**0**	**– 0.25**
Drain current, I_D in mA	**8**	**8.2**	**7.5**

Determine (i) a.c. drain resistance (ii) transconductance and (iii) amplification factor.

Solution: (i) The drain current increases from 8.0 mA to 8.2 mA when drain-source voltage is increased from 5 V to 12 V, keeping gate-source voltage constant at 0.V.

$$\therefore \text{ A.C. drain resistance, } r_d = \frac{\Delta V_{DS}}{\Delta I_D} = \frac{(12-5)\text{V}}{(8.2-8)\text{mA}} = \frac{7}{0.2\times10^{-3}} = 35 \text{ k}\Omega \quad \textbf{Ans.}$$

(ii) The drain current changes from 8.2 mA to 7.5 mA when gate-source voltage is changed from 0 to – 0.25 V, keeping drain-source voltage constant at 12 V.

$$\therefore \text{ Transconductance, } g_m = \frac{\Delta I_D}{\Delta V_{GS}} = \frac{(7.5-8.2)\text{mA}}{(-0.25-0)\text{V}} = 2.8 \text{ mA/V} = 2.8 \text{ mS} \quad \textbf{Ans.}$$

(iii) Amplification factor, $\mu = r_d \times g_m = 35 \times 1{,}000 \times 2.8 \times 10^{-3} = 98$ **Ans.**

EXAMPLE 6.30 **A JFET has V_P = – 4.5 V, I_{DSS} = 10 mA and I_{DS} = 2.5 mA. Determine the trans-conductance. (A.M.I.E. Examination, 1996)**

Solution: Drain-source saturation current,

$$I_{DSS} = 10 \text{ mA}$$

Pinch-off voltage, $V_P = -4.5$ V

Drain-source current, $I_{DS} = 2.5$ mA

From Shockley's equation, drain-source current,

$$I_{DS} = I_{DSS}\left[1-\frac{V_{GS}}{V_P}\right]^2 \quad ...(i)$$

or

$$V_{GS} = V_P\left[1-\sqrt{\frac{I_{DS}}{I_{DSS}}}\right] = -4.5\left[1-\sqrt{\frac{2.5}{10}}\right] = -2.25 \text{ V}$$

From equation (i), transconductance will be

$$g_m = \frac{-2I_{DSS}}{V_P}\left[1-\frac{V_{GS}}{V_P}\right] = \frac{-2\times10\times10^{-3}}{-4.5}\left[1-\frac{-2.25}{-4.5}\right] = 2.22 \text{ m A/V} \quad \textbf{Ans.}$$

EXAMPLE 6.31 A JFET has g_m = 10 mS, I_{DSS} = 10 mA. Calculate $V_{GS(OFF)}$.

(A.M.I.E. Examination, 1997)

Solution: Transconductance, $g_m = 10 \text{ mS} = 10 \times 10^{-3}$ S

Drain-source saturation current, $I_{DSS} = 10 \text{ mA} = 10 \times 10^{-6}$ A

$$V_{GS(OFF)} \to V_{GS} = V_P$$

$$g_m = g_{mo} = \frac{-2I_{DSS}}{V_P} = \frac{-2I_{DSS}}{V_{GS(OFF)}}$$

or
$$10 \times 10^{-3} = \frac{-2\times10\times10^{-6}}{V_{GS(OFF)}}$$

or
$$V_{GS(OFF)} = \frac{-2\times10\times10^{-6}}{10\times10^{-3}} = 2 \text{ mV} \quad \textbf{Ans.}$$

EXAMPLE 6.32 An N-channel JFET has I_{DSS} = 10 mA and V_P = – 4 V. Determine the minimum value of V_{DS} for pinch-off region and drain current I_D for V_{GS} = – 2 V in pinch-off region.

(A.M.I.E. Examination, 1996)

Solution: Pinch-off voltage, $V_P = -4$ V

Gate-source voltage, $V_{GS} = -2$ V

Drain-source saturation current, $I_{DSS} = 10 \text{ mA} = 10 \times 10^{-3}$ A

$$\text{Drain current, } I_D = I_{DSS}\left[1-\frac{V_{GS}}{V_P}\right]^2 = 10\times10^{-3}\left[1-\frac{-2}{-4}\right]^2 = 2.5 \text{ mA} \quad \textbf{Ans.}$$

The minimum value of V_{DS} for pinch-off region is equal to V_P. Thus the minimum value of V_{DS}

$$V_{DS(min)} = V_P = -4 \text{ V} \quad \textbf{Ans.}$$

EXAMLE 6.33 For an N-channel JFET

$$I_{DSS} = 8.7 \text{ mA}$$
$$V_P = -3 \text{ V}$$
$$V_{GS} = -1 \text{ V.}$$

Find the values of (i) I_D (ii) g_{mo} (iii) g_m.

Solution: Drain-source saturation current,

$$I_{DSS} = 8.7 \text{ mA} = 0.0087 \text{ A}$$

Pinch-off voltage, $V_{GS} = -1$ V

(i)
$$\text{Drain current, } I_D = I_{DSS}\left(1-\frac{V_{GS}}{V_P}\right)^2 = 0.0087\left(1-\frac{-1}{-3}\right)^2 = 3.8667 \text{ mA} \quad \textbf{Ans.}$$

(ii) Transconductance for $V_{GS} = 0$, is given by

$$g_{mo} = \frac{-2I_{DSS}}{V_P} = \frac{-2\times8.7 \text{ mA}}{-3} = 5.8 \text{ mA/V} = 5.8 \text{ mS} \quad \textbf{Ans.}$$

(iii) Now, transconductance, $g_m = g_{mo}\left(1-\dfrac{V_{GS}}{V_P}\right) = 5.8\left(1-\dfrac{-1}{-3}\right) = 3.867 \text{ mA/V} = 3.867 \text{ mS}$ **Ans.**

EXAMPLE 6.34 **The FET shown in figure 6.49 has the following parameters: I_{DSS} = 5.6 mA, V_P = 4 Volt. If V and V_0 are constant, then determine: (i) V_0 when $V_i = 0$ (ii) V_0 when V_i = 10 Volts (iii) V_i when $V_0 = 0$**

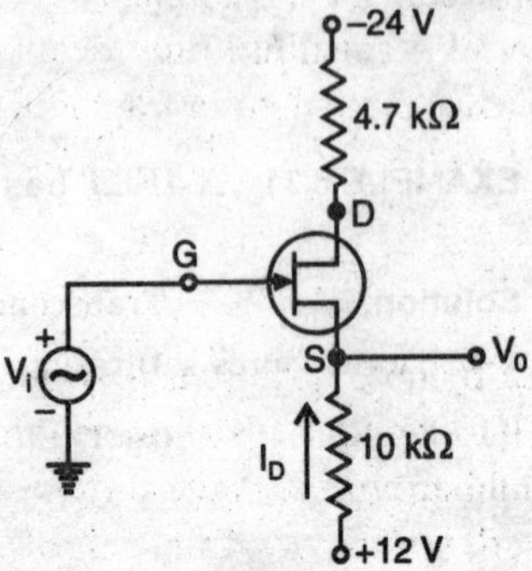

Fig. 6.49. Circuit for Example 6.34

Solution: Applying KVL to the gate source loop in figure 6.49, we have

$$12 = I_D \times 10 - V_{GS}$$

or $$V_{GS} = 10 \times I_D - 12$$

or $$I_D = \frac{V_{GS}+12}{10} \quad \text{...(i)}$$

Since, we do not know the value of I_D, therefore, let us find the value of I_D

$$I_D = I_{DSS}\left[1-\frac{V_{GS}}{V_P}\right]^2 = 5.6\left[1-\frac{V_{GS}}{4}\right]^2 \quad \text{...(ii)}$$

Now substituting V_{GS} from equation (i) into equation (ii), we get

$$I_D = 5.6\times10^{-3}\left[1-\frac{10I_D-12}{4}\right]^2 = 5.6\times10^{-3} - \frac{56}{4}I_D + 3\times5.6\times10^{-3}$$

Therefore, $$15\ I_D = 22.4 \times 10^{-3} \quad \text{or} \quad I_D = 1.49 \text{ mA} \quad \text{...(iii)}$$

Substituting the value of I_D into equation (i), we get

$$V_{GS} = (10 \times 1.49) - 12 = 2.9 \text{ Volts}$$

The output voltage will be

$$V_o = 12 - I_D R_S$$

Substituting values, we get

$$V_o = 12 - (1.49 \times 10) = -29 \text{ Volts} \quad \textbf{Ans.}$$

EXAMPLE 6.35 **The JFET used in figure 6.50 has V_P = – 2 V, I_{DSS} = 1.65 mA. It is desired to bias the circuit at I_D = 0.8 mA. If V_{DD} = 24 V and $r_d >> R_D$.**

Find: (i) V_{GS}, (ii) g_m, (iii) R_S.

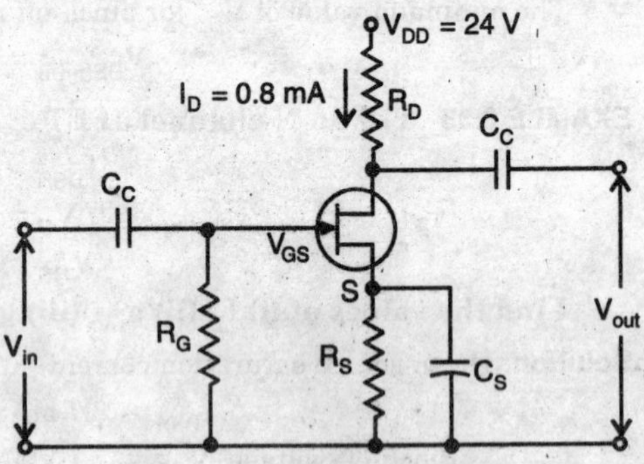

Fig. 6.50.

Solution: (i) We know that

$$I_D = I_{DSS}\left[1-\frac{V_{GS}}{V_P}\right]^2 \quad \text{...(i)}$$

or $$V_{GS} = V_P\left[1-\sqrt{\frac{I_D}{I_{DSS}}}\right]$$

Substituting the values, we get

$$V_{GS} = -2\left[1-\sqrt{\frac{0.8}{1.65}}\right] = -2\left[1-\frac{1}{\sqrt{2}}\right]$$

$$= +0.6074 \text{ Volts} \quad \textbf{Ans.}$$

(ii) We know that $$g_m = g_{mo}\left[1-\frac{V_{GS}}{V_P}\right] \quad \text{...(ii)}$$

But we do not know the value of g_{mo} so let us find it first as under:

$$g_{mo} = \frac{-2I_{DSS}}{V_P}$$

Substituting the values of I_{DSS} and V_P, we get

$$g_{mo} = \frac{-2\times1.65\times10^{-3}}{-2} = 1.65 \times 10^{-3} \quad \text{...(iii)}$$

Substituting this value into equation (iii), we get

$$g_m = 1.65\times10^{-3}\left[1-\frac{0.6074}{2}\right] = 1.15 \times 10^{-3} \text{ mho} \quad \textbf{Ans.}$$

(iii) Applying KVL to the gate to source circuit of figure 6.72, we get

$$0 = V_{GS} + I_D R_S$$

or $$R_S = \frac{-V_{GS}}{I_D}$$

Substituting the values, we get

$$R_S = \frac{0.6074}{0.8\times10^{-3}} = 759.25\ \Omega \quad \textbf{Ans.}$$

EXAMPLE 6.36 The maximum transconductance of a certain N-channel JFET is 9.4 mS. If I_{DSS} is 20 mA, what is the pinch off voltage ?

Solution: The maximum transconductance is given by

$$g_{mo} = \frac{-2I_{DSS}}{V_P} \quad \text{or} \quad V_P = \frac{-2I_{DSS}}{g_{mo}}$$

Substituting the values, we get

$$V_P = \frac{-2\times 20\times10^{-3}}{9.4\times10^{-3}}$$

or Pinch off voltage $V_P = -4.255$ Volts **Ans.**

EXAMPLE 6.37 The maximum transconductance of a certain N-channel JFET is 3.8×10^{-3} S. If I_{DSS} = 18 mA. What would be the pinch off voltage?

Solution: We know that, Maximum transconductance,

$$g_{mo} = \left|\frac{2I_{DSS}}{V_P}\right|$$

or $$V_P = \frac{2I_{DSS}}{g_{mo}} = \frac{2\times18\,\text{mA}}{9.8\times10^{-3}} = 3.6 \text{ Volts} \quad \textbf{Ans.}$$

EXAMPLE 6.38 The drain current in milliamperes of the enhancement type MOSFET shown in figure 6.51 is given by, $I_D = 0.3\,(V_{GS} - V_P)^2$ mA in the region $V_{DS} \geq V_{GS} - V_P$.

If V_P = + 4 V determine I_{DSQ}, V_{GSQ} and V_{DSQ}.

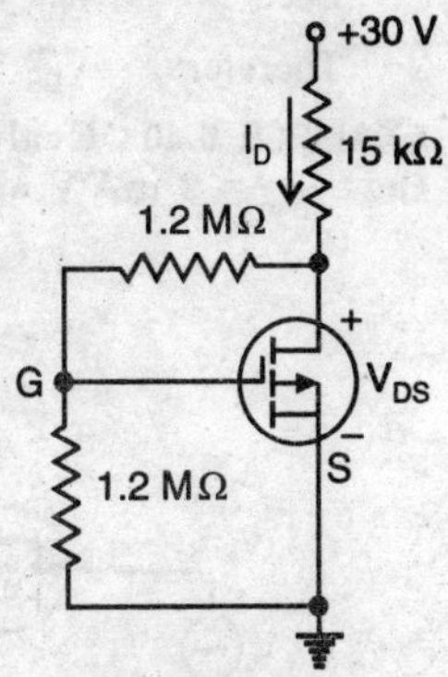

Fig. 6.51.

Solution: Since $I_G = 0$, therefore, we have

$$V_{GS} = \frac{1.2\ \text{M}\Omega}{(1.2\ \text{M}\Omega + 1.2\ \text{M}\Omega)} \times V_{DS} = 0.5\ V_{DS} \quad \text{...(i)}$$

Applying KVL to the drain circuit, we get

$$V_{DS} = 30 - 15 \times I_D \quad \text{...(ii)}$$

Substituting equation (i) into equation (ii), we get

$$V_{GS} = 0.5\,[30 - 15\,I_D] = 15 - 7.5I_D \quad \text{...(iii)}$$

Now given that $I_D = 0.3\,(V_{GS} - V_P)^2$, but $V_P = 4$

Substituting equation (iii) into last expression, we get

$$I_D = 0.3\,(15 - 7.5I_D - 4)^2 = 0.3\,[11 - 7.5I_D]^2$$

or $$I_D = 0.3\,[121 - 165\,I_D + 56.25\,I_D^2] = 16.875\,I_D^2 - 43.5\,I_D + 36.3$$

or $16.875\,I_D^2 - 50.5\,I_D + 36.3 = 0$

or $$I_D = \frac{50.5 \pm \sqrt{(50.5)^2 - 4(16.875 \times 36.3)}}{2 \times 16.875} = 1.2 \text{ mA}$$

Let us select, $I_D = 1.2$ mA **Ans.**

Also $V_{DS} = 30 - 15 \times 1.2 = 12$ Volt **Ans.**

$V_{GS} = 0.5 \times V_{DS} = 6$ Volt **Ans.**

EXAMPLE 6.39 Consider the following circuit (figure 6.52). Determine $I_D \cdot V_{GS}$ and V_{DS} for $|I_{DSS}| = 4$ mA, $V_P = 4$ V.

(U.P. Tech., Sem. Exam. 2005-06) (10 Marks)

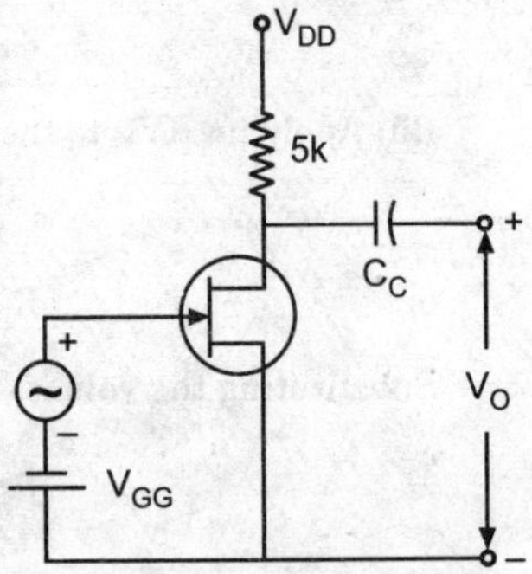

Fig. 6.52.

Solution: The given circuit is a CS amplifier with fixed bias circuit.

(i) Calculation of V_{GS}

For the DC analysis, the ac source acts as a short-circuit and capacitors are equivalent to open circuit.

Now, applying KVL to the gate loop of figure 6.53, we get

$$V_{GG} + V_{GS} = 0$$

or $$V_{GS} = -V_{GG}$$

Let $$V_{GG} = 2 \text{ V}$$

then, $$V_{GS} = -2 \text{ V}$$

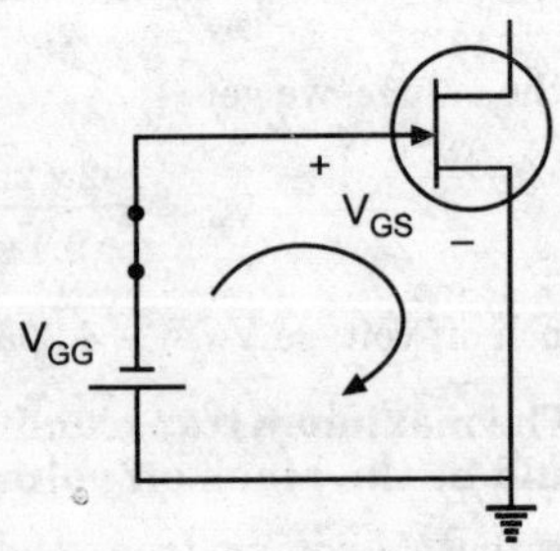

Fig. 6.53.

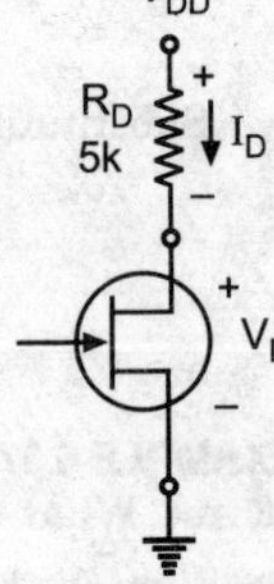

Fig. 6.54.

(ii) Calculation of I_D

We know that

$$I_D = I_{DSS}\left[1 - \frac{V_{GS}}{V_P}\right]^2$$

or $$I_D = 4\left[1 - \frac{(-2)}{(-4)}\right]^2 = 1 \text{ mA.} \quad \textbf{Ans.}$$

(iii) Calculation of V_{DS}

Applying KVL to the drain loop of figure 6.54, we obtain

$$V_{DD} = I_D R_D + V_{DS}$$

or $$V_{DS} = V_{DD} - I_D R_D$$

Let $$V_{DD} = 10 \text{ V}$$

Therefore, $V_{DS} = 10 - (1 \times 5) = 5$ V. **Ans.**

EXAMPLE 6.40 Evaluate the voltage gain and output resistance of the following circuit. Given that $g_m = 2$ mA/V and $r_d = 10$ k. (U.P. Tech., Sem. Exam. 2005-06) (10 marks)

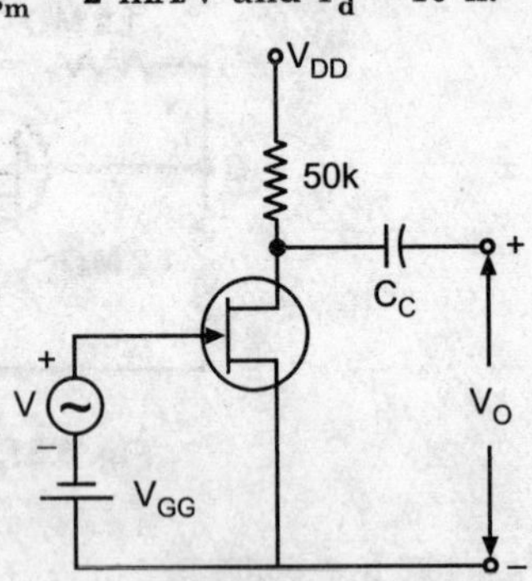

Fig. 6.55.

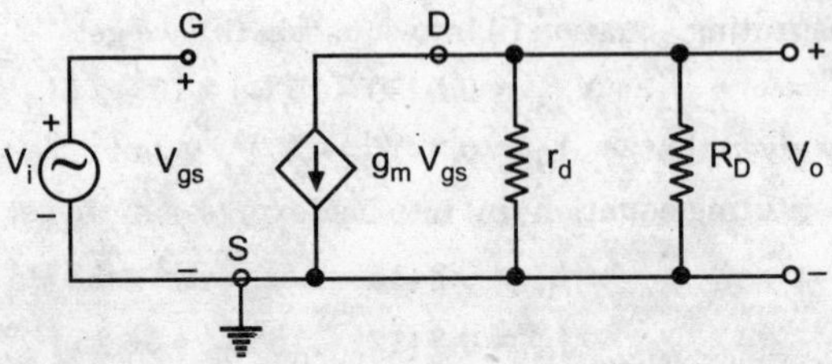

Fig. 6.56. Small signal equivalent circuit

Solution: First, let us draw the small signal equivalent circuit as shown in figure 6.56.

(i) Voltage gain

$$A_V = \frac{V_o}{V_i}$$

But $V_o = -g_m V_{gs} (r_d \parallel R_D)$ and $V_i = V_{gs}$

Therefore, $$A_V = \frac{-g_m V_{gs}(r_d \parallel R_D)}{V_{gs}} = -g_m V_{gs} (r_d \parallel R_D)$$

or $$A_V = -2 \times 10^{-3} [10\text{ k} \parallel 50\text{ k}) = -2 \times 10^{-3} \times 8.33\text{ k}\Omega = -16.67 \quad \textbf{Ans.}$$

(ii) Output Resistance

Let us draw the equivalent circuit with $V_i = 0$ as shown in figure 6.57. As $V_i = V_{gs} = 0$, $g_m V_{gs} = 0$, so the current source acts as an open circuit.

Hence $R_o = r_d \parallel R_D = 10\text{ k} \parallel 50\text{ k}$

or $$R_o = \frac{10 \times 50}{10 + 50} = 8.33\text{ k}\Omega. \quad \textbf{Ans.}$$

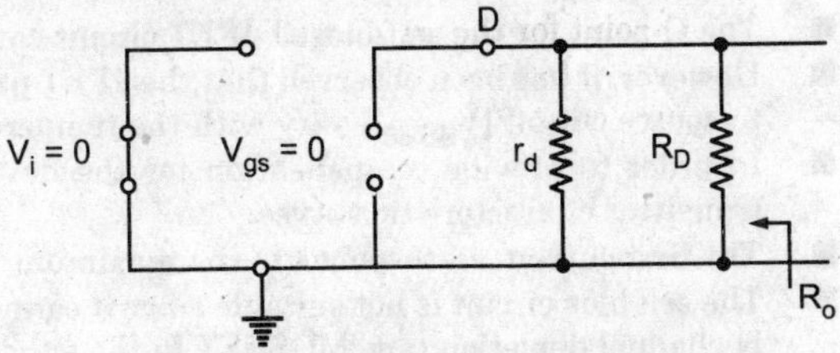

Fig. 6.57. Equivalent circuit to calculate R_o

EXAMPLE 6.41. Determine V_{GS}, I_D, V_{DS}, V_D, V_G and V_S for the following circuit. **(U.P. Tech., Sem. Exam. 2009-10) (10 marks)**

Solution :

(i) First, we calculate V_{GSQ}.

For a self bias circuit, $V_{GSQ} = -I_D R_S = -1\text{ k } I_D$

(ii) Then, we calculate I_D.

$$I_D = I_{DSS}\left[1 - \frac{V_{GSQ}}{V_P}\right]^2 = I_{DSS}\left[1 + \frac{I_D R_S}{V_P}\right]^2$$

$$I_D = 8\left[1 + \frac{1\text{ k } I_D}{-6}\right]^2 = \frac{8}{36}[6 - I_D]^2$$

Therefore, $$I_D = \frac{2}{9}(6 - I_D)^2$$

$$9 I_D = 2\left[36 + I_D^2 - 12 I_D\right] = 72 + 2I_D^2 - 24 I_D$$

or $$2I_D^2 - 33 I_D + 72 = 0$$

or $$I_D = \frac{+33 \pm \sqrt{(33)^2 - 4 \times 2 \times 72}}{4} = 13.91\text{ mA or } 2.59\text{ mA}$$

But, I_D cannot be higher than I_{DSS}, $\therefore I_D = 2.59$ mA **Ans.**

(iii) Then, we find V_{DS}.

$$V_{DS} = V_{DD} - I_D (R_D + R_S) = 20 - 4.225\,(3.3 + 1) = 1.8325\text{ V} \quad \textbf{Ans.}$$

(iv) Lastly, we find V_{GS}, V_D, V_S.

We have $V_{GS} = 1\text{ k} \times I_D = 1 \times 2.59 = 2.59$ V **Ans.**

$V_S = I_D R_S = 2.59 \times 1 = 2.59$ V **Ans.**

$V_D = V_S + V_{DS} = 2.59 + 1.8325 = 4.4225$ V **Ans.**

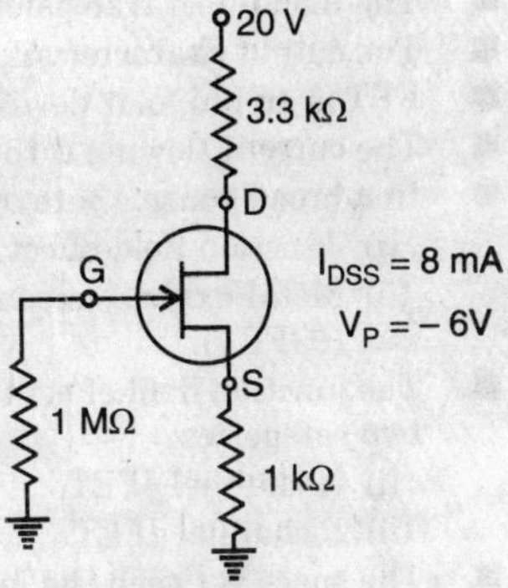

Fig. 6.58.

SUMMARY

- The biasing deals with setting a fixed level of current which should flow through the FET with a desired fixed voltage drop across the FET junctions.

- The commonly used biasing circuits for FET are as under:
  (i) Fixed-bias circuit.
  (ii) Self-bias circuit.
  (iii) Voltage divider bias circuit.
- The Q-point (i.e., quiescent or operating point or bias point) for a self-biased JFET is established by determining the value of drain current (I_D) for a desired value of gate-to-source voltage (V_{GS}) or vice-versa.
- In analytical method, we note down the values of the maximum drain current (I_{DSS}) and the gate-to-source cut-off voltage [$V_{GS(off)}$] from the data sheets of JFET supplied by the manufacturer.
- In graphical method, we draw a self-bias line such that it intersects the transfer characteristic curve near its mid-point.
- The Q-point for the self-biased JFET circuit can also be obtained graphically by using d.c. load line.
- However, it has been observed that the JFET parameters like maximum drain current (I_{DSS}) and the gate-to-source cut-off [$V_{GS(off)}$] vary with the temperature.
- In order to provide compensation for the device parameter, variation the manufacturers provide two transistor characteristic curves.
- The first current corresponds to the maximum values of I_{DSS} and $V_{GS\ (off)}$.
- The self bias circuit is not suitable since it cannot provide the positive V_{GS} required for the operation of an N-channel depletion type MOSFET in the enhancement region.
- The transfer characteristics of the enhancement type MOSFET are completely different from that of a JFET.
- The other most popular biasing arrangement for the enhancement type MOSFET is voltage divider biasing circuit.
- The field-effect transistor (FET), is a voltage-controlled device.
- The output characteristics of FET are controlled by the applied input voltage and not by the input current.
- FET is an unipolar device.
- The current flowing through it is only due to one type of charge particles, i.e., holes or electrons.
- In a broad sense, there are two main types of field-effect transistors as under:
  (i) Junction field-effect transistors (JFET).
  (ii) Metal oxide semiconductor field-effect transistor (MOSFET) or insulated gate field-effect transistor (IGFET).
- The junction field-effect transistor (JFET) may be divided, depending on their structure, into the following two categories:
  (i) N-channel JFET.
  (ii) P-channel JFET.
- The space between the junction (i.e., N-type regions) is known as channel.
- Whenever a voltage is applied across the drain and source terminals a current flows through the N-channel.
- The current carriers in JFET are the holes which flow through the P-type channel.
- This PN-junction is always reverse-bias in JFET operation.
- The reverse bias is applied by a voltage V_{GG} connected between the gate and source terminals.
- Whenever a PN-junction is reverse-biased, the electrons and holes diffuse across the junction and leave behind the positive ions on the N-side and negative ions on the P-side. The region containing these immobile ions is called depletion region.
- The reverse-bias across the gate-source junction of a JFET can also be achieved simply by applying a voltage across the drain and source terminals.
- In the presence of positive supply voltage V_{DD} (with gate open) the electrons flow from source to drain through the N-channel and constitutes a current known as drain current (I_D).
- The gate-to-source voltage (V_{GS}) at which the drain current is zero (or completely cut-off) is known as pinch off voltage.
- The value of pinch-off voltage V_P is negative for N-channel JFET. It depends upon
  (i) doping of the N and P regions of the device,
  (ii) width of the original channel structure.
- For a JFET there are two important characteristics as under:
  (i) V-I or drain characteristics
  (ii) Transfer Characteristics.

- The drain current in the pinch off region depends on the gate-to-source voltage and is given by the relation

$$I_D = I_{DSS}\left(1 - \frac{V_{GS}}{V_P}\right)^2$$

This relation is known as Shockley's equation. The pinch-off region is the normal operating region of JFET, when used as an amplifier.

- The drain-to-source voltage corresponding to point C is known as the breakdown voltage.
- Mathematically, the d.c drain resistance is expressed as

$$R_{DS} = \frac{V_{DS}}{I_D}$$

- The a.c drain resistance is expressed as

$$r_d = \frac{\Delta V_{DS}}{\Delta I_D}$$

It is, sometimes, written as r_{ds}, which indicates that it is the resistance from drain-to-source terminals. Since r_d is the output resistance of JFET, it can also be expressed as an output admittance y_{DS} (i.e., equal to $1/r_d$).

- The transconductance is expressed as

$$g_m = \frac{\Delta I_D}{\Delta V_{GS}}$$

The value of transconductance (g_m) is expressed in siemens (S) or mhos. We know that transfer characteristic of a JFET is a part of a parabola i.e., it is non-linear. Therefore, the value of g_m varies depending upon the location on the curve as set by the gate-to-source voltage (V_{GS}). It has a greater value near the top of the curve than near the bottom.

- The amplification factor is expressed as

$$\mu = \frac{\Delta V_{DS}}{\Delta V_{GS}}$$

- A field-effect transistor has a number of advantages over a bipolar junction transistor namely extremely high input resistance, less noisy, better thermal stability etc.
- In order to use the JFET as an amplifier, it is necessary to operate it in the "saturation region" of its drain characteristics. This region is very much similar to the active region of a BJT.

SHORT QUESTIONS WITH ANSWERS

Q.1. What is junction field-effect transistor (JFET)? Explain.

Ans. The junction field-effect transistor (JFET) may be divided, depending on their structure, into the following two categories:

(i) N-channel JFET.

(ii) P-channel JFET.

Q.2. What is depletion region in JFET? Expain.

Ans. Whenever a PN-junction is reverse-biased, the electrons and holes diffuse across the junction and leave behind the positive ions on the N-side and negative ions on the P-side. The region, containing these immobile ions, is called depletion region. As the reverse bias voltage across the junction is increased, the thickness of depletion region is also increased. If both the P-side and N-side of the junction are equally doped, the depletion region will extend equally in both the regions.

Q.3. What is pinch-off region? Explain.

Ans. This region has been shown by the curve BC. It is also known as the saturation region or constant current region. In this region, the drain current remains constant at its maximum value (i.e., I_{DSS}). The drain current in the pinch off region depends on the gate-to-source voltage and is given by the following relation:

$$I_D = I_{DSS}\left(1-\frac{V_{GS}}{V_P}\right)^2$$

This relation is known as Shockley's equation. The pinch-off region is the normal operating region of JFET, when used as an amplifier.

Q.4. Define breakdown region for JFET.

Ans. This region has been shown by the curve CD. In this region the drain current increases rapidly as the drain-to-source voltage is also increased. It happens because of the breakdown of gate-to-source junction due to the avalanche effect. The drain-to-source voltage corresponding to point C is known as the breakdown voltage.

Q.5. How MOSFET works as a resistor?

Ans. The MOSFETs have an important property that they can be used as a resistor capacitor, amplifier and a switch. This makes the design of electronic circuits very simple, since the entire circuit consists of only MOSFETs and no other component. The examples of such circuits are microprocessors and memory circuits.

Q.6. What is FET biasing? Explain.

Ans. We know the meaning and the need of biasing a device. As a matter of fact, the biasing deals with setting a fixed level of current which should flow through the FET with a desired fixed voltage drop across the FET junctions.

Q.7. What are the FET biasing circuits?

Ans. The commonly used biasing circuits for FET are as under:

(i) Fixed-bias circuit.

(ii) Self-bias circuit.

(iii) Voltage divider bias circuit.

Q.8. What do you mean by field-effect transistor amplifier?

Ans. The field-effect transistor (FET) has a capability to amplify a.c. signals like a bipolar junction transistor (BJT). Depending upon the types of configuration, the FET amplifiers may be classified under the following three heads:

(i) Common source amplifier

(ii) Common drain amplifier

(iii) Common gate amplifier.

Q.9. Why are field-effect transistors called unipolar transistors ?

Ans. In field-effect transistors, current conduction is by only one type of majority carriers (either by electrons or by holes) and, therefore are called unipolar transistors.

Q.10. Why the channel of a JFET is never completely closed at the drain end ?

Ans. If the channel is completely closed in JFET, then there will be no drain current, so there will be no voltage drop along the channel length and amount of reverse bias will become uniform and wedge-shaped depletion region will become rectangular one.

Q.11. How is drain current controlled in a JFET?

Ans. In a JFET, drain current is controlled by controlling the reverse bias given to its gate (i.e., V_{GS}).

Q.12. What is meant by drain characteristics of FETs ?

Ans. The curve drawn between drain current and drain-source voltage with gate-to-source voltage as the parameter is called the drain characteristic.

Q.13. What is meant by transfer characteristic of FETs ?

Ans. The curve drawn between drain current and gate-source voltage for a given value of drain-source voltage is called the transfer characteristic.

Q.14. What is pinch-off voltage in a JFET ?

Ans. The value of drain-source voltage at which channel is pinched-off (i.e., all the free charges from the channel get removed) is called the pinch-off voltage.

Q.15. What is meant by saturation region ?

Ans. The region of drain characteristic of a FET in which drain current remains fairly constant is called the saturation or pinch-off region.

Q.16. Why is input impedance of the FET very high ?

Ans. FET has very high input impedance because its input circuit (gate-to-source) is reverse biased and the input gate current is very small (of the order of few nano-amperes).

Q.17. What is the significant difference between the construction of an enhancement type MOSFET and a depletion type MOSFET ?

Ans. In the depletion type MOSFET, a channel is physically constructed and a current between drain and source is due to voltage applied across the drain-source terminals while in enhancement type construction, no channel is formed during its construction. Voltage is applied to the gate in case of enhancement type MOSFET, to develop a channel of charge carriers so that a current results when a voltage is applied across the drain-source terminals.

Q.18. Why E-MOSFET is called sometimes normally-off MOSFET ?

Ans. E-MOSFET operates with large positive gate voltages only and does not conduct when gate-source voltage $V_{GS} = 0$, so it is called normally-off MOSFET.

Q.19. Why MOSFETs are never connected or disconnected in the circuit when power is ON ?

Ans. If a MOSFET is connected or disconnected in circuit while the power is ON, transient voltages caused by inductive kickback and other effects may exceed $V_{GS(max)}$ and thus wipe out the MOSFET.

Q.20. Name the factors which make the JFET superior to BJT?

Ans. The high input impedance, low output impedance and low noise level make JFET far superior to the BJT.

Q.21. Explain FET is a voltage controlled device or current controlled device.

Ans. The voltage applied between gate and source (V_{GS}) controls the drain current I_D. Therefore, FET is a voltage controlled device. The transistor on the contrary is a current controlled device because the output current I_C is controlled by varying the base current I_B. The name field effect is derived from the fact that the current flow in FET is controlled by an electric field set by an external voltage.

Q.22. Explain FET is an unipolar device or a bipolar device.

Ans. FET is a unipolar device. This means that the current flowing through it is only due to one type of charge particles, holes or electrons. Transistor, on the other hand, is a bipolar device as holes and electrons both contribute to the flow of current. Just as there are NPN and PNP transistors, there are n-channel and p-channel field effect transistors. One of the most important characteristics of the FET is its very high input impedance. Typically, it is in the range of 1 to several megaohms, which is much higher than the input resistance of a BJT. FETs are more temperature stable as compared to the BJT and it requires less space than that for a BJT. Therefore, FETs are preferred in integrated circuits.

Q.23. Write the Advantages of JFET Over BJT

Ans. The FET enjoys several advantages over the conventional bipolar junction transistor (BJT). Some of them may be listed as under :

(i) It is a unipolar device so its operation depends on the flow of majority carriers only.

(ii) It is relatively immune to radiation.

(iii) FET has a very high input resistance (typically few megaohms).

(iv) FET is less noisy.

(v) It does not exhibit any offset voltage at zero drain current hence it can be used as an excellent signal chopper.

(vi) FET has a better thermal stability.

Q.24. Write Disadvantage of JFET.

Ans. The main disadvantage of the JFET is that it has a relatively small gain-bandwidth product as compared with that of a conventional transistor (i.e., BJT).

Q.25. Why are ordinary transistors called bipolar transistors ?

Ans. In ordinary transistors, current conduction is by both majority carriers and minority carriers (both by holes as well as electrons) and, therefore, these are called bipolar transistors.

Q.26. No current flows through the channel of a JFET when $V_{GS} = 0$ and $V_{DS} = 0$, though channel is fully-open. Why ?

Ans. When drain-source voltage is zero, there is no attracting potential at the drain, so no current flows inspite of the fact that the channel is fully open.

Q.27. What are the parameters that control the pinch-off voltage of JFET ?

Ans. Parameters that control the pinch-off voltage of JFET are electron charge, donor/acceptor concentration density, permittivity of channel material and half width of channel bar.

Q.28. Why FET is called a voltage controlled device ?

Ans. In a FET, drain current is controlled by the effect of the extension of the field associated with the depletion region developed by the reverse bias on the gate, so it is called a voltage controlled device.

Q.29. How does the FET behave (i) for small values of $|V_{DS}|$ and (ii) for large values of $|V_{DS}|$?

Ans. (i) FET behaves as an ordinary resistor for small value of $|V_{DS}|$ i.e. in ohmic region.

(ii) FET behaves as a constant current source for large values of $|V_{DS}|$ till breakdown occurs.

Q.30. How does the current vary with the gate voltage in the saturation region ?

Ans. Drain current decreases with the increase in gate-source bias $|V_{GS}|$. When VGS = 0; drain current $I_D = I_{DSS}$, drain-source saturation current and when $V_{GS} = V_p$; drain current $I_D = 0$.

Q.31. How does the transconductance vary with drain current ?

Ans. Transconductance g_m of a FET varies with the drain current I_D as given by the following equation

$$g_m = \frac{2}{|V_p|}\sqrt{I_{DS} \cdot I_{DSS}}$$

Q.32. What is meant by gate-source cut-off voltage ?

Ans. The gate-source bias voltage required to reduce the drain current to zero is designated the gate-source cut-off voltage V_{GS} (off)- it is equal to pinch-off voltage V_p.

Q.33. What is meant by drain-source saturation current I_{DSS} ?

Ans. The drain current in pinch-off or saturation region with zero gate-source voltage ($V_{GS} = 0$) is referred to the drain-source saturation current I_{DSS}.

Q.34. Why is input impedance of the FET very high ?

Ans. FET has very high input impedance because its input circuit (gate-to-source) is reverse bias and the input gate current is very small (of the order of few nanoamperes).

Q.35. What is the value of gate-source voltage V_{GS} that gives drain current of both N- and P-cha JFETs a zero temperature coefficient ?

Ans. $|vgs|$ $|V_P|-0.63V$.

Q.36. What is dynamic resistance of a JFET ?

Ans. The ratio of change in drain-source voltage to change in drain current at a given gate-source voltage is known as ac drain resistance or dynamic resistance r_d

i.e., $$r_d = \frac{\Delta V_{DS}}{\Delta I_D} \text{ at constant } V_{GS}.$$

Q.37. What is meant by transconductance with reference to JFET ?

Ans. The control that gate-source voltage has over the drain current is measured by the transconductance of a JFET. It may be defined as the ratio of change in drain current to the change in gate-source voltage for a given value of drain-source voltage i.e.

$$g_m = \frac{\Delta I_D}{\Delta V_{GS}} \text{ at constant } V_{DS}.$$

Q.38. What are the main disadvantages of JFETs ?

Ans. JFETs have the drawbacks of small gain, small gain-bandwidth product and gate leakage current as compared to MESFETs.

Q.39. How does MESFET differ from P-N junction FET ?

Ans. MESFET is the same basic device (JFET) except that the PN junction is replaced by a Schottky barrier rectifying junction. The PN junction FET is fabricated with PN junction where as the MESFET is fabricated with a Schottky barrier rectifying junction.

Q.40. Why MESFET is usually made from GaAs or other compound semiconductor materials rather than from silicon?

Ans. GaAs or InP semiconductors are used instead of silicon in MESFET fabrication because of higher mobility and carrier drift velocity. Furthermore, GaAs can be operated at higher temperature (and, therefore higher power levels).

Q.41. Enumerate the special features of MESFETs.

Ans. The MESFETs have special features of possibility of operation at high temperature (and, therefore, higher power levels); high operational speed, higher carrier drift velocities (about twice of silicon), higher electron mobility (about six times of silicon) and high frequency operation (because drift-time and capacitance may be reduced to minimum).

Q.42. What is the main drawback of MESFET ?

Ans. The main drawback of MESFET is a limited gate voltage swing and that is due to small turn-on voltage of the Schottky gate. But this limitation is not of much significance in small power circuits operating with a low supply voltage.

Q.43. Enumerate the advantages of MESFETs.

Ans. MESFETs have the advantages of higher gain, higher gain-bandwidth product and relatively smaller gate leakage currents, lower parasitic capacitances; Good heat dissipation for power devices; lower resistance and therefore, low voltage drop along the channel width; Lower parasitic resistances, larger transconductances and shorter electron transit time, and possibility of close geometrical tolerances and smaller channel lengths because MESFET fabrication does not require diffusion. Furthermore, the use of GaAs in MESFET fabrication provides higher mobility and higher carrier drift velocity. Higher electron mobility results in lower noise figures.

Q.44. Enumerate the applications of MESFETs.

Ans. MESFETs are employed in high performance circuits of communication, computers and military systems, microwave amplifiers, oscillators, switches and mixers.

Q.45. Give the advantages of HEMT over silicon MOSFET.

Ans. The advantages of HEMT over silicon MOSFET are the higher electron mobility and maximum electron velocity in AlGaAs/GaAs heterojunction compared with Si/SiO_2 interface. The high performance of the HEMT translates into an extremely high cut-off frequency, and devices with fast access times.

Q.46. Why HEMT or MODFET, is also referred to as a two-dimensional electron gas FET ?

Ans. The HEMT or MODFET, is also referred to as a two-dimensional electron gas FET (2-DEG FET, or TEGFET) to emphasize the fact that conduction along the channel occurs in a thin sheet of charge.

Q.47. Why HEMT is also referred to as SEDFET ?

Ans. HEMT is also referred to as a separately doped FET (SEDFET), to emphasize the fact that the doping occurs in a separate region from the channel.

Q.48. Why is MOSFET called sometimes IGFET ?

Ans. MOSFET is constructed with the gate terminal insulated from the channel so it is sometimes called insulated gate FET (or IGFET).

Q.49. How does MISFET differ from MOSFET ?

Ans. Metal-insulator semiconductor field-effect transistor (MISFET) is the more general term where insulator is not necessarily SiO_2 and semiconductor is not necessarily Si, where as in case of a MOSFET SiO_2 is necessarily an insulator and Si a semiconductor.

Q.50. What is the significant difference between the construction of an enhancement type MOSFET and a depletion type MOSFET ?

Ans. In the depletion type MOSFET a channel is physically constructed and a current between drain and source is due to voltage applied across the drain-source terminals while in enhancement type construction no channel is formed during its construction. Voltage is applied to the gate in case of enhancement type MOSFET, to develop a channel of charge carriers so that a current results when a voltage is applied across the drain-source terminals.

Q.51. Why E-MOSFET is called sometimes normally-off MOSFET ?

Ans. E-MOSFET operates with large positive gate voltages only and does not conduct when gate-source voltage $V_{GS} = 0$, so it is called normally-off MOSFET.

Q.52. What is meant by gate-to-source threshold voltage in E-MOSFET ?

Ans. The minimum value of gate-to-source voltage V_{GS} that is required to form the inversion layer is termed the gate-to-source threshold voltage V_{GST}.

Q.53. How is threshold voltage of the MOS transistor adjusted ?

Ans. Threshold voltage of a MOS transistor can be adjusted by the following three methods.

(i) If the silicon crystal orientation is $< 111 >$, it results in high threshold voltage. If the crystal has orientation of $< 100 >$, the value of threshold will go to half as compared to the previous one.

(ii) Instead of SiO_2, a layer of Si_3N_4 and SiO_2 is used. It makes the dielectric constant twice as that of SiO_2 alone. This causes V_{GST} to come down to half.

(iii) Instead of aluminium as gate electrode polycrystalline silicon doped with boron is used as the gate electrode. This causes reduction in contact potential between gate electrode and gate dielectric resulting in reduction of V_{GST}.

Q.54. Why MOSFETs are never connected or disconnected in the circuit when power is ON ?

Ans. If a MOSFET is connected or disconnected in circuit while the power is ON, transient voltages caused by inductive kickback and other effects may exceed $V_{GS\,(max)}$ and thus wipe out the MOSFET.

Q.55. Name the factors which make the JFET superior to BJT ?

Ans. The high input impedance, low output impedance and low noise level make JFET far superior the BJT.

REVIEW QUESTIONS

1. Illustrate with neat circuit diagrams the technique of biasing an enhancement MOSFET.
2. Explain giving necessary diagrams how JFET can be biased suitable to overcome the effect of unit to unit variation.
3. Explain briefly the technique of biasing a JFET for zero current drift with temperature.
4. Write notes on: Biasing techniques of MOSFET.
5. Explain any one method of biasing a single stage JFET amplifier.
6. With the help of transfer curve and bias line, explain the biasing techniques of N-channel FET.
7. With the help of neat diagrams explain the operation and drain characteristics of N-channel MOSFET.
8. Draw circuit diagram and describe the following applications of MOSFET:
   (i) Active load switching
   (ii) CMOS inverter.
9. Write a short note on: Depletion mode MOSFET.
10. Explain giving illustrative diagrams, how the pinch off condition occurs in a MOSFET.
11. Explain the difference between the operation of a JFET in ohmic and current saturation regions. Also explain the cut off condition.
12. What is thermal runway in BJT ? How can it be avoided? Is the thermal runway possible in JFET?
13. Draw the laboratory experimental setup for plotting output static characteristics of a JFET. Explain clearly, different operating regions in characteristics with proper reasoning.
14. Compare BJT and FET.
15. Define the parameters: Amplification factor, drain resistance, and mutual conductance for JFET and obtain the expression relating them.
16. Write a short note on MOSFET.
17. Compare JFET and MOSFET.

18. Give reason: FET behaves as a voltage controlled current source.
19. Sketch the drain characteristics of a JFET and explain its behaviour before and after pinch off.
20. Write a short note on: FET as VVR.
21. State true or false with reason.
    (i) The relationship between I_D and V_{GS} is non-linear.
    (ii) Gate current in FET is always neglected.
    (ii) In JFET, the P-N junction between gate and source is always kept in reverse biased condition.
    (iv) FET is a current controlled device.
22. What do you understand by "pinch off" voltage and cut-off voltage as applied to FET?
23. Draw the drain characteristics of depletion type MOSFET. Explain different operating regions in the characteristics with proper reasoning.
24. Draw the circuit diagram of JFET as an amplifier and explain its working.
25. Draw the a.c. equivalent circuit of JFET and mention important points.
26. Obtain the approximate equivalent circuit for JFET as an amplifier.
27. Draw the circuit diagram of JFET amplifier using fixed-bias configuration and explain the process.
28. What is the significant difference between the construction of an enhancement type MOSFET and a depletion type MOSFET ?
29. What is meant by gate-to-source threshold voltage in E-MOSFET ?
30. An n-type channel FET is superior to p-type channel FET because .... of electrons is greater than that of holes. **(U.P. Tech. First Semester Exam. 2008-09)(02 marks)**
31. Which one of the following has the highest input resistance?
    (*a*) npn transistor in CB configuation
    (*b*) pnp transistor in CE configuration
    (*c*) n-type channel JFET
    (*d*) p-type channel MOSFET **(U.P. Tech. First Semester Exam. 2008-09)(02 marks)**
32. How is an FET used as a voltage variable resistor? Explain.
    (U.P. Tech. First Semester Exam. 2008-09)(05 marks)
33. Sketch the structure of a p-channel depletion type MOSFET and explain its principle of operation with neat diagrams. Also sketch its V-I characteristics and circuit symbols for it.
    (U.P. Tech. First Semester Exam. 2008-09)(10 marks)
34. Define trans-conductance (g_m), output resistance (r_d) and gain of an FET. How these parameters are determined from graphically? **(U.P. Tech. Second Semester Exam. 2008-09)(06 marks)**
35. The operation of a JFET involves.
    (*a*) A flow of minority carriers
    (*b*) A flow of majority carriers
    (*c*) Recombination
    (*d*) Negative resistance
    (U.P. Tech. First Semester Exam. 2009-10)(02 marks)
36. Define the following :
    1. Drain to source saturation current of JFET
    2. Pinch off voltage of JFET
    3. Voltage controlled resistance of JFET **(U.P. Tech. First Semester Exam. 2009-10)(02 marks)**
37. What is the major difference between a bipolar and an unipolar device? Explain with example.
    (U.P. Tech. Second Semester Exam. 2009-10)(05 marks)
38. What are the advantage of FET over BJT? Explain.
    (U.P. Tech. Second Semester Exam. 2009-10)(05 marks)
39. Describe the construction and operation of a MOSFET in enhancement mode. Draw its characteristics and equivalent circuit of the device. **(U.P. Tech. Second Semester Exam. 2009-10)(10 marks)**
40. When the positive voltage on the gate of a p-channel JFET is increased, its drain current:
    (U.P. Tech. Second Semester Exam. 2009-10)(02 marks)
    (i) increases
    (ii) decreases
    (iii) remains the same
    (iv) none of the above
41. For common source FET amplifier with source resistance R_s, derive the expression for voltage gain input impedance and output impedance. **(U.P. Tech. Second Semester Exam. 2008-09)(06 marks)**

□□□

UNIT-V

Syllabus

Some Special Diodes: Photodiodes, Photo detectors, Solar Cell, Light Emitting Diodes, Semiconductor Lasers, Light Emitting Materials.
Tunnel Diode: Degenerate Semiconductors.
IMPATT Diode
The Transferred Electron Mechanism: The GUNN Diode
P-N-P-N diode, Semiconductor Controlled Rectifier (SCR), Bilateral Devices: DIAC, TRIAC, IGBT.

Above syllabus has been covered in following chapters :

Chapter 7

Photonic Devices

Syllabus

Photodiodes, Photodetectors, Solar cell, Light emitting diodes, Semiconductor lasers, Light emitting materials.

Chapter Outline

7.1. INTRODUCTION

As a matter of fact, with the development of science, man has made a lot of developments in the field of optics, from the early days of candle and now to the latest development like lasers. But the basic principles of each light source is same, that is-principle of conversion of electrical energy into light energy. In this chapter we shall discuss about different optoelectric devices. Optoelectronics basically, is the branch of electronics which deals with optics as well as electronics. Optics takes into account the effect of light and the electronics involved takes care of the processing part. The optoelectronic devices may be classified into following three categories as:

(i) Light emitters

(ii) Light sensors or detectors

(iii) Optical couplers or optoisolator.

7.2. CLASSIFICATION OF OPTOELECTRONIC DEVICES

The optoelectronic devices may be classified as light emitters, light detectors optocouplers or isolators.

(i) Light emitters

These devices give out light whenever excited electrically. Light emitters are also known as light sources.

(ii) Light detectors or sensors

These devices can convert light into some form of electrical energy, or they can modify some electrical parameters such as resistance.

(iii) Optocouplers or Optoisolators

These are formed by combining a light source and a light detector into one device. Optocouplers couple the signals electrically from one point to the other but isolate the two points physically. Figure 7.1 shows the classification of optoelectronic devices.

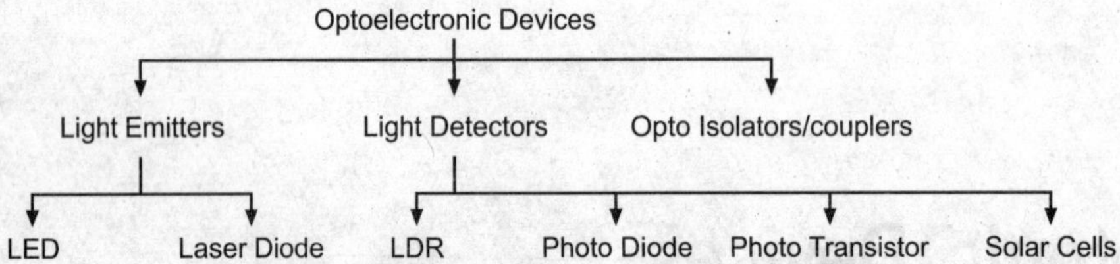

Fig. 7.1. Classification of optoelectronic devices

> **Important Point:** The optoelectronic devices may also be classified as photoemissive, photoconductive and photovoltaic devices.

7.3. PHOTOEMISSIVITY

(U.P. Tech, Sem. Exam; 2007-08, 2009-10)

1. Definition

The liberation of electrons from matter under the influence of light is called as the photoelectric effect. This was first observed by Hertz in 1887. If electrons are liberated from a metal surface under the influence of light then the phenomenon is known as **photoemissive effect.** This phenomenon is observed in the vacuum and gas filled **phototubes.** The experimental setup to understand the characteristics of the photoemissive effect is as shown in figure 7.2.

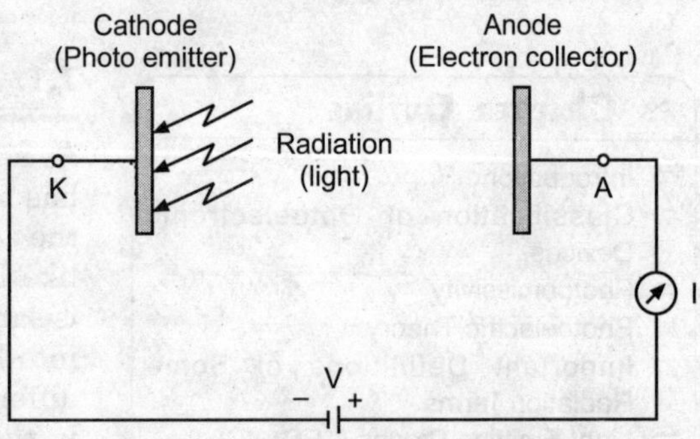

Fig. 7.2 Setup to measure the photoelectric current as a function of voltage V between photoemitter and photocollector

2. Important Characteristic

The important characteristics of the photoemissive effect may be listed as under:

(i) When the light is incident on the cathode which is made of some photo photosensitive material, it starts emitting electrons. The number of electrons emitted by the cathode is proportional to the intensity of incident light. This means that the photoelectric current is directly proportional to the intensity of light.

(ii) Keeping the intensity of incident light constant, if we increase the positive voltage V applied between the cathode and anode, then the number of electrons to the collector increase until saturation occurs.

(iii) The maximum energy of the photoelectrons emitted by the cathode is independent of light intensity but varies linearly with the frequency of incident light.

(iv) The photoelectric characteristics are practically independent of temperature, within a wide range of temperature.

(v) There is no time lag between the instant of exposure of the cathode to light and the instant of starting the emission.

3. Photoemissive sensors

(i) The basic phototubes, vacuum or gas-filled depend upon the emission of electrons from the photosensitive cathode, under the influence of light.

(ii) The cathode can be made to have different spectral responses.

(iii) One of the photoemissive device that is still very much used is the **photomultiplier tube**. It is used for photocurrent amplification. This tube is capable of multiplying th e photocurrent by 10 millions.

(iv) The photomultiplier tube operates on the principle of secondary emission. The operating voltages for such tubes are 1 kV to 5 kV.

(v) These tubes are very sensitive, they produce very low noise and low dark current. Therefore, they are used in radiation detectors or X-rays and gamma rays, in space exploration and laser communications.

7.3.1 Retarding Potential

When we increase the applied voltage, the electrons will start getting retarded. The value of applied voltage needed to fully retard the electrons to reduce the photocurrent to zero is known as the retarding potential and it is denoted by V_r.

Important Point: Here, it may be noted that the retarding potential is negative and has polarities opposite to those of applied voltage.

7.3.2. Plot of Photocurrent Versus Applied Voltage

1. Graph of Photocurrent (I) Versus Voltage (V)

Let us consider figure 7.3 which shows the graphical relation between the applied voltage and the photocurrent.

Important Point: Here, it may be noted that the frequency of the incident light is kept constant.

2. Accelerating Potential

When anode voltage is positive with respect to cathode, the voltage (V) is known as the **accelerating potential**. It is a positive potential. With increase in the accelerating potential the number of electrons reaching the anode increases upto a point of saturation. After the saturation point, the number of electrons reaching the anode will remain constant independent of the accelerating potential.

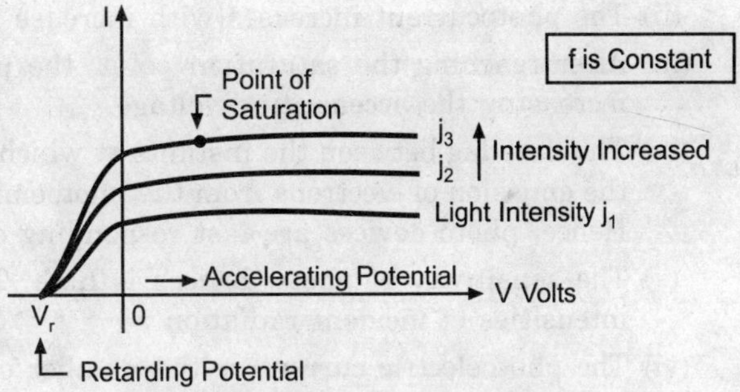

Fig. 7.3. A graph of photocurrent (I) versus voltage V at constant f

In figure 7.3 we have plotted a set of characteristics for different values of light intensities keeping the frequency of radiation constant.

Here, it may be noted that the retarding potential V_r for all the curves is same. It does not depend upon the intensity.

7.3.3. Graph of I Versus V at Constant Intensity, but Variable Frequency

(i) Now, let us observe the effect of frequency on the graph of I versus V while keeping the intensity of light constant.

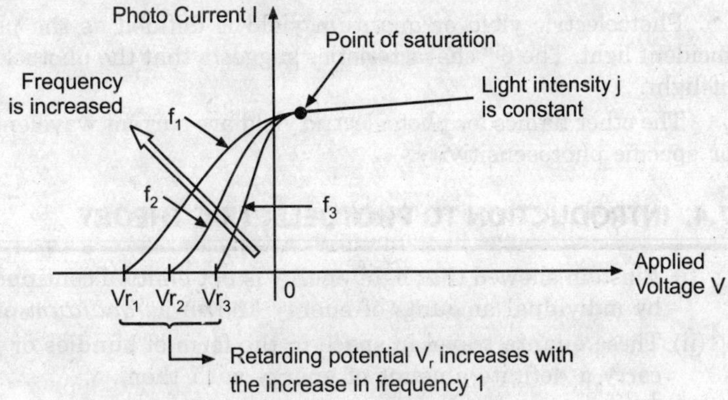

Fig. 7.4. Graph of photocurrent V_r applied voltage for constant intensity but variable frequency

(ii) Let us consider figure 7.4 which shows the effect of light of different frequencies focussed on the cathode at a constant intensity.

(iii) Figure 7.4 illustrates the relation between the retarding potential and frequency. It also shows that the magnitude of retarding potential V_r increases with increase in frequency. This is because of the fact that the maximum velocity V_{max} is higher for higher frequency and the retarding voltage required to reduce it to zero is also higher for higher frequency.

(iv) Hence, V_{max} and V_r both increase with the increase in frequency of incident light.

7.3.4. Relation between f and V_r

The relation between the retarding voltage and the frequency of incident radiation is a linear one as depicted in figure 7.5.

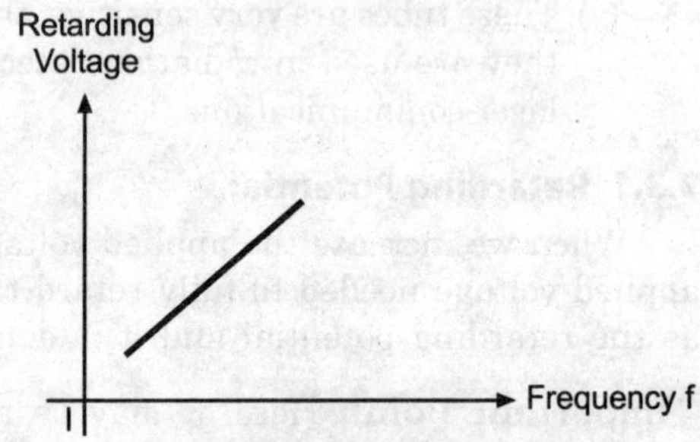

Fig. 7.5. Linear relation between retarding voltage

7.3.5. Characteristics of Photoemissivity

Few important characteristics of photoemissivity may be listed as under:

(i) The retarding voltage V_r does not depend on the intensity of the incident radiation but it is linearly dependent on the frequency of incident radiation.

(ii) The photocurrent increases with increase in intensity.

(iii) After reaching the saturation point, the photocurrent remains constant irrespective of increasing the accelerating voltage.

(iv) The time lag between the instants at which the light is focussed and the instant at which the emission of electrons from the photoemissive surface is very small (few nanoseconds). Hence, photo devices are fast responding devices.

(v) The sensitivity of photo devices is high. They can produce large phtocurrents for low intensities of incident radiation.

(vi) The photoelectric current or the number of liberated electrons in response to the same intensity of incident radiation, changes with the change in frequency of the incident radiation. In other words the number of liberated electrons for red incident light is different than that for the blue one. The photodevices are therefore said to selective.

7.3.6. Photoelectric Yield or Quantum Yield

Photoelectric yield or quantum yield is defined as the photocurrent in amperes per watt of incident light. The 6th characteristics suggests that the photoelectric yield depends on the frequency of light.

The other names for photoelectric yield are current wavelength characteristics, spectral response or specific photosensitivity.

7.4. INTRODUCTION TO PHOTOELECTRIC THEORY

(i) Einstein showed that light energy is not emitted continuously but it is emitted intermittently by individual amounts of energy known as *quantum* of energy.

(ii) These quanta travel in space in the form of bundles or packets known as photons. Photons carry a definite amount of energy with them.

(iii) These photons do not spread out in the form of waves. They are localized bits of radiant energy.

(iv) Hence, radiant energy is always emitted in the form of *bundles* or *packets* called as photons. And radiant energy is the energy which can be transmitted through space without the necessity of some transmitting agent.

(v) The energy contained by a photon is expressed as

$$E = hf \quad ...(7.1)$$

Here, E = Energy in eV

h = Plancks cosntant = 6.623×10^{-34} Joule/sec. or 4.137×10^{-13} eVs

and f = Frequency in hertz

(vi) Equation (7.1) illustrates that the energy contained by a photon is dependent directly on the frequency. Hence, with increase in frequency the energy contents will also increase.

(vii) But, light is not expressed in terms of its frequency but it is referred in terms of its wavelength λ.

Thus, we have

$$\lambda = \frac{c}{f} \quad ...(7.2)$$

Here, c is the speed of light, expressed in meter per second. If frequency f is expressed in Hz, then the wavelength λ will be in meters. But, wavelength of light is so small that "meter" is too big unit for it. Therefore, the wavelength is normally expressed in Å = where 1 A° = 1×10^{-10} meters.

7.4.1. Einstein's Equation

(i) It is essential to define the work function of a metal before discussing the Einsteins equation. The work function of a metal is the minimum amount of energy that has to be given to the fastest moving electron at 0° K so as to make the electron escape from the metal.

(ii) For a monochromatic light of frequency f incident on a metal, the velocity of emitted electron is given by the Einstein's equation as follows:

$$\frac{1}{2}mv^2 \le (hf - Uw) \quad ...(7.3)$$

(iii) Here, $1/2\ mv^2$ is the energy, Uw is the work function expressed in Joules. The corresponding energy Ew represents the work function in eV. Thus, Ew and Uw both represent the work function but in different units.

7.4.2. Threshold Wavelength (λ_c)

(i) We can define the critical frequency f_c as that frequency of the incident light which just causes the photoelectric emission to take place.

(ii) The corresponding velocity would be equal to zero. Hence, equation (7.3) gets modified to

$$0 = h_{fc} - Uw$$

Therefore,

$$f_c = \frac{Uw}{h} \quad ...(7.4)$$

(iii) The wavelength corresponding to the critical frequency (f_c) is known as the threshold wavelength (λ_c). The relation between λ_c and f_c is as under:

$$f_c\ \lambda_c = c \quad ...(7.5)$$

Here, c = velocity of light

(iv) Therefore,

$$\lambda_c = \frac{c}{f_c} = \frac{c}{Uw/h} = \frac{hc}{Uw} \quad ...(7.6)$$

7.4.3. Spectral Response

1. Definition

Spectral response may be defined as the graph of spectral sensitivity (on Y axis) plotted against the wavelength (on X axis).

2. Graph for Spectral Response

Figure 7.6 shows the spectral response curves for different materials such as Sodium (Na), Potassium (K). The spectral response does not exist for frequencies below f_c. This means that the cutoff occurs at the point $f = f_c$.

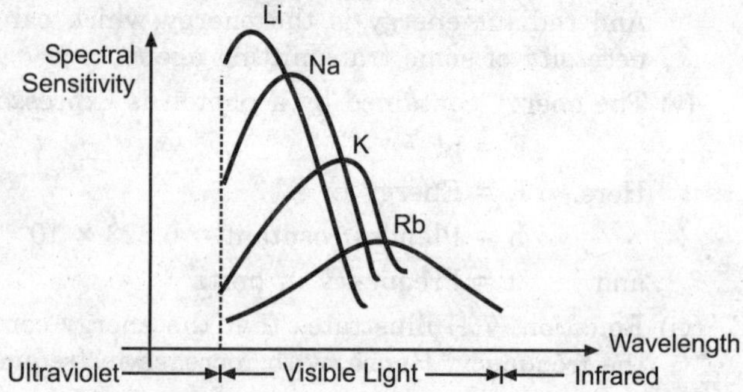

Fig. 7.6. Spectral response of various materials

3. Basic Aspects

(i) As the frequency is increased above f_c the energy of the incident photons (hf) will increase and some electrons in the levels the maximum energy state are permitted to escape.

(ii) Therefore, the response increases with increase in frequency i.e. with decrease in wavelength. Then there comes the point of maximum response.

(iii) The maximum point occurs at different wavelengths for different metals. If the energy of the light is U Joules and the number of photons in the light beam is (Uhf).

(iv) Thus, as f increases, hf increases, so number of photons (U/hf) decrease and the photocurrent decreases. This is the explanation for reduction in response with increase in f or decrease in λ.

4. Importance of spectral response

From the spectral response, we can understand the range of frequency or wavelengths for which we get a high response. This range is different for every material as depicted in figure 7.6. Hence, we can select the material which may suit the frequency range of our applications.

7.5. FEW IMPORTANT DEFINITIONS OF SOME RADIATION TERMS

In this article let us define some important terms related to radiation.

1. Radiant flux

The power in watts of the light beam is given by the radiant flux. It is possible to obtain the radiant flux if the number of photons striking the surface and energy of each photon is known.

2. Irradiation

Irradiation can be defined as the total radiant power density incident on the receiving surface. The unit of irradiation is mW/cm^2.

3. Illumination

The concept of illumination is based upon the ability of the human eye to see different wavelengths of radiation. Figure 7.7 illustrates a plot of relative spectral response of the human eye and it is called as the standard luminosity curve. Figure 7.7 shows that the human

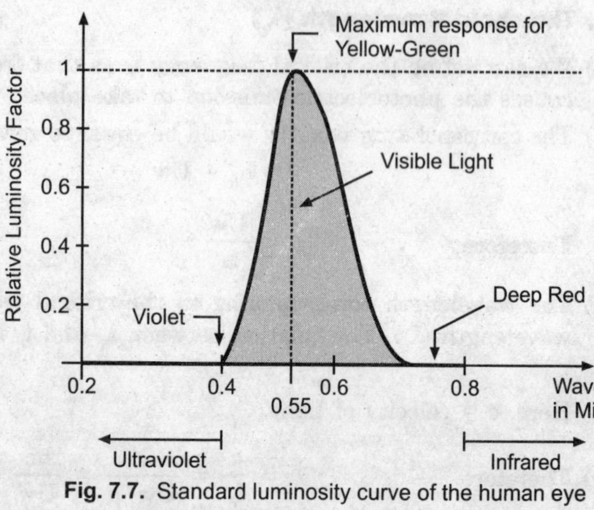

Fig. 7.7. Standard luminosity curve of the human eye

eye responds to light between 0.38 and 0.76 microns and it has the maximum response at a wavelength of 0.55 micron which corresponds to yellow green. A radiant power density which is weighted in proportion to the standard luminosity curve is called as illumination. The unit of luminous flux is called the lumen. It is defined so that 1 W of light at $\lambda = 0.5$ microns may be equal to 680 lumen.

Important Point: This means that 1 lumen = 0.0016 W for yellow-green light. From the standard luminosity curve we can observe that 1 W of light at $\lambda = 0.61$ microns is equal to 360 lumens. One candle power corresponds to a total flux of 4π lumens. One foot candle is defined as one lumen per square foot.

7.5.1. Electromagnetic Spectrum

1. Basic Aspect

The complete range of wavelength emitted by light is called as *spectrum*. According to the wave theory of light, it is in the form of electromagnetic waves. The spectrum shown in figure 7.8 covers both the visible and invisible regions.

2. Important Observations

The important observations from the electromagnetic spectrum may be listed as under:

(i) The energy content decreases with increase in wavelength.

(ii) The visible light occupies a very small portion of the complete electromagnetic spectrum. It generally covers the range from 380 μm to 780 μm.

(iii) The ultraviolet region is present at about 4000 Å, and at 8000 Å the infrared region will begin.

The optoelectronic devices will work over a limited range of the spectrum i.e. between the visible and infrared region.

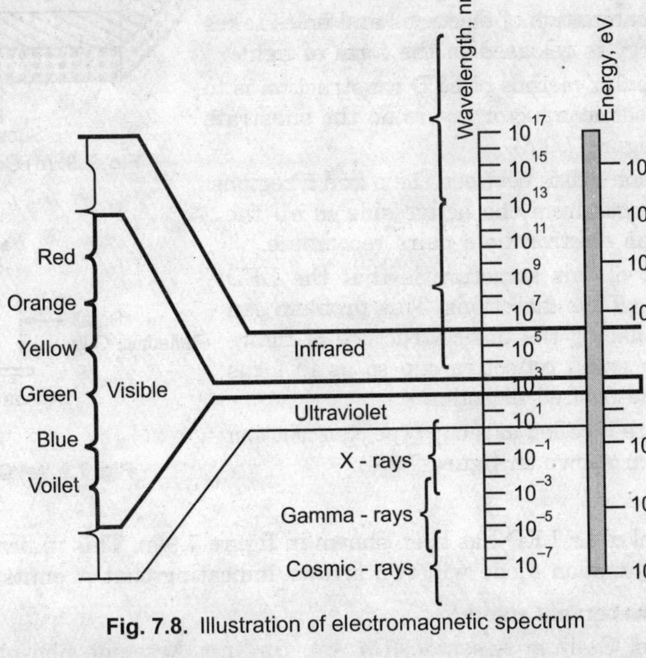

Fig. 7.8. Illustration of electromagnetic spectrum

7.6. LIGHT EMITTING DIODES (LEDs)

(U.P. Tech, Sem. Exam; 2010-11)

In this article, let us discuss about one of the important light sources which is the Light Emitting

Diode (LED). An LED emits light when electrical energy is applied to it. The construction and biasing of LED is as shown in figure 7.9(a) and (b).

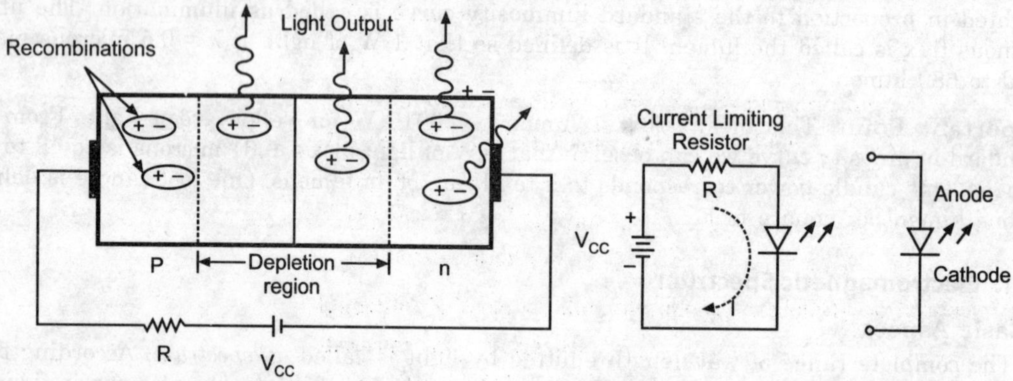

(a) Construction and Principle of Operations of a LED (b) LED Biasing (c) Circuit Symbol of LED

Fig. 7.9.

7.6.1. Construction of LED

1. Basic Aspects

(i) The construction of LED is same as that of a P-N junction semiconductor diode and the LED is operated in the forward biased region a shown in figure 7.9(b). R is the current limiting resistor.

(ii) The LED operates on the principle which states that when the recombination of electrons and holes takes place, an energy is released in the form of light.

(iii) One of the popular methos of LED construction is to deposit three semiconductor layers on the substrate as shown in figure 7.9(d).

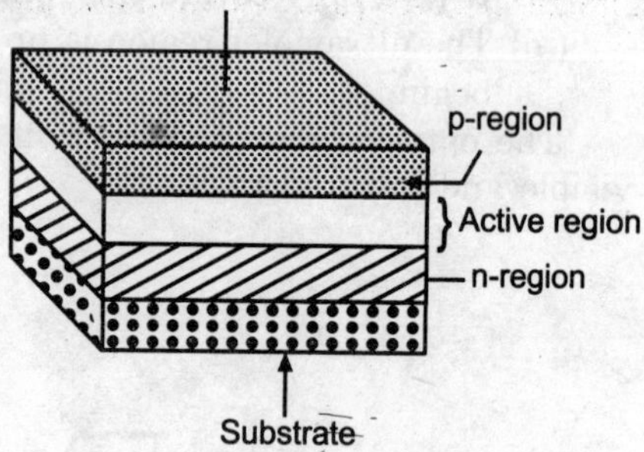

Fig. 7.9. (d) Construction of LED

(iv) The active region exists between the p and n regions. The light emerges from the active side in all the directions when electron hole pairs recombine.

(v) The drawback of this structure is that the LED emits light in all the directions. This problem can be solved by placing the basic structure of figure 7.9(d) inside a small reflective cup so as to focus the light in the desired direction.

(vi) Such a structure is called as a cup type construction and it has been shown in figure 7.9(e).

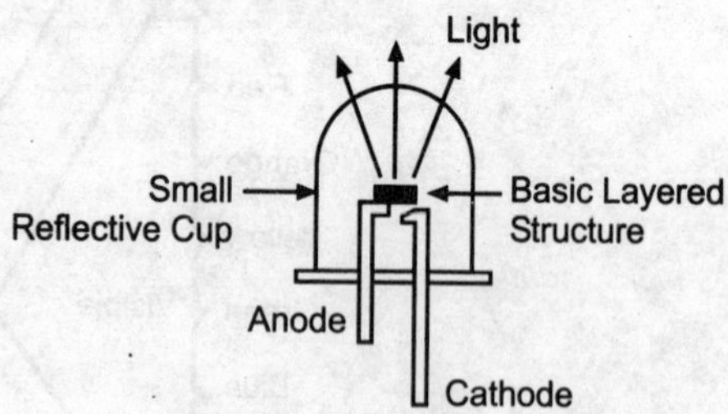

Fig. 7.9. (e) Cup type construction

2. Circuit Symbol

The circuit symbol of an LED has been shown in figure 7.9(c). This indicates that the symbol is identical to a P-N junction diode wih two arrows indicating that it emits light.

3. Semiconductor materials used

LEDs are made of Gallium Arsenide (Ga As), Gallium Arsenide phosphide (Ga As P) and Gallium phosphide (Ga P). Silicon and Germanium are not used since they are essentially heat producing materials and are very poor in producing light.

* Punishment is a sort of medicine. — **Aristotle**

7.6.2 Principle of LED Operation

1. Basic Aspects

(i) When the LED is forward biased, the electrons in the N-region will cross the junction and recombine with the holes in the P-type material.

(ii) These free electrons reside in the conduction band and hence at a higher energy level than the holes in the valence band.

(iii) When the recombination takes place, these electrons return back to the valence band which is at a lower energy level than the conduction band.

(iv) While returning back, the recombining electrons give away the excess energy in the form of light. This has been shown in figure 7.10. This process is known as electroluminescence. In this manner, an LED emits light. This is the principle of operational LED.

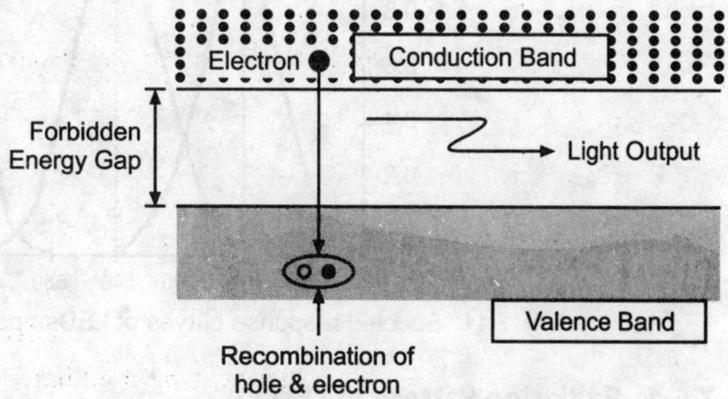

Fig. 7.10. Principle of operation of LED

2. Colour of the emitted light

The colour of the emitted light is decided by its wavelength. The wavelength of emitted light (and hence its colour) depends on the value of forbidden gap (E_G). Forbidden gap value depends on the material used. Hence, the colour of emitted light is dependent on the material used. The colours associated with different materials are as follows.

S.No.	Material used	Colour of the emitted light
1.	Gallium Arsenide (Ga As)	Infrared (IR)
2.	Ga As P	Red or Yellow
3.	Gallium phosphide (Ga P)	Red or Green

3. Silicon and Germanium are not used

(i) Silicon and Germanium are not used for LEDs because their forbidden gaps do not allow the lights emission in visible spectrum.

(ii) The commonly available LEDs emit green, red, yellow, orange and blue colours.

(iii) Some LEDs are capable of emitting infrared light which is not visible to our eyes.

(iv) Various impurities are added during the doping process to get precisely the required colour of the emitted light.

4. LED Biasing

(i) The voltage drop across a conducting LEDs is in the range of 1.2 V to 3.2 V, depending on the material used. This voltage is much higher than that across a conventional diode.

(ii) The current range of the LEDs available in market is 10 to 80 mA. LEDs have extremely low reverse breakdown voltage of the order of 3 to 10 V. Hence, they must not be subjected to high reverse voltages.

7.6.3. Spectral Response Curves of LED

The graph of output light (Y axis) versus the wavelength is called as the spectral responses curve of an LED. The shape and the wavelength at which maximum response is obtained varies depending on the material used as depicted in figure 7.11.

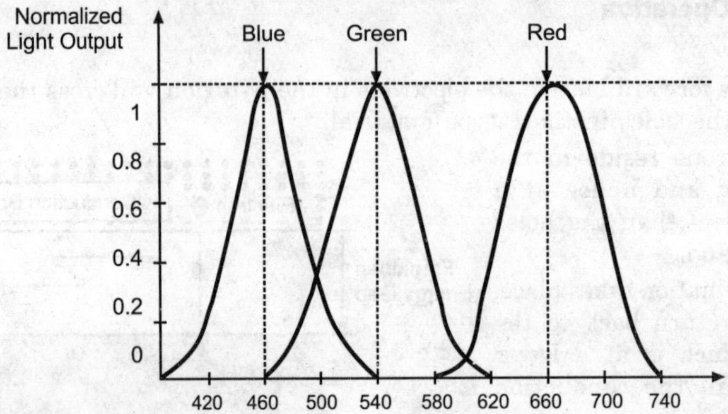

Fig. 7.11. Spectral response curves of LEDs operating in the visible portion of spectrum

7.6.4. Radiation Pattern of LED

Radiation pattern is a graph of light output plotted against angle as illustrated in figure 7.12. It gives as an idea about how directional the given light source is. The LED is a highly directional source since it radiates only in a particular direction. Most of the radiated energy is confined within 20° of the directional of maximum light.

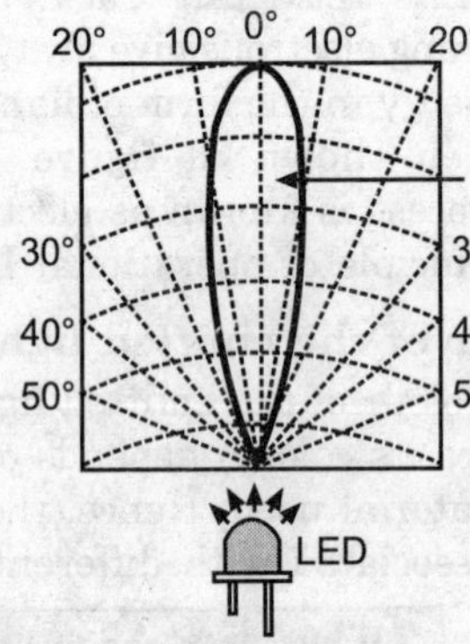

Fig. 7.12. Radiation pattern of LED

7.6.5. Output Characteristics of LED

The graph of figure 7.13(b) known as the output characteristics of the LED. It indicates the linear dependence of light output on the forward LED current. But, it is not possible to increase I_F indefinitely. Typically the maximum value of I_F is 80 mA.

How to drive an LED?

Some of the simplest LED driving circuits have been shown in figure 7.14. The output of a logic circuit may be either 0 or 1.

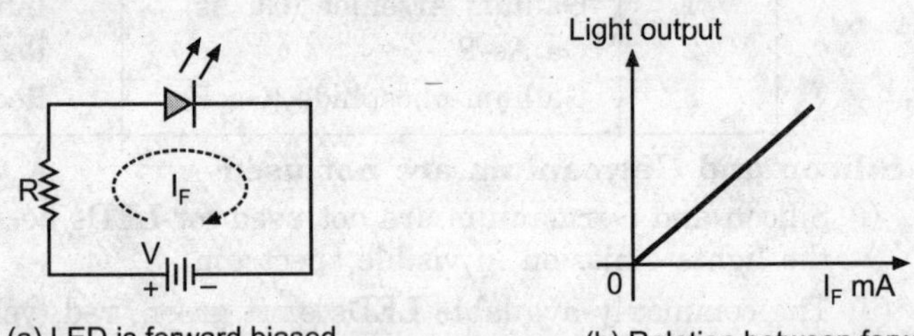

(a) LED is forward biased

(b) Relation between forward current and light output

Fig. 7.13.

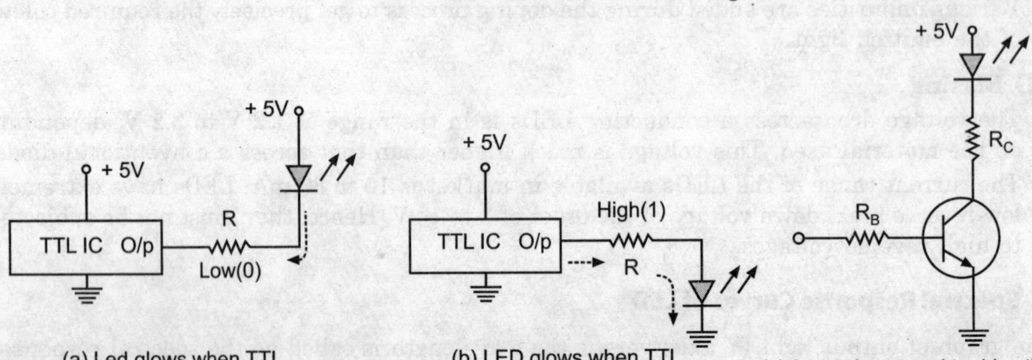

(a) Led glows when TTL output is low

(b) LED glows when TTL output is high

(c) A transistor used to drive the LED

Fig. 7.14. Driving arrangements of LED

Different circuit arrangements are required in order to turn on the LED when the output is 0 or 1. Figure 7.14(a) illustrates the arrangement where LED glows for a low output and figure 7.14(b) shows the arrangement to turn on the LED for a high output. It is quite possible to use a transistor to drive the LED as shown in figure 7.14(c). This arrangement reduces the loading on a logic circuit.

7.6.6. Important Advantages of LEDs

(i) LEDs are of small size and light weight. Therefore it is possible to pack a large number of LED, in a small space while manufacturing a display.

(ii) They are available in different spectral colours.

(iii) They have longer life as compared to the lamps.

(iv) The light emitted by a LED is proportional to the amount of current flowing through it. Hence we can control the current flowing through LEDs to vary their brightness as per the requirements of the application.

(v) They are suitable at high operating speeds as they take less than 1 μs to turn on or off.

(vi) LEDs can be easily interfaced with the other electronic circuits.

7.6.7. Drawbacks of LEDs

(i) Output power is affected by changes in temperature.

(ii) Overcurrent can damage it easily.

(iii) They need larger power for their operation.

(iv) Luminous efficiency of LEDs is low.

7.6.8 Important Applications of LEDs

Few typical applications of LED

(i) In the optocouplers.

(ii) In the infrared remote controls.

(iii) As indicators in various electronic circuits.

(iv) In seven segment and alphanumeric displays.

EXAMPLE 7.1. A LED is to be used to indicate the ON/OFF status of a DC power supply having a dc output voltage of 5 volt. Draw the indicator circuit and specify the component values.

Solution : Let us assume that the ON state voltage drop across LED is 1.5 V. For adequate brightness, let the current flowing through the LED be 5 mA.

Therefore, we have $$R = \frac{5V - 1.5}{5\text{ mA}}$$

$$= 700\ \Omega$$ **Ans.**

Since this is not standard value, we use R = 680 Ω which is a standard value.

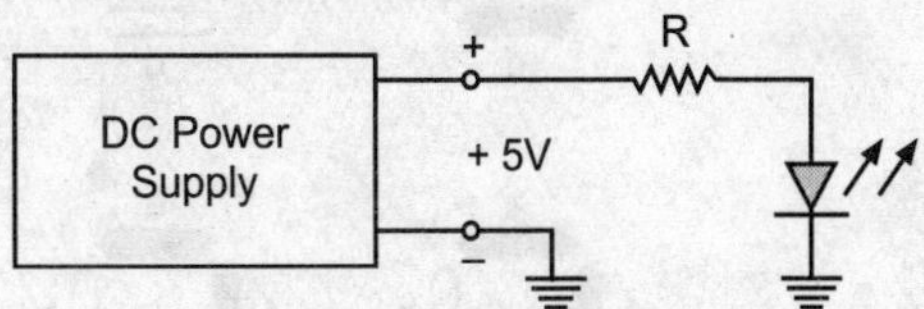

Fig. 7.15.

7.6.9 Performance Comparison of PN Junction Diode and LED

S.No	Parameter of comparsion	Conventional diode	LED
1.	Symbol		
2.	Material used	Silicon or Germanium	Gallium Arsenide
3.	Capacity to emit light	Cannot emit light	Can emit light when excited electrically
4.	Onstate voltage drop	0.7 V for silicon diode and 0.3 V for germanium diode	Ranges between 1.2 to 2 volts
5.	Reverse breakdown voltage	High	Very low
6.	Applications	Rectifier, clipper, clamper etc.	As a light source in optical fiber applications. As indicator, in 7 segment displays.

7.7 SEVEN SEGMENT DISPLAY

We know something about a seven segment display. Figure 7.16 shows the construction of a seven segment display. We can display any number from 0 to 9 by turning on various combinations of the segments as depicted in table 7.1.

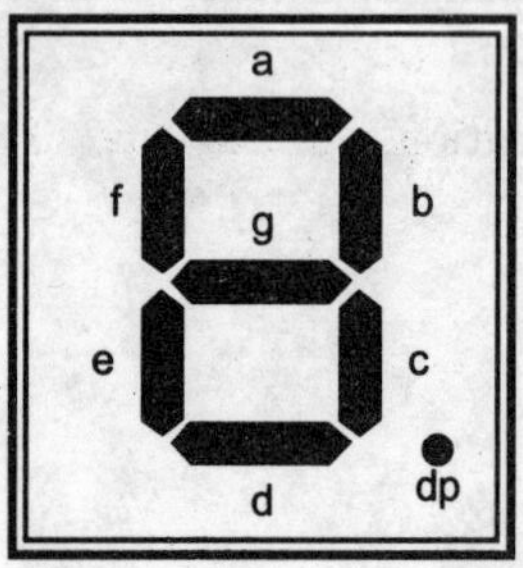

Note: Each segment (a to g and dp) is and LED in the shape of the segment

Fig. 7.16. Standard form of seven segment display

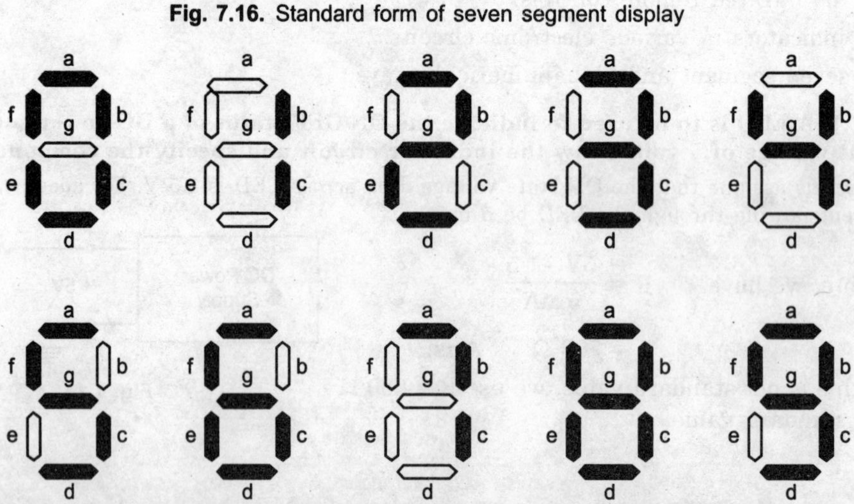

Fig. 7.17. Various digits displayed with seven segement display

In fact, each segment (a to f) is nothing but an LED in a segment form. Their connection leads are brought out, and by applying a forward voltage to the segments to be turned on we can display any number between 0 and 9. As an example if we want to display the number 7 then the segments a, b and c should be turned on and all other segments should be off. In a similar manner the other numbers can be displayed as depicted in figure 7.17.

Table 7.1

Segements							Number	Display
a	b	c	d	e	f	g		
ON	ON	ON	ON	ON	ON	–	(0)	0
–	ON	ON	–	–	–	–	(1)	1
ON	ON	–	ON	ON	–	ON	(2)	2
ON	ON	ON	ON	–	–	ON	(3)	3
–	ON	ON	–	–	ON	ON	(4)	4
ON	–	ON	ON	–	ON	ON	(5)	5
ON	–	ON	ON	ON	ON	ON	(6)	6
ON	ON	ON	–	–	–	–	(7)	7
ON	ON	ON	ON	ON	ON	ON	(8)	8
ON	ON	ON	ON	–	ON	ON	(9)	9

7.8. PHOTOCONDUCTIVITY

(U.P. Tech, Sem. Exam; 2009-10)

If radiation (light) falls on a semiconductor, its conductivity increases. This is known as the **photoconductive effect** and it may be explained as under:

Because of the radiant energy supplied to the semiconductor, the covalent bonds are broken and the electron hole pairs are generated. These increased current carriers increase the conductivity of the material and hence decrease the resistivity. Such a device is known as a **photoresistor or photoconductor**.

7.9. PHOTOCONDUCTIVE CELL

The photoconductive cell or a light dependent resistor (LDR) makes use of the principle of photoconductivity. It is a semiconductor device in which resistance is dependent on the intensity of incident light. The resistance of the LDR decreases with increase in the intensity of incident light. The construction of LDR has been shown in figure 7.18(a) and its symbol depicted shown in figure 7.18(b).

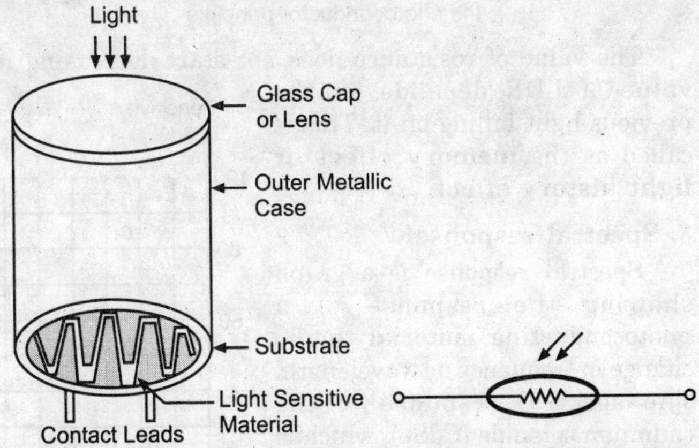

(a) Construction of Photoconductive Cell (b) Symbol of LDR

Fig. 7.18.

7.9.1. Construction of LDR

As depicted in figure 7.18(a), a light sensitive material such as cadmium sulphide (CdS) is deposited on a ceramic substrate. Then, the substrate alongwith the photosensitive layer of CdS is enclosed in a metal container. Contact leads are brought out for external connections as depicted in figure 7.18(a). Light is incident on the light sensitive material through the glass cap or lens which is added at the top of this assembly.

7.9.2. Principle of Opertion

As depicted in figure 7.18(c), when light is incident on the photosensitive material, the events take place in the following sequence:

(i) The incident photons collide with the atoms of the light sensitive semiconductor material and impart energy to them.

(ii) Because of this energy, the valence electrons will cross the forbidden energy gap and enter into the conduction band.

(iii) Because of more number of electrons entering the conduction band, the conductivity of the material increases and resistance decreases. Hence, the resistance of the photoconductive material decreases with increase in the intensity of light.

7.9.3. Variation of Resistance with Intensity of Light

1. Graph

Figure 7.20 shows the variation of resistance of LDR with change in intensity of light.

2. Dark resistance

It is the resistance offered by the LDR in the absence of light. Typical value of dark resistance is a few hundred k Ω. With increase in the light intensity, the resistance of LDR reduces significantly.

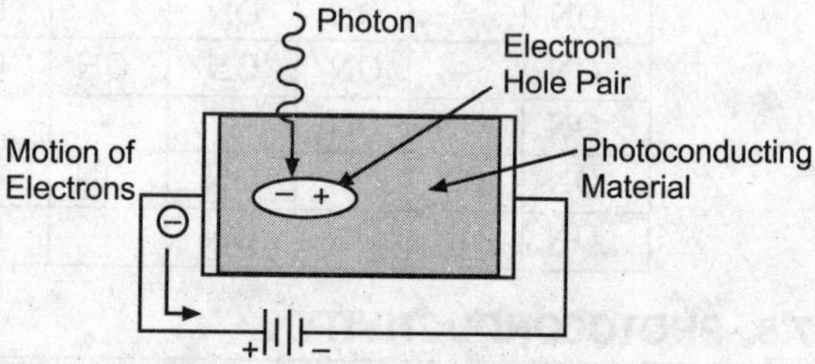

Fig. 7.18. (c) Principle of operation of a photoconductor

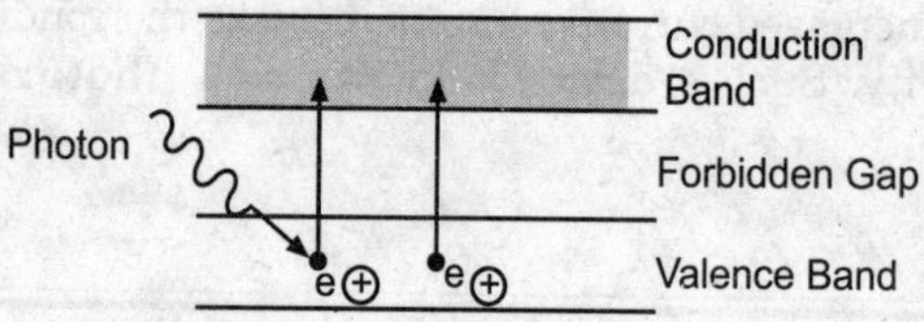

Fig. 7.19. Energy level diagram to explain the photoconductor principle

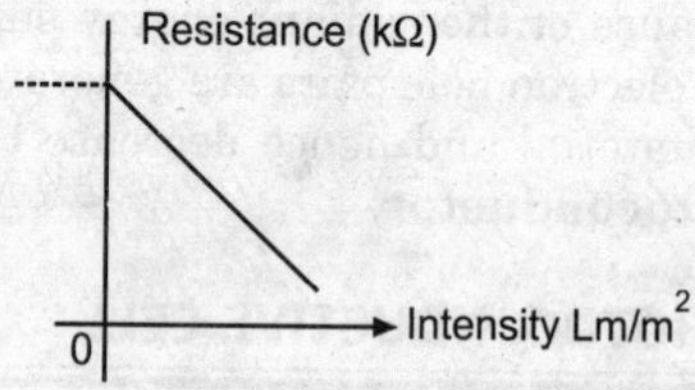

Fig. 7.20. Variation of resistance with intensity of light

The value of resistance does not start decreasing immediately. At any time, the resistance value of LDR depends on the previous light falling on it. This is called as the **memory effect or light history effect.**

3. Spectral response

Spectral response is a graph showing the response of a photoconducting material to the change in frequency or wavelength. The spectral response of the cadmium selenide (CdSe), which is a widely used material for LDRs, is as shown in figure 7.21.

The sensitivity of cadmium selenide is highest at a wavelength of about 7500 Å.

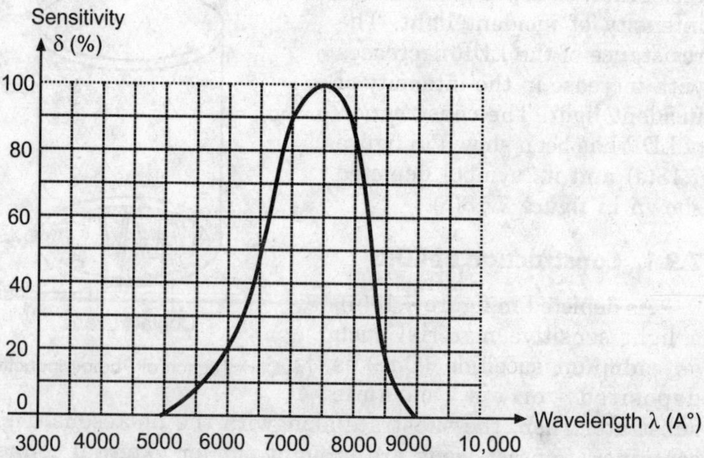

Fig. 7.21 Spectral response of a cadmium selenide

7.9.4. Applications of LDR

(i) Object counter

By using a light source and LDR detector it is possible to count the number of objects passing on a conveyor belt as depicted in figure 7.22.

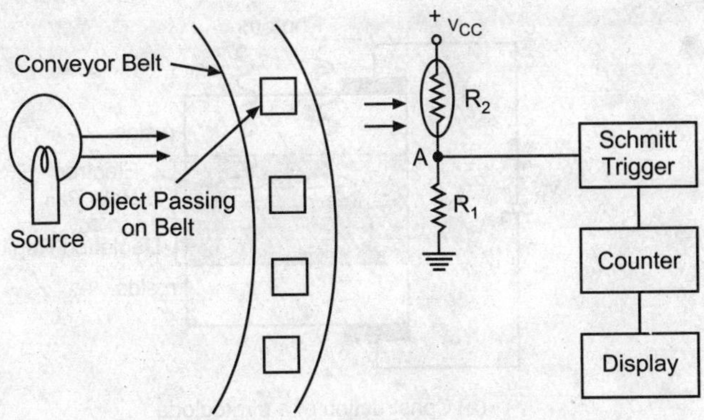

Fig. 7.22. Object counting scheme

(ii) Street light on-off system

LDR can be used in the automatic on-off switch circuit used for the street lights. This has been shown in figure 7.23. During the day time, the intensity of light falling on LDR is high, hence it offers a low resistance. The voltage across it is therefore not sufficient to turn on the transistor Q_1. The relay will remain unenergized and the lamp continues to be in the off state.

In the evening light intensity reduces. The resistance of LDR increases. Hence, voltage across it would be large enough to turn on the transistor. The relay would be energized and the N.O. contact gets closed to turn on the lamp.

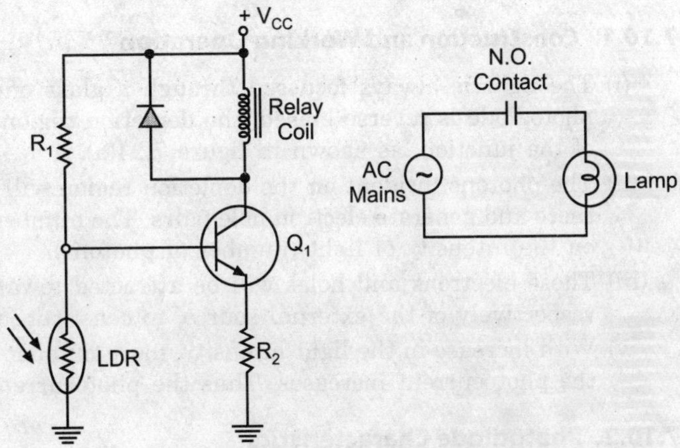

Fig. 7.23. Street light-on-off system

7.9.5. Few Advantages of LDR

(i) Simple to construct and operate
(ii) Good sensitivity
(iii) Low cost
(iv) Linear response

7.9.6. Drawbacks of LDR

(i) It has a light memory effect or light history effect (i.e., its resistance changes slowly).
(ii) It is a slow device. Its resistance does not change quickly.
(iii) It can work only in a small range of electromagnetic spectrum.
(iv) Resistance changes with temperature.
(v) Poor stability.

Important Point: The main drawback of the LDR is their slowness. This makes them unsuitable for the applications where light intensity changes rapidly. We need to use faster devices for such applications.

7.10. PHOTODIODE

(U.P. Tech, Sem. Exam; 2008-09, 2010-11)

The photodiode is a P-N junction semiconductor diode which is always operated in the reverse biased condition. The construction of a photodiode and its circuit symbols have been shown in figure 7.24 (a) and (b) respectively.

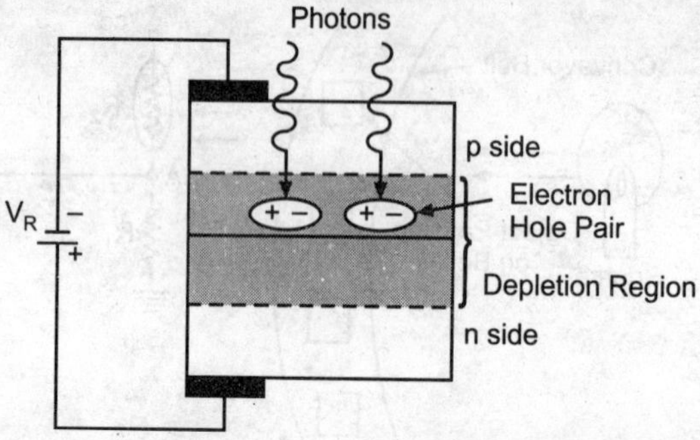

(a) Construction of a photodiode

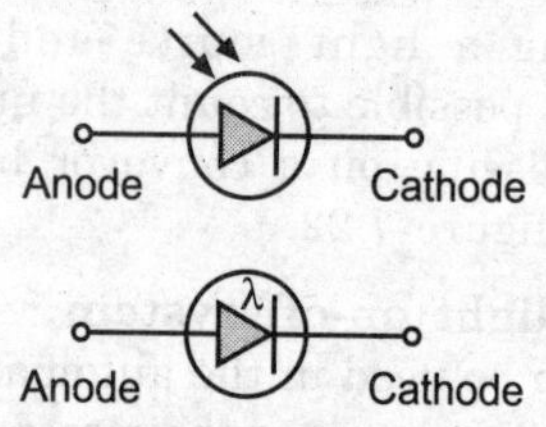

(b) Symbols of a photodiode

Fig. 7.24.

7.10.1. Construction and Working Operation

(i) The light is always focussed through a glass on the junction of the photodiode. As the photodiode is reverse biased, the depletion region is quite wide, penetrated on both sides of the junction, as shown in figure 7.24(a).

(ii) The photons incident on the depletion region will impart their energy to the ions present there and generate electron hole pairs. The number of electron hole pairs will be dependent on the intensity of light (number of photons).

(iii) These electrons and holes will be attracted towards the positive and negative terminals respectively of the external source, to constitute the photo current.

(iv) With increase in the light intensity, more number of electron hole pairs are generated and the photocurrent increases. Thus the photocurrent is proportional to the light intensity.

7.10.2. Photodiode Characteristics

1. Characteristics

The photodiode V-I characteristics have been shown in figure 7.25(a) and the variation of photocurrent with light intensity has been shown in figure 7.25(b).

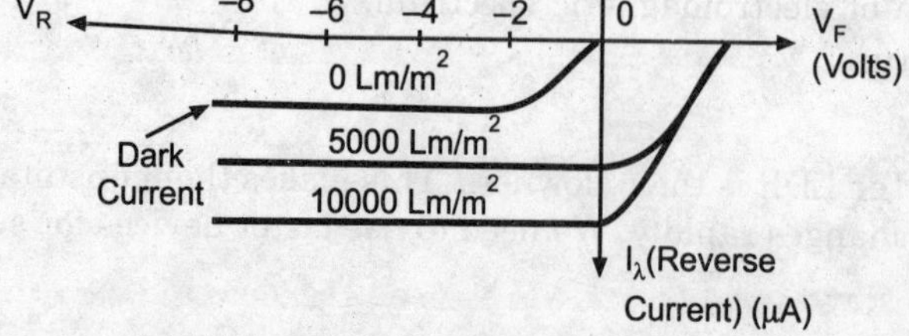

(a) V-I characteristics of photodiode

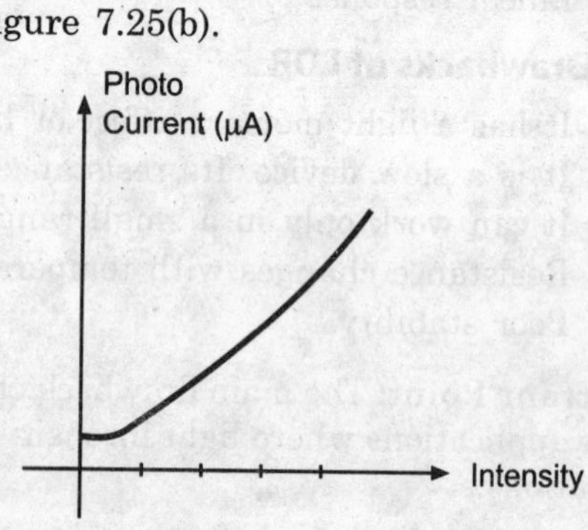

(b) Variation of photocurrent with intensity of light

Fig. 7.25.

2. Dark current

It is the current flowing through a photodiode in the absence of light. (See figure 7.25(a)). Dark current flows because of the thermally generated minority carriers, and hence increases with increase in temperature. The reverse current I_λ (photo current) depends only on the intensity of light incident on the junction. It is almost independent of the reverse voltage.

7.10.3. Advantages of Photodiode

1. High sensitivity

This means that a large change in the photocurrent for a small change in light intensity. High speed of operation as compared to LDR (Light Dependent Resistor).

7.10.4. Drawbacks of Photodiode

(i) Dark current increases with temperature.

(ii) Poor temperatures stability.

(iii) External bias voltage is essential for operation.

(iv) Amplification is required, as the output current is of small magnitude.

Why is the photodiode operated in reverse biased condition?

In the reverse biased condition, the only current flowing through the diode in the absence of light is the reverse saturation current which is very small in magnitude. Therefore, the change in diode current due to the light incident on it is significant (photocurrent is in μA). If the diode was forward biased then the forward current in the absence of light would be in mA and the change in forward current due to light will not be even noticeable. Thus, the photodiode is operated in the reverse biased condition.

7.10.5. An Important Application of Photodiode

A popular application of the photodiode is in object counting system shown in figure 7.26. This system is used for counting the number of objects that are passing on a conveyor belt. A light beam is focussed continuously on the photodiode and the reverse photocurrent is flowing through it. As soon as an object passes and interrupts the beam of light, the photocurrent flowing through the photodiode will reduce to zero. This will be counted by a counting mechanism to count the number of objects.

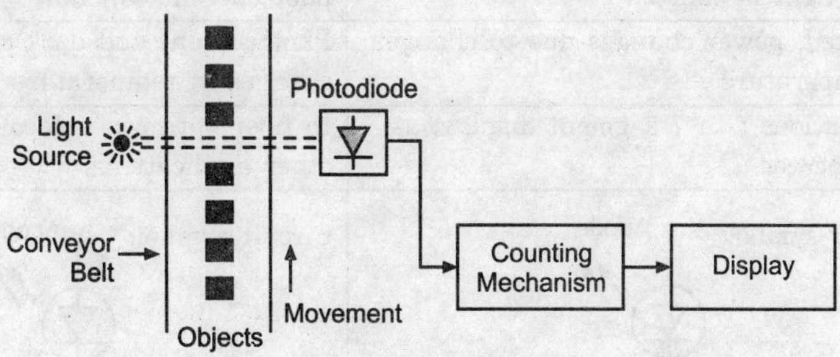

Fig. 7.26 Object counting system using photodiode

Important Point: In general the applications of photodiode are same as those of an LDR. In addition they can be used in optical measurement system and fiber optic communication.

7.10.6. Some Other Applications

(i) In the cameras for sensing the light intensity.

(ii) In the fiber optic receiver.

(iii) In light intensity meters.

7.10.7. Performance Comparison of LDR and Photodiode

Table 7.2. Comparison of LDR and Photodiode

S.No.	Photoresistor	Photodiode
1.	Large active area	Small active area
2.	More noise present	Less noise
3.	Light history effect is present	No such effect is present
4.	Poor spectral response	Good spectral response
5.	Amplification not needed	Amplification needed
6.	Low dark current	High dark current
7.	Reacts slowly to changes in light intensity	Reacts quickly to change in light intensity
8.	Applications in light meter, street light on-off system etc	Fiber optic communication, switching applications

7.10.8. Performance Comparison of LED and Photodiode

S.No.	LED	Photodioide
1.	It is a light emitting device	It is a light detecting device
2.	Electrical energy is converted into light	Electric current proportional to light intensity is produced
3.	LED is always forward biased	It is always reverse biased
4.	GaAs or GaP or GaAsP are the materials used	Silicon is used
5.	Due to recombination of electrons and holes, light is emitted	Due to generation of electron hole pairs, the photocurrent will flow
6.	Radiated, power changes due to changes in temperature	Photocurrent and dark current vary due to changes in temperature
7.	Applications : In 7 segment display, as an indicator	In fiber optics in optocouplers and many. other applications
8.	Circuit symbol: 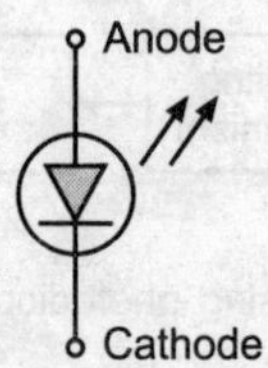	Circuit symbol: 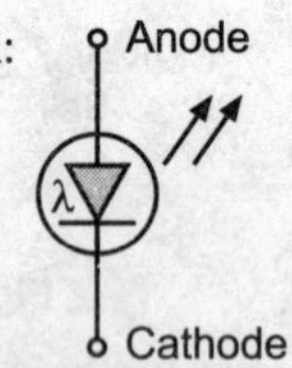

7.11. PIN DIODES

The P-I-N diode is a special type of diode. An intrinsic semiconductor layer (i-layer) is sandwitched between heavily p^+ layer and a thin n^+ layer. Therefore, the name P-I-N diode. This diode is used at microwave frequencies i.e. frequencies above 1 GHz. The construction of PIN diode has been shown in figure 7.27(a).

The intrtinsic i-layer is actually a high resistivity p-type silicon layer. This region is also known as π region. Figure 7.27(a) illustrates the pictorial view of the PIN or p-π-N diode. It shows that when this diode is open circuited, electrons will flow from i-region to p^+ region to recombine with the express number of holes there and holes will flow from i-region to n^+ region to recombine with the excess number of electrons in the n^+ region.

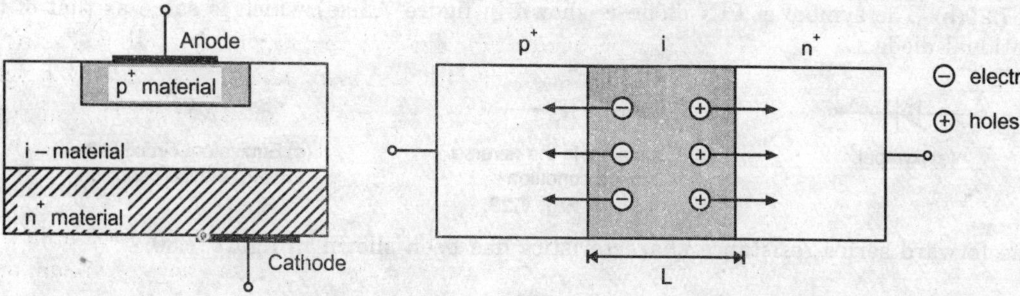

(a) Construction of PIN diode

(b) Pictorial representation of PIN diode

Fg. 7.27.

7.11.1. Working Operation

Let us discuss the working operation in the forward as well as in the reverse biased conditions.

1. Working Opertion when Reverse Biased

When a PIN diode is reverse biased as shown in figure 7.28(a) the electrons, and holes move in opposite directions to *sweep free* the intrinsic material off electrons and holes. It is important to note that the length "L" of the depletion region in PIN diodes is approximately same as that of the intrinsic i-region. Thus, the length of the depletion region remains approximately constant, independent of the externally applied reverse voltage. Therefore, the reverse capacitance or transition capacitance of a PIN diode remains approximately constant, independent of the bias. The variation of C_T with respect to reverse voltage is shown in figure 7.28(b). The breakdown voltages of PIN diodes are in the range of 100 to 1000 V.

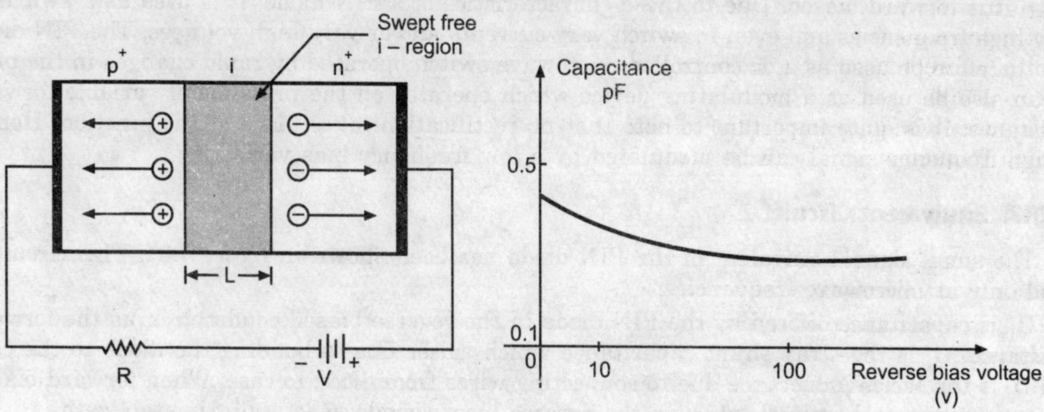

(a) Reverse biased PIN dipde

(b) Variation of diode capacitance with reverse bias voltage

Fig. 7.28.

2. Working Operation when Forward Biased

When a PIN diode is forward biased, holes from the p layer will diffuse into the i-layer. They will attract equal number of holes from the n-side. This double injection of charged particles **will** reduce the resistivity of the i-region. In other words its conductivity increases. Thus depending on the forward voltage, conductivity of the i-layer will keep changing. This is known as **conductivity modulation** of the PIN diode. The forward current flows due to both electrons and holes. In the forward biased condition, the voltage drop across the i-region is very small due to the **conductivity modulation**. In short, the PIN diode acts like a nearly constant capacitance when it is reverse biased whereas it acts as a current controlled variable resistance as shown in figure 7.29(b). The symbol of PIN diode is shown in figure 7.29(a) which is same as that of the conventional diode.

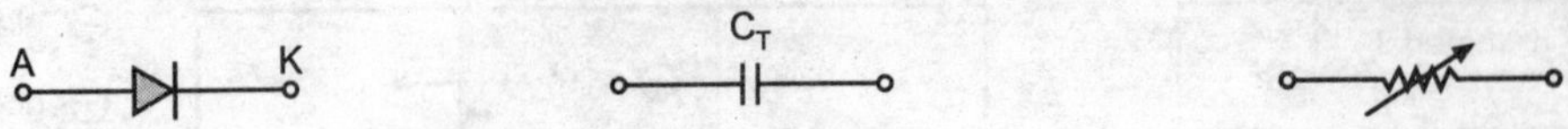

(a) Symbol (b) Equivalent in the reverse biased condition (c) Equivalent circuit in the forward biased condition

Fig. 7.29.

The forward series resistance characteristics has been shown in figure 7.30.

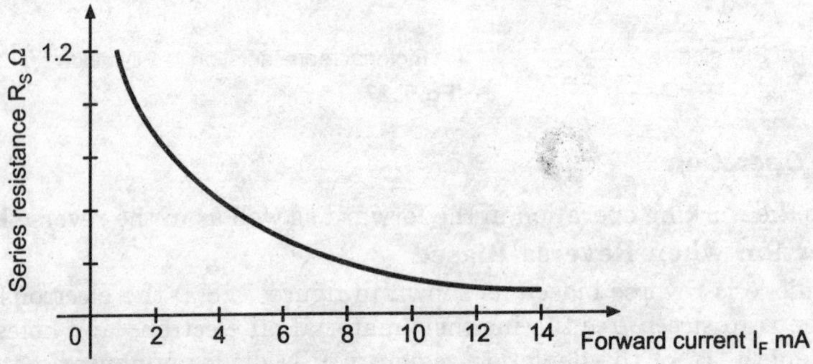

Fig. 7.30. Variation in series resistance with forward current

7.11.2. PIN Diode as a Switch

PIN diodes are used at very high frequencies (higher than 1 GHz) because even at these frequencies, the diode has a very high impedance when reverse biased and very low impedance when it is forward biased. Due to these characteristics of a PIN diode, it is used as a switch at very high frequencies and even to switch vary currents and/or vary high voltages. The PIN diode can therefore be used as a dc controlled microwave switch operated by rapid changes in the bias. It can also be used as a modulating device which operates on the principle of variable forward resistance. It is quite important to note that no rectification takes place at the junction. Hence, a high frequency signal can be modulated by a low frequency bias variation.

7.11.3. Equivalent Circuit

The small signal equivalent of the PIN diode has been shown in figure 7.31. This circuit is valid only at microwave frequencies.

C_T is capacitance offered by the PIN diode in the reverse biased condition. r_d is the forward resistance, C_s is the stray shunt capacitance which arises due to bonding the diode to the case and L_s is the series inductance due to connecting wires from diode to case. When forward biased C_T is approximately infinite while in the reverse biased mode r_d is approximately zero.

Important Advantages

(i) High breakdown voltage (100 to 1000 V)

(ii) High current/power handling capacity.

(iii) Low on state voltage drop, hence low power dissipation.

(iv) Can operate at very high frequencies (beyond 1 GHz).

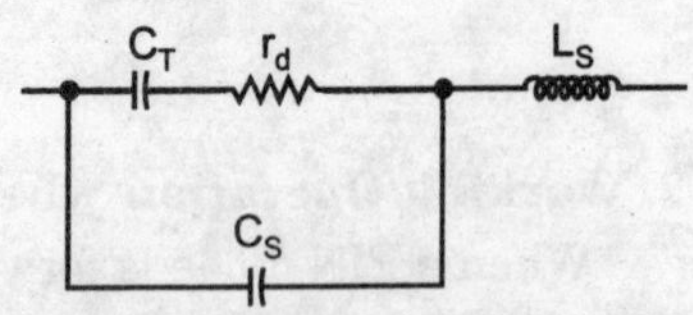

Fig. 7.31. PIN diode equivalent circuit at microwave frequency

7.11.4. Important Applications

(i) PIN diode may be used as an amplitude modulator at microwave frequencies beyond 1 GHz.

(ii) It can be used as a dc controlled microwave switch for switching high voltage/current at microwave frequencies.

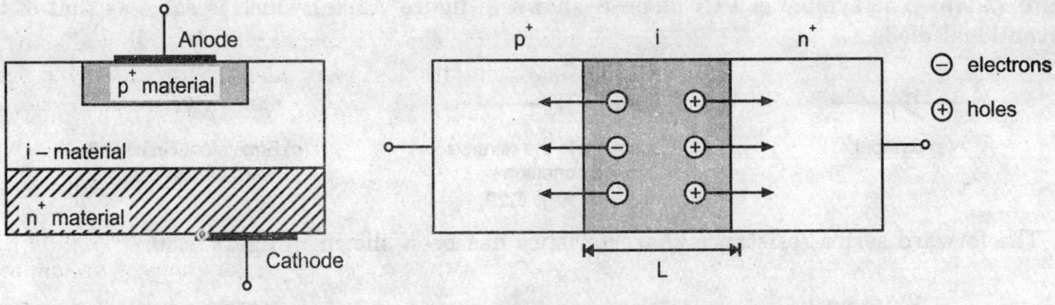

(a) Construction of PIN diode (b) Pictorial representation of PIN diode

Fg. 7.27.

7.11.1. Working Operation

Let us discuss the working operation in the forward as well as in the reverse biased conditions.

1. Working Opertion when Reverse Biased

When a PIN diode is reverse biased as shown in figure 7.28(a) the electrons, and holes move in opposite directions to *sweep free* the intrinsic material off electrons and holes. It is important to note that the length "L" of the depletion region in PIN diodes is approximately same as that of the intrinsic i-region. Thus, the length of the depletion region remains approximately constant, independent of the externally applied reverse voltage. Therefore, the reverse capacitance or transition capacitance of a PIN diode remains approximately constant, independent of the bias. The variation of C_T with respect to reverse voltage is shown in figure 7.28(b). The breakdown voltages of PIN diodes are in the range of 100 to 1000 V.

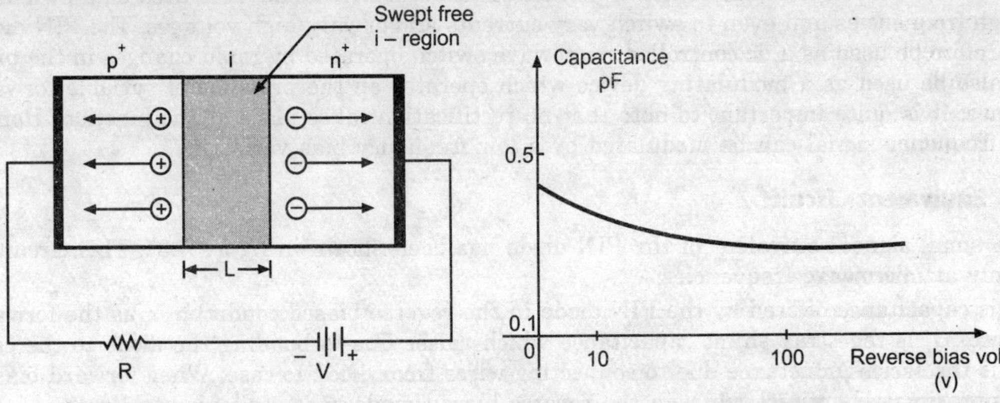

(a) Reverse biased PIN dipde (b) Variation of diode capacitance with reverse bias voltage

Fig. 7.28.

2. Working Operation when Forward Biased

When a PIN diode is forward biased, holes from the p layer will diffuse into the i-layer. They will attract equal number of holes from the n-side. This double injection of charged particles will reduce the resistivity of the i-region. In other words its conductivity increases. Thus depending on the forward voltage, conductivity of the i-layer will keep changing. This is known as **conductivity modulation** of the PIN diode. The forward current flows due to both electrons and holes. In the forward biased condition, the voltage drop across the i-region is very small due to the **conductivity modulation**. In short, the PIN diode acts like a nearly constant capacitance when it is reverse biased whereas it acts as a current controlled variable resistance as shown in figure 7.29(b). The symbol of PIN diode is shown in figure 7.29(a) which is same as that of the conventional diode.

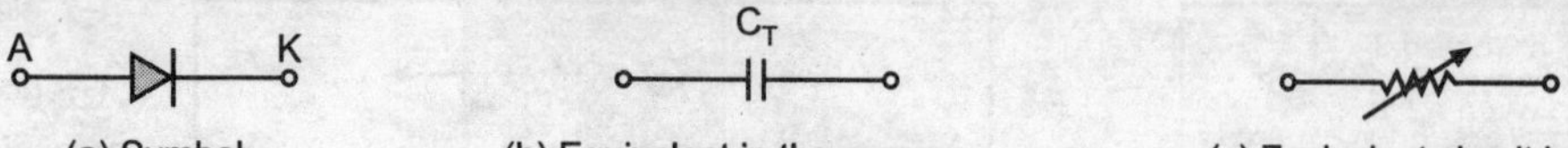

(a) Symbol (b) Equivalent in the reverse biased condition (c) Equivalent circuit in the forward biased condition

Fig. 7.29.

The forward series resistance characteristics has been shown in figure 7.30.

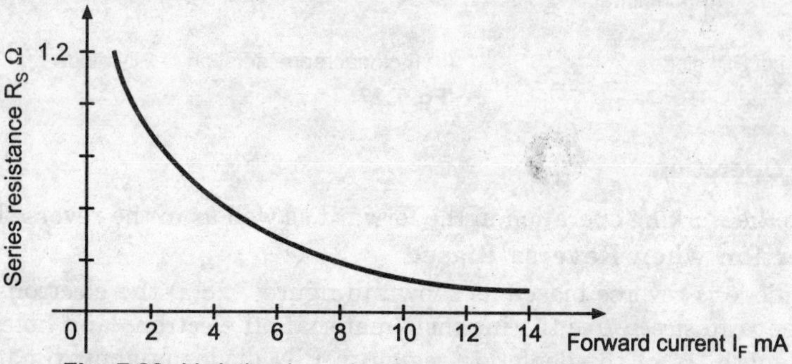

Fig. 7.30. Variation in series resistance with forward current

7.11.2. PIN Diode as a Switch

PIN diodes are used at very high frequencies (higher than 1 GHz) because even at these frequencies, the diode has a very high impedance when reverse biased and very low impedance when it is forward biased. Due to these characteristics of a PIN diode, it is used as a switch at very high frequencies and even to switch vary currents and/or vary high voltages. The PIN diode can therefore be used as a dc controlled microwave switch operated by rapid changes in the bias. It can also be used as a modulating device which operates on the principle of variable forward resistance. It is quite important to note that no rectification takes place at the junction. Hence, a high frequency signal can be modulated by a low frequency bias variation.

7.11.3. Equivalent Circuit

The small signal equivalent of the PIN diode has been shown in figure 7.31. This circuit is valid only at microwave frequencies.

C_T is capacitance offered by the PIN diode in the reverse biased condition. r_d is the forward resistance, C_s is the stray shunt capacitance which arises due to bonding the diode to the case and L_s is the series inductance due to connecting wires from diode to case. When forward biased C_T is approximately infinite while in the reverse biased mode r_d is approximately zero.

Important Advantages

(i) High breakdown voltage (100 to 1000 V)

(ii) High current/power handling capacity.

(iii) Low on state voltage drop, hence low power dissipation.

(iv) Can operate at very high frequencies (beyond 1 GHz).

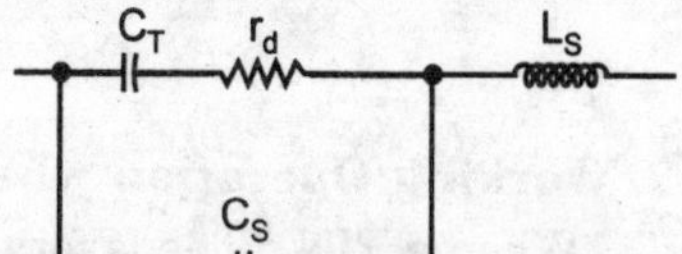

Fig. 7.31. PIN diode equivalent circuit at microwave frequency

7.11.4. Important Applications

(i) PIN diode may be used as an amplitude modulator at microwave frequencies beyond 1 GHz.

(ii) It can be used as a dc controlled microwave switch for switching high voltage/current at microwave frequencies.

7.11.5. Peculiarities of a PIN Diode

(i) Used as an intrinsic semiconductor layer between P and N layers.

(ii) It is equivalent to a constant capacitance when reverse biased and a variable resistance in the forward biased condition.

(iii) Operating frequency is beyond 1 GHz.

(iv) Switching time: 8 to 10 ns.

(v) Breakdown voltage: 100 to 1000 V.

(vi) Power dissipation: 10 W.

(vii) It can be used as switch at microwave frequencies.

7.12. PIN PHOTODIODE

(i) In several photodetector applications, the speed of response is important, so the prompt photocurrent generated in the space charge region is the only photocurrent of interest.

(ii) To increase the photodetector sensitivity, the depletion region width should be made as large as possible. It may be achieved in PIN photodiode. The PIN diode consists of a P region and an n region separated by an intrinsic region.

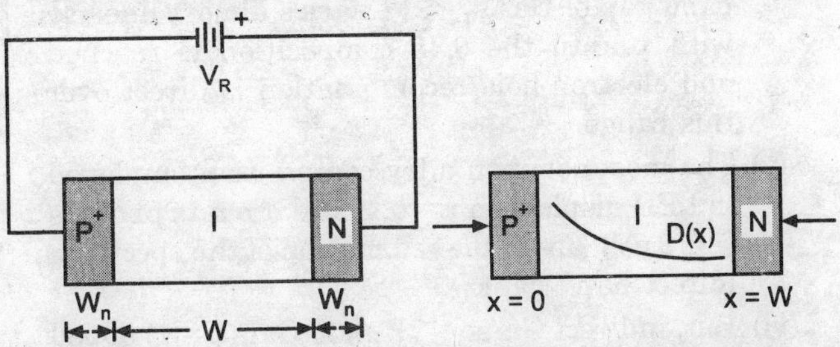

(a) A reverse biased PIN photodiode

(b) Geometry showing non-uniform photon absorption

Fig. 7.32.

(iii) A sketch of a PIN diode is shown in figure 7.32(a). The intrinsic region width W is much larger the space charge width of a normal pn junction. When a reverse bias is applied to the PIN diode, the space charge region extends completely through the intrinsic region.

(iv) Let us assume that a photon flux ϕ_0 is incident on the p^+ region. If we assume that the p^+ region width W_p is very thin, then the photon flux, as a function of distance, in the intrinsic region is

$\phi(x) = \phi_0 \, e^{-\alpha x}$, where α is the photon absorption coefficient.

(v) The non-linear photon absorption is depicted in figure 7.32(b). The photocurrent density generated in the intrinsic region may be found as under:

$$J_L = e \int_0^W G_L \, dx = e \int_0^W \phi_0 \, e^{-\alpha x} \, dx = e\phi_0 \cdot (1 - e^{-\alpha w}) \quad \text{...(7.7)}$$

(vi) Where G_L be the generation rate of excess carriers, W is the space charge width. The above mentioned equation (7.7) assumes that there is no electron-hole recombination within the space charge region and also that each photon absorbed creates the one electron-hole pair.

Important Point: The worthnoting point is that the prompt photocurrent density of a PIN photodiode will be larger than that of a regular photodiode because in case of a PIN diode space charge region is wider.

7.13 LIGHT EMITTING MATERIALS

(i) The band gaps of different binary compound semiconductors are provided in previous chapter. A large variation exists in photon energy between the infrared band gaps. (InSb of 0.18 eV) to the ultraviolet band gap (GaN of 3.4 eV).

(ii) In fact, use of ternary and quaternary compounds can increase the number of available energy significantly.

(iii) As an example, of the variation in photon energy available from the compound semiconductors is the ternary alloy gallium arsenide-phosphide as shown in figure 7.33.

(iv) As the percentage of As decreases and increase in percentage of P, the resulting band gap varies from the indirect 1.43 eV gap of GaAs (infrared) to the indirect 2.26 eV gap of GaP (green). The band gap of $GaAs_{1-x}P_x$ varies almost linearly with x until the 0.45 composition is reached and electron-hole recombination is direct over this range.

(v) The most common alloy composition employed in LED displays in x = 0.4, and it emits photons (~ 1.9 eV) are in the red portion of the spectrum (direct band gap).

(vi) But, indirect $GaAs_{1-x}P_x$ and GaP(x = 1) doped with N_2 can be employed in LED's to provide light in the yellow to green spectrum, this is due to the fact that nitrogen impurity binds electron very lightly.

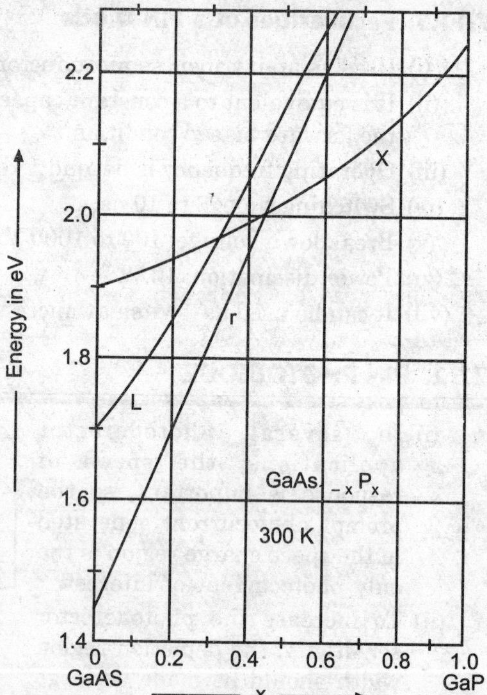

Fig. 7.33. Conduction band energies as a function of alloy Composition for $GaAs_{1-x}P_x$

(vii) In several applications, light from a laser or an LED need not visible to the eye. Infrared emitters like InP, GaAs and mixeds alloys of suchs compounds are well suitable for optical communication systems. As an example, we can apply a laser or LED in conjunction with a photodiodes or other photosensitive device for transmission of information optically between locations.

(viii) Just for instance, a semiconductor laser-photodetector arrangement may be employed in a compact disc system for reading digital information from the spinning disc.

(ix) A large number of available compound III-IV semiconductor materials have proved useful to provide for increased number of applications using semiconductor lasers and LEDs with visible and infrared wavelengths. In addition to the AlGaAs and GaAsP systems, the InAlGaP system is useful for yellow and green wavelengths and a GaN is a strong emitter in the blue.

(x) Even more wavelengths will be accessible GaN and related materials are increasingly employeds in LEDs and LASERS. With traditional doping methods, crystal defects tend to compensate the doping impurities such that N-type (Zns, ZnSe, Cds, CdSe) or p-type (ZnTe) can be obtained. The availability of blue light from GaN is particularly important in extending III-V semiconductor light emission across the entire visible spectrum.

7.14. PHOTOTRANSISTOR

A photodiode cannot provide any amplification action. A phototransistor can provide the internal current multiplication and can generate a larger output signal. The light is allowed to fall on the collector base junction of the transistor. The photocurrent produced there acts as the base current for the transistor and gets multiplied β times. The construction and circuit symbol of a phototransistor have been shown in figure 7.34(a) and (b) respectively.

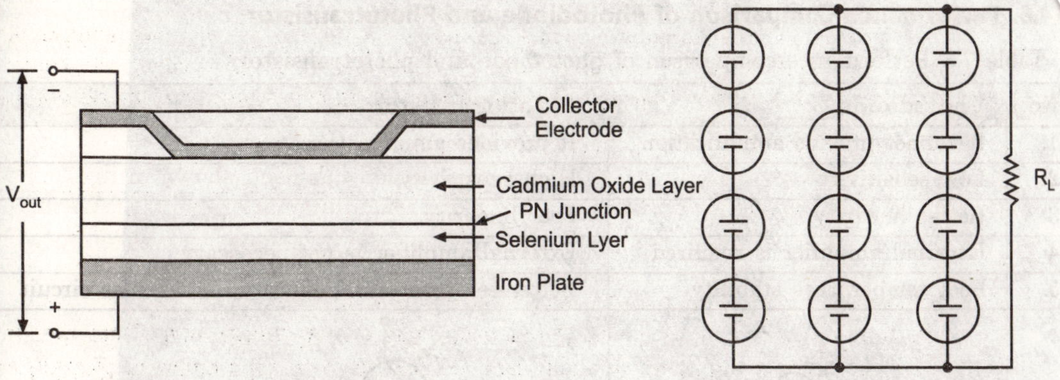

(a) Construction of Selenium photovoltaic cell (b) Array of Photovoltaic cells

Fig. 7.39.

(vii) The number of electron hole pairs is proportional to the intensity of incident light. The electrons and holes are separated out by the depletion region potential of the p-n junction.

7.16.1. Solar Cell: Construction and Working Operation (U.P. Tech, Sem. Exam; 2009-10, 2011-12)

(i) A solar cell is a P-N junction device with no voltage directly applied across the junction. In other words we can say that it is essentially a large photodiode designed to operate solely as a *photovoltaic* device and give as much output power as possible.

(ii) The solar cell converts solar energy into electrical energy.

(iii) The construction of a solar cell is depicted in figure 7.40(a) and the simplified diagram depicted in figure 7.40(b) can be used to explain the operation of the solar cell.

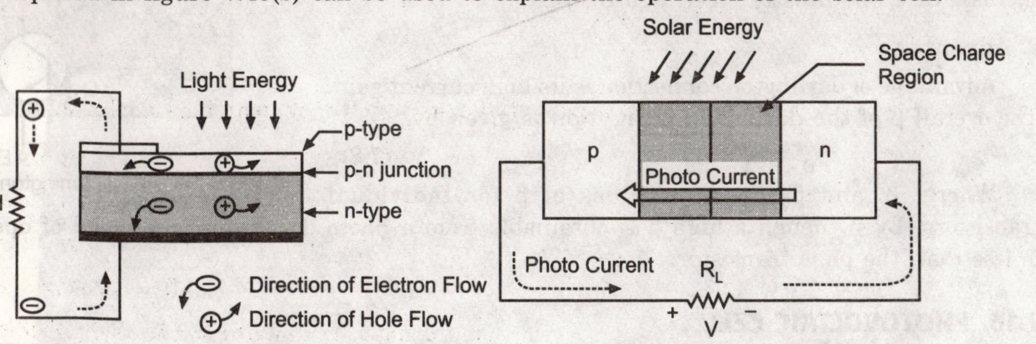

(a) Construction of a Solar Cell (b) Simplified Diagram

Fig. 7.40.

(iv) When the light hits the space charge region, electrons and holes are generated due to the photons striking the valence electrons and imparting energy to them. The optically generated electron-hole pairs are quickly separated and swept outside the space charge region (depletion region) by the electric field.

(v) These electrons and holes flow to constitute the photocurrent as shown in figure 7.40(b).

(vi) This photocurrent produces a voltage V across the load resistance R_L. Thus

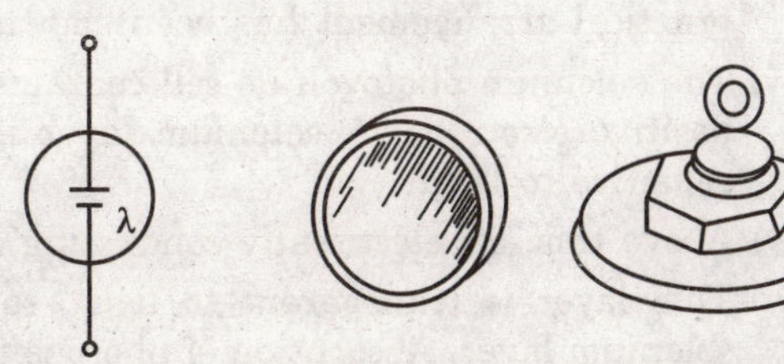

(a) Symbol of a solar cell (b) Construction of a solar cell

Fig. 7.41.

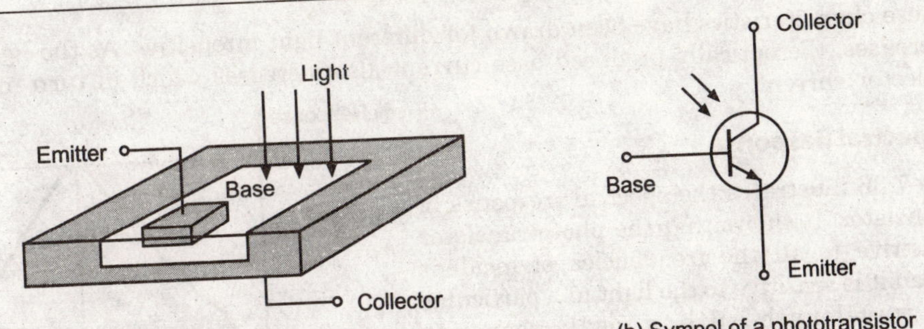

(a) Construction of a phototransistor (b) Sympol of a phototransistor

Fig. 7.34.

7.14.1. Construction and Biasing

1. Basic Aspects

(i) The construction has been shown in figure 7.34(a). The base collector region of the photo transistor is responsible for the generation of photo base current. Hence, in the construction of the transistor, the base collector area is kept as large as possible.

(ii) The base terminal is somtimes not brought out and the phototransistor acts as a two terminal device. However, if the base terminal is brought out, then it acts as a conventional transistor. The base terminal is generally left open or connected to grounds through a large value resistor. The biasing of the phototransistor is otherwise same as that of a conventional transistor.

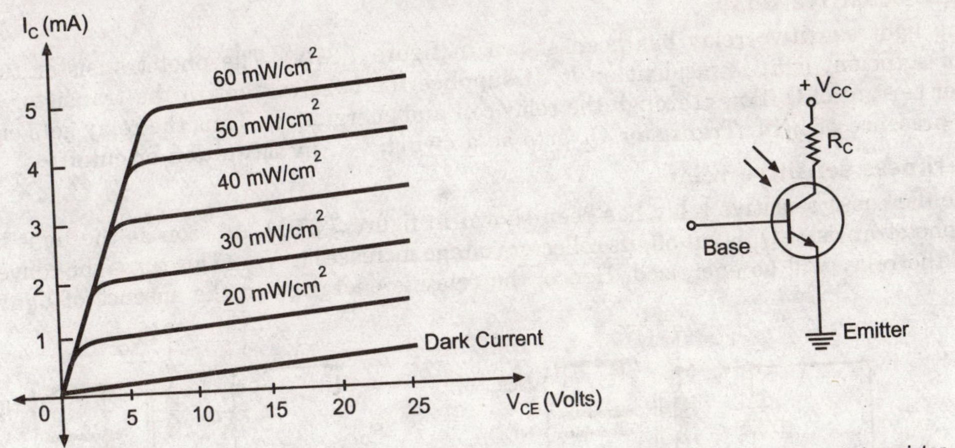

(a) Output characteristics of a phototransistor (b) Biasing of a phototransistor

Fig. 7.35.

2. Dark current

(i) In absence of light, a small thermally generated current I_{CEO} flows through the device. It is in the nA range.

(ii) The dark current should be as small as possible. In presence of light, the base current produced (I_λ) is directly proportional to the intensity of light. The collector current in β times base current.

(iii) The phototransistor is biased depicted in figure 7.35(b). The output characteristics of a photo transistor are as shown in figure 7.35(a). They are very much similar to those of conventional transistor.

(iv) These characteristics have been drawn for different light intensities. As the light intensity increases, the optically produced base current also increases which in turn increases the collector current.

7.14.2. Spectral Response

Figure 7.36 illustrates the spectral response of a phototransistor. It shows that the phototransistor is not sensitive to all the frequencies of incident light. Instead it is sensitive to the light in a particular range of wavelength with a peak response corresponding to a particular wavelengths.

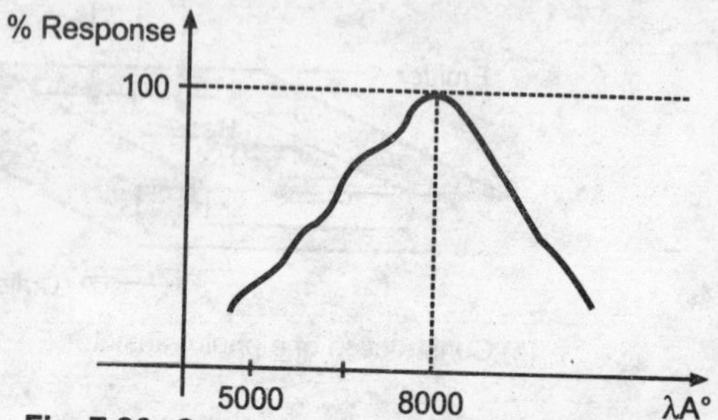

Fig. 7.36. Spectral response of a phototransistor

7.14.3. Advantages of a Phototransistor

Photocurrent gets internally multiplied by β, so no additional amplifier is required. High sensitivity of the order of 100 times that of a photodiode. Good switching speed. No memory effect.

7.14.4. Drawbacks of a Phototransistor

Phototransistor is not a fast as a conventional transistor due to sluggishness of the photoconducting material. Poor linearity. Phototransistors are temperature sensitive devices. Signal to dark current ratio is poor. This is because the dark current also gets multiplied by of the transistor. External voltage source is needed for operation.

7.14.5. Applications of Phototransistor

(i) Light sensitive relay

The light sensitive relay has been shown in figure 7.37(a). The phototransistor turns ON when a sufficient light is incident on it. It supplies the base current to the transistor Q_2. The collector current of Q_2 flows through the relay coil and energizes it. Thus the relay gets energized in the presence of light. Transistor Q_2 acts as a switch i.e., in saturation or cutoff.

(ii) Darkness sensitive relay

The darkness sensitive relay has been shown in figure 7.37(b). As soon as the light incident on the phototransistor Q_1 is cut-off, its collector voltage increases to V_{CC}. This turns the conventional Q_2 ON the relay will be energized. Hence, the relay is energized in the absence of light.

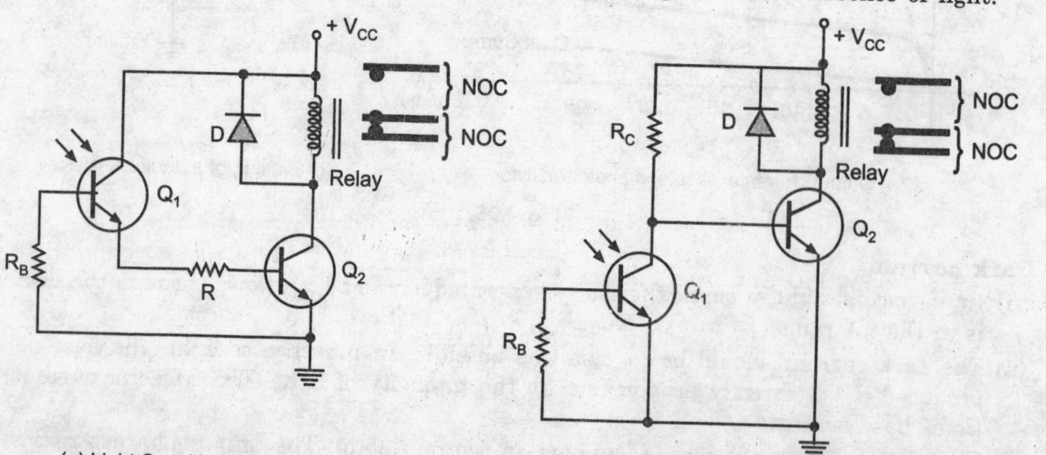

Fig. 7.37.

7.14.6. Performance Comparison of Photodiode and Phototransistor

Table 7.3 Performance comparison of photodiode and phototransistor

S.No.	Photodioide	Ph ototransistor
1.	It cannot provide amplification	It provides amplification
2.	Low sensitivity	High sensitivity
3.	Better linearity	Poor linearity
4.	External amplifier is required	External amplifier is not necessary
5.	Poor temperature stability	Better temperature stability with self bi
6.	Symbol: 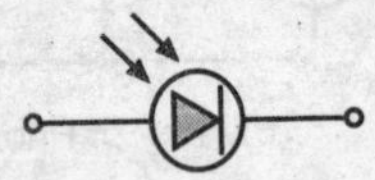	Symbol: 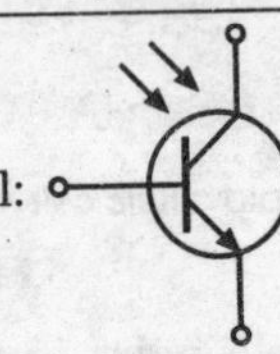

7.15. PHOTODARLINGTON

1. Circuit Symbol

The circuit symbol for a photo darlington is as depicted in figure 7.38. A photo transistor is connected in the darlington connection with a conventional transistor as shown. In the so called darlington connection the collectors of both the transistors are connected together and the emitter of Q_1 is connected to the base of Q_2 as illustrated in figure 7.38.

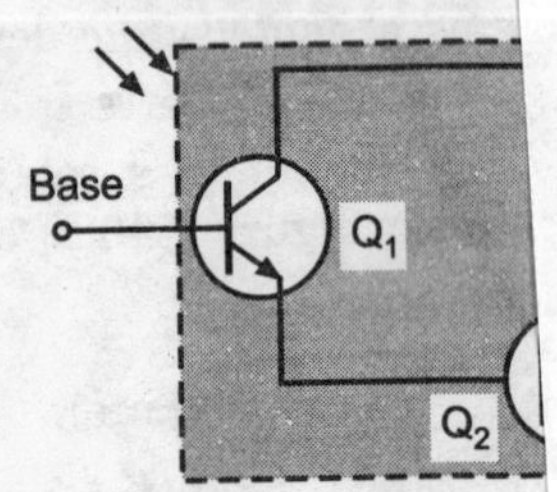

Fig. 7.38. Photo darlin

2. Advantage

Advantage of darlington connection is its high current gain. The overall β of the darlington connection is given by,

$$\beta_D = \beta_1 \times \beta_2 \qquad ...(7.8)$$

Where, β_1 and β_2 are the values of β for individual transistors. Even though a high β is obtainable from a photo darlington, its speed of is less than the phototransistor.

7.16. PHOTOVOLTAIC CELL

(i) A photovoltaic cell is an altogether different device from all the devices discu now. The photovoltaic cell generates voltage proportional to the intensity of incid

(ii) The photovoltaic cell operates on the principle of photovoltaic effect.

(iii) The construction of a selenium photovoltaic cell has been shown in figure 7.39(a) practical arrangement has been shown in figure 7.39(b).

(iv) The selenium photovoltaic cell contains a base plate made from or steel, which positive electrode. A selenium layer is placed above the base plate. Selenium sensitive to light.

(v) Above this, an electrically conducting cadmium oxide layer is deposited.

(vi) This layer is transparent to light, so the photons incident on the cell will re selenium layer. Absorption of photons takes place at the selenium layer and elect pairs are created.

solar cell supplies power. Solar cells are usually fabricated from silicon, but they can be made GaAs or other trivalent or pentavalent materials. The symbol and construction of a solar cell is shown in figure 7.41.

7.16.2. Output Characteristics of a Solar Cell

(i) The typical output characteristics of a power photocell has been shown in figure 7.42(a). Let us consider the characteristics when the incident illumination is 100 mW/cm^2.

(ii) If the cell is short circuited, then the output current is 50 mA, however, the output voltage is zero. Therefore, the output power is zero.

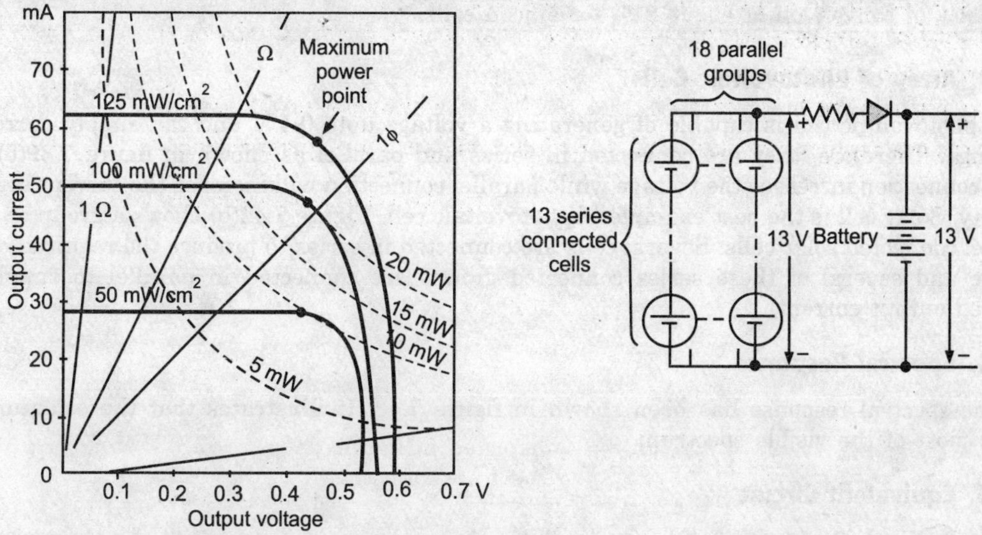

(a) Typical output characteristics of a solar cell (b)

Fig. 7.42.

(iii) Now if the cell is open circuited, then the output voltage is 0.55 V and the output current is zero.

(iv) Therefore, the output power is zero. For maximum output power, the cell should be operated on the knee of the characteristics, as depicted in figure 7.42 (a).

(v) Like other devices, the output power should be derated at increased temperatures.

Parameters of solar cell: (Consider figure 7.42 (c))

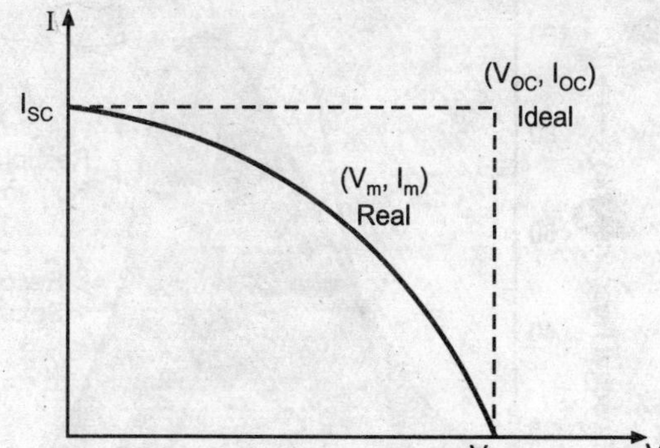

Fig. 7.42 (c)

1. The open circuit voltage V_{OC}

It is the output voltage from the solar cell when the load impedance is very high. (i.e. $R_L \rightarrow \infty$).

2. The short circuit current I_{SC}

It is the current output when the load impedance is very small (i.e. $R_L \rightarrow 0$).

3. The fill factor (F.F.)

The product of open circuit V_{OC} and short circuit current I_{SC} gives the ideal power output from the cell.

Therefore, $$P_{ideal} = I_{SC} \times V_{OC}$$

However, the actual maximum power output is less than P_{ideal} and is given by the product of maximum power current I_m and maximum power voltage V_m.

Therefore, $$P_{max} = I_m \times V_m$$

The fill factor of the solar cell is defined as the ratio of the maximum useful power to ideal power.

$$F.F = \frac{I_m \times V_m}{I_{SC} \times V_{OC}}$$

Important Point: Usually the voltage developed by every cell is of the order of 0.6 V and the efficiency of conversion of about 21% for silicon cells.

7.16.3. Array of Photovoltaic Cells

A photovoltaic cell is capable of generating a voltage upto 0.4 V and can supply current in μA range. Therefore, they are connected in series and parallel as shown in figure 7.39(b). The series connection increases the voltage while parallel connection will increase the current sourcing capacity. Solar cell is the best example of photovoltaic cell. Figure 7.42(b) shows a group of series parallel connected solar cells. Several cells are connected in series to produce the required output voltage and several of these series connected groups are connected in parallel to supply the required output current.

7.16.4. Spectral Response

The spectral response has been shown in figure 7.43. It illustrates that the selenium cell covers most of the visible spectrum.

7.16.5. Equivalent Circuit

The equivalent circuit of selenium cell has been shown in figure 7.44. In the equivalent circuit, E represents the photoelectric source, the voltage of which is dependent on the intensity of incident light. C is the effective capacitance R represents the equivalent resistance of the potential barrier. R_S is the series resistance and R_L is the load resistance.

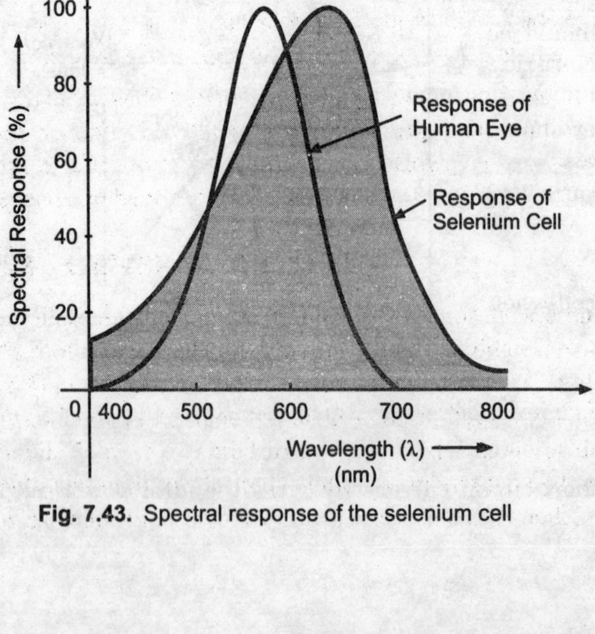

Fig. 7.43. Spectral response of the selenium cell

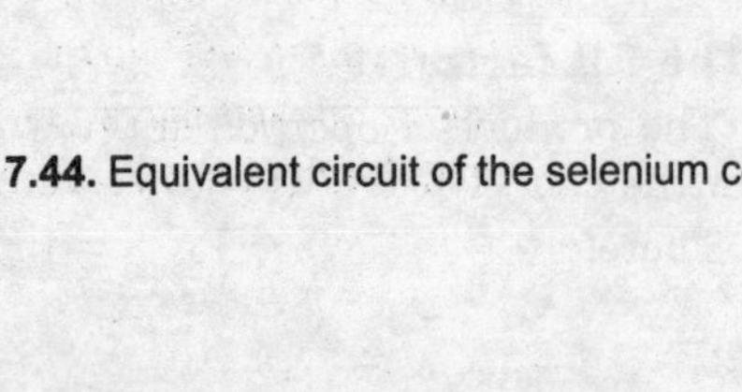

Fig. 7.44. Equivalent circuit of the selenium cell

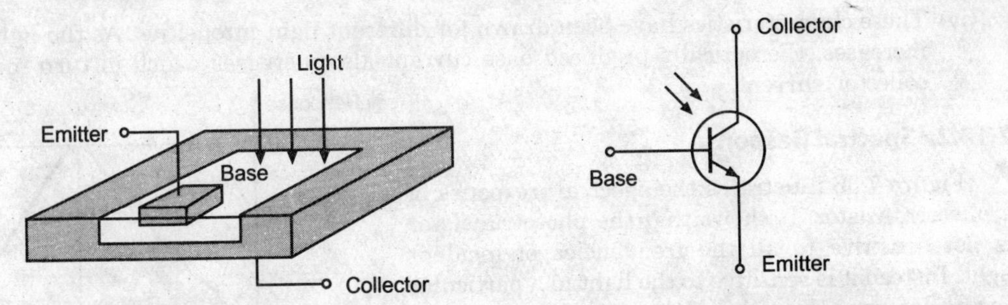

(a) Construction of a phototransistor

(b) Sympol of a phototransistor

Fig. 7.34.

7.14.1. Construction and Biasing

1. Basic Aspects

(i) The construction has been shown in figure 7.34(a). The base collector region of the photo transistor is responsible for the generation of photo base current. Hence, in the construction of the transistor, the base collector area is kept as large as possible.

(ii) The base terminal is somtimes not brought out and the phototransistor acts as a two terminal device. However, if the base terminal is brought out, then it acts as a conventional transistor. The base terminal is generally left open or connected to grounds through a large value resistor. The biasing of the phototransistor is otherwise same as that of a conventional transistor.

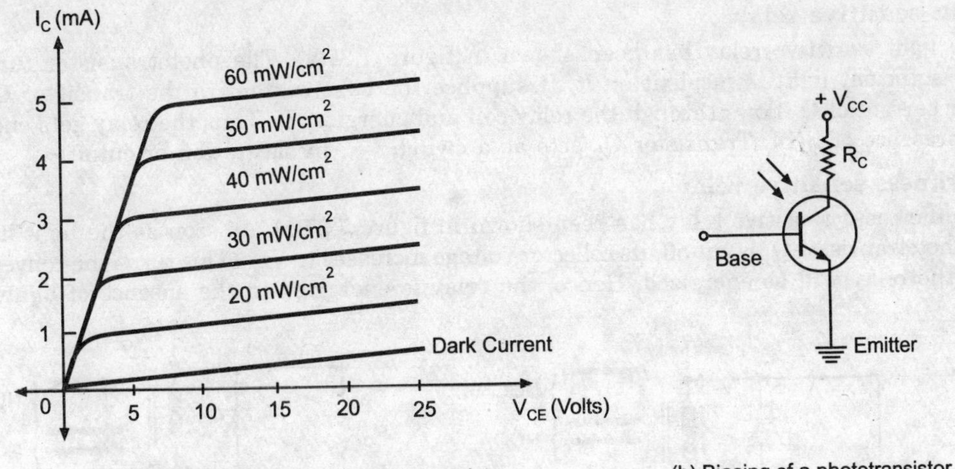

(a) Output characteristics of a phototransistor

(b) Biasing of a phototransistor

Fig. 7.35.

2. Dark current

(i) In absence of light, a small thermally generated current I_{CEO} flows through the device. It is in the nA range.

(ii) The dark current should be as small as possible. In presence of light, the base current produced (I_λ) is directly proportional to the intensity of light. The collector current in β times base current.

(iii) The phototransistor is biased depicted in figure 7.35(b). The output characteristics of a photo transistor are as shown in figure 7.35(a). They are very much similar to those of conventional transistor.

(iv) These characteristics have been drawn for different light intensities. As the light intensity increases, the optically produced base current also increases which in turn increases the collector current.

7.14.2. Spectral Response

Figure 7.36 illustrates the spectral response of a phototransistor. It shows that the phototransistor is not sensitive to all the frequencies of incident light. Instead it is sensitive to the light in a particular range of wavelength with a peak response corresponding to a particular wavelengths.

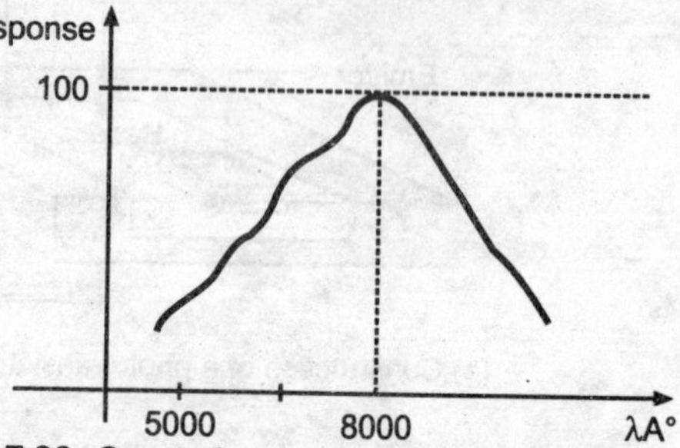

Fig. 7.36. Spectral response of a phototransistor

7.14.3. Advantages of a Phototransistor

Photocurrent gets internally multiplied by β, so no additional amplifier is required. High sensitivity of the order of 100 times that of a photodiode. Good switching speed. No memory effect.

7.14.4. Drawbacks of a Phototransistor

Phototransistor is not a fast as a conventional transistor due to sluggishness of the photoconducting material. Poor linearity. Phototransistors are temperature sensitive devices. Signal to dark current ratio is poor. This is because the dark current also gets multiplied by of the transistor. External voltage source is needed for operation.

7.14.5. Applications of Phototransistor

(i) Light sensitive relay

The light sensitive relay has been shown in figure 7.37(a). The phototransistor turns ON when a sufficient light is incident on it. It supplies the base current to the transistor Q_2. The collector current of Q_2 flows through the relay coil and energizes it. Thus the relay gets energized in the presence of light. Transistor Q_2 acts as a switch i.e., in saturation or cutoff.

(ii) Darkness sensitive relay

The darkness sensitive relay has been shown in figure 7.37(b). As soon as the light incident on the phototransistor Q_1 is cut-off, its collector voltage increases to V_{CC}. This turns the conventional Q_2 ON the relay will be energized. Hence, the relay is energized in the absence of light.

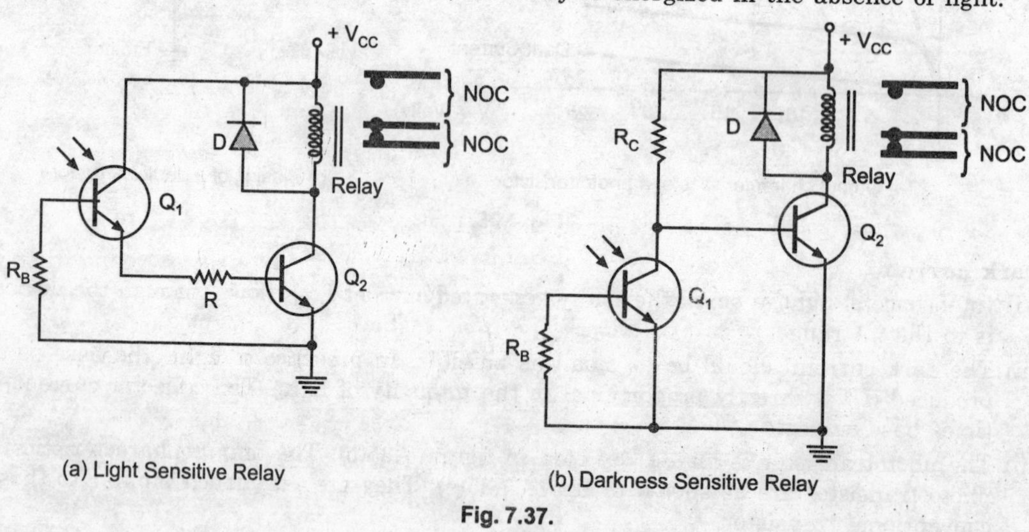

(a) Light Sensitive Relay

(b) Darkness Sensitive Relay

Fig. 7.37.

7.14.6. Performance Comparison of Photodiode and Phototransistor

Table 7.3 Performance comparison of photodiode and phototransistor

S.No.	Photodioide	Ph ototransistor
1.	It cannot provide amplification	It provides amplification
2.	Low sensitivity	High sensitivity
3.	Better linearity	Poor linearity
4.	External amplifier is required	External amplifier is not necessary
5.	Poor temperature stability	Better temperature stability with self bias circuit
6.	Symbol: 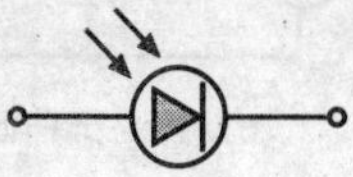	Symbol: 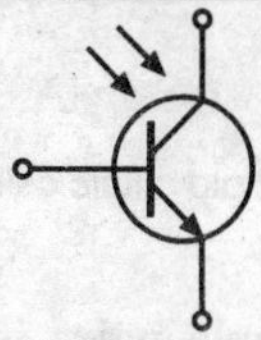

7.15. PHOTODARLINGTON

1. Circuit Symbol

The circuit symbol for a photo darlington is as depicted in figure 7.38. A photo transistor is connected in the darlington connection with a conventional transistor as shown. In the so called darlington connection the collectors of both the transistors are connected together and the emitter of Q_1 is connected to the base of Q_2 as illustrated in figure 7.38.

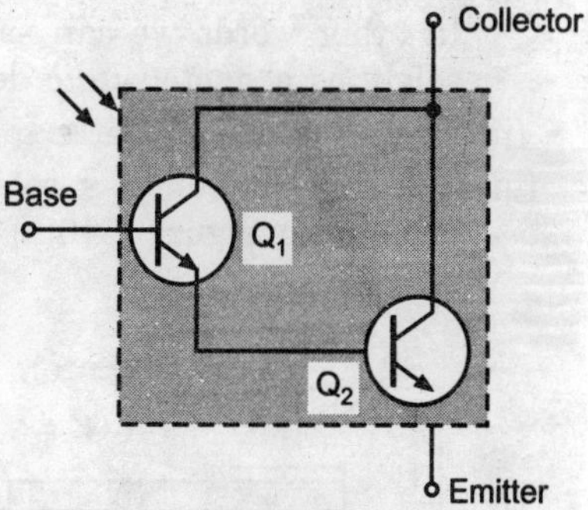

Fig. 7.38. Photo darlington

2. Advantage

Advantage of darlington connection is its high current gain. The overall β of the darlington connection is given by,

$$\beta_D = \beta_1 \times \beta_2 \qquad ...(7.8)$$

Where, β_1 and β_2 are the values of β for individual transistors. Even though a high β is obtainable from a photo darlington, its speed of operation is less than the phototransistor.

7.16. PHOTOVOLTAIC CELL

(i) A photovoltaic cell is an altogether different device from all the devices discussed upto now. The photovoltaic cell generates voltage proportional to the intensity of incident light.

(ii) The photovoltaic cell operates on the principle of photovoltaic effect.

(iii) The construction of a selenium photovoltaic cell has been shown in figure 7.39(a) and their practical arrangement has been shown in figure 7.39(b).

(iv) The selenium photovoltaic cell contains a base plate made from or steel, which acts as a positive electrode. A selenium layer is placed above the base plate. Selenium layer is sensitive to light.

(v) Above this, an electrically conducting cadmium oxide layer is deposited.

(vi) This layer is transparent to light, so the photons incident on the cell will reach the selenium layer. Absorption of photons takes place at the selenium layer and electron hole pairs are created.

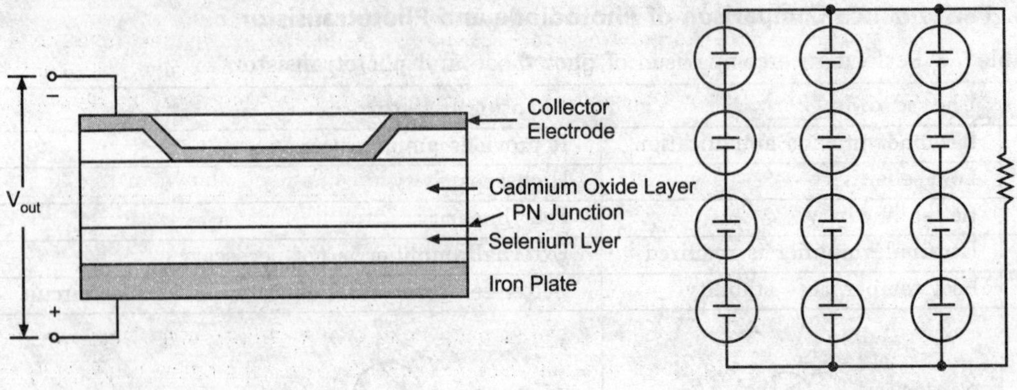

(a) Construction of Selenium photovoltaic cell (b) Array of Photovoltaic cells

Fig. 7.39.

(vii) The number of electron hole pairs is proportional to the intensity of incident light. The electrons and holes are separated out by the depletion region potential of the p-n junction.

7.16.1. Solar Cell: Construction and Working Operation (U.P. Tech, Sem. Exam; 2009-10, 2011-12)

(i) A solar cell is a P-N junction device with no voltage directly applied across the junction. In other words we can say that it is essentially a large photodiode designed to operate solely as a *photovoltaic* device and give as much output power as possible.

(ii) The solar cell converts solar energy into electrical energy.

(iii) The construction of a solar cell is depicted in figure 7.40(a) and the simplified diagram depicted in figure 7.40(b) can be used to explain the operation of the solar cell.

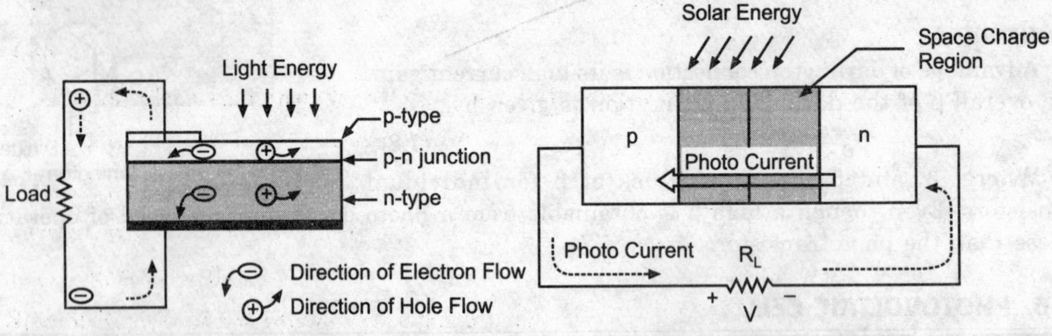

(a) Construction of a Solar Cell (b) Simplified Diagram

Fig. 7.40.

(iv) When the light hits the space charge region, electrons and holes are generated due to the photons striking the valence electrons and imparting energy to them. The optically generated electron-hole pairs are quickly separated and swept outside the space charge region (depletion region) by the electric field.

(v) These electrons and holes flow to constitute the photocurrent as shown in figure 7.40(b).

(vi) This photocurrent produces a voltage V across the load resistance R_L. Thus

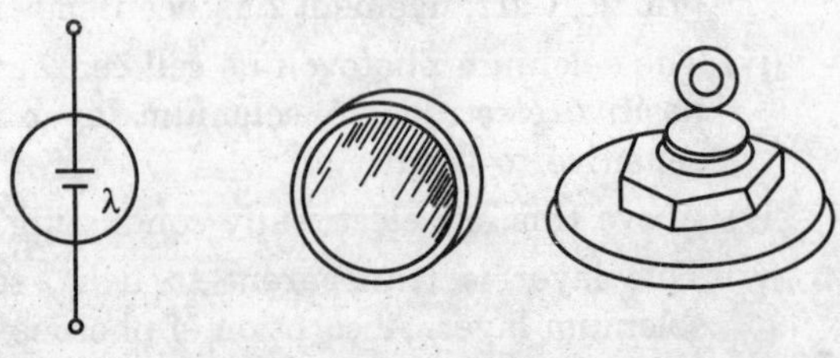

(a) Symbol of a solar cell (b) Construction of a solar cell

Fig. 7.41.

solar cell supplies power. Solar cells are usually fabricated from silicon, but they can be made GaAs or other trivalent or pentavalent materials. The symbol and construction of a solar cell is shown in figure 7.41.

7.16.2. Output Characteristics of a Solar Cell

(i) The typical output characteristics of a power photocell has been shown in figure 7.42(a). Let us consider the characteristics when the incident illumination is 100 mW/cm^2.

(ii) If the cell is short circuited, then the output current is 50 mA, however, the output voltage is zero. Therefore, the output power is zero.

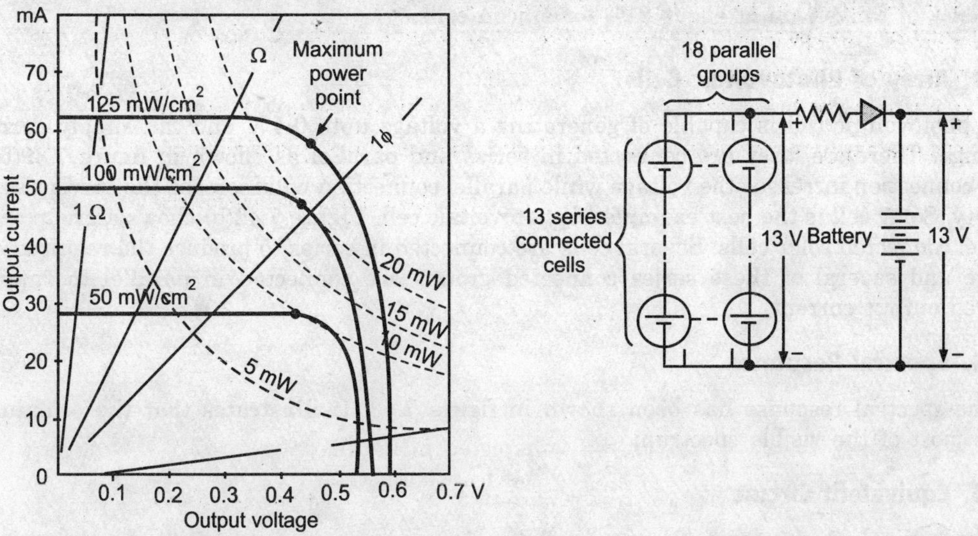

(a) Typical output characteristics of a solar cell (b)

Fig. 7.42.

(iii) Now if the cell is open circuited, then the output voltage is 0.55 V and the output current is zero.

(iv) Therefore, the output power is zero. For maximum output power, the cell should be operated on the knee of the characteristics, as depicted in figure 7.42 (a).

(v) Like other devices, the output power should be derated at increased temperatures.

Parameters of solar cell: (Consider figure 7.42 (c))

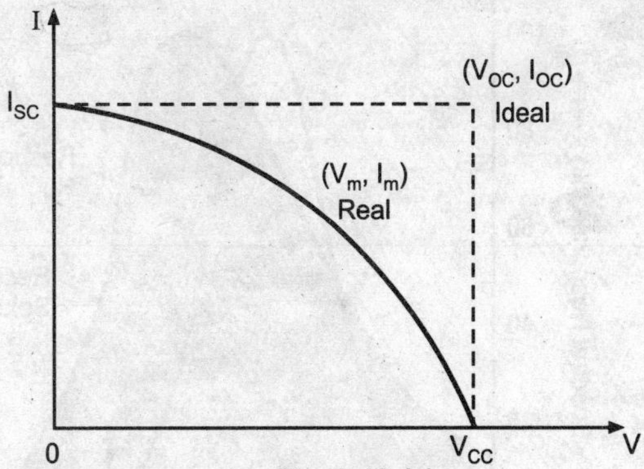

Fig. 7.42 (c)

1. The open circuit voltage V_{OC}

It is the output voltage from the solar cell when the load impedance is very high. (i.e. $R_L \to \infty$).

2. The short circuit current I_{SC}

It is the current output when the load impedance is very small (i.e. $R_L \to 0$).

3. The fill factor (F.F.)

The product of open circuit V_{OC} and short circuit current I_{SC} gives the ideal power output from the cell.

Therefore, $$P_{ideal} = I_{SC} \times V_{OC}$$

However, the actual maximum power output is less than P_{ideal} and is given by the product of maximum power current I_m and maximum power voltage V_m.

Therefore, $$P_{max} = I_m \times V_m$$

The fill factor of the solar cell is defined as the ratio of the maximum useful power to ideal power.

$$F.F = \frac{I_m \times V_m}{I_{SC} \times V_{OC}}$$

Important Point: Usually the voltage developed by every cell is of the order of 0.6 V and the efficiency of conversion of about 21% for silicon cells.

7.16.3. Array of Photovoltaic Cells

A photovoltaic cell is capable of generating a voltage upto 0.4 V and can supply current in μA range. Therefore, they are connected in series and parallel as shown in figure 7.39(b). The series connection increases the voltage while parallel connection will increase the current sourcing capacity. Solar cell is the best example of photovoltaic cell. Figure 7.42(b) shows a group of series parallel connected solar cells. Several cells are connected in series to produce the required output voltage and several of these series connected groups are connected in parallel to supply the required output current.

7.16.4. Spectral Response

The spectral response has been shown in figure 7.43. It illustrates that the selenium cell covers most of the visible spectrum.

7.16.5. Equivalent Circuit

The equivalent circuit of selenium cell has been shown in figure 7.44. In the equivalent circuit, E represents the photoelectric source, the voltage of which is dependent on the intensity of incident light. C is the effective capacitance R represents the equivalent resistance of the potential barrier. R_S is the series resistance and R_L is the load resistance.

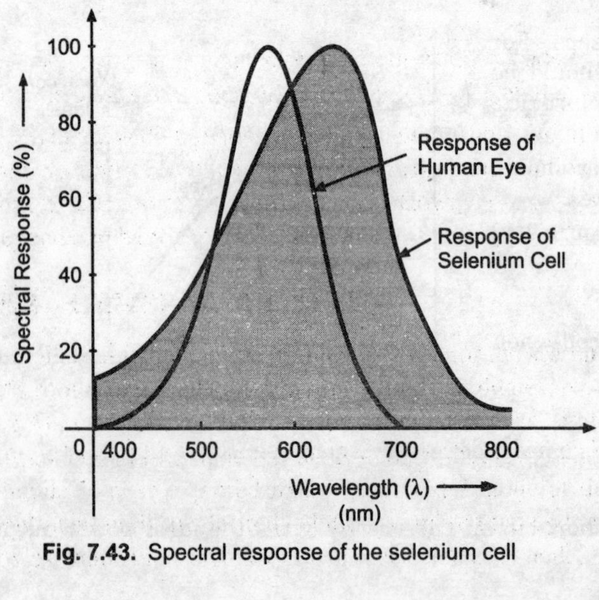

Fig. 7.43. Spectral response of the selenium cell

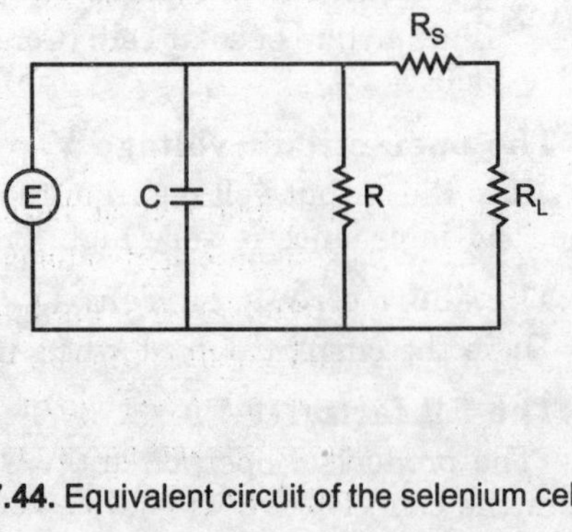

Fig. 7.44. Equivalent circuit of the selenium cell

7.16.6. Advantages of Solar Cells

Solar cells respond very well over a wide range of incident wavelength. No need of external dc source for operation. Can produce an adequately large photocurrent. It is robust and does not get damaged easily.

7.16.7. Drawbacks

(i) Slow operation. The solar cells cannot change their output rapidly if the light intensity changes rapidly.
(ii) Solar cells are temperature sensitive.
(iii) Low output voltage and current.
(iv) Solar panels are expensive.
(v) Conversion efficiency is low.

7.16.8. Important Applications of Solar Cells

Solar cells are used to power the electronic circuits used in satellites and space vehicles. Power supply to calculators. For powering the cars run on solar energy. Typically, a car requires about $8m^2$ of solar cell arrays that can produce 800 W of power on a sunny day at noon. For charging the batteries. To supply power to street lights. Solar panels are used for water pumps, home lighting etc.

7.17. OPTOCOUPLERS / ISOLATORS

1. Optocoupler or optoisolator is a combination of light source and light detector in the same package. These are used to couple signal from one point to the other optically, by providing a complete electrical isolation between them.
2. This kind of isolation is provided between a low power control circuit and high power output circuit, to protect the control circuit.
3. Depending upon the type of light sources and detectors used we can get a variety of optocouplers. They are as under:
   (i) LED – LDR optocoupler
   (ii) LED – photodiode optocoupler
   (iii) LED – phototransistor optocoupler.

Before we discuss them, let us discuss about the important characteristics of an optocoupler.

7.17.1. Characteristics of Optocoupler

Some of the important characteristics of optocoupler may be listed as under:
(i) Current transfer ratio (CTR)
(ii) Isolation voltage
(iii) Response time
(iv) Common mode rejection

(i) Current transfer ratio (CTR)

It is defined as the ratio of output collector current (I_C) to the input forward current (I_F).

$$CTR = \frac{I_C}{I_F} \times 100\%$$

CTR should be as large as possible. Its value depends on the devices used as source and detector.

(ii) Isolation voltage between input and output

It is the maximum voltage which can exist differentially between the input and output without affecting the electrical isolation. Isolation voltage is specified in K V_{rms} with a relative humidity of 40 to 60%. Isolation voltage should be as high as possible.

(iii) Response time

Response time indicates how fast an optocoupler can change its state. Response time largely depends on the detector transistor, input current and load resistance.

(iv) Common mode rejection

Even though the optocouplers are electrically isolated for dc and low frequency signals, an impulsive input signal (the signal which changes suddenly) can give rise to a displacement current $I_d = C_f \frac{dv}{dt}$. This current can flow between input and output due to the stray capacitance C_f existing between input and output. This will allow the noise to appear in the output. Common mode rejection should be as high as possible.

7.17.2. Types of Optocouplers

LED-Photodiode Optocoupler

The LED-photodiode optocoupler has been shown in figure 7.45(a). Here, the infrared LED works as a light source and photodiode is used as a detector. The advantage of using the photodiode is its high linearity. When the pulse at the input goes high, the LED turns ON. It emits light. This light is focussed on the photodiode. In response to this light, the photocurrent will start flowing through the photodiode. As soon as the input pulse reduces to zero, the LED turns OFF and the photocurrent through the photodiode reduces to zero. Hence, the pulse at the input is coupled to the output side.

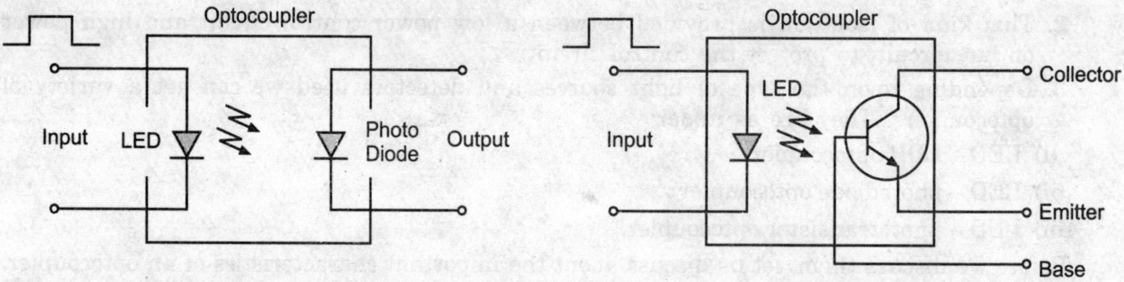

(a) LED-photodiode optocoupler (b) LED-phototransistor optocoupler

Fig. 7.45.

2. LED-Phototransistor Optocoupler

The LED-phototransistor optocoupler has been shown in figure 7.45(b). An infrared LED acts as the light source and these phototransistor acts as a photodetector. This is the most populariy used optocoupler, because it does not need any additional amplification. When the pulse at the input goes high, the LED turns ON. The light emitted by the LED is focussed on the CB junction of the phototransistor. In response to this light photocurrent starts flowing which acts as base current for the phototransistor. The collector current of phototransistor starts flowing. As soon as the input pulse reduces to zero the LED turns OFF and the collector current of phototransistor reduces to zero. Thus, the pulse at the input is optically coupled to the output side.

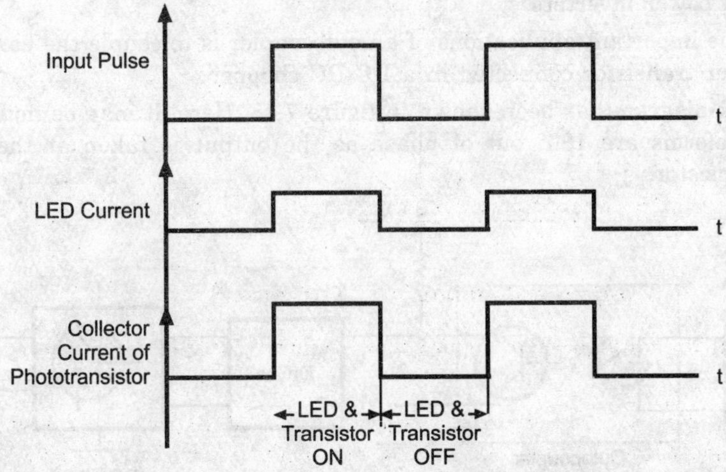

Fig. 7.46. Waveforms of LED-phototransistor optocoupler

3. Optoisolator Circuit

Figure 7.47 illustrates the typical connection diagram of an optoisolator. The DC input voltage V_1 forward biases the LED and the resistance R_1 is used for limiting the forward LED current. Due to the forward LED current, the LED emits light which is focussed on the base collector junction of the internal phototransistor. Proportional to the light intensity, there is a proportional output current produced. R_2 limits the output current.

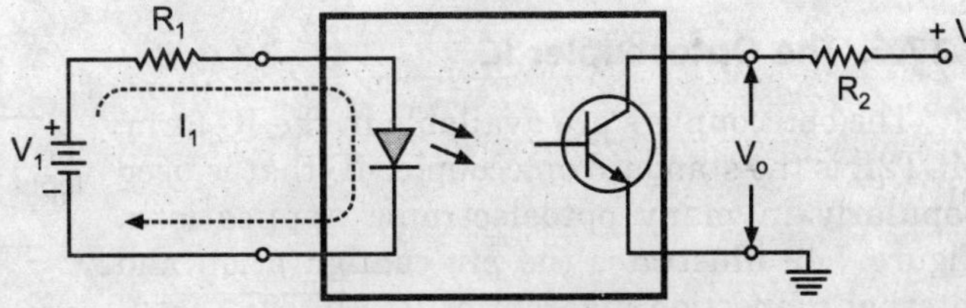

Fig. 7.47. Optoisolator circuit

7.17.3. Some Advantages of Optocouplers

(i) Control circuits are well protected due to electrical isolation.
(ii) Wideband signal transmission is possible.
(iii) Due to unidirectional signal transfer, noise from the output side does not get coupled to the input side.
(iv) Interfacing with logic circuits is easily possible.
(v) It is small size and light weight device.

7.17.4. Drawbacks

(i) Slow speed
(ii) Possibility of signal coupling for high dV/dt signals.

7.17.5. Some Important Applications

(i) Optocouplers are used basically to isolate low power circuits from high power circuits.
(ii) At the same time the control signals are coupled from the control circuits to the high power circuits.
(iii) Some of such applications are:
 (i) AC to DC converters used for DC motor speed control
 (ii) High power choppers

(iii) High power inverters

(iv) One of the important applications of an optocoupler is to couple the base driving signals to a power transistor connected in a DC-DC chopper.

(v) The block diagram has been shown in figure 7.48. Here, it may be noted that the input and waveforms are 180° out of phase as the output is taken at the collector of the phototransistor.

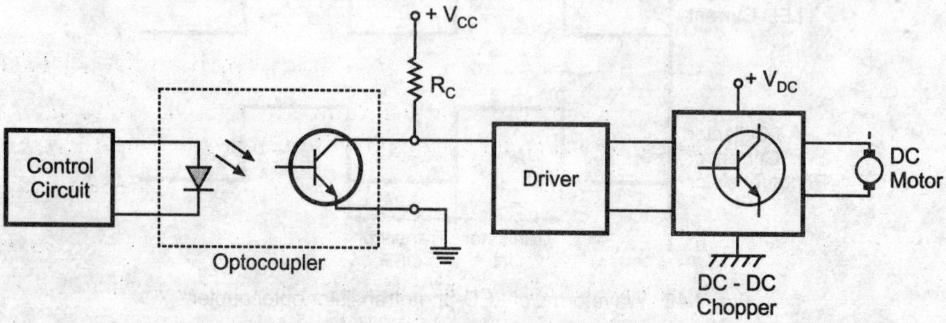

Fig. 7.48. Application of an optoisolator

7.17.6. The Optocoupler IC

The optocouplers are available in the IC form. MCT2E is the standard optocoupler IC that is used popularly in many optoelectronic applications. Figure 7.49 illustrates the pin configuration and internal connection diagram of IC MCT2E.

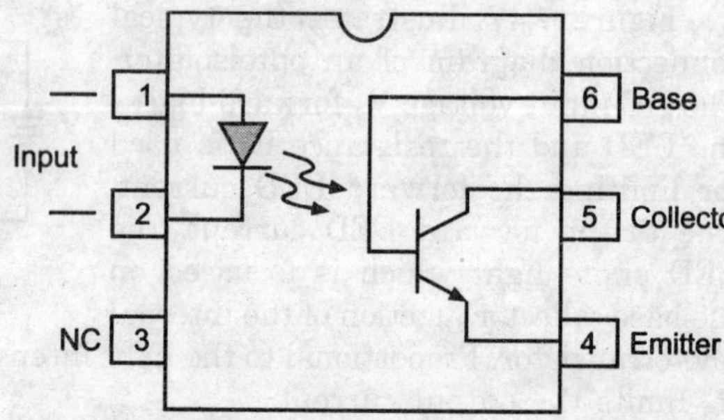

Fig. 7.49. Pin configuration and internal connecetion of IC MCT2E

7.18. LASER DIODE

(U.P. Tech, Sem. Exam; 2008-09)

The term LASER stands for light amplification by stimulated emission of radiation. Here, the word *light* represents the electromagnetic radiation which covers the range from ultraviolet region to the infrared region. The three basic steps in the generation of a LASER may be listed as under:

1. Absorption
2. Spontaneous emission
3. Stimulated emission

1. Absorption

When light is incident on a photo sensitive material, electrons absorb energy from the incident photons and move to a higher energy level as depicted in figure 7.50(a). This process is known as absorption.

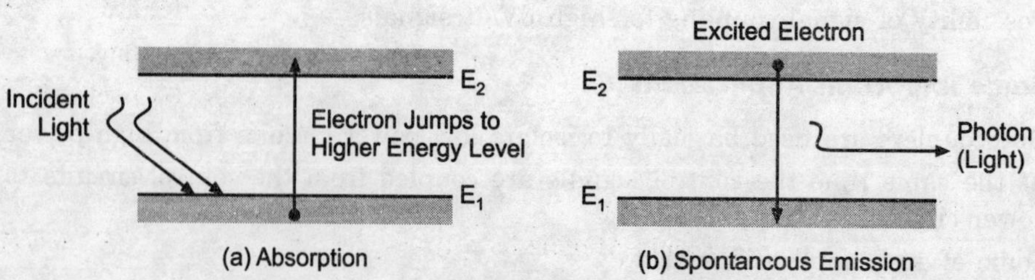

Fig. 7.50.

2. Spontaneous emission

(i) After absorption electrons are present in the higher energy level. These are called as *excited* electrons. If no light is incident on them, then they can not stay for a longer time at this higher energy level.

(ii) After a short time the *excited* electrons will fall back to their original energy level.

(iii) While doing so, it will radiate the energy, it had absorbed during the absorption process. Hence, this energy is radiated in the form of electromagnetic radiation (light).

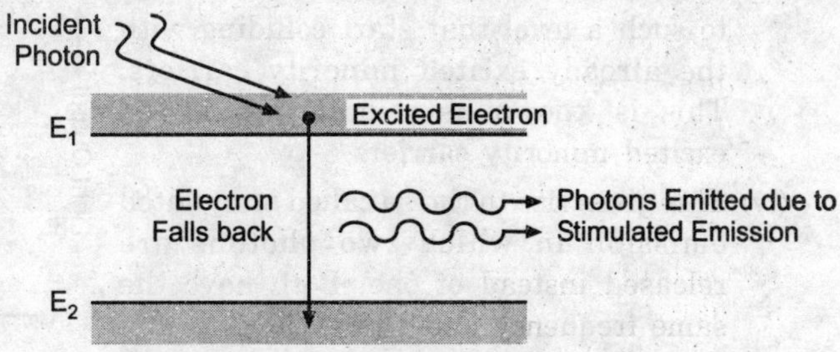

Fig. 7.51. Process of stimulated emission

(iv) This process is know as spontaneous emission and it has been shown in figure 7.50(b).

3. Stimulated emission

(i) This is the most important process in the formation of a LASER. In this process the *amplification* of light takes place.

(ii) Let us consider an excited electron present at a higher energy level E_2 as shown in figure 7.51.

(iii) If the incident light strikes this electron, when it is in the excited state and if there is no chance for this electron to go higher energy level then this electron will fall back to its original energy level and will emit light.

(iv) Here, we get two photons emitted instead of one in the spontaneous emission. Hence, one incident photon causes the emission of two photons and light amplification takes place.

(v) The two emitted photons are in phase and travel in the same direction. This process is known as as *stimulated emission* and it has been illustrated in figure 7.51. For the stimulated emission to take place, it is necessary to have a large number of excited electrons. The process in which the number of excited atoms is made larger than the electrons present at the ground state (low energy level) is known as **population inversion**.

7.18.1. Construction of a Laser Diode

The construction of a laser diode has been shown in figure 7.52. This diode is used to generate a LASER beam. The semiconductor material used for the P and N sides is **Gallium Arsenide**. The laser diode is operated in the forward biased condition. The LASER produced by the P-N junction diode is called semiconductor Laser.

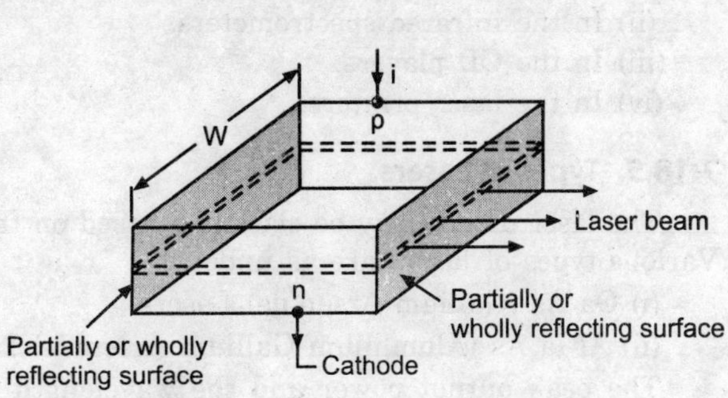

Fig. 7.52. Construction of semiconductor laser diode

Working Operation

Semiconductor laser is a special type of laser. The laser action takes place within a semiconductor diode junction which is of the same type used for LEDs. The steps in generation of semiconductor laser may be listed as under:

(i) When current is passed through a diode junction light is emitted due to *spontaneous emission* as discussed earlier preceeding section.

(ii) When a critical current level is exceeded, the population of minority carriers on either sides of the junction increases. This increases the number of recombinations of electrons and holes and thus, the number of photons emitted.

(iii) Hence, the density of photons increases to such a level that start colliding with the already excited minority carriers. This is known as stimulation of the excited minority carriers.

(iv) This gives rise to the so called stimulated emission in which two photons are released instead of one. Both have the same frequency and energy level.

(v) The laser action of the semiconductor diode can be enhanced by placing a reflecting surface on each side of the junction as depicted in figure 7.52.

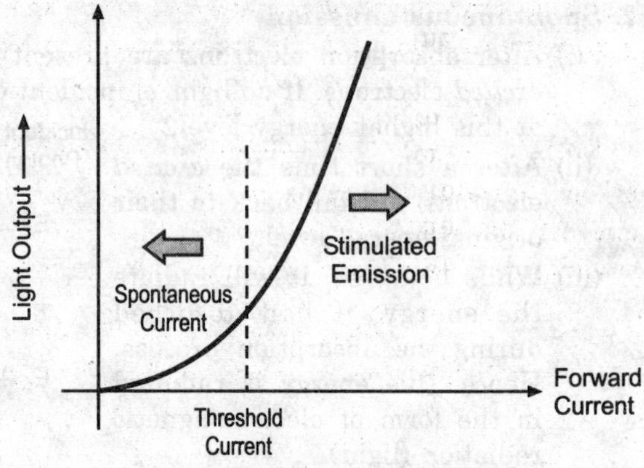

Fig. 7.53. Light output Vs forward current characteristics of a laser diode

Important Point: It is important to note that the stimulated emission will take place only when the forward diode current exceeds a minimum level called *threshold level.* The light outputs characteristics of a laser diode has been shown in figure 7.52.

7.18.2. Some Advantages of Laser Diode

Laser diode is the cheapest and smallest device to produce LASER output. They may be easily fabricated in arrays. They have high efficiency (typically 50% to 60%).

7.18.3. Drawbacks of Laser Diode

(i) Divergence of laser beam is more.
(ii) Heating problems are critical

7.18.4. Some Applications of Laser Diode

(i) As lights source in fiber optic communication.
(ii) In the infrared spectrometers.
(iii) In the CD players.
(iv) In the laser printers.

7.18.5. Types of Lasers

The laser diodes may be classified based on the material used for forming the P-N junction. Various types of lasers are as under:

(i) Ga As (Gallium Arsenide) Lasers
(ii) AlGa As (Aluminium Gallium Arsenide) Laser

The peak output power and the wavelength corresponding to the peak output power also varies depending on the material used.

SOLVED NUMERICAL PROBLEMS

EXAMPLE 7.2. Find out the open-circuit voltage of a silicon P-N junction solar cell at 300°K with the following parameters:

$N_A = 7.5 \times 10^{24}$ atoms/m^3, $N_D = 1.5 \times 10^{22}$ atoms/m^3, $D_0 = 25 \times 10^{-4}$ m^2/s
$D_h = 1 \times 10^{-3}$ m^2/s $\quad \tau_{eo} = 500$ ns, $\quad \tau_{ho} = 100$ ns, $\quad n_i = 1.5 \times 10^{16}$/m^3
and photo current density = 12.5 mA/cm^2

Solution: We know that electron diffusion length is given by $L_e = \sqrt{D_e \tau_{eo}}$

$$L_e = \sqrt{25\times10^{-4}\times500\times10^{-9}} = 35.36 \text{ μm}$$

We know that, hole diffusion length is given by $L_h = \sqrt{D_h \tau_{ho}} = \sqrt{1\times10^{-3}\times100\times10^{-9}} = 10.0 \text{ μm}$

$$J_s = e\, n_i^2\left(\frac{D_e}{L_e N_A}+\frac{D_h}{L_h N_D}\right)$$

$$= 1.6\times10^{-19}\times(1.5\times10^{16})^2\left[\frac{25\times10^{-4}}{35.36\times10^{-6}\times7.5\times10^{24}}+\frac{1\times10^{-3}}{10\times10^{-6}\times1.5\times10^{22}}\right]$$

$$= 2.4\times10^{-7} \text{ A/m}^2 = 2.4\times10^{-11} \text{ A/cm}^2$$

Also, open circuit voltage is given by

$$V_{OC} = V_T \log_e\left(1+\frac{I_\lambda}{I_s}\right) = 26 \text{ mv} \log_e\left(1+\frac{12.5\times10^{-3}}{2.4\times10^{-4}}\right)$$

$$= 521.84 \text{ mv} = 0.522 \text{ V} \quad \textbf{Ans.}$$

EXAMPLE 7.3. Find out photocurrent density in a silicon PIN photodiode with an intrinsic region width of W = 25 μm assuming the photon flux of 10^{21} m^{-2} s^{-1} and absorption coefficient $\alpha = 1\times10^5$ m^{-1}.

Solution : Photon flux $\phi_0 = 1\times10^{21}$ m^{-2} s^{-1}

Absorption coefficient $\alpha = 1\times10^5$ m^{-1}

The generation rate of EHP at the front edge of the intrinsic region

$$G_{L1} = \alpha\,\phi_0 = 1\times10^5\times1\times10^{21} = 1\times10^{26} \text{ m}^{-3}\text{ s}^{-1}$$

and the generation rate at the back edge of the intrinsic region

$$G_{L2} = \alpha\,\phi_0\, e^{-\alpha w} = 1\times10^5\times1\times10^{21}\, e^{-105\times25\times10-6}$$

$$= 0.082\times10^{26} \text{ m}^{-3}\text{ s}^{-1}$$

Hence, the generation rate in non-uniform is the intrinsic region and the density of photocurrent will then,

$$J_L = e\,\phi_0\,(1-e^{-\alpha w})$$

$$= 1.6\times10^{-19}\times10^{21}\,(1\, e^{-105\times25\times10-6})$$

$$= 146.9 \text{ A/m}^2 = 14.69 \text{ mA/cm}^2 \quad \textbf{Ans.}$$

SUMMARY

- The optoelectronic devices may be classified into following three categories as:
  (i) Light emitters
  (ii) Light sensors or detectors
  (iii) Optical couplers or optoisolator.
- The optoelectronic devices may be classified as light emitters, light detectors optocouplers or isolators.
- These devices give out light when excited electrically. Light emitters are also known as light sources.
- These devices convert light into some form of electrical energy, or they can modify some electrical parameters such as resistance.
- These are formed by combining a light source and a light detector into one device. Optocouplers will couple the signals electrically from one point to the other but isolate the two points physically.
- The liberation of electrons from matter under the influence of light is called as the photoelectric effect. This was first observed by Hertz in 1887. If electrons are liberated from a metal surface under the influence of light then the phenomenon is known as photoemissive effect. This phenomenon is observed in the vacuum and gas filled phototubes.

- The important characteristics of the photoemissive effect may be listed as under:
  (i) When the light is incident on the cathode which is made of some photo photosensitive material, it starts emitting electrons. The number of electrons emitted by the cathode is proportional to the intensity of incident light. This means that the photoelectric current is directly proportional to the intensity of light.
  (ii) Keeping the intensity of incident light constant, if we increase the positive voltage V applied between the cathode and anode, then the number of electrons to the collector increase until saturation occurs.
  (iii) The maximum energy of the photoelectrons emitted by the cathode is independent of light intensity but varies linearly with the frequency of incident light.
  (iv) The photoelectric characteristics are practically independent of temperature, within a wide range of temperature.
  (v) There is no time lag between the instant of exposure of the cathode to light and the instant of starting the emission.
- Photoelectric yield or quantum yield is defined as the photocurrent in amperes per watt of incident light. The 6th characteristics suggests that the photoelectric yield depends on the frequency of light. The other names for photoelectric yield are current wavelength characteristics, spectral response or specific photosensitivity.
- LEDs are made of Gallium Arsenide (Ga As), Gallium Arsenide phosphide (Ga As P) and Gallium phosphide (Ga P). Silicon and Germanium are not used since they are essentially heat producing materials and are very poor in producing light.
- The colour of the emitted light is decided by its wavelength. The wavelength of emitted light (and hence its colour) depends on the value of forbidden gap (E_G). Forbidden gap value depends on the material used. Hence, the colour of emitted light is dependent on the material used.
- (i) LEDs are of small size and light weight. Therefore it is possible to pack a large number of LED, in a small space while manufacturing a display.
  (ii) They are available in different spectral colours.
  (iii) They have longer life as compared to the lamps.
  (iv) The light emitted by a LED is proportional to the amount of current flowing through it. Hence we can control the current flowing through LEDs to vary their brightness as per the requirements of the application.
  (v) They are suitable at high operating speeds as they take less than 1 μs to turn on or off.
  (vi) LEDs can be easily interfaced with the other electronic circuits.
- (i) Output power is affected by changes in temperature.
  (ii) Overcurrent can damage it easily.
  (iii) They need larger power for their operation.
  (iv) Luminous efficiency of LEDs is low.
- If radiation (light) falls on a semiconductor, its conductivity increases. This is known as the photoconductive effect and it may be explained as under: Because of the radiant energy supplied to the semiconductor, the covalent bonds are broken and the electron hole pairs are generated. These increased current carriers increase the conductivity of the material and hence decrease the resistivity. Such a device is known as a photoresistor or photoconductor.
- The photoconductive cell or a light dependent resistor (LDR) makes use of the principle of photoconductivity. It is a semiconductor device in which resistance is dependent on the intensity of incident light.
- The main drawback of the LDR is their slowness. This makes them unsuitable for the applications where light intensity changes rapidly. We need to use faster devices for such applications.
- This means, a large change in the photocurrent for a small change in light intensity. High speed of operation as compared to LDR (Light Dependent Resistor).
- PIN diodes are used at very high frequencies (higher than 1 GHz) because even at these frequencies, the diode has a very high impedance when reverse biased and very low impedance when it is

forward biased. Due to these characteristics of a PIN diode, it is used as a switch at very high frequencies and even to switch vary currents and/or vary high voltages.

- A photodiode cannot provide any amplification action. A phototransistor can provide the internal current multiplication and can generate a larger output signal. The light is allowed to fall on the collector base junction of the transistor.
- Photocurrent gets internally multiplied by β, so no additional amplifier is required. High sensitivity of the order of 100 times that of a photodiode. Good switching speed. No memory effect.
- Phototransistor is not a fast as a conventional transistor due to sluggishness of the photoconducting material. Poor linearity. Phototransistors are temperature sensitive devices. Signal to dark current ratio is poor. This is because the dark current also gets multiplied by of the transistor. External voltage source is needed for operation.
- Optocoupler or optoisolator is a combination of light source and light detector in the same package. They are used to couple signal from one point to the other optically, by providing a complete electrical isolation between them. This kind of isolation is provided between a low power control circuit and high power output circuit, to protect the control circuit. Depending on the type of light sources and detectors used we can get a variety of optocouplers.
- Some of the important characteristics of optocoupler may be listed as under:
  (i) Current transfer ratio (CTR)
  (ii) Isolation voltage
  (iii) Response time
  (iv) Common mode rejection
- The optocouplers are available in the IC form. MCT2E is the standard optocoupler IC that is used popularly in many optoelectronic applications.
- The term LASER stands for light amplification by stimulated emission of radiation. Here, the word *light* represents the electromagnetic radiation which covers the range from ultraviolet region to the infrared region.
- The three basic steps in the generation of a LASER may be listed as under:
  1. Absorption
  2. Spontaneous emission
  3. Stimulated emission

SHORT QUESTIONS WITH ANSWERS

Q.1. What are photonic devices ?

Ans. Photonic devices are those in which the basic particle of light (the photone) plays an important role. Photodetectors, solar cells and LASERS etc. are photonic devices.

Q.2. What are photodetectors ?

Ans. Photodetectors are the semiconductor devices that can detect optical signals through electronic processes.

Q.3. Enlist the three electronic processes involved in photodetection.

Ans. Three electronic processes involved in photodetection are (i) carrier generation by incident light (ii) carrier transport and/or multiplication and (Hi) interaction of current with the external circuit to give the output signal.

Q.4. Name some devices used for photodetection.

Ans. The best known devices used for photodetection are LDRs, photodiodes and phototransistors.

Q.5. What is photodiode ?

Ans. Photodiode is a two-terminal semiconductor P-N junction device and is designed to operate with reverse bias.

Q.6. How the current is reduced to zero in a photodiode ?

Ans. The reverse saturation current I_0 flowing through a photodiode is reduced to zero by applying forward bias voltage of magnitude equal to barrier potential.

Q.7. Photodiode is a photovoltaic device or a photoconductive device or both ?

Ans. Photodiode is a photovoltaic device as well as photoconductive device. When it is operated with reverse bias, it is photoconductive device and when operated without the reverse bias, it photovoltaic device.

Q.8. What is PIN photodiode ?

Ans. PIN photodiode consists of a P-region and N-region separated by an intrinsic region where width of intrinsic region is much larger than the space charge width of an ordinary PN junction.

Q.9. How does avalanche photodiode differ from PIN photodiode ?

Ans. The avalanche photodiode is similar to PIN photodiode except that it is operated at high reverse-bias voltages so that impact ionization occurs.

Q.10. How does a phototransistor differ from an ordinary BJT ?

Ans. Phototransistor is similar to an ordinary BJT except that no base terminal is provided. Instead of a base current, the input to the transistor is in the form of light.

Q.11. What are photovoltaic or solar cells ?

Ans. Photovoltaic or solar cells are semiconductor junction devices used to convert radiation energy into electrical energy. These cells generate a voltage proportional to electromagnetic radiation intensity and are called the photovoltaic cells because of their voltage generating capability.

Q.12. What is meant by LED ?

Ans. LED stands for light-emitting diode.

Q.13. What precautions are required to be observed in the use of LEDs ?

Ans. LEDs should neither be reverse-biased nor operated near their maximum current rating. The leads of a LED should never be bent closer than about 2 mm from the encapsulation.

Q.14. What does LASER stand ?

Ans. LASER stands for light amplification by stimulated emission of radiation.

Q.15. List the advantages of laser devices.

Ans. Laser devices are small in size, highly efficient, operate at low power inputs and their outputs are adjustable by varying the junction currents.

Q.16. What are heterojunction lasers ?

Ans. Heterojunction lasers are the multiple layer PN junction lasers with more than one junction and can provide still greater efficiencies and more power.

Q.17. Why II-VI compounds are not used for the fabrication of semiconductor lasers though they are generally very efficient light emitters ?

Ans. Because junctions are difficult to form.

Q.18. Why semiconductors with indirect band gaps are not used for the fabrication of semiconductor lasers ?

Ans. Because semiconductors with indirect band gaps are not sufficiently efficient light emitters for practical laser fabrication.

REVIEW QUESTIONS

1. Silicon material is used for construction of photodiode. Give reason.
2. Phototransistor needs no base biasing. Give reason.
3. Photodiode is used in the reverse biased condition. Give reason.
4. Give reason. Ge material is not used for the construction of photodiodes.

5. Define the term optocoupler. Explain important characteristics of an optocoupler. State the advantages of optocoupling.
6. Write notes on: Application of opto-devices. Explain applications of one or two optical devices.
7. Define and explain the following terms in case of photodetectors (any four):
   (i) Responsivity (R)
   (ii) Quantum efficiency (η)
   (iii) Noise equivalent power (NEP)
   (iv) Directivity (D)
   (v) Dark current (I_D)
   (vi) Spectral response
8. Differentiate between: LED and conventional diode.
9. Differentiate between LED and photodiode.
10. Write detailed note on: Photoconductive sensors.
11. Draw the light output characteristics of a laser diode.
12. State advantages, disadvantages and applications of a laser diode.
13. Draw the sketch indicating construction, sketch the characteristics and give the area of application in respect of the following device - photovoltaic device.
14. Write explanatory note on opto devices and their applications. Also mention the peculiarities of each device.
15. Explain construction, principle of operation and characteristics of phototransistor. What is dark current?
16. Explain the operation of optocoupler/isolator. Give any one application in brief.
17. With the help of neat diagram explain the operation of a laser diode.
18. Write short notes on the following:
   (i) Photodetector
   (ii) LDR
   (iii) PIN photodiode
   (iv) Solar cells
   (v) LEDs
   (vi) Optical fibers
   (vii) Semiconductor lasers
   (viii) Light emitting materials
19. Describe monomode fiber, multimode fiber, step index fiber and graded index fiber.
20. Compare a phototransistor and photodiode. Name some of the applications of phototransistor.
21. What is LED? Give its principle of working construction, circuit symbol, merits demerits and applications.
22. Briefly discuss the use of LEDs as display devices in instrumentation.
23. What is an LDR? Give its basic construction symbol and characteristics. How a LDR can be used as an overlight detector?
24. What is a photodiode? Give its biasing arrangement and construction. Explain its construction and operation.
25. What is optical fiber? Explain how optical fibers are classified. Define angle of acceptance and numerical aperture.
26. Explain the construction and working of a photovoltaic cell. Give its circuit symbol, voltage and current characteristics. Explain also its merits and applications.

□□□

Chapter 8

Negative Conductance Microwave Devices

Syllabus

Tunnel Diode: Degenerate Semiconductors.
IMPATT Diode
The Transferred Electron Mechanism: The GUNN Diode

Chapter Outline

- Introduction
- Tunnel Diode
- Degenerate and Nondegenerate Semiconductors
- IMPATT Diode
- Gunn Diode

8.1. INTRODUCTION

As a matter of fact, semiconductor instability may be affected by negative conductance and this instability is effectively used in semiconductors for high frequency applications. There are number of devices that are useful in microwave circuits like the varactor and the specially designed high-frequency transistor, but they can provide amplification and other functions at microwave frequencies upto 10^5 MHz. This is because of transit time and other effects. To perform electronic functions like switching and conversion to dc power to microwave power, others devices are required to be specially designed.

Basically popular negative conductance devices are as under:

(i) Tunnel diodes

(ii) Transit time diodes

(iii) Gunn diodes

All the diodes mentioned above may be designed to operate in a negative conductance mode to provide amplification or oscillation at microwave frequencies in a proper circuit. Further, negative resistance devices may be voltage controllable type or current controllable type.

8.2. INTRODUCTION TO TUNNEL DIODE

(U.P. Tech, Sem. Exam; 2007-08, 2009-10, 2011-12)

The conventional P-N junction diode has an impurity

concentration of about 1 part in 10^3. This amount of doping results in the width of depletion region of the order of 5 microns (5×10^{-4} cm). The depletion region results in the potential barrier at the junction. This potential barrier opposes the flow of carriers. If the impurity concentration (doping level) is increased to a great extent say 1 part in 10^3 then the device characteristics are completely changed. This new diode is known as **Tunnel Diode**. It was first announced in 1958 by Esaki.

8.2.1. Working Operation of Tunnel Diode

The operation of a tunnel diode is based upon a special characteristics known as the negative resistance. The semiconductor materials used for constructing the tunnel diodes are Germanium or Gallium Arsenide. Construction of a tunnel diode quite similar to that of a P-N junction diode but both P and N sides are very heavily doped, than the conventional rectifier diodes. The heavy doping of both the sides of junction results in extremely narrow depletion region when the tunnel diode is reverse biased. This has been shown in the tunnel diode characteristics of figure 8.2(b).

8.2.2. The Tunneling Phenomenon

The width of the depletion region is inversely proportional to the square root of impurity concentration (doping level). Hence, with increase in the impurity concentration, the depletion region width reduces from 5 microns to less than 100 Å (10^{-6} cm). Theoretically, the energy of a particle should be atleast equal to the height of the potential energy barrier if it is to move from one side of the junction to the other. However, the thickness of the depletion region of this diode is so small that the schrodinger equation indicates that there is a large probability that an electron can penetrate through this barrier. This quantum mechanical behaviour is called as tunneling and hence the name of the high impurity density PN junction is known as a **Tunnel Diode**.

8.2.3. Construction of Tunnel Diode

The material used for the construction of a tunnel diode is Germanium or Gallium Arsenide. Figrue 8.1 illustrates the basic construction of a tunnel diode.

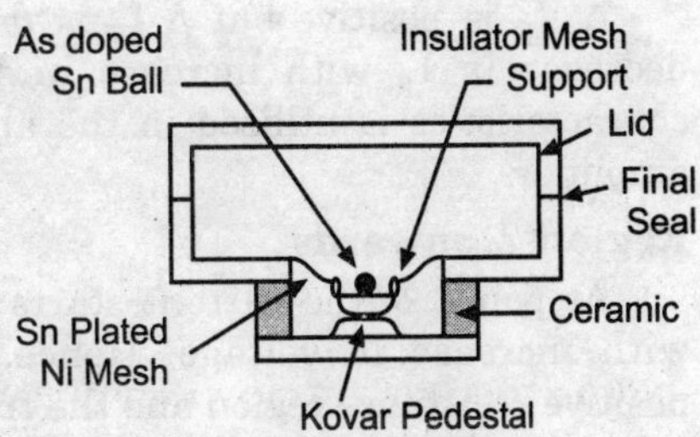

Fig. 8.1. Construction of a tunnel diode

8.2.4. Tunnel Diode Characteristics

The voltampere characteristics of a tunnel diode has been shown in figure 8.2(b) and its symbols have been shown in figure 8.2(a).

1. Reverse characteristics

(i) Due to heavy doping of p and n sides, the depletion region is extremely narrow when the tunnel diode is reverse biased.

(ii) Therefore the reverse blocking capacity of the junction is lost and reverse current will start flowing as soon as a very small reverse voltage is applied.

(iii) Hence the tunnel diode allows the conduction to take place for all the reverse voltages. There is no breakdown effect as observed in the conventional rectifier diode.

(iv) Thus, we can not use the tunnel diode as a rectifier.

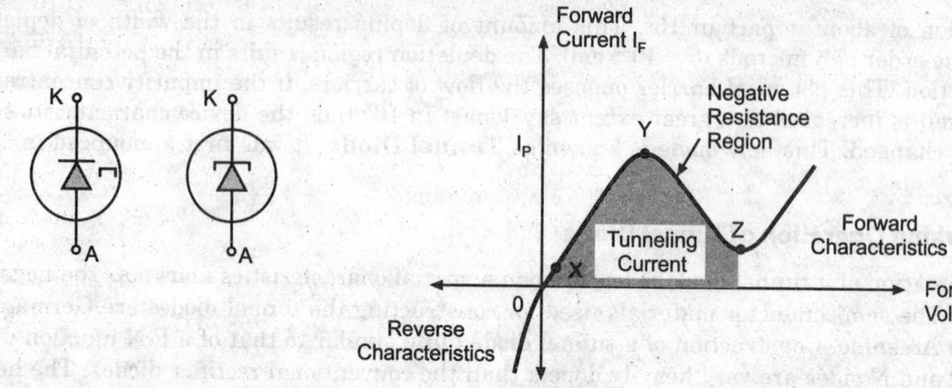

(a) Circuit symbol of tunnel diode (b) Volt-Ampere characteristics of a tunnel diode

Fig. 8.2.

2. Forward Characteristics

Forward characteristics may be divided into three regions namely X to Y, Y to Z and onwards.

Region X to Y

In this region the forward voltage V_F is extremely small. However, heavy conduction take places in this region because electrons *tunnel* through the pn junction. The *tunnelling* takes place as a result of heavy doping.

Region Y to Z

In this region, at point Y the forward voltage begins to develop a barrier and the forward current starts decreasing inspite of continued increase in V_F. This region is called as the negative resistance region. The diode resistance in the region Y to Z is mathematically expressed as under:

$$R_F = \frac{\Delta V_F}{\Delta I_F} \qquad ...(8.1)$$

ΔV_F is positive but ΔI_F is negative. Therefore, R_F is known as negative resistance. The decrease in I_F with increase in V_F is opposite to the Ohms law. The negative resistance characteristics is utilized in the applications of tunnel diode such as oscillator and microwave amplifier.

Region Z onwards

At point, Z, the current starts increasing with increase in voltage. Hence, this is a positive resistance region and the tunnel diode acts as the conventional diode.

8.2.5. Detail Characteristics of a Tunnel Diode

(Important)

The detailed characteristics of the tunnel diode has been shown in figure 8.3. The forward characteristics of a conventional semiconductor diode has been shown by dotted lines for the sake of comparison.

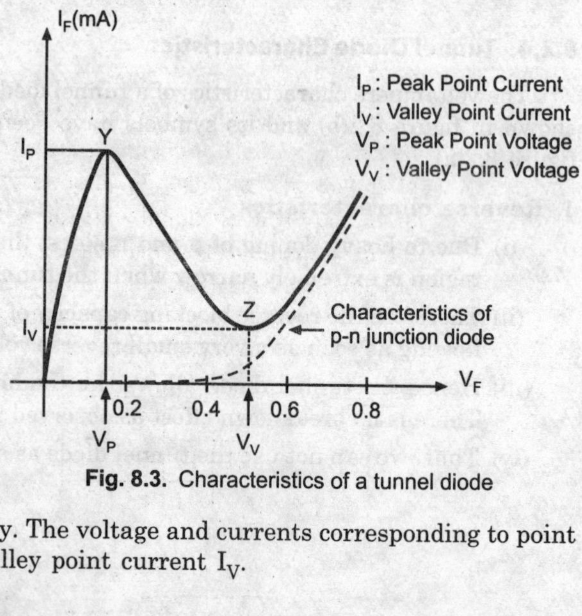

Fig. 8.3. Characteristics of a tunnel diode

The maximum current and voltage corresponding to point Y are known as the peak current I_p and peak point voltage V_p respectively. The voltage and currents corresponding to point Z are called as valley point voltage V_V and valley point current I_V.

8.2.6. Equivalent Circuit

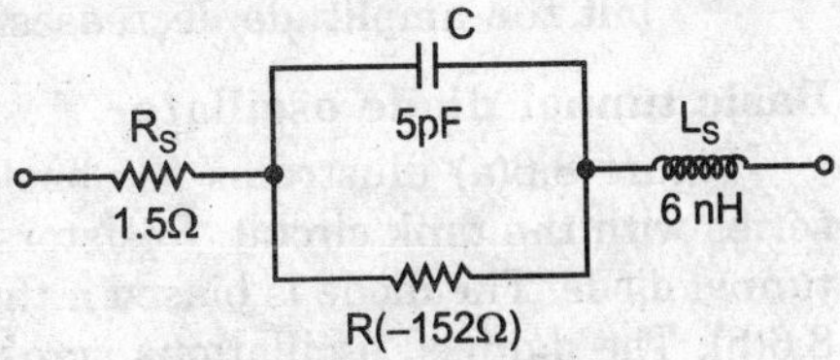

Fig. 8.4. Equivalent circuit of a tunnel diode

Equivalent circuit of a tunnel diode in the negative resistance region has been shown in figure 8.4. The values of various parameters in figure 8.4 correspond to 1N2939 tunnel diode. R_s is because of the leads, ohmic contact at the lead semiconductor junction and the semiconductor itself.

Inductor L_s is mainly due to the terminal leads. C is the junction diffusion capacitance and R is the negative resistance.

8.2.7. Important Specifications

Few important specifications of tunnel diode Ge 1N2939 have been as given in Table 8.1.

Table 8.1. Specification of a tunnel diode Ge 1N2939

S.No	Specifications	Value		
		Minimum	Typical	Maximum
1.	Forward current		5 mA	
2.	Reverse current		10 mA	
3.	Reverse voltage (I_R = 1 mA)			30 mV
4.	V_P	50	60	65 mV
5.	I_P	0.9	1.0	1.1 mA
6.	V_V		350 mV	
7.	I_V		0.1	0.14 mA

8.2.8. Peculiarities of a Tunnel Diode

(i) Heavy doping of P and N sides
(ii) Extermely narrow depletion region
(iii) No reverse blocking capability, so can't be used as rectifier
(iv) Very low cut-in voltage. Conducts heavily at very small forward voltages
(v) Exhibits a negative resistance characteristics when sufficiently forward biased
(vi) Useful for very high speed applications

8.2.9. Few Applications of Tunnel Diode as an Oscillator

(i) Figure 8.5 (a) shows a parallel resonant circuit which is also called as the tank circuit.
(ii) If the switch SW is closed momentarily, then a current pulse is applied to the tank circuit.
(iii) The capacitor charges and discharges sinusoidally to produce damped oscillations across the tank circuit as shown in figure 8.5(b).

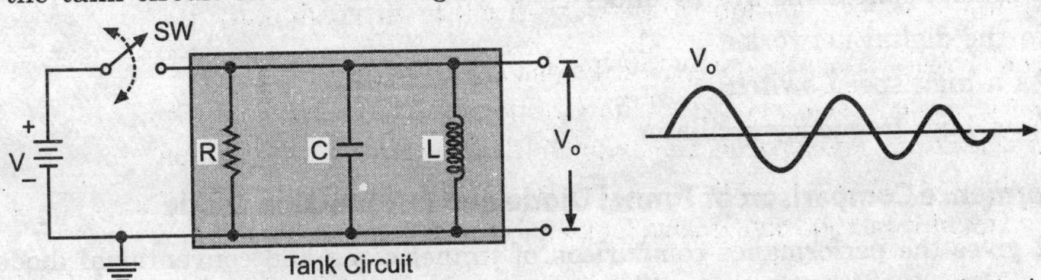

(a) Tank circuit

(b) Damped oscillations across the tank circuit

Fig. 8.5.

(iv) The frequency of the damped oscillations is fixed and it is decided by the RLC components but the amplitude decreases as the time progresses.

Basic tunnel diode oscillator

Figure 8.6(a) illustrates the basic tunnel diode oscillator. The tunnel diode is connected in series with the tank circuit. Resistors R_1 and R_2 are the biasing resistors. They forward bias the tunnel diode. The diode is biased a the center of the negative resistance region as shown in figure 8.6(b). The damped oscillations across the tank circuit occur due to the energy lost in the tank circuit resistance. The negative resistance characteristics of the tunnel diode counteracts the positive resistance characteristics of the tank circuit resistance.

This results in sustained oscillations at the output of the oscillator as depicted in figure 8.6(a).

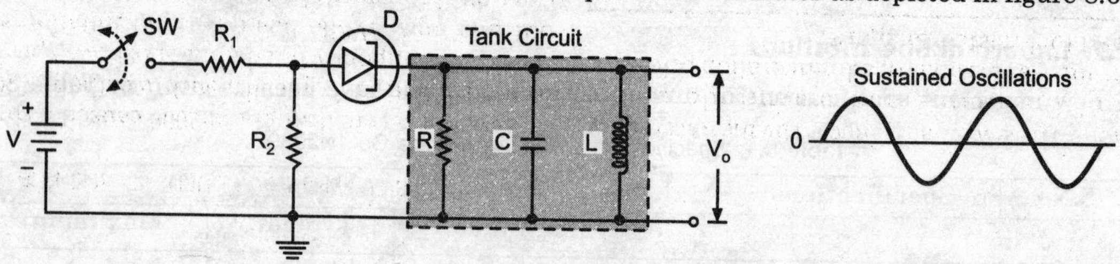

(a) Basic Tunnel Diode Oscillator

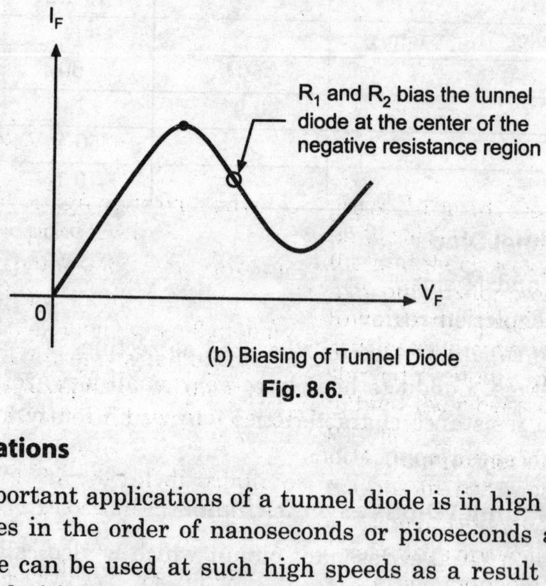

(b) Biasing of Tunnel Diode

Fig. 8.6.

8.2.10. Other Applications

1. One of the important applications of a tunnel diode is in high speed computers where the switching times in the order of nanoseconds or picoseconds are desirable.
2. A tunnel diode can be used at such high speeds as a result of the electrons **punching through** at velocities that far exceed those in conventional diodes.
3. The punching through takes place due to narrow depletion region.
4. Some other applications are as under :
   (i) In the digital networks
   (ii) As a high speed switch
   (iii) As a high frequency oscillator

8.2.11. Performance Comparison of Tunnel Diode and P-N junction Diode

Table 8.2 gives the performance comparison of tunnel diode and conventional diode.

Table 8.2. Performance Comparison of tunnel diode and conventional diode

S.No.	Parameter of comparison	Conventional diode	Tunnel diode
1.	Cut-in voltage	0.6 to 0.7 V for silicon	Less than 0.1 volts
2.	Breakdown	Large reverse breakdown voltage. Breakdown due to avalanche or zener effect	Breakdown does not take place
3.	Reverse blocking capacity	Conventional diode can block high reverse voltage	Can not block the reverse voltage
4.	Doping levels	Moderate	Very high
5.	Width of depletion region	Wider than that in tunnel diode	Narrower than that in the conventional diode
6.	Characteristics of negative resistance	Not applicable	The region between V_p and V_V exhibits resistance
7.	Symbol		
8.	Characteristics	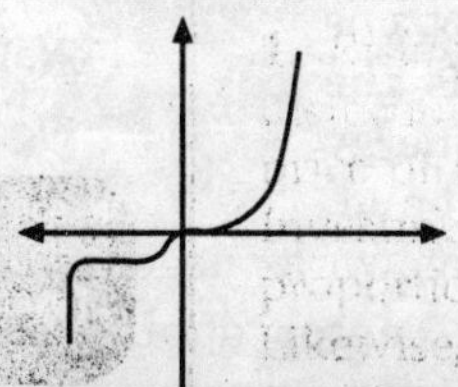	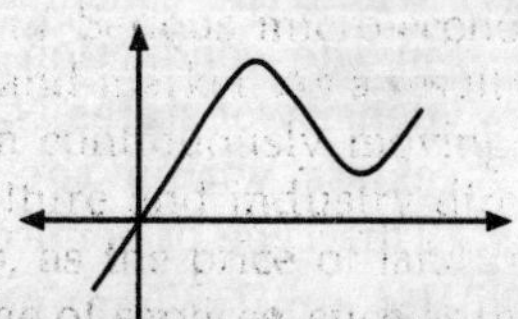
9.	Applications	Rectifiers, clipping and clamping circuits voltage multipliers	Oscillator, high speed computers, microwave amplifiers
10.	Speed of operation	Very slow. Suitable for operation at mains frequency	Very fast due to the tunneling effect
11.	Material	Silicon or Germanium	Gallium Arsenide

8.2.12. Some Advantages of Tunnel Diode

(i) Low cost.

(ii) The peak point voltage and current (V_p and I_p) are not dependent on temperature.

(iii) Tunnel diode needs a small number of external components and a dc power supply for its operation.

(iv) It consumes low power.

(v) It produces low noise.

(vi) It is a high speed device and suitable for the high frequency applications.

8.2.13. Drawbacks

Some of the may be listed as under:

(i) No isolation between the input and output.

(ii) Low output voltage swing. Hence, amplification is needed.

8.3. DEGENERATE AND NONDEGENERATE SEMICONDUCTORS

(i) In previous discussion of adding dopant atoms to a semiconductor, we have already assumed that the concentration of dopant atoms added is small when compared to the density to host semiconductor atoms.

(ii) The small number of impurity atoms are spread far apart so that there is no interaction between donor electrons, as an example N-type semiconductor materials. We have assumed that the impurities introduce discrete, noninteracting donor energy states in the N-type semiconductor and discrete, noninteracting acceptor states in the P-type semiconductor. These types of semiconductors are known as non-degenerate semiconductors.

(iii) When the impurity concentration increases, the distance between the impurity atoms decrease and a point shall be reached when donor electrons will begin to interact with each other. When this occurs, the single discrete donor energy will split into a band of energies.

(iv) When the donor concentration further increases, the band of donor states widens and may overlap the bottom of the conduction band. This overlap occurs when the donor concentration becomes comparable with the effective density of states.

(v) When the concentration of electrons in the conduction band exceeds the density of states N_C, the fermi energy exists within the conduction band. This type of semiconductor is known as degenerate N-type semiconductor.

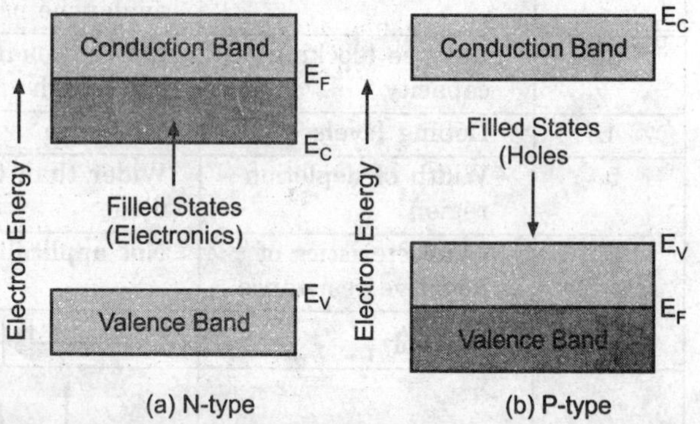

Fig. 8.7. Energy band diagram for degenerately doped

(vi) In similar manner, as the acceptor doping concentration increases in a P-type semiconductor, the discrete acceptor energy states will split into a band of energies and may overlap the top of the valence band.

(vii) The Fermi energy exists in the valence band if the concentration of holes exceeds the density of states Nv. This type of semiconductor is known as degenerate P-type semiconductor. Energy band diagrams of degenerate N-type and degenerate P-type semiconductor are shown in figure 8.7.

(viii) The energy states below E_F are almost filled up with electrons and energy states above E_F are mostly empty. In the below degenerate N-type semiconductor, the states between E_F and E_C are mostly filled with electrons, so the electron concentration in the conduction band is very large. In a similar manner, in the degenerate P-type semiconductor, the energy states between E_v and E_F are mostly empty, so that hole concentration in the valence band is very large.

8.4. IMPATT DIODE

(U.P. Tech, Sem. Exam; 2010-11, 2011-12)

Many important devices for high-frequency applications use the instabilities which occur in semiconductors. An important type of instability involves negative conductance. In this article we shall discuss two most commonly used negative conductance devices:

(i) IMPATT diodes, which depend upon a combination of impact ionisation and transit time effects.

(ii) Gunn diodes, which depend upon the transfer of electrons from a high-mobility states to a low-mobility states.

Each is a two-terminal device which can be operated in a negative conductance mode to provide amplification or oscillation at microwave frequencies in a proper circuit. The original suggestion for a microwave devices of the IMPATT types was made by W.T. Read and involved an $n^+ - p - i - p^+$ structure such as that depicted in figure 8.8. Although IMPATT operation can be obtained in simpler structures, the Read diode is best suited for illustration of the basic principles. The devices consists of essentially two parts:

(i) $n^+ - p$ region at which avalanche multiplication occurs.

(ii) intrinsic region i through which generated holes must drift in moving to the p^+ contact.

Similar devices can be built in the $p+ - n - i - n^+$ configuration, in which electrons resulting from avalanche multiplication drift through the i region. This device operates in a negative conductance mode, when the a.c. component of current is negative over a portion of the cycle during which the ac voltage is positive and vice-versa. The negative conductance occurs because of two processes, causing the current to lag behind the voltage in time:

(i) delay due to the avalanche process.

(ii) further delay due to transit time of carriers across the drift region.

When the sum of these delay time is approximately one-half cycle of the operating frequency, negative conductance occurs and the device can be used for oscillation and amplification. Taking into consideration another point of view, the a-c conductance is negative if the a-c component of carrier flow drifts opposite to the influence to the a-c electric field. As an example, with d-c reverse bias on the device shown in figure 8.8 holes drift from left to right i.e. in the direction of the field. Now, if we superimpose an a-c voltage such that Electric field decreases during the negative half cycle, we would expect the drift of holes to decrease also. However, in IMPATT diode operation the drift of holes through the i region actually increases while the a-c field is decreasing.

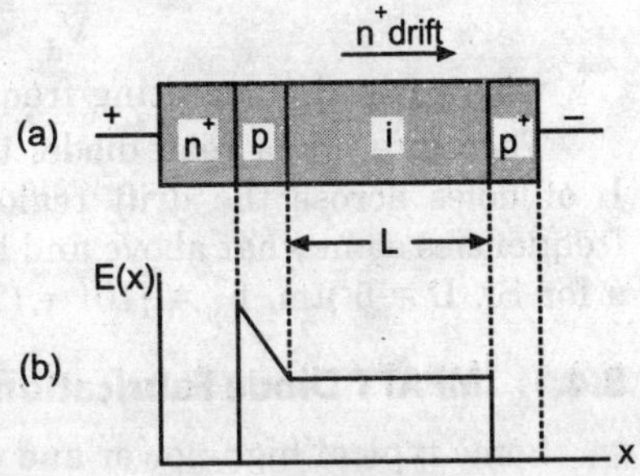

Fig. 8.8.

To demonstrate this effect, let us consider the effects of avalanche and drift for various points in the applied voltage cycle as shown in figure 8.9. For the simplicity of discussion, we shall assume that the P region is very narrow and that all the avalanche multiplication takes place in a thin region near the n^+-p junction. Also we shall assume that field in the narrow p region has uniform value. When the d-c bias is such that the critical field for avalanche E_a is just met in the n^+ - p space charge region as shown in figure 8.9(a)., avalanche multiplication begins at t = 0. Electrons generatd in the avalanche move to the n^+ region and holes enter the intrinsic(i) drift region, we assume that the device is mounted in a resonant microwave circuit so that an a-c signal can be maintained at a given frequency. As the applied a-c voltage goes positive, more and more holes are generated in the avalanche region. In fact, the pulse of holes (represented by dotted line) generated by the multiplication process continues to grow as long as the electric field in above E_a as shown in figure 8.9(b). It can be shown that the particle current due to avalanche increases exponentially with time while the field is above the critical value. The important result of this growth is that

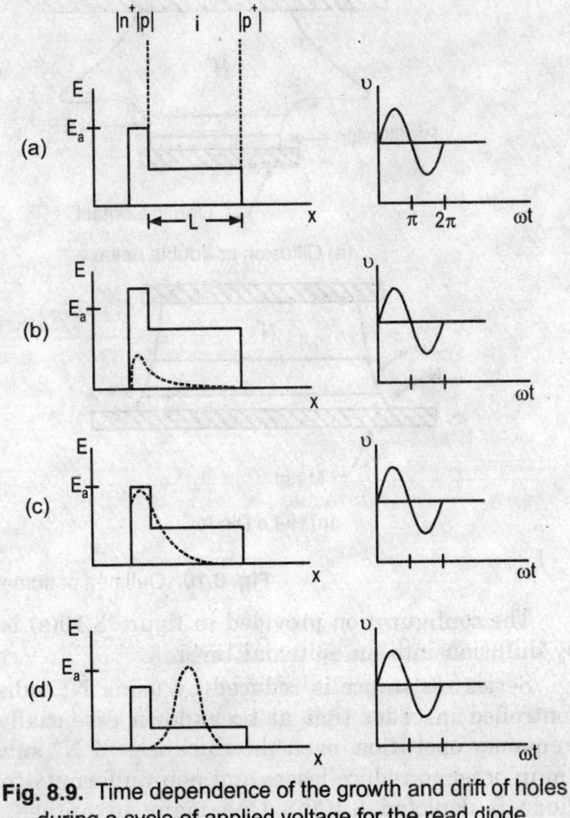

Fig. 8.9. Time dependence of the growth and drift of holes during a cycle of applied voltage for the read diode, (a) ωt = 0, (b) ωt = π, (c) ωt = π, (d) ωt = 3π/2. The hole pulse is sketched as a dotted line on the field diagram

the hole pulse reaches its peak value, not at $\pi/2$, when the voltage is maximum, but at π as shown in figure 8.9(c). So, there is a phase delay of $\pi/2$ inherent in the avalanche process itself.

A further delay is provided by the drift region, once the avalanche multiplication slope ($\omega t > \pi$) the pulse of holes simply drifts towards the P^+ contact as shown in figure 8.9(d). But during this period, the a-c terminal voltage is negative. Hence, the dynamic conductance is negative and energy is supplied to the a-c field. If the length of the drift region is chosen properly, the pulse of holes is collected at the p^+ contact just as the voltage cycle is completed and the cycle then repeats itself. The pulse will drift through the length L of the intrinsic(i) region during the negative half cycle if we choose the transit time to be one half the oscillation period i.e.,

$$\frac{L}{V_d} = \frac{1}{2}\cdot\frac{1}{f} \quad \text{or} \quad f = \frac{V_d}{2L} \qquad \text{...(8.2)}$$

where f is the operating frequency and V_d is the drift velocity of holes.

Therefore, for a read diode, the optimum frequency is one half the inverse transit time V_d/L of holes across the drift region. Negative resistance is exhibited by an IMPATT diode for frequencies somewhat above and below this optimum frequency 10^{10} Hz. (We assume $V_d = 10^7$ cm/s for Si, L = 5 μm, $f_{op} = [10^7 \div (2 \times 5 \times 10^{-4}) = 10^{10}$ Hz] for exact 180° phase delay.

8.4.1. IMPATT Diode Fabrication

Some typical high-power and efficient IMPATT diode designs have been shown in figure 8.10.

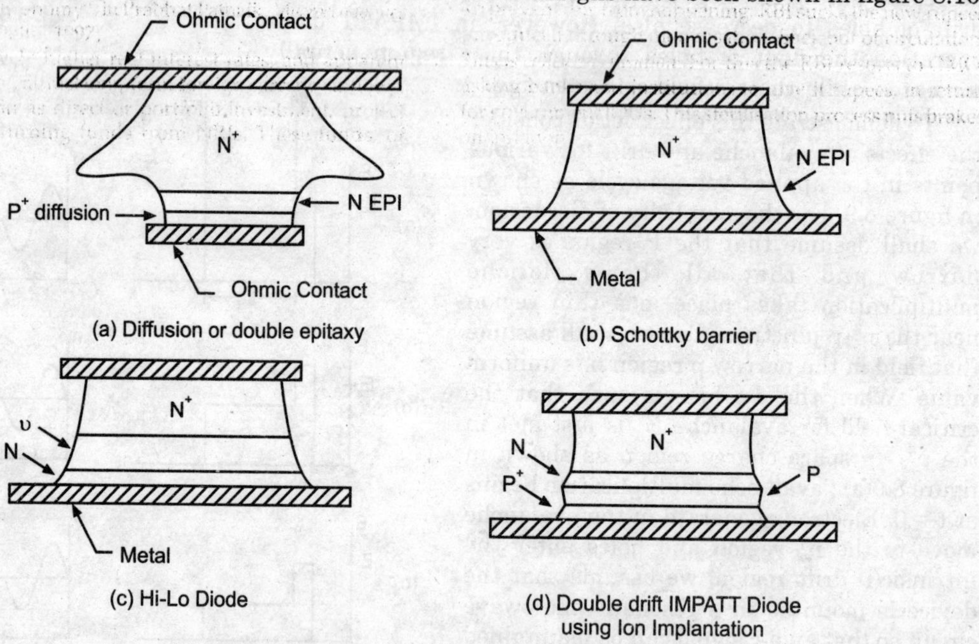

Fig. 8.10. Outlines of some IMPATT diode designs

The configuration provided in figure 8.10(a) is formed by using a double epitaxial process or by diffusion into an epitaxial layer.

Series resistance is reduced by using N^+ substrate. The thickness of epitaxial layer is also controlled in order that at breakdown essentially no unswept epitaxial layer is left. For high frequency operation, even the thickness of N^+ substrate should be reduced to the order of a few μm in order to reduce losses and non-uniformity from the skin effect. A schottky barrier IMPATT diode is depicted 8.10(b). One main advantage is that the maximum field is at the metal-semiconductor interface, and the heat produced can be easily conducted away from the metal

* Punishment is a sort of medicine. —**Aristotle**

contact. As shown in figure 8.10(b), the device can be formed in a truncated cone shape in order that the high field singularity at the periphery is removed and uniform breakdown can be achieved in the device.

One more advantage is that the device can be fabricate at relatively low temperature, in order that the original high-quality epitaxial layer can be preserved. The Schottky-barrier approach can be employed in modified Read diode structures, as shown in figure 8.10(c) by replacing p^+ layer by metal contact. As the Schottky barrier is basically a majority carrier device, the minority-carrier storage effect that occurs in the original Read structure, is eliminated and efficiency is improved.

A dramatic increase in microwave frequency in comparison to the diode with constant doping profile is observed in modified Read diodes. Another new device configuration in shown in figure 8.10(d). It uses an ion implanation technique. For dopants like B and P, the ion penetration depth is of the order of 0.5 μm per 100 keV, therefore, 1 μm depth can be easily obtained from ion implantation equipment with energy sources a few hundred keV.

These energy ranges are useful in making millimetre wave IMPATT diodes. Power output per unit area and impedance per unit aera are both approximately doubled, hence such structure increased power output with improved efficiency. The fabricated diode is usually mounted in a microwave package, which is shown in figure 8.11. In both cases the IMPATT diode is mounted with its diffused side or metal-electrode side in contact with a copper or diamond heat sink so as to conduct away the heat produced at the junction easily.

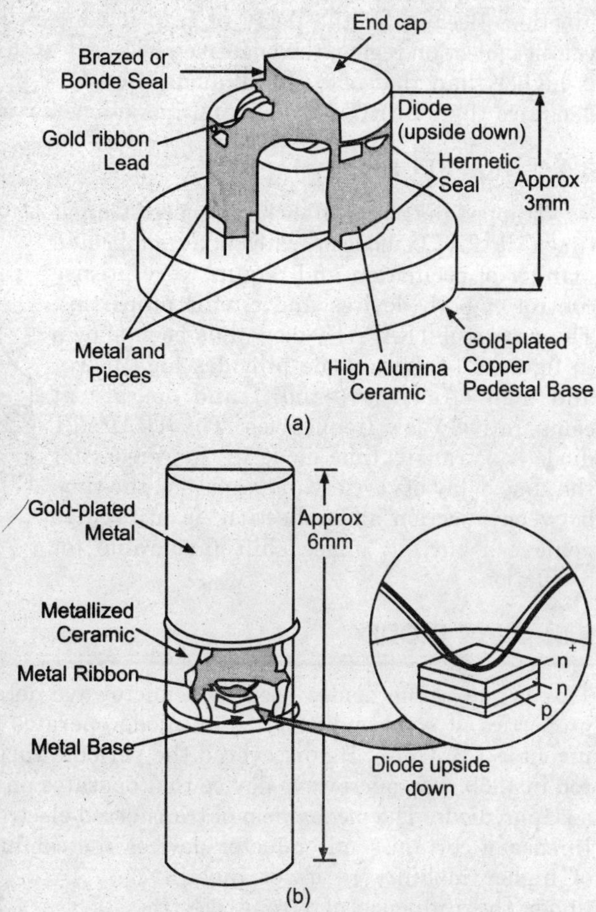

Fig. 8.11. Two microwave packages with IMPATT diodes mounted

8.4.2. BARITT Diode

Basically the barrier-inject transit time (BARITT) diode is a back to back pair diodes biased into reach-through condition. It belongs to the transit-time microwave diode family. In BARITT diode microwave oscillation occurs due to thermionic injection and diffusion of minority carriers across the forward-biased barrier and a $3\pi/2$ transit angle of the injected carriers traversing the drift region. These diodes provides low power efficiency but provides good signal to noise performance and hence useful as local oscillators in microwave receivers. The design of BARITT diode has been shown in figure 8.12. It has small signal negative resistance, hence it provides self starting oscillation.

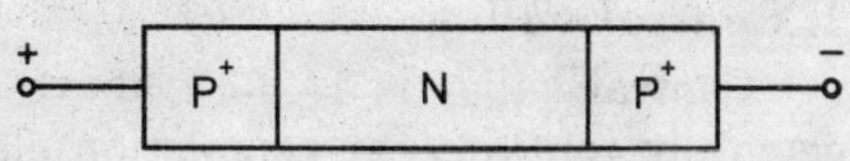

Fig. 8.12.

8.4.3. DOVETT Diode

The double velocity transit time (DOVETT) diode is quite similar to the BARITT diode, except that the velocity of carriers near the injection contact is much less than that near the collection contact. A heterostructure DOVETT diode is designed and shown in figure 8.13. The device is biased, such that P-N anisotype heterojunction is completely punched through and large current can be injected ove the forward biased n^+-p junction. Because of the delay of transit time in the low velocity injection region, the negative resistance of this diode is higher than that of BARITT diode. DOVETT diode may be operated at a higher d-c current densities than BARITT diodes and has efficiency above 25%.

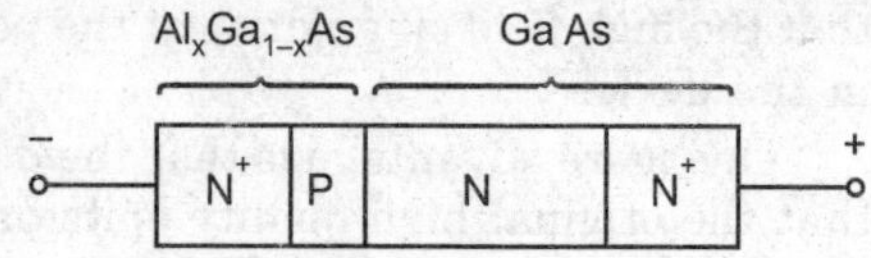

Fig. 8.13.

8.4.4. TRAPATT Diode

Trapped plasma avalanche triggered transit time (TRAPATT) diode operates in a complicated manner of oscillation and require very precise control of both devices and circuit properties. The design of TRAPATT diode has been shown in figure 8.14. This diode provides high power and high efficiency (> 50%) and operate at comparatively low frequencies. The TRAPATT diode is a transit time diode in the sense that the time delay of carriers in transit i.e. the time between injection and collection, is utilized to achieve a current phase shift favourable for oscillation.

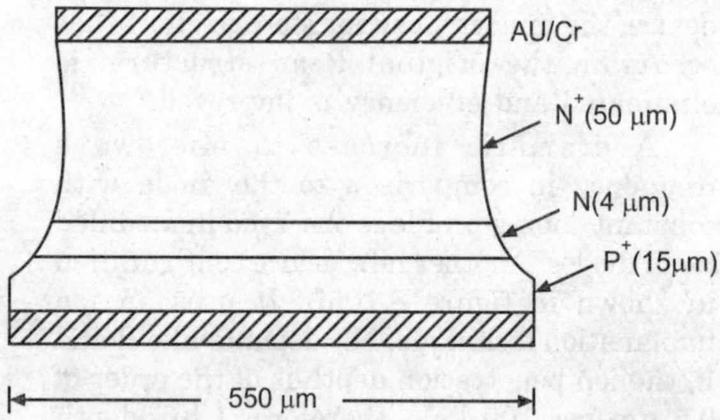

Fig. 8.14.

8.5. GUNN DIODE

(U.P. Tech, Sem. Exam; 2009-10, 2010-11)

This is one of the diodes used as a microwave detector. It uses the avalanche and transit time properties of semiconductor. Gunn diode operates on the *Gunn effect* which is named after its inventor J.B. Gunn. He discovered the periodic variation of current in a GaAs (gallium arsenide) rod in 1963. The microwave device that operates on the principle of transfer of electrons is known as Gunn diode. The mechanism of transferred electrons is related to the conduction band electrons. In case of certain semiconductor devices the conduction band electrons get shifted from a state of higher mobility to lower mobility, under the influence of strong electric field.

8.5.1. Working Operation

1. Basic Aspect

Let us consider the energy diagram of GaAs semiconductor shown in figure 8.15. The conduction band of GaAs is divided into two levels namely central valley and satellite valley.

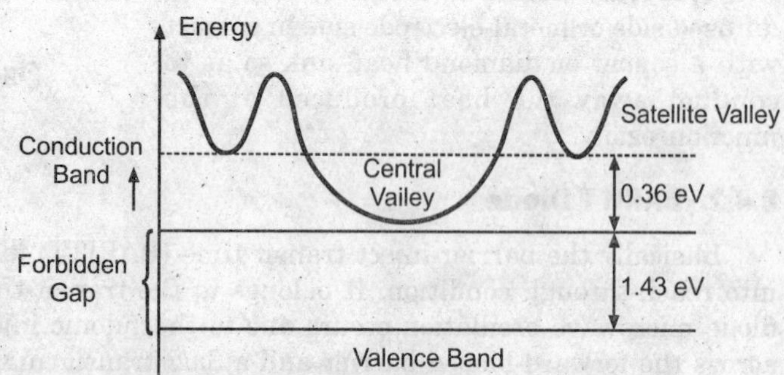

Fig. 8.15. Band structure of GaAs

2. Central valley

Central valley is 1.43 eV above the valence band. Its height is 0.36 eV. The central valley generally contains the conduction electrons.

3. Satellite valley

The satellite valley is generally at higher energy level than the central valley. In each energy band the electrons have following properties:

(i) Density (ii) Effective mass (iii) Mobility

The effective mass and mobility of electrons are related to the valleys as under:

Upper valley → Higher effective mass, lower mobility
Lower valley → Small effective mass, higher mobility

Keeping all this in mind, now we shall observe the mechanism of electron transfer

4. Mechanism of electron transfer

We know that relation between mobility, drift velocity and electric field is as under

$$v = \mu E \quad ...(8.3)$$

The current density J is expressed as

$$J = n.q.v = n \cdot q \cdot \mu \cdot E \quad ...(8.4)$$

At room temperature when the applied electric field intensity is low (E is small), then almost all the electrons are present in the lower valley i.e. at the bottom of the conduction band. Now, if the electric field intensity is increased then more photons are generated. They will strike the electrons present at the bottom of the conduction band. This collision will transfer energy from photons to electrons. Hence, kinetic energy of these electrons increases and they start moving in the upward direction. This is known as electron transfer. If the electric field intensity is increased further the electrons move to the satellite valley. The effective mass of electrons will increase as they move towards the satellite valley but the mobility reduces. With further increase in electric field intensity the energy of electrons will increase but velocity decreases. Hence, the current density J decreases. Hence, in the range of increasing electric field intensity we get a negative differential coefficient (–dJ / dE). Let μ_1 be the mobility of electrons in the lower valley and μ_2 be the mobility in the upper valley. Then, the mean mobility will be given by,

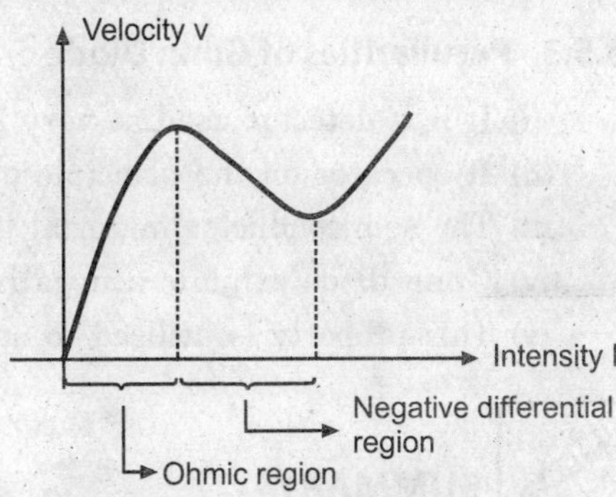

Fig. 8.16. Velocity versus electric field intensity

$$\bar{\mu} = \frac{n_1 \mu_1 + n_2 \mu_2}{(n_1 + n_2)} \quad ...(8.5)$$

Also, the current density is given by

$$J = ((n_1 + n_2)\bar{\mu}\, qE \quad ...(8.6)$$

Hence, the conductivity of the GaAs rod is given by

$$\sigma = q(\mu_1 n_1 + \mu_2 n_2) \quad ...(8.7)$$

5. Velocity Vs Electric field Curve

The velocity versus electric field curve is shown in figure 8.16.

The characteristics is divided into following two regions:

(i) Ohmic region

(ii) Negative differential

(i) Ohmic region

In this region, the electrons are present in the central valley. With increase in applied electric field, the velocity of electrons increases.

(ii) Negative differential resistance region

In this region, the velocity decreases with increase in electric field intensity. This is known as negative differential resistance region.

6. Modes of Operation

(i) The modes of operation of a Gunn diode are as may be listed as under:

(i) The Gunn mode

(ii) Stable amplifier mode

(iii) Limited space charge accumulation mode (LSA).

(ii) Depending on the applications the gunn diode is biased in one of the modes stated above.

(iii) The V-I characteristics of a gunn diode has been shown in figure 8.17. The diode is biased to operate in the negative resistance region if it is to be operated as oscillator. The bias voltage is adjusted in such a way that the diode is biased at the center of the negative resistance region.

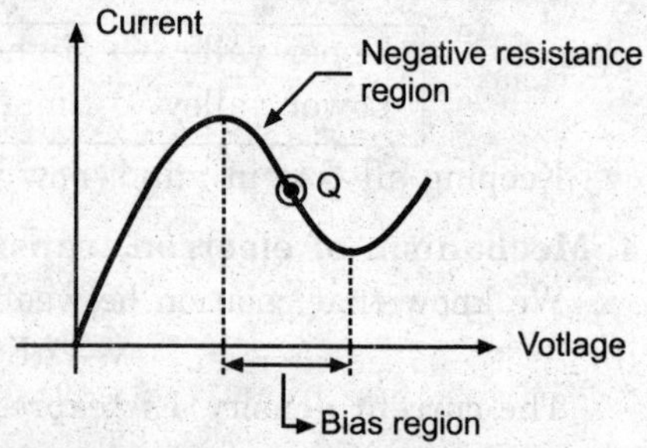

Fig. 8.17. V-I characteristics of Gunn diode

8.5.2. Some Applications

(i) Gunn diodes are generally used in power oscillators in microwave instruments.

(ii) It may be used as a pump source.

(iii) It may be used as low power transmitter of electromagnetic waves at microwaves frequencies.

8.5.3. Peculiarities of Gunn Diode

(i) It is a detector used at very high frequencies (microwave frequencies).

(ii) It operates on the principle of transfer of electrons.

(iii) The semiconductor material used in Gallium Arsenide.

(iv) Gunn diode exhibits a negative resistance property.

(v) This property is utilized to operate the diode as an oscillator at microwave frequency.

SUMMARY

- A number of devices which are extensively employed in pulse and switching circuitry are most conveniently characterised in terms of volt-ampere curve which displays, over a limited range, a negative incremental resistance. Negative resistance may result in two ways. Some devices exhibit a characteristic in part of which as the voltage increases, the current drops thereby resulting in a negative resistance while some other devices may exhibit a characteristics in part of which decrease in voltage causes increase in current thereby resulting in a negative resistance. Negative resistance devices may be voltage controllable type or controllable type.
- A tunnel diode is a high-conductivity two terminal heavily (1000 times higher than a conventional junction diode) doped P-N junction diode. It can conduct in reverse as well as forward direction but is usually used in forward biased mode.

  A tunnel diodes characteristics exhibit a negative resistance region and it may be used as an ultra-fast switching device or in ultra-high frequency oscillators.
- In relatively pure semiconductors, no charge transport can occur within the donor or acceptor themselves because the few impurity atoms are so widely spaced throughout the sample. But when doping levels are increased, the impurities are very close to each other in the structure leading to interaction within the donors or acceptors. In case the concentration of electrons in the conduction band is greater than the effective density of states, the Fermi legvel is no longer withiin the band gap but extends into the conduction band and when this occurs, the material is referred to as degenerate N-type. A similar reaction occurs in a degenerate P-type material where the concentration of acceptors is high and the Fermi level extends to the valence band.

- The impact ionization and transit time properties of the semiconductor structures are used to develop negative resistance at microwave frequencies

  Transit time devices can convert dc to microwave as signals with high efficiency and are very useful in the generation of microwave power for many applications.
- TED devices are microwave devices and are extensively employed as local oscillators and power amplifiers in the range of 1 to 100 GHz. These are used in RADARs, intrusion alarms, and microwave test instruments.
- IMPATT diode stands for impact Avalanche and Transit Time diode. Any device exhibiting dynamic negative resistance for dc will also exhibit it for ac. When alternating voltage is applied, the current will increase with the decrease in applied voltage. Thus, negative resistance may be redefined as that property of the device which causes the current through it to be 180° out of phase with the voltage applied across it.
- The transfer of conduction electrons from a high-mobility energy valley to low-mobility, higher-energy satellite valleys is called the *transferred-electron effect*. Although GaAs and InP ae the most common materials for TEDs, transferred-electron effects have also been observed in CDTe, ZnSe, GaAsP, and other materials.
- The Gunn diodes are named after J B Gunn, who first demonstrated one of the forms of oscillation, are microwave devices which operate by the transferred electron mechanism, These devices are called diodes only because they are two-terminal devices, through no P-N junction is involved.

  However the Gunn diode and other related devices use bulk instabilities which do not require junctions.

  Power dissipation is very important in Gunn devices because considerable heating (roughly 10,000 kW/cm^3) and the increase in temperature results in change of devices characteristics.

SHORT QUESTIONS WITH ANSWERS

Q.1. What is tunnel diode?

Ans. Tunnel diode is a highly conductivity two-terminal P-N junction doped heavily (about 1,000 times) higher than a normal diode).

Q.2. What is tunneling?

Ans. The mechanism of conduction in a semiconductor diode in which charge carriers (possessing very little energy) punch through a barrier directly instead of climbing over it is called tunneling.

Q.3. What are the applications of tunnel diodes?

Ans. Tunnel diodes can be used as amplifiers, oscillators or switching devices, being an exclusive high-frequency component because of its very fast response of inputs.

Q.4. What are degenerate devices?

Ans. Degenerate devices are the devices in which the doping levels are increased largely, the impurities are close to each other leading to interaction between donors and acceptors.

Q.5. How does negative resistance develop in transit time devices?

Ans. The negative resistance occurs because of two procsses, causing the current to lag behind the voltage in time: (i) delay due to the avalanche process and (ii) a further delay due to the transit time of the carriers across the drift region.

If the sum of above delay times is approximately one half cycle of the operating frequency, negative resistance occurs.

Q.6. Name a few transit time devices.

Ans. Transit time devices include (i) IMPATT (ii) BARITT (iii) DOVETT and (iv) TRAPATT diodes

Q.7. What are the applications of transit time devices?

Ans. Transit time devices can convert dc to microwave signals with high efficiency and are very useful in the generation of microwave power for many applications.

Q.8. What is transferred-electron effect?

Ans. The transfer of conduction electrons from a high-mobility energy valley to low-mobility higher energy satellite valleys is called the tranferred electron effect.

Q.9. Name the materials most commonly used for TEDs?

Ans. GaAs and InP are the materials most commonly used for TEDs.

Q.10. What are Gunn diode?

Ans. The Gunn diodes are named after J.B. Gunn, who first demonstrated one of the forms of oscillation, and microwave devices which operate by the transferred electron mechanism. These devices are called diodes only because they are two terminal devices, though no P-N junction is involved. However, the Gunn diode and other related devices use bulk instabilities which do not require junction.

Q.11. Why power dissipation is very important to Gunn devices?

Ans. Power dissipation is very important in Gunn devices because considerable heating (roughly 10,000 kW/cm^3) and the increase in temperature results in change of device characteristics.

Q.12. Name the bulk NDR devices.

Ans. Bulk NDR devices ae of two types: viz, (i) current controlled NDR (S-shaped) and voltage-controlled NDR (N-shaped).

REVIEW QUESTIONS

1. What are the basic constructional differences between a conventional P-N junction diode and a tunnel diode? Explain briefly the operation of a tunnel diode. **(U.P. Tech. Sem. Exam; 2006-07)**
2. How does a tunnel diode after from a conventional P-N junction diode? With the help of its energy-band diagram explain the principle of operation in forward and reverse bias conditions. Sketch its V-I characteristics and mark negative resistance on it.
3. Draw the V-I characteristics of tunnel diode. Briefly describe the mechanism of junction breakdown.
4. What is tunneling? In what mode the tunnel diodes are usually operated? Give the construction and V-I characteristics of tunnel diodes.
5. Obtain I-V characteristics of a tunnel diode with special reference to its negative resistance starting from equilibrium condition to forward and reverse bias conditions.
6. Explain the tunnel diode in detail. What are its applications? **(U.P. Tech. Sem. Exam; 2009-10)**
7. Describe briefly the principle of operation of an IMPATT diode. **(U.P. Tech. Sem. Exam; 2006-07)**
8. What is IMPATT diode? Give its construction. Give outlines of some IMPATT diode designs.
9. What are transferred electron devices and where they are used? What is transferred electron effect? Name the materials in which transferred electron effect can be observed. Draw the velocity field characteristics for GaAs at different temperatures.
10. What is Gunn diode? Why is it called a diode? How it can be fabricated? Explain formation and drift of space charge domains.
11. What is the basic difference between GUNN diode and the IMPATT diode? **(U.P. Tech. Sem. Exam; 2005-06)**
12. Write short notes on the following:
    (i) Tunnel diode (ii) Degenerate semiconductors
    (iii) IMPATT diode (iv) Transferred electron devices
    (v) Transit time device (vi) Gunn diode **(U.P. Tech. Sem. Exam; 2006-07)**

□□□

Chapter 9

Some Special Power Devices

Syllabus

P-N-P-N diode, Semiconductor Controlled Rectifier (SCR), Bilateral Devices: DIAC, TRIAC, IGBT.

Chapter Outline

- ⇨ Introduction
- ⇨ Classification of Thyristors
- ⇨ Applications of Thyristors
- ⇨ Silicon Controlled Rectifier (SCR)
- ⇨ Basic Structure of SCR
- ⇨ I-V Characteristics
- ⇨ Working Operation of SCR
- ⇨ Transistor Analogy: Two Transistor Equivalent Circuit
- ⇨ Dynamic Characteristics of SCR
- ⇨ P-N-P-N or Schottky Diode
- ⇨ Bidirectional Triode SCR (Triacs)
- ⇨ Diac
- ⇨ Power Mosfet
- ⇨ The Characteristics of (GTO)
- ⇨ Advantages and Disadvantages of GTO Over the Conventional Thyristor (SCR)
- ⇨ Insulated Gate Bipolar Transistor
- ⇨ Basic Structure of IGBT
- ⇨ I-V Characteristics
- ⇨ Types of IGBT
- ⇨ Principle of Operation of IGBT
- ⇨ Equivalent Circuit of IGBT
- ⇨ Protection Circuits for IGBT
- ⇨ Merits and Demerits and Applications of IGBT

9.1. INTRODUCTION

As a matter of fact, thyristors is a general term which includes devices such as SCR, Triac, GTO, LASCR, RCT, SCS, ... and many more. Thus, thyristors is a family of devices. SCR i.e. silicon controlled rectifier is the most popular device of the thyristor family.

9.2. CLASSIFICATION OF THYRISTORS

Figure 9.1 illustrates the classification of thyristors.

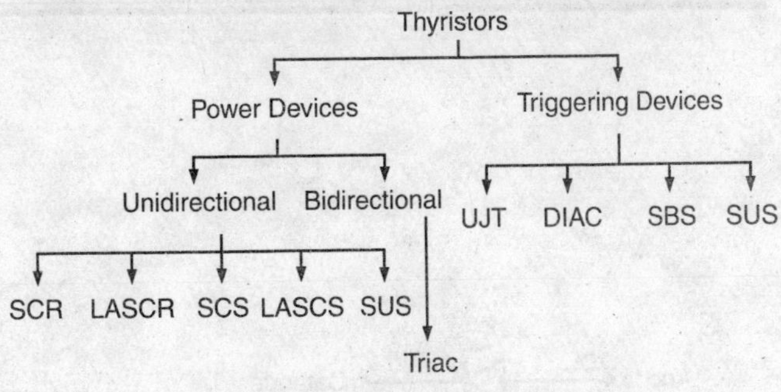

Fig. 9.1. Classification of thyristors.

9.3. APPLICATIONS OF THYRISTORS

Some of the typical applications of thyristors may be listed as under :

1. Temperature controllers
2. Level controllers

3. Intensity controllers
4. Speed control of AC and DC motors
5. Fan regulator
6. Battery charger
7. Power inverters
8. DC to DC choppers
9. Static switches
10. Cyclo converters
11. Controlled rectifiers (only SCRs)
12. HVDC transmission
13. AC voltage stabilizers.

9.4 SILICON CONTROLLED RECTIFIER (SCR)*

SCR is one of the oldest type of solid state power device. It was invented in 1957 by the General Electric Research Laboratories. SCRs have the highest power handling capacity of all the power semiconductor devices. They have a four layer construction with three user accessible terminals. SCR is a latching type device that can be turned on by the control terminal (gate) but once turned on, the gate loses control on it (i.e. it cannot be turned off by the gate).

9.4.1. Important Features of SCR

(i) It is a latching type device.
(ii) It can handle a very large power.
(iii) It is a current controlled device, because the gate current controls SCR.
(iv) It acts as an open or closed switch.
(v) The on state voltage drop is very low.
(vi) It can handle thousands of amperes of current.

9.5 BASIC STRUCTURE OF SCR

(U.P. Tech, Sem. Exam; 2009-10)

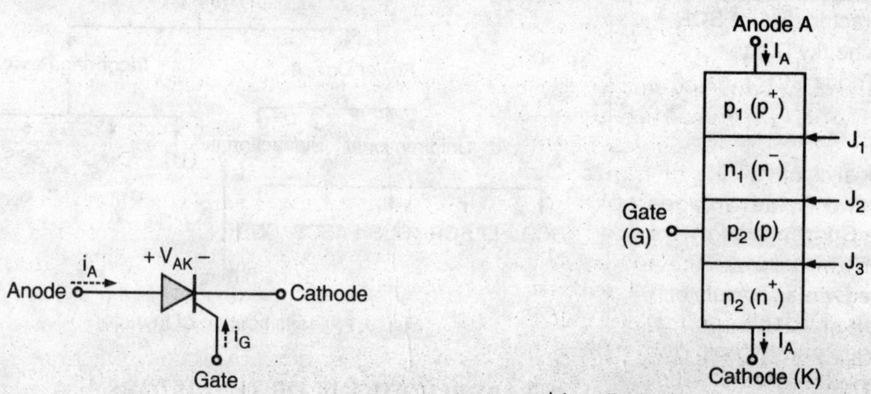

(a) Circuit symbol for SCR

p^+ heavily doped layer p moderately doped layer
n^- lightly doped layer n^+ heavily doped layer

(b) Basic structure of SCR

Fig. 9.2. Structural details of SCR.

* Write a short note about construction of SCR.

The basic structure of SCR has been shown in figure 9.2(b). It is a four layer pnpn device, with three terminals brought out for the user, namely Anode, Cathode and Gate. The gate terminal is the control terminal that can turn on the device whenever required. The symbol of SCR has been shown in figure 9.2 (a). It is essentially the symbol of a rectifying diode with a third control terminal, the gate added to it. The direction of forward anode current, voltage across the thyristor and direction of conventional gate current have been shown in figure 9.2(a).

Number of Junctions

As observed from figure 9.2(b), there are three junctions J_1, J_2 and J_3. In order to turn on SCR, the anode must be at a higher positive potential than cathode. This means that SCR must be forward biased.

Current Directions

The directions of the anode and gate current in figure 9.2(a) are conventional directions. It clearly indicates that SCR is a unidirectional device and that the gate current can be only positive. The gate current can flow only in one direction i.e. into the gate terminal.

9.5.1. Gate Cathode Layouts

Plan views of two different gate and cathode layouts are as shown in figure 9.3(a) and figure 9.3(b). The localised gate cathode structure is for smaller SCRs whereas the distributed gate structure is for larger diameter SCRs.

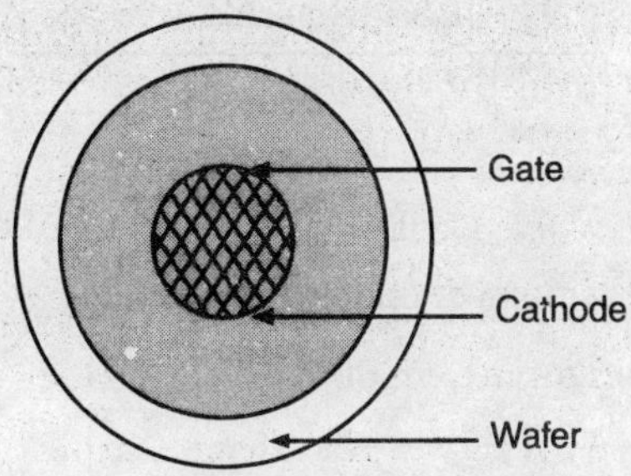

(a) Gate cathode layout for the localised gate electrode

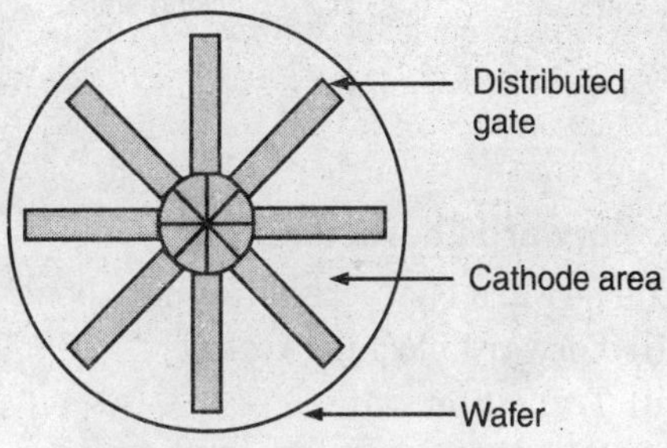

(b) Gate cathode area layout for distributed gate electrode

Fig. 9.3. Structure details of an SCR.

9.6. I-V CHARACTERISTICS*

(U.P. Tech. Sem. Exam., 2005-07)

The I-V characteristics of SCR is a graph of anode current i_A on y-axis and anode to cathode voltage plotted on the x-axis as shown in figure 9.3. The I-V characteristics can be split into two parts namely the forward characteristics and reverse characteristics. The characteristics in the reverse direction (anode to cathode voltage negative) is similar to a reverse biased diode. For small reverse voltage, a small reverse leakage current flows until the avalanche breakdown takes place at reverse breakover voltage V_{BR}.

As soon as the reverse breakdown takes place due to avalanche breakdown, a large current flows through SCR whereas the voltage across the device remains constant. It is dangerous to operate SCR in the reverse breakdown state because it may get damaged due to overheating.

---

* (i) Explain the V-I characteristics of SCR. **(U.P. Tech. Sem. Exam., June 2008)**
(ii) Draw the V-I characteristics of SCR and define all related terms. **(U.P. Tech. Sem. Exam., June 2009)**
(iii) Explain V-I characteristics of an SCR. **(U.P. Tech. Sem. Exam., Dec, 2007)**

The region from 0 volts upto V_{BR} volts in which the SCR is reverse biased and non-conducting is called as reverse blocking state.

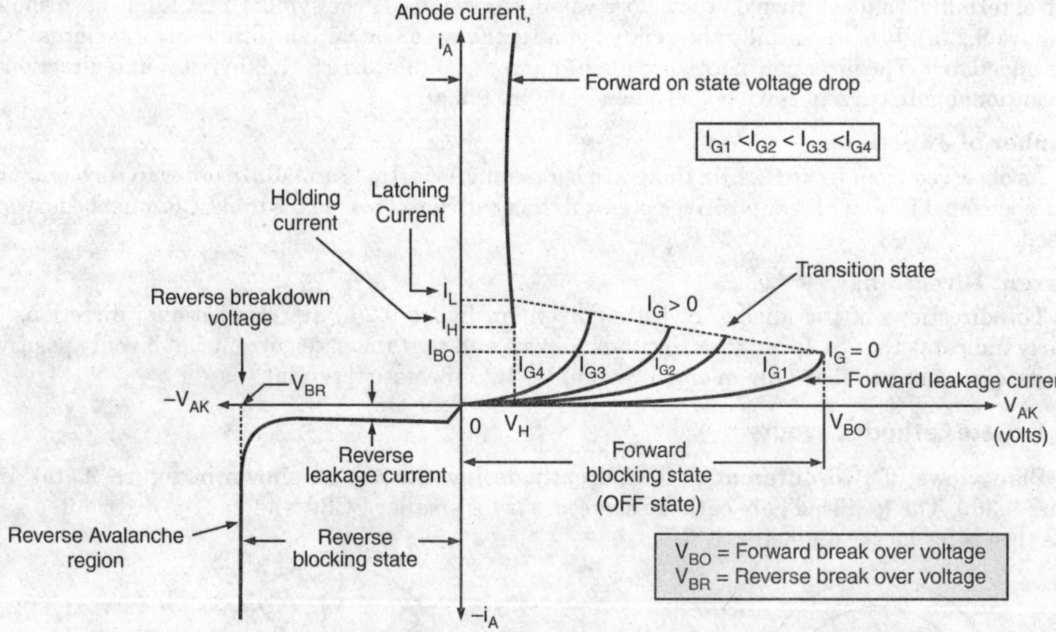

Fig. 9.4. I-V characteristics of SCR.

9.6.1. Forward Characteristics*

The forward characteristics may be divided into three regions of operation as under :

(i) Forward blocking state.

(ii) Transition state.

(iii) ON state.

(i) Forward Blocking State

This is the high voltage low current mode of operation in which SCR is in the OFF state. The current through it is Forward Leakage Current. This current flows due to thermally generated minority carriers.

(ii) Transition State

SCR remains in the blocking state as long as the forward anode to cathode voltage is less than the breakover voltage. As soon as V_{AK} becomes greater than the breakover value, the anode current i_A increases sharply to a high value and the voltage across SCR reduces sharply to a low value (on state voltage). The switching of SCR from off state to on state and vice versa takes place in a short time. This change over state is called as transition state. As this is an unstable state, it is shown by dotted line in figure 9.4.

(iii) The Low Voltage High Current Mode or the ON State

The voltage across SCR is low in this state. In the on state a high power SCR can conduct average currents as large as 2000 amp with on state voltage drop of a few volts.

9.6.2. Unstable State

These two stable starts or modes are connected together by an unstable mode (called as the

* Sletch the V-I characteristics of an SCR and explain its working in detail.

(U.P. Tech. Sem. Exam., 2007-08)

transition mode) of operation that appears as a negative resistance on the I-V characteristics. (dotted line in figure 9.4).

The important voltage and current terms in the I-V characteristics may be listed as under :

(i) The forward breakover voltage (V_{BO})

This is the maximum forward voltage that can be applied between anode and cathode, without initiating forward conduction. This voltage is defined for a zero gate current. In short, this is the maximum forward voltage across SCR in its off state.

(ii) The forward leakage current

This is the small current flowing in the forward direction in the off state of the device. This current is generated due to the minority current carriers and therefore is dependent on the operating temperature.

(iii) The holding current (I_H)

It represents the minimum current that can flow through SCR and still hold it in the on state. The accompanying voltage is termed as holding voltage V_H. If the forward anode current is reduced below holding current, SCR will be turned off. The holding current is defined for zero gate current ($I_G = 0$).

(iv) ON state voltage

It may be noted that the voltage across SCR in its on state is very low as compared to the off state voltage.It will be proved later on that the on state voltage is equal to drop across one junction (i.e. the order of 1 to 1.5 volts).

(v) Latching current (I_L)

It is minimum anode current that must flow through SCR to latch it into the on state. The latching current is higher than the holding current. The latching current is important when SCR is being turned on.

9.6.3. Difference Between Holding and Latching Currents

(i) The latching current is important only at the time of SCR turn on, whereas holding current is important only at the time of SCR turn off.

(ii) If the anode current goes below the holding current then the conducting SCR turns off. The holding current is defined for $I_G = 0$, whereas if anode current at the time of turn on is higher than latching current then only SCR will latch into its on state.

(iii) If $I_A < I_L$ then SCR does not latch. It will remain on as long as I_G is being supplied. As soon as I_G is removed, SCR will turn off.

(iv) Latching current is always higher than the holding current.

(v) The values of I_L and I_H are constant. They do not depend on the gate current magnitude.

9.6.4. Effect of Gate Current

Let us consider figure 9.4. It shows that as we increase the positive gate current from I_{G1} to I_{G4}, the breakover voltage of SCR reduces progressively. Thus, SCR will turn on at lower anode to cathode voltages as we increase the gate current.

9.6.5. How to Turn off SCR ?

The gate current can turn on SCR but once turned on, the gate loses control. Even if we reduce the gate current to zero, SCR continues to conduct. We can turn off a conducting SCR by :

(i) Reducing its anode current below the holding current or

(ii) Applying a reverse voltage across the SCR.

The experimental setup for plotting SCR characteristics has been shown in figure 9.5.

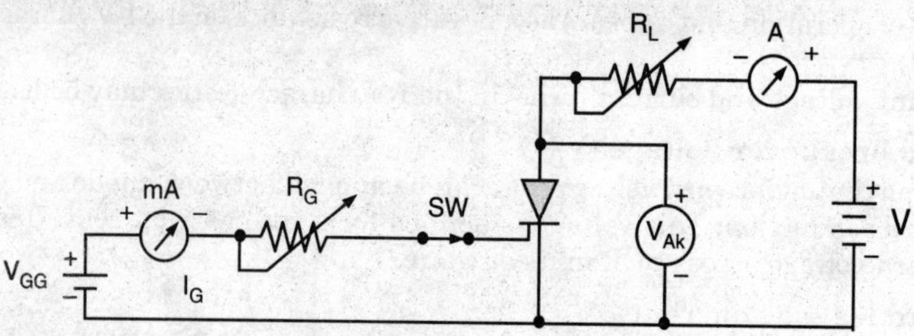

Fig. 9.5. Experimental setup to plot the I-V characteristics of SCR.

9.7. WORKING OPERATION OF SCR

Working operation of SCR can be explained under two different operating conditions. These two operating conditions are as under:

9.7.1. Working Operation Without Gate Current

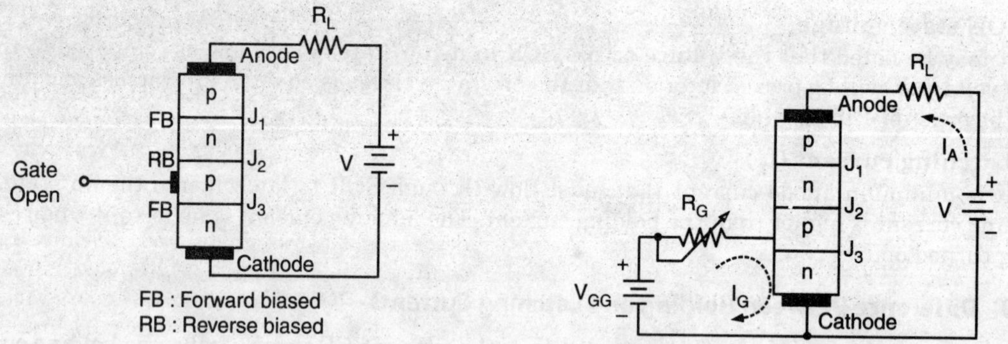

(a) Working operation of SCR without gate current (b) Working operation of SCR with gate current

Fig. 9.6.

Let us consider figure 9.6(a). The gate terminal of SCR is left open so that $I_G = 0$. SCR is forward, biased by applying a positive voltage to anode with respect to cathode. Out of the three junctions, the junctions J_1 and J_3 are forward biased and junction J_2 is reverse biased. Therefore, current does not flow through the SCR. As the anode to cathode voltage is increased, the voltage across junction J_2 increases. At a certain voltage, this junction will breakdown and SCR will start conducting. This voltage is called as forward breakover voltage.

9.7.2. Operation With Gate Current

Let us consider figure 9.6(b). It illustrates that SCR is forward biased as before and the gate cathode junction also is forward biased using an external power source. The gate current I_G starts flowing which can be adjusted by the resistance R_G. As explained earlier, the value of gate current decides breakover voltage of SCR. As I_G increases, breakover voltage will decrease i.e., SCR will turn on at lower voltages.

9.7.3. Reverse Blocking Capacity

In the reverse blocking state, anode is biased negative with respect to cathode. Therefore, the junctions J_1 and J_3 in figure 9.7(a) are reverse biased and junction J_2 is forward biased. Out of the two reverse biased junction J_1 and J_3, the lower junction J_3 has a low breakdown voltage due to the heavy doping on both sides of it. Hence, the reverse blocking capacity depends entirely on junction J_1. The reverse blocking capacity of junction J_1 is usually decided by the length of n^- (n_1) region.

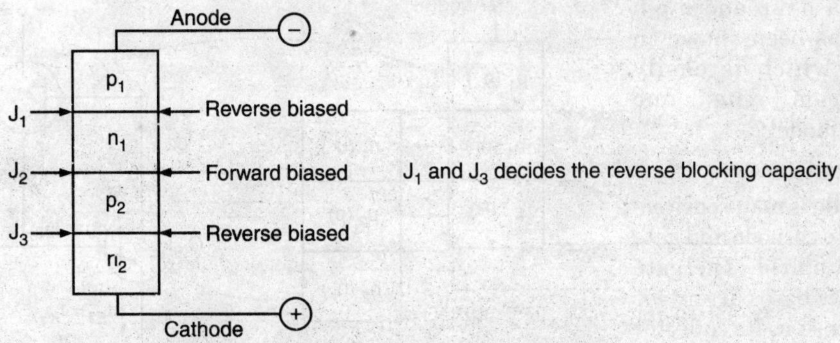

Fig. 9.7. Illustration of SCR in reverse blocking state.

9.7.4. Forward Blocking Capacity of SCR

In the forward blocking state, the thyristor has to block a forward voltage because anode is biased positive with respect to cathode. Therefore, the junctions J_1 and J_3 are forward biased, and junction J_2 is reverse biased. The junction J_2 has to block the entire anode to cathode voltage. In order to achieve this, the n^- layer (n_1) in figure 9.7(b) is lightly doped and made wide enough.

The depletion region of the reverse biased J_2 junction appears in this layer (n_1 layer). Thus, this region decides the forward blocking voltage.

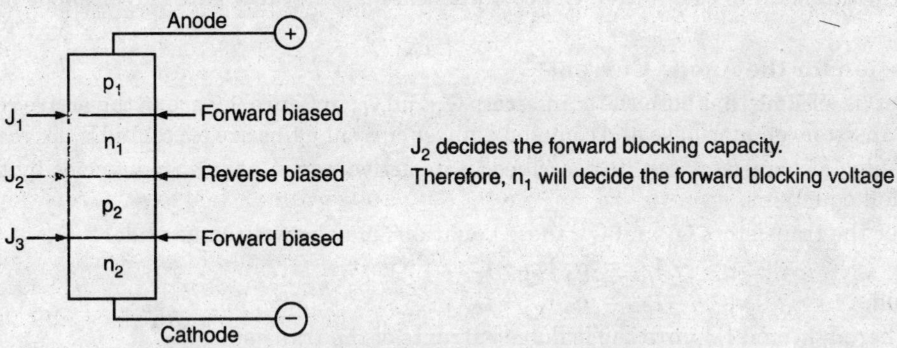

Fig. 9.8. Illustration of SCR in forward blocking state.

Why reverse voltage rating is higher than forward voltage rating?

The reverse voltage rating of the device is the sum of ratings of J_1 and J_3, while the forward voltage rating is the rating of J_2 junction. Since, the p_2 layer (gate) is lightly doped and narrower than p_1 (anode) layer the voltage blocking capacity of J_1 is larger than that of J_2. Also, this means rating of ($J_1 + J_3$) is even larger than that of J_2. Hence, for a thyristor the reverse voltage rating is usually greater than its forward voltage rating. Therefore, the forward voltage is used as a criteria while designing thyristorised circuits. The forward and reverse blocking voltages are dependent on the junction temperature. They reduce with increase in junction temperature beyond 150°C. Therefore, most manufacturers specify the maximum junction temperature for their thyristors at 125°C.

9.8. TRANSISTOR ANALOGY: TWO TRANSISTOR EQUIVALENT CIRCUIT*

1. Equivalent Circuit

The working operation of SCR can be explained with the help of the one dimensional model as shown in figure 9.9. The device is considered to be an ideal device. A low frequency equivalent circuit

* Briefly describe two transistor analogy. **(U.P. Tech. Sem. Exam., 2007-08, 2010-11)**

made up of a p-n-p and n-p-n transistor has been shown in figure 9.9, which is easily derived from the one dimensional model.

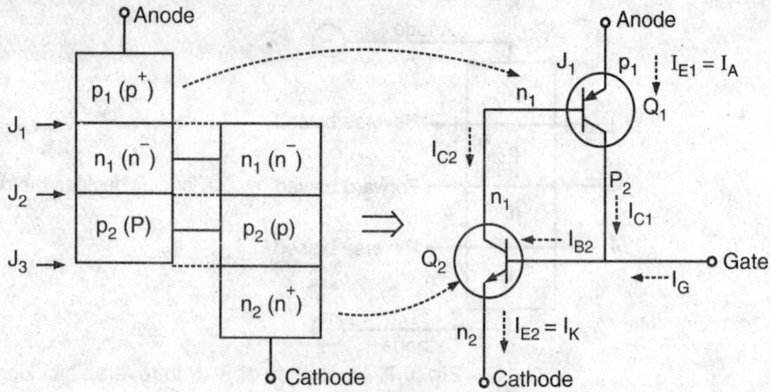

Fig. 9.9. Two transistor equivalent circuit of a thyristor.

2. SCR Operation

The anode is made positive with respect to cathode and gate current is supplied. The gate current acts as base current for Q_2 and it turns on. Its collector current $I_{C2} = \beta_2 I_G$ starts flowing. But, collector current of Q_2 acts as base current for Q_1 and Q_1 starts conducting. Its collector current $I_{C1} = \beta_1 I_{B1} = \beta_1 \beta_2 I_G$ now starts flowing. The collector current of Q_1 acts as base current of Q_2 and collector current of Q_2 in turn acts as base current of Q_1. Due to this type of connection, the current multiplication takes place which is called as cumulative current multiplication and both the transistors will saturate. The saturated transistors are equivalent to closed switches. Hence, the SCR will get latched into on state and acts as a closed switch. Even if the gate current is reduced to zero, the two internal transistors will supply base current to each other and hence will remain in saturation. Hence once located, removal of gate current does not turn off the SCR.

3. Expression for the Anode Current

(i) Let us assume that both the transistors Q_1 and Q_2 in figure 9.9 are in the active region. The transistor common base and common emitter current gains are α_1, α_2 and β_1, β_2 respectively.

(ii) A bipolar transistor can be described in its active region at low frequencies by the Ebers-Moll equations.

For the transistors Q_1 and Q_2, these equations may be written as under :

$$I_{C1} = \alpha_1 I_{E1} + I_{CO1} \qquad \text{...(9.8)}$$

and

$$I_{C2} = \alpha_2 I_{E2} + I_{CO2} \qquad \text{...(9.9)}$$

where I_{CO1} and I_{CO2} are the leakage currents of the transistors

and

$$\alpha = \frac{\beta}{(1+\beta)} \qquad \text{...(9.10)}$$

(iii) If we note that $I_A = I_{E1}$ and that $I_K = I_{E2} = I_A + I_G$ and setting the sum of all the currents into one of the transistors to zero, then, we have

From figure 9.9, we have

$$I_{E2} = I_{C2} + I_{B2} \qquad \text{...(9.11)}$$

(iv) But $I_{E2} = I_A + I_G$

therefore,

$$I_A + I_G = I_{C2} + I_{B2} \qquad \text{...(9.12)}$$

(v) But $I_{E2} = I_{C1} + I_G$

therefore,

$$I_A + I_G = I_{C2} + I_{C1} + I_G \qquad \text{...(9.13)}$$

But

$$I_{C2} = \alpha_2 I_{E2} + I_{CO2} \text{ and } I_{C1} = \alpha_1 I_{E1} + I_{CO1}$$

therefore,

$$I_A = \alpha_2 I_{E2} + I_{CO2} + \alpha_1 I_{E1} + I_{CO1} \qquad \text{...(9.14)}$$

(vi) But

$$I_{E2} = I_A + I_G \text{ and } I_{E1} = I_A$$

Substituting all these values, equation (9.14) can be written as under :

$$I_A = \alpha_2 (I_A + I_G) + I_{CO2} + \alpha_1 I_A + I_{CO1} \qquad \text{...(9.15)}$$

or

$$I_A - \alpha_2 I_A - \alpha_1 I_A = \alpha_2 I_G + I_{CO1} + I_{CO2}$$

Therefore, we have $$I_A = \left[\frac{\alpha_2 I_G + I_{CO1} + I_{CO2}}{1-(\alpha_1+\alpha_2)}\right] \quad ...(9.16)$$

(vii) In the blocking state (off state), the sum of $(\alpha_1 + \alpha_2)$ is much less than unity, so anode current I_A is quite small, a few microamperes for low current devices and to a few milliamperes for the high current SCR.

4. How does SCR Turn On?

In equation (9.16), as long as the values of α_1 and α_2 are small, the anode current I_A will remain low and SCR will be in the off state. If $(\alpha_1 + \alpha_2)$ approaches unity, the denominator of equation (9.16) will be very small and SCR will be at the breakdown point, ready to enter into the transition state. The current gains of the two transistors Q_1 and Q_2 would be high (greater than one) and the collector current of one transistor acts as base current for the other one and gets multiplied by its current gain β. This current then again acts as base current of the first transistor and gets multiplied by its current gain. This cumulative current multiplication action would soon increase the anode current to such a value that SCR will enter into its ON state. This is the turn on process of SCR.

5. How do the Values of α Increases?

The key to the turn on process is to understand how the values of α_1 and α_2 of two transistors Q_1 and Q_2 increase from the small values in blocking state to the higher values required to turn SCR on. As the forward anode-cathode voltage is increased, the depletion region of junction J_2 grows into the n_1 (n^-) layer. This layer is the base for the upper p-n-p transistor Q_1. Due to the growth of depletion region of J_2 into n_1 layer the effective thickness of the base region of p_1 n_1 p_2 transistor (Q_1) reduces and hence α_1 i.e. α of Q_1 increases. Similarly, the extension of the depletion region layer J_2 into the P_2 layer (the base of the n-p-n transistor Q_2) will cause increase in the value of α_2.

9.8.1. SCR as a Switch

The SCR has only two states of operation i.e. it can be either fully ON or completely OFF. There is no active region of operation like transistor. Therefore, SCR is equivalent to a switch, as shown in figure 9.10. Since it cannot operate in the active region, it cannot be used as an amplifier.

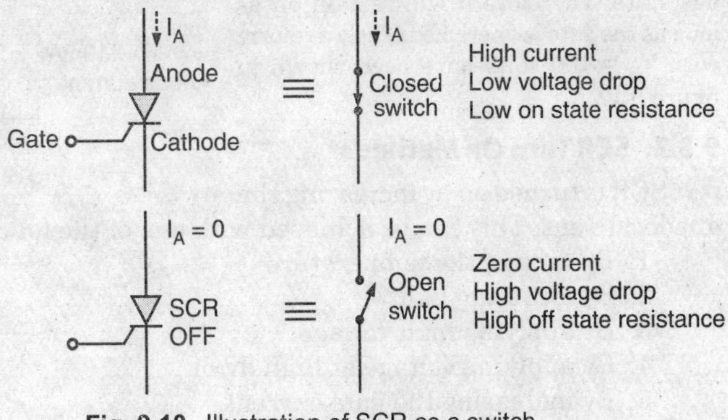

Fig. 9.10. Illustration of SCR as a switch

EXAMPLE 9.1. An SCR, having average forward current rating of 25 Amp, holding and latching current of 20 mA is connected in series with a 10 kΩ resistor across a DC supply of V volts. At t = 0, a periodic pulse train at a frequency of 5 kHz with 50% duty cycle, varying from 0 to 5 Volts, capable of causing conduction in the device is applied to the SCR gate for 1 ms. Sketch the gate voltage, anode voltage and anode current waveforms by neglecting the turn on and turn off times of the SCR, for a time interval t = 0 to t = 1.2 ms, if (i) V = 300 Volts, (ii) V = 100 Volts.

Solution: The circuit diagram from the given description has been shown in figure 9.11.

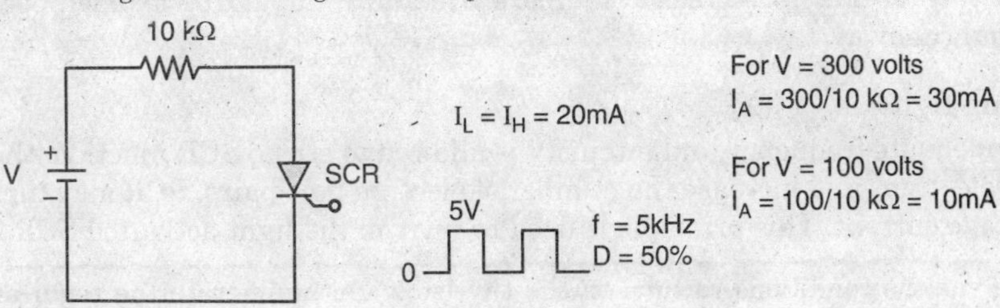

Fig. 9.11.

Case I : Waveforms with V = 300 Volts

If V = 300 volts, then the anode current is I_A = 300 V/10 kΩ = 30 mA. Since this current is higher than the latching current, the SCR is latched as soon as the first gate pulse is applied. The waveforms with V = 300 volts have been shown in figure 9.12.

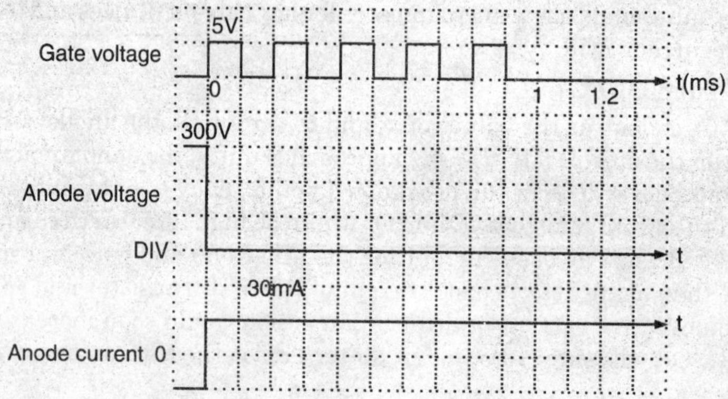

Fig. 9.12. Waveform with V = 300 volts.

Case II : Waveforms with V = 100 Volts

If V = 100 volts, then the anode current is given by

$$I_A = \frac{100\ \text{V}}{10\ \text{k}\Omega} = 10\ \text{mA}$$

Since this current is small that the latching current, the SCR can not latch into ON state. Therefore it will remain on as long as the gate is energized. The waveform with V_S = 100 volts have been shown in figure 9.13.

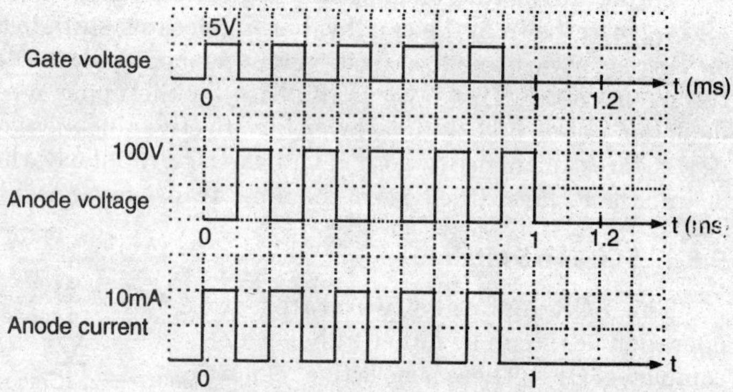

Fig. 9.13. Waveforms with V = 100 volts

9.8.2. SCR Turn On Methods*

SCR is turned on by increasing the anode current. This can be achieved with one of the following methods :

1. By raising the temperature
2. By focussing light
3. By applying high voltage
4. By applying voltage at high dv/dt
5. By increasing the gate current.

1. By raising the temperature

If the temperature of SCR increases, there is an increase in the number of electron-hole pairs. This will increase the leakage current. Therefore, the value of α_1 and α_2 will increase. If $(\alpha_1 + \alpha_2)$ tends to unity, then SCR may be turned on.

> **Note :** SCR turn on due to increased temperature and it should be avoided because it may cause thermal runaway.

2. By focussing light

If light of adequate frequency and intensity is allowed to strike SCR junction, then the photons will strike the electrons and increase the number of electron hole pairs. SCR may turn on due to the increased leakage current. This principle is used to turn on the light activated SCR (LASCR).

---

* Mention the various conditions to turn-on the thyristor and enumerate the requirements of ideal firing circuit.

3. By applying high voltage

SCR will turn on if the forward anode to cathode voltage is greater than the forward breakdown voltage V_{BO}, because then a sufficient leakage current will flow to turn on SCR. We should avoid this type of turn on because it may prove to be destructive.

4. By applying voltage at high dv/dt

We have already explained how a high dv/dt may result in the accidental turn on of a SCR. It may even damage SCR. Hence, SCR should be protected against high dv/dt. Care should be taken to limit the applied dv/dt below the value specified by the manufacturer.

5. By increasing the gate current

As discussed earlier, the breakdown voltage will decrease with increase in the gate current.

(i) The gate signal should be removed, once SCR is turned on. The gate signal should not be applied continuously in order to avoid gate power dissipation.

(ii) Gate current should not be applied when SCR is reverse biased, because then SCR may fail due to increased leakage current.

(iii) The width of the gate triggering pulse should be longer than the time taken by the anode current to the holding current I_h. This pulse width should be longer than the turn on time t_{on} of thyristor.

9.8.3. Salient Features of Gate Turn ON

The gate turn on has following salient features :

(i) Less power consumption.

(ii) It is possible to turn on the SCR precisely at the desired value of anode to cathode voltage.

(iii) If the supply voltage is AC, then it is possible to adjust the firing angle α to the desired value.

(iv) Triggering circuits are easy to design.

(v) Turn on gain is very high. Turn on gain is the ratio of anode current to the gate current.

(vi) Pulsed triggering circuit provide greater flexibility.

9.9. DYNAMIC CHARACTERISTICS OF SCR

The I-V characteristics (static characteristics) of SCR, discussed earlier, can provide all the information about the behaviour of thyristor in the forward blocking, reverse blocking and on state. However, it cannot provide any information about the transition state. Therefore, to know about SCR behaviour at the time of turn on and the turn off, we must understand the dynamic characteristics of switching characteristics of SCR.

The dynamic characteristics can be further classified into two categories as under :

(i) Turn on characteristics

(ii) Turn off characteristics

The turn on characteristics is useful in knowing the behaviour of SCR at the time of turn on whereas the turn off characteristics is useful to understand its behaviour at the time of turn off.

9.9.1. Turn On Characteristics of SCR*

SCR is turned on by applying a current pulse of specified magnitude and duration to the gate of the device. The gate current is applied at t = 0 as shown in figure 9.14. The waveforms for the anode current and anode to cathode voltage at the time of turn on have been shown in figure 9.14. The

* Draw and explain the behaviour of switching characteristics of a thyristor during its turn-on and turn-off process..

forward anode current increases at a constant rate di_F/dt. The total time taken by SCR to turn on completely can be subdivided into three times intervals. They are as under:

(i) The turn on delay time (t_d)

(ii) The rise time (t_r), and,

(iii) The spreading time (t_{ps})

Therefore, we have

$$t_{on} = t_d + t_r + t_{ps} \quad \text{...(9.17)}$$

(i) Turn on delay time $t_{d\,(on)}$

During the turn on delay time $t_{d\,(on)}$, SCR appears to be in the blocking (off) state. The voltage across SCR is constant and high and current through it is zero as shown in figure 9.14. But as the gate current has already been applied, excess carriers are injected into the p_2 layer (the base of n-p-n transistor in the equivalent circuit). This increases the sum $(\alpha_1 + \alpha_2)$. This sum finally reaches unity and SCR turns on, due to regeneration. The anode current begins to increase and this indicates the end of turn on delay time and start of rise time interval.

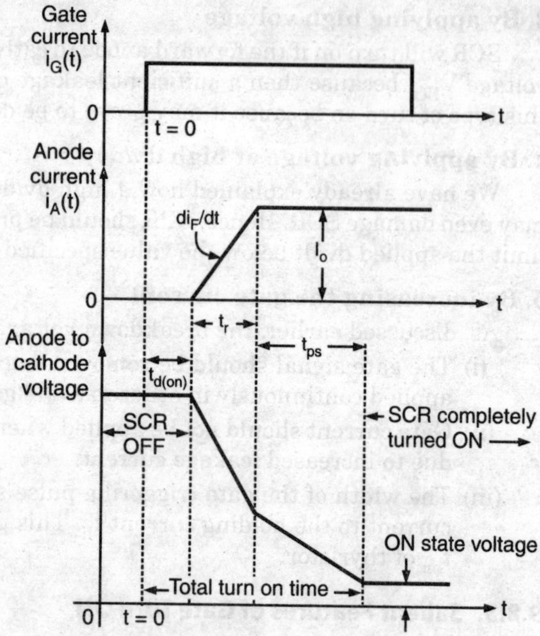

Fig. 9.14. SCR voltage and current waveforms at the time of turn on.

(ii) Rise time interval (t_r)

During the rise time interval (t_r), the following events take place :

A large excess carrier density or plasma is built up near the gate region. This then spreads laterally until the entire cross sectional area of the SCR is filled up with excess carrier density. Simultaneously, the carrier injection begins from p_1 (anode) layer to n_1 layer and the number of carriers injected begins to grow. The anode current rises at the rate di_F/dt. This rate is large enough so that the anode current reaches its steady state value (I_0) within a short time. The rise time interval con. s (t_r) to an end as soon as the anode current reaches its steady state value (I_0). The anode to cathode voltage begins to drop rapidly during the rise time interval as shown in figure 9.14. Even after the rise time interval is over the plasma will continue to spread over the lateral area of SCR until SCR is completely shorted out by the large number of excess carriers.

(iii) Plasma spreading interval (t_{ps})

The time required for the plasma to spread from the region around the gate to the entire device cross-section is the Plasma Spreading Time (t_{ps}). The rate of voltage drop during the plasma spreading time is slower than that in the rise time interval because t_{ps} is larger than t_r and because most of the voltage drop occurs during the rise time interval as shown in figure 9.15.

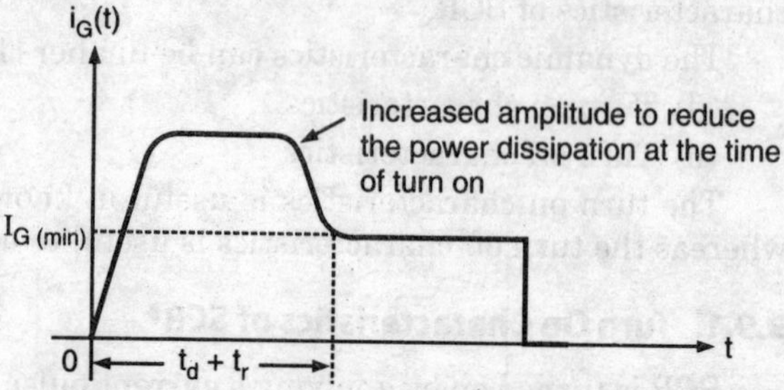

Fig. 9.15. Gate current with an initial large value.

9.9.2. Precautions to Avoid SCR Damage

(i) In order to avoid SCR damage or failure, the rate of anode current rise di_F/dt should be kept less than the maximum value given in SCR specification sheet. The failure of the device may take place due to excessive power dissipation that takes place during the time interval t_r.

(ii) This power dissipation can be reduced by applying large values of gate current during the $t_{d\,(on)}$ and t_r intervals. For this reason, the gate current applied is often large at the start of the turn on interval and is gradually reduced as time proceeds, as shown in figure 9.15.

9.9.3. Turn Off Process

Once SCR is turned on, it gets latched into its on state and acts like a closed switch. The gate terminal loses control over SCR i.e. even though the gate pulses are removed SCR remains in the on state. A conducting SCR can be turned off with the help of an external circuit, by reducing the anode current below its holding current for a minimum specified time. During this time, the stored charge inside the device will be removed due to internal recombination and carrier sweep out. Due to this, the transistors Q_1 and Q_2 in the equivalent circuit of figure 9.12 come out of saturation and enter into active region. The values of α_1 and α_2 will decrease, the regeneration will decrease and the conducting SCR will turn off. The minimum time required for the recombination and carrier sweep out should be provided before the device is turned on again. This is essential for the successful turn off of SCR.

9.9.4. Turn Off Characteristics of SCR

The turn off characteristics using the forced commutation has been shown in figure 9.16.

The successful turn off a conducting SCR requires that it should be reverse biased by the external circuit for a minimum time period. The process to turn off SCR is also called as commutation. The turn on time of SCR is appreciably shorter than the turn off time. The turn off waveforms of SCR have been shown in figure 9.16.

(i) The current through the conducting SCR starts decreasing at $t = 0$ at a fixed rate di_R/dt. This rate of change of current is decided by the external circuit. As the current decreases, the excess carrier in the four regions of SCR decrease from steady state values, due to the internal recombination and carrier sweep out.

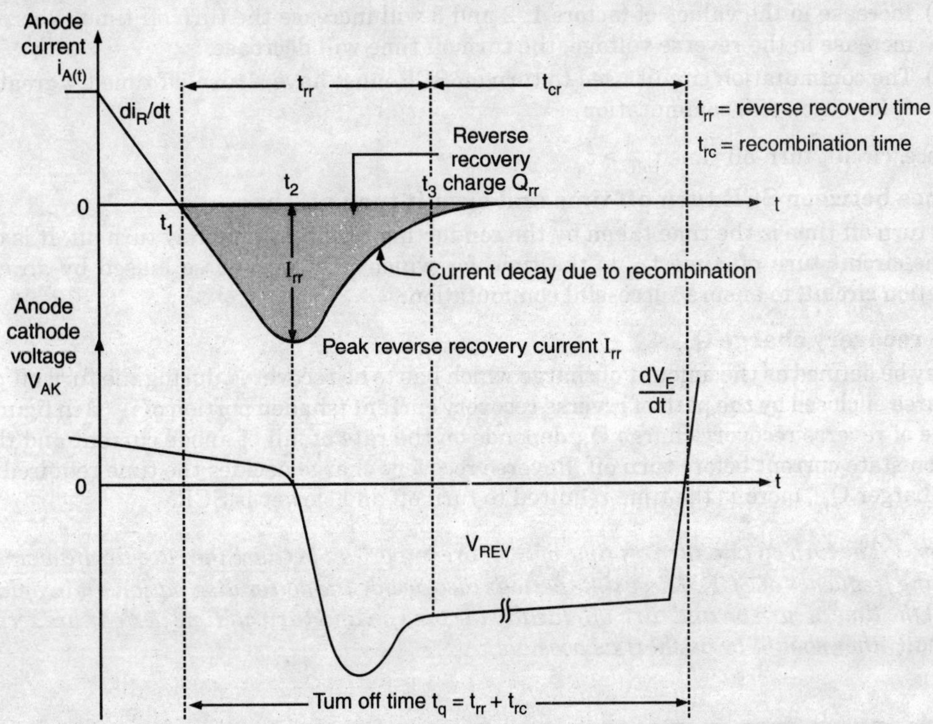

Fig. 9.16. SCR voltage and current waveforms during turn off.

(ii) Due to the two outer junctions J_1 and J_3, the turn off characteristics is similar to that of a diode. The peak reverse recovery current is denoted by I_{rr} and its magnitude is much higher than the normal reverse blocking current I_r.

(iii) Reverse recovery time t_{rr}

The reverse recovery time t_{rr} is the time required for the minority carriers to recombine with the opposite charges and to be neutralized. It is the time from zero cross over point of SCR current to 25% of the peak reverse recovery current I_{rr}.

(iv) Recombination time t_{rc}

This is the time required for the recombination of excess charges in the middle p-n junction J_2. A negative reverse voltage will reduce this time.

9.9.5. Turn Off Time (t_q)*

(i) Device manufacturers specify a turn off time t_q for their SCRs, which represents the minimum time SCR should remain in the reverse blocking mode (i.e. reverse biased) before any forward voltage is reapplied.

(ii) This means that for successful commutation (turn off) of SCR, it is essential to reverse bias it for a period equal to or greater than (t_q) before reapplying a forward voltage.

(iii) As shown in figure 9.16, the turn off time (t_q) of SCR is sum of reverse recovery period (t_{rr}) and the recombination time (t_{rc}) and is of the order of 40 μs to 200 μs. Thus, we have

$$t_q = t_{rr} + t_{rc} \qquad \text{...(9.18)}$$

The turn off time of SCR is dependent upon following factors:

(i) The current through SCR (I_G)
(ii) di/dt at the time of commutation (turn off)
(iii) The junction temperature
(iv) Magnitude of reverse voltage.

(a) Increase in the values of factors 1, 2 and 3 will increase the turn off time, whereas with increase in the reverse voltage, the turn off time will decrease.

(b) The commutation circuit used to turn off SCR must have a turn off time t_{off} greater than t_q for successful commutation.

Hence, circuit turn-off time $t_{off} > t_q$ ...(9.19)

Difference between SCR turn off time and circuit turn off time

SCR turn off time is the time taken by the conducting SCR to completely turn off. It is denoted by t_q. The circuit turn off time t_{off} is the time for which SCR is reverse biased by an external commutation circuit, to ensure successful commutation.

Reverse recovery charge Q_{rr}

It may be defined as the amount of charge which has to be recovered during the turn off process. It is the area enclosed by the path of reverse recovery current (shaded portion of i_A (t) in figure 9.16). The value of reverse recovery charge Q_{rr} depends on the rate of fall of anode current and the peak value of on-state current before turn off. Reverse recovery charge decides the time required to turn off SCR. Larger Q_{rr}, more is the time required to turn off and slower is SCR.

Note: *The turn on and turn off time of SCR are important because they decide the maximum switching frequency of SCR. These time periods also decide the power dissipation taking place in SCR at the time of turn on and turn off. Ideally the turn on and turn off time periods are zero and practically they should be as short as possible.*

---

* Define the turn-on and turn-off time of an SCR. **(U.P. Tech. Sem. Exam., 2008-09, 2010,-11)**

EXAMPLE 9.2. Determine the peak reverse recovery current for SCR, if the reverse recovery time t_{rr} = 10 μ sec and reverse recovery charge Q_{rr} = 150 μC.

Solution: Given that $t_{rr} = 10\ \mu s$, $Q_{rr} = 150\ \mu C$

We know that peak reverse recovery current = $\dfrac{2 \times \text{Reverse recovery charge}}{\text{Reverse recovery time}}$

Therefore,
$$I_{rr} = \frac{2 \times Q_{rr}}{t_{rr}} = \frac{2 \times 150 \times 10^{-6}}{10 \times 10^{-6}}$$

Solving, we get I_{rr} = **30 A** **Ans.**

9.9.6. Turn Off Time and Effect of Reapplied dV_F /dt

Let us consider figure 9.16, which illustrates that even at instant t_3, substantial excess carriers remain in the n_1 and p_2 layers of a thyristor.

If a forward voltage is reapplied which increases at a rate dV_F/dt as shown in figure 9.17, a pulse of forward current (called as forward recovery current) will flow through the SCR. This current flows due to the remaining excess carriers. This forward recovery current may act as a pulse of gate current and therefore may turn on the thyristor even though the turn on is not desired. This is known as accidental turn on of thyristors. The accidental turn on of the thyristors is unwanted and can be avoided by taking the following measures :

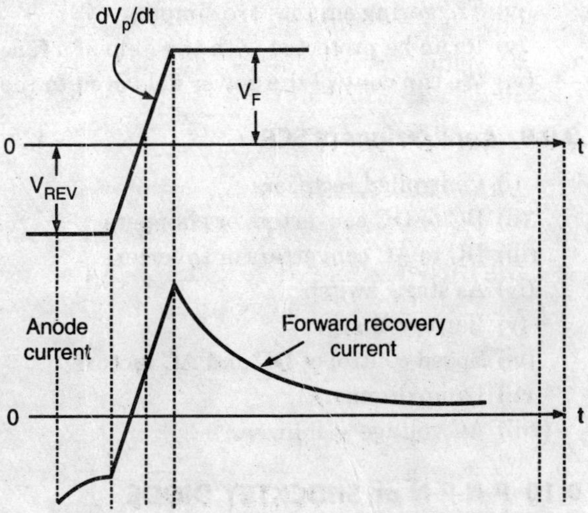

Fig. 9.17. Forward recovery current resulting from reapplication of a forward voltage across the thyristor before the end of the specified recovery time.

(i) A thyristor must remain reverse biased for a minimum time equal to (t_q) specified by the device manufacturer. No forward voltage be reapplied during this time period.

(ii) The value of the rate of increase of the forward voltage dV_F/dt should be kept below the maximum value specified by the manufacturer. This is because larger the value of dV_F/dt is, larger will be the peak forward recovery current and hence more will be the possibility of accidental turn on.

(iii) The maximum value of reapplied dV_F/dt is decided by the value of space charge capacitance of junction J_2, because if the current through this capacitance due to dV_F/dt exceeds, the breakover current I_{BO}, accidental turn on will take place.

Types of SCRs

Depending on the turn off time of the SCR, they may be classified into following two categories:

(i) **Converter grade SCRs**

These SCRs have a large turn off time so they are suitable only for the low frequency applications like rectification.

(ii) **Inverter grade SCRs**

These SCRs have shorter turn off time. Hence, they can operate at higher frequencies, and hence are suitable for the high frequency applications like inverters.

9.9.7. Drawbacks of SCR

(i) It can conduct only in one direction. Hence, it can control power only during one half cycle of ac.

(ii) It can turn on accidentally due to high dv/dt of the source voltage.
(iii) It is not easy to turn off the conducting SCR. We have to use special circuits called commutation circuits to turn off SCR.
(iv) SCR can not be used at high frequencies. The maximum frequency of its operation is 400 Hz.
(v) Gate current can not be negative.

9.9.8. Advantages of SCR

(i) It can handle large voltages, currents and power.
(ii) The voltage drop across conducting SCR is small. This will reduce the power dissipation in the SCR.
(iii) Easy to turn on.
(iv) Triggering circuits are simple.
(v) It can be protected with the help of a fuse.
(vi) We can control the power delivered to the load.

9.9.9. Applications of SCR

(i) Controlled rectifiers
(ii) DC to DC converters or choppers
(iii) DC to AC converters or inverters
(iv) As static switch
(v) Battery charger
(vi) Speed control of DC and AC motors
(vii) Lamp dimmers
(viii) AC voltage stabilizers.

9.10 P-N-P-N or SHOCKTEY DIODE

(U.P. Tech, Sem. Exam; 2010-11, 2011-12)

The four layer diode is the basic device developed before the SCR (silicon controlled rectifier). Many important concepts about SCR are introduced through the P-N-P-N diode.

The four layer diode is a special type of diode. The construction of the four layer diode is as shown in figure 9.18(a) and its circuit symbol has been shown in figure 9.18(b).

9.10.1. Construction and Operation

(i) The four layer diode consists of four layers of silicon doped alternately with p and n type impurities as shown in figure 9.18(a). Because of this construction, this device is called as p-n-p-n diode. It is a two terminal device with the terminals called as anode and cathode.

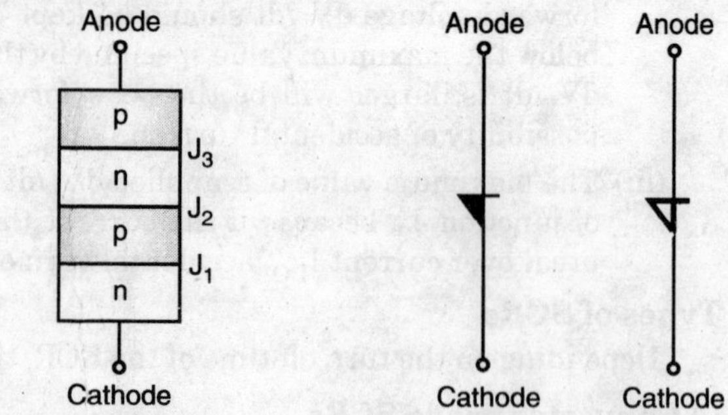

(a) Construction (b) Circuit symbols

Fig. 9.18. Four-layer diode

1. Working operation when forward biased

(ii) Due to the four alternate layers, there are three junctions. When an external voltage is applied to make the anode positive with respect to cathode, the junctions J_1 and J_3 are forward biased and the centre junction J_2 is reverse biased.
(iii) The externally imposed voltage appears principally across the reverse biased junction J_2. Therefore, a very small current will flow through the diode.

(iv) As we increase the external voltage, the diode current increases slowly. This will continue upto a voltage called as the firing voltage or breakover voltage V_{BO} is reached.

(v) As soon as the breakover voltage is reached the diode current increases sharply and the voltage across the device decreases sharply.
This happens because the center junction J_2 breaks down at the breakover voltage. Thus, at the breakover point the p-n-p-n diode switches from its OFF state (blocking state) to its ON state. This has been shown in the V-I characteristics in figure 9.20.

2. Working operation when reverse biased

(i) When reverse biased, the p-n-p-n diode acts like a typical p-n junction diode. The current flowing through the device is very small leakage current. If the reverse voltage reaches the reverse breakdown voltage then the device breaks down and a large current flows through it as shown in figure 9.20.

3. Resistance offered by the diode

The four layer diode acts basically as a switch. Therefore, it has two stable states. When reverse biased, it acts like an open switch and offers a resistance as high as 100 MΩ. When forward biased, it is equivalent to a closed switch and offers a resistance less than 10 Ω.

9.10.2. Equivalent Circuit of the Four-layer Diode

The two transistor equivalent circuit of a four-layer diode has been shown in figure 9.19 :

(i) Figure 9.19(a) illustrates that a p-n-p-n diode has been split into two parts which are displaced mechanically from each other but connected electrically.

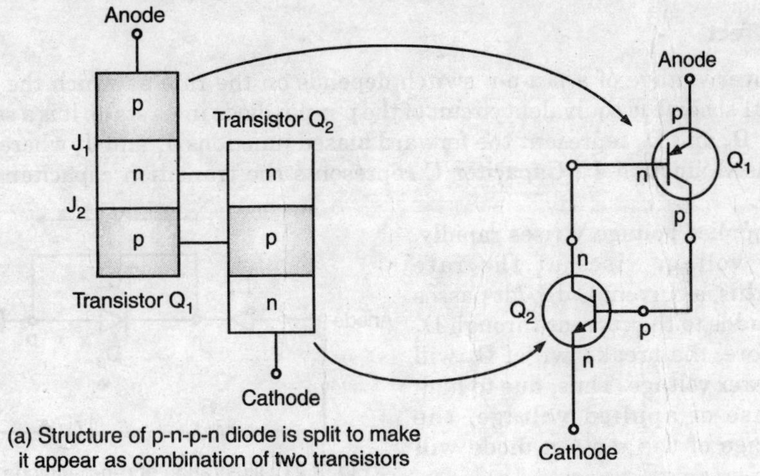

(a) Structure of p-n-p-n diode is split to make it appear as combination of two transistors

(b) Equivalent circuit

Fig. 9.19. Equivalent circuit of a p-n-p-n diode.

(ii) This splitting illustrates that the four layer diode can be viewed as two transistor connected back to back. Transistor Q_1 is a p-n-p transistor and Q_2 is an n-p-n transistor.

(iii) The equivalent circuit has been shown in figure 9.19(b), which shows that the base of Q_1 is same as collector of Q_2 and base of Q_2 is same as the collector of Q_1.

9.10.3. V-I Characteristics of the Four-Layer Diode

The volt-ampere characteristics of p-n-p-n diode has been shown in figure 9.20.

1. Reverse Characteristics

When the voltage is applied in the reverse direction, the junctions J_1 and J_3 are reverse-biased. At a sufficiently large reverse voltage, breakdown will occur at these junctions. This voltage is termed as reverse avalanche voltage V_{RA}.

2. Forward Characteristics

(i) When the four-layer diode is forward biased, a small forward current flows until the voltage reaches the breakover voltage V_{BO}. The corresponding current is I_{BO}. The diode now switches to ON state from it OFF state or blocking state.

(ii) Once the diode enters into the ON state of saturation region, it gets *latched* there.

(iii) If the forward voltage is reduced, the diode remains ON, as long as the forward current is less than the holding current I_H. It is the minimum current required to hold the diode in its ON state.

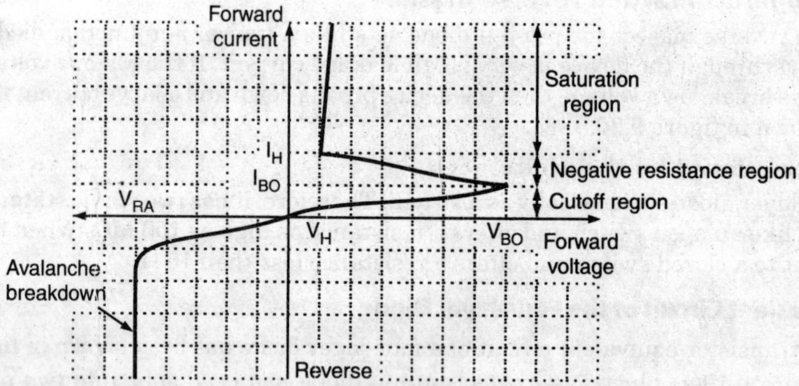

Fig. 9.20. Volt-ampere (V-I) characteristics of the p-n-p-n diode.

9.10.4. Rate Effect

The breakover voltage of a p-n-p-n switch depends on the rate at which the required voltage rises. Figure 9.21 shows the equivalent circuit of the p-n-p-n diode in its state. It is a series combination of three diodes. D_1 and D_3 represent the forward biased junctions J_1 and J_3 whereas D_2 represents the reverse biased junction J_2. Capacitor C represents the transition capacitance of the reverse biased junction.

When the applied voltage V rises rapidly, the capacitor voltage rises at the rate dV_C/dt. Due to this, a current C dV_C/dt passes through C and adds to the current through D_1 and D_3. Therefore, the breakdown of D_2 will take place at lower voltage. Thus, due to high rate of increase of applied voltage, the breakover voltage of the p-n-p-n diode will reduce.

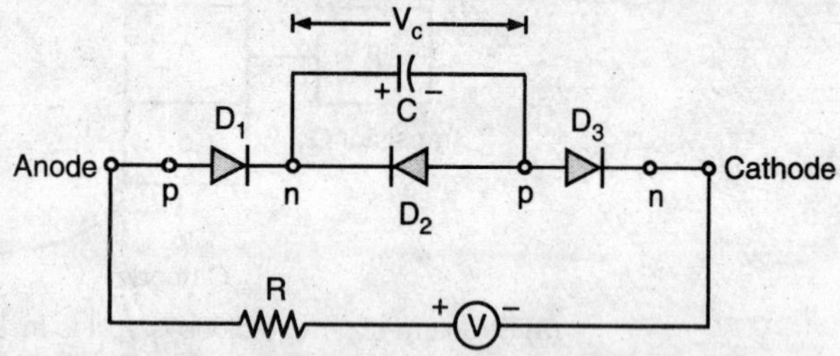

Fig. 9.21. Illustration of p-n-p-n diode in the off state.

9.10.5. Peculiarities of Shockley Diode

It has four P-N-P-N layers with three junctions. It is a undirectional device and allows the current to flow in only one direction i.e. from anode to cathode. It is equivalent to two transistors connected in base to collector manner. It basically acts as a switch. Its resistance in the forward direction is of the order of 10 Ω and the reverse resistance is typically 100 MΩ. Once turned on this, diode gets latched into ON state. This diode can be turned on by increasing the forward voltage beyond the breakover voltage or by applying a voltage of high dV/dt across the device.

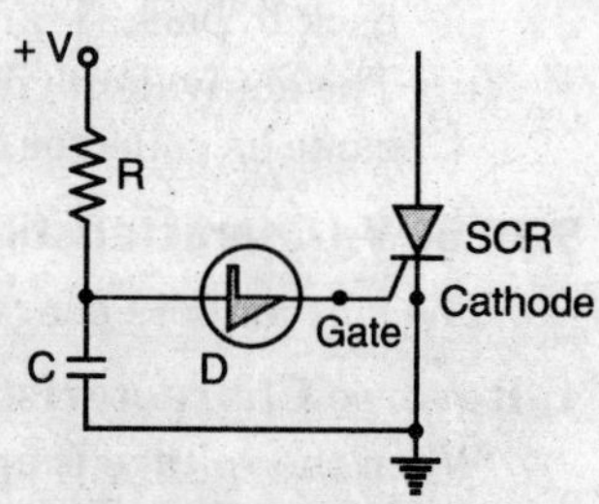

Fig. 9.22. Shockley diode used to trigger SCR

9.10.6. Applications of Shockley Diode

One common application of the shockley diode is shown in figure 9.22, where it is used to trigger the SCR. When the circuit is turned on, the capacitor voltage begins to rise towards the supply voltage. As soon as V_C reaches the breakdown voltage of Shockley diode, it will be turned on. The capacitor discharges through the diode and gate-cathode junction of SCR. This current acts as gate current for the SCR and turns it ON.

9.11. BIDIRECTIONAL TRIODE SCR (TRIACS) (U.P. Tech, Sem. Exam; 2010-11)

A TRIAC is a bidirectional device *i.e.* a device which allows current to flow through it in either directions.

Construction Details of a Triac

In the SCR family, after the SCR, Triac is the most widely used device for power control. Triacs with large voltage and current ratings are now available in market. Triac is a three terminal device, with the terminals named as main terminals 1, 2 (MT_1 and MT_2) and gate, out of which gate is the control terminal. Triac is a bidirectional device, i.e. the current can flow in both the directions through it. (Note that SCR is an unidirectional device). The basic structure of triac has been shown in figure 9.23(a). Its working operation is equivalent to two SCRs connected in antiparallel. The two main terminals are designated MT_2 and MT_1 (main terminal 2 and main terminal 1). The gate is near MT_1. When the gate is open, the triac will block both the polarities of the voltage across MT_1 and MT_2 if the magnitude of voltage is less than the breakover voltage of the device as shown by the characteristics of the triac in figure 9.24. This means that the triac will remain in the OFF state.

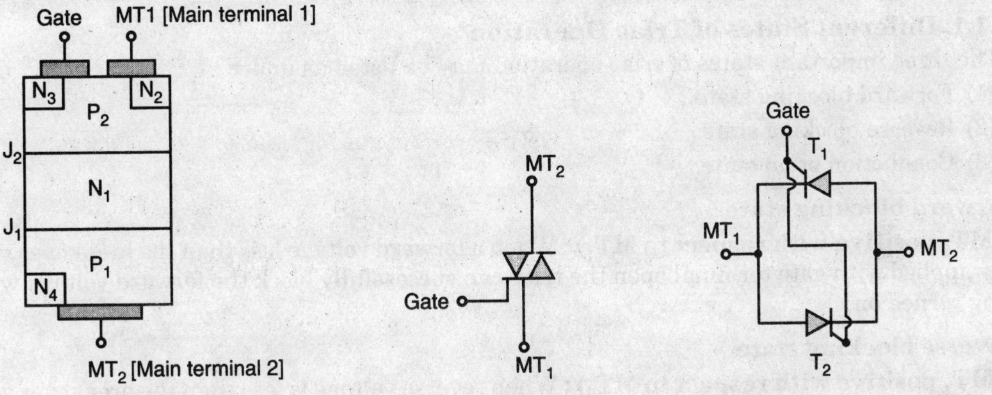

(a) Triac structure (b) Triac symbol (c) Triac is equivalent to two SCRs connected back to back

Fig. 9.23.

The characteristics of the triac have been shown in figure 9.24 and they are similar to those of an SCR both in blocking and conducting states. The only difference is that the SCR conducts only in the forward direction (anode cathode) whereas the triac conducts in both the directions. The other difference in the working operation is in the triggering mechanism. The triac can be turned on by applying either a positive or negative voltage to the gate with respect to terminal MT_1, whereas the SCR can be triggered only by a positive gate signal.

9.11.1. V-I Characteristics of Triac (Static Characteristics)

As observed from figure 9.24, the characteristics of triac is same as that of two back to back connected SCRs. The effect of gate current is also same i.e. with increase in the gate current, the breakdown voltage decreases. In triac, the gate current can be positive or negative whereas in SCR, the gate current can be only positive.

The triac characteristics can be divided into following three regions of operation :

(i) Block state or off state
(ii) Transition or unstable state
(iii) Conduction state or on state

Depending on the polarity of voltage applied between its MT_2 and MT_1 terminals, it will operate in either the first quadrant or the third quadrant as shown in figure 9.24.

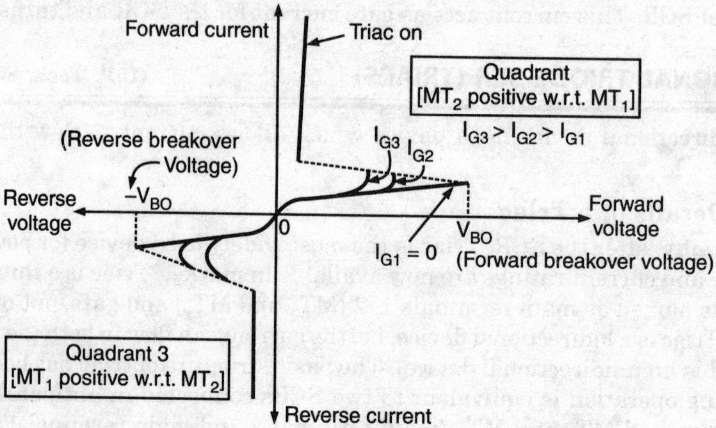

Fig. 9.24. Static I-V characteristics of triac.

MT_2 positive w.r.t. MT_1 : Operation is in the first quadrant.
MT_2 negative w.r.t. MT_1 : Operation is in the third quadrant.

9.11.1.1. Different States of Triac Operation*

The three important states of triac operation may be listed as under :

(1) Forward blocking state.
(2) Reverse blocking state.
(3) Conduction or on state.

1. Forward blocking state

(MT_2 positive with respect to MT_1) : When a forward voltage less than the breakover voltage V_{BO} is applied with gate terminal open the triac can successfully block the forward voltage without getting turned on.

2. Reverse blocking state

(MT_1 positive with respect to MT_2) : When reverse voltage is less than the breakover voltage V_{BO} with gate open circuited the triac will block the reverse voltage without getting turned on.

3. Conduction or ON State

The triac is equivalent to two SCRs connected back to back. Therefore it is a bidirectional device that can conduct in positive as well as negative half cycles of the supply voltage. The gate current can be positive or negative. The forward and reverse breakover voltage reduces with increase in the gate current. Depending on the polarity of supply voltage and the polarity of the gate current, the triac can operate in four different modes of operation as under:

Modes of Operation

I^+, I^-, III^+ and III^- are the four modes of operation of a TRIAC where I or III represents the quadrant of operation and (+) and (–) signs indicate the direction of gate current. The sensitivity of a mode is defined as the minimum gate current required to turn on a triac. The sensitivity of I^+ mode is the highest whereas that of III^+ mode is the lowest.

9.11.1.2. Operating Modes of Triac

Depending on the biasing of triac and the polarity of gate voltage (or polarity of gate drive), the triac can operate in any one of the following four modes of operation :

* Explain the different modes of operation of a triac. **(U.P. Tech., Sem. Exam., 2007-08)**

1. Mode I or I^+ Mode of Operation

In this mode of operation terminal MT_2 is positive with respect to MT_1 and the triac can be turned on by applying a positive voltage between gate and MT_1 as shown in figure 9.25(a). The gate current is positive and the triac operates in the first quadrant of its I-V characteristic. Therefore this is known as I^+ mode. In this mode triac behaves like a conventional SCR with four layers P_1, N_1, P_2, N_2 and the cathode connected to layer N_2 [Fig. 9.25(b)]. This is the recommended mode of triggering if the operation is in the first quadrant. In this mode, the triac acts as shorted emitter SCR as shown in the equivalent circuit of figure 9.25(c).

The device is most sensitive in this mode than other modes of operation. The sensitivity is defined as the amount of gate current required to turn on the device. Lower the gate current is higher is the sensitivity. The currents shown in figure 9.25 are the conventional currents.

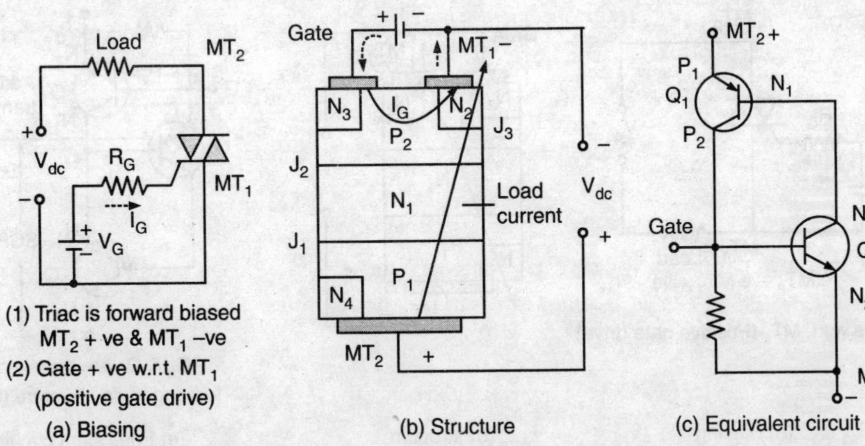

Fig. 9.25. Biasing conditions, structure and equivalent circuit of a triac in the first model i.e. I^+ mode of operation.

2. Mode II or I^- Mode of Operation (Junction Gate)

In this mode, terminal MT_2 is positive and MT_1 is negative and the triac is turned on by applying a negative gate voltage with respect to MT_1 as shown in figure 9.26(a). The gate current is thus negative but the triac operates in the first quadrant of its I-V characteristics. Therefore this is known as I^- mode of operation. The initial conduction of current is through the structure $P_1\ N_1\ P_2\ N_3$ [see figure 9.26(b)] which acts like a pilot or auxiliary SCR. Finally the conduction takes place through the structure $P_1\ N_1\ P_2\ N_2$ which is the main SCR.

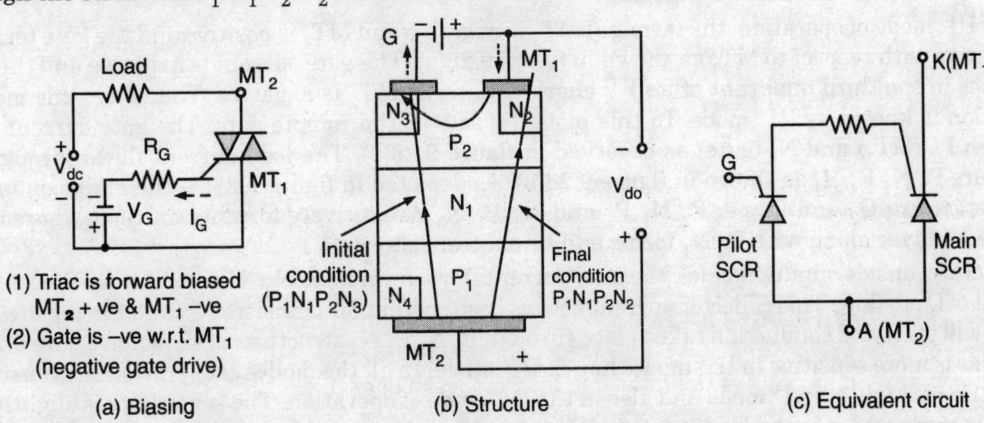

Fig. 9.26. Biasing conditions, structure and equivalent circuit of a triac in I^- mode of operation

The anode current of the pilot SCR acts as gate current for the main SCR. The device is less sensitive to the gate current in this mode than I^+ mode and needs more gate current to turn on the triac. This is also known as the *Junction gate operation*. Since it involves high switching losses, this type of gate drive is not used normally.

3. Mode III or III^+ Mode of Operation (Remote Gate)

In this mode, the terminal MT_2 is negative and MT_1 is positive and the triac is turned on by applying a positive gate voltage with respect to MT_1 as shown in figure 9.27(a). The gate current is positive and the device operates in the third quadrant of its I-V characteristic as MT_2 is negative. Therefore this mode of operation is known as III^+ mode. In this mode the triac is switched by the remote gate operation. The gate current flows from P_2 (gate) to N_2 (MT_1) terminal as shown in figure 9.27(b).

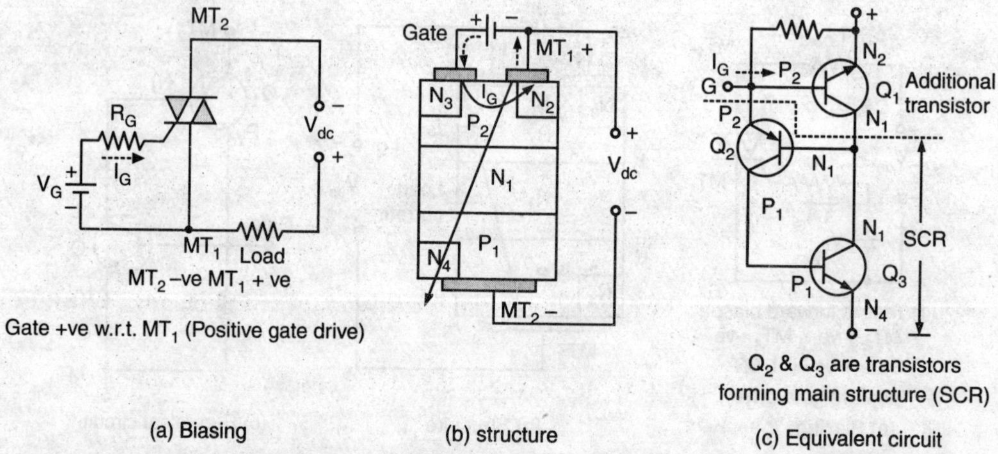

Fig. 9.27. Biasing, structure and equivalent circuit of triac operating in the III^+ mode of operation.

As observed from figure 9.27(b), the gate terminal is not the part of the SCR formed by Q_2 and Q_3, but still it turns the SCR on. Therefore, this is called the remote gate operation. The two transistors P_2, N_1, P_1 and N_1, P_1, N_4 form the main $P_2\ N_1\ P_1\ N_4$ (SCR) structure whereas the remote N_2 layer alongwith P_2N_1 forms an additional transistor [see Fig. 9.27(c)]. Since, the gate (P_2) is positive w.r.t. MT_1 (N_2 layer), the collector current of Q_1 acts as base current for Q_2 transistor. This will initiate the regenerative action in the main structure $P_2\ N_1\ P_1\ N_4$ and the triac starts conducting [figure 9.27(c)]. The triac is less sensitive in this mode.

4. Mode IV or III^- Mode of Operation

In III^- mode of operation, the terminal MT_2 is negative and MT_1 is positive and the gate terminal is negative with respect to MT_1 as shown in figure 9.28(a). The gate current is negative and the triac operates in the third quadrant of its I-V characteristic as MT_2 is negative. Therefore, this mode of operation is known as III^- mode. In this mode N_3 acts as the remote gate. The gate current flows through P_2 (MT_1) and N_3 (gate) as observed in figure 9.28(b). The load current flows through the structure $P_2\ N_1\ P_1\ N_4$ as shown in figure 9.28(b). As depicted in figure 9.28(c), the main conducting transistors are Q_2 and Q_3 i.e. $P_2\ N_1\ P_1$ and $N_1\ P_1\ N_4$ respectively forming an SCR whereas the remote N_3 layer along with $P_2\ N_1$ forms additional transistor.

Let us consider figure 9.28(c), the gate current flows from P_2 to N_3. This will cause the collector current of Q_1 to flow. This collector current acts as base current for transistor Q_2 and the regenerative action will start and conduction takes place through $P_2\ N_1\ P_1\ N_4$ structure as shown in figure 9.28(b). The triac is more sensitive in III^- mode than in III^+ mode. In all the modes of operation the sensitivity of the triac is highest in I^+ mode and also in the III^- mode of operation. The sensitivity is slightly less in the I^- mode and it is the lowest in the III^+ mode of operation.

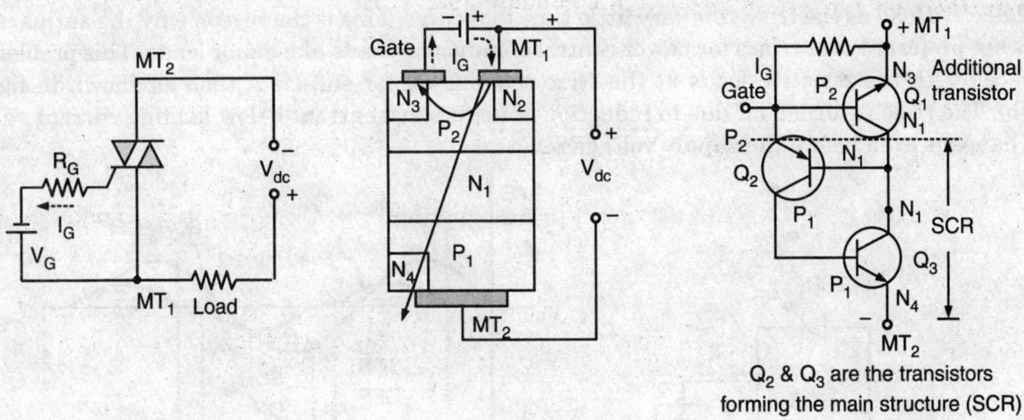

(a) Biasing (b) Structure (c) Equivalent circuit

Fig. 9.28. Biasing structure and equivalent circuit of triac operating in III$^-$ mode.

9.11.2. Commutation of Triac (Triac Turn Off)

A triac is equivalent to a pair of SCRs connected in antiparallel form. The output voltage waveform in an ac circuit with a triac is identical to that with antiparallel SCRs used. But, there is one important difference, between operation of triac and antiparallel SCRs. In antiparallel SCRs, each SCR has entire half cycle for its turn off, but the triac must turn off during the short interval while the load current is passing through zero. For the resistive load, it is possible to achieve, because the time available for the triac to turn off extends from the instant at which the triac current reduces below the holding current, until the reapplied voltage exceeds the value of the line voltage which is required to allow the latching current as shown in figure 9.29. But, with the inductive loads the triac can be commutated easily. Figure 9.29 shows the triac voltage and current waveforms for typical inductive load. Due to the inductive nature of the load, at the point of triac turn off, a large voltage is induced in the inductive load. Due to this, the dv/dt overshoot is observed in the triac voltage waveform. Due to this high dv/dt, the triac may get turned on again immediately after it is turned off. Therefore, to avoid this malfunctioning, the RC snubber circuit must be connected across the triac.

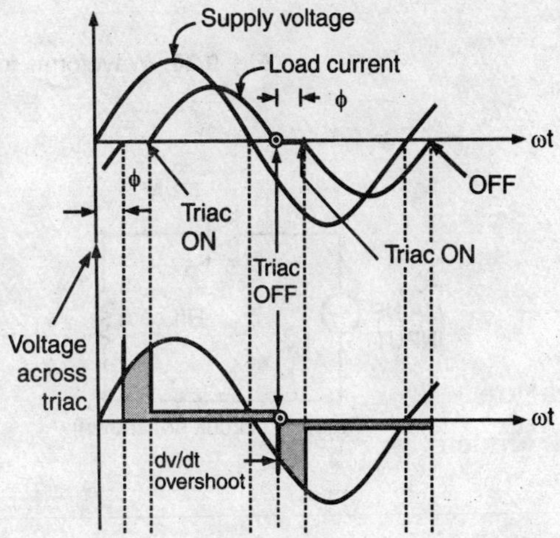

Fig. 9.29. Inductive load waveforms.

9.11.3. Why Antiparallel SCRs are Preferred Over Triac?

If the triac is replaced by two SCRs in antiparallel, they are capable of tolerating a greater dv/dt. The waveforms for the inductive load with antiparallel SCRs have been shown in figure 9.30(a). As observed from the waveform, each SCR conducts in one direction and as long as one of them is conducting the other will be in the off state due to a reverse voltage across it. Since the SCR which is turned on at given instant is initially off for a sufficiently long time with reverse voltage across it, the maximum permissible dv/dt equals to the static dv/dt capability of SCR. However, in case of triac, the same device is conducting in both the directions and so the dv/dt is equal to the reapplied dv/dt rating which is less than the static dv/dt rating. Due to this, the triac may get turned on accidentally due to high dv/dt, even without the gate pulse applied to it. This possibility is more with the inductive load as the triac gets very little time to recover. This is the reason why the antiparallel SCRs are preferred over triacs for power control of inductive loads like motor loads. This problem is not faced with the resistive loads at the triac remains off for sufficient time as shown in figure 9.30(b). The triac is turned off due to reduction of its forward current below holding current value. This happens even before the supply voltage reversal.

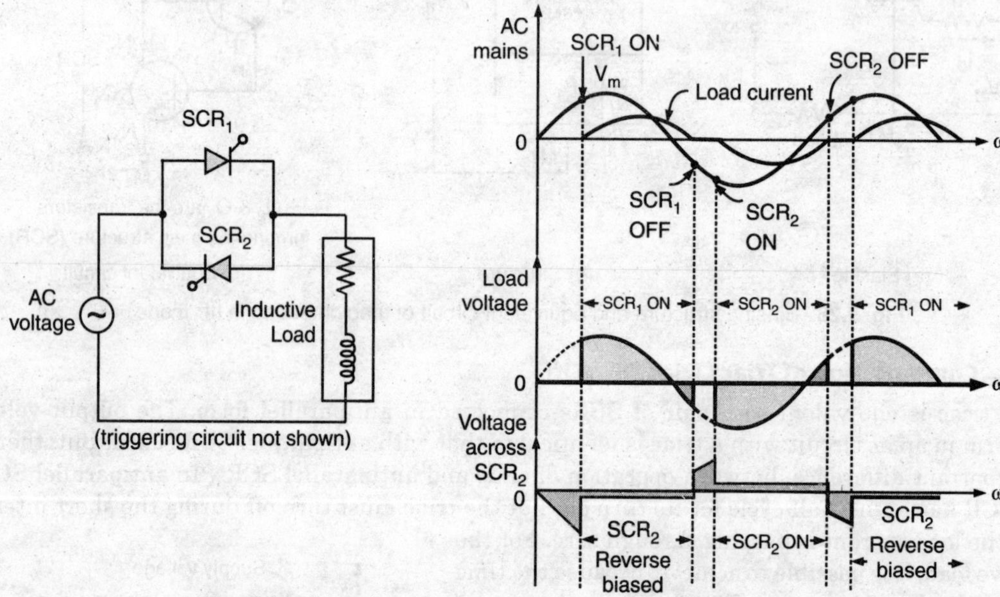

Fig. 9.30. Waveforms for inductive load with antiparallel SCRs

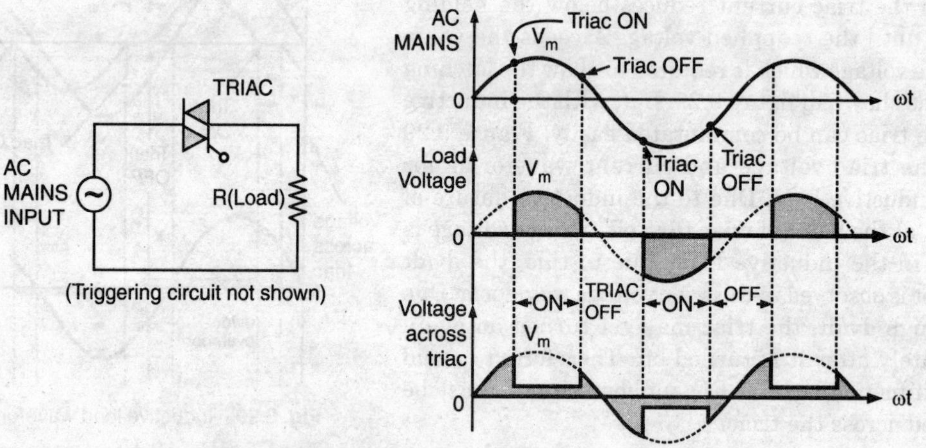

Fig. 9.31. Power control to resistive load using triac

9.11.4. Snubber for Triac

In order to avoid the accidental turn on of the triac due to high dv/dt, a RC snubber circuit is connected across the triac. The snubber capacitor reduces the rate of rise of voltage and the resistor will limit the discharge current of capacitor when the triac is turned on. The snubber for triac is identical to that of the SCR.

9.11.5. Salient Features of Triac

(i) It is a bi-directional device. So, we can control the power delivered to load in both the half cycles of ac supply.

(ii) It is equivalent to two SCRs connected back to back.
(iii) We can turn it on by using a positive as well as negative gate current.
(iv) It is more suitable for resistive loads.
(v) It is more economical than SCRs, since inside the same package, we get two SCRs connected back to back.
(vi) It can control power delivered to ac loads such as fan motor.
(vii) It is not necessary to use a protection diode across the Triac.
(viii) We can use a single heat sink.

9.11.6. Drawbacks of Triac

(i) We can not use it as a controlled rectifier.
(ii) Low dv/dt rating than SCR. So possibility of accidental turn on is higher than that of SCR.
(iii) Low di/dt rating.
(iv) Suitable for resistive loads only. Not suitable for controlling power to highly inductive loads.
(v) Its power rating is lower than that of SCR.
(vi) Triggering circuits require to be designed more carefully.

9.11.7. Applications of Triac

(i) As a static switch
(ii) In the AC voltage stabilizers
(iii) Fan speed regulator
(iv) Lamp dimmer
(v) Flasher circuit
(vi) Proximity detector
(vii) Temperature controller

9.12 DIAC

(U.P. Tech, Sem. Exam; 2008-09)

Diac is one of the most important devices used to trigger the Triac. It is sometimes called as a Triac without gate terminal.

9.12.1. Construction Details of Diac

The diac is basically a two terminal device. It is a parallel inverse combination of semiconductor layers that permits triggering in either direction. The diac is used extensively as a triggering device for the triac circuits. The arrangement of a semiconductor layers has been shown in figure 9.32(a). The two terminals accessible to the user are MT_1 and MT_2 (main terminals one and two)

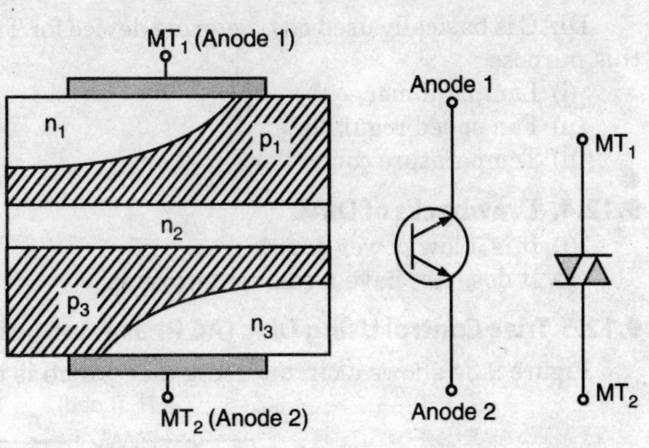

(a) Structure (b) Diac symbols

Fig. 9.32.

When MT_1 is positive with respect to MT_2 the conduction takes place through the structure formed by the layers $p_1 - n_2 - p_2 - n_3$ which is equivalent to an SCR. Similarly, when MT_2 is positive with respect to MT_1 the conduction takes place through the structure formed by the layers $p_2 - n_2 - p_1 - n_1$ which is another SCR. This reflects in the equivalent circuit of Diac as shown in figure 9.32 (b), which shows two SCRs back to back. Figure 9.32 shows the circuit symbols for the diac. The arrows indicate that the diac can conduct in both the directions. The other symbol of diac is identical to that of a triac without gate terminal.

9.12.2. I - V Characteristics of a Diac

The static characteristic of diac has been shown in figure 9.33. The diac resembles a bipolar transistor with no base connection. The device can be turned on by either positive or negative half cycle of an ac voltage. For the positive cycle of ac voltage, if the applied voltage is less than the forward breakover voltage, a very small current called as the leakage current, flows through the device. This current is produced due to the drift of electrons and the holes at the depletion region and is not sufficient to cause conduction. Hence the diac remains practically in the non-conducting mode called the blocking state. As soon as the applied voltage reaches the breakover voltage, the device starts conducting. The current through the device starts increasing and the voltage across it starts decreasing. This region is the conduction state. The breakover voltage for the diac remains unchanged as the gate terminal is absent here. The characteristics of the diac is very much similar to that of a triac.

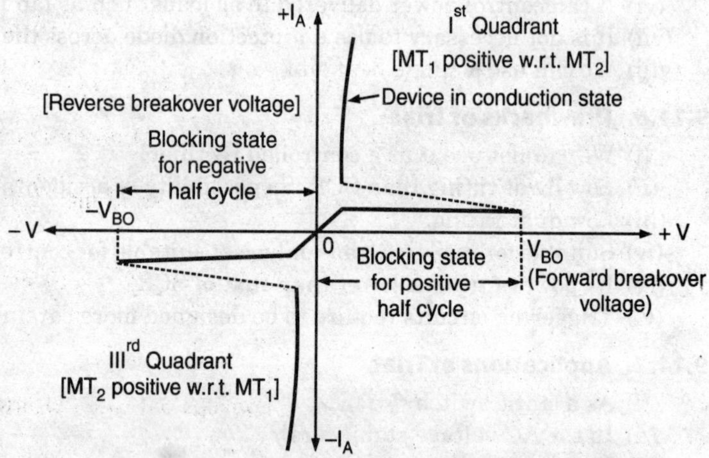

Fig. 9.33. Characteristic of diac.

For the positive half cycle, the characteristic is obtained in the first quadrant and similar characteristic for the negative half cycle is obtained in the third quadrant as shown in figure 9.33. The breakover voltage of a diac ranges between 28 volts to 42 volts. The typical turn on time ranges from 50 to 500 m sec. The turn off time is much longer (upto about 100 n sec). Diacs have a power handling capacity between 300 mW to 1 W.

9.12.3. Diac Applications

DIAC is basically used as triggering device for Triac. The following control systems are DIAC for this purpose :

(i) Lamp dimmer
(ii) Fan speed regulator
(iii) Temperature controller

9.12.4. Drawbacks of DIAC

(i) It is a low power device.
(ii) It does not have a control terminal.

9.12.5. Triac Control Using Diac (AC Phase Control)

Figure 9.34 shows a circuit using diac which is used to trigger (turn on) the triac.

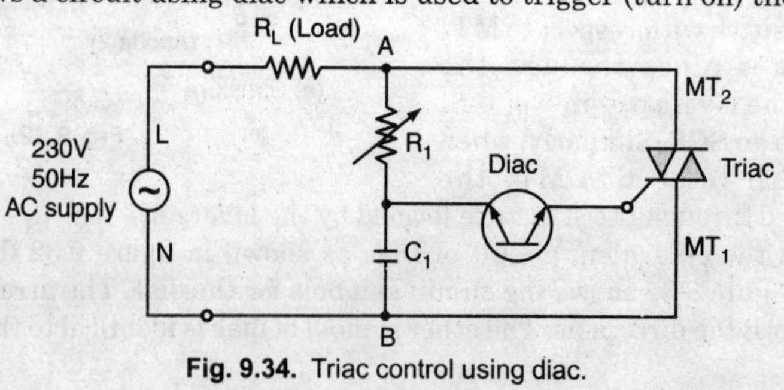

Fig. 9.34. Triac control using diac.

Working Operation in Positive Half Cycle

In the positive half cycle of the input ac supply voltage, the live point (L) is positive with respect to the neutral point (N). The charging current for the capacitor C_1 flows through R_1 as shown in figure 9.35(a). As soon as the voltage across C_1 reaches the breakover voltage of the diac, it is turned on to supply gate current for the triac and the triac will be turned on.

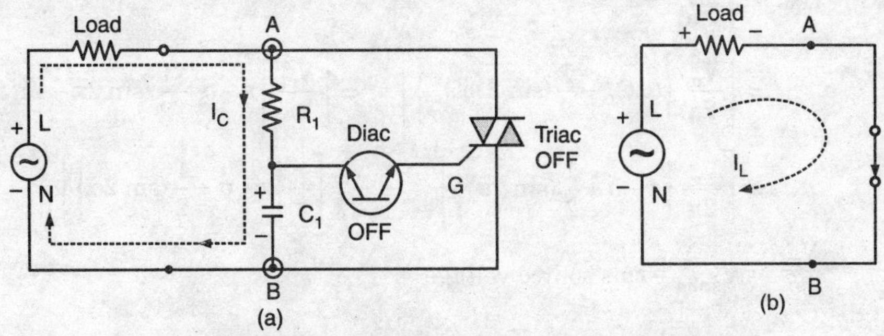

Fig. 9.35. Equivalent circuits for positive half cycle.

The conducting triac is equivalent to a closed switch. Hence, the R_1 C_1 and diac circuit is short circuited as shown in figure 9.35(b). The load voltage is equal to the instantaneous supply voltage. In the negative half cycle of the ac supply voltage the live point (L) is negative with respect to neutral (N). The triac and diac both are in the off state. The charging current for the capacitor C_1 flows through R_1 as shown in figure 9.36(a). Here, it may be noted that the voltage on C_1 is now negative. As soon as this voltage reaches the breakover voltage of the diac, it is turned on and supplies gate current to the triac. The triac is then turned on. The points A and B are short circuited through the triac as shown in figure 9.36(b). The load current reverses its direction and the voltage across the load will be negative equal to the instantaneous ac supply voltage.

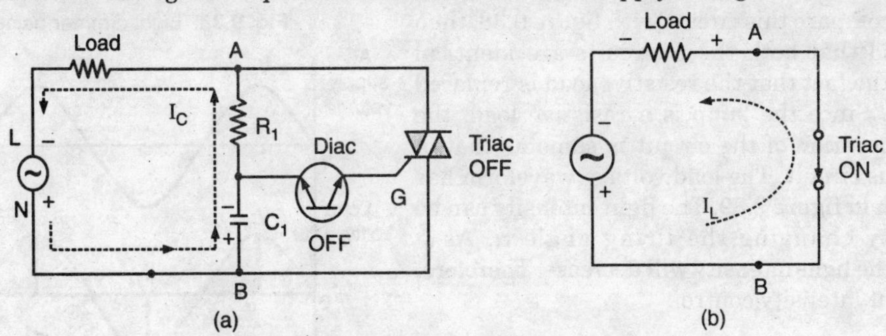

Fig. 9.36. Equivalent circuits for negative half cycle.

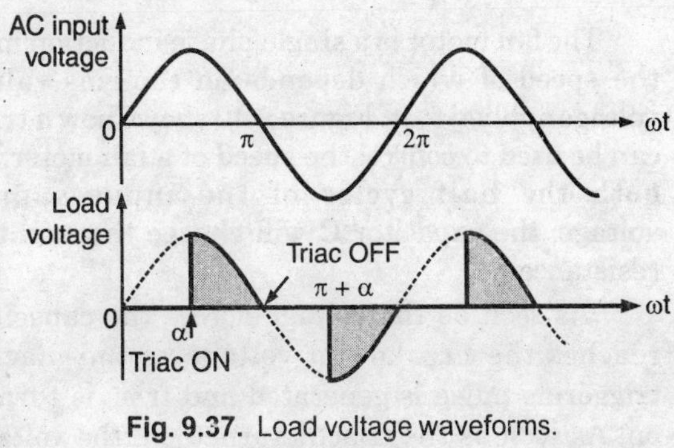

Fig. 9.37. Load voltage waveforms.

Thus, diac being a bi-directional device can turn on the triac in both the half cycles of input ac supply. The capacitor C_1 must be a non-polarized capacitor, being capable of charging to positive as well as negative voltages. The firing angles (α) i.e. the instant at which the triac is turned on in the positive as well as negative half cycles of the ac supply voltage can be controlled by making the resistance R_1 of figure 9.37, variable. This variable resistance would then decide the charging rate of capacitor C_1 and hence firing angle (α) in each half cycle. As the control circuit operates directly on AC supply, it is automatically synchronized with the supply. The load voltage waveforms for a resistive load are as shown in figure 9.39.

Expression for V_{Lrms}

Let us obtain the expression for the rms value of load voltage waveform. We consider the load voltage waveform of figure 9.37.

We have
$$V_{Lrms} = \left[\frac{1}{\pi}\int_{\alpha}^{\pi} V_m^2 \sin^2 \omega t \, d\,\omega t\right]^{-1/2} = \left[\frac{V_m^2}{\pi}\int_{\alpha}^{\pi} \frac{1-\cos 2\,\omega t}{2} d\,\omega t\right]^{1/2}$$

$$= \left[\frac{V_m^2}{2\pi}\left\{(\omega t)_{\alpha}^{\pi} - \frac{1}{2}(\sin 2\,\omega t)_{\alpha}^{\pi}\right\}\right]^{1/2} = \left[\frac{V_m^2}{2\pi}\left\{\pi - \alpha - \frac{1}{2}(\sin 2\pi - \sin 2\alpha)\right\}\right]^{1/2}$$

$$= \left[\frac{V_m^2}{2\pi}\left\{\pi - \alpha + \frac{1}{2}(\sin 2\alpha)\right\}\right]^{1/2} = \frac{V_m}{\sqrt{2}}\left[\frac{1}{\pi}\left\{\pi - \alpha + \frac{1}{2}(\sin 2\alpha\right\}\right]^{1/2}$$

But,
$$\frac{V_m}{\sqrt{2}} = V_{srms} = \text{rms source voltage}$$

Therefore,
$$V_{Lrms} = V_{srms}\left[\frac{1}{\pi}\left\{\pi - \alpha + \frac{1}{2}(\sin 2\alpha\right\}\right]^{1/2} \quad ...(9.20)$$

By controlling the firing angle, it is possible to change the value of RMS load voltage. We can use this fact to control the intensity of lamp or speed of a fan.

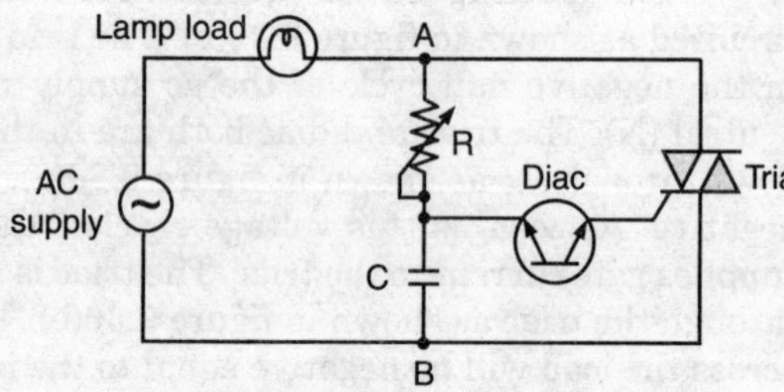

Fig. 9.38. Light dimmer using triac

9.12.6. Light Dimmer Using Triac

The circuit diagram for the light dimmer using triac has been shown in figure 9.38.

If we compare this circuit with figure 9.38, then we can find that both these circuits are identical except for the fact that the resistive load is replaced by a lamp. Since the lamp is a resistive load, the working operation of the circuit is same as that of the previous circuit. The load voltage waveform has been shown in figure 9.39. The light intensity can be changed by changing the firing angle α. As α increases, the light intensity will decrease. Therefore, R acts as an intensity control.

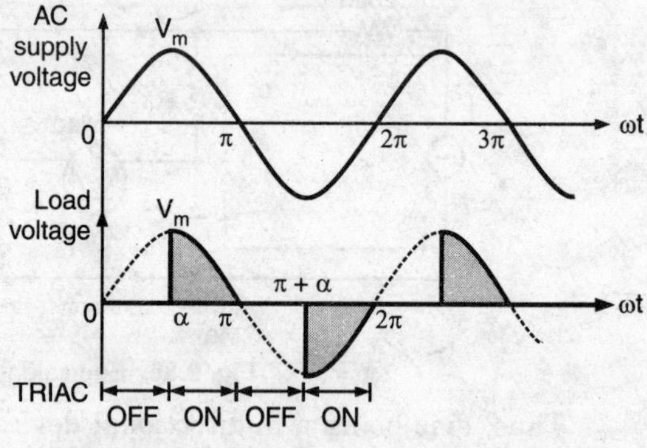

Fig. 9.39. Waveforms for the light dimmer.

9.12.7. Fan Regulator Using Triac

The fan motor is a single phase induction motor, the speed of which depends on the rms value of voltage applied to it. Figure 9.40 shows how a triac can be used to control the speed of a fan motor. In both the half cycles of the input supply voltage, the capacitor C will charge through the resistance R.

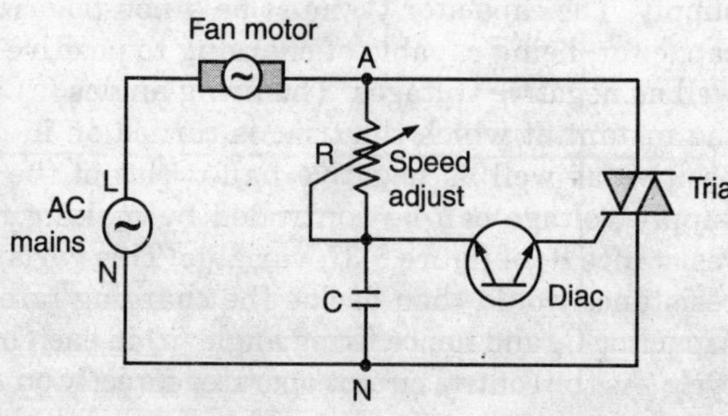

Fig. 9.40. Fan regulator using triac.

As soon as the voltage across the capacitor reaches the breakdown voltage of the diac, a triggering pulse is generated and triac is turned on. As soon as the triac is turned on, the voltage between points A and N will drop down to on state voltage drop of the triac. The voltage on C is negligibly small. The capacitor can charge again after the triac is turned off. The charging rate of the capacitor C is decided by the variable resistance R. If value of R is low, the capacitor will charge to the diac breakdown voltage in less time, that is the firing angle will be small, the rms voltage applied to the motor will be higher and the fan speed increases. With increase in the value of R, the capacitor charges slowly and thus increasing the firing angle (α) of the triac and in turn reducing the speed. In this way, the fan speed can be regulated by controlling the firing angle of the triac in both the half cycles.

9.12.8. Performance Comparison of SCR and Triac **(Important)**

Table 9.1. Performance Comparison of SCR and Triac

S. No.	SCR	Triac
1.	SCR is a unidirectional device.	Triac is bi-directional device.
2.	The gate current can be only positive.	The gate current can be positive or negative.
3.	SCR can operate in only one quadrant of V-I characteristics.	Triac can operate in two quadrants of V-I characteristics.
4.	Antiparallel SCRs are preferred over-triac for power control of inductive loads. They can be successfully used for resistive loads as well.	Triac is not suitable for power control of inductive loads but suitable for the resistive loads.
5.	UJT is used for triggering SCR.	Diac is used for triggering triac.
6.	For both SCRs and triac the gate terminal is used to turn on the device however it looses control as soon as the device is turned on.	
7.	RC snubber circuits are necessary for both these devices to avoid the accidental firing due high dv/dt.	
8.	Symbol 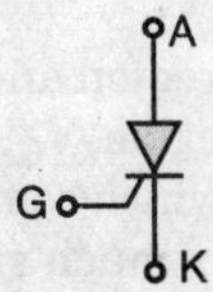	Symbol 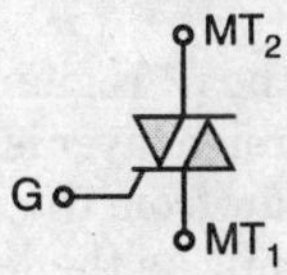

9.12.9. Performance Comparison of Diac and Triac

Table 9.2. Performance Comparison of Diac and Triac

S. No.	Diac	Triac
1.	Symbol : 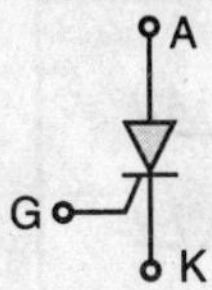	Symbol : 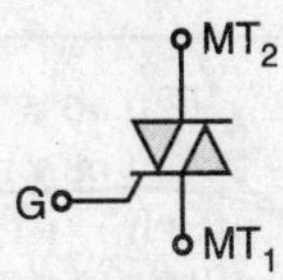
2.	It is a two terminal device.	It is a three terminal device.
3.	Breakover voltage cannot be controlled.	Breakover voltage can be controlled by adjusting the gate current.
4.	Diac is a low power device.	Triac is a high power device.
5.	It is used as a triggering device for the triac.	It is used in applications like fan control, light dimmer etc.

9.13. POWER MOSFET

A power MOSFET is a voltage-controlled device and needs only a small input current. In this circuit, the control signal is applied to a metal gate electrode that is separated from the semiconductor surface by an intervening insulator, namely, silicon dioxide (SiO_2). The control signal needed is essentially a bias voltage with no significant steady-state gate current flow in either the on-state or the off-state.

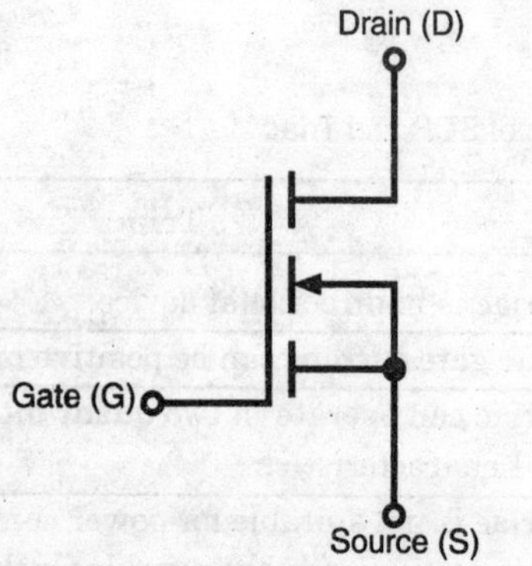

(a) Circuit Symbol of Power MOSFET

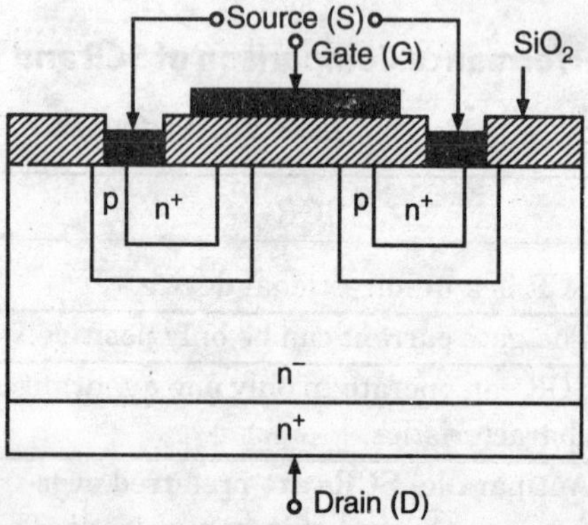

(b) Cross-sectional View of Double-diffused MOS (DMOS) Transistor

Fig. 9.41.

In a power MOSFET the current flow is vertical and the whole area of the bottom surface collects the majority carriers. This means that the current conduction area is greatly increased while the conduction length is very much reduced and thus much higher average current can be conducted. The circuit symbol of power MOSFET is shown in figure 9.41 (a). A cross-sectional view of a double-diffused MOS (DMOS) transistor is shown in figure 9.41 (b). This is most commonly used power MOSFET.

In figure 9.41 (b), the n^+ region of source shifts on the top of p-type region and so channel width is much wider. An inversion layer is formed and enhanced in the p-type region, $V_{GS} > V_{TH}$ (threshold). The majority carrier flow from the inversion layer in vertical direction which constitutes the length of the channel ($\simeq$ 1 μ m). As the length to width ratio of the channel is much smaller in a DMOS transistor, its transconductance is about two or three orders of magnitude larger than that for a lateral device. Further, the provision of a high resistivity n^- epitaxial layer above n^+ layer in contact with drain gives very much values for drain to source breakdown voltage.

The characteristics of DMOS transistor are similar to those of an ordinary MOSFET except for higher current and voltage levels as shown in figure 9.42. (a) and 9.42 (b), respectively.

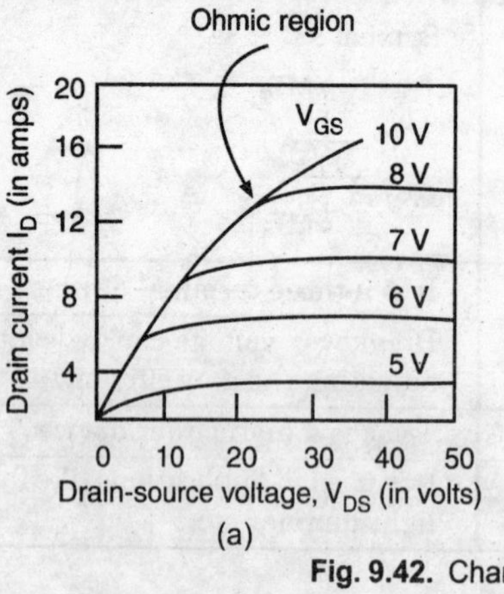

(a)

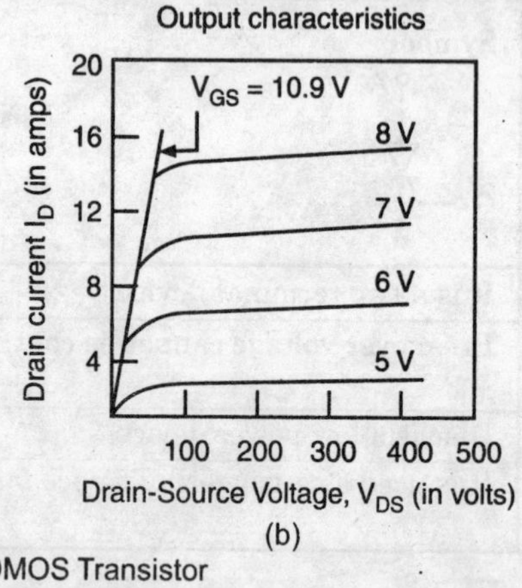

(b)

Fig. 9.42. Characteristics of DMOS Transistor

The power MOSFET is a unipolar device. The current conduction occurs through transport of majority carriers in the drift region without the presence of minority carrier injection needed for BJT operation. The power MOSFET having operating frequencies are well above 100 kHz. The switching time of power MOSFET is in order of 50 n-sec to 100 n-sec and generate many kilowatts of power at frequencies upto 500 kHz. The power MOSFET greatly simplifies the gate drive circuitry and reduces the cost of the power electronics.

9.14. IMPORTANT CHARACTERISTICS OF A POWER MOSFET

Important characteristics of a power MOSFET are mentioned below :

(i) Transfer characteristics I_D versus V_{GS} is almost linear beyond a small value of current (compared to rated drain current).

(ii) Extremely low noise figure is used.

(iii) The switching is very fast because there is no minority carrier storage.

(iv) The power MOSFETs are available with VTH ranging for 0.8 V to 2V, so as to be compatible with TTL logic.

(v) The input impedance is very high and so extremely small input power is needed.

Therefore, power MOSFET can be directly driven from CMOS logic gates. The resultant power gain is extremely high.

(vi) Power MOSFETs can be used for high frequency circuits because of low capacitance between the gate and drain electrodes.

9.15. GATE TURN-OFF (GTO) THYRISTOR

(U.P. Tech, Sem. Exam; 2008-09)

The GTO is a three terminal device with anode, cathode and gate terminals. The different circuit symbols of GTO are shown in figure 9.43 (a) and 9.43 (b), respectively. The two way arrow convention on the gate lead distinguishes the GTO from the conventional thyristor. Figure 9.44 shows the two transistor analogy of GTO.

Like the thyristor, the GTO can be turned-on by a short-duration gate current pulse and once in the on-state, the GTO may stay on without any further gate current. However, unlike the thyristor, the GTO can be turned-off by applying a negative gate-cathode voltage, therefore, causing a sufficiently large negative gate current to flow.

The reduction in gain of the p-n-p transistor can be achieved by the diffusion of gold or other heavy metal to reduce carrier lifetime. Therefore, the gold-doped GTO retains its reverse blocking capability but has a high on-state voltage drop. The large GTOs also have an interdigitated gate-cathode structure in which the cathode emitter consists of many parallel connected n-type fingers diffused into the p-type gate region.

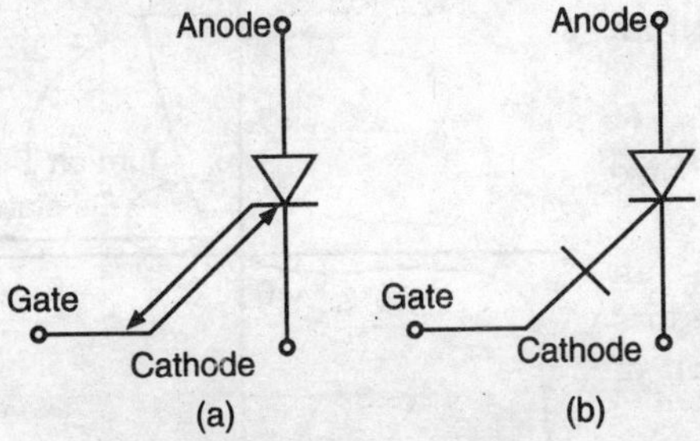

Fig. 9.43. Circuit Symbols of GTO

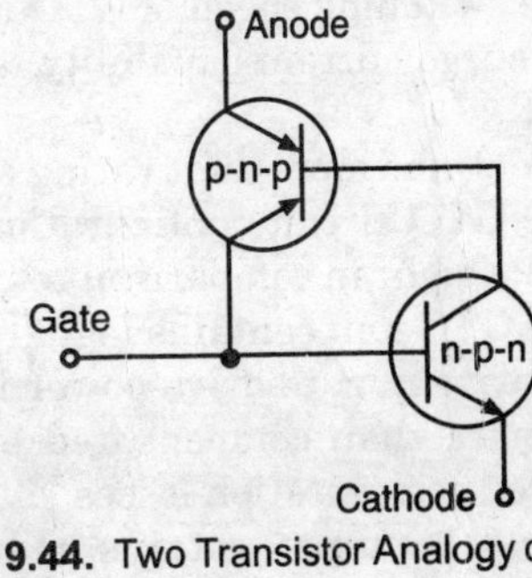

Fig. 9.44. Two Transistor Analogy of GTO

The GTO is a controllable switch in the same category as MOSFETs and BJTs, it turn-off switching transient is different. The on-state voltage of a GTO is slightly higher than those of thyristors. The GTO switching speeds are in the range of a few microseconds to 25 μ-sec. Because of their capability to handle large voltages and large currents, the GTO is used when a switch is required for high voltages and large currents in a switching frequency range of a few hundred hertz to 10 kHz. The GTOs are available with symmetric or a symmetric voltage blocking capabilities. The use of asymmetrical GTOs needs the connection of a diode in series with each GTO to gain the reverse blocking capability, whereas symmetrical GTOs have the ability to block a reverse voltage. The n-base is doped with a heavy metal to reduce the turn-off time in symmetrical GTOs. The asymmetrical GTOs offer more stable temperature characteristics and lower on-state voltage compared to symmetrical GTOs.

9.16. CHARACTERISTICS OF GTO

In order to realize the gate turn-off capability of the GTO, the basic design trade-offs are necessary. Because of low internal regeneration, there is an increase in latching and holding current levels, Also, there is an increase in the on-state voltage drop and the associated power loss.

In a GTO with a shorted anode emitter, the reverse voltage rating is appreciably less compared to the forward blocking voltage. However, the GTO retains many of the merits of the thyristor and has a faster switching speed.

Generally, the GTO has the high blocking voltage and large current capability that are characteristics of thyristor devices. A wide range of GTO devices is now available and a single device can be obtained with a present power capability of 4.5 k V and 3500 A.

GTO inverter has a number of merits over the thyristor inverter. Particularly, the GTO circuit has about 60% of the size and weight of the thyristor unit and has a higher efficiency because the increase in gate-drive power and on-state power loss is more compared to compensated by the elimination of forced communication losses.

The current-voltage (I-V) characteristics of a GTO in the forward direction is identical to that of thyristor. However, in the reverse direction, the GTO has virtually non-blocking capacity because of the anode short structure. The I-V characteristics of GTO is shown in figure 9.45.

For large power GTOs, the latching current is several amperes (here 2A) as compared to 100 mA to 500 mA for thyristors of the same rating. If gate current is not able to turn-on the GTO, it behaves like a high voltage, low gain transistor with considerable anode current.

9.17. ADVANTAGES AND DISADVANTAGES OF GTO OVER THE CONVENTIONAL THYRISTOR (SCR)

Advantages

Advantages of GTO over the conventional thyristor (SCR) are given as under :

(i) The switching speed of GTO is faster.

(ii) Its surge current capability is comparable with an SCR.

(iii) It contains more di/dt rating at turn-on.

(iv) The GTO circuit configuration contains lower size and weight in comparison to SCR circuit unit.

(v) The GTO unit contains higher efficiency because an increase in gate-drive power loss and on-state loss is more than compensated by the elimination of forced commutation losses.

(vi) The GTO unit contains reduced acoustical and electromagnetic noise because of elimination of commutation chokes.

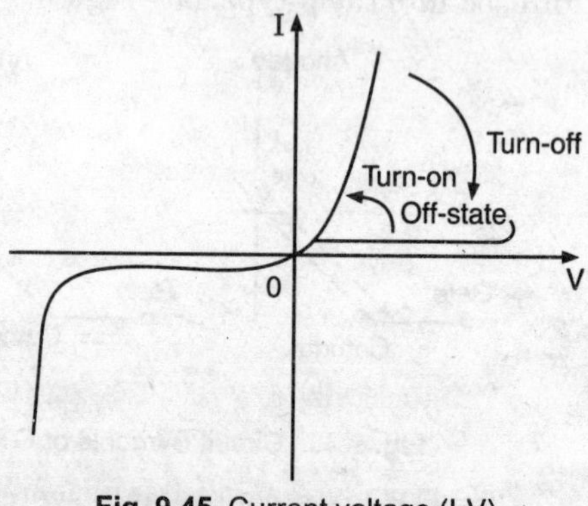

Fig. 9.45. Current voltage (I-V) Characteristics of GTO

Disadvantages

Disadvantages of GTO over the conventional thyristor (SCR) are given as under :

(i) GTO has more magnitude of the latching and holding currents.

(ii) The GTO has more on-state voltage drop and the associated loss.

(iii) Because of the multicathode structure of GTO, triggering gate current is higher than that needed for a conventional SCR.

(iv) It has more gate drive circuit losses.

(v) Its reverse-voltage blocking capability is less than its forward- voltage blocking capability. But, in the concern of inverter circuits, this is not a disadvantage.

9.18. INSULATED GATE BIPOLAR TRANSISTOR (IGBT)* (U.P. Tech, Sem. Exam; 2008-09, 2011-12)

9.18.1. Introduction

The power BJT has advantage of low on state power dissipation, but it cannot be switched at faster rates due to longer turn-off time. The power MOSFET can be switched at much higher frequency but has a drawback of higher on state power loss. These observations have led to attempts to combine BJTs and MOSFETs monolithically on the same silicon wafer to develop a new device that will have the best qualities of both, BJT and MOSFET. The new device is known as insulated gate bipolar transistor (IGBT) which is finding use in many applications. Other names to this device are GEMFET, COMFET (conductivity modulated field effect transistor), IGT (insulated gate transistor) and bipolar mode MOSFET or bipolar MOS transistor.

9.18.2. Salient Features of IGBT

The important salient features of IGBT may be listed as under

(i) Low on state voltage drop.

(ii) Low on state power loss.

(iii) It has a higher switching frequency than that of a power BJT.

(iv) IGBT has the best qualities of BJT and MOSFET.

9.19 BASIC STRUCTURE OF IGBT

The vertically oriented structure of IGBT has been shown in figure 9.46(a). Like all other devices, IGBT also uses the vertically oriented structure in order to maximize the area available for the current flow. This will reduce the resistance offered to the current flow and hence the on state power loss taking place in the device. The IGBT also uses highly inter digitated gate-source structure in order to reduce the possibility of source/emitter current crowding. The doping levels used in different layers of IGBT are similar to those used in the comparable layers of the vertical MOSFET structures except for the body region. The main difference in the structure of IGBT as compared to that of a MOSFET is the existence of p^+ layer that forms the drain of the IGBT. This device also uses the n^- type drain drift layer which improves its breakdown voltage capacity. This is same as that in case of power MOSFETs. It is also possible to make a p channel IGBT by changing the doping type in each of the layers of the device. The n^+ buffer layer is not essential for the operation of the IGBT and some IGBTs are made without it*.

This layer improves the operation of IGBT in two important aspects :

(i) It reduces the on state voltage drop across the device and

(ii) It shortens the turn-off time. But, the drawback is that presence of the buffer layer greatly reduces the reverse blocking capacity of the IGBT.

* Write short note on insulated gate bipolar transistor (IGBT). ((U.P. Tech., Sem. Exam., 2008-09)

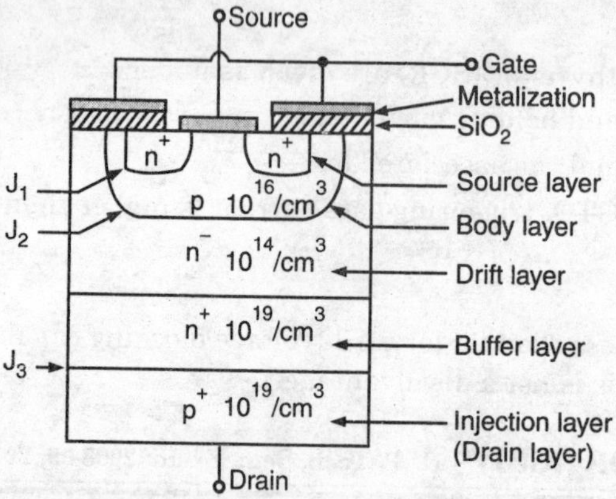

(a) Vertically oriented structure of IGBT with doping densities

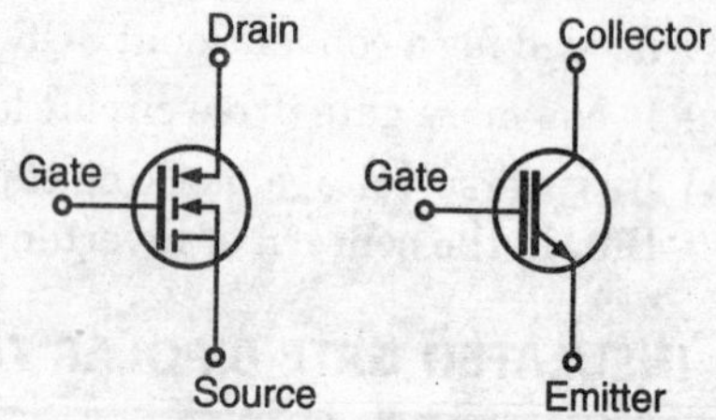

(b) and (c) Circuit symbols for the n-channel IGBT

Fig. 9.46.

9.19.1. Circuit Symbol for IGBT

The circuit symbols for an n channel IGBT have been shown in figure 9.46(b) and (c). As observed from figure 9.46(b) and (c), IGBT is a three terminal device, gate being the control terminals. The directions of the arrowheads will reverse in a p channel IGBT. The symbol shows in figure 9.46(b) is essentially the same as that used for the n channel power MOSFET, but with the addition of an arrowhead in the drain lead pointing into the body of the device indicating an injection contact. The symbol shown in figure 9.46(c) is used if IGBT is considered to be basically a BJT with MOSFET gate input. This device has a collector and emitter rather than a drain and source. This symbol indicates that the IGBT has output characteristics similar to power BJT and input characteristics similar to the power MOSFET.

9.20. I-V CHARACTERISTICS

The I-V characteristics of IGBT has been shown in figure 9.47. In the forward direction, they are similar to those of the logic level bipolar transistor, the only difference here is that the controlling parameter is the gate to source voltage and the parameter being controlled is the drain current. Thus, IGBT is a voltage controlled device with an insulated gate. The IGBT possesses all the advantages of MOSFET due to the insulated gate. It also has all the advantages of the BJT due to bipolar conduction.

As observed from figure 9.47, the drain current (or the collector current) increases with increase in the voltage between gate and source (V_{GS}). Also, it may be noted that the gate to source voltage V_{GS} is positive. B V_{DSS} is the forward breakdown voltage. This the voltage at which the avalanche breakdown takes place. At this point the voltage across the device and current through it both are high. Therefore, the power dissipated in the device will be very large and will damage it. The device must be therefore operated below this voltage.

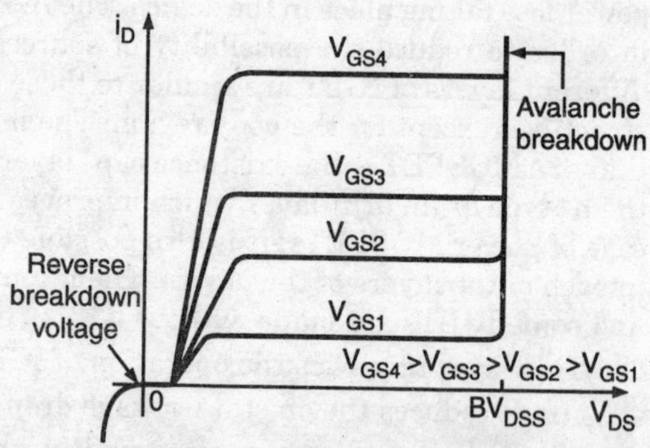

Fig. 9.47. IGBT I-V characteristics

* Sometimes termed symmetric IGBTs whereas those with the buffer layer are termed asymmetric IGBTs.

9.21 TYPES OF IGBTS

Depending on whether the n^+ buffer layer has been included in the structure of IGBT or not, they can be classified into following two categories :

(i) Non-punch through IGBT (n^+ layer is absent).

(ii) Punch through IGBT (n^+ layer present).

The non-punch through type IGBTs have a symmetrical blocking capacity. This means that without breaking down they can block high positive and negative voltages (V_{DS}) successfully. The punch through IGBTs have an asymmetrical blocking capacity. This means that they can block positive V_{DS} successively but cannot block negative voltage successfully.

Effect of n^+ Buffer Layer

Depending on whether the n^+ buffer layer has been included in the structure of IGBT or not, they can be classified into following two categories :

(i) Non punch through IGBT (n^+ layer is absent)

(ii) Punch through IGBT (n^+ layer is present).

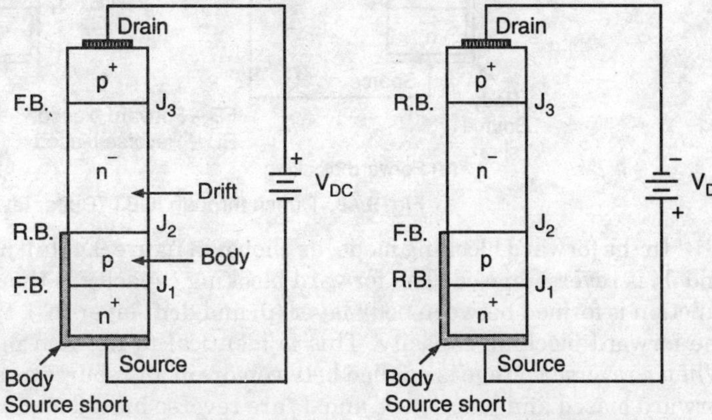

Fig. 9.48. Non punch through IGBT (Buffer layer absent)

9.21.1. Non Punch Through IGBT

The structure of non punch through IGBT has been shown in figure 9.48. The n^+ buffer layer is absent in the structure.

In the forward blocking mode, as shown in figure 9.48(a), the drain (or collector) is positive with respect to source (emitter). Therefore the junction J_1 and J_3 are forward biased and hence are useless in deciding the blocking capability of the device. The junction J_2 is reverse biased and hence decides the forward blocking capability of IGBT. The junction J_2 is formed between moderately doped p (body) layer and lightly doped n^- (drift) layer. The forward blocking capacity is decided by the thickness on n^- layer. When a reverse voltage is applied between drain and source, as shown in figure 9.48(b), junction J_2 is forward biased and junctions J_1 and J_3 are reverse biased and the reverse blocking capacity is decided by them. Since the source later (n^+) is heavily doped and the body layer (P layer) is thin and moderately doped, the junction J_1 does not have capacity to block voltage. However the junction J_3 will block all the reverse voltage due to the presence of n^- drift layer on its one side. Since the drift layer is designed to block large voltages, the reverse blocking voltage of the non punch through structure will be comparable to the forward blocking voltage. In this way, the non punch through IGBT has a symmetrical blocking capacity and they can be used in rectifier circuits.

9.21.2. Punch Through IGBT

The structure of punch through IGBT has been shown in figure 9.49. The n^+ buffer layer has been included between the p^+ drain injection layer and the n^- drift layer.

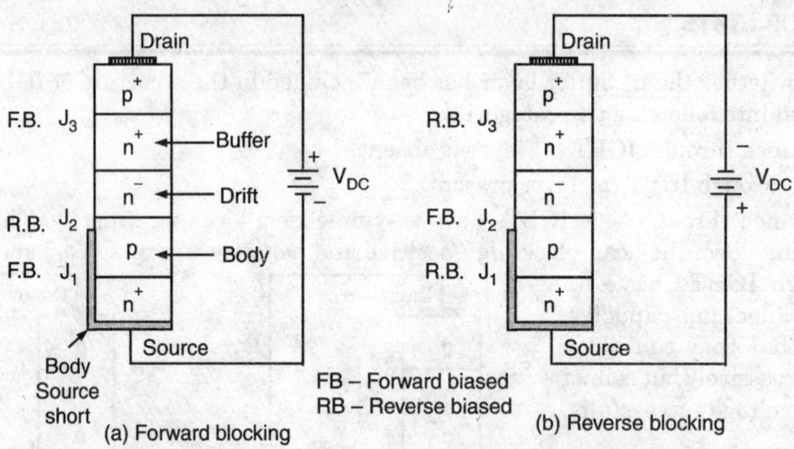

Fig. 9.49. Punch through IGBT (Buffer layer included)

In the forward blocking mode, as shown in figure 9.49(a), junctions J_1 and J_3 are forward biased and J_2 is reverse biased. The forward blocking capacity is therefore decided by junction J_2. As this junction is formed between body layer (p) and drift layer (n^-), the thickness of drift layer will decide the forward blocking capacity. This is identical to the non punch through structure of Fig. 9.49. When a reverse voltage is applied between drain and source as shown in figure 9.49(b), junction J_2 is forward biased and junction J_1 and J_3 are reverse biased and will have to block the reverse voltage. But J_1 cannot block large voltages as explained earlier and now due to the presence of n^+ buffer layer J_3 also cannot block a large reverse voltage. Therefore, the device will breakdown at low reverse voltages. The punch through structure will thus have an asymmetric blocking capacity.

The advantage of n^+ layer is faster turn off the IGBT. The punch through structure of IGBT is more popular.

9.22 PRINCIPLE OF OPERATION OF IGBT

The principle of operation of IGBT is similar to that of a MOSFET. The operation can be divided into two parts :

(i) Creation of the inversion layer and

(ii) Conductivity modulation

(i) Creation of Inversion Layer

The operation of IGBT is based on the principle of creation of inversion layer for the power MOSFET. In IGBTs also, when the gate to source voltage V_{GS} is greater than V_{GS} (threshold), n type inversion layer is created beneath the SiO_2 (oxide) layer as shown in figure 9.50.

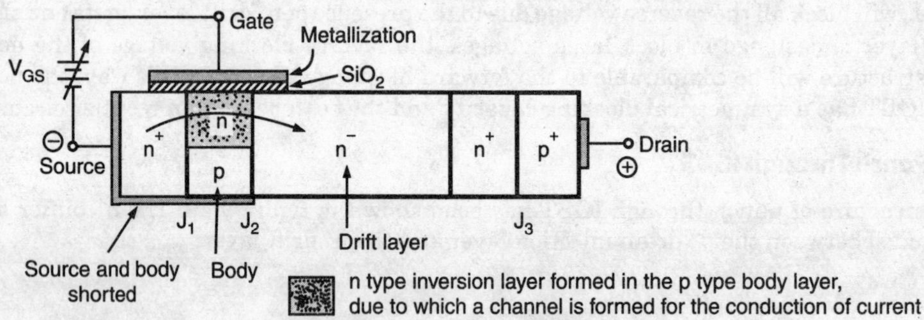

Fig. 9.50. Creation of an inversion layer.

Due to the formation of n type of layer in the p type body layer, a channel is formed. ($n^+ n n^-$) which helps to establish the current as shown in figure 9.51. The only difference between the MOSFET and IGBT is that there is no conductivity modulation of drift layer in MOSFET. Therefore, the on state resistance $R_{DS(on)}$ and hence the on state power loss is very high in MOSFET. In IGBT, however the conductivity modulation takes place as explained earlier.

(ii) Conductivity Modulation

In IGBT, the conductivity modulation of the n^- drift layer takes place. The effect of conductivity modulation is reduction in the on state resistance and hence the on state power loss. Therefore, the on state losses in IGBT are less than that in MOSFET. The conductivity modulation in the n^- drift layer can be explained with the help of figure 9.51. Due to the application of forward voltage between drain (collector) and source (emitter) the junction J_3 is forward biased. Due to the creation of inversion layer, electrons from the source are injected into the n^- drift layer via the $n^+ p n^-$ channel. (Refer to Figure 9.50). Since the junction J_3 is already forward biased, it will inject holes in the n^+ buffer layer. The electrons injected in the n^- drift layer create a space charge which will attract holes from the n^+ buffer layer which were injected by the p^+ layer. In this way, double injection takes place into the n^- drift region from both sides as shown in figure 9.51. This increases the conductivity of the drift region and reduces the resistance to its minimum. In this way, the conductivity modulation reduces the on state voltage across the IGBT.

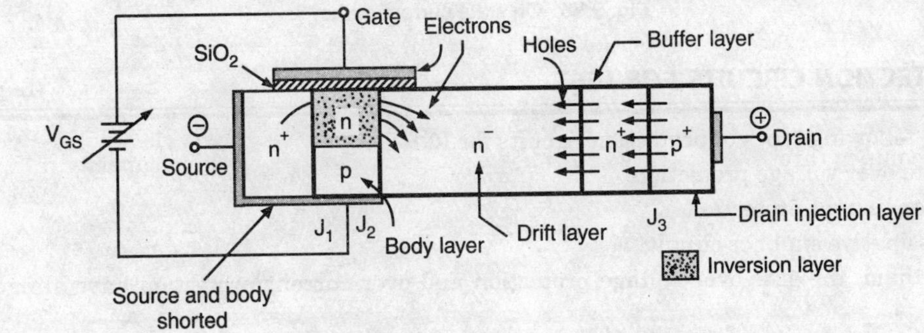

Fig. 9.51. Conductivity modulation in IGBT

9.23. EQUIVALENT CIRCUIT OF IGBT

The complete equivalent circuit of the IGBT has been shown in figure 9.52. Since the IGBT has a pnpn structure, in the equivalent circuit a parasitic thyristor will appear. This parasitic thyristor has been represented by the two transistors Q_1 and Q_2 in the equivalent circuit. The main collector (drain) current is carried by the pnp transistor Q_1. The input side of the IGBT is like MOSFET so the MOSFET has been included in the equivalent circuit. The two resistances that appear in the equivalent circuit of IGBT are the drift region resistance and the body spreading resistance R_B. The body spreading resistance R_B is due to the Lateral component of current flowing through the IGBT structure as shown in figure 9.53. The drift region resistance is considerably large (and has to be included in the equivalent circuit) as the drift layer is very lightly doped and therefore has low conductivity.

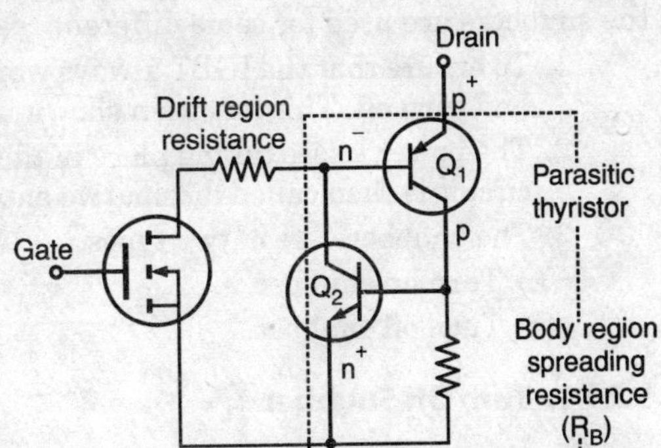

Fig. 9.52. Equivalent circuit of IGBT.

As shown in figure 9.53 a part of the total current crosses the p layer (body) laterally to reach the source. The resistance offered by the body layer to this component of current is considerably larger due to longer path it has to travel through the body layer.

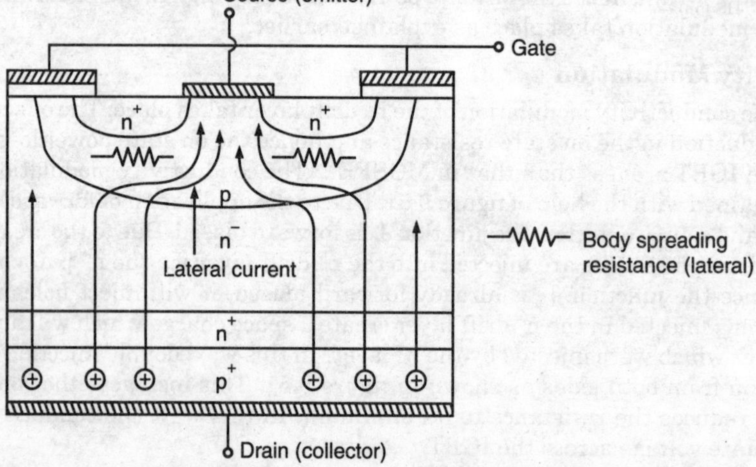

Fig. 9.53. On state current paths.

9.24 PROTECTION CIRCUITS FOR IGBT

(Important)

We can use following types of protection circuits for IGBT :

(i) Gate over voltage protection.

(ii) Over current protection.

(iii) Dissipative snubber circuit.

Out of them, the gate over voltage protection and over current protection have already been discussed.

9.24.1. Snubber Circuits for IGBT

The snubber circuits are used for the protection of almost all the power devices. In the thyristors the snubber is used to avoid the false triggering due to high dv/dt. But, in the power BJT and IGBT, the snubbers are used for some different reasons. They are :

1. To ensure that the IGBT always works within the safe operating area at the time of turn on and turn off. This has been shown in figure 9.54.
2. The power losses taking place in the IGBT are diverted to the snubber circuit. The snubber circuit is then called dissipative snubber.
3. The snubbers are of two types :
   (i) Turn on snubber
   (ii) Turn off snubber.

9.24.2. Turn Off Snubber

Figure 9.54 shows a typical turn off snubber. The snubber circuits will reduce the power losses in the IGBT during switching (the total switching loss does not change but a part of it is absorbed by the snubber). This will protect the device from the switching stresses of high voltage and current.

1. Working Operation

The snubber circuit provides another path for the load current during the transistor turn off. As the IGBT is turned off voltage across it starts increasing which will forward bias diode D and connect the snubber capacitor C across the IGBT. The capacitor begins to charge. As this capacitor appears directly across the IGBT, it will force the voltage across it to rise slowly as shown in figure 9.54. Thus, the voltage transition from low to high is delayed. The size of C determines the rate of voltage rise across the transistor.

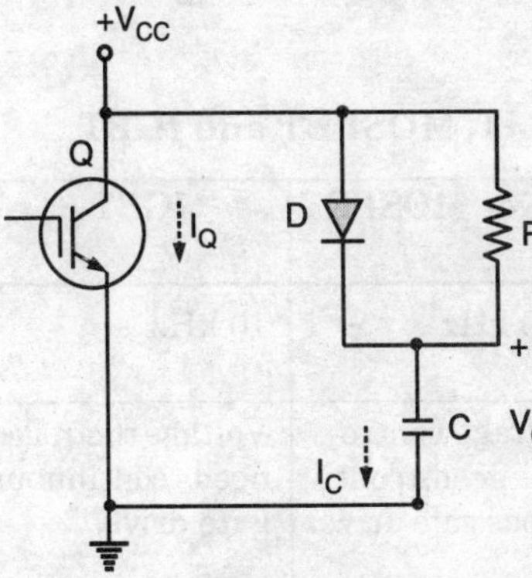

Fig. 9.54.

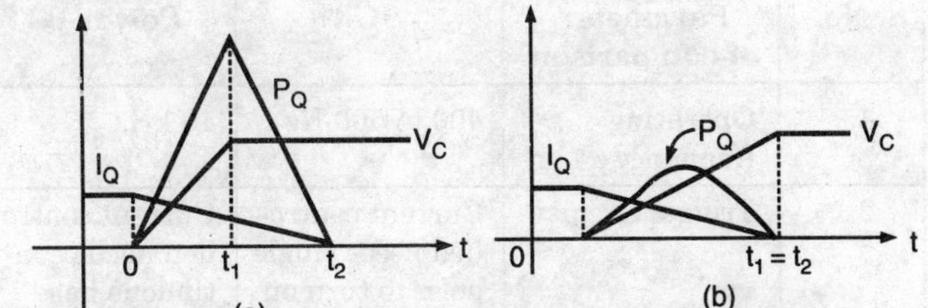

(a) (b)

Fig. 9.55. Voltage current and power waveforms at the time of transistor turn off.

2. Effects of the value of C

As seen from the figure 9.55(a) a small capacitor results in voltage reached V_{CC} before the IGBT current reaches zero, whereas shown in figure 9.55(b) a large C results in longer times for the voltage to reach V_{CC}. Note that the energy absorbed by the IGBT (area under the instantaneous power curve) during turn off reduces with increase in the value of C.

3. Selection of R

When the IGBT turns on the charged capacitor will discharge through it. The resistance R is connected to limit the discharge current below a safe value. But the R should not be too large or the capacitor will not discharge before the next time the IGBT turns off. Typically

$$t_{ON} > 5\ RC \qquad \text{...(9.21)}$$

or

$$R < t_{ON}/5\ C \qquad \text{...(9.22)}$$

The energy stored by the capacitor is,

$$W = \frac{1}{2} CV_{CC}^2 \qquad \text{...(9.23)}$$

This energy is transformed to the resistance *R* during the on time of the IGBT. The total power for the IGBT turn off is the sum of IGBT and snubber powers. Equation (9.23) shows that power absorbed by the snubber capacitor increases with increase in the value of C and therefore large C reduces the power loss in the IGBT, but at the expense of power loss in the snubber resistor. A snubber can alter the trajectory of SOA and prevent the operation outside the FBSOA and RBSOA. This is as shown in figure 9.56.

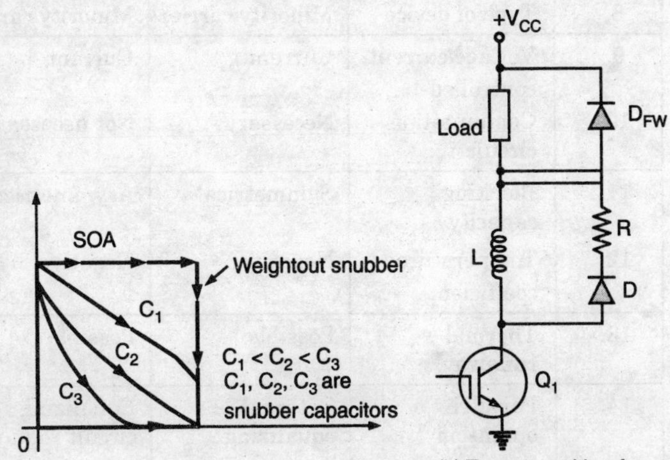

(a) Effect of snubber

(b) Turn on snubber for power transistor

Fig. 9.56.

9.24.3. Turn on Snubber

The turn on snubber protects the device from simultaneous high voltage and current during the turn on. The turn on snubber will reduce the power loss at the time of turn on. The turn on snubber is as shown in figure 9.56. If the turn off snubber is used along with the turn on snubber then the resistor and diode shown in figure 9.56 need not be used. The energy stored by the turn on snubber inductor can be transferred to the turn off snubber without R and D.

9.24.4. Performance Comparison of SCR, BJT, MOSFET and IGBT

Table 9.3. Performance comparison between the SCR, BJT, MOSFET and IGBT

S. No.	Parameter of comparison	SCR	Power BJT	Power MOSFET	IGBT
1.	Operating frequency	400 to 500 Hz	10 kHz	100 kHz	10 kHz
2.	Trigger circuit	Current controlled needs single pulse to turn on.	Current controlled needs continuous base drive.	Voltage controlled needs continuous gate drive.	Voltage controlled needs continuous gate drive.
3.	On state voltage drop.	< 2 volts.	< 2 volts.	4-5 volts.	3 volts.
4.	Snubbers	Necessary (unpolarised)	Necessary (polarised)	Snubber can be eliminated. If used a polarised snubber is used.	Snubber can be eliminated. If used a polarised snubbers is used.
5.	Maximum VI	10 kV/5000 Amp	2 kV/1000 A	600 V/200 Amp.	150 V/400 Amp.
6.	Applications	DC motor drives, inverters, rectifiers.	UPS, SMPS static VAR motor control SMPS	AC motor control, SMPS	SMPS, BLDC drives AC motor control UPS.
7.	Symbol				
8.	Type of device	Minority carrier	Minority carrier	Majority carrier	Minority carrier
9.	Voltage/current controlled	Current	Current	Voltage	Voltage
10.	Commutation circuit	Necessary	Not necessary	Not necessary	Not necessary
11.	Blocking capacity	Symmetrical	Asymmetrical	Asymmetrical	Asymmetrical
12.	Temperature coefficient	Negative	Negative	Positive	Flat
13.	Thermal runaway	Possible	Possible	Not present	Not present
14.	Parallel operation	External equalizing circuit is necessary.	Equalizing circuit required	Easy to parallel	Easy to parallel

9.25. MERITS AND DEMERITS AND APPLICATIONS OF IGBT

9.25.1. Salient Features of IGBT

(i) Easy to turn on and off.
(ii) It is a voltage controlled device. Therefore simpler driver circuit.
(iii) Low on state voltage drop. Therefore low on state power dissipation.
(iv) Switching frequency higher than that of a power BJT.
(v) It does not need snubber circuit for its protection.
(vi) It has a flat temperature coefficient of resistivity.

9.25.2. Drawbacks of IGBT

(i) It has an asymmetric blocking capacity. It cannot block high reverse voltages.
(ii) Switching frequency is not as high as that of a power MOSFET.
(iii) Problem of latch ups.
(iv) Excessive power dissipation can take place at the time of turn-off due to the "current tail" present in the turn-off characteristics.

9.25.3. Applications of IGBT

Some of the important applications of IGBT are :
(i) Switching mode power supplies (SMPS)
(ii) UPS systems
(iii) AC motor controllers
(iv) Choppers
(v) Inverters

9.25.4. Salient Features of IGBT over BJT

(i) It is a voltage controlled device so that gate driving in easy.
(ii) It can switch at higher frequency that BJT. Typically f_{max} = 20 kHz.
(iii) IGBTs can be easily connected in parallel.
(iv) Second breakdown does not take place.
(v) IGBTs do not need snubber circuits for their protection.

9.25.5. Advantages of IGBT over MOSFET

(i) On state voltage drop across IGBT is less than that across MOSFET.
(ii) Low on state power dissipation.
(iii) IGBTs can handle larger power than MOSFETs.

SUMMARY

- As a matter of fact, thyristors is a general term which includes devices such as SCR, Triac, GTO, LASCR, RCT, SCS, ... and many more. Thus, thyristors is a family of devices. SCR i.e. silicon controlled rectifier is the most popular device of the thyristor family.
- Some of the typical applications of thyristors may be listed as under :
  1. Temperature controllers
  2. Level controllers
  3. Intensity controllers
  4. Speed control of AC and DC motors
  5. Fan regulator
  6. Battery charger

7. Power inverters
8. DC to DC choppers
9. Static switches
10. Cyclo converters
11. Controlled rectifiers (only SCRs)
12. HVDC transmission
13. AC voltage stabilizers.

- SCR is one of the oldest type of solid state power device. It was invented in 1957 by the General Electric Research Laboratories. SCRs have the highest power handling capacity of all the power semiconductor devices. They have a four layer construction with three user accessible terminals. SCR is a latching type device that can be turned on by the control terminal (gate) but once turned on, the gate loses control on it (i.e. it cannot be turned off by the gate).
- Important Features of SCR
  (i) It is a latching type device.
  (ii) It can handle a very large power.
  (iii) It is a current controlled device, because the gate current controls SCR.
  (iv) It acts as an open or closed switch.
  (v) The on state voltage drop is very low.
  (vi) It can handle thousands of amperes of current.
- The forward characteristics may be divided into three regions of operation as under :
  (i) Forward blocking state.
  (ii) Transition state.
  (iii) ON state.
- The voltage across SCR is low in this state. In the on state a high power SCR can conduct average currents as large as 2000 amp with on state voltage drop of a few volts.
- It may be noted that the voltage across SCR in its on state is very low as compared to the off state voltage.It will be proved later on that the on state voltage is equal to drop across one junction (*i.e.* of the order of 1 to 1.5 volts).
- It is minimum anode current that must flow through SCR to latch it into the on state. The latching current is higher than the holding current. The latching current is important when SCR is being turned on.
- The gate current can turn on SCR but once turned on, the gate loses control. Even if we reduce the gate current to zero, SCR continues to conduct. We can turn off a conducting SCR by :
  (i) Reducing its anode current below the holding current or
  (ii) Applying a reverse voltage across the SCR.
- SCR is turned on by increasing the anode current. This can be achieved with one of the following methods :
  1. By raising the temperature
  2. By focussing light
  3. By applying high voltage
  4. By applying voltage at high dv/dt
  5. By increasing the gate current.
- The gate turn on has following salient features:
  (i) Less power consumption.
  (ii) It is possible to turn on the SCR precisely at the desired value of anode to cathode voltage.
  (iii) If the supply voltage is AC, then it is possible to adjust the firing angle α to the desired value.
  (iv) Triggering circuits are easy to design.
  (v) Turn on gain is very high. Turn on gain is the ratio of anode current to the gate current.
  (vi) Pulsed triggering circuit provide greater flexibility.
- **Drawbacks of SCR**
  (i) It can conduct only in one direction. Hence, it can control power only during one half cycle of ac.
  (ii) It can turn on accidentally due to high dv/dt of the source voltage.
  (iii) It is not easy to turn off the conducting SCR. We have to use special circuits called commutation circuits to turn off SCR.

(iv) SCR can not be used at high frequencies. The maximum frequency of its operation is 400 Hz.
(v) Gate current can not be negative.

- **Advantages of SCR**

(i) It can handle large voltages, currents and power.
(ii) The voltage drop across conducting SCR is small. This will reduce the power dissipation in the SCR.
(iii) Easy to turn on.
(iv) Triggering circuits are simple.
(v) It can be protected with the help of a fuse.
(vi) We can control the power delivered to the load.

- **Applications of SCR**

(i) Controlled rectifiers
(ii) DC to DC converters or choppers
(iii) DC to AC converters or inverters
(iv) As static switch
(v) Battery charger
(vi) Speed control of DC and AC motors
(vii) Lamp dimmers
(viii) AC voltage stabilizers.

- The three important states of triac operation may be listed as under :

(1) Forward blocking state.
(2) Reverse blocking state.
(3) Conduction or on state.

- **Salient Features of Triac**

(i) It is a bi-directional device. So, we can control the power delivered to load in both the half cycles of ac supply.
(ii) It is equivalent to two SCRs connected back to back.
(iii) We can turn it on by using a positive as well as negative gate current.
(iv) It is more suitable for resistive loads.
(v) It is more economical than SCRs, since inside the same package, we get two SCRs connected back to back.
(vi) It can control power delivered to ac loads such as fan motor.
(vii) It is not necessary to use a protection diode across the Triac.
(viii) We can use a single heat sink.

- **Drawbacks of Triac**

(i) We can not use it as a controlled rectifier.
(ii) Low dv/dt rating than SCR. So possibility of accidental turn on is higher than that of SCR.
(iii) Low di/dt rating.
(iv) Suitable for resistive loads only. Not suitable for controlling power to highly inductive loads.
(v) Its power rating is lower than that of SCR.
(vi) Triggering circuits require to be designed more carefully.

- **Applications of Triac**

(i) As a static switch
(ii) In the AC voltage stabilizers
(iii) Fan speed regulator
(iv) Lamp dimmer
(v) Flasher circuit
(vi) Proximity detector
(vii) Temperature controller

- Diac is one of the most important devices used to trigger the Triac. It is sometimes called as a Triac without gate terminal.

- DIAC is basically used as triggering device for Triac. The following control systems are DIAC for this purpose:
  (i) Lamp dimmer
  (ii) Fan speed regulator
  (iii) Temperature controller
- **Drawbacks of DIAC**
  (i) It is a low power device.
  (ii) It does not have a control terminal.
- The power BJT has advantage of low on state power dissipation, but it cannot be switched at faster rates due to longer turn-off time. The power MOSFET can be switched at much higher frequency but has a drawback of higher on state power loss. These observations have led to attempts to combine BJTs and MOSFETs monolithically on the same silicon wafer to develop a new device that will have the best qualities of both, BJT and MOSFET. The new device is known as insulated gate bipolar transistor (IGBT) which is finding use in many applications. Other names to this device are GEMFET, COMFET (conductivity modulated field effect transistor), IGT (insulated gate transistor) and bipolar mode MOSFET or bipolar MOS transistor.
- **Salient Features of IGBT**
  The important salient features of IGBT may be listed as under
  (i)Low on state voltage drop.
  (ii)Low on state power loss.
  (iii) It has a higher switching frequency than that of a power BJT.
  (iv)IGBT has the best qualities of BJT and MOSFET.
- Power devices include power diodes (general purpose diodes, fast recovery diodes and Schottky diodes), power transistors [BJTs, MOSFETs (N-channel) SIT and IGBT] and thyristors (SCR, Triac, GTO, SITH and MCT).
- The silicon controlled rectifier (SCR) is the most widely used thyristor. The SCR may be thought of as a PNP and an NPN transistor interconnected, with regenerative feedback that is initiated by a gate current pulse. It can switch very large currents on and off. On application of forward voltage to SCR, a small forward leakage current flows. When the applied voltage exceeds the breakover voltage, the device turns on and a large current flows. The SCR may be triggered into a conducting state by a pulse gate current, but the gate loses control of anode current after firing. An SCR may be turned off by reducing the current below the holding value or by reducing the anode-to-cathode voltage to zero. Gate triggering of an SCR requires the application of gate voltage and current that must lie in the preferred area of the graphical gate characteristics is to ensure reliable switching. SCR has turn-on time of 1-4 μsec and turn-off time of 10-300 μsec. An SCR has a number of voltage, current and power ratings.These ratings should not be exceeded. The applications of SCRs include power control, switching, zero voltage switching, overvoltage protection, pulse circuits, battery charging regulator, heater control circuit, emergency lighting system etc.
- The four-layer diode, also called a Shockley diode, is the simplest thyristor. It is essentially a low-current SCR without a gate.
- Silicon controlled switch (SCS), like an SCR, is a unilateral, four-layer, three junction P-N-P-N silicon device, with four electrodes namely cathode, cathode gate, anode gate and anode. It is a low power device compared with the SCR. A few of the more common areas of application of SCS include a variety of computer circuits (such as counters, registers, and timing circuits) voltage sensors, pulse generators, oscillators etc.

SHORT QUESTIONS WITH ANSWERS

Q.1. Write the important features of SCR.

Ans. Important Features of SCR

(i) It is a latching type device.
(ii) It can handle a very large power.
(iii) It is a current controlled device, because the gate current controls SCR.

(iv) It acts as an open or closed switch.
(v) The on state voltage drop is very low.
(vi) It can handle thousands of amperes of current.

Q.2. Write the difference between holding and latching currents.

Ans. Difference Between Holding and Latching Currents

(i) The latching current is important only at the time of SCR turn on, whereas holding current is important only at the time of SCR turn off.

(ii) If the anode current goes below the holding current then the conducting SCR turns off. The holding current is defined for $I_G = 0$, whereas if anode current at the time of turn on is higher than latching current then only SCR will latch into its on state.

(iii) If $I_A < I_L$ then SCR does not latch. It will remain on as long as I_G is being supplied. As soon as I_G is removed, SCR will turn off.

(iv) Latching current is always higher than the holding current.

(v) The values of I_L and I_H are constant. They do not depend on the gate current magnitude.

Q.3. What are SCR turn on methods?

Ans. SCR is turned on by increasing the anode current. This can be achieved with one of the following methods :

1. By raising the temperature
2. By focusing light
3. By applying high voltage
4. By applying voltage at high dv/dt
5. By increasing the gate current.

Q.4. What are the salient features of Gate Turn ON?

Ans. The gate turn on has following salient features :

(i) Less power consumption.
(ii) It is possible to turn on the SCR precisely at the desired value of anode to cathode voltage.
(iii) If the supply voltage is AC, then it is possible to adjust the firing angle α to the desired value.
(iv) Triggering circuits are easy to design.
(v) Turn on gain is very high. Turn on gain is the ratio of anode current to the gate current.
(vi) Pulsed triggering circuit provide greater flexibility.

Q.5. What should be the pracautions to Avoid SCR damage?

Ans. Precautions to Avoid SCR Damage

(i) In order to avoid SCR damage or failure, the rate of anode current rise di_F/dt should be kept less than the maximum value given in SCR specification sheet. The failure of the device may take place due to excessive power dissipation that takes place during the time interval t_r.

(ii) This power dissipation can be reduced by applying large values of gate current during the $t_{d\,(on)}$ and t_r intervals. For this reason, the gate current applied is often large at the start of the turn on interval and is gradually reduced as time proceeds.

Q.6. Write the Drawbacks of SCR.

Ans. Drawbacks of SCR

(i) It can conduct only in one direction. Hence, it can control power only during one half cycle of ac.
(ii) It can turn on accidentally due to high dv/dt of the source voltage.
(iii) It is not easy to turn off the conducting SCR. We have to use special circuits called commutation circuits to turn off SCR.
(iv) SCR can not be used at high frequencies. The maximum frequency of its operation is 400 Hz.
(v) Gate current can not be negative.

Q.7. Write the advantages of SCR.

Ans. Advantages of SCR

(i) It can handle large voltages, currents and power.
(ii) The voltage drop across conducting SCR is small. This will reduce the power dissipation in the SCR.
(iii) Easy to turn on.
(iv) Triggering circuits are simple.
(v) It can be protected with the help of a fuse.
(vi) We can control the power delivered to the load.

Q.8. Write the applications of SCR?

Ans. Applications of SCR

(i) Contrlled rectifiers
(ii) DC to DC converters or choppers
(iii) DC to AC converters or inverters
(iv) As static switch
(v) Battery charger
(vi) Speed control of DC and AC motors
(vii) Lamp dimmers
(viii) AC voltage stabilizers.

Q.9. Write the salient features of Triac.

Ans. Salient Features of Triac

(i) It is a bi-directional device. So, we can control the power delivered to load in both the half cycles of ac supply.
(ii) It is equivalent to two SCRs connected back to back.
(iii) We can turn it on by using a positive as well as negative gate current.
(iv) It is more suitable for resistive loads.
(v) It is more economical than SCRs, since inside the same package, we get two SCRs connected back to back.
(vi) It can control power delivered to ac loads such as fan motor.
(vii) It is not necessary to use a protection diode across the Triac.
(viii) We can use a single heat sink.

Q.10. Write the drawbacks of Triac.

Ans. Drawbacks of Triac

(i) We can not use it as a controlled rectifier.
(ii) Low dv/dt rating than SCR. So possibility of accidental turn on is higher than that of SCR.
(iii) Low di/dt rating.
(iv) Suitable for resistive loads only. Not suitable for controlling power to highly inductive loads.
(v) Its power rating is lower than that of SCR.
(vi) Triggering circuits require to be designed more carefully.

Q.11. Write the applications of Triac.

Ans. Applications of Triac

(i) As a static switch
(ii) In the AC voltage stabilizers
(iii) Fan speed regulator
(iv) Lamp dimmer
(v) Flasher circuit
(vi) Proximity detector

Q.12. What is meant by Diac.

Ans. Diac is one of the most important devices used to trigger the Triac. It is sometimes called as a Triac without gate terminal.

Q.13. How is current limited in conducting state of an SCR ?

Ans. The current in conducting state of an SCR is controlled by external impedance.

Q.14. What is the effect of negative gate current on a normal SCR ?

Ans. Negative gate current will increase the holding current of the SCR.

Q.15. What is false triggering ?

Ans. False triggering is unintended turn-on of an SCR either through gate due to noise pick-up or excessive anode voltage.

Q.16. What are the factors on which 'turn-off time' of an SCR depends ?

Ans. Turn-off time of an SCR depends upon doping densities, recombination time, junction temperature, on-state current, rate of decay of on-state current.

Q.17. How the forced turn-off of an SCR is different from natural turn-off ?

Ans. In forced turn-off of an SCR current is brought below the holding current by an additional delicated circuit employing energy storage elements while in natural turn-off, it is the natural voltage/ current variation.

Q.18. Why 'pulse triggering* is preferred and when does it fail ?

Ans. Pulse triggering is preferred as it is effective to turn on an SCR keeping the device gate dissipation low and ensures a fast turn-on keeping the $\frac{di}{dt}$ stress on the device low.

Pulse triggering shall fail if the pulse is of short duration and the load circuit contains large inductance. With highly inductive load, the device current builds up slowly. If the pulse duration is small and the anode current in not able to build up to latching current, I_L, the device turns off on removal of gate pulse.

Q.19. Why is it necessary to keep supply voltage much less than breakover voltage of an SCR ?

Ans. The supply voltage is kept much less than breakover voltage of an SCR otherwise the device will get damaged.

Q.20. What is meant by breakover voltage of an SCR ?

Ans. The minimum forward voltage, with gate open, at which SCR starts conducting heavily {i.e. turns on) is called the breakover voltage.

Q.21. What is maximum on-state voltage in reference to an SCR ?

Ans. Maximum on-state voltage is the maximum value of the voltage appearing across the SCR during conduction.

Q.22. What is meant by forward current rating of an SCR ?

Ans. The maximum value of anode current, that an SCR can handle safely (without any damage) is called the forward current rating of the SCR.

Q.23. What is holding current in an SCR ?

Ans. Holding current is the minimum on-state current required to keep the SCR in conducting state without any gate drive.

Q.24. Define latching current of an SCR.

Ans. Latching current is the minimum device current, which must be attained by the device, before the gate drive is removed.

Q.25. Give the typical values of current ratings of SCR.

Ans. Forward breakover current—less than a few hundred microamperes. Peak forward current 30 A to over 100 A. Holding current—a few mA to few hundred milliamperes.

Q.26. How Shockley diode is switched on ?

Ans. Shockley diode has no trigger inputs, the only way to switch the device on is to increase the anode-to-cathode voltage to the switching voltage equivalent to SCR forward breakover voltage.

Q.27. What is silicon controlled switch ?

Ans. Silicon controlled switch (SCS), like the SCR, is a unilateral, four-layer, three junction P-N-P-N silicon device with four electrodes namely cathode, cathode gate, anode gate and anode.

Q.28. What are bilateral devices ?

Ans. Bilateral devices are those devices which switch symmetrically with forward and reverse bia Triac and diac fall under this category.

Q.29. What is triac?

Ans. The triac is a three terminal, four layer bidirectional semiconductor device. It incorporates SCRs connected in inverse parallel with a common gate terminal in a single chip. Triac is abbreviation for a triode ac switch.

Q.30. What is diac?

Ans. A diac is a P-N-P-N structured four-layer, two-terminal semiconductor device. This is just like a triac but without gate terminal. Diac is an abbreviation for a diode ac switch.

Q.31.What is IGBT?

Ans. IGBT (insulated gate bipolar junction transistor) is a new high conductance MOS controlled power switch and incorporates some of the best features of power MOSFETs and power transistors.

Q.32. Give the applications of IGBTs.

Ans. IGBTs are finding increasing applications in medium power applications such as dc and ac motor drives, power supplies, solid-state relays and contactors.

REVIEW QUESTIONS

1. Draw the two transistor model of SCR and explain it.
2. Why is reverse breakdown voltage greater than forward breakdown voltage in SCR ?
3. Explain why the circuit recovery time should be greater than the turn off time t_q of a SCR ?
4. What is the significance of di/dt rating of SCR? How do you protect it against high di/dt ?
5. Define latching current and explain how is it different from holding current.
6. Write a note on : Current ratings of SCR.
7. Explain the significance of $\int i^2\, t$ or $I^2\, t$ rating of SCR.
8. Explain the following SCR ratings specifications .
   (i) Turn-off time (ii) Latching current,
   (iii) dv/dt.
9. Write a short note on : Protection circuits of SCR.
10. Explain the use of Triac in the phase control applications.
11. Define thyristor and name different thyristor devices.
12. The temperature coefficient of IGBT is
    (a) positive (b) negative (c) flat.
13. Draw the static characteristics of SCR and explain different regions of operation.
14. Explain the turn-off characteristics of SCR giving typical values of turn off time for normal and inverter grade SCRs.
15. Why snubber circuit is required ? Explain the snubber circuit used for SCR.
16. Explain why SCR is developed by using silicon and not germanium.
17. Draw the structure, symbol and V-I characteristics of a Diac.
18. Explain any one application of the Diac.
19. Draw and explain the V-I characteristics of a Triac.
20. Explain role of thyristors in industrial systems.
21. Give the classification of thyristor devices.
22. Explain the basic differences between a triac and thyristor ? Draw and explain I-V characteristic of a triac.
23. What is the origin of the name of diac. Explain the construction, operation and characteristics of a diac. Discuss its applications.
24. Draw the physical structure and equivalent circuit of the IGBT. Explain its special features. Draw its I-V characteristics.
25. Write short notes on : *(i)* TRIAC, *(ii)* DIAC, *(iii)* IGBT, *(iv)* SCR
26. Draw the structure of an IGBT and explain the function of each layer.
27. Explain the operation of thyristor in detail with its V-I characteristics. Also discuss its two transistor model.
28. What are the differences between a transistor and a thyristor? Explain the operation of a thyristor with the help of its characteristics.
29. Discuss the construction and explain the characteristic curve of an SCR showing how forward voltage depends on the value of gate current.
30. Draw the block and symbolic representation for SCR. State its application. Sketch its V-I characteristics for different gate currents and indicate there upon holding current, latching current, and breakover voltage.
31. Explain the basic construction of a silicon controlled rectifier (SCR). Explain its principle of operation using two transistor models. Explain the turn-off mechanism of SCR. Give a circuit explain your answer.
32. Give the two transistor analogous circuit of an SCR and explain its operation. Sketch its V-I characteristics. Mark 'hold current' and 'latch current' on it and explain their importance in SC power control circuits.
33. What are the normal methods of turn-on of an SCR ? Also state the schemes of commutation.
34. What are the advantages and disadvantages of an SCR switch over a mechanical switch ?

35. Describe the characteristics and applications of SCR and P-N-P-N diode.
36. Why a four-layer diode is called the Shockley diode ? Give its basic structure, equivalent transistor circuit and schematic symbol. Give its breakover characteristic and explain.
37. What is silicon controlled switch (SCS) ? How is it different from an SCR ? Give its basic structure, equivalent transistor circuit and schematic symbol. Give its advantages over an SCR and applications.
38. Why triac is called a double ended SCR ? Explain the construction and operation of a triac with the help of a neat diagram.
39. With the help of V-I characteristics show the four triggering modes of triac.
40. With neat diagram explain the constructional features of a triac. Sketch its V-I characteristics and explain its operation. How does it differ from an SCR ? Mention its applications.
41. Explain the working of TRIAC alongwith its V-I characteristics.
42. Explain the construction of diac. Draw a circuit showing use of a diac to trigger a triac.
43. What is the origin of the name diac? Explain the construction, operation and characteristics of a diac. Discuss same applications of diac.
44. Explain the construction and working of (i) TRIAC and (ii) IGBT.
45. Explain the characteristics of IGBT.
46. What is IGBT ? Draw its equivalent circuit and enumerate its special features.
(U.P. Technical Univ. 2007-08)
47. What is IGBT ? Give the operation and characteristics of the device. What precautions are taken while handling IGBTs ?
48. Draw the physical structure and equivalent circuit of the IGBT. Explain the special features. Draw its I-V characteristics.
49. Write short notes on the following:
    (i) Thyristor (ii) SCR
    (iii) TRIAC (iv) DIAC and TRIAC
    (v) DIAC (vi) IGBT

OBJECTIVE TYPE QUESTIONS

1. An IGBT structure is obtained
   (a) by combining a MOSFET and BJT.
   (b) by adding an insulated gate to the power transistor and by adding a p^+ layer.
   (c) none of the above.
2. The SOA of IGBT is better than that of a power transistor because
   (a) it is a majority carrier device.
   (b) it is a minority carrier device.
   (c) second breakdown is absent due to its flat temperature coefficient of resistivity.
3. The maximum operating frequency of an IGBT is approximately.
   (a) 10 kHz (b) 50 kHz,
   (c) 100 kHz
4. The on state voltage drop across IGBT is
   (a) less than that across a MOSFET (b) greater than that across MOSFET
   (c) equal to that of a MOSFET.
5. The reduction in the on state voltage drop in IGBT takes place due to
   (a) added p^+ layer in the IGBT structure. (b) conductivity modulation.
   (c) the n^- drift layer.
6. IGBT can be protected against a large load current by
   (a) using a fast acting fuse (b) using a crowbar circuit
   (c) removing its gate driving voltage.

7. The non punch through IGBT has a
   (a) symmetrical blocking capacity
   (b) asymmetrical blocking capacity
   (c) no blocking capacity at all.
8. The blocking capacity of a punch through IGBT is
   (a) symmetrical
   (b) asymmetrical
   (c) none of the above.
9. The gate-source voltage to the IGBT should be sufficiently large in order to
   (a) avoid latch up in IGBT
   (b) to reduce the on state voltage drop
   (c) to reduce the turn off time.

Answers:

1. (b) 2. (c) 3. (a) 4. (a) 5. (b) 6. (c) 7. (a)
8. (b) 9. (b)

❑❑❑

PREVIOUS YEAR QUESTION PAPERS

B.TECH
THIRD SEMESTER THEORY EXAMINATION, 2009-10
FUNDAMENTALS OF ELECTRONICS DEVICES

Time: 3 hour **Total Marks : 100**

Note: *Attempt all the questions. All questions carry equal marks.*

1. **Attempt any four parts of the following:**
   (a) Derive the formula for Density and lattice constant.
   (b) What is atomic radius and atomic packing factor?
   (c) What do you mean by mobility of carrier? How does it depend on temperature and doping concentration?
   (d) Explain the basic crystal growth techniques.
   (e) What is Hall effect and Hall angle? show that $QH = \tan^{-1}(\mu_n B_z)$ where symbols have their usual meaning
   (f) In an intrinsic semiconductor the effective mass of the electron is 0.07 m_0 and that of the hole is 0.4 m_0, where m_0 is the rest mass of the electron. Calculate the intrinsic concentration of change carriers at 300°K given by $E_g = 0.7$ eV.

2. **Attempt any four parts of the following:**
   (a) Find the diffusion coefficients of electrons and holes of a silicon single crystal at 27°C if the mobilities of electrons and holes are 0.17 and 0.025 $m^2 V^{-1} \rho^{-1}$ respectively at 27°C.
   (b) Design a 5 μM cds photoconductor with 10 MΩ dark resistance, 0.5 cm square. Assume $\tau = 10^{-6}$ s and $N_d = 10^{14}$ cm^{-3}.
   (c) Obtain relationship between photoconductivity and mobility of carriers.
   (d) Define mobility of a change carrier. Show that

$$\frac{\mu}{D} = \frac{e}{KT}$$

   (e) Using the concept of diffusion and drift of carriers establish Einsteins relation and modified from of Ohm's law.
   (f) Define
      (i) Photoconductivity
      (ii) Luminescense.

3. **Attempt any two parts of the following:**
   (a) What do you mean by zener diode? Explain how zener diode? Explain how zener diode maintains constant voltage across the load.
   (b) Consider a Ge diode with $N_D = N_A$ and will impurity concentration of 8×10^{14} cm^3. Assume $\eta_i = 2 \times 10^{13}/cm^3$. At room temperature of 300 K calculate the height of the potential barriers under open circuits conditions. Assume Bolt constant $K = 8.61 \times 10^{-5}$ eV/k and electron $\rho = 1.6 \times 10^{-19}$ C. Describe the Schotky diode detail.
   (c) What do you mean by reverse bias break down? Calculate the thermal equilibrium electron and hole concentration in a compensated P-type silicon semiconductor at 300° K in which $N_A = 10^{16}$ cm^{-3}, $N_D = 3 \times 10^{15}$ cm^{-3}
   $\eta_i = 1.6 \times 10^{10}$ cm^{-3}

4. **Attempt any two parts of the following:**

   (a) Prove that the transconductance g_m of JFET is given by $gm = \frac{2}{|V_P|}\sqrt{I_{DS} I_{DSS}}$

(b) What is a MOSFET? How many types of MOSFETs are there? Point out the basic difference between the FET and BJT?

(c) For the N-channel zero biased D-MOSFET circuit of figure below calculate the V_{DS} if J_{DSS} = 10 mA and $V_{GS\ (off)} = -6V$.

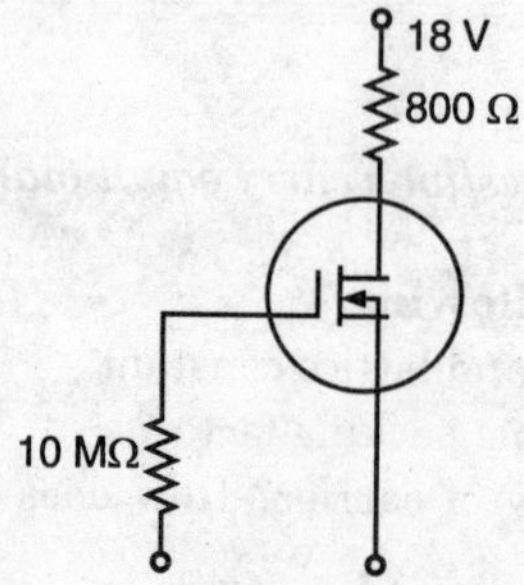

Fig. 1

5. Attempt any two parts of the following:

(a) What is a photodetector? Describe the working of solar cell. What are the applications of these devices?

(b) Explain the tunnel diode in detail. What are its applications?

(c) Describe the characteristics and application of SCR and P-N-P-N diode.

□□□

B.TECH
THIRD SEMESTER THEORY EXAMINATION, 2010-11
FUNDAMENTALS OF ELECTRONICS DEVICES

Time: 3 hour *Total Marks : 100*

Note : (*i*) Attempt **ALL** questions.
(*ii*) All questions carry equal marks.

1. Attempt any Four parts: **(4×5=20)**

(a) With a suitable sketch describe briefly the lattice structure of GaAs.

(b) What is Miller Indices? Show the (643) plane and the [643] direction in a cubic crystal lattice.

(c) Classify semiconductor according to band structure. Explain with examples.

(d) What do you mean by effective mass of carriers? How does it depend on energy bands?

(e) Explain briefly why the temperature coefficient of the resistivity of a doped semiconductor is negative at low and high temperature but positive at intermediate temperature.

(f) Calculate the Fermi level position in Si containing 10^{16} Phosphorus atoms/cm^3 at 100 K assuming 50% of the impurities are ionized at this temperature. Also calculate the equilibrium electrons and holes concentrations.

2. Attempt any Two parts: **(2×10=20)**

(a) What are the direct and indirect recombination? Derive an expression for minority carrier life time.

(b) What is the physical significance of diffusion length? How is it related with mobility of the carrier Explain.

(c) What is quasi-fermi levels? Explain.
A Ge sample with 10^{17} Phosphorous atoms/cm^3 is optically excited at 300 K such that gop = 10^{20} EHP/cm^3-sec and $\tau_n = \tau_p = 10$ μsec. What is the separation of the quasi-Fermi levels? Draw a band diagram showing the result. Given that intrinsic carrier concentration for Ge at room temperature is 2.5×10^{13} per cm^3.

3. Attempt any Two parts : **(2×10=20)**

(a) What is diffusion potential? Assuming equilibrium condition at a step junction, derive an expression for diffusion potential.

(b) What do you mean by minority carrier injection and minority carrier extraction? Derive an expression for total current through the diode for either forward or reverse bias.

(c) Describe the physical mechanism for p-n junction breakdown. Draw a circuit which uses a breakdown diode to regulate the voltage across a load. Explain its operation.

4. Attempt any Two parts: **(2×10=20)**

(a) Enumerate the special features of MESFET. Explain its working and state the difference between normally-on and normally-off MESFETs.

(b) What are the advantages and disadvantages of FET over BJT? Describe the construction and operation of MOSFET with suitable diagra.

(c) Explain how a Bipolar Junction Transistor can be used as a switch. Explain the factor which determine the switching speed of BJT.

5. Attempt any Two parts : **(2×10=20)**

(a) What is transferred electron effect? Describe a device based on this effect with suitable diagram in detail. Also draw its characteristics.

(b) What is photo-detector? Explain the operation of a p-i-n photodetector. What are the suitable material for it? How can be be made more sensitive to low level intensity of light?

(c) With a suitable diagram describe the operation and characteristics of a four-layer diode. Explain the various triggering mechanism of this diode.

□□□

B.TECH

THIRD SEMESTER THEORY EXAMINATION, 2011-12

FUNDAMENTALS OF ELECTRONICS DEVICES

Time: 3 hour **Total Marks : 100**

Note: *Attempt all the questions. All questions carry equal marks.*

1. Attempt any four parts of the following: (5×4=20)

(a) State the fermi Dirac energy distribution function, show the fermi level position in p-type semiconductor at 0, 300 and 400 K.

(b) Explain the various energy bands in solids. Based on these bands, distinguish between insulator, conductor and semiconductors.

(c) For a semiconductor, $\mu_n = \mu_p = 1000$ cm^2/vsec and $N_c = N_v = 10^{19}$ cm^{-3}. If the conductivity of the intrinsic semiconductor at 300 K is 4×10^{-6} $(\Omega\text{–cm})^{-1}$. What is the conductivity at 600 K?

(d) Explain the effects of temperature on doping and mobility

(e) In a semiconductor at room temperature, the intrinsic carrier concentration and intrinsic resistivity are 1.5×10^{16}/m^3 and 2×10^3 Ωm respectively. It is converted into an extrinsic semiconductor with a doping concentration of 10^{20}/m^3 for the extrinsic semiconductor. Calculate for the extrinsic minority carrier conc. e^- mobility and resistivity of doped semiconductor, minority carrier concentration when temperature is increased to a value at which intrinsic carrier concentration n_i is doubled $\mu_n = \mu_p$.

(f) How many times the conductivity increases because of doping? Assume total number of atoms in 4.421×10^{21} per cm^3 in Ge, $n_i = 2.5 \times 10^{13}$ atoms/cm^3, $\mu_n = 3800$ cm^2/vsec, $\mu_p = 1800$ cm^2/vsec.

2. Attempt any two parts of the following: (10×2=20)

(a) What is luminescence? Explain excitation and recombination of photoluminescence with trap level for electrons.

(b) What is photoconductivity? Explain how the photocurrent is proportional to the lifetime and inversely proportional to transit time of carrier.

(c) Draw the spectral response of Si and Ge and relationship between wavelength and E_g (Energy Gap) for Si and Ge at room temperature.

3. Attempt any two parts of the following: (10×2=20)

(a) For the regulator circuit shown below, calculate Regulated voltage, Load Resistance unregulated voltage, P_z, R_z, $I_z = 10\%$ I_L. Also explain Zener and Avalanche Breakdown.

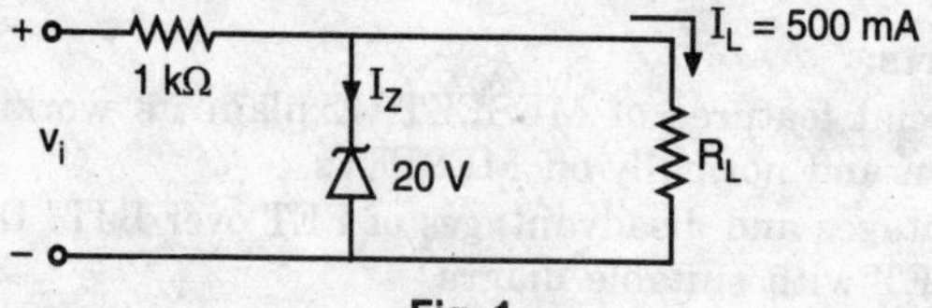

Fig. 1

(b) Explain the formation of built in potential across a p-n junction diode without the application of an external bias voltage. How is this built in potential modified by the application of forward and reverse bias voltage that leads to a net current across the diode?

(c) Write short notes on any two of the following :

(i) Transition and diffusion capacitance of a junction diode.

(ii) Varactor diode

(iii) Storage and transition times of p-n junction diode.

4. Attempt any two parts of the following: **(10×2=20)**

(a) Compare CB, CE and CC configurations of a transistor. Prove that $I_C = \beta I_B + (1 + \beta)I_{CO}$. In figure, if $V_{CB} = 0.5$ V, find I_Q :

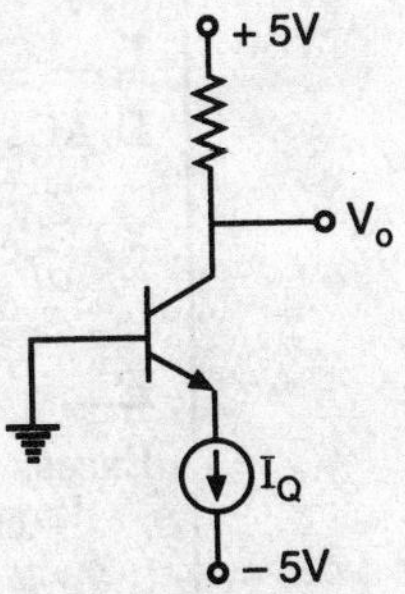

Fig. 2.

(b) Explain the construction and working of any two :

(i) Depletion type MOSFET

(ii) Enhancement type MOSFET

(iii) Junction Field Effect Transistor (JFET)

(c) Enumerate the special feature of MESFET. Explain its working and discuss difference in its characteristics from the MOSFET.

5. Attempt any four parts of the following: **(5×4=20)**

(a) What is an LDR? Give its basic construction symbol and characteristics. How an LDR can be used as a detector?

(b) Explain the working principle of Photovoltaic cell. Give its circuit symbol, voltage-current characteristics.

(c) What is a photodetector? Describe the working of Solar Cell?

(d) Define light emitting materials. What determines the emission of colour of LED?

(e) What is the basic difference between GUNN diode and IMPATT diode?

(f) What is tunneling phenomenon? Explain the V-I characteristics of Tunnel diode.

□□□

INDEX

T....

X....

□□□

Win Prizes

ATTENTION READERS

Win Prizes

We request for your frank assessment regarding various aspects of the book as given below :

FUNDAMENTALS OF ELECTRONICS DEVICES

BY: Dr. SANJAY SHARMA

Please Fill up the Sheet in Neat and Clean Writing

(i) What topics of your syllabus, important for your examination point of view, are not covered in this book ?

(ii) What are the chapters wherein the treatment of the subject matter is not systematic or organised or updated ?

(iii) Have you come across misprints/mistakes/factual inaccuracies in the book? Please specify the chapters and the page numbers.

(iv) What other books on the same subject have you found/heard better than the present book? Please specify in terms of price and quality.

(v) Further suggestions and comments for the improvement of this book.

(PS: If need be please attach a separate sheet to write your views)

OTHER DETAILS

(i) Who recommended you this book ? (Please tick (✓) in the box near the option relevant to you.

Teacher ☐ Friends ☐ Bookseller ☐

....

....

....

....

(ii) Name of the recommending teacher, his designation and address :

....

....

....

....

(iii) Name and address of the bookseller you purchased the book from :

....

....

....

....

(iv) Name and address of your institution (Please mention the University or Board as the case may be)

....

....

....

....

(v) Your name and complete postal address :

....

....

....

....

(vi) Course you are enrolled in :

....

....

....

....

The best assessment will be awarded each month: The award will be in the form of our publications as decided by the Editorial Board.

Please mail the filled up coupon at your earliest to :

S.K. KATARIA & SONS®
4760-61/23, Ansari Road,
Daryaganj, New Delhi-110002
☎: +91-11-23243489, 23269384
E-mail : katariabooks@yahoo.com
katariabook@gmail.com